# PLANETARY NEBULAE

INTERNATIONAL ASTRONOMICAL UNION
UNION ASTRONOMIQUE INTERNATIONALE

# PLANETARY NEBULAE

PROCEEDINGS OF THE 155TH SYMPOSIUM OF THE
INTERNATIONAL ASTRONOMICAL UNION,
HELD IN INNSBRUCK, AUSTRIA, JULY 13-17, 1992

EDITED BY

R. WEINBERGER
*Institut für Astronomie der Universität, Innsbruck, Austria*

and

A. ACKER
*Observatoire Astronomique de Strasbourg, Strasbourg, France*

SPRINGER-SCIENCE+BUSINESS MEDIA, B.V

Library of Congress Cataloging-in-Publication Data

```
International Astronomical Union. Symposium (155th : 1992 :
  Innsbruck, Austria)
    Planetary nebulae : proceedings of the 155 Symposium of the
  International Astronomical Union, held in Innsbruck, Austria, July
  13-17, 1992 / edited by R. Weinberger and A. Acker.
      p.   cm.
    "International Astronomical Union symposia, volume 155."
    Includes index.
    ISBN 978-0-7923-2440-9 (pkb.)
    1. Planetary nebulae--Congresses.   I. Weinberger, R. (Ronald)
  II. Acker, Agnès.  III. Title.
  QB855.5.I67  1992
  523.1'135--dc20                                          93-21552
```

ISBN 978-0-7923-2440-9     ISBN 978-94-011-2088-3 (eBook)
DOI 10.1007/978-94-011-2088-3

---

*Printed on acid-free paper*

*All Rights Reserved*
© *1993 Springer Science+Business Media Dordrecht*
*Originally published by Kluwer Academic Publishers in 1993*

No part of the material protected by this copyright notice may be reproduced or utilized in any form or by any means, electronic or mechanical including photocopying, recording or by any information storage and retrieval system, without written permission from the publisher.

# TABLE OF CONTENTS

| | |
|---|---:|
| List of Participants | xix |
| Foreword | xxxvii |
| Introductory Review: **Planetary Nebulae** – L.H. ALLER | 1 |

## I. THE OBSERVATIONAL PN DATABASE ......... 9

| | |
|---|---:|
| **Classification Criteria and Databases** – B. STENHOLM | 11 |
| **Observational Parameters: Definitions and Limits** – J.H. LUTZ | 19 |
| **Results from Space** – S.R. HEAP | 23 |
| Strasbourg-ESO Catalogue of Galactic Planetary Nebulae – A. ACKER, F. OCHSENBEIN | 33 |
| Four New Evolved Planetary Nebulae – S. TAMURA, R. WEINBERGER | 34 |
| New Planetary Nebulae – J. MARCOUT | 35 |
| New and Misclassified Planetary Nebulae – L. KOHOUTEK | 36 |
| Three Possible Planetary Nebulae from Near-IR Observations – P. PERSI, A. MARENZI, A. PREITE-MARTINEZ, M. FERRARI-TONIOLO | 37 |
| Planetary and Proto-Planetary Nebulae which are Strong $25\mu m$ Emitters – R.D. WOLSTENCROFT, M.A. READ, S.M. SCARROTT, C.J. LONSDALE, Q.A. PARKER | 38 |
| The Southern Deep Near Infrared Survey (DENIS). A Prospect of PN Exploration – S. KIMESWENGER, C. KIENEL | 39 |
| Radio Continuum Observations of Southern Planetary Nebulae Candidates – G.C. VAN DE STEENE, S.R. POTTASCH | 40 |
| The Photographic Observations of Some Planetary Nebulae – X. HAO | 41 |
| CCD Spectrophotometry of Planetary Nebulae at Wendelstein Observatory – M.M. ROTH, R.P. KUDRITZKI, R.H. MÉNDEZ | 42 |
| Planetary Nebulae as a Research Field – W. SAURER, R. WEINBERGER | 43 |
| A Clustering Method Applied to the Analysis of Planetary Nebulae – O.S. YATSYK | 44 |

## II. HIGHLIGHTS ON THE NUCLEI ......... 45

| | |
|---|---:|
| **Model Atmospheres of Central Stars of PN** – R.P. KUDRITZKI, R.H. MÉNDEZ | 47 |
| **Observed Mass Loss from Central Stars of Planetary Nebulae** – M. PERINOTTO | 57 |
| **Temperatures of Central Stars of Planetary Nebulae** – A. PREITE-MARTINEZ | 65 |
| **Atomic Data – a Bibliography** – K. BUTLER | 73 |
| Luminosities of Central Stars of PN in the Galactic Bulge – R.H. MÉNDEZ | 81 |
| Unified Model Atmosphere Studies of Central Stars of Planetary Nebulae – R. GABLER, A. GABLER, R.H. MÉNDEZ, R.P. KUDRITZKI | 82 |

| | |
|---|---|
| Metal Line Blanketed Non-LTE Model Atmospheres for Central Stars of Planetary Nebulae – K. WERNER, S. DREIZLER | 83 |
| Non-LTE Spectra of Iron Group Elements for CSPN – S.R. BECKER, K. BUTLER | 84 |
| The Properties of Planetary Nebulae Nuclei: Stellar Winds – L. BIANCHI, G. DE FRANCESCO | 85 |
| Spectroscopic Properties of the Nucleus of NGC 6826 – B. ALTNER, S.R. HEAP, I. HUBENY | 86 |
| Spectral Analyses of WC-Type Central Stars – W.R. HAMANN, L. KOESTERKE | 87 |
| Central Stars of Old Planetary Nebulae – R. NAPIWOTZKI | 88 |
| Infrared Emission-Lines and the Stellar Temperature – S.M. VIEGAS, R. GRUENWALD | 89 |
| Ultraviolet and Optical Spectra of Central Stars of Halo Planetary Nebulae – M. PEÑA, S. TORRES-PEIMBERT, M.T. RUIZ | 90 |
| A High Resolution Far-UV Spectral Atlas of CSPNs and Hot White Dwarfs – R.W. TWEEDY | 91 |
| The Wind Temperature and C/He and O/He Ratios of the WC10 Central Star CPD-56°8032 – M.J. BARLOW, P.J. STOREY | 92 |
| Mass Loss in Two Low Temperature Central Stars of Planetary Nebulae – A. MODIGLIANI, M. PERINOTTO, P. PATRIARCHI | 93 |
| Stark Broadening Parameters of C IV Lines for Stellar Plasma Research – M.S. DIMITRIJEVIC, S. SAHAL-BRECHOT | 94 |
| E2 and M1 Transition Probabilities in Ions of the Nitrogen Isoelectronic Sequence Calculated Using MBPT – G. GAIGALAS, R. KISIELIUS, G. MERKELIS, M. VILKAS | 95 |
| MBPT Results for $\Delta n=0$ Electric Dipole Transitions – G. GAIGALAS, R. KISIELIUS, G. MERKELIS, Z. RUDZIKAS, M. VILKAS | 96 |

## III. HIGHLIGHTS ON THE NEBULAE ............... 97

| | |
|---|---|
| **Energy Distribution of Planetary Nebulae (UV to Radio)** – C.Y. ZHANG | 99 |
| **Distances of Planetary Nebulae** – Y. TERZIAN | 109 |
| **Advances in Numerical Simulations of Gaseous Nebulae** – G.F. FERLAND | 123 |
| **Evolution of PN Morphologies: Concepts, Models and Observations** – B. BALICK | 131 |
| **Ring Nebulae around PN Nuclei and Massive Stars** – Y.-H. CHU | 139 |
| **The Neutral Envelopes of Planetary Nebulae: Molecules and H I** – P.J. HUGGINS | 147 |
| **Photodissociation Regions and Planetary Nebulae** – A.G.G.M. TIELENS | 155 |
| **Dust in Planetary Nebulae and in Post-AGB Objects** – M.J. BARLOW | 163 |

On the Distances to Galactic Planetary Nebulae – C.Y. ZHANG .......... 173
A Comparison of Nebular Distance Scales – M. SAMLAND, J. KÖPPEN,
    A. ACKER, B. STENHOLM ......................................... 174
Trigonometric Parallaxes of Planetary Nebulae – J.R. PIER, H.C. HARRIS,
    C.C. DAHN, D.G. MONET ........................................ 175
Infrared Excess (IRE) as an Indicator of PN Distance – G. JACOBY ...... 176
Interstellar Extinction of Planetary Nebulae – G. STASINSKA,
    R. TYLENDA, A. ACKER, B. STENHOLM ......................... 177
Interstellar Reddening Towards S188, HW4 and We1-6 – W. SAURER ... 178
The Shklovsky Paradox – D. BUCKLEY, S.E. SCHNEIDER,
    D. VAN BLERKOM ............................................... 179
Reddening Distances for Planetary Nebulae from Broad Band BVIc
    CCD Imaging – D.L. POLLACCO, G. RAMSAY ..................... 180
The Ionization Structure of Planetary Nebulae – T. BARKER ............ 181
CCD Photometry of NGC 2453 – D.C.V. MALLIK, R. SAGER,
    A.K. PATI ......................................................... 182
Electron Density and Nitrogen Abundances from FIR Lines –
    R.H. RUBIN, S.W.J. COLGAN, E.F. ERICKSON, M.R. HAAS,
    S.D. LORD, J.P. SIMPSON ....................................... 183
Filling Factors and Ionized Masses of Planetary Nebulae – F.R. BOFFI,
    L. STANGHELLINI ............................................... 184
Mean Electron Densities, Distances and Filling Factors for Galactic
    Planetary Nebulae – R.L. KINGSBURGH, M.J. BARLOW .......... 185
The Ionization and Thermal Structure of NGC 2392 and NGC 3242 –
    X.-W. LIU, J. DANZIGER ....................................... 186
Self-Consistent Photoionization Models of Planetary Nebulae Luminescence –
    V.V. GOLOVATY, Yu.F. MALKOV ................................ 187
Temperature Fluctuations in PN – R. GRUENWALD, S.M. VIEGAS ..... 188
Radiation Charge Exchange and Radiation Ion-Atom Recombination as a
    Source of Continual E-M Radiation from Astrophysical Plasma –
    A.A. MIHAJLOV, M.S. DIMITRIJEVIC, A.M. ERMOLAEV ....... 189
Observations of the Bowen Fluorescence Mechanism and Charge Transfer
    in Planetary Nebulae I. – X.-W. LIU, J. DANZIGER ................. 190
Bowen Resonance Fluorescence Lines of OIII in Planetary Nebulae –
    C.R. O'DELL, C.O. MILLER ..................................... 191
Dielectronic Recombination in the Gaseous Nebulae as a Cooling Process –
    A.F. KHOLTYGIN ................................................ 192
High Dispersion Spectra of Bright Planetary Nebulae – S. HYUNG,
    L.H. ALLER ...................................................... 193
Imaging Spectrophotometry of the Ring Nebula – N.J. LAME,
    R.W. POGGE ..................................................... 194
Planetary Nebulae with the Strong [NII] Emission Lines –
    L.N. KONDRATJEVA ............................................ 195
Probable Type I Planetary Nebulae – H. MORENO, A. GUTIERREZ-
    MORENO, G. CORTES ........................................... 196

Extended X-Ray Emission from Planetary Nebulae – H.C. KREYSING,
C. DIESCH, J. ZWEIGLE, R. STAUBERT, M. GREWING ........ 197
Spatial Variations in UV-Optical Lines across the Ring Nebula –
R.J. DUFOUR, R. QUIGLEY ............................................ 198
Collimating Discs and Bipolar Flows in SH 2-71 – L. CUESTA,
J.P. PHILLIPS ............................................................. 199
M4-18: The Low Excitation PN around a WC11 Star –
R. SURENDIRANATH, N. KAMESWARA RAO ................... 200
Extended Nebulae around WC11 Stars: IRAS 17514-1555 –
D.L. POLLACCO ......................................................... 201
Kinematics of the Planetary Nebula Hb 5. A Progress Report – P. PISMIS,
M. MANTEIGA, A. MAMPASO, G. CRUZ-GONZALEZ ............ 202
A Symmetric Jet-Like Structure in the Planetary Nebula FG-1 –
J.A. LÓPEZ, M. TAPIA, M. ROTH .................................. 203
Monochromatic CCD Images of Three Planetary Nebulae –
W.A. FEIBELMAN ....................................................... 204
High-Resolution CCD Imagery of NGC 6537 and NGC 7027 –
S.R. HEAP ................................................................ 205
Optical Imagery of NGC 6302 – J. BOHIGAS .......................... 206
A Twodimensional Ionisation Model of NGC 2440 – M. BÄSSGEN,
C. DIESCH, M. GREWING ............................................ 207
Extended Structures in the Planetary Nebulae He2-111 and He2-119 –
J.A. LÓPEZ, M. TAPIA, M. ROTH .................................. 208
CCD Imaging of Planetary Nebula Halos – K.B. KWITTER,
Y.-H. CHU, R.A. DOWNES ........................................... 209
Near- and Mid-Infrared Imaging of the PN IC 418 – J.L. HORA,
L.K. DEUTSCH, W.F. HOFFMANN, G.G. FAZIO,
K. SHIVANANDAN ..................................................... 210
Mid-IR (8-13$\mu$m) Images of Planetary Nebulae – M. MEIXNER,
J.F. ARENS, J.G. JERNIGAN, J.R. BALL, C.J. SKINNER ........ 211
Imaging of Magellanic Cloud Planetary Nebulae with the Hubble Space
Telescope – M.A. DOPITA, S.J. MEATHERINGHAM, P.R. WOOD,
H.C. FORD, R.C. BOHLIN, T.P. STECHER, S. MARAN,
J.P. HARRINGTON ..................................................... 212
Hubble Space Telescope Images of Four Magellanic Cloud Planetary
Nebulae – M.J. BARLOW, J.C. BLADES, S. OSMER, THE FAINT
OBJECT CAMERA I.D.T. ............................................. 213
H$\alpha$ Morphological Classification of Planetary Nebulae – H.E. SCHWARZ,
R.L.M. CORRADI, L. STANGHELLINI ........................... 214
A PC-Based Quicklook-Program for PN Images – M. BÄSSGEN,
M. BREMER .............................................................. 215
Point Symmetry in Planetary Nebulae – R.L.M. CORRADI,
H.E. SCHWARZ, L. STANGHELLINI ............................... 216
Spectrophotometry and Multicolour Imagery of the Planetary Nebula
around the P Cygni Star AG Carinae – C. ROSSI, A. ALTAMORE,
R.D.D. COSTA, A. DAMINELI NETO, J.A. DE FREITAS PACHECO,

A. CASSATELLA, A.R. MARENZI, P. PERSI,
    V.F. POLCARO, R. VIOTTI ......................................... 217
An Iterative Method for the Reconstruction of Two-Dimensional Density
    Distributions – M. BREMER, M. GREWING ....................... 218
Ring Nebulae Around Population I WR Stars: Is Their Origin Similar
    to the PNe? – T.A. LOZINSKAYA, M.A. DOPITA, Y.-H. CHU ..... 219
Photodissociation Regions in Planetary Nebulae – V. ESCALANTE,
    A. GÓNGORA-T. ..................................................... 220
A Model of the Chemistry in the Neutral Shell of a Planetary Nebula –
    S.N. GOULDSWORTHY, D.R. FLOWER .......................... 221
CO Line Emission in Planetary and Proto-Planetary Nebulae:
    The Molecular Envelope – G. SILVESTRO, I. PORRO .............. 222
CO Interferometric Maps of CIT 6 and CRL 618 – M. MEIXNER,
    W.J. WELCH ........................................................ 223
The Spatio-Kinematic Structure of the CO Envelopes in Evolved Planetary
    Nebulae – R. BACHILLER, P.J. HUGGINS, P. COX, T. FORVEILLE . 224
High Resolution Observations of CO in PNe – K.M. SHIBATA, S. DEGUCHI,
    T. KASUGA, S. TAMURA, N. HIRANO, O. KAMEYA ........... 225
1.6–1.75 and 3.1–3.75 $\mu$m Spectrum of Hb 5 – A. MAGAZZÙ,
    G. STRAZZULLA .................................................... 226
Chemistry in the Molecular Envelope of NGC 7027 – P. COX,
    R. BACHILLER, P.J. HUGGINS, A.OMONT, S. GUILLOTEAU .. 227
Axisymmetric Dust-Shells in Planetary Nebulae – W. HOPFENSITZ,
    M. GREWING ....................................................... 228
Molecular-Line Observations of the Remnant AGB Envelopes Around
    Planetary Nebulae – R. SAHAI, A. WOOTTEN, R.E.S. CLEGG .... 229
NGC 7027: New 7.8 – 20.0 $\mu$m Array Camera and $H\alpha/H\beta$ CCD
    Image Analysis of Dust, PAH and Ionized Gas Distribution –
    D.Y. GEZARI, M.D. THORNLEY, S.R. HEAP, S.N. SHORE,
    F. VAROSI, S.J. MEATHERINGHAM, S.P. MARAN ............. 230
Polarized Line Profiles in Planetary Nebulae – J.R. WALSH,
    R.E.S. CLEGG ...................................................... 231
Millimetre and Submillimetre Continuum Observations of Planetary
    Nebulae – M.G. HOARE, P.F. ROCHE, R.E.S. CLEGG ............ 232

## IV. PLANETARY NEBULAE CONNECTION: EVOLUTION FROM THE AGB ................. 233

**The Third Dredge-Up: Status and Problems –**
    J.C. LATTANZIO .................................................... 235
**Carbon- and Oxygen-Rich Progenitors of Planetary Nebulae –**
    H.J. HABING, J.A.D.L. BLOMMAERT ........................... 243
**Planetary Nebulae from Miras?** – P.A. WHITELOCK,
    M.W. FEAST ........................................................ 251
**Evolution from the AGB: Variability** – D.D. SASSELOV ............ 259
**Proto-Planetary Nebulae** – S. KWOK ................................ 263

**Post-AGB Candidates** – L.B.F.M. WATERS, K.C. SAHU .............. 271
**Planetary Nebulae with Binary Nuclei** – M. LIVIO ................ 279
**Thermal Pulses and Planetary Nebula Ejection** – P.R. WOOD,
  E. VASSILIADIS ..................................................... 291
**Basic Problems of Planetary Nebula Gas Dynamics** –
  J.E. DYSON ......................................................... 299
**Interaction of Planetary Nebulae with the Interstellar Medium** –
  K.J. BORKOWSKI ..................................................... 307
**Spindles, Spheres and a Few Jets: The Radiation Gasdynamics
  of Planetary Nebulae** – A. FRANK .................................. 311
**Dynamical Structures of Planetary Nebulae – Models Against
  Observations** – H. MARTEN, K. GESICKI, R. SZCZERBA ....... 315

Mass-Losing AGB Stars in the Magellanic Clouds – N. REID ............. 319
$10\mu m$ Images of AGB Stars & Supergiants – C.J. SKINNER,
  G. HAWKINS, M.M. MEIXNER, J.G. JERNIGAN, J.F. ARENS .... 320
Mid-IR Spectra of AGB and Post-AGB Stars – C.J. SKINNER,
  M.J. BARLOW, K. JUSTTANONT, R.J. SYLVESTER ............. 321
Comparative Analysis Miras/PPN – D. BARTHES,
  M.O. MENNESSIER, F. GLEIZES, A. LÈBRE ...................... 322
Mechanisms for Radio Continuum Emission of Long-Period Variable Stars –
  G.M. RUDNITSKIJ .................................................... 323
History of the Light Curves and Molecular Maser Emission of the Miras
  U Ori and R Leo – I.L. ANDRONOV, L.S. KUDASHKINA,
  G.M. RUDNITSKIJ .................................................... 323
On the Possible Relationship between the Photometric Parameters
  of the AGB Stars and their Evolutionary Status –
  L.S. KUDASHKINA, I.L. ANDRONOV ............................. 324
Spline Fits: Modelling the Observations – I.L. ANDRONOV .............. 325
Near IR-Photometry of Semiregular Variables – F. KERSCHBAUM,
  J. HRON ............................................................ 326
Space Distribution of Short Period Mira Variables – J. HRON .......... 327
The Galactic Distribution of O-Rich AGB Stars – B.W. JIANG, J.Y. HU . 328
Ice Mantle Formation in the Envelopes of OH/IR Stars –
  S.B. CHARNLEY, R.G. SMITH ..................................... 329
First Results of a Near IR Monitoring Program of OH/IR Stars –
  P. GARCÍA-LARIO, D. ENGELS, A. MANCHADO ................ 330
Planetary and Proto-Planetary Nebulae in the IRAS Two-Colour Diagram –
  P. GARCÍA-LARIO, A. MANCHADO, S.R. POTTASCH .......... 331
A New Evolutionary Interpretation of the IRAS Two-Colour Diagram –
  P. GARCÍA-LARIO, A. MANCHADO, S.R. POTTASCH .......... 332
A Systematic Study of IRAS Selected Proto-Planetary Nebula Candidates –
  J.Y. HU, B.W. JIANG, T. DE JONG, S. SLIJKHUIS .............. 333
FI Lyr: A Candidate Binary System Consisting of Carbon-Rich and
  Oxygen-Rich Companions – J.J. WANG, J.Y. HU, X. ZHOU ...... 334

Near-Infrared Spectroscopy of Proto-Planetary Nebulae –
B.J. HRIVNAK, S. KWOK, T.R. GEBALLE ....................... 335
High Resolution Optical Imaging of Proto-Planetary Nebulae –
P.P. LANGILL, S. KWOK, B.J. HRIVNAK ....................... 336
On the Evolution of Proto-Planetary Nebulae – J.Y. HU, B.W. JIANG,
S. SLIJKHUIS ..................................................... 337
The Molecular Features in the Optical Spectra of the Proto-Planetary
Nebulae – J.Y. HU ................................................ 338
Optical Spectroscopy of Six Carbon-Rich Proto-Planetary Nebulae –
B.J. HRIVNAK .................................................... 339
Near-Infrared Imaging of Proto-Planetary Nebulae – R.E.S. CLEGG,
N.A. WALTON, M.J. BARLOW ................................... 340
Mid-Infrared Spectroscopy of Four 21μm Emission Band Sources –
K. JUSTTANONT, M.J. BARLOW, C.J. SKINNER ............... 341
UKIRT CGS3 Observations of New IRAS 21 Micron Sources – S. KWOK,
B.J. HRIVNAK, T.R. GEBALLE, P.L. LANGILL .................. 342
Mid-IR (8–13μm) Images of Proto-Planetary Nebulae – M. MEIXNER,
J.F. ARENS, J.G. JERNIGAN, C.J. SKINNER, G. HAWKINS ..... 343
The Morphology of MID-Infrared UIR Feature Emission in the PPN
M 2-9 and IRAS 21282+5050 – L.K. DEUTSCH, J.L. HORA,
W.F. HOFFMANN, G.G. FAZIO, K. SHIVANANDAN .............. 344
Visual Extinction and Physical Conditions in the Bipolar Nebula M2-9 –
A. RIERA ......................................................... 345
Complex Motions in Planetary Nebulae – V. ICKE ...................... 346
The Nature of the High Velocity Flow in CRL 618 – R. NERI,
M. GUÉLIN, S. GUILLOTEAU, R. LUCAS, S. GARCIA-BURILLO,
J. CERNICHARO ................................................. 347
High Velocity Outflows in IRAS 17423-1755 – A. RIERA, P. GARCÍA-
LARIO, A. MANCHADO, S.R. POTTASCH ...................... 348
Rotation-Pulsation Coupling in the Bipolar Preplanetary Nebula, V Hya –
M. MORRIS, C. BARNBAUM ..................................... 349
Influence of the Stellar Winds on the Post-AGB Evolution – R. SZCZERBA   350
Axially Symmetric Dynamics of PNe – L. WANG ....................... 351
The Chemical Composition of Post AGB Stars – M. PARTHASARATHY,
P. GARCÍA-LARIO, S.R. POTTASCH ............................ 352
UBVRI Polarization Measurements of Post AGB Stars –
M. PARTHASARATHY, S.K. JAIN ................................ 353
High-Resolution Radial Velocity and Hα Study of Proto-Planetary Nebulae –
B.J. HRIVNAK, A.W. WOODSWORTH ........................... 354
Hα Profiles of Selected Candidates for Proto-Planetary Nebulae –
S. TAMURA ....................................................... 355
LSIV -12° 111 – A Newly Emerging Halo Planetary Nebula – E.S. CONLON,
P.L. DUFTON, F.P. KEENAN, R.J.H. McCAUSLAND ............ 356
A Very Rapid-Evolving Young Planetary Nebula – A. MANCHADO,
P. GARCÍA-LARIO, K.C. SAHU, S.R. POTTASCH ............... 357

Search for the Young Planetary Nebulae. Preliminary Results –
L.N. KONDRATJEVA ............................................. 358
About the Suspected Very Young PN IRAS 17516-2525 – H.U. KÄUFL,
L. STANGHELLINI ............................................... 359
A Spectroscopic Search for Hot (B-Type) Post-AGB Stars –
E.S. CONLON .................................................... 360
New Calculations of Thermal Pulses and s-Process Nucleosynthesis in
AGB Stars – M. BUSSO, A. CHIEFFI, R. GALLINO,
M. LIMONGI, C.M. RAITERI, O. STRANIERO .................. 361
Evolutionary Properties of Post-AGB and Post-EAGB Stars –
M. LIMONGI, A. TORNAMBE, M. CASTELLANI ............... 362
Evolution of a Dust Shell along a Stellar Post-AGB Track –
H. MARTEN, R. SZCZERBA, T. BLÖCKER ...................... 363
Dust Driven Mass Loss from Pulsating AGB-Stars – E.A. DORFI,
M.U. FEUCHTINGER, S. HÖFNER .............................. 364
Is There a Connection Between Thermal Pulses and PNe Halos: an Approach
to an Answer – A. FRANK, B. BALICK, W. VAN DER VEEN ..... 365
Linear Pulsation Periods of the Post-AGB Stars – M. TAKEUTI,
R. TAKANO, S. TAMURA ........................................ 366
Effects of New Opacity on the Post-AGB Evolution – M. KATO,
I. HACHISU ..................................................... 367
Angular Momentum Loss in Post-Main Sequence Stellar Evolution
through the PN Stage – M. VILLATA .......................... 368
Modelling PN Formation from Hydrodynamics and Radiation –
G. MELLEMA ................................................... 369
Evolution of Planetary Nebulae Envelopes: an Empirical Approach –
V.V. GOLOVATY, Yu.F. MALKOV .............................. 370
Numerical Study of the Shaping of Planetary Nebulae –
I.V. IGUMENSHCHEV ........................................... 371
Deprojection of Planetary Nebula Images – K. VOLK, D.A. LEAHY ..... 372
Twodimensional Axialsymmetrical Hydrodynamical Simulations of
PN-Evolution – J. ZWEIGLE, M. BREMER, M. GREWING ........ 373
Spherically Symmetric Kinematic Modelling of Planetary Nebulae –
C. DIESCH, M. GREWING ...................................... 374
Shock Modelling and High Resolution Spectroscopy of NGC 6905 –
L. CUESTA, J.P. PHILLIPS ..................................... 375
A Modelling of Expansion Velocities of Planetary Nebulae –
K. GESICKI, R. SZCZERBA ..................................... 376
Echelle Measurements of the Expansion Velocities of the Faint Giant Haloes
of Planetary Nebulae – M. BRYCE, J. MEABURN, J.R. WALSH .... 377
Kinematical Studies of Planetary Nebulae Using Taurus+CCD –
K.C. SAHU, J.R. WALSH, N.A. WALTON, S.R. POTTASCH ...... 378
High-Dispersion Spectroscopy of IC 351 and NGC 3242, Planetaries
with High Internal Motion – Y. YADOUMARU, S. TAMURA ........ 379
Interaction of Planetary Nebulae with Prenebulae Debris – J. FIERRO ... 380

The Magnetic Fields in the Envelopes of Proto-Planetary Nebulae –
   J.Y. HU ........................................................ 381
On Bipolar Jet Formation in Planetary Nebulae – G. PASCOLI ........... 382
Stripping of a Planetary Nebula from the Globular Cluster M22 –
   K.J. BORKOWSKI, J.P. HARRINGTON, Z. TSVETANOV ......... 383
A Detailed Study of the Galactic Planetary Nebula G 258-15.7 –
   P. LEISY, M. DENNEFELD ........................................ 384
Shock Modelling of the Bipolar Outflow Source NGC 6537 –
   L. CUESTA, J.P. PHILLIPS ........................................ 385
The Dust in the Hydrogen-Poor Ejecta of Abell 30 – J.P. HARRINGTON,
   K.J. BORKOWSKI, W.P. BLAIR, J. BREGMAN ................... 386
The Central Region of the Planetary Nebula A58 – D.L. POLLACCO,
   P.W. HILL, R.E.S. CLEGG ........................................ 387
Morphology & Kinematics of the 'Born-Again' Planetary Abell 78 –
   R.E.S. CLEGG, M.N. DEVANEY, A.P. DOEL, C.N. DUNLOP,
   J.V. MAJOR, R.M. MYERS, R.M. SHARPLES ..................... 388
The Formation of Single and Binary Nuclei of Planetary Nebulae –
   L.R. YUNGELSON, A.V. TUTUKOV ............................... 389
A Spectroscopic Study of Binary Star Planetary Nebulae – J.R. WALSH,
   N.A. WALTON, S.R. POTTASCH ................................. 390
The Peculiar Light Variation of the Planetary Nebula NGC 2346 –
   X.-L. HAO ...................................................... 391
New Eclipsing Phenomena of the Central Star in NGC 2346 –
   R. COSTERO, M. PEÑA, W.J. SCHUSTER, M. TAPIA,
   J. ECHEVARRIA, J. FIERRO ...................................... 392
Observational Studies of Close Binary Central Stars of Planetary Nebulae:
   HFG 1 and A 63 – H.L. MALASAN, A. YAMASAKI ................ 393
Imaging and Spectroscopy of Abell 63 (UU Sge) – N.A. WALTON,
   J.R. WALSH, S.R. POTTASCH .................................... 394
New Light on UU Sagittae – S.A. BELL, D.L. POLLACCO .............. 395
Precataclysmic Binaries in the Centre of Planetary Nebulae –
   G. JASNIEWICZ, A. ACKER ...................................... 396
The Abell 35-Type Planetary Nuclei – H.E. BOND, R. CIARDULLO,
   M.G. MEAKES .................................................. 397
The IUE Ultraviolet Spectrum of PC 11 – M. PARTHASARATHY,
   S.R. POTTASCH, J. CLAVEL ..................................... 398
On the Photometric Behaviour of the Central Star of the Planetary
   Nebula Sh2-71 – J. JURCSIK ..................................... 399
On Some Links Between Symbiotic Stars and Planetary Nebulae –
   L. LEEDJÄRV ................................................... 400
Is There Any Connection Between Planetary Nebulae and Symbiotic
   Stars? – M. FRIEDJUNG .......................................... 401
Elemental Abundances in Symbiotic Stars – H.M. SCHMID,
   H. NUSSBAUMER ............................................... 402
Diagnostic Diagrams for Planetary Nebulae and Symbiotic Stars –
   A. GUTIERREZ-MORENO, H. MORENO, G. CORTES ............ 403

On the Dereddening of Symbiotic Stars – D. RAYKOVA,
B. RAYTCHEV .................................................... 404
ROSAT Observations of Symbiotic Stars – K.F. BICKERT,
R.E. STENCEL, R. LUTHARDT ................................. 405
The Active Phase of the Hot Component of Z Andromedae –
T. FERNANDEZ-CASTRO, R. GONZALEZ-RIESTRA,
A. CASSATELLA, A.R. TAYLOR, E.R. SEAQUIST ............ 406
BZ Camelopardalis = 0623+71: The Cataclysmic Variable Inside a
Bow-Shock Nebula – N.M. SHAKHOVSKOY, Y.S. EFIMOV,
I.L. ANDRONOV, S.V. KOLESNIKOV ........................... 407
Theoretical Light Curves of Recurrent Novae – M. KATO ........ 408
The Environs of Supernova Precursors – O.A. TSIOPA ........... 409
Circumstellar Nebular Lines in the Optical Spectrum of SN 1987A –
I. KHAN, H.W. DUERBECK .................................... 410
SN 1987A Deconvolved by MIM – H. GRATL, J. PFLEIDERER ...... 411

## V. PLANETARY NEBULAE CONNECTION: EVOLUTION TO WHITE DWARFS .............. 413

**Evolutionary Tracks** – D. SCHÖNBERNER ......................... 415
**Diagrams for Observational Testing of Evolution of
Planetary Nebula Nuclei** – R. TYLENDA ....................... 423
**The Evolution of the Planetary Nebulae in the Magallanic
Clouds and the Galactic Bulge** – M.A. DOPITA ................ 433
**White Dwarf Central Stars** – J.W. LIEBERT ..................... 443
**On the Relation of Core Mass with Chemical Composition in PN** –
S.R. POTTASCH ................................................. 449
**Simulations of a Population of Planetary Nebulae** –
G. STASINSKA, R. TYLENDA .................................. 461
**Hydrogen and Helium Burning Evolutionary Tracks** –
P.R. WOOD, E. VASSILIADIS ................................... 465
**Further Models of Planetary Nebula Spectral Evolution** –
K. VOLK ........................................................ 469
**Synthetic P-AGB Evolution** – L. STANGHELLINI, A. RENZINI ..... 473

Further Models of Planetary Nebula Spectral Evolution – K. VOLK ....... 477
Evolution of 1-5 $M_\odot$ Stars with Mass Loss – E. VASSILIADIS,
P.R. WOOD ..................................................... 478
On the Fading of AGB Remnants – T. BLÖCKER, D. SCHÖNBERNER . 479
Planetary Nebula Evolution Traced by Distance-Independent Parameters –
C.Y. ZHANG, S. KWOK ........................................ 480
Excitation Class of Nebulae as an Evolution Criterion –
G.A. GURZADYAN, A.G. EGIKYAN ............................ 481
Morphology and Evolution of Planetary Nebulae – L. STANGHELLINI,
R.L.M. CORRADI, H.E. SCHWARZ ............................ 482
Influence of the Stellar Winds on the Evolution of the Planetary
Nebula Nuclei – S.K. GÓRNY .................................. 483

Detection of Evolution of the Nucleus of NGC 2392 – S.R. HEAP ......... 484
A Search for Optical-UV Fading of Central Stars – B. ALTNER,
S.R. HEAP ...................................................................... 485
The Central Stars of He 2-131 and He 2-138: Photometric Variations –
R.G. HUTTON, R.H. MÉNDEZ ........................................ 486
Time-Resolved CCD-Photometry of Planetary Nebula Nuclei –
M.M. ROTH, T. SOFFNER, W. MITSCH ........................... 487
Variable Spectra of IC 4997 and NGC 6572 – S. HYUNG,
L.H. ALLER, W.A. FEIBELMAN ...................................... 488
A Search for Pulsations in O VI Planetary Nuclei – H.E. BOND,
R. CIARDULLO ............................................................. 489
Global Photometric Campaigns on Pulsating Planetary Nuclei –
R. CIARDULLO, H.E. BOND ........................................... 490
Photoelectric Photometry of Five PNNi – R. SILVOTTI,
C. BARTOLINI, F.R. BOFFI, G. COSENTINO, A. GUARNIERI,
A. PICCIONI, L. STANGHELLINI ..................................... 491
O VI Central Stars of Planetary Nebulae: NGC 2371 –
L. STANGHELLINI, J.B. KALER, R.A. SHAW .................... 492
Precision Asteroseismology of Pre-White Dwarfs and PN Central Stars –
S.D. KAWALER, P.A. BRADLEY ...................................... 493
A Spectacular Mass-Loss Event of the Central Star of Longmore 4 –
K. WERNER, W.-R. HAMANN, U. HEBER, R. NAPIWOTZKI,
T. RAUCH, U. WESSOLOWSKI ........................................ 494
Discovery of a Planetary Nebula Associated with the White Dwarf GD 561 –
R. NAPIWOTZKI, D. SCHÖNBERNER ............................... 495
A New PG1159-Type Central Star Discovered in the ROSAT XRT
All Sky Survey: Non-LTE Analysis of X-Ray and Optical Spectra –
K. WERNER, C. MOTCH, M. PAKULL .............................. 496
ROSAT Studies of the Composition and Structure of DA White Dwarf
Atmospheres – C.J. DIAMOND, M.A. BARSTOW, A.E. SANSOM,
M.C. MARSH, S.R. ROSEN, T.A. FLEMING, D. KOESTER,
D.S. FINLEY, J.B. HOLBERG, K. KIDDER ....................... 497
HST FOS Observations of KPD0005+5106: A Subluminous WN-WC
Descendant with Ongoing Mass Outflow? – E.M. SION,
R.A. DOWNES ............................................................... 498
HST Observations of the Nuclei of EGB 6 (0950+139) and Abell 58
(V605 Aql) – H.E. BOND, M.G. MEAKES, J.W. LIEBERT,
A. RENZINI .................................................................. 499
The Low Luminosity Central Star of the PN ESO166-21 – M.T. RUIZ,
M. PEÑA, S. TORRES-PEIMBERT ................................... 500

**VI. PLANETARY NEBULAE IN GALACTIC SYSTEMS** .... 501

**Luminosity Functions of Planetary Nebulae** – G. JACOBY,
R. CIARDULLO .............................................................. 503

Why are Planetary Nebulae Poor Distance Indicators? –
G.A. TAMMANN .................................................... 515
Planetary Nebula Birth Rates in the Galaxy and Other Galaxies –
M. PEIMBERT ...................................................... 523
Planetary Nebulae and Halo Dynamics in Early Type Galaxies –
X. HUI, H.C. FORD ................................................. 533
Dynamics of AGB Stars and Planetary Nebulae in the Galaxy –
H. DEJONGHE ...................................................... 541
PN Abundances in Different Galactic Systems – R.E.S. CLEGG ... 549
How to Model the Chemical Evolution of Galaxies –
J. KÖPPEN ......................................................... 557

Distribution of Planetary Nebulae Perpendicular to the Disk –
D.C.V. MALLIK, S. CHATTERJEE ................................... 567
Kinematics of Disk Planetary Nebulae – W.J. MACIEL, C.M. DUTRA ... 568
Spectrophotometry and Kinematics of the Newly Discovered PN in the
Outer Field of the LMC – M.A. DOPITA, E. VASSILIADIS,
D.H. MORGAN ...................................................... 569
The Radial Velocities of Planetary Nebulae in NGC 3379 –
R. CIARDULLO, G. JACOBY .......................................... 570
UV and Optical Abundances for a Sample of Southern Galactic
Planetary Nebulae – R.L. KINGSBURGH, M.J. BARLOW ............ 571
Chemical Enrichment and Central Star Properties – C.Y. ZHANG ........ 572
Determination of Element Abundances in Planetary Nebulae from
Recombination Line Spectra – A.A. NIKITIN, A.F. KHOLTYGIN,
A.A. SAPAR, T.KH. FEKLISTOVA .................................. 573
Clumps in the Planetary Nebulae – A.F. KHOLTYGIN,
T.KH. FEKLISTOVA ................................................. 574
The Chemical Features of Galactic Planetary Nebulae – P.R. AMNUEL ... 575
O, S, Ar from Planetary Nebulae Data and the Chemical Evolution of the
Galactic Disk – J.A. DE FREITAS PACHECO ....................... 576
Evolution of Radial Abundances Gradients from Planetary Nebulae –
W.J. MACIEL, J. KÖPPEN ........................................... 577
Dependance of the Metallicity of Planetary Nebulae with the Galactic
Height Above the Disk – F. CUISINIER, A. ACKER, J. KÖPPEN ... 578
Chemical Composition of Planetary Nebulae: A New Determination –
V.V. GOLOVATY, Yu.F. MALKOV .................................. 579
Galactic Ba Enrichment from TP-AGB Stars – C.M. RAITERI,
M. BUSSO, F. MATTEUCCI, R. GALLINO ......................... 580
Chemical Abundances in Galactic Bulge PN – N.A. WALTON,
M.J. BARLOW, R.E.S. CLEGG ...................................... 581
A Reanalysis of C/O Ratios in Planetary Nebulae – C. ROLA,
G. STASINSKA ..................................................... 582
The Helium-to-Metals Enrichment Ratio in Planetary Nebulae –
C.M.L. CHIAPPINI, W.J. MACIEL ................................. 583

Spectrophotometry of Selected Planetary Nebulae of Type I in the
    Magellanic Clouds – S. TORRES-PEIMBERT, M. PEIMBERT,
    M.T. RUITZ, M. PEÑA ............................................. 584
Synthetic AGB Evolution in the LMC: The Abundances of PN –
    M. GROENEWEGEN, T. DE JONG ............................... 585
Chemical Abundances of Planetary Nebulae in the LMC –
    J.A. DE FREITAS PACHECO, R.D.D. COSTA ..................... 586

**The Evolution of Planetary Nebulae, Their Precursors and**
    **Their Progeny – A Commentary** – I. IBEN, JR. ................. 587

Author Index ........................................................... 597
Object Index ........................................................... 601

# LIST OF PARTICIPANTS

| | |
|---|---|
| ACKER, Agnes | Observatoire de Strasbourg, 11, rue de l'Universite', F-67000 Strasbourg, FRANCE |
| ALLER, Lawrence H. | Department of Astronomy, University of California, Los Angeles, CA 90024-1562, USA |
| ALTNER, Bruce | Applied Research Corporation, 8201 Corporate Drive, Suite 1120, Landover, MD 20785, USA |
| AMNUEL, P.R. | Department of Physics and Astronomy, Tel-Aviv University, Ramat-Aviv, 69978 Tel-Aviv, ISRAEL |
| ANDRONOV, Ivan L. | Astronomical Observatory of the Odessa State University, T.G. Shevchenko Park, Odessa 270014, UKRAINIA |
| ARINGER, Bernhard | Institut für Astronomie, Türkenschanzstr. 17, A-1180 Wien, AUSTRIA |
| ASHLEY, Michael | School of Physics, University of New South Wales, Kensington P.O. Box 1 2033, AUSTRALIA |
| BACHILLER, Rafael | Centro Astronomico, de Yebes (OAN-IGN), Apartado 148, E-19080 Guadalajara, SPAIN |
| BALICK, Bruce | Astronomy Department, FM-20, University of Washington, Seattle, Washington 98195, USA |
| BARKER, Timothy | Department of Physics and Astronomy, Wheaton College, Norton, MA 02766, USA |
| BARLOW, Michael J. | Department of Physics and Astronomy, University College London, Gower Street, London WCIE 6BT, UK |
| BÄSSGEN, Martin | Astronomisches Institut der Universität, Waldhäuserstraße 64, D-7400 Tübingen, GERMANY |
| BÄSSGEN, Gabriele | Astronomisches Institut der Universität, Waldhäuserstraße 64, D-7400 Tübingen, GERMANY |
| BECKER, Sylvia | Institut für Astronomie und Astrophysik der Universität, D-8000 München 80, GERMANY |
| BELL, Steve | Department of Physics and Astronomy, North Haugh, St. Andrews, Fife, KY16 9SS, UK |
| BENGER, Werner | Institut für Astronomie, Universität Innsbruck, Technikerstr. 25, A-6020 Innsbruck, AUSTRIA |
| BIANCHI, Luciana | Osservatorio Astronomico, Strada Osservatorio 20, I-10025 Pino Torinese, ITALY |
| BICKERT, Klaus | MPI für Extraterrestrische Physik, Giessenbachstraße, D-8046 Garching, GERMANY |
| BLÖCKER, Thomas | Institut für Theoretische Physik und Sternwarte der Universität Kiel, Olshausenstr. 40, D-2300 Kiel, GERMANY |
| BOFFI, Francesca R. | Department of Physics and Astronomy, 440 West Brooks, Norman, OK 73019, USA |

BOND, Howard E.    Space Telescope, Science Institute, Homewood Campus, 3700 San Martin Dr., Baltimore, MD 21218, USA

BOHIGAS, Joaquin    Observatorio Astronomico, P.O. Box 439027, San Diego CA 92143-9027, USA

BORKOWSKI, K. J.    Department of Astronomy, University of Maryland, College Park Campus, College Park, Maryland 20742, USA

BRATSCHITSCH, R.    Institut für Astronomie, Universität Innsbruck, Technikerstr. 25, A-6020 Innsbruck, AUSTRIA

BREMER, Michael    Astronomisches Institut der Universität, Waldhäuserstraße 64, D-7400 Tübingen, GERMANY

BRUHWEILER, F.    Astrophysics Program, Department of Physics, Catholic University of America, Washington, DC 20064, USA

BRYCE, Myfanwy    University of Manchester, Department of Astronomy, Manchester M13 9PL, UK

BUCKLEY, David    Department of Physics, East Stroudsburg University, East Stroudsburg, PA 18301-2999, USA

BUTLER, Keith    Institut für Astronomie und Astrophysik der Universität, D-8000 München 80, GERMANY

CALOI, Vittoria    Istituto di Astrofisica, Spaziale, C.P. 67, 00044 Frascati, Roma, ITALY

CASSATELLA, Angelo    Istituto di Astrofisica, Spaziale, C.P. 67, 00044 Frascati, Roma, ITALY

CHARNLEY, Steven    Space Science Division, Theoretical Studies Branch, NASA Ames Research Center, Moffet Field, CA 94035, USA

CHU, You-Hua    Astronomy Department, University of Illinois, 1002 W. Green St, Urbana, IL 61801, USA

CIARDULLO, Robin    Dept. of Astronomy and Astrophysics, Pennsylvania State University, 525 Davey Laboratory, University Park, PA 16802, USA

CLEGG, Robin E.S.    Royal Greenwich Observatory, Madingley Road, Cambridge, CB3 0EZ, UK

CONLON, Elizabeth    Department of Pure and Applied Physics, University of Belfast, Belfast BT7LNN, UK

CORRADI, Romano L.M.    ESO, Cassilla 19001, Santiago 19, CHILE

COSTERO, Rafael    P.O. Box 439027, San Diego, CA 92143, USA

COX, Pierre    MPI für Radioastronomie, Auf dem Hügel 69, D-5300 Bonn, GERMANY

CUESTA, Luis    Instituto de Astrofisica, de Canarias, Via Lactea, S/N, La Laguna, Tenerife, E-38200, SPAIN

CUISINIER, Francois    Observatoire de Strasbourg, 11, rue de L'Universite, F-67000 Strasbourg, FRANCE

| | |
|---|---|
| DE FREITAS PACHECO, J.A. | Instituto Astronomico e Geofisico USP, Aven. Miguel Stefano 4200-04301, CP 9638, Sao Paolo, BRAZIL |
| DEJONGHE, Herwig | Rijksuniversiteit te Gent, Sterrenkundig Observatorium, Krijgslaan 281, B-9000 Gent, BELGIUM |
| DELUEG, Klaus | Institut für Astronomie, Universität Innsbruck, Technikerstr. 25, A-6020 Innsbruck, AUSTRIA |
| DENNEFELD, Michel | Institut d'Astrophysique, 98 bis, Bvd. Arago, F-75014 Paris, FRANCE |
| DEUTSCH, L.K. | NASA/Ames Research Ctr., MS 245-6, Moffet Field, CA 94035, USA |
| DIAMOND, C. J. | Space Research Group, Schools of Physics and Space Research, University Birmingham, Birmingham/Edgbaston B15 2TT, UK |
| DIESCH, C. | Astronomisches Institut der Universität Tübingen, Waldhäuserstraße 64, D-7400 Tübingen, GERMANY |
| DOPITA, Michael A. | Mount Stromlo and Siding Spring Observatories, Australian National University, Private Bag, Weston Creek, P.O. ACT 2611, AUSTRALIA |
| DORFI, Ernst | Institut für Astronomie, Türkenschanzstr. 17, A-1180 Wien, AUSTRIA |
| DREIZLER, Stefan | Altenburger Str. 10, D-8600 Bamberg, GERMANY |
| DUFOUR, Reginald | Department of Space Physics and Astronomy, Rice University, P.O. Box 1892, Houston, TX 77251, USA |
| DYSON, John E. | Department of Astronomy, The University, Manchester M13 9PL, UK |
| ESCALANTE, Vladimir | Instituto de Astronomia, UNAM, Apdo. Postal 70-264, Mexico, DF 04510, MEXICO |
| FEAST, Michael W. | Institute of Astronomy, University of Cambridge, Madingley Road, Cambridge CB3 0HA, England, UK |
| FEKLISTOVA, T. Kh. | W. Struve Astrophysical Observ., Estonian Academy of Sciences, Tartu, Toravere 202444, ESTONIA |
| FELDMEIER, Achim | Schreiner Str. 1, D-8000 München 80, GERMANY |
| FERLAND, G.J. | Cerro Tololo Inter-American Observatory, Casilla 603, La Serena, CHILE |
| FERNANDEZ-CASTRO, Telmo | Planetario de Madrid, Parque Tierno Galvan, 28045 Madrid, SPAIN |
| FERRARI-D'OCCHIEPPO, K. | Innstraße 17, A-6020 Innsbruck, AUSTRIA |
| FIERRO, Julieta | Instituto de Astronomia, UNAM, Apdo. Postal 70-264, 04510 Mexico D.F., MEXICO |
| FRANK, Adam | Astronomy Department, FM-20, University of Washington, Seattle, WA 98195, USA |
| FRIEDJUNG, Michael | Institut d'Astrophysique, 98 bis, Boulevard Arago, F-75014 Paris, FRANCE |

| | |
|---|---|
| GABLER, Rudolf | Institut für Astronomie und Astrophysik der Universität, Scheinerstraße 1, D-8000 München 80, GERMANY |
| GAIGALAS, Gediminas | Institute of Theoretical Physics and Astronomy, A. Gostauto 12, Vilnius, 232 600, LITHUANIA |
| GARCIA-LARIO, Pedro | Laboratorio de Astrofisica, Espacial y Fisica Fundamental, LAEFF, Estacion de Villa franca del Castillo, Aptdo. de Correos 50727, SPAIN |
| GESICKI, Krysttof | Copernicus Astronomical, Center, Chopina 12/18, 87-100 Torun, POLAND |
| GOLOVATYJ, V. V. | Astronomical Observatory of L'vov State University, Ul Lomonosova 8, L'vov, UKRAINIA |
| GONZALEZ-RIESTRA, R. | ESA IUE Observatory, P.O. Box 50727, 28080 Madrid, SPAIN |
| GORNY, Slawomir | Copernicus Astronomical Center, Chopina 12/18, 87-100 Torun, POLAND |
| GOULDSWORTHY, S. | Physics Department, University of Durham, South Road, Durham DH1 3LE, England, UK |
| GREWING, Michael | IRAM, F-38405 St. Martin d' Heres, FRANCE and Astronomisches Institut der Universität, Waldhäuserstraße 64, D-7400 Tübingen, GERMANY |
| GROENEWEGEN, M. | Astronomical Institute, University of Amsterdam, Kruislaan 403, 1098 SJ Amsterdam, THE NETHERLANDS |
| GRUENWALD, Ruth B. | Instituto Astronomico e Geofisico, Universidade de Sao Paulo, Avenida Miguel Stefano 4200, BR 04301 Sao Paulo (SP), BRAZIL |
| GUERRERO, M. A. | Instituto de Astrofisica de Canarias, Tenerife, 38200-La Laguna, SPAIN |
| GURZADYAN, G.A. | 375002 Erevan, Str. Moscovian 33, Flat 32, Erevan, ARMENIA |
| GUTIERREZ-MORENO, A. | Dept. de Astronomia, Universidad de Chile, Casilla 36 - D, Santiago, CHILE |
| HABING, Harm J. | Sterrewacht Leiden, P.O. Box 9513, NL-2300 Ra Leiden, THE NETHERLANDS |
| HACHISU, Izumi | Dept. of Earth Sciences and Astronomy, College of Arts and Science, The University of Tokyo, Komaba 3-8-1, Meguro-ku, JAPAN |
| HAMANN, Wolf-Rainer | Institut für Theoretische Physik und Sternwarte der Universität, Olshausenstr., D-2300 Kiel, GERMANY |
| HANDLER, Gerald | Institut für Astronomie, Türkenschanzstr. 17, A-1180 Wien, AUSTRIA |
| HARRINGTON, J. P. | Astronomy Program, University of Maryland, College Park, MD 20742, USA |

| | |
|---|---|
| HARTL, Herbert | Institut für Astronomie, Universität Innsbruck, Technikerstr. 25, A-6020 Innsbruck, AUSTRIA |
| HARVEY, Paul M. | Observatoire de Grenoble, CERMO, BP 53 X, 38041 Grenoble Cedex, FRANCE |
| HEAP, Sara R. | NASA / GSFC, Laboratory for Astronomy and Solar Physics, Greenbelt, Maryland 20771, USA |
| HEBER, Ulrich | Institut für Theoretische Physik und Sternwarte der Universität Kiel, Olshausenstrasse 40-60, D-2300 Kiel, GERMANY |
| HERRERO, Artemio | Institut für Astronomie und Astrophysik, Universitätssternwarte, Scheinerstr.1, D-8000 München 80, GERMANY |
| HILL, Phil | Department of Physics and Astronomy, North Haugh, St. Andrews, Fife, KY16 9SS, UK |
| HOPFENSITZ, Wolfgang | Institut für Astronomie, Waldhäuserstr. 64, D-7400 Tübingen, GERMANY |
| HORA, Joseph | Institute of Astronomy, 2680 Woodlawn Drive, Honolulu, HI 96822-1839, USA |
| HOUZIAUX, Leo | 15 Avenue des Tilleuls, B-4000 Liege, BELGIUM |
| HRIVNAK, Bruce J. | Dept. of Physics and Astronomy, Valparaiso University, Valparaiso, IN 46383, USA |
| HRON, Josef | Institut für Astronomie, Türkenschanzstr. 17, A-1180 Wien, AUSTRIA |
| HU, Jing-Yao | Department of Physics and Astronomy, University of Calgary, 2500 University Drive N.W., Calgary, Alberta, T2N 1N4, CANADA |
| HUGGINS, Patrick J. | Physics Department, New York University, 4 Washington Place, New York 10003, USA |
| HUI, Xiaohui | Astronomy Department, 105 - 24, Caltech, Pasadena, CA 91125, USA |
| IBEN, JR., Icko | University of Illinois, Department of Astronomy, Astronomy Building, 1002 West Springfield Avenue, Urbana, IL 61801, USA |
| ICKE, Vincent | Sterrewacht Leiden, Postbus 9513, 2300 RA Leiden, THE NETHERLANDS |
| IGUMENSHCHEV, I. V. | Academy of Sciences, Astronomical Council, Ul. Pyatnitskaya 48, SU-109017 Moscow, RUSSIA |
| IIJIMA, Takashi | Universita' di Padova, Osservatorio Astrofisico, (Asiago Astrop. Observatory), I-36012 Asiago, ITALY |
| JACOBY, George H. | Kitt Peak Nat. Observatory, Nat. Optical Astron. Observ., P.O. Box 26732, Tucson, AZ 85726-6732, USA |
| JURCSIK, Johanna | Konkoly Observatory, Hungarian Academy of Sciences, P.O. Box 67, H-1525 Budapest, HUNGARY |
| JUSTTANONT, Kay | Department of Physics and Astronomy, University College London, Gower street, London, WC1E 6BT, UK |

| | |
|---|---|
| KATO, Mariko | Department of Astronomy, Keio University, Hiyoshi 4-4-1, Kouhoku-ku, Yokohama 223, JAPAN |
| KAWALER, Steven D. | Dept. of Physics and Astron., Iowa State University, Ames, IA 50011, USA |
| KERSCHBAUM, Franz | Institut für Astronomie, Türkenschanzstr. 17, A-1180 Wien, AUSTRIA |
| KHAN, I. | Fachbereich Physik F11.03, Universität Wuppertal, Gaußstr.20, D-W5600 Wuppertal 1, GERMANY |
| KHOLTYGIN, A. F. | St. Petersburg University, Astronomical Observatory, Bibliotechnaya Pl.2, 198904, St. Petergof, RUSSIA |
| KIMESWENGER, Stefan | Institut für Astronomie, Universität Innsbruck, Technikerstr. 25, A-6020 Innsbruck, AUSTRIA |
| KINGSBURGH, Robin | Department of Physics and Astronomy, University College London, Gower Street, London, WC1E 6BT, UK |
| KOHOUTEK, Lubos | Hamburger Sternwarte, Gojenbergsweg 112, D-2050 Hamburg, GERMANY |
| KONDRAT'EVA, L. N. | Astrophysical Institute, Kazakh Academy of Sciences, Kamenskoye Plato, Alma Ata, KAZAKH |
| KÖPPEN, Joachim | Institut für Theoretische Physik und Sternwarte, Universität Kiel, Olshausenstr. 40-60, D-2300 Kiel, GERMANY |
| KREYSING, H.-C. | Universität Tübingen, Astronomisches Institut, Waldhäuserstr. 64, D-7400 Tübingen, GERMANY |
| KUCZAWSKA, Ewa | Mt. Suhora Observatory, Pedagogical University, ul. Podchorazych 2, 30082 Cracow, POLAND |
| KUDASHKINA, Larisa S. | Astronomical Observatory of the Odessa State University, T.G. Sherchenko Park, 270014 Odessa, UKRAINIA |
| KUDRITZKI, Rolf-Peter | Institut für Astronomie und Astrophysik der Universität München, Scheinerstrasse 1, D-8000 München 80, GERMANY |
| KWITTER, Karen B. | Thomson Physics Laboratory, Williams College, Williamstown, Massachussetts 01267, USA |
| KWOK, Sun | Department of Physics and Astronomy, The University of Calgary, Alberta, Canada T2N 1N4, CANADA |
| LAME, Nancy Jo | The Ohio State University, Department of Astronomy, 174 W. 18th Ave., Columbus OH 43210, USA |
| LANGILL, Philip P. | Department of Physics and Astronomy, The University of Calgary, Calgary Alberta, CA T2N 1N4, CANADA |
| LATTANZIO, John | Monash University, Clayton, Victoria 3168, AUSTRALIA |
| LEEDJÄRV, Laurits | W.Struve Astrophysical Observ., Estonian Academy of Sciences, Tartu, Toravere 202444, ESTONIA |
| LEISY, Pierre | 17 Avenue d'Alembert, F-92160 Antony, FRANCE |

| | |
|---|---|
| LERCHER, Georg | Institut für Astronomie, Universität Innsbruck, Technikerstr. 25, A-6020 Innsbruck, AUSTRIA |
| LEWIS, Brian | 120 L St. Ramey, Aquadilla, PRCC604, PUERTO RICO |
| LIEBERT, James | Steward Observatory, University of Arizona, Tucson, Arizona 85721, USA |
| LIU, Xiaowei | European Southern Observatory, Karl-Schwarzschild-Str. 2, D-8046 Garching, GERMANY |
| LIVIO, Mario | Space Telescope Science Institute, 3700 San Martin Drive, Baltimore, MD 21218, USA |
| LOPEZ, J.A. | P.O. Box 439027, San Diego, CA 92143-9027, USA |
| LOZINSKAYA, T.A. | Sternberg Astronomical Institute, Universitetskij Prospekt 13, SU-119899, Moscow, RUSSIA |
| LUTZ, Julie | Washington State University, Department of Pure and Applied Mathematics, Pullman, WA 99164-3113, USA |
| M.L.CHIAPPINI, Cristina | Instituto Astronomico e Geofisico, Universidade de Sao Paulo, Departamento de Astronomia, C.P. 9638 Sao Paulo 01065, BRAZIL |
| MACIEL, Walter J. | Instituto Astronomico e Geofisico da USP, Departamento de Astronomia, Av. Miguel Stefano 4200, 04301 Sao Paulo, BRAZIL |
| MALASAN, H. L. | Institute of Astronomy, Faculty of Science, University of Tokyo, Mitaka, Tokyo 181, JAPAN |
| MALLIK, Dipankar C.V. | Indian Institute, of Astrophysics, Koramangala, Bangalore 560 034, INDIA |
| MANCHADO, Arturo | Instituto de Astrofisica de Canarias, 38200 La Laguna, Tenerife, Islas Canarias, SPAIN |
| MARTEN, Holger | Institut für Theoretische Physik und Sternwarte der Universität Kiel, Olshausenstr. 40, D-2300 Kiel, GERMANY |
| MEIXNER, Margret | Department of Astronomy, University of California, Berkley, CA 94720, USA |
| MELLEMA, Garrelt | Sterrewacht Leiden, University of Leiden, P.O. BOX 9513, 2300 RA, Leiden, THE NETHERLANDS |
| MENDEZ, Roberto H. | Institut für Astronomie und Astrophysik der Universität München, Scheinerstr. 1, D-8000 München 80, GERMANY |
| MENNESSIER, M.-O. | GRAAL (cc072), Univ. Montpellier II, F-34095 Montpellier Cedex 5, FRANCE |
| MITSCH, Wolfgang | Universitäts-Sternwarte, Scheinerstraße 1, D-8000 München 80, GERMANY |
| MORENO, Hugo | Departamento de Astronomia, Universidad de Chile, Casilla 36-D, Santiago, CHILE |
| MORRIS, Mark | Astronomy Department, University of California, Los Angeles, 8979, MSB/ UCLA, Los Angeles, Ca 90024, USA |

| | |
|---|---|
| MÜRSET, Urs | Institut für Astronomie, ETH - Zentrum, CH-8092 Zürich, SWITZERLAND |
| NAPIWOTZKI, Ralf | Institut für Theoretische Physik und Sternwarte der Universität Kiel, Olshausenstr. 40, D-2300 Kiel, GERMANY |
| NERI, Roberto | IRAM, Domaine Universitaire, 300 rue de la Piscine, 38406 St-Martin-d'Heres, FRANCE |
| O'DELL, C.R. | Department of Space Physics and Astronomy, Rice University, P.O. Box 1892, Houston, TX 77251, USA |
| OLSON, Randy W. | University of Wisconsin, Dept. of Physics and Astronomy, Stevens Point, WI 54481, USA |
| OUDMAIJER, Rene | Kapteyn Astronomical Institute, P.O. BOX 800, 9700 AV Groningen, THE NETHERLANDS |
| PARTHASARATHY, M. | Indian Institute of Astrophysics, Sarjapur Road, Bangalore 560034, INDIA |
| PASCOLI, Gianni | Universite de Picardie, Departement de Physique, 33 rue Saint-Leu, F-80039 Amiens Cedex, FRANCE |
| PEIMBERT, Manuel | Instituto de Astronomia, UNAM, Apartado Postal 70-264, Mexico 04510, D.F., MEXICO |
| PENA, Miriam | Instituto de Astronomia, UNAM, Apartado Postal 70-264, Mexico 04510, D.F., MEXICO |
| PERINOTTO, Mario | Dipartimento di Astronomia e Scienza dello Spazio, Largo E. Fermi 5, 50125 Firenze, ITALY |
| PERSI, Paolo | Istituto Astrofisica Spaziale, CNR, CP 67 00044 Frascati, ITALY |
| PETRINI, Daniel | Observatoire de la, Cote D' Azur, B.P. 139-06003, Nice Cedex, FRANCE |
| PFITSCHER, Kurt | Institut für Astronomie, Universität Innsbruck, Technikerstr. 25, A-6020, AUSTRIA |
| PFLEIDERER, Jörg | Institut für Astronomie, Universität Innsbruck, Technikerstr. 25, A-6020, AUSTRIA |
| PFLEIDERER, Mircea | Institut für Astronomie, Universität Innsbruck, Technikerstr. 25, A-6020, AUSTRIA |
| PIER, Jeff | U.S. Naval Observatory, P.O. Box 1149, Flagstaff, AZ 86002-1149, USA |
| PISMIS, Paris | Instituto de Astronomia, UNAM, Apdo. Postal 70-264, 04510 Mexico D.F., MEXICO |
| POLLACCO, Don | Department of Physics and Astronomy, North Haugh, St. Andrews, Fife, KY16 9SS, UK |
| POTTASCH, Stuart R. | Kapteyn Laboratorium, Postbus 800, 9700 AV Groningen, THE NETHERLANDS |
| PREITE-MARTINEZ, A. | Istituto di Astrofis. Spaziale, C.P. 67, I-00044 Frascati, ITALY |
| RAITERI, Claudia Maria | Osservatorio Astronomico di Torino, Strada Osservatorio 20, 10025 Pino Torinese, ITALY |

| | |
|---|---|
| RAYTCHEV, Blagoy | 28 Bvd. D'Anvers,, 67000 Strasbourg, FRANCE |
| REID, Neill | 105 - 24, CalTech, Pasadena, CA 91125, USA |
| RIERA, Angels | Department d'Astronomia i Meteorologia, Av. Diagonal 647, 08028 Barcelona, SPAIN |
| ROLA, Claudia | DAEC, Observatoire de Meudon, F-92195 Meudon Principal Cedex, FRANCE |
| ROSSI, Corinne | Istituto Astronomico, Universita La Sapenzia, Via G.M. Lancisi 29, 00161 Roma, ITALY |
| ROTH, Miguel | Las Campanas Observatory, Casilla 601, La Serena, CHILE |
| ROTH, Martin | Institut für Astronomie und Astrophysik der, Universität München, Scheinerstrasse 1, D-8000 München 80, GERMANY |
| RUBIN, Robert H. | NASA Ames Reseach Center, MS 245-6, Moffett Field, CA 94035, USA |
| RUDNITSKIJ, Georgij | Sternberg Astronomical Institute, Universitetskij Prospekt 13, SU-119899, Moscow, RUSSIA |
| RUDZIKAS, Zenonas B. | Institute of Physics and Astronomy, A. Gostauto 12, Vilnius, 232 600, LITHUANIA |
| RUECKERT, H.-J. | Kristeneben 38, A-6094 Axams, AUSTRIA |
| RUIZ, Maria Teresa | Departamento de Astronomia, Universidad de Chile, Casilla 36-D, Santiago, CHILE |
| SAHAI, Raghvendra | JET Propulsion Lab. (Radio Astronomy Group), Oak Grove Drive, Pasadena, CA, USA |
| SAHU, Kailash C. | Kapteyn Laboratory, P.O. BOX 800, 9700 AV Groningen, THE NETHERLANDS |
| SALPETER, Edwin | 308 Newman Laboratory, Cornell University, Ithaca, N.Y. 14853, USA |
| SAMLAND, Markus | Institut f. Theoretische Physik und Sternwarte der Universität, Olshausenstr. 40, D-2300 Kiel, GERMANY |
| SASSELOV, Dimitar D. | Harvard - Smithsonian Center for Astrophysics, 60 Garden Street, Cambridge, MA 02138, USA |
| SAURER, Walter | Institut für Astronomie, Universität Innsbruck, Technikerstr. 25, A-6020 Innsbruck, AUSTRIA |
| SCHMID, Hans Martin | Institute of Astronomy, ETH Zentrum, CH-8092, Zürich, SWITZERLAND |
| SCHNELL, Anneliese | Institut für Astronomie, Türkenschanzstr. 17, A-1180 Wien, AUSTRIA |
| SCHOBER, Hans Josef | Institut für Astronomie, Universität Graz, Universitätsplatz 5, A-8010 Graz, AUSTRIA |
| SCHÖNBERNER, Detlef | Institut für Theoretische Physik und Sternwarte, Universität Kiel, Olshausenstr. 40-60, D-2300 Kiel, GERMANY |
| SCHULTHEIS, Matthias | Institut für Astronomie, Türkenschanzstr. 17, A-1180 Wien, AUSTRIA |

| | |
|---|---|
| SCHWARZ, Hugo E. | E.S.O., Casilla 19001, Santiago 19, CHILE |
| SEEBERGER, Robert | Institut für Astronomie, Universität Innsbruck, Technikerstr. 25, A-6020 Innsbruck, AUSTRIA |
| SELVELLI, Pierluigi | Osservatorio Astronomico di Trieste, Via Tiepolo 11, I-3431 Trieste, ITALY |
| SHIBATA, K. M. | Nobeyama Radio Observatory, Nobeyama, Minamimaki, Minamisaku, Nagano 384-13, JAPAN |
| SILVESTRO, Giovanni | Istituto di Fisica Generale, Universita' di Torino, via P.Giuria 1, Torino, ITALY |
| SILVOTTI, Roberto | Dipartimento di Astronomia, Casella Postale 596, 40100 Bologna, ITALY |
| SION, Edward | Villanova University, Villanova, PA 19085, USA |
| SIVAGNANAM, Philippe | Observatoire de Meudon (DERAD), F-92195 Meudon Principal Cedex, FRANCE |
| SKINNER, Chris | Lawrence Livermore National Laboratory, University of California, P.O. Box 808 L-413, CA 94550, USA |
| SOFFNER, Till | Institut für Astronomie, Univ.- Sternwarte München, Scheinerstr.1, D-8000 München 80, GERMANY |
| STANGHELLINI, Letizia | Osservatorio di Bologne, Via Zamboni 33, I-40126 Bologne, ITALY |
| STASINSKA, Grazyna | DAEC, Observatoire de Paris, Section de Meudon, F-92195 Meudon Cedex, FRANCE |
| STENHOLM, Björn | Lund Observatory, Box 43, S-22100 Lund, SWEDEN |
| STRAZZULLA, Giovanni | Osservatorio Astrofisico and Istituto di Astronomia, Citta Universitaria, I-95125 Catania, ITALY |
| SURENDIRANATH, R. | Indian Institute of Astrophysics, Sarjapur Road, Bangalore 560034, INDIA |
| SZCZERBA, Ryszard | Copernicus Astronomical, Center, Chopina 12/18, 87-100 Torun, POLAND |
| TAMMANN, Gustav A. | Astronomisches Institut der Universität Basel, Venusstr. 7, CH-4102 Binningen, SWITZERLAND |
| TAMURA, Shin'ichi | Astronomical Institute, Faculty of Science, Tohoku University, Aobaku Aramaki Aza Aoba, Sendai 980, JAPAN |
| TAPIA, Mauricio | Instituto de Astronomia, UNAM, P.O. Box 73 027, San Ysidro CA 92073, USA |
| TERZAN, Agop | Observatoire de Lyon, F-69561 St.-Genis-Laval, FRANCE |
| TERZIAN, Yervant | Department of Astronomy, Space Sciences Building, Cornell University, Ithaca, NY 14853-6801, USA |
| TIELENS, A. | Sterrewacht Leiden, P.O. Box 9513, NL- 2300 RA Leiden, THE NETHERLANDS |

| | |
|---|---|
| TORRES-PEIMBERT, S. | Instituto de Astronomia, Universidad National, Autonoma de Mexico, Apartado Postal 70-264, Mexico 04510 D.F., MEXICO |
| TSIOPA, Olga | Main astronomical observatory of the Russian Academy of Sciences, St. Petersburg, Pulkovo, RUSSIA |
| TWEEDY, Richard | Steward Observatory, University of Arizona, 2nd and Cherry Avenue, Tucson, AZ 85721, USA |
| TYLENDA, R. | Laboratory of Astrophysics, Copernicus Astronomical Center, Chopina 12/18, P-87-100 Torun, POLAND |
| VAN DE STEENE, Griet | Kapteyn Laboratory, PO. Box 800, NL-9700 AV Groningen, THE NETHERLANDS |
| VILLATA, Massimo | Istituto di Fisica, Generale - Universita di Torino, Via P. Giuria 1, I-10125 Torino, ITALY |
| VOLK, Kevin | Department of Physics and Astronomy, University of Calgary, 2500 University Drive N.W., Calgary, T2N 1N4, CANADA |
| WALSH, Jeremy R. | STECF, European Southern Observatory, Karl-Schwarzschild-Str. 2, D-8046 Garching, GERMANY |
| WALTON, N.A. | University College London, Department of Physics and Astronomy, Gower Street, London WCIE 6BT, UK |
| WANG, Lifan | Department of Astronomy, The University of Manchester, M139PL, UK |
| WEIDEMANN, Volker | Institut für Theoretische Physik und Sternwarte, der Universität, Olshausenstraße 40-60, D-2300 Kiel, GERMANY |
| WEINBERGER, Ronald | Institut für Astronomie, Universität Innsbruck, Technikerstr. 25, A-6020, AUSTRIA |
| WERNER, Klaus | Institut für Theoretische Physik und Sternwarte der Universität Kiel, Olshausenstrasse 40, D-2300 Kiel, GERMANY |
| WHITELOCK, P. A. | South African Astronomical Observatory, P.O. Box 9, Observatory, Cape 7935, SOUTH AFRICA |
| WOLSTENCROFT, R.D. | Royal Observatory, Blackford Hill, Edinburgh EH9 3HJ, UK |
| WOOD, P.R. | Mount Stromlo and Siding Spring Observatories, Australian National University, Private Bag, Weston Creek P.O., Canberra, ACT 2611, AUSTRALIA |
| WSZOLEK, Bogdan | Jagiellonian University, Astronomical Observatory, ul. Orla 171, Pl-30-244 Krakow, POLAND |
| YADOUMARU, Yasushi | Astronomical Institute, Faculty of Science, Tohoku University, Aobaku Aramaki Aza Aoba, Sendai 980, JAPAN |
| YAMASAKI, Atsuma | Department of Geoscience, National Defense Academy, Hashirizu, Yokosuka, Kanagawa 239, JAPAN |

| | |
|---|---|
| YATSYK, Olga S. | Astronomical Observatory of L'vov State University, Ul Lomonosova 8, L'vov, UKRAINIA |
| YUNGELSON Lev.R. | Institute of Astronomy, 48 Pyatnitskaya Str., 109017, Moscow, RUSSIA |
| ZHANG, Cheng-Yue | Department of Astronomy, University of Texas, RLM 15 308, Austin TX 78712, USA |
| ZWEIGLE, Jürgen | Astronomisches Institut der Universität, Waldhäuserstraße 64, D-7400 Tübingen, GERMANY |

A photograph of those participants who had attended the 1st IAU Symposium on "Planetary Nebulae", held in Tatranska Lomnica in 1967. From left to right: M. Perinotto, V. Weidemann, M.W. Feast, E. Salpeter, Y. Terzian, C.R. O'Dell, S.R. Pottasch, L. Kohoutek, L.H. Aller, G.A. Gurzadyan.

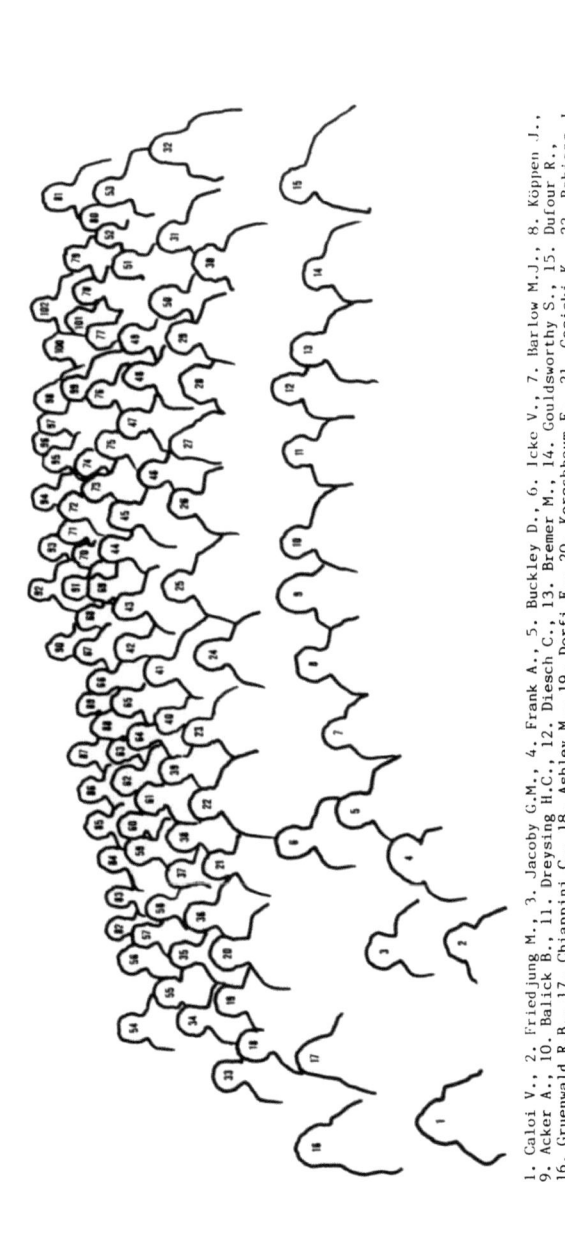

1. Caloi V., 2. Friedjung M., 3. Jacoby G.M., 4. Frank A., 5. Buckley D., 6. Icke V., 7. Barlow M.J., 8. Köppen J.,
9. Acker A., 10. Balick B., 11. Dreysing H.C., 12. Diesch C., 13. Bremer M., 14. Gouldsworthy S., 15. Dufour R.,
16. Gruenwald R.B., 17. Chiappini C., 18. Ashley M., 19. Dorfi E., 20. Kerschbaum F., 21. Gesicki K., 22. Bohigas J.,
23. Dyson J.E., 24. Kwitter K.B., 25. Iben I. jr., 26. Butler K., 27. Becker S., 28. Gonzalez-Riestra R.,
29. De Freitas Pacheco J.A., 30. Feklistova T.K., 31. Golovatyj V., 32. Fierro J., 33. Chu Y.H., 34. Kwok S.,
35. Igumenshchev I.V., 36. Hron J., 37. Hui X., 38. Gorny S., 39. Bachiller R., 40. Guerrero M.A., 41. Bässgen M.,
42. Iijima T., 43. Grewing M., 44. Barker T., 45. Hopfensitz W., 46. Houziaux L., 47. Amnuel P.R., 48. Diamond C.J.,
49. Kholtygin A.F., 50. Kondrat'eva L.N., 51. Kuczawska E., 52. Garcia-Lario P., 53. Pascoli G., 54. Gurzadyan G.A.,
55. Jurcsik J., 56. Kato M., 57. Huggins P.J., 58. Khan I., 59. Aringer B., 60. Borkowski K.J., 61. Dopita M.A.,
62. Handler G., 63. Groenewegen M., 64. Hill P., 65. Bell S., 66. Bond H.E., 67. Hora J., 68. Kawaler S.D.,
69. Bickert K., 70. Ciardullo R., 71. ..., 72. Blöcker T., 73. Ferland G.J., 74. Herrero A., 75. Harrington J.P.,
76. Heber U., 77. Dreizler S., 78. Hamann W.R., 79. Escalante V., 80. Kudashkina L.S., 81. Andronov I.L.,
82. Hrivnak B.J., 83. Kingsburgh R., 84. Aller L.H., 85. Hachisu I., 86. Kohoutek L., 87. Cuisinier F., 88. Deutsch L.K.,
89. Clegg R.E.S., 90. Dennefeld M., 91. Altner B., 92. Gaigalas G., 93. Harvey P.M., 94. Hu J.Y., 95. Feast M.W.,
96. ..., 97. Corradi R.L.M., 98. Conlon E., 99. Fernandez-Castro T., 100. Bianchi L., 101. Boffi F.R., 102. Habing H.J.

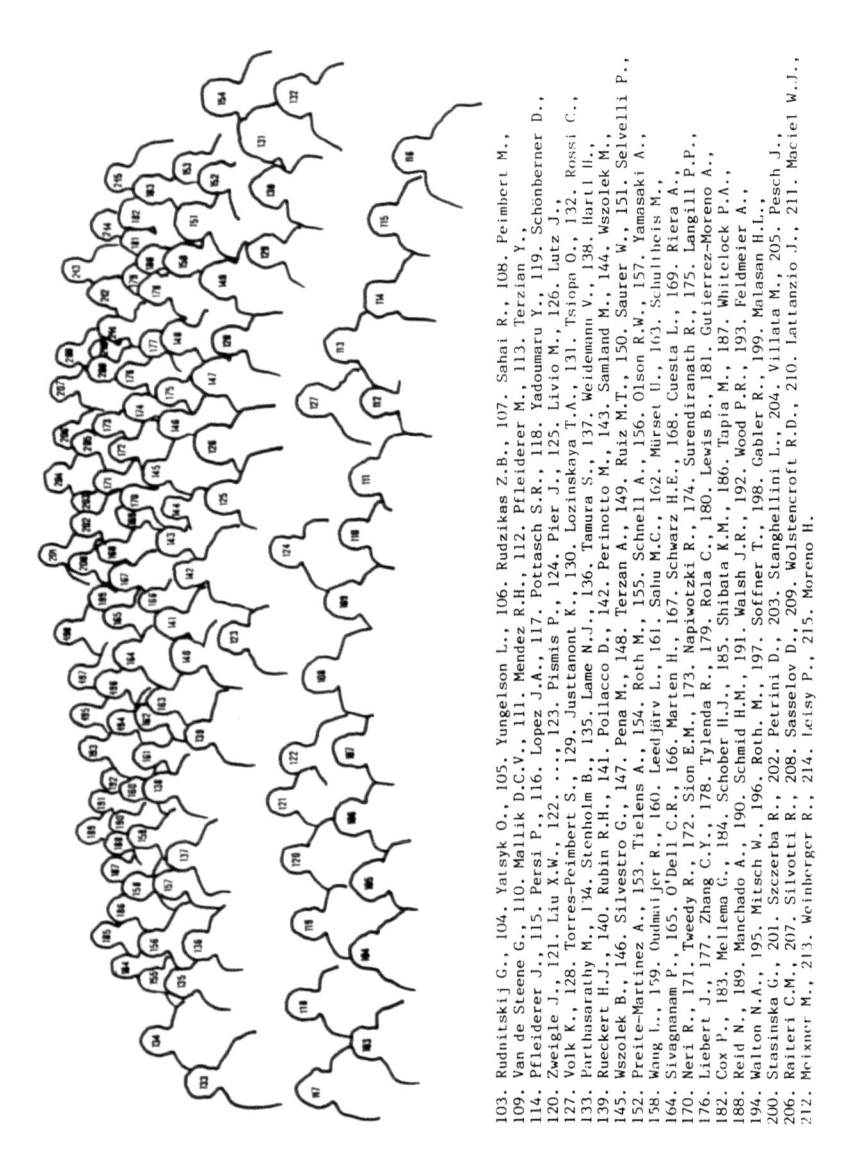

103. Rudnitskij G., 104. Yatsyk O., 105. Yungelson L., 106. Rudzikas Z.B., 107. Sahai R., 108. Peimbert M.,
109. Van de Steene G., 110. Mallik D.C.V., 111. Mendez R.H., 112. Pfleiderer M., 113. Terzian Y.,
114. Pfleiderer J., 115. Persi P., 116. Lopez J.A., 117. Pottasch S.R., 118. Yadoumaru Y., 119. Schönberner D.,
120. Zweigl J., 121. Liu X.W., 122. ...,123. Pismis P., 124. Pier J., 125. Livio M., 126. Lutz J.,
127. Volk K., 128. Torres-Peimbert S., 129. Justtanont K., 130. Lozinskaya T.A., 131. Tsiopa O., 132. Rossi C.,
133. Parthasarathy M., 134. Stenholm B., 135. Lame N.J., 136. Tamura S., 137. Weidemann V., 138. Hartl H.,
139. Rueckert H.J., 140. Rubin R.H., 141. Pollacco D., 142. Perinotto M., 143. Samland M., 144. Wszolek M.,
145. Wszolek B., 146. Silvestro G., 147. Pena M., 148. Terzan A., 149. Ruiz M.T., 150. Saurer W., 151. Selvelli P.,
152. Preite-Martinez A., 153. Tielens A., 154. Roth M., 155. Schnell A., 156. Olson R.W., 157. Yamasaki A.,
158. Wang L., 159. Oudmaijer R., 160. Leedjärv L., 161. Sahu M.C., 162. Mürset U., 163. Schultheis M.,
164. Sivagnanam P., 165. O'Dell C.R., 166. Marten H., 167. Schwarz H.E., 168. Cuesta L., 169. Riera A.,
170. Neri R., 171. Tweedy R., 172. Sion E.M., 173. Napiwotzki R., 174. Surendiranath R., 175. Langill P.P.,
176. Liebert J., 177. Zhang C.Y., 178. Tylenda R., 179. Rola C., 180. Lewis B., 181. Gutierrez-Moreno A.,
182. Cox P., 183. Mellema G., 184. Schober H.J., 185. Shibata K.M., 186. Tapia M., 187. Whitelock P.A.,
188. Reid N., 189. Manchado A., 190. Schmid H.M., 191. Walsh J.R., 192. Wood P.R., 193. Feldmeier A.,
194. Walton N.A., 195. Mitsch W., 196. Roth M., 197. Sofner T., 198. Gabler R., 199. Malasan H.L.,
200. Stasinska G., 201. Szczerba R., 202. Petrini D., 203. Stanghellini L., 204. Villata M., 205. Pesch J.,
206. Raiteri C.M., 207. Silvotti R., 208. Sasselov D., 209. Wolstencroft R.D., 210. Lattanzio J., 211. Maciel W.J.,
212. Meixner M., 213. Weinberger R., 214. Leisy P., 215. Moreno H.

# FOREWORD

The working group on Planetary Nebulae as subgroup of the IAU Commission 34 selected the Highlands of Europe for hosting the IAU symposium 155 devoted to the most beautiful and the most interesting – though small and unsubstantial – objects in our universe.
The Institute of Astronomy of the Innsbruck University spent a lot of time organizing the meeting with an exemplary rigour and boundless generosity, with a lot of refinements, and with particular effort for our colleagues from Eastern Europe, whose borders have recently been opened. The scientific organisation was built through many oral and written discussions with the members of the *Scientific Organizing Committee*, with whom I had the unique opportunity to work.

During the last years, an impressive number of papers were devoted to the planetary nebulae, the record being held in Europe by Pottasch and his collaborators: 75 papers appeared between 1986 and 1990, followed by Jacoby (36), Kwok (35), Clegg (31), Kaler (30). Thanks to all these works, and to those of all other specialists throughout the world, extensive surveys of planetary nebulae (PN) in our Galaxy and in the Local Group have been performed, both in the optical and in the radio domains. A number of new objects were discovered, mainly proto–planetary nebulae, through their IRAS colors. These studies allow, first, to give a better **classification** in Hydrogen–rich, Helium–rich, etc, PN and in better defined types of central stars.

Second, due to their intense emission lines, the PN are traceable throughout large distances, and should be characterized as **"stellar populations"** having specific (chemical, spatial, kinematical) properties: disc objects, halo objects, bulge objects,... , related to models of dynamical and chemical evolution of galaxies.

On the other hand, a large set of observations and theoretical work was devoted to the **evolution** of PN: OH/IR stars, Mira variables, carbon stars as proto–PN, and with white dwarfs as nuclei of old PN. The understanding of the formation of PN requires multi–wavelength observations. Large infrared molecular CO emissions and continuum excesses have been observed in PN, especially in young PN. Optical CCD imaging of PN revealed faint outer haloes which are probably remnants of the AGB winds. From the comparison of observed HR diagrams and the theoretical tracks, a mass distribution of the stars is deduced, depending strongly on the adopted theoretical tracks. A controversial comparison of these theories is needed.

First, the field of research usually referred to as "Planetary Nebulae" is not isolated from other fields; on the contrary, it represents one of the most interesting intersections between different areas, subjects, methodologies in the astrophysical world (e.g.: stellar evolution, stellar atmospheres, stellar winds, asteroseismology and thermal evolution, binaries; galactic populations and evolution,...). In addition, PN are *per se*, unrivalled "physical laboratories" (physical processes in nebular gas, magnetohydrodynamics, photoionisation models, atomic and molecular data, dust, etc...).

A second, equally important theme will be the PN "evolutionary connection". Still open problems remain like those concerning evolutionary scenarios from the Asymptotic Giant Branch (timescales, dredge-ups, link to OH/IR stars, physics and dynamics of envelopes,...) down to white dwarfs (composition, instabilities, birth-rates,...), and like those concerning the mechanism of PN formation, including the evolution of the central star, the interaction of an aspherical superwind with a spherical AGB wind, and interactions with the interstellar medium.

During the symposium, we brought advances on those uncleared paths, and found, on our way, new unsolved riddles, such as PG1159–pulsators, or proto–PN shells with water, Polyclic Aromatic Hydrocarbons, and other exotic constituents.

We are indebted to the SOC and the LOC members who made their contribution to the success of the symposium. Thanks also to the Strasbourg Observatory for its help. Special thanks go to Dr Kimeswenger for his editorial assistance. The generous sponsorship of several organisations made it possible to support travel and subsistance for numerous participants.

Agnès ACKER, chairperson of the SOC
Observatoire de Strasbourg, France

**Scientific Organizing Committee (SOC):** A. Acker (France, chairperson), R.E.S. Clegg (UK), M.A. Dopita (Australia), H.J. Habing (Neth.), I. Iben (USA), G. Jacoby (USA), J. Kaler (USA), L.N. Kondratjeva (Kazakh.), S. Kwok (Can.), D.C.V. Mallik (India), M. Peimbert (Mex.), S.R. Pottasch (Neth.), A. Preite-Martinez (Italy), D. Schönberner (Ger.), Y. Terzian (USA), R. Tylenda (Pol.), and R. Weinberger (Austria).

**Local Organizing Committee (LOC):** H. Hartl (chairperson), R.Weinberger, A. Hörtnagel, S. Kimeswenger, K. Pfitscher, J. Pfleiderer, P. Riedlsperger, W. Saurer,and R. Seeberger

The editors would like to thank Drs. A. Frank, B. Balick, V. Icke, and G. Mellema for providing the cover picture.

# PLANETARY NEBULAE, AN INTRODUCTORY REVIEW

L.H. ALLER

*University of California, Los Angeles, CA 90024, USA*

The time is long past when we could regard planetary nebulae (PN) as semi-static or even steady state features, anchored in the phase in which we happen to catch them. We must regard them as evanescent phenomena, celestial flowers caught between bud and final loss of petals. Throughout most stages of its existence, a star can be regarded as essentially an equilibrium structure, but a PN and its central star evolve on a time scale comparable with human history.

Two separate developments provide impetus for PN research, a vastly expanded observational data base and important theoretical insights aided and abetted by powerful computers. In addition to optical observations, we have extensive radio-frequency (rf), infrared (IR), and ultraviolet (UV) data. Some 350 PN have been measured in the rf range, a thousand have been detected by IRAS, and photometric data exist in 4 bands for ~300 of these, while some 200 PN have been studied with the IUE.

Observed stellar properties can be mostly summarized by a position, and a high resolution spectral scan, but a PN requires a two dimensional map in several radiations, e.g., [OII], [OIII], Hβ, [NII] and [SII] as obtained, e.g. by Jacoby et al for NGC 40, and NGC 6826. Then we can obtain values of $T_e$ and $N_e$, N(O+)/N(H+), etc. averaged along a pencil taken through the nebula, and make comparisons with models. The photometric information should be supplemented with velocity measurements, such as those obtained with a Fabry-Perot etalon. The quantity of data involved is tremendous, but the effort is justified for a few objects amenable to detailed theoretical modeling.

New IR and rf techniques supply crucial structural information. With VLA, one can construct continuum isophotic maps, free of effects of interstellar extinction (ISE). Furthermore, by comparing maps from different epochs, we can get nebular angular expansion rates. Spectroscopic measurements give the expansion velocity in km/sec at least in the observers direction; and if one can allow for an increase in the size of the H+ zone as the ionization front invades the neutral zone, the PN's distance can be found. This principle, originally applied to optical data by W. Liller has been used by Masson (1989) to obtain distances and structures of NGC 7027, BD +30 3639 and NGC 6572. Rapid advances in IR technology make it possible to secure high resolution images for many PN. With a 256×256 array and a 2'x2' field one is limited mostly by seeing. We can compare, e.g., images in $H_2$ with atomic H images from the nearby Brγ line. Also, we hope to get CIII, CIV, and other images in bright PN with the Hubble Space Telescope. Krishna et al (1989) measured X-rays from 8PN - or more accurately from their central stars or "nuclei" (PNN) with the EXOSAT satellite.

Examples of some recent work are: photometric optical data on Hβ, Hα + [NII], [OIII] and B and V magnitudes for their (PNN) by Shaw and Kaler (1989) for 134 southern objects and measurements of Hβ fluxes in 460 southern PN by Acker et al's (1989). These data are especially valuable for objects of small angular size. Much effort has been devoted to compact and presumably young PN by a number of observers who have obtained optical, IR, and r.f. observations. Zhang and Kwok (1991) compiled energy distributions from 100A to 100 μm for 66 PN. They attempt to separate contributions of nebular continua, lines and dust. Image tube scanners (ITS) and other devices have been used to extend studies of PN spectra to the S. Hemisphere and Magellanic Clouds. Development of high dispersion echelle spectroscopy provides us with high resolution data.

**The Persistent Problem of Planetary Nebular Distances.**

A most unsatisfactory feature of the PN problem is our poor knowledge of their distances. The Liller-Masson method promises rich dividends but workers in the field are few. VLA observations are tedious, time-consuming and require great care and patience. For a few precious objects, the spectroscopic parallax method of Kudritzki and his associates can be used, and distance for a few other PNs that are members of binaries or clusters are also obtainable; but for most PNs, many years will be needed to get the necessary VLA observations. Statistical methods have proved disappointing. The Shklovsky method which postulates a uniform mass for all fully ionised PN is confronted by an actual mass range from 0.3 to 1.3 solar masses for PN in the Magellanic Clouds (Dopita & Meatheringham 1991). The Shklovsky method gives a poor determination of the distance of the galactic bulge (Pottasch & Zijlstra 1992).

**Evolution and Structure.**

The picture of the transformation of an evolving asymptotic giant branch (AGB) star to a PN gradually becomes better defined, but many details and ambiguities remain. As a star ascends the AGB, the mass loss rate increases. Radiation pressure drives the grains away and the mass ejection rate becomes especially large for long period variables with periods greater than 400 days, since radiation pressure drives grains from an extended atmosphere more easily than from a normal one. The mass loss problem appears more tractable for carbon stars. The grains seem to be amorphous rather than graphite. Jura (1991) estimates that in the solar neighborhood half of all stars in the range of 1 to 5 solar masses become C stars, with mass-loss rates between $10^{-7}$ and $10^{-5}$ solar masses/yr. A larger fraction of metal-poor stars evolve into carbon stars before they become PN than is the case for metal-rich stars.

If the dying star is sufficiently massive, it becomes surrounded by a thick, warm, dusty envelope. It emits copiously in the IR; hence, the great value of IRAS data. At this stage, the post AGB star has stopped pulsating. The dust cloud is moving into space and what still can be called the "star" migrates to the blue side of the HR diagram. When the envelope mass has fallen to about 0.001 that of the sun, a large mass loss rate can no longer be sustained. The stellar core gradually becomes uncovered and $T(*)$ rises from 5000°K to 30,000°K. The dusty envelope is exposed to high frequency radiation, molecules dissociate, and the gas becomes ionized. All this time the escaping envelope is vanishing into space; and if the stellar spot-light is not turned on fast enough, there will be no PN. If we don't quibble over too many details, this picture is probably generally acceptable. What really happens from the late AGB phase through the proto-PN phase until we get a dusty, shiny, new PN is open to much discussion. The IRAS data have supplied many clues but many questions remain.

Infra-red C stars may evolve unobtrusively into C-rich PN. The OH-IR stars and those with strong 9.7 $\mu$m silicate absorption features are expected to produce O-rich PN, but once in a while there is a surprise. Zhang and Kwok (1990) found 6 young PNs that showed both O-rich and C-rich dust features. One of these, SwSt 1 whose PNN exhibits a WC type spectrum, is an O-rich PN. This might be a transition phase where a C-rich photosphere emerges after the O-rich layers escape from the star. Among long-known PN we recall NGC 40 and BD +30° 3639 with WC nuclei and N-rich envelopes with no trace of N in the central star spectrum.

An important clue to the ejection process must be the fact that the dominant symmetry in PN is elliptical or bipolar, a symmetry that can be traced to the central stars. Properties of bipolar pre-PNs are discussed by Mark Morris (1990). They have a thick toroidal ring, so that the outflowing wind, whose velocity depends on latitude, is strongly directed towards the poles. This polar stream (with an average velocity of 100-200 km/sec) moves about ten times as fast as one would expect for a spherical, symmetrical shell driven by radiation pressure on dust grains. The polar streams often exhibit masers and some have ansae along the polar axis, presumably the effect of polar streaming on the AGB envelope. These objects are clearly the antecedents of bipolar and maybe other kinds of PNs with their characteristic topologies and bright line ansae (e.g., Balick 1987).

Bipolar phenomena are common in young PN like Vy2-2 (Miranda & Solf 1991). One striking example is IC 4406. There is a dense, cool, dusty, molecular torus mapped in CO (Sahai et al 1991). The equatorial expansion velocity is about 15 km/sec and four times larger in higher latitudes. The mass of the molecular cloud is at least 0.16 $M(\odot)$ and may be much larger. It is a young object with a post-AGB age of less than 10,000 years. Pascoli (1990) has discussed the morphology of bipolar PNs and has set up a classification scheme somewhat similar to that of Balick (1987) and of Sabbadin (1986).

Although bipolar PN possibly may be explained most easily as ejecta from close binaries, nebular statistics do not favor such a hypothesis. There are too many bipolars. Bond and Livio (1990) suggest that PN produced by binaries are most likely ejected at the end of a common envelope phase when the primary star in a wide binary system expands and engulfs the companion, reducing the orbit of the latter. No close binary PN shows double or multiple shells, although half of the planetaries with binary nuclei show butterfly or elliptical features with moderate to extreme density contrasts.

In the currently accepted model, basically due to Kwok, there is first the leisurely red giant wind with a velocity of about 5 km/sec and mass loss rate from $10^{-8}$ to $10^{-9}$ $M(\odot)$/yr which is followed by the superwind with a velocity of 10km/sec and a mass loss rate between $10^{-4}$ and $10^{-6}$ $M(\odot)$/yr. This phenomenon occurs at the OH-IR star phase and results in expulsion of the AGB envelope. Then, as the hot core is exposed, the very fast wind (v ~ 2000km/sec) appears with a mass loss rate of $10^{-7}$ to $10^{-8}$ solar masses per year. This wind can compress the previously ejected shell. Giant PN haloes are often interpreted as the remnants of a giant red envelope as buffeted by the superwind.

In the scenario of Kahn and Breitschwerdt (1990), the H+ gas accelerates the sluggishly moving neutral shell producing Rayleigh Taylor instabilities that can cause fragmentation of the shell. They conclude that this model can explain such phenomena as the bright inner ring, low ionization knots, and faint surrounding halo. The knots are regarded as transient features in a dynamical evolution. Much work remains to be done to really understand the vast differences between the smoothness of the Owl Nebula, NGC 3587, and NGC 7293 with its numerous small (~200 astr. units) ansae, presumably cold blobs (Dyson et al 1989). Some PNs show extremely intricate morphologies and kinematics. An example is Abell 78, whose chemical composition is a function of the distance from the PNN. The outflow is non-isotropic and collimated (Pismis 1989).

Fast winds from the central star were suggested 60 years ago by the Wolf-Rayet profiles in the spectra of NGC 40 and BD +30° 3639, but meaningful measurements were possible only with the IUE. Actual mass loss measurements depend on fundamental stellar parameters and are difficult to make (e.g., Cerruti Sola & Perinotto 1989). They give loss rates of 1.4 to 6.3 x $10^{-9}$ $M(\odot)$/yr for a sample of 5 well-observed PN. Sometimes, as in the PNN of Abell 78 (Kaler et al 1988), where V ~ 2700 km/sec, $dM/dt = 4 \times 10^{-9}$ $M(\odot)$/yr, the pattern is quite complex; there exists evidence for rebound shocks.

Can these winds affect the physical state as well as the dynamics of the PN? Apparently, they can. Analyses of the wispy giant halo of NGC 6543 by conventional diagnostics suggested an unlikely chemical composition difference between this envelope and the main ring. The discordance is removed if one supposes that a small fraction of the wind energy is expended exciting the halo (Middlemass et al 1989).

Very high excitation NGC 6302 which shows lines of 5 and 6-fold ionized silicon presents an even more dramatic case. Although a PNN with an unlikely temperature of 450,000° could explain the general level of excitation, Lame and Ferland (1991) found that a strictly photo-ionization model would not work in detail, while ionization in a post shock gas at T = 2,000,000°K plus radiation from a PNN at T = 200,000°K could explain the entire spectrum. Could coronal lines be observed in this PN?

**Theoretical Nebular Models**

Improving knowledge of nebular evolution and morphology now suggests we build better theoretical models. Heretofore, most PN models have been static structures although some workers, most notably Tylenda and Harrington, have investigated effects of time-dependent changes. Bobrowsky and Zipoy (1989) calculated spherically symmetrical kinematical models with a brisk stellar wind in which a molecular shell is exposed to a star that is turned on over a period of 100 to 1000 years. They treat it as a hydrodynamical problem with equations of motion, momentum and energy conservation, and dissipation of energy by the excitation of forbidden lines, etc. all taken into account. Energy input is primarily by photoionization of H although the influence of a stellar wind is also considered. Aware of advances in IR spectroscopy, they include $H_2$ emission predictions, as well as those of familiar ionic lines. Comparison with observations will require high-quality data. Observationally, one can estimate v(r) from line profiles of H, HeI, HeII, OIII, etc. by Sabbadin's method.

An attempt to explain the PN forms has been made by Icke et al (1992). They start with an intermediate mass star in its final evolutionary stages where it exhibits a slow, dense aspherical wind which is responsible for the toroidal ring. This stage is followed by a fast spherically symmetrical wind from the core. An extremely complicated hydrodynamical code is required to execute the calculations. Their attempt to explain the evolving shapes and morphologies - especially elliptical and butterfly formations, but not ansae - seems to be successful. Effects of different density distributions are most dramatic. The Icke et al theory needs to be extended to include radiative effects and to predict the spectrum. With a sufficiently powerful computer, the task is horrendously difficult but not intractable.

**Central Stars of Planetary Nebulae**

The fundamental parameters required for PNN are their temperatures (which require spectrophotometric measurements) and their luminosities (which require a knowledge of their distances). Magellanic Cloud data, when they can be acquired, are thus particularly useful. The chief limitation is imposed by object faintness and PN size.

The pattern of evolution seems well established by the work of Paczynski, of Schönberner, and of Wood and Faulkner. During the hydrogen burning phase, the initial evolution from the post-AGB stage is at constant luminosity. When the envelope becomes too small, H burning suddenly ceases and the main energy source becomes gravitational contraction. The luminosity now declines and the star begins to cool towards the white dwarf (WD) stage. Stars that burn He in their shells fade more slowly. The rate of evolution depends on the core mass. Below 0.57 $M(\odot)$, a small decrement in the mass results in a large decrement in L. Very old PN, such as NGC 7293, may contain degenerate, essentially WD stars.

Theory gives an evolutionary time scale. If the PN distance is known, then from the measured rate of expansion in km/sec and the present angular size of the nebula one can deduce a dynamical age. McCarthy et al (1990) found significant discrepancies, e.g., the kinematical age of NGC 7009 was 7000 years, while its evolutionary age was only 1740 years. For a few PN, the kinematical age was less than the evolutionary age. They suggested that the discrepancy could be explained by having a small amount of residual envelope material remaining after the superwind mass-loss phase.

Accurate distances are necessary to resolve the problem of PNN masses. Most of them are generally regarded as falling in the mass range 0.55 to 0.60, although Kaler, Shaw and Kwitter (1990) find a much wider distribution with about 40% of their sample having masses in excess of 0.7 and 15% in excess of 0.8 M(o). From a sample of PNN for which they could get T(eff), log g, and spectroscopic parallaxes, Napiwotzki and Schönberner found masses in the rage 0.55 to 0.60 M(o) for stars for which Kaler et al had obtained much higher values.

Considerable justification appears to exist for Zanstra temperatures when they are properly interpreted. Jacoby and Kaler (1989), concerned with investigations of optically thick PN with a view to applications to galaxy distance determinations, measured Hβ and HeII 4686 fluxes to establish the UV fluxes for PNN shortward of 912A and 228 A respectively. You assume a V magnitude and get $T_z(H)$ and $T_z(He^+)$ as a function of V, defining a V (cross-over) such that $T_z(H) = T_z(He^+)$. The method can be calibrated on galactic PN for which the actual V(PNN) can be measured. If the PNN has a dense wind or powerful corona, the HeII Zanstra temperature could be falsified, at least in principle. For galactic PN, T(eff) can be found from model nebular parameters required to reproduce the observed spectrum.

**Molecules and Dust**

Molecular gas provides the link connecting AGB stars, post-AGB stars, proto-PN, and PN themselves. They supply important probes to evolutionary history and clues to complex morphologies, enabling us to investigate nebular stratification and dense clumps ($10^4$ to $10^5$ atoms/cm$^3$) which cast shadows protecting fragile molecules from destruction by UV radiation. A PN starts out as a molecular cloud, typically between 0.1 and 1 solar mass, and of non-uniform density. It becomes exposed to the intense UV radiation of a defunct star, which destroys by photoionization most of the molecules present in the proto-PN. The molecular cloud erodes away as it is penetrated by UV quanta. In dense clumps where the density of $H_2$ molecules can attain $10^4/cm^3$, fragments like CN and HCN can persist (Howe et al 1992). CO is the hardiest of molecules. It is found in a number of mostly young, disk population PN with high N/O ratios and massive progenitors (Huggins & Healy 1989).

All PN with strong $H_2$ emission are N-rich objects with equatorial toroidal structures and faint bipolar extensions (Webster et al 1988). In NGC 2440 where T(★) ~ 350,000, V(wind) ~ 2000 km/sec, $dM/dt \sim 3 \times 10^{-7}$ M(o)/yr, the v = 1-0 S(1) line of $H_2$ which is strong in the position of the two intense clumps (Reay et al 1988) appears to be excited by winds. In most PN, excited $H_2$ is distributed the same as in the Hα + [NII] images. Emitting molecules lie in or close to the transition region between the neutral and ionized H gas. Generally it is believed that vibrational and rotational transitions of $H_2$ arise from thermal emission from a gas shocked by a brisk stellar wind. However, Dinerstein et al (1988) found strong lines in Hb12 originating from v = 1, 2, and 3 vibrational levels. These must be produced by fluorescence, not by thermal excitation. The dust/gas mass ratio which varies from PN to PN averages much less than in the ISM. Lenzuni et al (1989) suggest dust grains evolve during a PN's evolution, their size and total mass declining as the nebular radius increases. Perhaps they are ablated by the wind. Hydrogenated amorphous carbon grains appear to contribute to a diffuse emission band extending from 5500-8200A (Furton and Witt 1992).

**Physical Processes**

Progress in the interpretation of PN spectra depends on availability of accurate atomic parameters whereby we can investigate detailed physical processes, establish nebular diagnostics, determine abundances and examine evolutionary developments. Many references are listed by Aller (1990) and in the excellent review article by Peimbert

(1990). The great importance of the primordial helium abundance has stimulated much work on the He spectrum. The metastability of the $2^3$ S and $2^1$ S levels, which behave somewhat like pseudo ground levels, mandate attention to many collisional and radiative effects: Peimbert & Torres-Peimbert (1987), Clegg & Harrington (1989).

Accurate electronic collision strengths and A-values have long been required for nebular diagnostics; but in turn, PN data provides checks on these very same atomic data. Important contributions have been made by Seaton and his associates in London and by the Belfast group, including F.P. Keenan, C. T. Johnson, A. E. Kingston, V. M. Burke, D. J. Lennon, K. M. Aggarwal, E. Barrett, and C. D. McKeith.

The OIII Bowen fluorescent mechanism seems to play a role in many astrophysical sources: the solar chromosphere, interacting binaries, the x-ray source Sco X-1, and active galactic nuclei; but the lines are studied in PN with a minimum of interference. Studies by O'Dell et al (1992) and others show that charge exchange with OIV can be important and emphasize the need for detailed quantitative theoretical and observational work.

On the other hand, NIII lines, such as 4097, 4103, 4634, and 4640, long popularly attributed to the NIII Bowen fluorescent mechanism cannot thus arise (Kastner & Bhatia 1991). Ferland (1992) finds that direct continuum fluorescence gives an intensity prediction for 4634 in good agreement with the observations; this line may serve as a direct probe of the stellar continuum in the 374A region, thereby providing a check on model PNN atmospheres.

**Chemical Abundances**

The results of many investigations indicate the following: (1) There is a large scatter in the C,N,O abundances among galactic disk PNs. The N-rich (Peimbert's type I) constitute a distinct class, but there also exists C-poor objects where the C/O ratio < 0.1 of the solar value. Some C-rich PN also exist. (2) The disk <O/H> is about a half the solar value. In some PN, such as NGC 6537, the (p,O) cycle may convert O to N. Some of the discordance may be blamed on grain formations or $T_e$ fluctuations. (3) He and N are enhanced in Type I PN and C in many objects. Ne and heavier elements were not affected by nucleogenesis in PN progenitors. Some elements, such as Ca and Fe, are tied up in grains. (4) The Ne/O ratio is remarkably stable, ~0.17 by numbers. (5) The N/O ratio is constant up to a core mass M(PNN) of about 0.65. Thereafter, it seems to rise with core mass for awhile or at least with T(PNN). N was produced more copiously in massive stars. (6) Comparison of C/H, O/H, and N/H ratios in PNs with red giants show many facets of similarity but some differences suggesting that the PN material may come from subsurface layers of AGB stars. (7) The CNO cycle and He burning produce most of the C and N in PN. O, largely, and all heavier elements come from the interstellar medium at the epoch of formation of the progenitor stars; the lower the metallicity and mass of the progenitor stars, the lower the N abundance.

Broad surveys of abundances carried out by R. B. C. Henry (1989), Perinotto (1991), and by Henry et al (1989), and by Dopita and Meatheringham (1991) in the Magellanic Clouds show that the C/H, N/H, O/H. and He/H correlations and anti-correlations follow expectations of nucleogenesis theory, at least qualitatively, although improvements in theories of element building are needed.

These broad statistical surveys are invaluable in establishing global patterns, but at some stage we will need more accurate data for individual PN's for which specific nuclear scenarios can be worked out. On the one hand, we need accurate spectrophotometric data over a wide wavelength range and on the other, fairly sophisticated models calculated eventually for an evolving PN. For NGC 7027, I compare 3 abundance estimates as obtained by Keyes et al (1990), Middlemass (1991) and Hyung with similar observational data. We normalize to N(H) = 10,000.

| el. | He | C | N | O | Ne | Mg | S | Cl | Ar |
|---|---|---|---|---|---|---|---|---|---|
| K. | 1100 | 6.9 | 1.3 | 3.1 | 1.0 | 0.26 | 0.071 | 0.002 | 0.021 |
| M. | 1000 | 13.0 | 1.9 | 5.5 | 1.1 | 0.21 | 0.086 |  | 0.025 |
| H. | 1000 | 8.5 | 1.8 | 4.6 | 0.9 | 0.20 | 0.10 | 0.0028 | 0.025 |

For some elements the discordances are still too large. Care must used with the observations. Fluxes of stronger lines often pertain to the integrated PN image, but for weaker lines we often employ a pencil beam through the image (Aller 1990).

## Planetary Nebulae in the Galactic Center and in Other Galaxies

Great emphasis has been placed on PN in the galactic center, in the Magellanic Clouds and other galaxies. We give up spatial resolution, accept small fluxes, but have the advantage of dealing with a sample at a known distance, thus avoiding the worst problem in galactic PN research.

Extensive studies by Pottasch and collaborators and by Stasinska et al (1991) suggest that the total number of galactic bulge PNs is about 600-700, <M(PNN)> = 0.593. Pottasch (1990) finds the distance to the galactic center to be in harmony with that found by other methods, somewhere near 7.5 to 7.8 kpc.

Various observers have carefully studied PNs in the Magellanic Clouds: Dopita and Meatherington measured PN masses, constructed HR diagrams for PNN and obtained nebular parameters and chemical compositions. Most of the PNN seem to have been caught on the H burning excursion to high temperatures, their T(eff) agreeing well with T(Zanstra). They find a huge range in PN masses. One of the most interesting objects in the SMC is the X-ray source, N67, investigated by Qinde Wang (1991).

The greatest of PN paradoxes is that although distances of individual local objects are hard to get, these nebulae yet serve as one of the most useful of standard candles for determining the distances of galaxies more remote than 10 Mpc. The reason is that the PN luminosity function is remarkably stable from galaxy to galaxy. The method is insensitive to the color or Hubble type of the host galaxy (Jacoby et al 1990), although metallicity may have a small effect (Ciardullo and Jacoby 1992). With modern methods special filters to isolate the [OIII] lines, etc., a group of PN can be measured in a galaxy, such as M81, and a luminosity function constructed and compared with that of a nearby galaxy of known distance, such as M31.

Thus planetary nebulae have paid off in a big way. Once regarded only as useful objects for visitors nights and for testing physical theories of atomic processes, they have emerged as a window of the last days of an ordinary star, as probes for stellar nucleosynthesis processes, and lastly, as yardsticks for measuring the nearby universe. The preparation of this summary was aided in part by National Science Foundation grant NSF AST 90-14133 to UCLA.

## REFERENCES

Acker, A., Stenholm, B., Tylenda, R. (1989), Astr.& Ap.Suppl., **77**, 487.
Aller, L.H. 1990, PASP, **102**, 1097.
Balick, B. 1987, Astron.J., **94**, 671.
Bobrowsky, M. & Zipoy, D.M. 1989, Ap.J., **347**, 307.
Bond, H.E. & Livio, M. 1990, Ap.J., **355**, 568.
Ciardullo, R. & Jacoby, G. 1992, Ap.J., **388**, 208.
Clegg, R.E.S. & Harrington, J.P. 1989, **239**, 869.

Dinerstein, H., Lester, D., Carr, J., & Harvey, P. 1988, Ap.J., **327**, L27.
Dopita, M. & Meatheringham, S. 1991, Ap.J., **367**, 115.
Dyson, J.E., Hartquist, T.W., Pettini, M. & Smith, L.J. 1989, MNRAS, **241**, 625.
Ferland, G.V. 1992, Ap.J., **389**, L63.
Furton, D.G. & Witt, A.N. 1992, Ap.J., **386**, 587.
Henry, R.B.C. 1990, Ap.J., **356**, 229.
Henry, R.B.C., Liebert, J., & Boronson, T.A. 1989, Ap.J., **339**, 812.
Howe, D.A., Millar, T.J. & Williams, D.A. 1992, MNRAS, **255**, 217.
Huggins, P.J. & Healy, A.P. 1989, Ap.J., **346**, 201.
Icke, V., Balick, B., & Frank, A. 1992, Astr.& Ap., **253**, 224.
Jacoby, G., & Kaler, J.B. 1989, Astron.J., **98**, 1662; Ap.J. **345**, 871.
Jacoby, G., Ciardullo, L., & Ford, H., 1990, Ap.J., **356**, 332.
Jacoby, G., Walker, A.R., & Ciardullo, R., 1990, Ap.J., **365**, 471.
Jura, M., 1992, Astr.Ap.Reviews, **2**, 227.
Kahn, F.D. & Breitschwerdt 1990, MNRAS, **242**, 505; **244**, 521, 526.
Kaler, J.B., Feibelman, W.A., & Henrich, H.F. 1988, Ap.J., **324**, 528.
Kaler, J.B., Shaw, R.A., & Kwitter, K.B. 1990, Ap.J., **359**, 392.
Kastner, S. & Bhatia, A.E. 1991, Ap.J., **381**, L59.
Keyes, C.D., Aller, L.H., & Feibelman, W.A. 1990, PASP, **103**, 59.
Krishna, M., Apparao, V., & Tarafdar, S.P. 1989, Ap.J., **344**, 826.
Lame, N.J. & Ferland, G.J. 1991, Ap.J., **367**, 268.
Lenzuni, P., Natta, A., & Panagia, N., 1989, Ap.J., **345**, 306.
Masson, C.R. 1989, Ap.J., **346**, 243.
McCarthy, J.K., Mould, J.R., Mendez, R.H., Kudritzki, R.P., Husfeld, D., Herrero, A. & Groth, H. 1990, Ap.J., **351**, 230.
Middlemass, D., 1990, MNRAS, **244**, 294.
Middlemass, D., Clegg, R.E.S. & Walsh, J.R. 1989, MNRAS, **239**, 1.
Miranda, L.F. & Solf, J. 1991, Astr.& Ap., **252**, 331.
Morris, M. 1990, From Miras to Planetary Nebulae, ed. M. Mennessier, France, Editions Frontieres, p. 520.
Napiwotzki, R., & Schönberner D. 1990, Toulouse White Dwarfc Workshop.
O'Dell, C.R. & Miller, C.O. 1992, Ap.J. **390**, 219.
Pascoli, G. 1990, Astr.& Ap., **232**, 184.
Peimbert, M. 1990, Rep.Prog.Phys., **53**, 159.
Peimbert, M., Torres-Peimbert, S., 1987, Rev.Mex.Astr.Astrofis., **15**, 117.
Perinotto, M. 1991, Ap.J.Suppl., **76**, 687.
Pismis, P. 1989, MNRAS, **237**, 611.
Pottasch, S.R. 1990, Astr.& Ap., **236**, 231.
Pottasch, S.R. & Zijlstra, A.A. 1992, Astr.& Ap., **256**, 251.
Reay, N.K., Walton, N.A., & Atherton, P.D. 1988, MNRAS, **232**, 615.
Sabbadin, F. 1986, Astr.& Ap., **160**, 31.
Sahal, R., Wootten, A., Schwarz, H.E., & Clegg, R.E.S. 1991, Astr.& Ap., **251**, 560.
Shaw, R.A. & Kaler, J.B. 1989, Ap.J.Suppl., **69**, 495.
Stasinska, G., Fresneau, A., Gameiro, G., Acker, A. 1991, Astr.& Ap., **252**, 762.
Wang, Qinde 1991, MNRAS, **253**, 47P.
Webster, B.L., Payne, P.N., Storgy, J. & Dopita, M. 1988, MNRAS, **235**, 533.
Zhang, C.Y. & Kwok, S. 1990, Astr.& Ap., **237**, 479.
Zhang, C.Y. & Kwok, S. 1991, Astr.& Ap., **250**, 179.

# I. THE OBSERVATIONAL PN DATABASE

L.H. ALLER
Introductory review

B. STENHOLM

J.H. LUTZ

S.R. HEAP

# CLASSIFICATION CRITERIA AND DATABASES

BJÖRN STENHOLM

*Lund Observatory, Box 43, S-221 00 Lund, Sweden*

**Abstract**

In this review lecture the increase of fundamental data for planetary nebulae is shortly reflected. Special attention is given to the new general catalogue of galactic planetary nebulae, and selection criteria for the entries are summarised. Some information on planetary nebula data in the Magellanic Clouds is also given.

## 1 Introduction

According to modern language, the term 'planetary nebula' is misleading, the objects of this kind have nothing to do with planets at all. But the view through the telescope by Antoine Darquier at the end of the 18th century, 'the fading disc of a planet' (for the Ring Nebula), was a hint that at least this object had a distinct morphology. Almost a century later, William Huggins spotted his spectroscope towards the heavenly bodies, noting that 'planetary nebulae' had an emission line spectrum of certain character. Since then, morphology and spectroscopy have gone hand in hand in the exploration of the ever-increasing number of planetary nebulae, combined with more modern observational approaches. Among them there is a wild variety of forms, and even the spectral appearance varies. The designation 'planetary nebula', however, still remains.

Although numerous listings and catalogues have appeared during the years, e.g. the most famouos of them all, the *Catalogue of Galactic Planetary Nebulae*, Perek and Kohoutek (1967)[2] , there is still reason to ask what a PN really is and what it looks like.

The ideal planetary nebula consists of two parts: 1) a spherical cloud or shell of gas (and dust) centred around and originating in 2) the central star, which once has expelled the gas shell during the late stages of its evolution. The central star, now a hot star of small diameter, is radiating mainly in the ultraviolet, exciting the gas which reradiates most of the energy through a number of emission lines. An ideal PN has thus a well-defined form or morphology, and the nebular component has a well-defined set of spectral lines.

In reality, morphology and radiation pattern can vary widely between individual objects. The morphology is affected by different processes where only little is known: the ejection mechanism, the rotation of the ejecting star, the nebular mass and the interaction between the ejected gas and the interstellar medium. Moreover, the morphology is practically unknown for a large fraction of the population, i.e. those PN which have just a "stellar" appearance because of large distance or compactness. On the other hand, the nebular spectrum is affected by the radiation properties of the central star as well as by the physical properties of the gas shell itself and its composition.

---

[1] The e-mail address of the author is bjorn@astro.lu.se.internet
[2] A fourth supplement to this catalogue is given as a poster at this symposium by L. Kohoutek.

This was the problem for the authors of the new *Strasbourg-ESO Catalogue of Planetary Nebulae*, Acker *et al.* (1992), SESO Cat, which is just released. Their selection criteria is given below in some detail.

## 2  How to tell a cat from a meat loaf

The problem may seem inappropriate, but at least in some applications it is important to know whether a population is homogenouos or not, i.e. if the objects in the sample represent for example the same evolutionary status or just happen to have important features in common. Therefore, along with the compilation of the new catalogue, the authors decided to perform spectroscopic observations of all possible PN candidates. This was also done. Optical spectroscopy in the region of approximately 400 – 740 nm was gathered for about 1 450 objects during the years 1985 – 1991. Armed with this information, along with other relevant data and a good portion of courage, the authors divided the PN candidate population in three parts: 1) true PN, 2) possible PN and 3) objects which clearly are something else than PN. The main criteria for selection into the three groups were in short:

1. **True planetary nebulae** *(1 143 objects)*

    1. Objects with properties close to the ideal PN mentioned above, preferably with an identified central star.
    2. Objects with stellar or nearly stellar appearance showing a spectrum close to the ideal one. If an IRAS observation exists, the object should be in the appropriate region of the IRAS two-colour diagramme.
    3. Objects with non-ideal spectra, showing low excitation, sometimes with an infrared excess and molecular emission, being young PN.

2. **Possible planetary nebulae** *(347 objects)*

    1. Objects without indentification, which therefore cannot be observed correctly.
    2. Objects proposed as planetary nebulae but for which no informative spectrum could be obtained because of faintness or high reddening.
    3. Objects of general unclear status with well-known spectra but possibly related to PN or objects which could evolve into PN.

3. **Objects not being planetary nebulae** *(330 objects)*

    1. Stellar objects showing a continuous spectrum, or showing a stellar continuum with emission lines not typical for PN.
    2. Objects showing a considerable redshift.
    3. Objects showing low-excited PN spectra but being in the wrong place in the IRAS two-colour diagramme.
    4. Non-existing objects.

The objects in the first category can be regarded as a very homogenous sample of PN in the Galaxy. The items here are really PN up to the highest degree of probability. A rough analysis of this population is given in the SESO Cat, such as galactic distribution, distributions of angular diameters, radial and expansion velocities etc.

Regarding the second category it was the ambition of the authors to keep the number as close to zero as possible. However, there is still a considerable amount of objects present. It is obviuos, that the above cited selection criteria is not sufficiently covering, and that there are a number of objects for which observational information is difficult to obtain. It is, in other words, still difficult to tell cats from meat loaves, at least when they are too distant!

In the third category we find objects which clearly are something else than planetary nebulae, namely:

| | |
|---|---|
| Galaxies | 50 |
| H II regions | 35 |
| Symbiotic stars | 70 |
| M stars | 35 |
| Other emission line stars | 45 |
| Objects without detectable emission | 95 |

Among the galaxies, one fourth of them shows highly red-shifted emission lines. The symbiotic stars are classified following the characteristic emission at 683 nm in most cases. A number of objects seem to be just ordinary M stars. The reason for this is obviously that the continuum between the TiO bands has been confused with emission lines on objective prism plates. There are also other kinds of emission line stars, such as Wolf-Rayet stars and Be stars, in the group. There are also many candidates which did not show any detectable (line) emission. They are called "non-existing objects" above. A number of them are known as plate faults, the remaining are thought to be artifacts of similar character.

## 3 The planetary nebula database

### 3.1 Planetary nebulae in the Galaxy

Although the number of PN in the Galaxy may be estimated to several thousand, maybe ten thousand or more, only a fraction of these can actually be observed. If we believe that the new catalogue is representative for our present knowledge of the galactic population, which contains about 1 500 true or possible PN, it is therefore completely clear that the majority of PN is unobservable, at least in the visual domain. Further observations of the 'possible' PN in the catalogue will clarify the situation for them. Still more objects will be found, following classical methods of discovery, but analyses of the IRAS observations seem to be more promising. Preite-Martinez (1988) studied the IRAS Point Source Catalogue finding 340 objects as possible PN, mainly due to their infrared colours. Some of these objects have later been confirmed in radio and optical searches. A method of discovery combining infrared (IRAS) and radio observations is outlined in Pottasch et al. (1988) presenting also 36 new PN around the direction of the galactic centre. Additional 48 PN were presented in a subsequent paper, Ratag et al. (1991), and a third paper dealt with radio continuum observations of IRAS PN sources, Ratag and Pottasch (1991). Their work is summarised in Pottasch et al. (1990), which also gives an interesting comparison of the distributions of 3 000 unidentified IRAS sources with PN colours and the SESO Cat sample of (optically) known PN.

Fundamental radio data from the VLA are given in Zijlstra et al. (1989) and Aaquist and Kwok (1990). The first paper presents radio fluxes, diameters and positions for about 300 PN. The second paper gives radio data at 6 cm for 174 objects, most of them being optically unresolved. Combined with IRAS data this gives the total infrared flux, dust temperature and infrared excess.

Spectroscopic line intensity data are also given in some of the above mentioned papers. The SESO Cat contains line data for lines of major importance as obtained in the spectroscopic survey, on which the selection of cat objects rests. Line data from this survey are also given in Acker et al. (1989b and 1991). Data from the survey have been used in a number of papers to derive nebular and stellar parameters, e.g.:

- Central star $B$ and $V$ magnitudes, Tylenda et al. (1991b), $\approx$ 350 objects.
- Absolute H$\beta$ fluxes, Acker et al. (1991), 880 objects.
- Extinction constants, Tylenda et al. (1992), $\approx$ 900 objects.
- Zanstra temperatures of central stars, Gleizes et al. (1989), 94 objects.
- Energy-balance (Stoy) temperatures of central stars, Preite-Martinez et al, (1989), 388 objects, (1991), 184 objects.
- Chemical compositions and galactic gradients, Köppen et al. (1991), 86 objects.
- Masses for PN nuclei in the galactic bulge, Tylenda et al. (1991a), 100 objects.

Shaw and Kaler (1989) studied a sample of 145 PN in the southern sky, measuring continuum and line fluxes, determining stellar and nebular parameters. Stanghellini and Kaler (1989) calculated electron densities for 146 objects using a large sample of forbidden lines. Another work concerning fundamental data is an up-dating of Kaler's (1976) catalogue of relative line intensities by Kaler and Browning (1992). Cahn et al. are publishing a catalogue of H$\beta$ fluxes, 468.6 nm intensities, extinctions and radio fluxes.

In the ultraviolet spectral region, we note the appearance of an IUE catalogue of PN compiled by Feibelman et al. (1988). In two subsequent papers, Feibelman and Bruhweiler (1990) and Feibelman et al. (1991), altogether 26 central stars are studied based on IUE observations.

Zhang and Kwok (1991) make use of observations in a wide spectral range, between 0.1 and 100 $\mu$m, to map the spectral distribution for 66 compact PN. They fit model curves to the observed data and derive stellar temperature, gas and dust temperature and interstellar extinction.

There is at present no good method to compute individual distances for most PN. However, distance estimations are necessary for derivation of certain stellar and nebular physical parameters. To avoid this difficulty with the whole sample of galactic PN, one has instead selected PN close to the direction of the galactic centre, assuming then that that sample really is situated in the galactic bulge, i.e. at the common distance of $\approx$ 8 kpc. Such a sample was the subject for the thesis by Ratag (1991), which discusses several aspects of PN situated within about 20 degrees of the galactic centre. In total, 110 PN are studied. Abundances for the most important elements are derived. The results are discussed in relation to stellar and galactic evolution.

Acker et al. (1991a), Stasińska et al. (1991) and Tylenda et al. (1991a) also treat a sample of galactic bulge PN. In their case the number of objects is 275 PN. In the first paper the scene is set, presenting the data together with derived basic parameters. In the second paper some statistical properties of the nebular shells are given, including a distribution of the derived Shklovsky distances. The third paper is devoted to the central stars and its mass distribution. In a recent paper, Pottasch and Zijlstra (1992) attack the previous investigators because their calculation of Shklovsky distances was based on optical diameters. However, Tylenda et al. (1991a) discuss already consequences of errors in their adopted nebular model.

Also Webster (1988) treats a sample of 65 luminous PN towards the galactic centre with spectroscopic observations of her own. She derives abundances for most of them and discusses the mass distribution and its history in the galactic bulge. Pottasch and Acker (1989) investigate a similar sample with respect to the central stars and their evolution.

## 3.2 Planetary nebulae in the Magellanic Clouds

The problem of distances is also solved for populations of PN in other galaxies than our own. In this case, it is correct to say that the PN sample is situated at the same distance as seen from us. Moreover, the reddening caused by galactic extinction is small. This makes it possible, following Peimbert (1990), to study properties of the PN themselves, e.g. luminosity functions, envelope masses and progenitor masses, as well as properties of the galaxies as systems such as stellar death rate, production rate of the interstellar medium and chemical evolution. However, due to the general large distances to other galaxies, selection effects are introduced as only the brightest part of the complete population can be reached. Information on morphology is also scarce or non-existent.

The method can be applied to the galaxies nearest to us, the Magellanic Clouds, situated at a well-known distance. The PN populations in these galaxies have been surveyed spectroscopically by Boroson and Liebert (1989) who studied the Jacoby sample (68 objects) at Las Campanas Observatory, by Meatheringham and Dopita (1991a, 1991b) and by Vassiliadis et al, (1992), who studied mainly the Sanduleak-MacConnell-Philip sample (130 objects) at Siding Spring Observatory. Some of these objects turned out to be other kinds of objects than PN, but the majority is now regarded as true PN. In a poster at this symposium, Dopita and Vassiliadis present additional spectrophotometric results from the Clouds , as well as an analysis of the whole samples of PN in these galaxies.

The Siding Spring group compared their line intensities with earlier determinations finding excellent or very good agreement in most cases. In their last paper, Vassiliadis et al. (1992), aimed at the faintest objects in their sample, the agreement is less good, giving slightly lower intensities than the comparison values. The Las Campanas group do not make any comparison with previous observations, but a comparison with some objects which also have been observed by the Siding Spring group, gives a good agreement too, at least for brighter lines.

From the line intensities calculated from the Las Campanas observations, several physical parameters were derived and presented in the original paper, Boroson and Liebert (1989), as well as in Henry et al. (1989).

In a paper by Morgan and Good (1992), 86 new PN candidates in the Large Magellanic Cloud are presented, discovered on objective prism plates from the UK 1.2 m Schmidt telescope. This fruitful search, which almost doubles the number of known PN in this galaxy, continues. Dopita and Vassiliadis present in a poster at this symposium new PN from a recent objective prism survey.

In another poster, Walton, Barlow and Clegg present results from spectroscopic investigations of more than 50 central stars of PN in the Clouds. They derive a number of stellar parameters, using the known distance to the sample.

## 4 Final words and future prospects

When I was at the beginning of my career in astronomy, I was told by a senior astronomer, that there was once some kind of belief that some day it would be possible to stop observing. The time which was referred to was the beginning of this century, when a number of large telescopes were being constructed. When these big machines had been in action for some years, the whole sky would have been surveyed, the telescopes could be dismantled or turned into museum pieces, and the astronomers could be kept busy by analysing the observations only.

Almost a century later we know that this scenario never occurred and never will occur. The situation today is the opposite. Almost any observation in any wavelength region promotes further observations with higher resolution or higher photon rate. The period we now have in the rear mirror is a period when we have experienced big efforts to increase the number of objects of interest and to increase the quality of information about the objects. The period we have in front of us contains a number of important points. New optical telescopes with a light-collecting area bigger than ever are being constructed, and telescopes with mirror diameter up to 25 m are planned. New astronomical satellites will soon orbit the earth. As one of the authors of the new PN catalogue, I also hope that that work will inspire and spur observers for years to come. However, when data quality and quantity increase, even this new catalogue will be obsolete. Let us therefore try to make it obsolete as soon as possible! This can, however, only be done with *more and better observations*.

## 5 References

Aaquist, O.B., Kwok, S.: 1990, *Astron. Astrophys. Suppl. Ser.* **84**, 229.
Acker, A., Stenholm, B., Tylenda, R.: 1989a, *Astron. Astrophys. Suppl. Ser.* **77**, 487.
Acker, A., Köppen, J., Stenholm, B., Jasniewicz, G.: 1989b, *Astron. Astrophys. Suppl. Ser.* **80**, 201.
Acker, A., Köppen, J., Stenholm, B., Raytchev, B.: 1991a, *Astron. Astrophys. Suppl. Ser.* **89**, 237.
Acker, A., Stenholm, B., Tylenda, R., Raytchev, B.: 1991b, *Astron. Astrophys. Suppl. Ser.* **90**, 89.
Acker, A., Ochsenbein, F., Stenholm, B., Tylenda, R., Marcout, J. Schohn, C.: 1992, *Strasbourg-ESO Catalogue of Galactic Planetary Nebulae*, ESO Publication.
Boroson, T.A., Liebert, J.: 1989, *Astrophys. J.* **339**, 844.
Cahn, J.H., Kaler, J.B., Stanghellini, L.: 1992, *Astron. Astrophys. Suppl. Ser.* **94**, 399.

Feibelman, W.A., Oliversen, N.A., Nichols-Bohlin, J., Garhart, M.P.: 1988, *NASA Ref. Publ.* 1203.
Feibelman, W.A., Bruhweiler, F.C.: 1990, *Astrophys. J.* **357**, 548.
Feibelman, W.A., Bruhweiler, F.C., Johansson, S.: 1991, *Astrophys. J.* **373**, 649.
Gleizes, F., Acker, A., Stenholm, B.: 1989, *Astron. Astrophys.* **222**, 237.
Morgan, D.H., Good, A.R.: 1992, *Astron. Astrophys. Suppl. Ser.* **92**, 571.
Henry, R.B.C., Liebert, J., Boroson, T.A.: 1989, *Astrophys. J.* **339**, 872.
Kaler, J.B.: 1976, *Astrophys. J. Suppl. Ser.* **31**, 517.
Kaler, J.B., Browning, P.: 1992, (private communication).
Köppen, J., Acker, A., Stenholm, B.: 1991, *Astron. Astrophys.* , **248**, 197.
Meatheringham, S.J., Dopita, M.A.: 1991a, *Astrophys. J. Suppl. Ser.* **75**, 407.
Meatheringham, S.J., Dopita,.M.A.: 1991b, *Astrophys. J. Suppl. Ser.* **76**, 1085.
Morgan, D.H., Good, A.R.: 1992, *Astron. Astrophys. Suppl. Ser.* **92**, 571.
Peimbert, M.: 1990, *Rep. Prog. Phys.* **53**, 1559.
Perek, L., Kohoutek, L.: 1967, *Catalogue of Galactic Planetary Nebulae*, Academia, Praha.
Pottasch, S.R., Acker, A.: 1989, *Astron. Astrophys.* **221**, 123.
Pottasch, S.R., Zijlstra, A.A.: 1992, *Astron. Astrophys.* **256**, 251.
Pottasch, S.R., Ratag, M.A., Olling, R.: 1990, 'Newly discovered young planetary nebulae', in *From Miras to planetary nebulae: Which path for stellar evolution?* Eds. M.O. Mennessier, A. Omont, Editions Frontières, Gif sur Yvette, p. 381.
Pottasch, S.R., Bignell, C., Olling, R., Zijlstra, A.A.: 1988, *Astron. Astrophys.* **205**, 248.
Preite-Martinez, A.: 1988, *Astron. Astrophys. Suppl. Ser.* **76**, 317
Preite-Martinez, A., Acker, A., Köppen, J., Stenholm, B.: 1989, *Astron. Astrophys. Suppl. Ser.* **81**, 309.
Preite-Martinez, A., Acker, A., Köppen, J., Stenholm, B.: 1991, *Astron. Astrophys. Suppl. Ser.* **88**, 121.
Ratag, M.A.: 1991, *A Study of Galactic Bulge Planetary Nebulae*, Thesis, Rijksuniversiteit Groningen.
Ratag, M.A., Pottasch, S.R.: 1991, *Astron. Astrophys. Suppl. Ser.* **91**, 481.
Ratag, M.A., Pottasch, S.R., Zijlstra, A.A., Menzies, J.: 1990, *Astron. Astrophys.* **233**, 181.
Shaw, R.A., Kaler, J.B.: 1989, *Astrophys. J. Suppl. Ser.* **69**, 495.
Stanghellini, L., Kaler, J.B.: 1989, *Astrophys. J.* **343**, 811.
Stasińska, G., Tylenda, R., Acker, A., Stenholm, B.: 1991, *Astron. Astrophys.* **247**, 173.
Tylenda, R., Stasińska, G., Acker, A., Stenholm, B.: 1991a, *Astron. Astrophys.* **246**, 221.
Tylenda, R., Acker, A., Stenholm, B., Gleizes, F., Raytchev, B.: 1991b, *Astron. Astrophys. Suppl. Ser.* **89**, 77.
Tylenda, R., Acker, A., Stenholm, B., Köppen, J.: 1992, *Astron. Astrophys. Suppl. Ser.* (in press).
Vassiliadis, E., Dopita, M.A., Morgan, D.H., Bell, J.F.: 1992, *Astrophys. J. Suppl. Ser.* (submitted).
Webster, B.L.: 1988, *Mon. Not. R. astr. Soc.* **230**, 377.
Zhang, C.Y., Kwok, S.: 1991, *Astron. Astrophys.* **250**, 179.
Zijlstra, A.A., Pottasch, S.R., Bignell, C.: 1989, *Astron. Astrophys. Suppl. Ser.* **79**, 329.

J. SCHIESTL, H. HARTL

# OBSERVATIONAL PARAMETERS: DEFINITIONS AND LIMITS

J.H. LUTZ

*Program in Astronomy, Washington State University, Pullman, Washington USA*

This review paper gives an overview of many of the observational topics that will be covered during the symposium. Trends and new developments in planetary nebula observations will be highlighted.

## 1. Introduction

The definition of a "planetary nebula" (PN) is clear enough as long as the terms are kept fairly general. A PN is a star that ejects some material while evolving from the red giant to the white dwarf stage. However, when we try to get more specific about particular sub-phases of the evolution, different observers adopt quite diverse terminologies. The following adjectives have been used to describe groups of planetary nebulae in the recent literature: proto-planetary nebulae, young pre-planetary nebulae, young PN, compact PN, evolved PN, large PN and higherly evolved PN. Presumably this list is in some rough order from the beginning to the end of the PN stage, but this is not entirely clear.

Further complicating the matter are several types of objects that are thought to be related, at least in part, to the PN stage. These include OH/IR stars, some IRAS sources, Mira variable stars, symbiotic stars, and calaclysmic variables. All but the cataclysmic variables are associated with the early stages of PN evolution.

Observational questions such as "what objects are representative of the earliest stage at which we can detect the nebula" will be the subject of much debate during this conference. Also, the extent to which objects such as OH/IR stars represent a stage in the evolution of all PN remains to be studied.

We are in a fascinating time with regard to the observations of PN. New hardware such as CCD's, IR detectors and space-borne telescopes have greatly expanded the capacity to obtain data. On the other hand, some fundamental data (such as distances) remain surprisingly elusive.

## 2. How Many and Where?

PN are spreading out. The number quoted for scale height in textbooks and previous symposia is about 150 pc. However, Zilstra and Pottasch (1991) find that the scale height is somewhat larger: 250±50 pc assuming an exponential model of the galactic disk or 190±40 pc assuming an isothermal disk. Their estimate for the total number of PN in the disk is 23,000±6000.

Accurate distances to large numbers of individual PN remain elusive. Most studies continue to rely upon the relatively crude statistical methods. However, a new application of a method that was used a number of years ago to find distances to individual PN from optical images shows great promise. This is the expansion parallax as measured from radio maps obtained at different epochs. Masson (1989) used VLA maps from three different epochs of observation to calculate a distance of

880±150 pc for NGC 7027. Seaquist (1991) was able to put a limit on the distance of $Vy2-2$ by measuring the radio expansion parallax.

## 3. Temperatures, Densities, Abundances

CCD's have made possible a vast expansion of the databases on PN physical parameters. Still, the typical PN with visible shells have electron temperatures of about $1-2 \times 10^4 K$ and electron densities of about $10^2-10^7$ cm$^{-3}$.

Authors tend to confine their studies to a particular physical parameter or to some identifiable subset of PN. For example, Stanghellini and Kaler (1989) studied electron densities derived from [O II], [Cl III], [S II] and [Ar IV] in 146 PN. Rowlands (1989) investigated electron temperatures in the high excitation zones of PN by using [Ne V] lines. Kaler (1990) concentrated on the characteristics of 75 large PN. Henry (1990) included 192 PN from the Milky Way Galaxy, the LMC, the SMC and M31 in a comprehensive study of abundance patterns. He concluded that nitrogen is produced via different thermonuclear cycles in different groups of PN.

Connected intimately to the determination of physical parameters is the issue of whether a particular PN is density bounded or ionization bounded. Zahng and Kwok (1991) analyzed spectra from 0.1 to 100 microns for 66 compact PN and concluded that about 90% of them were ionization bounded. Further pursuit of their broad-based approach will lead to progress on the issue of whether or not most PN have neutral material surrounding the ionized regions.

Studies of PN in the galactic bulge have addressed a variety of topics Webster (1988) provided the first large scale investigation of the masses and abundances of bulge PN. Most recently Ratag, et al (1992) studied 120 bulge PN. Dopita, et al (1990) did high resolution imaging of bulge PN, achieving a resolution 0.35 arcsec by using photon counting and image reconstruction techniques.

A review of the considerable literature that has accumulated recently on extragalactic PN will be left to those speaking on the subject later in the conference.

## 4. Molecules, Infrared, Dust

At previous IAU symposia on PN, studies of molecules were scarce and the number of objects detected was low. What a change we see now! In particular, there has been an extension of the CO surveys such that detections are occurring frequently. Huggins and Healey (1989) surveyed 100 PN and detected CO in 19 of them. It is now possible to estimate both the molecular mass and the ionized mass of a nebula, and sometimes the molecular mass is considerably larger. For example, Gomez, et al (1992) found that the molecular mass of the young PN M3-28 is 0.2 solar masses, whereas the ionized mass is 0.03 solar masses. In evolved PN the ionized mass dominates (Healey and Huggins 1990).

Other molecular studies include the discovery of ammonia in the proto-planetary nebulae CRL618 and CRL2688 (Martin-Pintado and Bachiller 1992). Many more detections can be expected as millimeter and infrared detectors improve.

In the infrared, IRAS spectra have been a powerful tool for studying dust in PN and related objects. Volk and Cohen (1990) were able to characterize the properties of dust emission in 170 PN by using the IRAS low resolution database. Extended red emission due to dust was discovered in bright PN by Furton and Witt (1992). Polarization studies (Johnson and Jones 1991), are proving to be a valuable tool in relating the characteristics of PN dust shells to those of related objects like Miras, OH/IR stars and carbon stars.

## 5. Shell Velocities, Morphologies and Masses

PN shell expansion velocities are typically from about 10-100 km s$^{-1}$ for the bulk of the material. However, much higher velocities in some regions have been found for objects such as He2-111 Meaburn and Walsh (1989) and NGC7139 (Walton et al 1990).

Regarding morphologies, there have been many CCD images produced of PN and a staggering amount of theory generated to explain their shapes (Frank et al 1990, Soker 1992). It remains to be decided whether stellar winds, binary nuclei or a combination of both are responsible for the shapes of PN.

Very large and faint PN halos are more common than previously believed. Thanks to the large dynamic range of CCD's, the physical properties of large halos can be studied. The masses derived for these halos are impressively large. Manchado and Pottasch (1989) find that the outer halo of NGC6543 has a mass of 1.1 solar masses, whereas the bright PN shell has only 0.07 solar masses.

## 6. Central Stars

There are now many more high quality observations of central star magnitudes than there were five years ago, particularly for PN observable in the southern skies (Jacoby and Kaler 1988, Shaw and Kaler 1989, Tylenda, et al 1989). Temperatures derived from the Zanstra and energy balance method range from 25,000 to 200,000 K (Jacoby and Kaler 1988, Preite-Martinez et al 1991).

A new view of central star spectra and magnitudes is provided by x-ray observations (Apparao and Tarafdar 1989).

## 7. Conclusions

Much of the progress in understanding PN and their relationship to other types of objects has come from expanding the observational database. Advances in CCD's, infrared arrays and millimeter-wave technology have made possible studies that could not have been done five years ago. As has been the case so many times, the new data suggest as many problems and avenues of inquiry as they overcome.

### References

Apparao, K.M.V. and Tarafdar, S.P.: 1989, *Astrophys. J.*, 344, 826.
Dopita, M.A., Henry, J.P., Tuohy, I.R., Webster, B.L., Roberts, E.H., Byun, Y.-I., Cowie, L.L., and Sungaila, A.: 1990, *Astrophys. J.*, 365, 640.
Frank, A., Balick, B., and Riley, J.: 1990, *Astron. J.*, 100, 1903.

Furton, D.G. and Witt, A.N.: 1992, *Astrophys. J.*, 386,587.
Gomez, Y., Rodriguez, L., and Garay, G: 1992 *Astron. Astrophys.*, 258,469.
Healey, A.P. and Huggins, P.J.: 1990, *Astron. J.*, 100, 511.
Henry, R.B.C.: 1990, *Astrophys. J.*, 356, 229.
Huggins, P.J. and Healey, A.P.: 1989, *Astrophys. J.*, 98, 1662.
Jacoby, G.H. and Kaler, J.B.: 1988, *Astron. J.*, 98, 1662.
Johnson, J.J. and Jones, T.J.: 1991, *Astron. J.*, 101, 1735.
Kaler, J.B.: 1990, *Astrophys. J.*, 359, 392.
Kaler, J.B. and Jacoby, G.H.: 1989, *Astrophys. J.*, 345, 871.
Manchado, A. and Pottasch, S.R.: 1989, *Astron. Astrophys.*, 222, 219.
Martin-Pintado, J. and Bachiller, R.: 1992, *Astrophys. J. Lett.*, 391, 693.
Masson, C.R.: 1989, *Astrophys. J.*, 336, 294.
Meaburn, J. and Walsh, J.R.: 1989, *Astron. Astrophys.*, 223, 277.
Preite-Martinez, A., Acker, A., Koppen, J., and Stenholm, B.: 1991, *Astron. Astrophys. Suppl.*, 88, 121.
Ratag, M.A., Pottasch, S.R., Dennefeld, M., and Menzies, J.W.: 1992, *Astron. Astrophys.*, 255, 255.
Rowlands, N., Houck, J.R., Herter, T., Gull, G.E., and Skrutskie, M.F.: 1989, *Astrophys. J.*, 341, 901.
Seaquist, E.R.: 1991, *Astron. J.*, 101, 2141.
Shaw, R.A. and Kaler, J.B.: 1989, *Astrophys. J. Suppl.*, 69, 495.
Soker, N.: 1992, *Astrophys. J.*, 389, 628.
Stanghellini, L. and Kaler, J.B.: 1989, *Astrophys. J.*, 343, 811.
Tylenda, R., Acker, A., Gleizes, F., and Stenholm, B.: 1989, *Astron. Astrophys. Suppl.*, 77, 39.
Volk, K. and Cohen, M.: 1990, *Astron. J.*, 100, 485.
Walton, N.A., Walsh, J.R., and Sahu, K.C.: 1990, *Astron. Astrophys.*, 230, 445.
Webster, B.L.: 1988, *Monthly Notices Roy. Astron. Soc.*, 230, 377.
Zahng, C.Y. and Kwok, S.: 1991, *Astron. Astrophys.*, 250, 179.
Zijlstra, A.A. and Pottasch, S.R.: 1991, *Astron. Astrophys.*, 243, 478.

# RESULTS FROM SPACE

SARA R. HEAP

Laboratory for Astronomy and Solar Physics, Goddard Space Flight Center,
Greenbelt MD, U.S.A.

**Abstract.** Space astronomy has made major and ever increasing contributions to planetary nebula research. Three astronomical satellites — ROSAT, *Hubble*, and EUVE — have been launched since our last meeting five years ago. In addition, SpaceLab experiments flying on the NASA Shuttle have now observed a planetary nebula. After fourteen years, the IUE satellite is still going strong, and IRAS data continue to provide new results on planetaries and their antecedents.

With such a large volume of space data and a broad range in research topics, it is impossible to describe all the results from these instruments. Fortunately, other reviews at this conference by Perinotto (IUE observations of stellar winds) and Zhang (broadband flux distributions) will cover some of these topics. I will limit this review to five topics: (1) the first far-UV spectrum of a planetary, (2) new observations concerning the interaction of a stellar wind and the nebula, (3) the first high-resolution pictures of planetaries made by *Hubble*, (4) new observational evidence on the masses of planetary nuclei, and (5) recent advances in UV spectroscopy of central stars.

**Keywords**: Planetary nebulae; Stars: atmospheres, early-type, mass-loss, evolution of; UV radiation, X-rays

## First Far-UV Spectrum of a Planetary

The far-UV spectral window (from 912 Å to Lyman α at 1216 Å), has special importance for planetaries because it contains: *i)* the O VI resonance doublet, which was detected by *Copernicus* in the winds of virtually all O and WR stars, and *ii)* the Werner absorption bands of molecular hydrogen. In late 1990, the Hopkins Ultraviolet Telescope (HUT) on the SpaceLab mission, ASTRO, obtained a far-UV spectrum of the central star of NGC 1535. Bowers *et al.* (1992) find that the O VI λλ1032,1038 doublet is the strongest wind feature in the spectrum of this hot star ($T_{eff}$=70,000 °K; Méndez *et al.* 1988). The HUT spectrum also shows strong $H_2$ absorption below 1110 Å. As yet, Bowers and his colleagues have not been able to determine the relative contributions of the interstellar medium and the nebula to the observed $H_2$ absorption. However, they note that if all the detected $H_2$ is in a thin shell around NGC 1535, then an upper limit of 0.007 $M_\odot$ of $H_2$ is derived.

## New Observational Evidence Of Wind - Nebula Interactions

Kwok *et al.*'s (1978) interacting wind model leads to two observable consequences: (1) acceleration of the nebula by the high-velocity stellar wind from the hot central star, and (2) production of X-ray emission from the nebula in the collision of the high-velocity wind with the low-velocity nebula. New evidence for both have recently been obtained by space observatories.

*Acceleration of the nebula by the wind.* Figure 1, based on data from Patriarchi and Perinotto (PP;1991), shows the relation between the terminal velocity of the wind and the expansion velocity of the nebula. While there is admittedly a large amount of scatter, it is lessened when the composition of the stellar atmosphere is taken into account. Apparently, carbon-rich central stars (■) -- both Wolf-Rayet stars and hot, C-rich subdwarfs like the nucleus of NGC 7094 -- have more rapidly expanding nebulae than do O or sdO-type nuclei (o) with the same wind velocity. A theoretical $V_\infty$-$V_{exp}$ relation can be computed from the equation of conservation of energy between the fast wind and the expanding nebula at time, $t$,

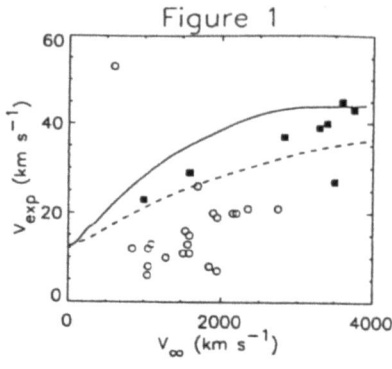

Figure 1

$$\tfrac{1}{2} f \int_0^t \dot{M}(t') V_\infty^2(t') dt' = \tfrac{1}{2} M_{neb} V_{exp}^2(t) \qquad (1)$$

where $f$ is the efficiency of converting the kinetic energy of the wind to the nebula. The "luminosity" of the wind, $1/2\,\dot{m}\,V_\infty^2$, can be computed from radiatively driven wind theory (Kudritzki *et al.* 1989) as a function of stellar mass, temperature, and luminosity. The stellar parameters, in turn, are given by evolutionary models. Figure 1 also plots the nebular expansion velocity computed with Blöcker and Schönberner's (B&S;1990) $m=0.836$ (dashes) and $m=0.605$ (line) models and an assumed AGB wind velocity, $V_{exp}=10$ km s$^{-1}$, a constant conversion efficiency, $f=0.3$, and nebular mass, $m_{neb}=0.2$. For a given $V_\infty$, slowly expanding nebulae are associated with more massive stars. Thus, some of the apparent scatter in Figure 1 may be indicative of a dispersion in stellar masses.

I have used B&S's models since only these models take into account the effects of a stellar wind and have the proper initial mass - final mass relation. Figure 2 shows some of the relevant characteristics of their $m=0.836$ (dashes) and $m=0.605$ models (line). While a star evolves at constant *bolometric* luminosity on the HR diagram (panel *a*), it fades in the UV-optical region of the spectrum (panel *b*). The course of evolution also causes the terminal wind velocity to increase while the rate of mass-loss decreases (panel *c*). The wind luminosity shows a steep rise as the star evolves along the horizontal track followed by a tapering off as the star becomes a white dwarf (panel *d*). B&S found that after completing its horizontal track, the $m=0.836$ model actually evolved more slowly than did the $m=0.605$ model. Consequently, the wind luminosity of the $m=0.836$ models remains higher than the $m=0.605$ model, even after $10^4$ years (panel *d*). This surprising result is consistent with Dopita and Meatheringham's (1990) inference that nebulae with massive nuclei undergo a rapid acceleration as the star

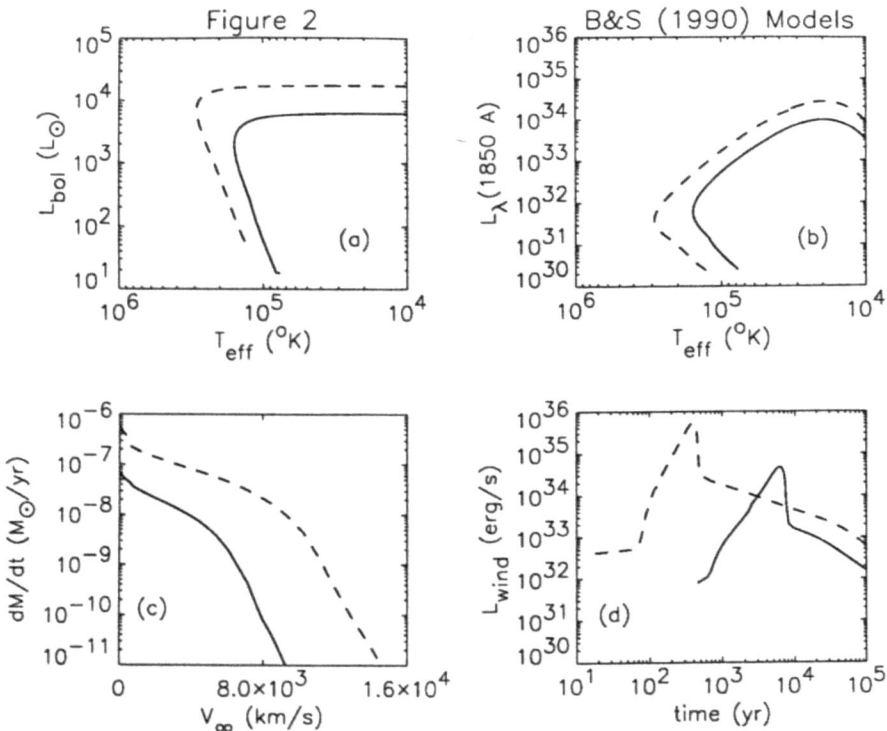

evolves along the horizontal track and continue to be accelerated as the central star fades.

*Extended X-ray emission from planetary nebulae.* The first results from the ROSAT all-sky survey have come in, and for the first time, X-ray emission from six planetary *nebulae* (not the stars) has been detected. According to Kreysing *et al.* (these proceedings; 1992), six planetary nebulae, BD +30°3639, NGC 6543, NGC 6853, A 12, NGC 4361, and LoTr 5, were detected by ROSAT. Perhaps these detections should not be surprising, given the ubiquity and strength of the winds from planetary nuclei -- up to $10^{35}$ erg s$^{-1}$ (panel *d* ). What is puzzling is why these six nebulae and not others were seen.

## High-Resolution Pictures Of Planetaries From *Hubble*

Despite its problems, *Hubble* is starting to fulfill its long-awaited potential for PN research (*c.f.* Barlow 1989). After image-restoration with the Maximum Entropy Method or the Richardson-Lucy method, structure as small as 0".06 can be discerned in *Hubble* pictures. Two international groups, one led by Blades and Barlow (these proceedings; 1992) using the Faint Object Camera, and the other led by Dopita (these proceedings) using the Planetary Camera, have obtained [O III]-filter pictures of planetary nebulae in the Magellanic Clouds. Both groups appear to have been surprised that some nebulae were much smaller than expected, and consequently, their nebular expansion ages were much shorter than their presumed stellar evolutionary ages.

*Hubble* pictures of planetaries in the galaxy are being used to detect faint central stars embedded in bright nebulae and to discern finer detail in the

**Figure 3. NGC 2440 as Viewed by the Planetary Camera.** The 20-minute observation was broken into two 10-minute exposures, so as to identify and remove cosmic-ray hits. The filter employed was the F517N filter, which transmits continuum light in the spectral range, 5130-5210 Å, as well as N I $\lambda\lambda$5198,5200 line emission.

nebulae. As an example, Figure 3 shows a picture of NGC 2440 taken by the Planetary Camera with the F5170N filter, which transmits continuum light from the central star and the emission lines of [N I] $\lambda\lambda 5198,5200$ doublet from the nebula. For the first time, the central star is clearly visible. The Zanstra temperature of the central star, $T_z = 200,000$ °K, derived from the *Hubble* images is in accord with previous ground-based measurements using image-subtraction techniques (Heap & Hintzen 1990).

### The Continuing Saga Of Central-Star Masses

At the last meeting in Mexico City, the Munich group announced a new, spectroscopic method of determining the masses of planetary nuclei (Kudritzki and Méndez 1989, Méndez *et al.* 1988) which is independent of distance and nebular parameters. By comparing the effective temperature and gravity with evolutionary tracks plotted on a $T_{eff}$-log g plane, they were able to derive the mass of a central star. Later, Pauldrach *et al.* (1988) showed that it was possible also to derive the stellar mass from the terminal velocity of the wind as measured on IUE spectra. Since the last meeting, the Munich group lowered its mass-estimates upon finding that the gravities might be underestimated due to filling in of the hydrogen absorption lines by nebular emission and/or wind emission ((McCarthy *et al.* 1990, Gabler *et al.* 1989). Even so, these estimates of stellar mass are still often higher than others (Table 1), and they show a disturbing correlation with temperature.

Since the last meeting, two new distance-independent methods of estimating stellar mass have been developed. One method (Heap, these proceedings; 1992) identifies massive central stars on the basis of UV fading by the central star over the lifetime of IUE. According to Blöcker and Schönberner's (1990) models, a star evolving along the high-luminosity, horizontal track increases in temperature at a rate,

$$dT_{eff}/dt = 4080 \, m^{10.63} \text{ °K per year.} \tag{2}$$

As the temperature increases at constant bolometric luminosity, the UV-optical luminosity decreases as shown in Panel 2b, since a greater fraction of the flux goes into the unobservable extreme-UV. The higher the mass, the faster the fading. Altner and Heap (these proceedings) used the IUE to look for signs of optical fading in planetary nuclei. In fact, they did detect UV-optical fading in just those central stars that Méndez *et al.* (1988) had identified as the most massive in their sample ($m$(spec)$\geq$0.77). However, their optical fading masses, listed as $m(L_{uv})$ in Table 1, are significantly lower than the spectroscopic masses.

The other new distance-independent method makes use of nebular fluxes measured by IRAS and radio telescopes. Zhang and Kwok (these proceedings; 1992) calculate what stellar mass would allow a star to evolve to the observed Zanstra temperature in the time that the nebula has faded to its observed surface brightness. Their results listed under the column $M(T,SB)$, are also shown in Table 1.

Comparing masses in Table 1 is a little like comparing distances (see Terzian, these proceedings): there is no clear best method. Furthermore, there is sometimes agreement among the methods (usually with the hot subdwarfs), and sometimes, strong disagreement (usually for the Of-type stars. See other papers in these proceedings on the Of-type nucleus of NGC 6826 by Gabler *et al.*, Becker & Butler, Altner *et al.*, and the nucleus of NGC 2392 by Heap, Altner & Heap). Nevertheless, there is some convergence in mass estimates compared to five years ago.

## Advances in UV Spectroscopy

With the steady accumulation of UV spectra of planetary nuclei from the IUE has come the realization of the importance of the iron-group elements (Dean and Bruhweiler 1985, Schönberner and Drilling 1985, Hubeny *et al.* 1991). Absorption lines of ionized iron and nickel dominate the UV spectra of all central-star spectra on the horizontal track of the HR diagram *i.e.* not subject to gravitational settling. According to Dreizler and Werner (1992; these proceedings), not only do the iron-group lines affect the appearance of the UV spectrum: they also affect the structure of the stellar atmosphere, so as to produce deeper and broader H and He line profiles. Because the profiles of H and He lines form the observational basis for determining temperature, gravity and helium abundance, their finding has profound implications for future spectroscopic analyses.

The quality of UV spectra has undergone a dramatic boost with the new spectra obtained by the GHRS and FOS spectrographs on the Hubble Space Telescope. As an example, Figure 4 (next page) shows a 10-Å segment of the UV spectrum of the sdO star, BD +75°325, obtained by the GHRS in June 1992. With a resolving power of 15 km s$^{-1}$, this GHRS spectrum has twice the resolution of IUE, and its high signal-to-noise, S/N=70, is totally unattainable by IUE. While not a planetary nucleus, BD +75°325 represents an important test-case for atmospheric modeling. According to Hubeny (1992), who is analyzing the spectrum, NLTE line-blanketed models will be necessary to match the observed spectrum.

With its rich UV spectrum and very sharp lines, the spectrum of BD +75°325 is also an excellent test-case for UV line-lists and associated atomic data and spectral synthesis codes. In anticipation of high-resolution GHRS spectra, Becker and Butler (these proceedings; 1992) made NLTE calculations of the formation of Fe V lines. They found that a full non-LTE treatment of line formation is necessary to interpret the Fe V spectrum observed in O and sdO-type central stars.

In a recent survey of planetary nuclei and hot subdwarfs, Feibelman and Bruhweiler (1990) claimed the detection of Fe VII in the UV spectrum of BD +75°325 as well as in the spectra of planetary nuclei with temperatures as low as 35,000 °K. In the case of BD +75°325, the GHRS spectrum confirms the presence

## Comparison of Stellar Mass Estimates

| ID | $T_{eff}$ | M(spec) | $M(V_\infty)$ | $M(\dot{L}_{UV})$ | M(T, SB) | M(OFD) |
|---|---|---|---|---|---|---|
| | (1) | (2) | (3) | (4) | (5) | (6) |
| A 36 | 95x10³ | 0.61 | -- | | | 0.59 |
| NGC 7293 | 90 | 0.55 | -- | | | 0.66 |
| NGC 7009 | 82 | 0.69 | 0.64 | | 0.63 | 0.66 |
| NGC 4361 | 80 | 0.55 | -- | | 0.59 | 0.62 |
| NGC 1360 | 75 | 0.55 | -- | | 0.56 | 0.58 |
| NGC 3242 | 75 | 0.68 | 0.63 | | 0.61 | 0.65 |
| NGC 1535 | 70 | 0.67 | 0.61 | | 0.61 | 0.63 |
| IC 3568 | 51 | | 0.57 | | 0.61 | 0.55 |
| NGC 6891 | 50 | 0.77 | 0.63 | | 0.61 | 0.58 |
| NGC 6210 | 50 | 0.77 | 0.55 | | 0.63 | 0.66 |
| NGC 6826 | 50 | 0.81 | 0.63 | | 0.59 | 0.59 |
| NGC 6629 | 47 | 0.77 | | | 0.61 | |
| NGC 2392 | 47 | 0.77 | 0.8 | 0.71 | 0.59 | 0.65 |
| IC 4593 | 36 | | 0.68 | 0.72 | | 0.59 |
| IC 418 | 36 | 0.82 | 0.65 | 0.73 | 0.68 | 0.66 |
| Hu 2-1 | 33 | 0.67 | | | 0.68 | 0.67 |
| He 2-108 | 33 | 0.87 | 0.67 | | 0.58 | |
| He 2-131 | 33 | | 0.75 | 0.71 | 0.68 | |
| He 2-138 | 27 | 0.87 | | 0.74 | 0.61 | 0.67 |

2) Mass and $T_{eff}$ derived from non-LTE spectroscopic analysis by Méndez et al. (1988) and McCarthy et al. (1990).
3) Calculated mass assumes wind parameters, α=0.709, δ=0.052, β=1.0; $T_{eff}$=0.9 x that listed here (to account for wind blanketing); $V_\infty$ = 0.85 $V_{edge}$, where $V_{edge}$ is taken from Heap (1986), Pauldrach et al. (1988), and Patriarchi and Perinotto (1991).
4) Optical fading mass from Altner and Heap (these proceedings)
5) Mass derived from $T_{eff}$ and nebular surface brightness, from Zhang and Kwok (1992).
6) Mass based on position of star on an optical fading diagram (OFD); Heap and Augensen (1987), but see also Weidemann's (1989) corrections, which result in lower masses.

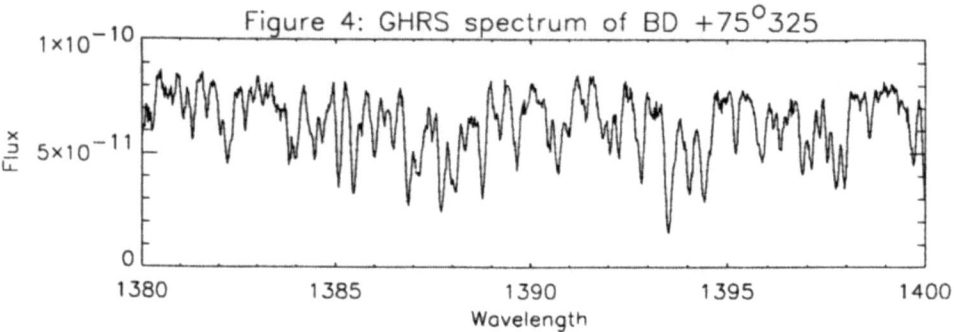

Figure 4: GHRS spectrum of BD +75°325

of an absorption feature near the Fe VII laboratory wavelength, but theoretical calculations by Hubeny indicate that the observed features are in reality due to Ni V, not Fe VII lines. For planetary nuclei in general, Tweedy (1992) disputes many of F&B's detections and identifications, and he maintains that Fe VII lines are present only in stars as least as hot as 70,000 °K.

The UV spectrum has importance not only for estimating abundances but also for determining the fundamental properties of very hot central stars, where the He I/He II ionization balance cannot be used to derive effective temperature. For example, Werner and Koesterke (1992) determined the properties of the nucleus of Abell 78 through a comprehensive analysis of its UV-optical spectrum. They found that one of the best handles on temperature was the ratio of strengths of O V $\lambda1371$/O VI $\lambda\lambda3811,3834$. As another example, Tweedy and Napiwotzki (1992), obtained an approximate value of the effective temperature of the central star of the nearby, large planetary, S 216, from the ratio of strengths of Fe VI to Fe VII lines in the UV spectrum. Clearly, while UV spectroscopy is still in its infancy, it is a field of great potential.

## References

Barlow, M.J. 1989, in *Planetary Nebulae* (IAU Symp. No. 131),
    ed. S. Torres-Peimbert, (Kluwer: Dordrecht), p. 319
Becker, S. & Butler, K. 1992, *Astr. & Ap.*, in press
Blades, J.C., Barlow, M.J. *et al.* 1992, preprint
Blöcker, T. & Schönberner, D. 1990, *Astr. & Ap.*, 240, L11
Bowers, C. *et al.* 1992, in preparation
Dean, C. & Bruhweiler, F. 1985, *Ap.J. Suppl.*, 57, 133
Dopita, M. *et al.* 1992, preprint
Dopita, M. & Meatheringham, S. 1990, *Ap.J.* 357, 14
Dreizler, S. & Werner, K. 1992, in *The Atmospheres of Early-Type Stars*,
    ed. U. Heber, C.S. Jeffery, (Springer-Verlag: Berlin), p. 436
Feibelman, W.A. & Bruhweiler, F.C. 1990, *Ap.J.*, 357, 548

Gabler, R. *et al.* 1989, *Astr. & Ap.*, 226, 162
Heap, S. R. 1986, ESO-SP-263, p. 291
Heap, S.R. 1992, *Ap. J.*, submitted
Heap, S. R. & Augensen, H. J. 1987, *Ap.J.*, 313, 268
Heap, S.R. & Hintzen, P. 1990, *Ap.J.*, 353, 200
Hubeny, I. *et al.* 1991, *Ap.J.*, 397, L33.
Hubeny, I. 1992, private communication
Kreysing, H.C. *et al.* 1992, *Astr. & Ap.*, in press.
Kudritzki, R.-P. *et al.* 1989, *Astr. & Ap.*, 219, 205
Kudritzki, R.-P. & Méndez, R. 1989, in *Planetary Nebulae* (IAU Symp. No. 131), ed. S. Torrest-Peimbert, (Kluwer: Dordrecht), p. 273
Kwok, S. *et al.* 1978, Ap.J., 219, L25
M<sup>c</sup>Carthy, J. *et al.* 1990, *Ap.J.*, 351, 230
Méndez, R. *et al.* 1988, *Astr. & Ap.*, 190, 113
Patriarchi, P. & Perinotto, M. 1991, *Astr. & Ap., Suppl.*, 91, 325
Pauldrach, A. *et al.* 1988, *Astr. & Ap.*, 207, 123
Schönberner, D. and Drilling, D. 1985, *Ap.J. Lett.*, 290, L49
Tweedy, R. 1992, *M.N.R.A.S*, in press.
Tweedy, R. & Napiwotski, R. 1992, M.N.R.A.S., in press
Weidemann, V. 1989, *Astr. & Ap.*, 213, 155
Werner, K. & Koesterke, L. 1992, in *The Atmospheres of Early-Type Stars*, ed. U. Heber, C.S. Jeffery, (Springer-Verlag: Berlin), 288
Zhang, C.Y. & Kwok, S. 1992, *Astr. & Ap. Suppl.*, submitted

R. WEINBERGER, A. ACKER, P.J. HUGGINS

# STRASBOURG-ESO CATALOGUE OF GALACTIC PLANETARY NEBULAE

ACKER AGNÈS and OCHSENBEIN FRANÇOIS
*Observatoire de Strasbourg, 11, rue de l'Université 67000 STRASBOURG, France*

A list of 1820 objects, each of them called at least once a planetary nebula, has been inspected; 1143 of them have been classified as true or probable planetary nebulae by the authors of the catalogue (Tables 1); 347 objects, the status of which is still unclear, were classified among the "possible" planetary nebulae (Table 2).; 330 objects have been rejected, on various grounds, from the community of planetary nebulae (Table 3).

The Catalogue consists of two parts. Part I includes the following chapters:

A  Explanations of the Catalogue
B  Tables
C  References of papers containing 20 objects or more
D  Finding Charts

A new designation system for planetary nebulae following the recommendations of IAU Commission 5 (*Astronomical Nomenclature*) was introduced here with the structure **PN** G$\ell\ell\ell.\ell \pm bb.b$ where **PN** means 'Planetary Nebula', **G** means 'Galactic coordinates' and $\ell\ell\ell.\ell \pm bb.b$ stands for the galactic longitude and latitude respectively, truncated to one decimal place only. Among the 1143 true PN, 846 come from the CGPN, 218 from the supplementary lists compiled by Kohoutek, and 79 were discovered recently. Of the 347 possible PN, 21 come from the CGPN and 82 from the supplementary lists. Of the 330 misclassified PN, 169 are from the CGPN and 101 from the supplementary lists.

Finding charts have been constructed for all *true* planetary nebulae in the catalogue. The identification was possible either from the finding charts provided by the discoverer, appearing mainly in the CGPN, or from information communicated by the observers in the framework of the spectroscopic survey.

In the main catalogue (Part II), the following data are given for each of the 1143 PNs: PN designation and names, discovery lists (cross-identifications), coordinates (1950 and 2000), diameters (optical and radio), radial and expansion velocities, relative line intensities, fluxes in $H\beta$, IR and radio, central star parameters, distances and notes, and bibliography (1065–1991).

Histograms illustrate the distributions of the galactic positions, the angular diameters, the radial (577) and expansion (284) velocities, and the fluxes in $H\beta$ (991), at 6cm (655), and in IR from IRAS (788).

# FOUR NEW EVOLVED PLANETARY NEBULAE

S. TAMURA

*Astronomical Institute, Tohoku University, Aoba-ku, Sendai 980, Japan*

and

R. WEINBERGER

*Institut für Astronomie, Universität Innsbruck, Technikerstr. 25, Austria*

While examining Palomar Observatory Sky Survey prints for various purposes, we came upon a number of hitherto uncatalogued nebulous objects, all of them of low surface brightness. Four of them are considered by us as new planetary nebula candidates due to their morphology. For the brightest one of them, spectroscopic observations were carried out with the Cassegrain spectrograph attached to the 74-inch telescope of the Okayama Astrophysical Observatory: this object (l = 65.49°, b = +3.18°) is clearly confirmed as a planetary nebula and obviously is in an advanced stage in its evolution; in Fig. 1, a spectrum of it is shown.

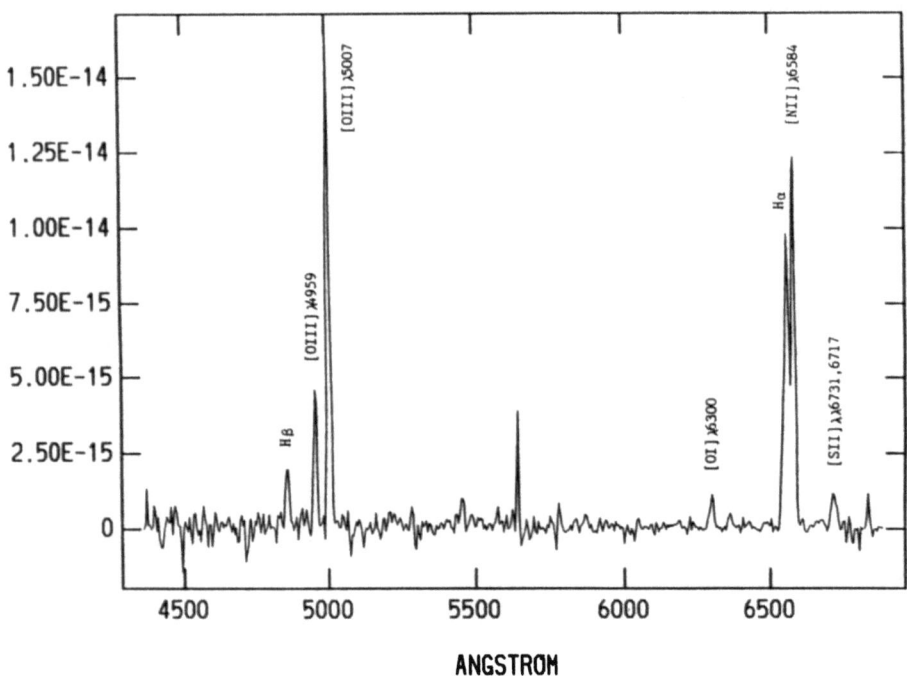

Fig. 1. The low-dispersion spectrum of the new PN at $\alpha=19^h41^m49^s$, $\delta=+30°06'49''$

# NEW PLANETARY NEBULAE

## J.MARCOUT

URA 1280 Observatoire de Strasbourg 11,rue de l'université 67000 STRASBOURG, France

As a complement to the "Strasbourg-ESO Catalogue of Galactic Planetary Nebulae" (Acker et al. 1992), we give the list of newly discovered planetary nebulae.

**New Planetary Nebulae discovered after the publication of the Strasbourg-ESO Catalogue of Galactic Planetary Nebulae**

| PN G | $\alpha_{2000}$ $\delta_{2000}$ | Name |
|---|---|---|
| PN? G000.6 + 00.6 | 17 44.7−28 03 | V744 Sgr; IRAS 17415-2801 OUDMAIJER et al. |
| PN? G004.1 + 11.2 | 17 14.1−19 22 | NSV 8355; IRAS 17112-1918 OUDMAIJER et al. |
| PN? G006.3 − 00.4 | 18 01.9−23 40 | AS 267; IRAS 17588-2340; MHα 359-71 OUDMAIJER et al. |
| PN? G009.3 + 05.7 | 17 45.2−17 56 | IRAS 17423-1755 RIERA et al (Optic. spectrosc.) |
| PN? G009.3 + 00.3 | 18 05.1−20 38 | HEN 1578; IRAS 18020-2037 OUDMAIJER et al |
| PN? G010.6 − 00.0 | 18 09.3−19 43 | BD -19 4866; HD 165970; MWC 909; IRAS 18063-1943; SAO 161116 OUDMAIJER et al |
| PN G065.4 + 03.1 | 19 43.8+30 14 | TaWe a TAMURA et al. (Optic. spectrosc.) |
| PN? G070.8 + 03.7 | 19 54.3+35 07 | TaWe b TAMURA et al. |
| PN? G071.5 + 04.9 | 19 50.7+36 20 | TaWe c TAMURA et al |
| PN? G080.3 − 10.4 | 21 17.1+34 12 | WMP a WERNER et al. |
| PN? G120.3 + 18.3 | 23 48.2+80 56 | SH 2-174 NAPIWOTZKI et al |
| PN? G208.9 − 07.8 | 06 16.2−00 00 | TaWe d TAMURA et al. |
| PN? G217.0 − 00.0 | 05 58.7−03 41 | IRAS 06562-0337 MANCHADO et al. (Optic. spectrosc.) |
| PN? G217.3 + 04.5 | 07 15.6−01 52 | IRAS 07131-0147; WSR a WOLSTENCROFT et al. |
| PN G260.1 + 00.2 | 08 37.4−40 38 | PM 1-41; IRAS 08355-4027; WSR b WOLSTENCROFT et al. (Optic. spectrosc.) |
| PN? G270.1 − 02.9 | 08 59.0−50 24 | IRAS 08574-5011; Vo a VOLK Private communication (Optic. spectrosc.) |
| PN? G293.7 + 01.1 | 11 36.4−60 21 | PM 1-58; IRAS 11339-6004; PMPF a PERSI et al. (JHK photom.) |
| PN? G298.7 + 13.2 | 12 26.5−49 24 | PM 1-66; IRAS 12238-4907; Vo b VOLK Private communication (Optic. spectrosc.) |
| PN? G300.5 − 01.7 | 12 29.1−64 34 | PM 1-67; IRAS 12262-6417; PMPF b PERSI et al. (JHK photom.) |
| PN? G300.8 − 00.7 | 12 33.1−63 34 | IRAS 12302-6317; PMPF c PERSI et al. (JHK photom.) |
| PN? G311.1 − 06.0 | 14 19.1−67 32 | PM 1-80; IRAS 14150-6718; Vo c VOLK Private communication (Optic. spectrosc.) |
| PN? G313.7 + 02.1 | 14 16.9−58 53 | PM 1-79; IRAS 14132-5839; Vo d VOLK Private communication (Optic. spectrosc.) |
| PN G354.4 + 04.0 | 17 15.9−31 22 | Te 233 ACKER et al.(1992 A&A) (Optic. spectrosc.) |

Fig. 1. New Planetary Nebulae Discovered after the publication of the Strasbourg-ESO Catalogue of Galactic Planetary Nebulae (1992)

# NEW AND MISCLASSIFIED PLANETARY NEBULAE

## L. KOHOUTEK
*Hamburg Observatory, Hamburg-Bergedorf, Germany*

This fourth supplementary list to the CGPN (Perek, Kohoutek, 1967) contains 81 new objects (Table 1) which were published mainly between 1987 and 1990. We did not include as new PN those objects, which are in a transition phase between AGB and PN (no emission lines), and possible post-PN namely objects having central stars on the evolutionary way to WD and without nebulae.The possible pre-PN are summarized in a separate incomplete list as an Appendix to Table 1.
In Table 2 it is suggested to remove 86 objects from the CGPN or from the previous supplementary lists of PN ( Paper I,II,III). In case of SS (symbiotic stars) we do not remove them from PN because their symbiotic characteristics do not exclude their simultaneous classification as PN. Even the red continuum including TiO bands cannot be a reason for removing such objects from the list of PN.We are convinced that there exist some SS (especially of type D), which can also be classified as PN. It seems that there exist emission line objects which can be called "symbiotic proto-PN". We neither removed some of those objects, for which the following statements were reported: (a) the IRAS spectrum does not correspond to PN, (b) small or compact H II regions, (c) H$\alpha$ emission line only, (d) no detected objects.
The number of new and misclassified PN differs strongly from that given by Acker and Stenholm (1990) which can be explained by the different classification criteria used (see Paper II and III).Inspite of the fact that the objects called planetaries very probably do not represent a uniform group, the decision about the question "what is or what is not a PN" would require not only a sufficient observational material, but also the correct understanding of the phenomenon "planetary nebula".

References:
Acker A., Stenholm B., 1990, Astron.Astrophys.Suppl.86, 219
Kohoutek L., 1978, IAU Symp.No.76 (ed.Y.Terzian), p.47. (Paper I)
Kohoutek L., 1983, IAU Symp.No.103 (ed.D.R.Flower),p.17. (Paper II)
Kohoutek L., 1989, IAU Symp.No.131 (ed.S.Torres-Peimbert), p.29. (Paper III)
Perek L., Kohoutek L., 1967, Catalogue of Galactic Planetary Nebulae, Academia Praha. (CGPN)

# THREE POSSIBLE PLANETARY NEBULAE FROM NEAR-IR OBSERVATIONS

P. PERSI, A. MARENZI, A. PREITE-MARTINEZ and M. FERRARI-TONIOLO
*Istituto di Astrofisica Spaziale, CP 67, 00044 Frascati, Italy*

A catalogue of 388 new possible planetary nebulae was selected by Preite-Martinez (1988) from the IRAS Point Source Catalogue. These unidentified sources have IRAS colours in the range $F(12)/F(25) \leq 0.35$ and $F(25)/F(60) \geq 0.35$, and are located in the proximity of the galactic equator ( $|b| \leq 15°$ ). In order to identify these IRAS sources we have undertaken a programme of near-IR observations using an InSb photometer and a near-IR camera. We report here on results relative to four of these sources.

A scanning at $2.2\mu$ within the positional error box of IRAS sources 12262−6417, 12302−6317, and 12316−6401 was made using the InSb photometer at the ESO 2.2m telescope, with an aperture of 8 arcsec. Photometry in the J, H, K, L, and M filters was obtained for each source.

The J-H, H-K colours of IRAS 12262−6417 and 12302−6317 show a strong IR excess, interpreted in terms of dust emission at a temperature of 1000−1500K. Only a background star was found in the field of IRAS 12316−6401. The IR energy distributions of IRAS 12262−6417 and 12302−6317 both show a hot as well as a cold component. A similar energy distribution has been observed in symbiotic stars, and in He 2-104 by Schwarz *et al.* (1989).

J, H, and K broad-band images of IRAS 11339−6004 were collected with the ESO Hg:Cd:Te 64×64 pixels array camera (IRAC-1) at the same telescope. The images were obtained with a scale of 0.8 arcsec/pix in beam-switching mode, and were calibrated using standard stars. Using IRAF packages we have derived J, H, K photometry of the four sources closer to the IRAS position. Only one of these show near-IR colours typical of an N-type planetary nebula (Whitelock, 1985; Persi *et al*, 1987).

The source that we associate with IRAS 11339−6004 shows an energy distribution dominated by nebular emission in the near-IR and by a cold component at $T_C = 228K$.

### References

Persi, P., Preite-Martinez, A., Ferrari-Toniolo, M., Spinoglio, L. 1987, *Planetary and protoplanetary nebulae: from IRAS to ISO*, ed. A. Preite-Martinez, Reidel Pub.Co., p.221
Preite-Martinez, A. 1989, *Astron. Astrophys. Suppl. Ser.* **76**, 317
Schwarz, H.E., Aspin, C., Lutz, J.H. 1989, *Astrophys. J. Lett.* **344**, L29
Whitelock, P.A. 1985, *Mon. Not. Royal Astron. Soc.* **213**, 59

# PLANETARY AND PROTO-PLANETARY NEBULAE WHICH ARE STRONG 25μm EMITTERS

R.D. WOLSTENCROFT and M.A. READ
*Royal Observatory, Edinburgh EH9 3HJ, UK*
and
S.M. SCARROTT
*Dept. of Physics, University of Durham, Durham DH1 3LE, UK*
and
C.J. LONSDALE
*IPAC, MS 100-22, Caltech, Pasadena, California 91125, USA*
and
Q.A. PARKER
*Anglo-Australian Observatory, Coonabarabran, NSW, Australia*

The IRAS Point Source Catalogue containing about 250,000 sources has yielded a large number of previously unknown planetary nebulae (PNe) and a smaller number of proto-planetary nebulae (PPNe). The spectral energy distributions for many of these objects peak at or close to 25μm. A program to optically identify sources in the complete IRAS Faint Source Database, which comprises about 750,000 sources, is currently under way (Wolstencroft et al. 1991), and in a related study Wolstencroft, Parker & Lonsdale are carrying out a spectroscopic survey of a small sample of the approximately 106,000 sources which either peak or are detected only at 25μm. So far spectra of 150 sources have been obtained: 3 of these sources are PNe and 1 is a PPNe. This suggests that this sample of 25μm emitters may be a rich source of new PNe and PPNe. In this note we discuss two of these four sources.

IRAS 08355-4027: this source is identified as a PNe based on its spectrum. It lies in the galactic plane and based on the observed H(alpha)/H(beta) is very heavily reddened with an extinction factor c=7.3 compared to the range cited by Osterbrock (1989) of 0.02 to 2.3 for published PNe values. A CCD image of the nebula shows a ring much brighter on one side with a funnel of emission connecting the star and the bright section of the ring. The nebula, which we have labelled the Stealth Bomber, is similar in shape to the PNe A35: Jacoby (1981) has suggested that the shape of A35 is due to the motion of the system relative to the local ISM. The HeII 4686A line strength of IRAS 08355-4027 is comparable with that of H(beta) implying a high excitation (E=88).

IRAS 07131-0147: this is a PNe with a classical bipolar structure. This object is described in detail by Scarrott et al. (1990).

### References

Jacoby, G.H. (1981) Astrophys.J. 244, 903-911.
Osterbrock, D.E. (1989) In 'Astrophysics of Gaseous Nebulae and Active Galactic Nuclei', Ch. 7, University Science, California.
Scarrott, S.M. et al. (1990) MNRAS 245, 484-492.
Wolstencroft, R.D. et al. (1991) In 'Digitised Optical Sky Surveys', Kluwer, Dordrecht, pp 471-483.

# THE SOUTHERN DEEP NEAR INFRARED SURVEY (DENIS)
# A PROSPECT OF PN EXPLORATION

STEFAN KIMESWENGER and CHRISTOPH KIENEL
*Institut für Astronomie der Leopold-Franzens Universität, Technikerstraße 25,*
*A-6020 Innsbruck, Austria*

**Abstract.** The near infrared light is important for the exploration of proto-planetary nebulae as well as for the planetary nebulae in early phases (Persi et. al. 1986, in Planetary and Proto-Planetary Nebulae: From IRAS to ISO, ed A.P.Martinez). Numerous work on the fluxes of the well known planetary nebulae was already done in the late 80's, but a sky survey will give a large sample of data to provide more detailed statistics.

The present work presents an overview of the data on planetary nebulae expected from the European project of a deep near infrared survey of the southern sky (**DENIS**) (IAP and DESPA Paris, Heidelberg, Leiden, IAC Tenerife, Grenoble, Lyon, Frascati, Innsbruck, Vienna) in the I, J and K band with a limiting magnitude of 14.5 to 15 for point sources and 17 mag arcsec$^{-1}$ for the surface brightness. The angular resolution for identification of non-point source objects will be about 5".

### The number of PNe in the survey

The problems of such calculations are the numerous assumptions on the space density, the average size and the average brightness of the objects. Assuming an average size of 0.2 to 0.5 pc (moderately bright surface) we attain up to 20 kpc as the distance of visibility of the PNe. Using a space density and radial scale factor determined by known objects we obtain a total number of PNe of approximately $2.10^4$ objects within this area. Assuming a 'determination efficiency' of 30% we get $6.10^3$ sources.

But due to the lack of extinction in these bands, the density estimates might have to be corrected heavily. Therefore the total number of candidates in the sample might increase by a factor of 10. Smaller sources might be filtered from the stellar sources by color-color diagrams. This might give an extra contribution to the sample.

### The IR data for already known PNe

Since the survey gives very accurate positions the crossidentification to objects of other catalogues will be possible almost 'online' while the survey is running. This will lead to numerous new data on the known objects. In particular statistics on hot dust properties in different stages of evolution can be done easily.

### The central stars of PNe

The IR photometry of the central stars is mainly of interest for a more accurate determination of the extinction of these objects and better estimates of the Plank curves of the photosphere.

Assuming an average absolute magnitude of $M_V \approx 5^m$ and an intrinsic color index similar to a O5V star (Scheffler & Elsässer, 1974) we get 1 kpc as a distance of visibility of the objects.

### References

Balick, B.: 1989 IAU Symp. **131**, 83
Mallik, D.C.V.: 1983 IAU Symp. **103**, 424
Scheffler, H., Elsässer, H.: 1974, Physik der Sterne und der Sonne, B.I. Verlag, Mannheim

# RADIO CONTINUUM OBSERVATIONS OF SOUTHERN PLANETARY NEBULAE CANDIDATES

G.C. VAN DE STEENE and S.R. POTTASCH

*Kapteyn Laboratorium, P.O. BOX 800, 9700 AV Groningen, The Netherlands*

In an attempt to find new Planetary Nebulae we have short radio continuum observations at 6 cm and 3 cm of 90 PN candidates with the Australian Compact Array. We selected the unidentified objects from the IRAS Point Source Catalogue on basis of their PN colors. Detection of radio continuum emission at the IRAS position almost certainly confirms that these objects are PN, because it indicates the presence of ionized gas. Therefore the 18 detected sources are considered new PN. Because of their high brightness temperatures, they are probably young PN.

Their radio and IR properties are remarkably similar to the known PN with stellar appearance studied by Aaquist and Kwok (1990). However, among our detections, PN with higher IR excess are much more abundant. When we compare the radio and IR properties of our discovered PN in the Galactic Plane with the ones in the Galactic bulge, found by Ratag and Pottasch (1990, 1991) using the same method and having approximately the same observational constraints in the radio observations, then the differences are more substantial. The new PN which they detected, have in general a lower brightness temperature and an even higher IR excess. Furthermore we also notice a decrease in IR excess for PN below -60 degrees in Galactic longitude as first mentioned by Zijlstra (1990). We appear to have found an indication for a different nature between Galactic bulge and Galactic plane PN.

## References

A. Aaquist, S. Kwok, 1990, A&A Suppl. Ser. 84, 229
M.A. Ratag, S.R. Pottasch, 1990, A&A 233, 181
M.A. Ratag, S.R. Pottasch, 1991, A&A Suppl. Ser. 91, 481
A.A. Zijlstra, 1990, A&A, 234, 387

# THE PHOTOGRAPHIC OBSERVATIONS OF SOME PLANETARY NEBULAE

XIANGLIANG HAO

*Beijing Astronomical Observatory, Beijing, China*

The first planetary nebula was discovered by Messier in 1794. But for some reasons it has not been studied detail for a long time, especially for the central star Of planetary nebula. The primary research for these objects showed that the lifetime of a planetary nebula is about $5\,10^4$ years, but in this period the luminosity of central star varies from 63 $L_\odot$ to nearly $3.5\,10^4 L_\odot$ and then decrease to 100 $L_\odot$ ;its temperature changes from $3.4\,10^4$ to $10^5 K$ and then begins to decrease (Seaton 1966). The radius of central stars also have fast varies in planetary nebula phase. For these reasons we consider that in the planetary nebula phase the activities of central star is very drastic and the result of these activities must cause some variation at the surface of central star witch may be detected on the earth, especially for the surface light variations. Some observers have been trying to find the luminosity variations in central stars. But until now no one has made systematical survey for these. Since the different authors used different instruments amd different processing methods at different places which may be caused a lot of uncertainty in the photometry of planetary nebulae and central stars. So it is hard to decide whether the differences between the authors or the essential variations of the objects is responsible of the observing differences. Therefore, we have selected over fifty planetary nebulae to observe for a long period at Beijing Observatory using the same instrument and the same processing method. From these observations we may determine the light variations and the brightness of the planetary nebulae and central stars more correctly.

Up to now we have taken over thousand photographic plates of planetary nebulae using the 40/200 cm double astrograph and others at Beijing Observatory since 1979. Some of these observations we have made processing and published in other papers.

### References

M.J. Seaton, 1966, Mon. Not. Roy. Astron. Soc. 132, 113

# CCD SPECTROPHOTOMETRY OF PLANETARY NEBULAE AT WENDELSTEIN OBSERVATORY

M. M. ROTH, R.-P. KUDRITZKI and R. H. MÉNDEZ

*Universitäts-Sternwarte München, Scheinerstr. 1, 8000 München 80, FRG*

During the past decade the achievements in the theory of stellar atmospheres of hot stars combined with improved spectrograph and detector technology at large telescopes have led to a significantly improved knowledge of PN nuclei properties (see Méndez et al. 1988, Kudritzki and Méndez 1989).

The spectroscopic determination of the fundamental stellar parameters effective temperature, gravity, Helium abundance, and derived quantities like mass, luminosity, or distance has the advantage of being independent of assumptions on the nebula as compared to more indirect methods which e.g. are based on the hypothesis of optical thickness to H–Lyman continuum radiation within the nebula (the effects of optically thin or "leaking" nebulae is discussed by Méndez et al., 1992).

Recently, improved NLTE–model–atmosphere calculations ("Unified Models") have become available (Gabler et al. 1989), now taking into account the effects of sphericity and the presence of stellar winds. The far UV flux distribution provided by these calculations enables one to investigate the ratio between stellar flux shortward of the H/He II absorption edges and the corresponding nebular emission line flux, thus comparing with the classical diagnostics for central star temperature (Zanstra method), see Gabler et al. 1991, A&A 245, 587.

The theoretical interest in this topic has motivated us to start a research program of developing a CCD–camera (specifically designed for interference filter spectrophotometry; see Roth 1990) and measuring PN emission line fluxes in the important lines of $H_\alpha$, $H_\beta$, $He\,II_{4686}$, and $[O\,III]_{5007}$. Global fluxes of the latter have become of interest since a comparison of published [O III] fluxes taken together with the spectroscopically derived distances of our sample of central stars yields a luminosity distribution which is consistent with the PNLF as determined by Jacoby and Ciardullo (Jacoby 1989, ApJ 339, 39; Ciardullo et al. 1989, ApJ 339, 53; and subsequent papers).

In view of the upcoming installations of 8m–class telescopes (VLT–ESO, SST–McDonald Obs.) we anticipate that central stars with visual magnitudes up to 16 will become accessible to spectroscopic analysis in the future. In addition to this we intend to conduct an extensive observing program at a local 0.8m telescope (Wendelstein Observatory) in order to check and supplement the existing data base of PN emission line fluxes making use of the CCD observing technique.

## References

Gabler R., Gabler A., Kudritzki R.-P., Puls J., Pauldrach A., 1989, A&A 226, 162
Kudritzki R.P., Méndez R.H., 1989, in: IAU Symp. 131, ed. S. Torres-Peimbert, Kluwer, p.273
Méndez R.H., Kudritzki R.-P., Herrero A., Husfeld D., Groth H.G., 1988, A&A 190, 113
Méndez R.H., Kudritzki R.-P., Herrero A., 1992, A&A 260, 329
Roth M.M., 1990, in: *CCD's in Astronomy*, ed G.H.Jacoby, ASP Conf. Series, p.380

# PLANETARY NEBULAE AS A RESEARCH FIELD

W. SAURER and R. WEINBERGER

*Institut für Astronomie, Universität Innsbruck, Technikerstr. 25, Austria*

The "Astronomy and Astrophysics Abstracts", edited twice a year by the "Astronomisches Recheninstitut" in Heidelberg, served as basis for the determination of some data concerning the development of planetary nebulae as a research field. From the numbered and unnumbered papers within the subject category 134 there it was, for example, possible to compare the development of the PN paper rate with that of the whole field of Astronomy; for the years 1986 to 1990, a list (including postal addresses) of all individuals (ca. 900!) who published at least one paper on PN was made. For these 5 years, we now know which scientist(s) published most, in how many countries research on PN is done, how the annual publication rate varies for a specific country etc. Below, we show two results of our statistics.

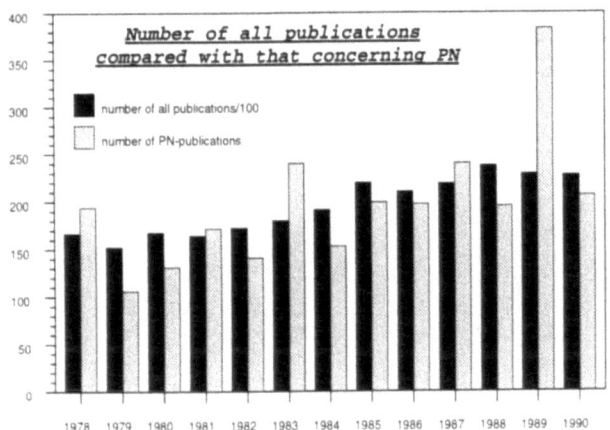

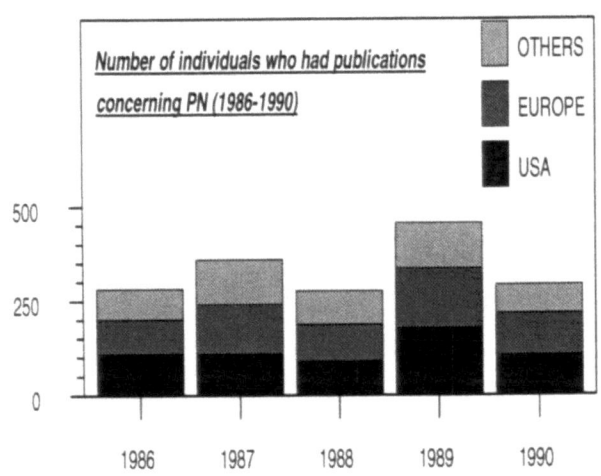

# A CLUSTERING METHOD APPLIED TO THE ANALYSIS OF PLANETARY NEBULAE

O.S.YATSYK

*Astronomical Observatory, L'viv State University Lomonosov St., 8, L'viv 290005*
*Ukraine*

Intensities of the observed spectral lines, radio fluxes and $H_\beta$ fluxes are used for the classification of planetary nebulae by centroid method of taxonomical analysis. Two variants of classification are proposed. The first one– in the three–dimensional space of relation of intensity of spectral lines He II $\lambda$ 4686/HeI $\lambda$ 4471, [OIII] $\lambda$ 4959+5007/[OII] $\lambda$ 3726+29, [OIII] $\lambda$ 4959+5007/[OIII] 4363. The second variants is the classification in the three-dimensional space with cooordinates being radio flux, $H_\beta$ flux and [OIII] $\lambda$ 4959+5007 intensity. The membership of classes (taxons) are presented. In the diagram joining the planetary nebula descriptors pairwise there are regions of complete (or predominant) of nebulae belonging to the same taxon, but there are also some regions of overlap. The corresponding taxons are not isolated but merge continuously into one another.

# II. HIGHLIGHTS ON THE NUCLEI

R.P. KUDRITZKI

M. PERINOTTO

A. PREITE-MARTINEZ

K. BUTLER

# MODEL ATMOSPHERES OF CENTRAL STARS OF PN

R.P. KUDRITZKI and R.H. MÉNDEZ
*University Observatory, Munich, Federal Republic of Germany*

ABSTRACT. We review the work done in this field since the time of the Mexico Symposium on PN, five years ago. Although substantial progress has been achieved, a lot of work is still needed, and we pay attention to some remaining problems. We briefly describe some recent results about stellar and nebular properties, obtained with NLTE model atmosphere techniques.

## 1. Introduction

Work on model atmospheres for central stars of planetary nebulae (CSPN) has two main motivations: (a) to obtain direct information about stellar properties by comparison with the observed stellar spectra, and (b) to provide a more reliable input, in particular more reliable stellar fluxes, for improved nebular modeling.

The most outstanding characteristic of CSPN is that they are very hot, with surface temperatures above 25000 K, and the very high energy density in the radiation fields of such hot stars produces two major difficulties: departures from local thermodynamic equilibrium (LTE) in the photosphere, and the presence of strong stellar winds. In Sections 2 and 3 of this review we describe recent progress and current limitations in the plane-parallel, hydrostatic NLTE treatment. Sections 4 and 5 deal in a similar way with NLTE models including winds. In Section 6 we describe some recent results which are relevant to the problem of the Galactic PN luminosity function.

## 2. "Classic" NLTE models (no metals, no winds)

At the Mexico Symposium we presented (Kudritzki and Méndez 1989; Méndez et al. 1988a,b) a large-scale effort to determine $T_{eff}$, log $g$ and photospheric He abundances of CSPN, by comparison of the observed stellar H and He absorption line profiles with theoretical profiles obtained from plane-parallel, hydrostatic, metal-free (H and He only) NLTE model atmospheres in radiative equilibrium. Subsequently, some more work has been done using the same kind of models (Heber et al. 1988; McCarthy et al. 1990, 1991; Méndez et al. 1990; Herrero et al. 1990; Napiwotzki 1992). This kind of "classic" NLTE analysis is suitable, in principle, for H-rich CSPN showing predominantly absorption-line spectra, if we take into account the corrections we will describe in Section 5. The determination of the basic atmospheric parameters, with typical uncertainties of ±10% in $T_{eff}$

and $\pm 0.2$ in log $g$, permits to place these CSPN in the log $g$ - log $T_{\text{eff}}$ diagram, leading to a better knowledge of their evolutionary properties and to the derivation of luminosities and spectroscopic distances. We will come back to these applications in Sections 5 and 6.

Napiwotzki (1992) has reported the existence of a "Balmer line problem" affecting temperature determinations in the case of very high gravity CSPN, already positioned on the white dwarf cooling tracks. The problem is that, for such high gravities, different Balmer lines indicate different temperatures. The reason for this problem is not clear; it might be attributed to deficiencies in the broadening theory or to metal line blanketing effects. For the moment, we would give more weight to temperatures derived from $H_\gamma$ or $H_\delta$, because, in the case of NGC 7293, they are in better agreement with the H and He II Zanstra temperatures, while $H_\beta$ and especially $H_\alpha$ give, according to Napiwotzki, much lower (clearly unrealistic) temperatures for this central star.

The available information about photospheric He abundances has been recently reviewed by Méndez (1991a). What about other elements? Given sufficiently high spectral resolution and signal-to-noise ratio, the visual spectra of CSPN show many useful lines of C, N, O, Mg, Al and Si, particularly at temperatures below 30000 K. If we are dealing with a H-rich CSPN, then a "classic" NLTE analysis, coupled to adequate NLTE line formation calculations, permits to obtain some information about photospheric abundances. One example of this kind of work is given by McCarthy et al. (1991), who made a determination of the O and Si photospheric abundances in the spectrum of CD-41 13967, the central star of He 3-1863. Microturbulent velocities of 10 and 20 km/s for Si III and O II, respectively, were adopted in order to obtain no dependence of abundance as a function of equivalent width. The O II microturbulent velocity may be a bit too high, perhaps indicating that more improvements are needed (see Section 4 below, and Kudritzki 1992).

The NLTE line formation calculations we have just mentioned are characterized by the assumption that the ion under consideration can be treated as a trace element, so that, given a good (NLTE) model atmosphere, the atmospheric structure can be kept fixed and we only need to solve a system of radiative transfer and statistical equilibrium equations restricted to the ion we are studying. Previous experience indicates that this method provides reliable abundances only if the model ion is sufficiently complex, including many energy levels and transitions: see e.g. Eber and Butler (1988) on C II; see also Becker and Butler (1988, O II; 1989, N II; 1990, Si II, III and IV).

Becker and Butler (1992) have recently completed a Fe V model ion, and have made NLTE line formation calculations aimed at reproducing the HST ultraviolet spectrum of the sdO star BD+75 325, resulting in a determination of the photospheric iron abundance. They have shown that NLTE effects in line formation must be taken into account. Model ions for Ni IV, V and VI have also been developed (Becker and Butler in preparation). This opens the way to the future determination of Fe and Ni photospheric abundances in CSPN. More comments about this project at the beginning of Section 4.

## 3. NLTE line blanketing

The CSPN show such a variety of spectral characteristics (see e.g. Méndez 1991a) that no single set of models can be adequate to study them all. Much work in the last 5 years has been devoted to remove the limitations of the "classic" NLTE models (no metals, no winds), in an effort to extend the NLTE analyses to all kinds of CSPN. In this Section we deal with recent progress in line blanketing problems.

Most remarkable has been the development, by the Kiel group, of plane-parallel, hydrostatic NLTE models for the quantitative analysis of some H-deficient, C-rich CSPN and related objects (the PG 1159 stars); see in particular Werner (1992). The analysis requires NLTE model atmospheres including the effect of line blanketing due not only to H and He, but also at least to C and O, because their abundances are so high that they cannot be treated as trace elements. Therefore it is necessary to introduce, in the model atmosphere code, the statistical equations for the populations of more than 100 atomic levels. This became possible in practice through the application of the so-called "Accelerated Lambda Iteration" method (Werner and Husfeld 1985).

Having obtained a reasonable representation of the atmospheric structure, it is still necessary to solve very complicated line formation problems, because most of the diagnostic lines arise from highly excited levels, thus requiring very complex model atoms. Besides, there is a severe blending problem, because all the ions that produce diagnostic lines have one-electron spectra. And finally, accurate broadening theories for the diagnostic C IV, N V, O VI lines are still lacking (although the situation is improving, at least for C IV; see the new Stark broadening calculations by Schöning 1992). Despite all these difficulties, Werner et al. have been able to obtain satisfactory line profile fits for a handful of PG 1159 stars, giving fundamental information about their $T_{\text{eff}}$, log $g$ and H, He, C, N and O abundances. Further work along this line will be of great help to understand the evolution of H-deficient CSPN.

One of the dreams of NLTE model makers is to be able to treat metal line blanketing in as much detail as in Kurucz's LTE models. Although that is not possible for the moment, some people are making good progress. Grigsby et al. (1992) have used NLTE metal line blanketed model atmospheres with solar abundances (see Anderson 1985, 1989) for the analysis of late O and early B young stars in clusters and associations; similar models can be used to study cool CSPN. Dreizler and Werner (1992) have constructed exploratory NLTE model atmospheres for hot post-AGB stars, featuring H, He and Fe with solar abundances. The NLTE iron line blanketing is treated in a statistical way, following Anderson (1989). As a first step, their models are blanketed by some tens of thousands of observed Fe III - Fe VII transitions. Further work will include all predicted transitions, which are roughly a factor 100 more numerous, as well as other iron-group elements.

Such NLTE metal line blanketing studies will make it possible to investigate, for example, if metal line blanketing can solve Napiwotzki's problem, or, in general, how relevant is metal line blanketing for the determination of $T_{\text{eff}}$, log $g$, energy distributions and photospheric abundances of CSPN. The first results for massive OB stars (Grigsby et al. 1992) would seem to indicate that the "classic" NLTE determinations of $T_{\text{eff}}$ and log $g$ are not wrong by more than their estimated error bars.

Kunze, Kudritzki and Puls (in preparation) are working on a hydrostatic code that includes NLTE opacity due to H, He, C, N, O, Ne, Mg, Al, Si, S, Ar in order to investigate the effects of metallicity on the emergent ionizing continuum flux between 228 and 911 Å. They have also used these models to investigate wind blanketing effects (Abbott and Hummer 1985), finding that wind blanketing does not significantly change the number of ionizing photons (see Kudritzki et al. 1991).

## 4. NLTE models including winds

Let us go back for a moment to the NLTE line formation problem for Ni and Fe. In the case of hot stars, the relevant ions of these elements (Ni IV - VI, Fe V - VII) produce many weak lines in the ultraviolet, giving us a chance to determine the photospheric Ni and Fe abundances. There is, however, a complication, recently described by Kudritzki (1992): for ions with so high atomic weights, the thermal velocities are almost one order of magnitude smaller than the sound speed. As a consequence, even a small, subsonic outflow velocity in the deep photosphere can shift the metal lines by a few thermal Doppler widths into the neighboring continuum, increasing their equivalent widths. Clearly, a hydrostatic NLTE line formation formulation would require the introduction of a spuriously high microturbulent velocity, in order to reproduce the resulting curve of growth. This is a good example of a case in which wind effects must be taken into account well below the sonic point. Becker and Butler (see their poster abstract in these proceedings) have reformulated the Fe V, Ni IV, Ni V NLTE line formation problem, in order to incorporate Kudritzki's (1992) analytical approximation of the aforementioned wind effect. They have tried to compare the resulting synthesized ultraviolet spectrum with the observed high-resolution IUE spectra of the central stars of IC 3568 and NGC 6826 (whose main atmospheric parameters had been improved by Méndez et al. 1992, see also Section 5 below). The value of this comparison is very limited because of the very low signal-to-noise ratio of both IUE SWP spectra. Hubble Space Telescope UV spectrograms, when available, will allow a reliable determination of photospheric Ni and Fe abundances.

One of the most encouraging results of the "classic" NLTE analysis of CSPN was that stars closer to the Eddington limit, in the log $g$ - log $T_{\text{eff}}$ diagram, show stronger winds (Méndez et al. 1988b, Pauldrach et al. 1988), as expected from the theory of radiatively driven winds. The Munich group (Pauldrach, Puls, Butler, the Gablers, Kudritzki) has devoted much effort along several years to improve this theory. Although it is not ready yet, it is increasingly used for spectral diagnostics in all kinds of hot stars with moderate winds. Just to remind the reader, the idea is to solve simultaneously the extended atmospheric structure and radiation driven wind hydrodynamics, treating the problem (including all opacities) fully in NLTE. A list of references is given by Kudritzki et al. (1992). A high degree of physical and numerical detail (e.g. the very complicated Grotrian diagrams always used to astonish the audience) is essential; otherwise the results are not satisfactory. The theory in its final form ought to provide masses, radii and distances for all kinds of hot stars with mass loss, given the following observational quantities determined from the stellar spectrum: wind terminal velocity, mass loss rate and effective temperature.

One aspect of this project is the development of so-called "unified" model atmospheres (Gabler et al. 1989; a preliminary report had been presented in our Mexico review). A brief description of unified models is also given in Gabler et al.'s poster abstract in these proceedings. The unified models permit to study the influence of radiatively driven winds and of atmospheric extension on the stellar energy distribution and on the profiles of diagnostic lines. In the remaining part of this Section we consider energy distributions, and in Section 5 we deal with diagnostic lines.

Concerning energy distributions: Gabler et al. (1989) have been able to reproduce the observed continuum IR excess of the famous O4f star $\zeta$ Puppis, implying that the unified models would be useful for modeling the IR spectral energy distribution of CSPN, with possible incidence upon studies of dust properties. Of more relevance to our knowledge of CSPN is the far-UV energy distribution, because it is directly related to the old problem of the "Zanstra discrepancy" between Tz(H) and Tz(HeII). As usual, we refer to Zanstra temperatures as calculated using blackbodies. Gabler et al. (1991) have used a grid of unified models to show that wind effects can substantially decrease the strong absorption edge shown by most plane-parallel, hydrostatic NLTE models at the He II Lyman limit, leading, in the most favorable cases, to an increment of more than two orders of magnitude in the extreme UV continuum flux of CSPN below 228 Å. This is an important step towards a model atmosphere able to explain simultaneously the stellar *and* the nebular spectrum. But it is not enough, because in several cases Tz(HeII) is higher than $T_{\text{eff}}$ as derived from a photospheric NLTE analysis, implying either that $T_{\text{eff}}$ must be increased or that the EUV flux produced by the unified models is still insufficient, sometimes by as much as one order of magnitude.

Therefore, some more work is needed to fully understand the difference between Tz(HeII) and $T_{\text{eff}}$. At the same time, photospheric NLTE studies have provided strong support (Méndez et al. 1992) to the idea that most of the difference between Tz(HeII) and Tz(H) can be attributed simply to optically thin PN giving too low values of Tz(H).

## 5. Improved diagnostics

Two of the fundamental motivations for the development of "unified" models are, first, to be able to extend the NLTE analysis technique to stars showing strong winds; second, to find out what systematic errors are made when the NLTE analysis is based on plane-parallel, hydrostatic models.

The first aspect is related to the problem of modeling those stellar lines which are more affected by wind and atmospheric extension. Gabler et al. (1989) were immediately able to reproduce, qualitatively at least, the strong observed stellar emission in lines like $H_\alpha$, $P_\alpha$, $B_\alpha$, He II $\lambda\lambda$ 1640, 4686, 10124. In particular, the most easily observable He II 4686 and probably also $H_\alpha$ are very promising mass loss rate indicators. What is needed at the present time is to implement a fully self-consistent code; current limitations are described by Gabler et al. (poster abstract in these proceedings).

A systematic comparison between "unified" versus plane-parallel, hydrostatic NLTE models is still in preparation. However, preliminary results have already shown that the unified models are indeed able to produce a much better fit to both absorption and emission lines for CSPN with Of spectra. Figures 1 to 3 show some examples. Although not very good, because some tuning is still required, these preliminary fits permit to estimate the size of the corrections needed for the values of $T_{\text{eff}}$ and log $g$ that had been determined using hydrostatic, plane-parallel NLTE models. The corrections are small, and in agreement with previous estimates (Méndez et al. 1992): the "hydrostatic" log $g$ must be increased by an amount which can be conveniently expressed as a function of the "hydrostatic mass" of the CSPN. The maximum correction (+0.2 in log $g$) must be applied for hydrostatic masses above 0.8 solar masses. For hydrostatic masses below 0.7 solar masses the correction to log $g$ is already smaller than +0.1. The temperature corrections, mostly caused by the increased gravity, are of the order of +1000 or +2000 K at temperatures between 30000 and 40000 K.

Méndez et al. (1992) have made preliminary corrections, where needed, to the gravities, masses and spectroscopic distances determined in previous studies. The corrected distances are between 5% and 20% smaller than derived previously (and several distances did not need any correction). The spectroscopic distances are still roughly comparable to (perhaps slightly smaller than) the distances of Cudworth (1974). See Fig. 5 in Méndez (1991b). More comments about distances in next Section.

A non-negligible fraction of all CSPN show very strong winds (Of-WR and WC objects, see e.g. Méndez 1991a). Most of them are H-deficient and show high He and C abundances. For such objects a photospheric NLTE analysis is not possible, because all the spectral

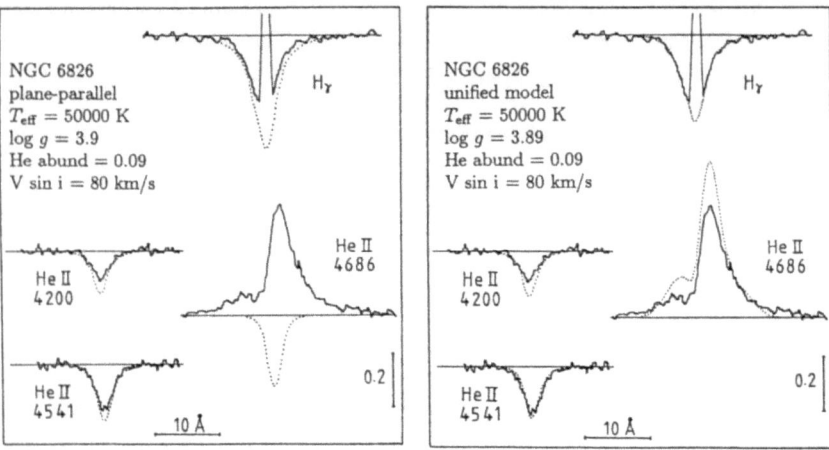

Figure 1. Comparison of profile fits obtained with plane-parallel, hydrostatic NLTE models versus fits obtained with "unified" models, for the central star of NGC 6826. The vertical segment indicates 0.2 times the intensity of the continuum. Wavelength increases to the right. Notice that a better "hydrostatic fit" to $H_\gamma$ would require a lower gravity.

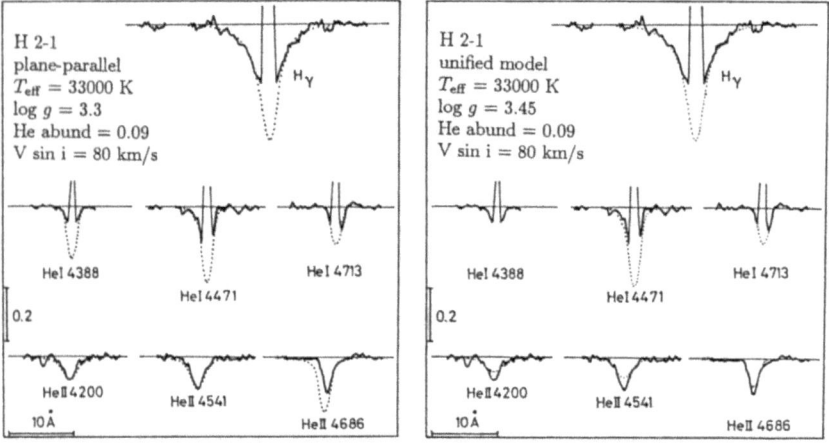

Figure 2. Same as Figure 1, for the central star of H 2-1. The best fit with unified models would require $\log g = 3.35$ and $T_{\text{eff}} = 34000$ or $35000$ K. At these temperatures and gravities the ratio between He I and He II absorptions is a very sensitive function of $T_{\text{eff}}$.

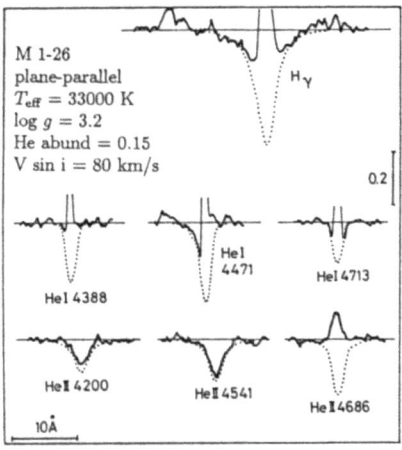
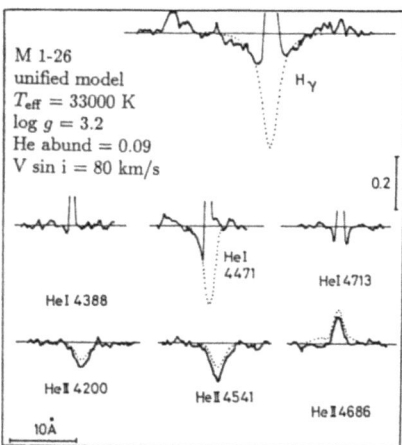

Figure 3. Same as Figures 1 and 2, for the central star of M 1-26. $H_\gamma$ is better fitted at log $g = 3.3$, and, again, a slightly higher $T_{\text{eff}}$ of about 34000 K would be needed to reproduce the observed ratio between He I and He II absorptions.

features are formed in the dense wind. A fully consistent hydrodynamical treatment of these expanding atmospheres is not available yet. However, there are good reasons to be optimistic. The semi-empirical NLTE modeling of Wolf-Rayet stars (for a recent review see Hamann 1992) is now well established as a useful tool. In particular, modeling of WC stars has been developed to such an extent (Hillier 1989, Hamann 1992) that the main spectral features are now reasonably well reproduced. We can expect that the empirical information collected through this kind of analyses will help to understand the physics of mass loss in these CSPN, opening the way to the direct determination of their basic properties.

## 6. Spectroscopic distances and the Galactic PN luminosity function

As we have seen, it is not possible yet to study all kinds of CSPN with NLTE model atmosphere techniques. However, many CSPN can be and have been analyzed in that way, giving important information for evolutionary discussions, and providing spectroscopic distances with estimated uncertainties of about 25% (the most recent data can be found in Méndez et al. 1992). During this Symposium we have detected some resistance to believe in the model atmosphere technique and the resulting stellar luminosities and distances. While we wait for more powerful methods of distance determination (the VLA expansion distances appear to be most promising) we think it is useful to emphasize that there is some independent evidence supporting the spectroscopic distances. First of all, a NLTE analysis of the hot post-AGB star ROB 162 in the globular cluster NGC 6397 (Heber and Kudritzki 1986) has given a distance of 2560 pc (Méndez et al. 1988b), in very good agreement with the cluster distance (2400 pc according to Alcaino and Liller 1980). Wind effects cannot change this result (no wind is detected in the spectrum of ROB 162), and the subsequent improvements made in the Stark broadening theory for He II 4686 (Schöning and Butler 1989), already incorporated in the CSPN work, would only slightly *decrease* the spectroscopic distance of ROB 162, bringing it in even better agreement with the cluster distance.

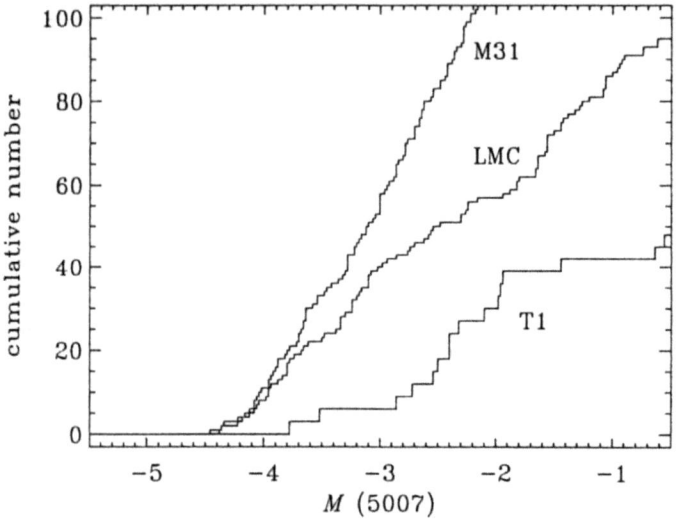

Figure 4. Cumulative PNLFs for M 31, LMC and the local sample (labeled T1), which consists of 24 PN for which spectroscopic distances are available (their central stars have been analyzed with NLTE models). Given the small amount of objects in the local sample, the corresponding cumulative numbers have been multiplied by a factor 3.

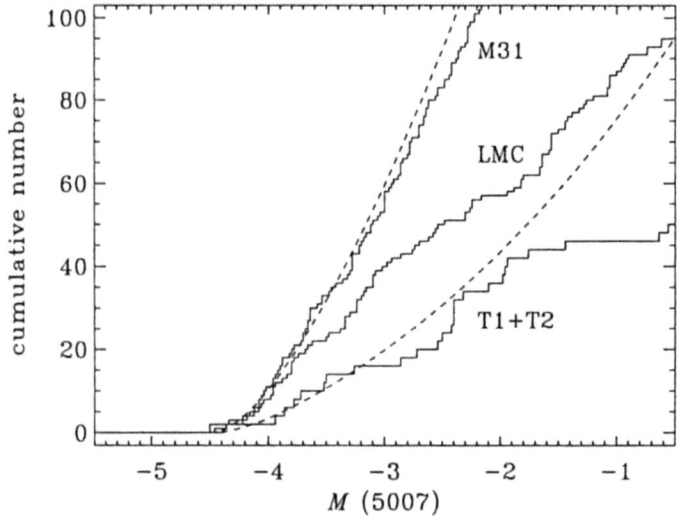

Figure 5. The same comparison, now using an extended local sample, labeled T1+T2 (see text), whose cumulative numbers have been multiplied by a factor 2. The dashed lines are analytical representations of the PNLF, derived from Formula (2) of Ciardullo et al. (1989). The three observed PNLFs have similar shapes and cutoffs.

Second, let us consider the extragalactic evidence. Careful surveys and measurements of PN in nearby galaxies like M 31 and the LMC (Ciardullo et al. 1989, Jacoby et al. 1990) have provided a good knowledge of the bright end of the PN luminosity function (PNLF). There is growing empirical evidence (Ciardullo et al. 1991, see also Jacoby's review in these proceedings) that this bright end of the PNLF is sufficiently invariant to be useful as a secondary standard candle for extragalactic distance determinations. Therefore we expect our Galaxy to show a similar PNLF. Are the spectroscopic distances consistent with this expectation? We have selected 24 PN whose central stars have been analyzed using NLTE models. Given the dereddened nebular fluxes at [O III] 5007 (which we call $I$), expressed in erg cm$^{-2}$ s$^{-1}$, we calculate the corresponding apparent magnitudes $m$, following Jacoby's (1989) definition: $m = -2.5 \log I - 13.74$. Knowing the spectroscopic distances, we can also obtain the corresponding absolute magnitudes $M$. Now we define the cumulative PNLF at any absolute magnitude $M$ as the number of PN with absolute magnitudes brighter than or equal to $M$. The resulting cumulative PNLF is shown in Figure 4, where it is compared with cumulative PNLFs calculated in the same way for the Large Magellanic Cloud (data taken from Jacoby et al. 1990; adopted distance 50 kpc) and for M 31 (data taken from Ciardullo et al. 1989; adopted distance 710 kpc).

In Figure 4 we find no obvious disagreement with the extragalactic cumulative PNLFs. The local cumulative PNLF does not reach as bright absolute magnitudes as the extragalactic ones, but this can be easily explained as a consequence of the small sample size; see e.g. the discussion in Sect. VII of Ciardullo et al. (1989). The important fact in our Figure 4 is that the situation would only become worse if we decreased the spectroscopic distances.

For further discussion we need to enlarge our local sample. For that purpose we have added 12 other local PN with what we consider to be sufficiently reliable distances, determined with the following methods: (a) VLA expansion distances (Masson 1989a, 1989b); (b) several distances derived from observations of H I absorption at 21 cm (Gathier et al. 1986a); (c) a few extinction distances (Gathier et al. 1986b). Full details will be given by Méndez, Kudritzki, Ciardullo and Jacoby (in preparation).

The improved local cumulative PNLF is shown in Figure 5. Although this cumulative PNLF must be still seriously affected by incompleteness, anyway it appears to be in reasonable agreement with the extragalactic ones. In the enlarged local sample we have now one very bright object: NGC 7027, with $M = -4.49$, right at the cutoff proposed by Ciardullo et al. (1989).

In summary, we find no reason to doubt that NLTE model atmosphere techniques, applied to CSPN with weak and even moderate winds (Of type), give reliable information about stellar temperatures, gravities, luminosities and distances. Given further theoretical and observational work, it may well be possible to promote the PNLF technique from a secondary to a primary extragalactic distance indicator.

**References**

Abbott D.C., Hummer D.G. 1985, ApJ 294, 286
Alcaino G., Liller W. 1980, AJ 85, 680
Anderson L.S. 1985, ApJ 298, 848
Anderson L.S. 1989, ApJ 339, 558
Becker S.R., Butler K. 1988, A&A 201, 232
Becker S.R., Butler K. 1989, A&A 209, 244

# OBSERVED MASS LOSS FROM CENTRAL STARS OF PLANETARY NEBULAE

M. PERINOTTO

*Dipartimento di Astronomia e Scienza dello Spazio, Universita' de Firenze*

ABSTRACT. In the Introduction we recall the mass loss history of a progenitor of a planetary nebula (PN). Then we concentrate on the status of knowledge of fast winds in central stars of planetary nebulae (CSPN) : the detection and statistics, the observed edge velocities, relationships of the edge velocities with other stellar or nebular parameters. We then summarize the methods used to derive the mass loss rates associated to the fast winds, and review the determinations of the "observed" mass loss rates. The comparison with predictions from the radiation driven theory (RDT) is then discussed as well as possible lines for future improvements.

## 1. Introduction

The general trend of the mass loss phenomenon along the life-time of a single star progenitor of a PN was clear already at the last IAU Symposium on PNe, five years ago, in Mexico city. Progress has been made meanwhile which helps to put in a better perspective some aspects of the subject, but much work needs still to be made before we reach a detailed quantitative understanding of the whole topic.

The wind phenomenon can be divided in various phases following the evolution of the star across the HR diagram. We can distinguish four main phases, each possibly with its own dominant mechanism for the production of the wind. Within each phase, some sub-phases might be identified from the specific stellar structure, which implies particular properties for the outher layers too, and consequently for the character of the wind. We have, in sequence: phase I, with sub-phases Ia (first Red Giant Branch), Ib (Horizontal Branch), Ic (early Asymptotic Giant Branch); phase II, with subphases IIa (pulsed Asymptotic Giant Branch), and IIb (transition from the AGB up to the end of the "superwind" phase; phase III, which represents the remaining transition up to the appearance of the optical nebula and phase IV, with sub-phases IVa (stellar luminosity approximately constant from the value at the end of the ABG phase up to 0.9 of it) and IVb (fading phase : stellar luminosity up to 0.1 of the final value on the AGB).

In phase I the wind velocity is around 10 km/s and the mass loss rate is believed to be represented by the Reimers (1975a,b) law : $dM/dt = \eta\, 4.0\, 10^{-13}\, LR/M$, where L,R and M are the usual stellar quantities in solar units, and $\dot{M}$ is in solar masses per year. The law has been obtained from some K and M giants and supergiants, with $\eta$ ranging from 0.3 to 3. Subsequent studies proved that that in Population II red giants $\eta$ can be set equal to $0.4 \pm 0.04$, while equally stringent limits are not easy to set for intermediate-mass stars (cf. Iben and Renzini, 1983) and then in general for the progenitors of PNe.

In phase II the wind becomes quite stronger, the so called "superwind", whose canonical $\dot{M}$ is of $10^{-4}$ Mo yr$^{-1}$ for a duration of 1000 years. These values are requested to match the nebular mass observed in the PN of well known distance K 648 in M 15 (Renzini, 1981). The well known OH/IR sources are believed to represent this phase (see Habing, these Proceedings).

To the subsequent phase III (transition typically from a stellar temperatures of T = 5000 K to T = 30,000 K), the Proto-PN objects (see Kwok, these Proceedings) and the optically bright Post-AGB stars (cf. Trams, 1992) belong.

Finally in phase IV, with the regular optically visible nebula coming out, the central star exhibits a wind much faster than in the previous phases : it is called the "fast" wind.

The importance of the winds in these various phases for the evolution of the central stars of planetary nebulae and for the relationship with the presently observed nebulae and with their history has been widely recognized. A great deal of work has been recently made and is currently pursed in the specific area of the hydrodynamics of the interaction of the various type of winds, with the aim of explaining the morphology in "groups" of PNe as well as in individual objects (see the review by Balick in these Proceedings). To this subject a precise knowledge of the fast wind parameters, in particular of the mass loss rate, is crucial.

## 2. The fast winds

The information on these winds in the CSPN come essentially from P Cygni profiles of resonance or excited lines of heavy ions observed in the UV range from the IUE satellite, with the addition of some P Cygni-like profiles observed in the optical range mainly in stars of W-R type.

2.1. DETECTION AND STATISTICS

A recent statistical study of the occurence of winds in CSPN has been made by Patriarchi and Perinotto (1991). Over a total of 159 objects examined, 81 showed a stellar continuum, thus giving the possibility to detect P Cygni profiles. Of these, 45 CSPN exhibited a fast wind, while 29 did not and 7 cases resulted to be doubtful. Thus about 60% of the examined CSPN did show a wind a 40% did not, to the level of detectability of the IUE spectra.

The occurence of the wind has been found to depend on the stellar gravity (Cerruti-Sola and Perinotto, 1985), in the sense that CSPN with gravity smaller than $\log g = 5.2$ (cgs units) almost always have a wind, while that is not so for the higher gravity CSPN, and on the luminosity (Pauldrach et al., 1988), so that the more a CSPN departs from the Eddigton luminosity, less frequent is the occurence of the wind.

2.2. EDGE VELOCITIES

A list of edge velocities in CSPN has been published by Patriarchi and Perinotto (1991), in connection with a study of the expansion velocities of the nebulae. To complete and update that list, the following values of Vedge are to be considered : 2700, 1700, 1550, 1350 Km/s in NGC 6210, 6826, 6891 and IC 4593, respectively, by Bianchi and De Francesco (these Proceedings), 3600 km/s in NGC 6905 (Heap, 1986), 1000-2000 km/s in He 2-99 (Kaler et al., 1989), 1800 km/s in Hu 2-1 (Cerruti-Sola and Perinotto, 1985), 2150 km/s in PK 75 +35 1 (Feibelman and Bruhweiler, 1989).

These values come from high resolution IUE spectra, a part from those of He 2-99 and Hu 2-1, where low resolution IUE spectra had to be used. Moreover there is a value of about 500 km/s in NGC 6302 by Meaburn and Walsh (1980) from optical data. In NGC 6826 there is another determination by Altner et al. (these Proceedings), equal to 1250 km/s, much lower than the above mentioned value, which is close to the one listed by Patriarchi and Perinotto, equal to 1600 km/s. This new determination differs from the previous ones quite more than the usual error from high resolution IUE spectra (about 100 km/s) and was obtained from the same high resolution IUE spectra used e.g. by Perinotto et al. (1989), who found Vedge = 1750 km/s and Vterminal = 1665 km/s. Looking back to the original profiles, in particular to the well defined profile of the CIV 1550 A line, we judge that a Vedge of 1560 km/s is in fact more realistic. This however is still unconfortably far from the 1250 km/s by Altner et al.. This object deserves then further attention.

The observed values of Vedge range from about 600 to 3600 km/s.

2.3. RELATIONSHIP OF Vedge WITH STELLAR AND NEBULAR PARAMETERS

An important, almost linear, relationship between the edge velocity of the stellar winds and the temperature of the central stars was found by Heap (1986). We have added 9 new objects to the 18 shown in her Fig.3. The new objects follow well the trend of the previous ones, so that the best fit obtained by Heap is confirmed by the new observations. Her interpretation of the relationship is thus strengthened : stars of similar mass (0.6 solar masses) evolve in the HR diagram at constant luminosity. The radius then decreases, the gravity increases and then also the escape velocity increases. That implies, according to the prediction of the radiation driven theory, an higher velocity of the wind. As a consequence, this relationship supports the radiation driven theory explanation for the fast winds in CSPN.

A similar relationship exists also for the O-B stars of Population I, where the RDT is firmly considered to provide the dominant mechanism for the production of the winds.

Another, almost linear, relationship has been found by Patriarchi and Perinotto (1991). That is between the edge velocity of the stellar winds and the expansion velocity of the associated optical nebula. Although the scatter is large, the authors argued that the relationship is consistent with the multiple-wind interacting theory, in the adiabatic case. Basically a larger Vedge implies an higher temperature of the shocked region and therefore a larger pressure on the previously ejected matter (see also the review by Heap in these Proceedings).

## 3. The mass loss rate

We quicly recall the methods used to derive the mass loss rates associated with the fast winds in CSPN, review the determinations made after the Mexico city

Symposium, focussing on the "observed" values, compare the observed values with those predicted by the radiation driven theory and present some suggestions for future work.

3.1. METHODS TO DERIVE THE MASS LOSS RATE

One can distinguish between "theoretical" and "semi-empirical" methods. In both cases the calculated P Cygni profiles must match the observed ones.

In the first type the mechanism(s) responsible for the production of the wind is (are) specified, while that is not so in the second type of methods. The mass loss rates coming from the last methods are called "observed" $\dot{M}$.

The theoretical methods are clearly more satisfactory and would give the best answer, provided one is able to include all the necessary physics. The dominant mechanism is recognized to be the pressure by radiation on the heavy ions. The status of art in this area is reviewed by Kudritzki (these Proceedings).

In the semi-empirical methods the velocity law is assumed and used as a parameter to fit the observations. The various semi-empirical methods then differ in the physical description of the phenomenon.

Basically the following approximations and approaches enter: 1a) Sobolev approximation for the opacity, or 1b) No Sobolev approximation; 2a) transfer equation in the observer's frame or 2b) in the comoving frame with the fluid; 3a) two-levels atom or 3b) multi-level atom; 4a) interaction among the components of multiplets not handled or rougly handled or 4b) correctly handled; 5a) behaviour of the opacity across the fluid assumed (and treated as a parameter) or 5b) occupation numbers derived from statistical equilibrium calculations; 6a) trend of the temperature across the wind assumed or 6b) derived from the solution of the energy balance; 7a) turbulence in the wind not considered or 7b) handled in some way; 8a) role of the underlying photospheric lines in the transfer considered or 8b) not taken into account; 9a) source function purely radiative or 9b) able to account for collisional effects; 10b) use of the full shape of the observed profile in the comparison with the observations or 10a) of some quantity derived from the profile, as the first moment of the flux distribution (which can be obtained also from spectra of resolution close to the instrumental resolution).

It is evident that a "b" condition specifies a superior method. The methods actually used to derive "observed" $\dot{M}$ in CSPN have been summarized by Perinotto (1989) : i) the escape probability method (EP), and the same with improvements by Olson (EPm), ii) the first moment of the flux distribution method (FM), and the same with modifications by Surdej (FMm), iii) the comoving frame method (CF) and iv) the SEI method (Sobolev plus exact integration). To these we must now add : v) the comoving frame method with statistical equilibrium equations across the wind solved in multi-level atom (CFS), used by Werner and Koesterke (1992).

Relative to the (a,b) conditions above illustrated, the various methods use "a" conditions except for the following "b" conditions. Method EP: 10; method FM : none and in addition the opacity law is not treated as a parameter but is fixed to some 'average' behaviour; method CF : 1,2,4,10; method SEI : 4,7,8,9,10 and as far as 1 is considered, something of quality between a and b. In the CFS method essentially all conditions are of type "b".

To the discussion of the merit of the various methods by Perinotto (1989), we add here the following. The EP method has been overcome by the SEI method (which has been widely distributed and is available upon request). The FM (or FMm) method may still be useful when spectra of resolution comparable with the instrumental resolution are available, but is of limited values because both the velocity and the opacity laws are assumed equal to some average shapes. The CF method is good, particularly for the regions of the wind closer to the star, but it has various limitations as is clear from its "a" conditions. At present a rather good method is the SEI method, while the most powerful one is the CFS method.

3.2. OBSERVED MASS LOSS RATES

In a review on the same subject at the Symposium of Mexico city, (Perinotto, 1989), I reported the determinations of the observed mass loss rates obtained by that time.

They referred to 20 CSPN studied with low or high resolution IUE spectra using the methods mentioned in the previous Section. They ranged between $2.10^{-6}$ and $5.10^{-11}$ Mo yr$^{-1}$, but values by different authors in the same objects did scatter by 1-2 order of magnitude, due to the different methods and to the different values of the fundamental stellar parameters adopted.

Since then the following studies have been made : A reanalysis by Hutsemekers and Surdej (1989) with their FMm method of 14 CSPN previously studied with low resolution IUE spectra and the FM method by Cerruti-Sola and Perinotto (1985); a study of five CSPN with the SEI method by Cerruti-Sola and Perinotto (1989) plus studies of the following individual objects : A78 by Kaler et al. (1988) with the FM method and by Werner and Koesterke (1992) with the CFS method; PK 75 +35 1 by Feibelman and Bruhweiler (1989) with the FM method and NGC 40 by Bianchi (1992) with the EPm method plus another approach based on the optical helium lines of WR stars.

These determinations are not of the same accuracy for what explained before.

To these one should add the preliminary values of $\dot{M}$ presented at this meeting of IC 2149 and Tc 1 by Modigliani et al. (these Proceedings) and of NGC 6210, 6826, 6891, IC 4593 by Bianchi and De Francesco (these Proceedings), in both cases with the SEI method.

We report in Table 1 the objects that we feel have at present the best observed $\dot{M}$ : they have been studied by Perinotto et al. (1989) and by Cerruti-Sola and Perinotto (1989) with the SEI method and by Werner and Koesterke with the CFS method.

| Object | T/1000(K) | R/R☉ | Velocity (km/s) | | $dM/dt (M_\odot\ yr^{-1})$ | | |
|---|---|---|---|---|---|---|---|
| | | | V(term.) | V(turb.) | Obs. | RDT | Reim. |
| NGC 1535 | 77 | 0.55 | 1900 | 95 | 1.4e-9 | 9.9e-9 | 3.4e-9 |
| 6210 | 90 | 0.29 | 2180 | 110 | 2.2e-9 | 2.9e-9 | 1.0e-9 |
| 6543 | 60 | 0.70 | 1900 | 15 | 4.0e-8 | 4.0e-9 | 2.6e-9 |
| 6826 | 45 | 1.45 | 1750 | 90 | 6.4e-8 | 7.5e-9 | 7.3e-9 |
| 7009 | 88 | 0.45 | 2770 | 30 | 2.8e-9 | 1.3e-8 | 3.2e-9 |
| IC 418 | 37 | 2.8 | 940 | 45 | 6.3e-9 | 2.2e-8 | 2.3e-8 |
| IC 4593 | 35 | 2.2 | 1000 | 100 | 4.2e-8 | 5.0e-9 | 8.7e-9 |
| A 78 | 115 | 0.24 | 3700 | – | 2.5e-8 | 7.7e-9 | 1.3e-9 |

The accuracy of the observed $\dot{M}$ is estimated to be within factors 3-5. That of the values from the radiation driven theory, as calculated here (see the next Section), is not easy to judge, but should be higher, particularly if the correct mass of the individual objects is used.

3.3. COMPARISON WITH PREDICTION BY THE RADIATION DRIVEN THEORY

In Table 1 we also report the values predicted by the radiation driven theory, calculated with the analytical formulation by Kudritzki et al. (1989) using the latest prescriptions of the force multiplier parameters for CSPN by Kudritzki et al. (1992), assuming the same mass of 0.6 Mo for all the central stars.

The calculated and the observed values are to within a factor of 10, apparently with no systematic effects. In the last column of Table 1 we give the values of $\dot{M}$ resulting from the Reimers formula (with $\eta = 1$) (see Introduction). These numbers are close to within a small factor to those from the radiation driven theory. That is surprising, considering the origin of the Reimers parametrization.

## 4. Conclusions and suggestions for future work

From the comparison made in the previous Section and considering the errors of the methods used to obtain the mass loss rates of fast winds in CSPN, one might conclude that the semi-empirical methods have been useful to prove the substantial correctness of the theoretical methods, based on the pressure of radiation on the heavy ions. And one might also conclude to be advisable for the future to rely only upon theoretical determinations of the mass loss rates.

We believe instead that there are reasons that justify the opportunity to continue also the work with the semi-empirical methods. First the correctness of the theoretical methods has to be proved on the full range of the fundamental stellar parameters, second it is well known that radiative winds are highly unstable (cf. Castor, 1991).

That means on one side that slightly different physical conditions in the subsonic-transonic regions of the wind, may produce rather different mass loss rates in stars which, generally speaking, are very similar. And, on another side, that at different epocs the mass loss rate of the same star might be different.

Moreover the present level of the radiation driven theory in hot stars needs still improvements (cf. Lamers, 1991), which are under way (see Kudritzki, these

Proceedings).

Then we believe that semi-empirical methods remain important for the understanding of the winds in CSPN and for our knowledge of their mass loss rates.

Improvements for a better measurement of this quantity include : a) a search for the variability in the winds; b) a better determination of the stellar radius (then of the distance, e.g. from the coming Hipparcos data), of the stellar temperature (via better atmospheric models and space experiments in the EUV spectra range) and of the chemical composition in the wind; c) a more precise definition of the observed p Cygni profiles and of the stellar continuum, properly accountign for the numerous spectral features; d) the use of other wind signatures (in addition to the UV P Cygni profiles) as optical P Cygni profiles and hopefully of the free-free wind continuum in the radio and infrared spectral ranges.

Clearly we need to observe at a sufficient spectral resolution more objects than those observable with IUE, and therefore we need to use the Hubble Space Telescope.

## REFERENCES

Bianchi,L.: 1992, *Astron. Astrophys.* **253**, 447.
Castor,J.I.: 1991, in *Stellar Atmospheres: Beyond Classical Models*, eds. L.Crivellari,I.Hubeny,D.G.Hummer, Kluwer Academic Publ., Dordrecht,p. 221.
Cerruti-Sola,M.,Perinotto,M.: 1985, *Astrophys. J.* **291**, 237.
Cerruti-Sola,M.,Perinotto,M.: 1989, *Astrophys. J.* **345**, 339.
Feibelman,W.A.,Bruhweiler,F.C.: 1989, *Astrophys. J.* **347**, 901.
Heap,S.: 1986, in *New Insights in Astrophysics*, ESA SP-263, p. 291.
Hutsemekers,D.,Surdej,J.: 1989, *Astron. Astrophys.* **219**, 237.
Iben,I.Jr.,Renzini,A.: 1983, *Ann. Rev. Astron. Astrophys.* **21**, 271.
Kaler,J.B.,Feibelman,W.A.,Henrichs,H.F.: 1988, *Astrophys. J.* **324**, 528.
Kaler,J.B.,Shaw,R.A.,Feibelman,W.A.,Lutz,J.H.: 1989, *Astrophys. J. Suppl.* **70**, 213.
Kudritzki,R.P.,Pauldrach,A.W.A.,Puls,J.,Abbott,D.C.: 1989, *Astron. Astrophys.* **219**, 205.
Kudritzki,R.P.,Hummer,D.G.,Pauldrach,A.W.A.,Puls,J., Najarro, F.,Imhoff,J.: 1992, *Astron. Astrophys.* **257**, 655.
Lamers,H.J.G.L.M.: 1991, in *Stellar Atmospheres: Beyond Classical Models*, NATO ASI Ser. Conference, eds. L.Crivellari, I.Hubeny, D.G.Hummer, Kluwer Academic Publ., Dordrecht,p. 311.
Meaburn,J.,Walsh,J.R.: 1980, *Mon. Not. R. astr. Soc.* **191**, 5p.
Patriarchi,P.,Perinotto,M.: 1991, *Astron. Astrophys. Suppl.* **91**, 325.
Pauldrach,A.W.A.,Puls,J.,Kudritzki.R.P.,Mendez,R.H.,Heap,S.R.: 1988, *Astron. Astrophys.* **207**, 123.
Perinotto,M.: in *IAU Symp. N. 131, Planetary Nebulae*, ed. S.Torres-Peimbert, Reidel Publ., Dordrecht,p. 293.
Perinotto,M.,Cerruti-Sola,M.,Lamers,H.J.G.L.M.: 1989, *Astrophys. J.* **337**, 382.
Reimers,D.: 1975a, *Mem. Soc. R. Sci. Liege, 6 Ser.*, **8**, 369.

Reimers,D.: 1975b, in *Problems in Stellar Atmospheres and Envelopes*, eds. B.Baschek,W.H.Kegel,G.Traving, Springer Publ., p. 229.
Renzini,A.: 1981, in *Physical Processes in Red Giants*, eds. I.Iben Jr., A.Renzini, Reidel Publ., Dordrecht,p. 165.
Trams,N.R.: 1992, *Thesis*, Utrecht, SRON Laboratorium.
Werner,K.,Koesterke,L.: 1992, in *The Atmospheres of the Early-Type Stars*, eds. U.Heber,C.S.Jeffery, Springer-Verlag, Berlin,p. 288.

# TEMPERATURES OF CENTRAL STARS OF PLANETARY NEBULAE

A. PREITE-MARTINEZ

*Istituto di Astrofisica Spaziale, CP 67, 00044 Frascati, Italy*

and

*Observatoire de Strasbourg, 11 rue de l'Université, 67000 Strasbourg, France*

The temperature of Central Stars of Planetary Nebulae (CSPN) is one of the most important parameters for a better understanding of their evolution, their properties and those of the surrounding photoionized nebula. It is an elusive quantity to derive though, because CSPN are generally faint objects and their spectrum is often heavily contaminated by nebular continuum emission.

In the following I will review the different methods used for the determination of the temperature of CSPN, the advances made in this difficult field, and discuss some of the problems arising when comparing temperatures derived with different methods.

## 1. Methods

There are essentially five methods to derive the temperature T of CSPN, plus nebular models that can be used to get some information on T, plus two scales of calibrated temperatures that can be used as "extrema ratio", e.g. when all the above methods are inapplicable.

### 1.1. ZANSTRA

The Zanstra method (Zanstra, 1931) is quite simple and effective: the number of ionizing photons emitted by the central star can be counted measuring the flux in a single hydrogen recombination line (usually H-beta). A temperature for the central star can be derived if the continuum flux of the star in a given frequency band can also be measured. This procedure yelds the so-called hydrogen Zanstra temperature, $T_Z(H)$. Similarly, one can use a HeII recombination line to count the number of ionizing photons shortward of the $He^+$ threshold, to derive $T_Z(HeII)$.

Its principal disadvantage is that the continuum of the central star should be observable, and that the nebular spectrum is good enough to give elements for the calculation of the nebular spectrum that has to be subtracted from the observed stellar continuum. Additional problems can arise if the reddening is (very) high or the angular size of the nebula is (very) large.

The basic assumption of the method is that the nebular gas is absorbing all the ionizing photons emitted by the central star (nebula optically thick in all directions).

Moreover, the shape of the ionizing continuum has to be selected "a priori": usually that of a blackbody.

The Zanstra method has basically not changed in the last 60 years. Advances in the last five years concerns the number of new applications of the method : Gathier and Pottasch, 1989; Gleizes et al., 1989; Kaler et al., 1989; Pottasch and Acker, 1989; Jacoby and Kaler, 1990; Kaler et al., 1990; Kaler et al., 1991; Tylenda et al., 1991; Kaler and Jacoby, 1991. Apparao and Tarafdar (1989) detected a number of PN with EXOSAT in the energy band 0.05 − 2.0KeV, and, in an interesting variation of the method, derived X-Ray Zanstra temperatures for 8 CSPN. They report a better agreement with $T_Z(HeII)$ than with $T_Z(H)$.

## 1.2. ENERGY BALANCE

The energy-balance (or Stoy) method is based on the assumption that thermal equilibrium holds in the ionized nebular gas, e.g. that energy losses and energy gains exactly balance and define the thermal content of the electron gas in the nebula. The energy-balance (EB) method was also suggested 60 years ago, by Stoy (1933) for a hydrogen nebula. The method was extended by Kaler (1976) to include helium in low-excitation nebulae, and revised and extended by Preite-Martinez and Pottasch (PMP,1983).

The advantage of the EB method is that only the nebular emission spectrum is needed. This is because the energy gain per photoionization (depending on the shape of ionizing continuum of the cantral star) is related to the intensity of all collisionally excited (CE) lines in units of the H-beta intensity. Measuring the central star is not required. The main disadvantage is that strong CE lines can fall in unobserved (or unobservable) regions of the spectrum. A possible way of estimating the intensity of unseen CE lines was already suggested by PMP. Recently Preite-Martinez et al. (1989, 1991) extended this correction scheme to the case of spectra taken in a very limited wavelength region. The correction is rather rough, yet it is the only available. As for the Zanstra method, the spectral distribution of the ionizing star is usually assumed to be that of a black-body.

A nice feature of the EB method is that the derived temperatures are rather insensitive to the optical depth of the nebula, in particular for low-temperature central stars (T < 50 − 60.000 K). Nonetheless, the fact that the application of the EB method results in values of the temperature depending on the assumed optical depth status of the nebula, has cast some doubt on the reliability of the method. To remove this drawback, Köppen and Preite-Martinez (1991) have recently thoroughly revised the EB method : with the addition of the observational parameter HeII 4686 / $H_\beta$ they derive simultaneously the temperature of the central star and the optical depth of the nebula.

## 1.3. IONIC ABUNDANCE RATIOS

With this method one essentially deduces the shape of the stellar ionizing continuum from the ratio of two consecutive stages of ionization of a given element. Many ratios corresponding to different ionization thresholds are necessary to define the shape of the stellar spectrum down to very short wavelengths. A very good nebular spectrum is the only requirement. After the works of Köppen and Tarafdar (1978) and Natta et al.(1980), no advances have been made.

It is worth noting though that in its simplest form, using the amount of $He^{++}$ relative to $H^+$, the method was suggested by Ambartsumyan (1932) and is often referred to as "cross-over" method (Kaler and Jacoby 1989).

## 1.4. SPECTROSCOPIC ANALYSIS

The method (Mendez et al. 1988) consists of fitting stellar photospheric H and He line profiles with theoretical profiles computed from non-LTE models (atmosphere and line formation). The analysis of the observed spectral lines requires sofisticated models, still in refinement. The application of the method is unfortunately limited to relatively right CSPN. Substancial improvements have been made in recent years in spectral resolution, signal-to-noise ratio, sky and nebula subtraction, as well as in non-LTE model atmospheres and line formation codes ( Mendez et al. 1988b; Mendez et al. 1990; Gabler et al. 1991; Mendez et al. 1992). All this reflects in the increased reliability of the method.

## 1.5. FIT TO CONTINUUM

The temperature of the observed central star is derived fitting a model for the continuum emission to the observed continuum. The UV wavelength range is used because it is there that the nebular continuum emission is less contaminating the stellar continuum (Bianchi et al. 1986; Bianchi and Grewing 1987). On the other hand, the UV band of hot stars falls in the Rayleigh-Jeans tail of the spectral distribution, reducing the sensitivity of the method at high temperatures. An interesting advance was made recently by Grewing and Neri (1990), who presented a variation of the classical method making use of extinction-independent UV colour indices.

## 1.6. CALIBRATED SCALES

In addition to the above mentioned methods, one can in principle calibrate an observational quantity against temperatures derived for large samples of CSPN. This approach has been taken by Zijlstra and Pottasch (1989) who tried to calibrate the excitation class of the nebula against the EB temperature of the central star. The fit is fear only for medium excitation nebulae. Kaler and Jacoby (1991) and Mendez (1992) used the observed strength of the OIII 5007 line to find an agreement among the various ways of determining temperatures of central star. Their fits are quite good, in particular at low-temperatures, and can be reliably used for statistical studies of low-excitation nebulae.

## 2. Open problems and Discussion

In the recent past reviewed here, the application of the above methods has led to the determination of the temperature (or of different temperatures referring to the same star) for almost 1000 CSPN. A compilation of all these temperatures has been produced by the author, and its content will be used in the following to discuss the problems that arise when comparing different determinations for the same object.

Because the agreement between temperatures derived with different methods if far from being satisfactory, let's first recall which are the open problems, and use the welth of derived data to try to get some clue on the origin of these problems. Of course a method cannot be better than the assumptions on which it is based, and produce results of better quality than that of the input data. Nonetheless, the answers we are looking for can help us in improving our assumptions, revise methods, and search for more reliable observational data.

The oldest question is: which is the origin of the Zanstra discrepancy? Or, why so many central stars show a $T_Z(H)$ lower than $T_Z(HeII)$? In Fig.1 the Hydrogen versus Helium II Zanstra temperatures are plotted for 133 CSPN with both determinations of the temperature. A well known result shows up clearly: $T_Z(H)$ tends to be lower than

$T_Z(\text{HeII})$ for $T_Z(\text{H}) < 100.000$K, almost in agreement above. The ratio $T_Z(\text{HeII})/T_Z(\text{H})$ (called Zanstra Ratio, ZR) can reach values >2 at low-T, and it is almost 1 at higher temperatures. Values of the ZR<1 are also present, although not statistically very significant. Another point to note is that in the low-T region there is a continuous distribution between the highest ZR values and 1.

We have at least three possible reasons for such a behaviour: (i) we are not counting all the stellar ionizing photons in the H-Lyman continuum, while we are counting all those emitted in the He-Lyman continuum, due to a low nebular optical depth; (ii) our other basic assumption concerning the shape of the far UV continuum emission of the central star is invalid, e.g. the star is not radiating as a black-body; (iii) errors in the determination of the magnitude of the central star. We can immediately rule out reason (iii) as an explanation of the general appearance of Fig.1. The pattern is too well defined to be due to random errors (although they certainly introduce noise); systematic errors (bright stars too brigth, faint stars too faint) could produce such a trend, but how to explain high values of ZR derived from high quality data?

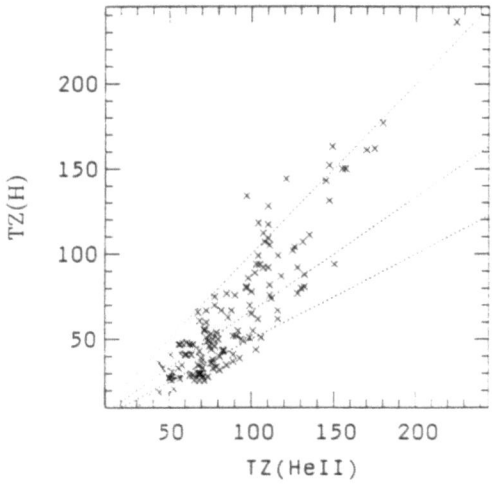

**Figure 1.** $T_Z(\text{H})$ is plotted versus $T_Z(\text{HeII})$ for 133 CSPN with both determinations of the Zanstra temperature. Dotted lines indicate Zanstra Ratios of 1, 1.5, and 2, clock-wise.

An optical depth effect (i) can easily explain the observed behaviour. In principle, departures from a black-body spectral distribution, alone or coupled with an optical depth effect, could also account for the Zanstra discrepancy. In practice, take the points in Fig.1 around, say, $T_Z(\text{HeII}) = 70.000$K. We should interpret the range in Zanstra Ratio as due to a continuously changing in the shape of the ionizing continuum of the CSPN. To explain that we need an additional stellar parameter, perhaps gravity. So the adoption of explanation (ii) raises two more questions: (a) is gravity (keeping T constant) able to influence the shape of the He-Lyman continuum of such an amount as to explain the observed spread in ZR? And (b) do we really observe such a systematic gradient of gravity in Fig.1? Waiting for the answers, we cannot rule out that non-blackbody effects are at work, but certainly the possibility of an optical depth effect gains some strength.

If we now compare Zanstra temperatures with results of direct spectroscopy of the central star, we find two features: $T_Z(H)$ is almost always lower than $T_M$ especially for $T_M > 40.000K$, and $T_Z(HeII)$ is reasonably well correlated with $T_M$ although frequently higher than $T_M$. The simplest explanation is that $T_Z(H)$ is too low because the nebula is not counting all the ionizing photons: indeed, from the results of the new EB method (Preite-Martinez et al. in preparation) we find that, of the six CSPN for which $\tau$ and $T_M$ are both available, all with $T_Z(H) < T_M$, five are surrounded by very thin nebulae (IC 2448, IC 4593, NGC 1535, NGC 2392, and NGC 7009). A low optical depth cannot be the only mechanism at work in these cases, beacause it cannot explain why the helium Zanstra temperature is often higher than $T_M$. Besides, large errors on stellar magnitudes are improbable, because these CSPN are brigth and well observed.

A point to bear in mind though is that while $T_M$ actually measures the effective temperature of the star assuming that it is radiating as predicted by model atmospheres, the Zanstra temperature could be considered as an effective temperature only if the star was indeed radiating as postulated by the Zanstra method (black-body). If this is not the case, the two methods measure different things, and it should not be surprising that they produce different results.

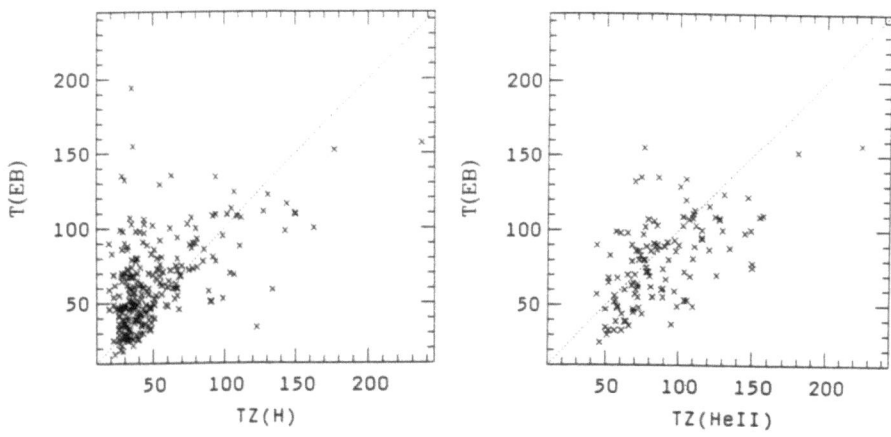

**Figure 2.** $T_{EB}$ is plotted versus $T_Z(H)$ (left panel) and $T_Z(HeII)$ (right panel).

An interesting comparison with $T_M$ can be made for another method, also using direct spectroscopic observation of the central star (UV colours). The temperatures $T_C$ so derived are in reasonable agreement with $T_M$ only in about half of the cases. If $T_C$ is compared with $T_Z(H)$, a similar fraction of coincidences if found, the remaining half of the cases being equally spread above or below $T_Z(H)$. A better agreement with $T_Z(HeII)$ is found, but again with a tendency for $T_Z(HeII)$ to be higher than $T_C$. In this case may well be that the major source of disagreement is due to observational errors: indeed the five most discrepant objects are all PN with small angular diameter, with a continuum emission that can contaminate the stellar continuum (Grewing and Neri, 1990).

We can now move on to discuss the results of the Energy-Balance method. In Fig.2 we plot $T_{EB}$ versus $T_Z(H)$ and $T_Z(HeII)$. The agreement with Zanstra temperatures is very poor, then we can probably derive important clues. Comparing the two panels, it is clear that EB temperatures are on the average understimated by 15–20%, at least the lowest ones. The most probable reason is that $T_{EB}$ come mostly from the application of the method to a large sample of nebulae observed in a restricted wavelength region (Preite-Martinez et al. 1989, 1991). The correction scheme to account for unobserved CE lines probably underestimates the amount of CE cooling in the nebula. Alternatively, or in addition to this, most nebular spectra did not allow the determination of the key parameters necessary to feed the correction scheme (electron temperature and/or density, and ionic abundances of the most important coolants). Again, the quality of the results reflects that of the input data.

Once this effect is taken in due account, we are left with the result that $T_{EB}$ is always $\geq T_Z(H)$, with very few exceptions. Now, both methods make use of the blackbody assumption, but while the Zanstra method assumes complete optical thickness for the nebula, $T_{EB}$ is little affected by the optical depth. So what we see in Fig.2 (left panel) is a direct test of the Zanstra's assumption on the optical thickness of the nebulae. The result of the test is that most nebulae are probably optically thin in the H-Lyman continuum.

From the right panel of Fig.2, we get an indirect confirmation because $T_{EB}$ and $T_Z(HeII)$ roughly correlate, although with a large spread. There could also be an indication that al least part of the scatter is due to observational errors in deriving the magnitude of the central star.

A direct confirmation of our interpretation of Fig.2 comes from Preite-Martinez et al. (in preparation) as the result of the application of the new EB method developed by Köppen and Preite-Martinez (1991). In Fig.3 we plot their new $T_{EB2}$ values against $T_Z(H)$, and in the left panel only the points corresponding to optically thin nebulae (optical depth < 3): thin nebulae fall exactly where they should, and the correlation between $T_{EB}$ and $T_Z(H)$ is much better once the optically thin nebulae are removed.

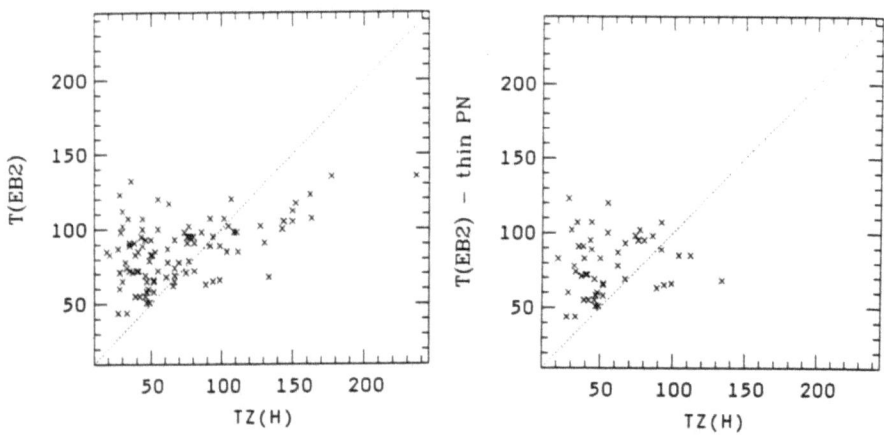

**Figure 3.** $T_{EB2}$ is plotted versus $T_Z(H)$ (left panel). In the right panel only nebulae with $\tau < 3$ are plotted.

Another interesting comparison is between $T_{EB}$ and spectroscopically derived temperatures $T_M$. The agreement is fairly good at low temperatures, beside a few discrepant cases of some interest, and it worsens for $T_M$ above 60.000K. There $T_{EB}$ is often much higher than $T_M$. What we are actually testing here are two different variations of the assumption on the shape of the ionizing stellar continuum: on one side a blackbody is assumed (EB), the other method uses model atmospheres. Unless we decide "a priori" that one of the two hypothesis is the "true" one, a comparison of this kind can only tell us how to reconcile the results of the two different methods. Indeed, the use of model atmospheres can reduce the values derived with the EB method for hot stars.

Undoubtably, one of the most promising advances recently made in this field is the possibility of deriving the optical depth of the nebula with the new version of the EB method proposed by Köppen and Preite-Martinez (1991). The application of the method to about 360 objects by Preite-Martinez et al., still unpublished, will open the possibility of a direct test on one of the most criticized assumptions: that planetary nebulae are mostly optically thick in the H-Lyman continuum. We can anticipate some of the new results on the determination of the optical depth $\tau$ : half of the nebulae examined are optically thin ($\tau < 3$). In Table 1 we extracted from our compilation of recent results those nebulae with at least 4 different determinations of the temperature of the central star (both Zanstra temperatures plus two other, or $T_Z(H)$ only, plus three other temperatures). Although the sample is small, the trend is clear: to thin nebula always correspond a $T_Z(HeII) > T_Z(H)$. Conversely, the Zanstra Ratio ZR correlates very well with $\tau$ (beside M 3-6, whose $T_Z(H)$ is unreliable, and NGC 3242).

Table 1 - *CSPN with at least 4 different determinations of the temperature (both Zanstra plus 2 others, or $T_Z(H)$, $T_M$, $T_C$ and $T_{EB}$).*

| PK | Name | $T_Z(H)$ | $T_Z(HeII)$ | $T_M$ | $T_C$ | $T_{EB}$ | $T_{EB2}$ | Log($\tau$) | ZR |
|---|---|---|---|---|---|---|---|---|---|
| 25+40 1 | IC 4593 | 28a | 50 | 40 | 35 | 47a | 60 | 0.29 | 1.8 |
| 34+11 1 | NGC 6572 | 66 | - | 60 | 34 | 60 | - | - | |
| 37−34 1 | NGC 7009 | 60 | 84 | 82 | 51 | 73 | 78 | 0.70 | 1.4 |
| 54−12 1 | NGC 6891 | 35a | - | 50 | 31 | 52a | - | - | |
| 43+37 1 | NGC 6210 | 61a | 69a | - | 37 | 65a | 68 | 1.96 | 1.1 |
| 83+12 1 | NGC 6826 | 33 | - | 50 | 42 | 33a | - | - | |
| 123+34 1 | IC 3568 | 31 | 52 | 50 | 59 | 65a | - | - | 1.7 |
| 197+17 1 | NGC 2392 | 27 | 67 | 47 | 33 | 118 | 105 | 0.79 | 2.5 |
| 206−40 1 | NGC 1535 | 35 | 68 | 70 | 63 | 79 | 89 | 0.59 | 1.9 |
| 215−24 1 | IC 418 | 38 | - | 36 | 39 | 29a | - | - | |
| 254+ 5 1 | M 3-6 | 48: | 58 | - | 43 | 48 | 60 | 0.24 | 1.2: |
| 261+32 1 | NGC 3242 | 54 | 91 | 75 | 50 | 109 | 100 | 1.36 | 1.7 |
| 285−14 1 | IC 2448 | 43 | 83 | 65 | 72 | 91 | 95 | 0.06 | 1.9 |
| 316+ 8 1 | He 2-108 | 26a | 52 | 33 | - | 33a | - | - | 2.0 |
| 327+10 1 | NGC 5882 | 55a | 72a | - | 32 | 64a | 72 | 0.77 | 1.3 |
| 345− 8 1 | Tc 1 | 30a | - | 33 | 33 | 34 | - | - | |

a = average value; T in $10^3$K.

We have seen in the discussion above that comparing results of different methods is sometimes a good way of testing assumptions and the methods themselves. Model atmospheres are certainly to be improved, and the correction scheme that is used to estimate the total CE cooling also needs improvement or revision. Observational errors still play a non negligible role in all methods. Piece by piece the picture is getting clearer: a good fraction of planetary nebulae must be thin to explain most of the discrepancies between the results of different methods. According to Preite-Martinez and collaborators a determination of the optical depth is now available for a large sample of medium/high-excitation nebulae, and half of them are indeed thin. Departures from blackbody distribution cannot be ruled out, but have little statistical weight.

References

Ambartsumyan, V. A. 1932, *Pulkovo Obs. Circ.* Vol.8, No.4
Apparao, K. M. V., Tarafdar, S. P. 1989, *Astrophys. J.* **344**, 826
Bianchi, L., Cerrato, S., Grewing, M. 1986, *Astron. Astrophys.* **169**, 227
Bianchi, L., Grewing, M. 1987, *Astron. Astrophys.* **181**, 85
Gabler, R., Kudritzki, R. P., Mendez, R. H. 1991, *Astron. Astrophys.* **245**, 587
Gathier, R., Pottasch, S. R. 1989, *Astron. Astrophys.* **209**, 369
Gleizes, F., Acker, A., Stenholm, B. 1989, *Astron. Astrophys.* **222**, 237
Grewing, M., Neri, R. 1990, *Astron. Astrophys.* **236**, 223
Jacoby, G. H., Kaler, J. B. 1989, *Astron. J.* **98**, 1662
Kaler, J. B. 1976 *Astrophys. J.* **210**, 843
Kaler, J. B., Jacoby, G. H. 1989, *Astrophys. J.* **345**, 871
Kaler, J. B., Jacoby, G. H. 1991, *Astrophys. J.* **372**, 215
Kaler, J. B., Shaw, R. A., Feibelman, W. A., Lutz, J. H. 1989, *Astrophys. J. Suppl. Ser.* **70**, 213
Kaler, J. B., Shaw, R. A., Kwitter, K. B. 1990, *Astrophys. J.* **359**, 392
Kaler, J. B., Shaw, R. A., Feibelman, W. A., Imhoff, C. L. 1991, *Publ. Astron. Soc. Pac.* **103**, 67
Köppen, J., Tarafdar, S. P. 1978, *Astron. Astrophys.* **69**, 363
Köppen, J., Preite-Martinez, A. 1991, *Astron. Astrophys.* **248**, 191
Mendez, R. H., Kudritzki, R. P., Herrero, A., Husfeld, D., Groth, H. G. 1988, *Astron. Astrophys.* **190**, 113
Mendez, R. H., Groth, H. G., Husfeld, D., Kudritzki, R. P., Herrero, A. 1988b, *Astron. Astrophys.* **197**, L25
Mendez, R. H., Herrero, A., Manchado, A. 1990, *Astron. Astrophys.* **229**, 152
Mendez, R. H. 1992, *Astron. Astrophys.*, in press
Mendez, R. H., Kudritzki, R. P., Herrero, A. 1992, *Astron. Astrophys.*, in press
Natta, A., Pottasch, S. R., Preite-Martinez, A. 1980, *Astron. Astrophys.* **84**, 284
Pottasch, S. R., Acker, A. 1989, *Astron. Astrophys.* **221**, 123
Preite-Martinez, A., Pottasch, S. R. 1983, *Astron. Astrophys.* **126**, 31
Preite-Martinez, A., Köppen, J., Acker, A., Stenholm, B. 1989, *Astron. Astrophys. Suppl. Ser.* **81**, 309
Preite-Martinez, A., Köppen, J., Acker, A., Stenholm, B. 1991, *Astron. Astrophys. Suppl. Ser.* **88**, 121
Stoy, R. H. 1933, *Montly Notices Roy. Astron. Soc.* **93**, 588
Tylenda, R., Stasinska, G., Acker, A., Stenholm, B. 1991, *Astron. Astrophys.* **246**, 221
Zanstra, H. 1931, *Publ. Dom. Astroph. Obs.* **4**, 209
Zijlstra, A. A. and Pottasch, S. R. 1989 *Astron. Astrophys.* **216**, 245

# ATOMIC DATA – A BIBLIOGRAPHY

K. BUTLER

*Universitätssternwarte München, Scheinerstraße 1, D-W-8000 München 80*

Space does not allow for a discussion of methods here. The reader is referred to the reviews or critical evaluations in the relevant sections. The current bibliography is an attempt to update the evaluations/reviews through to June 1992. Note that the Opacity Project provides an enormous amount of LS-coupling radiative data (see the Opacity Workshop reference). There is to be a follow-up project, the IRON project to provide radiative data in intermediate coupling including electric quadrupole and magnetic dipole transitions AND collisional excitation data.

The following abbreviations are used for journals:

**AA** *Astron. Astrophys.* and **AAsup** *Astron. Astrophys. Suppl. Ser.*
**ADNDT** *At. Data Nucl. Data Tables*
**ApJ** *Astrophys. J.* and **ApJsup** *Astrophys. J. Suppl. Ser.*
**IAU103** *IAU Symposium No. 103*, Ed. Flower, D.R.
**JPB** *J. Phys. B: Atom. Molec. Phys.*
**JPCRD** *J. Phys. Chem. Ref. Data*
**JQSRT** *J. Quant. Spectrosc. Radiat. Transfer*
**MNRAS** *Mon. Not. R. astr. Soc.*
**PRA** *Phys. Rev. A*
**ZMLD** *J. de Phys., Vol. 1, Coll. 1, Suppl. JP II, 3*, ed. Zeippen, C.J. and Le Dourneuf, M.

## Energy Levels and Wavelengths

**Review** Wiese, W.L., 1991, in *ZMLD*, 287
**Review** Martin, W.C., 1992, in *Lecture Notes in Physics*, in press
**Na ions** Martin, W.C. and Zalubas, R., 1981, *JPCRD*, **10**, 153
**Mg ions** Martin, W.C. and Zalubas, R., 1980, *JPCRD*, **9**, 1
**Al ions** Martin, W.C. and Zalubas, R., 1979, *JPCRD*, **8**, 817
**Si ions** Martin, W.C. and Zalubas, R., 1983, *JPCRD*, **12**, 323
**P ions** Martin, W.C., Zalubas, R. and Musgrove, A., 1985, *JPCRD*, **14**, 751
**S ions** Martin, W.C., Zalubas, R. and Musgrove, A., 1990, *JPCRD*, **19**, 821
**K – Ni** Sugar, J. and Corliss, C., 1985, *JPCRD*, **14**, Suppl. 2.
**Mg ions** Kaufman, V. and Martin, W.C., 1991a, *JPCRD*, **20**, 83
**Al ions** Kaufman, V. and Martin, W.C., 1991b, *JPCRD*, **20**, 775

## Collisional Ionization

**Critical Evaluation** Itikawa, Y., 1992, *ADNDT*, **49**, 209
**Critical Evaluation** Bell, K.L., Gilbody, H.B., Hughes, J.G., Kingston, A.E., Smith, F.J., 1983, *JPCRD*, **12**, 891.
**Critical Evaluation** Lennon, M.A., Bell, K.L., Gilbody, H.B., Hughes, J.G., Kingston, A.E., Murray, M.J., Smith, F.J., 1988, *JPCRD*, **17**, 1285
**General** Arnaud, M. and Rothenflug, R., 1990, *AAsup*, **60**, 425
**Excited States** Golden, L.B. and Sampson, D.H., 1980, *JPB*, **13**, 2645.
**H and He** Jones, S., Madison, D.H. and Srvastava, M.K., 1991, *JPB*, **24**, 1899
**General** Freund, R.S., Wetzel, R.C., Shul, R.J. and Hayes, T.R., 1990, *PRA*, **41**, 3575
**Ca II** Badnell, N.R., Griffin, D.C. and Pindzola, M.S., 1991, *JPB*, **24**, L275

## Line Broadening

**Critical Evaluations** Konjević, N., Dimitrijević, M.S. and Wiese, W.L, 1984, *JPCRD*, **13**, 619 and 649
**Review** Dimitrijević, M.S. and Sahal-Bréchot, S., 1991, in *ZMLD*, 111
**C IV** Burke, V.M., 1992, *JPB*, **25**, in press
**C IV** Schöning, T, 1992, *AA*, in press and *JPB*, submitted
**General** Seaton, M.J., 1987/8, *JPB*, **20**, 6431 and *JPB*, **21**, 3033
**General** Dimitrijević, M.S. and Konjević, N., 1987, *AA*, **172**, 345
**Neutrals** Seaton, M.J., 1989, *JPB*, **22**, 3603
**H** Seaton, M.J., 1990, *JPB*, **23**, 3255
**Shifts** Wiese, W.L. and Konjević, N., 1992, *JQSRT*, **47**, 185
**He I** Dimitrijević, M.S. and Sahal-Bréchot, S., 1990, *AAsup*, **82**, 519
**Li I** Dimitrijević, M.S. and Sahal-Bréchot, S., 1991, *JQSRT*, **46**, 41
**C IV** Dimitrijević, M.S., Sahal-Bréchot, S. and Bommier, V., 1991, *AAsup*, **89**, 581
**O VI** Dimitrijević, M.S. and Sahal-Bréchot, S., 1992, *AAsup*, **93**, 359
**Mg I and Mg II** Djeniže, S., Srećković, A., Platiša, M., Labat, J., Konjević, R. and Purić, J., 1990, *JQSRT*, **44**, 405
**Si III** Djeniže, S., Srećković, A., Labat, J., Purić, J. and Platiša, M., 1992, *JPB*, **25**, 785
**Si III** Bakshi, V., 1990, *AIP Conf. Proc.*, **216**, 71
**Si IV** Djeniže, S., Srećković, A., Labat, J., Purić, J. and Platiša, M., 1992, *JPB*, **25**, 785
**Si IV** Dimitrijević, M.S., Sahal-Bréchot, S. and Bommier, V., 1991, *AAsup*, **89**, 591
**Na I** Dimitrijević, M.S. and Sahal-Bréchot, S., 1990, *JQSRT*, **44**, 421
**Ca I** Spielfiedel, A., Feautrier, N., Chambaud, G. and Lévy, B., 1991, *JPB*, **24**, 4711

## Opacities

**Workshop** Ed. Lynas-Gray, A.E., Mendoza, C. and Zeippen, C.J., 1992, *Rev. Mex. Astro. y Astrofísica*, **23**
**Tables** Rogers, F.J. and Iglesias, C.A., 1992, *ApJsup*, **79**, 507

## Photoionization

**Bibliography** Le Dourneuf, M., 1991, in *ZMLD*, 227
**H ions** Storey, P.J. and Hummer, D.G., 1991, *Comput. Phys. Commun*, **66**, 129
**Heavy systems** Kelly, H.P, 1990, *AIP Conf. Proc.*, **215**, 292
**Positive Ions** Manson, S.T., 1990, *AIP Conf. Proc.*, **205**, 189
**Li I and Li II** Hollauer, E. and Nascimento, M.A.C., 1990, *PRA*, **42**, 6608
**C seq** Luo, D. and Pradhan, A.K., 1989, *JPB*, **22**, 3377
**N I** Bell, K.L., Berrington, K.A. and Ramsbottom, C.A., 1991, *JPB*, **24**, 1209
**N I** Bell, K.L. and Berrington, K.A., 1991, *JPB*, **24**, 933
**O I** Butler, K. and Zeippen, C.J., 1990, *AA*, **234**, 569
**Al VII** Baliyan, K.S. and Kingston, A.E., 1991, *JPB*, **24**, 4743

## Recombination

**Review** Pindzola, M.S., Badnell, N.R. and Griffin, D.C., 1990, *AIP Conf. Proc.*, **215**, 659
**General fits** Péquignot, D., Petitjean, P. and Boisson, C., 1991, *AA*, 251, 680
**General** Smits, D.P., 1991, *MNRAS*, **248**, 193
**Scaling** McLaughlin, D.J. and Hahn, Y., 1991, *PRA*, **43**, 1313
**C&O ions** Griffin, D.C., 1989, *Phys. Scr.*, **T28**, 17
**C&O ions** Hahn, Y., 1989, *Phys. Scr.*, **T28**, 25
**C&O ions** Badnell, N.R., 1989, *Phys. Scr.*, **T28**, 33
**H Seq** Pindzola, M.S. and Badnell, N.R., 1990, *PRA*, **42**, 282
**H Seq-with dust** Hummer, D.G. and Storey, P.J., 1992, *MNRAS*, **254**, 277
**He, Li, Be and Ne Seq** Romanik, C.J., 1988, *ApJ*, **330**, 1022
**Li Seq** Dittner, P.F., Dutz, S., Miller, P.D. and Pepmiller, P.L., 1987, *PRA*, **35**, 3668
**Be Seq** McLaughlin, D.J., LaGattuta, K.J. and Hahn, Y., 1987, *JQSRT*, **37**, 47
**Be Seq** Badnell, N.R., 1987, *JPB*, **20**, 2081
**Be and B Seq** Badnell, N.R., Pindzola, M.S., Andersen, L.H., Bolko, J. and Schmidt, H.T., 1991, *JPB*, **24**, 4441
**B III** Pradhan, A.K., 1984, *PRA*, **30**, 2141
**B III** Griffin, D.C., Pindzola, M.S. and Bottcher, C., 1985, *PRA*, **31**, 568
**B III** Geltman, S., 1985, *JPB*, **18**, 1425
**Ar ions** Moussa, A.H. and Hahn, Y., 1990, *JQSRT*, **43**, 45
**C I** Escalante, V. and Victor, C.A., 1990, *ApJsup*, **73**, 513
**C IV** Pradhan, A.K., 1984, *PRA*, **30**, 2141
**C IV** Griffin, D.C., Pindzola, M.S. and Bottcher, C., 1985, *PRA*, **31**, 568
**C IV** Geltman, S., 1985, *JPB*, **18**, 1425
**C V** Hahn, Y. and Bellantone, R., 1989, *PRA*, **40**, 6117
**C V and C VI** Bellantone, R. and Hahn, Y., 1989, *PRA*, **40**, 6913
**C V** Badnell, N.R., Pindzola, M.S. and Griffin, D.C., 1990, *PRA*, **41**, 2422
**N II** Escalante, V. and Victor, C.A., 1990, *ApJsup*, **73**, 513
**N III** Nasser, I. and Hahn, Y., 1989, *PRA*, **39**, 401
**N V and F VII** Andersen, L.H., Pan, G.-Y., Schmidt, H.T., Pindzola, M.S. and Badnell, N.R., 1992, *PRA*, **45**, 6332

O I Chung Sunggi and Lin, C.C., 1991, *PRA*, **43**, 3433
O II Terao, M., Bell, K.L., Burke, P.G. and Hibbert, A., 1991, *JPB*, **24**, L321
O III Roszman, L.J., 1989, *Phys. Scr.*, **T28**, 36
O IV Nasser, I. and Hahn, Y., 1989, *PRA*, **39**, 401
O IV Janjusevic, M. and Hahn, Y., 1989, *PRA*, **40**, 5641
O VI Pradhan, A.K., 1984, *PRA*, **30**, 2141
O VI Griffin, D.C., Pindzola, M.S. and Bottcher, C., 1985, *PRA*, **31**, 568
O VI Roszman, L.J., 1989, *Phys. Scr.*, **T28**, 36
O VII Bellantone, R. and Hahn, Y., 1989, *PRA*, **40**, 6913
O VII Badnell, N.R., Pindzola, M.S. and Griffin, D.C., 1990, *PRA*, **41**, 2422
O VIII Bellantone, R. and Hahn, Y., 1989, *PRA*, **40**, 6913
F V Nasser, I. and Hahn, Y., 1989, *PRA*, **39**, 401
Ne Seq Dalhed, S., Nilsen, J. and Hagelstein, P., 1986, *PRA*, **33**, 264
Ne Seq Chen, M.H., 1986, *PRA*, **34**, 1073
Ne ions Nussbaumer, H. and Storey, P.J., 1987, *AAsup*, **69**, 123
Mg Seq Dube, M.P., Rasoanaivo, R. and Hahn, Y., 1985, *JQSRT*, **33**, 13
Mg Seq Dube, M.P. and LaGattuta, K.J., 1987, *JQSRT*, **38**, 311
Mg II Pradhan, A.K., 1984, *PRA*, **30**, 2141
Mg II Geltman, S., 1985, *JPB*, **18**, 1425
Mg, Al and Si ions Nussbaumer, H. and Storey, P.J., 1986, *AAsup*, **64**, 545
Cl VII Pradhan, A.K., 1984, *PRA*, **30**, 2141
Ca II Geltman, S., 1985, *JPB*, **18**, 1425

## Charge Exchange

Review Dalgarno, A., 1985, *Nucl. Instrum. & Methods. Phys. Res. Sect. B*, **B9**, 662
General Butler, S.E. and Dalgarno, A., 1980, *ApJ*, **241**, 838 and the program **LZRATE** Bienstock, S., 1983, *Comput. Phys. Commun.*, **29**, 333
H I Jouin, H. and Harel, C., 1991, *JPB*, **24**, 3219
H I Hoekstra, H.R., de Heer, F.J. and Morgenstern, R., 1991, *JPB*, **24**, 4025
H I and He II Jackson, D., Slim, H.A., Bransden, B.H. and Flower, D.R., 1992, *JPB*, **25**, L127
H I Errea, L.F., López, A., Méndez, L. and Riera, S., 1992, *JPB*, **25**, 811
He I Valiron, P., Gayet, R., McCarroll, R., Masnou-Seeuws, F. and Philippe, M., 1979, *JPB*, **12**, 53
He I Slim, H.A., Heck, E.L., Bransden, B.H. and Flower, D.R., 1991, *JPB*, **24**, 1683
C II, C III and C IV Butler, S.E., Heil, T.G. and Dalgarno, A., 1980, *ApJ*, **241**, 442
C IV Watson, W.D. and Christensen, R.B., 1979, *ApJ*, **231**, 627
C IV Errea, L.F., Herrero, B., Mendez, L., Mó, O. and Riera, A., 1991, *JPB*, **24**, 4049
C IV Opradolce, L., Benmeuraiem, L., McCarroll, R. and Piacentini, R.D., 1988, *JPB*, **21**, 503
C V Gargaud, M., McCarroll, R. and Valiron, P., 1987, *JPB*, **20**, 1555
N III and N IV Butler, S.E., Heil, T.G. and Dalgarno, A., 1980, *ApJ*, **241**, 442
N IV Watson, W.D. and Christensen, R.B., 1979, *ApJ*, **231**, 627
N IV and N VI Gargaud, M. and McCarroll, R., 1985, *JPB*, **18**, 463
O III and O IV Butler, S.E., Heil, T.G. and Dalgarno, A., 1980, *ApJ*, **241**, 442

O III Bacchus-Montabonel, M.C., Courbin, C. and McCarroll, R., 1991, *JPB*, **24**, 4409
O III Gargaud, M., McCarroll, R. and Opradolce, L., 1989, *AA*, **208**, 251
O VI Andersson, L.R., Gargaud, M. and McCarroll, R., 1991, *JPB*, **24**, 2073
Ne III and Ne IV Butler, S.E., Heil, T.G. and Dalgarno, A., 1980, *ApJ*, **241**, 442
Ne III Forster, H.C., Cooper, I.L., Dickinson, A.S., Flower, D.R. and Mendez, .L, 1991, *JPB*, **24**, 3433
Al III Gargaud, M., McCarroll, R., Lennon, M.A., Wilson, S.M., McCullough, R.W. and Gilbody, H.B., 1990, *JPB*, **23**, 505
Si II Gargaud, M., McCarroll, R. and Valiron, P., 1982, *AA*, **106**, 197
Si III McCarroll, R. and Valiron, P., 1976, *AA*, **53**, 83
Si V Opradolce, L., McCarroll, R. and Valiron, P., 1985, *AA*, **148**, 229
Ar VII Opradolce, L., Valiron, P. and McCarroll, R., 1983, *JPB*, **16**, 2017

## Transition Probabilities

**MBPT** Gaigalas *et al.*, This Meeting
**Bibliography** Biémont, E. and Zeippen, C.J., 1991, in *ZMLD*, 209
**Review** Wiese, W.L., 1991, in *ZMLD*, 287
**Review** Martin, W.C., 1992, in *Lecture Notes in Physics*, in press
**M1–Tables** Kaufman, V. and Sugar, J., 1986, *JPCRD*, **20**, 321
**Critical Evaluation** Mendoza, C., 1983, in *IAU103*, 143
**Critical Evaluation** Martin, W.C., Fuhr, J.R. and Wiese, W.L., 1988, *JPCRD*, **16**, Suppl. 3 and Fuhr, J.R., Martin, W.C. and Wiese, W.L., 1988, *JPCRD*, **17**, Suppl. 4
**Critical Evaluation** Fuhr, J.R. and Wiese, W.L., 1990, in *CRC Handbook of Chemistry and Physics, 71st Edition*, Ed. Lide, D.R.
**Critical Evaluation** Morton, D.C., 1991, *ApJsup*, **77**, 119
**Iron Group** Kurucz, R.L., 1992, in *Rev. Mex. Astro. y Astrofisica*, **23**, Ed. Lynas-Gray, A.E., Mendoza, C. and Zeippen, C.J.
**H ions** Storey, P.J. and Hummer, D.G., 1991, *Comput. Phys. Commun*, **66**, 129
**Be Seq** Trabert, E., 1990, *Phys. Scr.*, **41**, 675
**Be Seq** Eissner, W.B. and Tully, J., 1992, *AA*, **253**, 625
**Be I** Chang, T.N., 1990, *PRA*, **41**, 4922
**Be I** Chang, T.N. and Tang, X., 1990, *JQSRT*, **43**, 45
**C seq** Luo, D. and Pradhan, A.K., 1989, *JPB*, **22**, 3377
**C I** Goldbach, C., Martin, M. and Nollez, G, 1989, *AA*, **221**, 155
**C III, N IV and O V** Allard, N., Artru, M.-C., Lanz, T. and Le Dourneuf, M., 1990, *AAsup*, **84**, 563
**N Seq** Becker, S.R., Butler, K. and Zeippen, C.J., 1989, *AA*, **221**, 375
**N Seq** Curtis, L.J., Rudzikas, Z.B. and Ellis, D.G., 1991, *PRA*, **44**, 776
**N I** Vaeck, N. and Hansen, J.E, 1991, *JPB*, **24**, L469
**N I** Bell, K.L. and Berrington, K.A., 1991, *JPB*, **24**, 933
**N I** Zhu, Q., Bridges, J.M., Hahn, T. and Wiese, W.L., 1989, *PRA*, **40**, 3721
**N I** Hibbert, A., Biémont, E., Godefroid, M. and Vaeck, N., 1990, *AAsup*, **88**, 505
**N II** Bell, K.L., Ramsbottom, C.A. and Hibbert, A., 1992, *JPB*, **25**, 1735
**N IV** Chang, T.N. and Mu Yi, 1990, *JQSRT*, **44**, 413
**N IV** Laughlin, C, 1990, *Phys. Scr.*, **42**, 551

N IV Chang, T.N., *JQSRT*, **44**, 413
O Seq Chen, M.H., 1989, *PRA*, **40**, 4330
O I Bell, K.L., Hibbert, A., McLaughlin, B.M. and Higgins, K., 1991, *JPB*, **24**, 2665
O I Hibbert, A., Biémont, E., Godefroid, M. and Vaeck, N., 1991, *JPB*, **24**, 3943
O I Biémont, E., Hibbert, A., Godefroid, M., Vaeck, N. and Fawcett, B.C., 1991, *ApJ*, **375**, 818
O I Bell, K.L. and Hibbert, A., 1990, *JPB*, **23**, 2673
O I Tayal, S.S. and Henry, R.J.W., 1989, *PRA*, **39**, 4531
O III Aggarwal, K.M. and Hibbert, A., 1991, *JPB*, **24**, 4685
O III Aggarwal, K.M. and Hibbert, A., 1991, *JPB*, **24**, 3445
Ne II Burshtein, M.L. and Vujnović, V., 1991, *AA*, **247**, 252
Mg Seq Trabert, E., 1990, *Phys. Scr.*, **41**, 675
Mg Seq Chang, T.N. and Rong-qi Wang, 1991, *PRA*, **44**, 80
Mg I Chang, T.N. and Tang, X., 1990, *JQSRT*, **43**, 45
Mg I Beck, D.R. and Ziyong Cai, 1989, *Phys Lett. A*, **142**, 378
Mg I Chang, T.N., 1990, *PRA*, **41**, 4922
Mg IV Mohan, M. and Baluja, K.L, 1989, *Z. Phys. D: At. Mol. Clusters*, **14**, 135
Mg IV Mohan, M. and Hibbert, A., 1991, *Phys. Scr.*, **44**, 158
Al Seq Hjorth-Jensen, M., Aashamar, K., 1990, *Phys. Scr.*, **42**, 309
Al I Davidson, M.D., Volten, H. and Dönszelmann, A., 1990, *AA*, **238**, 452
Al II Serrâo, J.M.P., 1990, *JQSRT*, **45**, 121
Al V – VII Biémont, E., 1991, *At. Dat. Nuc. Tab.*, **48**, 1
Si III Serrâo, J.M.P., 1990, *JQSRT*, **45**, 349
Si VI Mohan, M. and Le Dourneuf, M., 1990, *PRA*, **41**, 2862
P Seq Curtis, L.J., Rudzikas, Z.B. and Ellis, D.G., 1991, *PRA*, **44**, 776
S I Doering, J.P., 1990, *J. Geophys. Res.*, **95**, 21313
S I Delalic, Z., Erman, P. and Kallne, E., 1990, *Phys. Scr.*, **42**, 540
S VII Hibbert, A., Mohan, M. and Le Dourneuf, M, 1992, *JPB*, **25**, 1107
S VIII Mohan, M. and Hibbert, A., 1991, *Phys. Scr.*, **44**, 158
Cl I Delalic, Z., Erman, P., Kallne, E. and Zastrow, K.-D., 1990, *JPB*, **23**, 2727
Cl I Ojha, P.C. and Hibbert, A., 1990, *Phys. Scr.*, **42**, 424
Cl VIII Hibbert, A., Mohan, M. and Le Dourneuf, M, 1992, *JPB*, **25**, 1107
Ar II Das, M.B. and Bhattacharya, R., 1991, *JPB*, **24**, 423
Ar II Lüdtke, T. and Helbig, V., 1990, *JQSRT*, **44**, 261
Ar IX Hibbert, A., Mohan, M. and Le Dourneuf, M, 1992, *JPB*, **25**, 1107
K seq Zilitis, V.A., 1989, *Opt. Spectrosc.*, **67**, 595
K seq Zeippen, C.J., 1990, *AA*, **229**, 248
Ca I Vaeck, N., Godefroid, M. and Hansen, J.E., 1991, *JPB*, **24**, 361
Ca II Guet, C. and Johnson, W.R., 1991, *PRA*, **44**, 1531
Ti II Savanov, I.S., Huovelin, J. and Tuominen, I., 1990, *AAsup*, **86**, 531
Fe I Bard, A., Kock, A. and Kock, M., 1991, *AA*, **248**, 315
Fe I O'Brian, T.R., Wickliffe, M.E., Lawler, J.E., Whaling, W. and Brault, J.W., 1991, *J. Opt. Soc. Am. B*, **8**, 1165
Fe I – IV Sawey, P.M.J. and Berrington, K.A., 1992, *JPB*, **25**, 1451
Fe II Heise, C. and Kock, M., 1990, *AA*, **230**, 244
Fe II Fawcett, B.C., 1987, *ADNDT*, **37**, 333
Fe III – VI Fawcett, B.C., 1989, *ADNDT*, **41**, 181

## Collisional Excitation

**Critical Evaluation** Itikawa, Y., 1992, *ADNDT*, **49**, 209
**Critical Evaluation** Pradhan, A.K. and Gallagher, J.W, 1992, *ADNDT*, in press and Gallagher, J.W. and Pradhan, A.K., 1985, *JILA Report No. 30*
**Critical Evaluation** Mendoza, C., 1983, in *IAU103*, 143
**General** Percival, I.C. and Richards, D., 1977, *JPB*, **8**, 1497
**van Regemorter?** Sampson, D.H. and Zhang, H.L., 1992, *PRA*, **45**, 1556
**H seq** Clark, R.E.H., 1990, *ApJ*, **354**, 382
**H I** Chang, E.S., Avrett, E.H. and Loeser, R., 1991, *AA*, **247**, 580
**H I** King, G.C., Trajnar, S. and McConkey, J.W., 1989, *Comments At. Mol. Phys.*, **23**, 229
**H I** Aggarwal, K.M., Berrington, K.A., Burke, P.G., Kingston, A.E. and Pathak, A., 1991, *JPB*, **24**, 1411
**H I** Fon, W.C., Aggarwal, K.M. and Ratnavelu, K., 1992, *JPB*, **25**, 2625
**He I** Fon, W.C., Berrington, K.A. and Kingston, A.E., 1991, *JPB*, **24**, 2161
**He I** Cartwright, D.C., Csanak, G, Trajmar, S. and Register, D.F., 1992, *PRA*, **45**, 1602
**He I** Sawey, P.M.J., Berrington, K.A., Burke, P.G. and Kingston, A.E., 1990, *JPB*, **23**, 4321
**He I** Scholz, T.T., Walters, H.R.J., Burke, P.G. and Scott, M.P., 1990, *MNRAS*, **242**, 692
**He I** Nakazaki, S, Berrington, K.A., Sakimoto, K. and Itikawa, Y., 1991, *JPB*, **24**, L27
**He II** Aggarwal, K.M., Berrington, K.A., Kingston, A.E. and Pathak, A., 1991, *JPB*, **24**, 1757
**He II** Aggarwal, K.M., Callaway, J., Kingston, A.E. and Unnikrishnan, K., 1992, *ApJsup*, **80**, 473
**Li Seq** Zhang, H.L., Sampson, D.H. and Fontes, C.J., 1990, *ADNDT*, **44**, 31
**Li I** Gien, T.T., 1987, *JPB*, **20**, 1337
**Li I** Tayal, S.S. and Tripath, A.N., 1984, *Can. J. Phys.*, **62**, 198
**Li II** Berrington, K.A. and Nakazaki, S., 1991, *JPB*, **24**, 1411
**Be I** Fon, W.C., Berrington, K.A., Burke, P.G., Burke, V.M. and Hibbert, A., 1992, *JPB*, **25**, 507
**Be II** Lengyel, V.I., Navrotsky, V.T. and Sabad, E.P., 1990, *JPB*, **23**, 2847
**B I** Nakazaki, S. and Berrington, K.A., 1991, *JPB*, **24**, 4263
**C I** Doering, J.P. and Dagdigian, P.J., 1989, *Chem. Phys. Lett.*, **154**, 234
**C I** Johnson, C.T., Burke, P.G. and Kingston, A.E., 1987, *JPB*, **20**, 2553
**C II** Luo, D. and Pradhan, A.K., 1990, *PRA*, **41**, 165 and Blum, R.D. and Pradhan, A.K., 1992, *ApJsup*, **80**, 425
**C IV** Burke, V.M., 1992, *JPB*, **25**, in press
**C VI** Aggarwal, K.M. and Kingston, A.E., 1991, *JPB*, **24**, 4583
**N I** Doering, J.P and Goembel, L., 1991, *J. Geophys. Res.*, **96**, 16021
**N III** Luo, D. and Pradhan, A.K., 1990, *PRA*, **41**, 165 and Blum, R.D. and Pradhan, A.K., 1992, *ApJsup*, **80**, 425
**O I** Williams, J.F. and Allen, L.J., 1989, *JPB*, **22**, 3529
**O I** Tayal, S.S. and Henry, R.J.W., 1989, *PRA*, **39**, 4531
**O I** Doering, J.P. and Gulcicek, E.E., 1989, *J. Geophys. Res.*, **94**, 2733
**O I** Laher, R.R. and Gilmore, F.R., 1990, *JPCRD*, **19**, 227
**O I - Review** Itikawa, Y. and Ichimura, A., 1990, *JPCRD*, **19**, 637

O I Mantas, G.P. and Carlson, H.C., 1991, *Geophys. Res. Lett.*, **18**, 159
O I Tayal, S.S., 1992, *JPB*, **25**, 2639
O III Aggarwal, K.M. and Hibbert, A., 1991, *JPB*, **24**, 3445
O IV Luo, D. and Pradhan, A.K., 1990, *PRA*, **41**, 165 and Blum, R.D. and Pradhan, A.K., 1992, *ApJsup*, **80**, 425
O V Kato, T., Lang, J. and Berrington, K.A., 1990, *ADNDT*, **44**, 133
F IV Conlon, E.S., Keenan, F.P. and Aggarwal, K.M., 1992, *Phys. Scr.*, **45**, 309
Na I Ganas, P.S., 1985, *J. Appl. Phys.*, **57**, 154
Na I Msezane, A.Z., 1988, *PRA*, **37**, 1787
Na I Msezane, A.Z., Handy, C.R., Mantica, G. and Lee, J., 1988, *PRA*, **38**, 1604
Na I Bielschowsky, C.E., Lucas, C.A. de Souza, G.G.B., and Noqueria, J.C., 1991, *PRA*, **43**, 5975
Na VI Conlon, E.S., Keenan, F.P. and Aggarwal, K.M., 1992, *Phys. Scr.*, **45**, 309
Ne Seq Chen, M.H. and Reed, K.J., 1989, *PRA*, **40**, 2292
Ne I Bubelev, V.E. and Grum-Grzhimailo, A.N., 1990, *Opt. Spectrosc.*, **69**, 178
Ne I Pilsof, N. and Blagoev, A., 1988, *JPB*, **21**, 639
Ne I Taylor, K.T., Clark, C.W. and Fon, W.C., 1985, *JPB*, **18**, 2967
Ne V Lennon, D.J. and Burke, V.M, 1991, *MNRAS*, **251**, 628
Ne VI Hayes, M.A., 1992, *JPB*, **25**, 2649
Mg I Clark, R.E.H., Csanak, G. and Abdallah, J., Jr., 1991, *PRA*, **44**, 2874
Mg I Meneses, G.D., Pagan, C.B. and Machado, L.E., 1990, *PRA*, **41**, 4740
Mg I Brunger, M.J., Riley, J.L., Scholten, R.E. and Teubner, P.J.O, 1988, *JPB*, **21**, 1639
Mg I Brunger, M.J., Riley, J.L., Scholten, R.E. and Teubner, P.J.O, 1989, *JPB*, **22**, 1431
Mg I McCarthy, I.E., Ratnavelu, K. and Zhou, Y., 1989, *JPB*, **22**, 2597
Mg II Lengyel, V.I., Navrotsky, V.T. and Sabad, E.P., 1990, *JPB*, **23**, 2847
Mg IV Mohan, M. and Baluja, K.L, 1989, *Z. Phys. D: At. Mol. Clusters*, **14**, 135
Mg VII Burgess, A., Mason, H.E. and Tully, J.A., 1991, *ApJ*, **376**, 803
Al VIII Conlon, E.S., Keenan, F.P. and Aggarwal, K.M., 1992, *Phys. Scr.*, **45**, 309
Si II Dufton, P. and Kingston, A.E., 1990, *MNRAS*, **248**, 827
Si III Dufton, P. and Kingston, A.E., 1989, *MNRAS*, **241**, 209
Si IV Whalin, E.K., Thompson, J.S., Dunn, G.H., Phaneuf, R.A., Gregory, D.C. and Smith, A.C.H., 1991, *Phys. Rev. Lett.*, **66**, 157
Si VI Mohan, M. and Le Dourneuf, M., 1990, *PRA*, **41**, 2862 and *AA*, **227**, 285
S I Ganas, P.S., 1990, *J. Chem. Phys.*, **92**, 2374
S II Ho, Y.K. and Henry, R.J.W., 1990, *ApJ*, **351**, 701
S VII Mohan, M., Hibbert, A., Berrington, K.A., and Burke, P.G., *JPB*, **23**, 989
Ar I Bubelev, V.E. and Grum-Grzhimailo, A.N., 1990, *Opt. Spectrosc.*, **69**, 178 and 1991, *JPB*, **24**, 2183
Ar I Subramanian, K.P. and Kumar, V., 1987, *JPB*, **20**, 5505
Ar I Bielschowsky, C.E., de Souza, G.G.B., Lucas, C.A. and Boechat-Roberty, H.M., 1988, *PRA*, **38**, 3405
Ar III Johnson, C.T. and Kingston, A.E., 1990, *JPB*, **23**, 3393
Cl I Ganas, P.S., 1988, *J. Appl. Phys.*, **63**, 277
Ca II Zatsarinny, O.I., Lendel, V.I. and Masalovich, E.A., 1989, *Opt. Spectrosc.*, **67**, 10
Fe III Berrington, K.A., Zeippen, C.J., Le Dourneuf, M., Eissner, W.B. and Burke, P.G., 1991, *JPB*, **24**, 3467
Fe VII Keenan, P. and Norrington, P.H., 1990, *ApJ*, **368**, 486

# LUMINOSITIES OF CENTRAL STARS OF PN IN THE GALACTIC BULGE

R. H. MÉNDEZ

*University Observatory, Munich, Federal Republic of Germany*

A calibration of the nebular intensity of [O III] 5007 (relative to $H_\beta$) in terms of stellar bolometric corrections is presented. This calibration (restricted to low- and medium-excitation PN) is based exclusively on the results of non-LTE model atmosphere analyses of absorption-line profiles in the spectra of bright central stars of PN (Méndez et al. 1988, A&A 190, 113 and subsequent papers). Knowing the bolometric corrections, the distance to the Galactic center, the amount of interstellar extinction and the visual apparent magnitudes (Tylenda et al. 1989, A&AS 77, 39) of central stars of PN in the direction of the Galactic center, luminosities for 20 of these central stars are obtained without any assumptions about nebular properties. The luminosities obtained with this method are substantially higher than those obtained for the same objects by Pottasch and Acker (1989, A&A 221, 123) and Tylenda et al. (1991, A&A 246, 221). The most probable explanation for this discrepancy is that luminosity estimates based on assumptions about nebular properties are too low, because many PN are not completely optically thick in the H Lyman continuum and/or because many ionizing photons in these PN are being absorbed by nebular dust. Which of the two effects is more important cannot be decided without solving the problem of the discrepancy between extinction determinations based on the Balmer decrement versus extinction determinations based on radio and $H_\beta$ fluxes (the radio-$H_\beta$ extinctions are too low).

The higher luminosities obtained here have removed the large discrepancy, reported by Pottasch and Acker, between nebular ages of a few thousand years and the stellar post-AGB ages of more than 100000 years that would be implied by the very low luminosities they obtained.

# UNIFIED MODEL ATMOSPHERE STUDIES OF CENTRAL STARS OF PLANETARY NEBULAE

R. GABLER, A. GABLER, R. H. MÉNDEZ and R. P. KUDRITZKI
*University Observatory, Munich, Federal Republic of Germany*

A first step in the accurate quantitative spectroscopic analysis of central stars of PN has been based on fitting the results of NLTE, hydrostatic, plane-parallel model atmosphere calculations to the observed H and He absorption-line profiles in high-resolution spectra of bright central stars (Méndez et al. 1988, A&A 190, 113 and subsequent papers). Such analyses have provided very useful determinations of the basic atmospheric parameters: $T_{\text{eff}}$, log $g$ and He abundance.

That kind of analysis is not suitable for central stars showing strong winds. In order to overcome this limitation, we are working on the development of "unified" model atmospheres (Gabler et al. 1989, A&A 226, 162; 1991, A&A 245, 587). These unified models make no artificial separation between a hydrostatic photosphere and a supersonically expanding stellar wind. Instead, a detailed radiatively driven wind code, giving the density structure and velocity field along the whole (sub- and supersonic) atmosphere, is combined with a NLTE model atmosphere code for spherical geometry, which gives the temperature structure. In this way, the whole atmospheric structure can, in principle (although not yet in practice), be described self-consistently as a function of three basic stellar parameters: $T_{\text{eff}}$, log $g$, and the stellar radius at which $T_{\text{eff}}$ and log $g$ are given (this stellar radius fixes the stellar luminosity and mass) plus the chemical composition. For the present calculations we have included only H and He.

At the present time the wind parameters $(k, \alpha, \delta)$ are not yet determined self-consistently but are instead adopted from other studies of the winds of hot massive stars like $\zeta$ Puppis. Our general idea is to use as input, if known, the value of $V_{\infty}$ derived from high-resolution UV spectroscopic studies, and to adjust $T_{\text{eff}}$, log $g$, He abundance and mass loss rate until a good fit is obtained to both absorption *and* emission line profiles in the central star spectra. This semi-empirical approach will be useful as a guideline for a future step in which we expect to achieve full self-consistency.

In this poster we compare plane-parallel and unified fits to the stellar line profiles in the spectra of several central stars showing weak and strong winds. Although not perfect, the fits permit to give an idea of the size of the corrections needed for the values of $T_{\text{eff}}$ and log $g$ that had been determined using hydrostatic, plane-parallel NLTE models. The corrections are small, and in agreement with previous estimates (Méndez et al. 1992, A&A 260, 329).

# METAL LINE BLANKETED NON-LTE MODEL ATMOSPHERES FOR CENTRAL STARS OF PLANETARY NEBULAE

K. WERNER
*Institut für Theoretische Physik und Sternwarte, Kiel, FRG*
and
S. DREIZLER
*Sternwarte der Universität Erlangen-Nürnberg, Bamberg, FRG*

Absorption lines of highly ionized iron group elements dominate the UV spectra of many hot stars. They were identified in several central stars as well as in sdO stars, in PG 1159 stars and in the hottest DO and DA white dwarfs. Due to the high effective temperatures atmospheric modelling including metal line blanketing has to be done under NLTE conditions. Adequate models have become available only very recently (Dreizler & Werner 1992). We present new NLTE model atmospheres blanketed by some 120 000 lines from iron group elements (Sc through Ni). We adopted Anderson's (1991) statistical approach along with an opacity sampling technique using our Accelerated Lambda Iteration (ALI) code. We generally found:

– Many strong metal lines cause substantial blocking at the flux maximum.
– The temperature in the continuum forming layers is increased by backwarming due to iron group lines; outer layers, cooled by CNO lines, are unaffected.
– The line profiles of H and He become deeper and broader, indicating that the neglect of NLTE metal line blanketing could cause the discrepancies encountered when fitting He II and H I lines in hot subdwarfs.

Anderson, L.S. 1991, NATO ASI series C, 341, 29
Dreizler, S., Werner, K. 1992, Lecture Notes in Physics, Springer, 401, 436

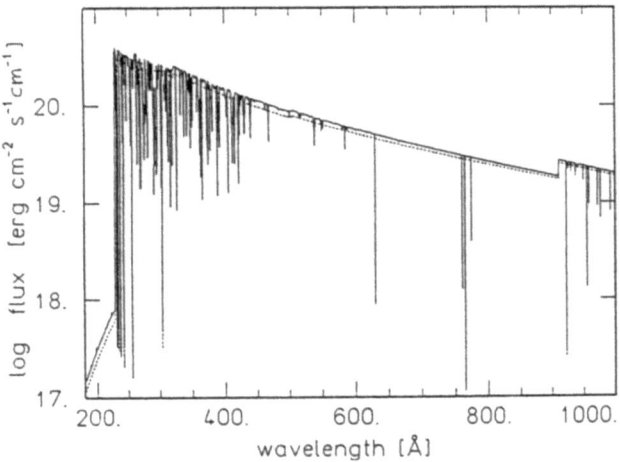

EUV model flux with (solid)/without metal line blanketing. $T_{eff}$=80kK, $\log g$ =7.5, abundances H/He/C/N/O/iron group= 1/98/0.3/0.3/0.1/0.1 by number

# NON-LTE SPECTRA OF IRON GROUP ELEMENTS FOR CSPN

S.R. BECKER and K. BUTLER
*Universitätssternwarte München, Scheinerstraße 1, 8000 München 80, Germany*

The calculations

UV spectra of CSPN are dominated by lines from ions of iron group elements. Recently we have developed a non-LTE model of the Fe V spectrum putting particular emphasis on this spectral region (Becker and Butler, 1992a). Lines of Ni IV and Ni V are major contributors to blends with the dominating Fe V lines. Meanwhile, non-LTE models for Ni IV, Ni V and Ni VI have been developed (Becker and Butler, 1992b) and the non-LTE model spectra composed of lines from all four ions can be compared to the observations.

In this paper we modelled the spectra of two CSPN (IC3568, NGC6826). IUE high resolution spectra were available and stellar parameters were taken from Mendez et al., 1990. The presence of a stellar wind has been accounted for according to the analytical stellar wind model by Kudritzki (1992a).

A resulting theoretical spectrum is shown in fig. 1 together with the observations. It is obvious from this that the major limitation for determination of abundances is currently given by the quality of the observations. HST spectra, when available, allow a reliable determination of the iron abundance (see Becker and Butler, 1992a, Kudritzki, 1992b).

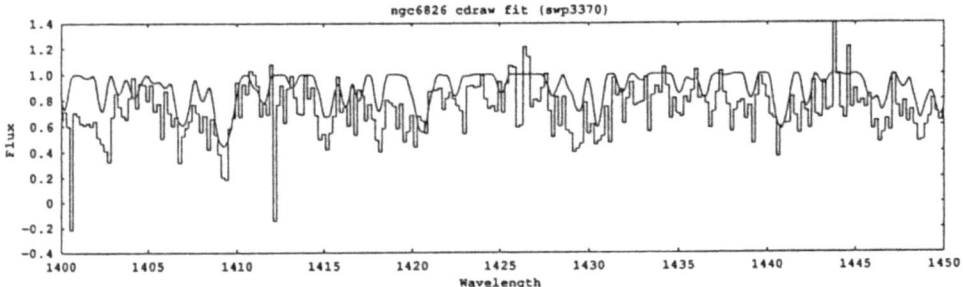

Figure 1: Comparison of a theoretical non-LTE spectrum with an IUE observation of NGC 6826.

REFERENCES

Becker, S.R. and Butler, K., 1992a, A&A, in press.
Becker, S.R. and Butler, K., 1992b, A&A, in preparation.
Kudritzki, R.P., 1992a, A&A in press.
Kudritzki, R.P., 1992b, this book.
Mendez, R.H., Herrero, A. and Manchado, A., 1990, A&A, **229**, 152.

# THE PROPERTIES OF PLANETARY NEBULAE NUCLEI: STELLAR WINDS

L. BIANCHI and G. DE FRANCESCO
*Osservatorio Astronomico di Torino, Strada Osservatorio, 20, 10025 Pino Torinese, Torino, Italy*

We present IUE observations of some nuclei of Planetary Nebulae. From these data we derive the stellar photospheric parameters ($T_{eff}$, $L_{bol}$, log g), and the wind characteristics (velocity, mass loss rate). $T_{eff}$, $R_*$, $L_{bol}$ are derived from UV low resolution spectra, combining optical and radio data, from Bianchi (1988) or from new IUE data, with the same method (fit of the UV continuum with model atmospheres for high gravity stars, after correcting for reddening and for the contribution of continuum emission by the nebular gas). P Cygni profiles from IUE high resolution spectra are fitted with the SEI method and $V_\infty$ is derived. The non-LTE ionisation in the wind and the mass loss rate are computed as in Bianchi et al. (1986). Details are given in a forthcoming paper. The results for a first group of objects are given in the Table below.

References
Bianchi, L., Cerrato, S., Grewing, M., 1986, Astron. Astrophys. 169, 227.
Bianchi, L., 1988, ESA SP-281, 2, 173.
Bianchi, L., Cascio, F., 1989, in "From Miras to Planetary Nebulae: Which Path for Stellar Evolution", eds. Mennessier and Omont, Editions Frontieres , p. 403.
Bianchi, L., 1991, Astron. Astrophys. 253, 447.
Bianchi, L., De Francesco, G., 1992, in preparation.
Heap, S., R., Augensen, H., J., 1987, Ap. J. 313, 268.

| Name | PK number | Sp.Type | $T_{eff}$ (K) | $R_*/R_\odot$ | $V_{edge}$ (km/s) | log $\dot{M}$ ($M_\odot$/yr) |
|---|---|---|---|---|---|---|
| NGC6210 | 43 +37 1 | O3 | 60000 | 0.19 | 2700 | -8.7 |
| NGC6826 | 83 +12 1 | O4f | 35000 | 1.49 | 1700 | -8.9 |
| NGC6891 | 54 -12 1 | O4f | 22500 | 1.91 | 1550 | -7.9 |
| IC4593 | 25 +40 1 | O7 | 30000 | 2.60 | 1350 | -8.1 |
| NGC40(a) | 120 + 9 1 | WC8 | 90000 | 1.7 | 1800 | -6 |
| NGC6543(b) | 96 +29 1 | Of-WR | 80000 | 0.64 | 1900 | -6.5 |

Note: Sp. Types are from: Heap et al., 1987.
(a) Results from Bianchi, 1991.
(b) Results from Bianchi et al., 1986.

# SPECTROSCOPIC PROPERTIES OF THE NUCLEUS OF NGC 6826

B. ALTNER
*Applied Research Corporation, Landover, MD USA*

S.R. HEAP
*Laboratory for Astronomy and Solar Physics*
*NASA Goddard Space Flight Center, Greenbelt, MD USA*

and

I. HUBENY
*Universities Space Research Association (USRA)*
*NASA Goddard Space Flight Center, Greenbelt, MD USA*

**Abstract.** We continue the trend of finding discrepancies in the values of central star masses derived from photospheric line profile fitting and other methods in our analyses of ground-based and UV spectra of the nucleus of NGC 6826.

Many apparent contradictions and uncertainties in the basic parameters of planetary nuclei cloud the general picture of the late stages of evolution of low- and intermediate-mass stars. The primary uncertainty is the mass distribution of the central stars. Different methods — non-LTE photospheric analyses (*e.g.*, Mendez et al. 1988, *Astr. Ap.*, **190**, 113), analyses of the wind spectrum (*e.g.*, Pauldrach et al. 1988 *Astr. Ap.*, **207**, 123), application of the Zanstra method, use of the optical fading diagram (*e.g.*, Heap and Augensen 1987 *Ap. J.*, **313**, 268) — yield different results for the stellar mass, and there is no clear consensus as to which method, if any, is the correct one (Pottasch 1989, in *"Planetary Nebulae", IAU Symp. No. 131*, p. 481).

We have used high-dispersion ultraviolet and visual spectra to derive the properties of the O3Iaf*-type central star of the planetary nebula NGC 6826. Based on comparisons to models within an extensive grid of non-LTE atmospheres we conclude that the star has a temperature of 50,000 K ±5000, and a log g ≈ 3.9, yielding a mass ≈ 0.82 $M_\odot$. However, both the strength of the He II λ4686 emission line and the terminal velocity of the stellar wind (as determined from the P Cygni profiles of the strong UV resonance lines), suggest a mass near 0.65 $M_\odot$. Application of our newest method of mass determination for central stars (optical-UV fading; see Altner and Heap, this volume) yields an upper limit of 0.64±0.1 $M_\odot$ for NGC 6826. Even lower masses are found from the optical fading diagram and the Zanstra temperature and luminosity. Therefore, this star, like many Of-type central stars, shows a discrepancy between the mass derived from photospheric analysis and that derived from other methods.

# SPECTRAL ANALYSES OF WC-TYPE CENTRAL STARS

W.-R. HAMANN, L. KOESTERKE

*Institut für Theoretische Physik und Sternwarte der Universität, Kiel, FRG*

Models have been developed in Kiel for massive (Pop. I) WR stars which account for multi-level non-LTE radiation transfer in spherically expanding atmospheres. The published (Koesterke et al. 1992) grid of models for WC composition (40% helium, 60% carbon by mass) can be applied to low-mass stars as well by means of the scaling properties of WR spectra (Hamann et al. 1992) and allow a rough guess of the parameters, while individual calculations are necessary for a detailed analysis and the determination of the chemical composition.

We have analyzed the spectrum of the central star of NGC 6751. The obtained parameters - still somewhat preliminary - are given in the Table and compared to the previously analyzed CPN of Abell 78 (Werner & Koesterke 1992) and Longmore 4 (cf. Werner et al., these proceedings). The luminosity has been adopted as $10^{3.7} L_\odot$ from evolutionary considerations for all three stars. The results reveal that the only essential difference between these three CPN concerns their mass-loss rate, which is exceptionally strong in NGC 6751. The Longmore 4 data refer to the state of high mass-loss observed fortuitously on Jan-27-1992, while in the quiet state Longmore 4 is of PG 1159 type with $\dot{M}$ being $10^{-8} M_\odot$ yr$^{-1}$ or less. Despite of their mass-loss all three stars resemble PG 1159 stars, indicating a close relationship.

The remarkable existence of nitrogen in NGC 6751 and Abell 78 has also its counterpart in a PG 1159 star (namely PG 1144+005, cf. Werner & Heber 1991); its explanation within the "born-again" scenario needs special assumptions on "mixing and burning" (cf. Schönberner & Blöcker 1992).

| CPN | NGC 6751 | Abell 78 | Longmore 4 |
|---|---|---|---|
| $T_* $ / kK | 105 | 115 | 120 |
| $\log \dot{M}/(M_\odot \text{ yr}^{-1})$ | -5.2 | -7.3 | -7.3 |
| $v_\infty/(\text{km s}^{-1})$ | 2000 | 3700 | 4000 |
| helium [% by mass] | 61.5 | 33 | 67 |
| carbon [% by mass] | 27 | 50 | 25 |
| nitrogen [% by mass] | 1.5 | 2 | 0 |
| oxygen [% by mass] | 10 | 15 | 8 |

Hamann W.-R., Leuenhagen U., Koesterke L., Wessolowski U.: 1992, A&A **255**, 200
Koesterke L., Hamann W.-R., Wessolowski U.: 1992, A&A (in press)
Schönberner D., Blöcker T.: 1992, in *Atmospheres of Early-type stars*, U. Heber and C.S. Jeffery (eds.), Lecture Notes in Physics 401, Springer, p. 305
Werner K., Heber U.:1991, A&A **247**, 476
Werner K., Koesterke L.: 1992, in *Atmospheres of Early-type stars*, U. Heber and C.S. Jeffery (eds.), Lecture Notes in Physics 401, Springer, p. 288

# CENTRAL STARS OF OLD PLANETARY NEBULAE

R. NAPIWOTZKI

*Institut für Theoretische Physik und Sternwarte der Universität, Olshausenstr. 40,*
*D-2300 Kiel, Germany*

During a program for the observation of central stars of old planetary nebulae 29 stars are classified until now. Most of them (22) are belonging to the hydrogen-rich sequence and resemble either high-gravity sdO stars or white dwarfs (14 DAOs, 3 DAs). 3 are hydrogen-deficient PG 1159 stars, and also 3 are hybrid type stars Further 3 CPN are close binaries, whose evolution has been certainly influenced by this circumstance. A complete list of so far observed objects is given in Napiwotzki (1992).

For the analysis we used model atmospheres containing H and He calculated with the sophisticated NLTE code of Werner (1986). We tried to determine temperature and gravity of the observed central stars in the usual way by fitting the Balmer lines. A consistent fit of the Balmer lines was not possible for a number of objects. Examples are S 216 and NGC 7293. In the case of S 216 we derived $T_{\text{eff}} = 50,000\,\text{K}$ from $H_\alpha$ and $T_{\text{eff}} = 90,000\,\text{K}$ from $H_\delta$, for NGC 7293 $T_{\text{eff}} = 55,000\,\text{K}$ ($H_\beta$) and $110,000\,\text{K}$ ($H_\delta$).

Possible reasons for the Balmer line discrepancy are (1) a change of the atmospheric structure caused by the line blanketing of iron and other metals, (2) possible errors in the line broadening theory (VCS) used for the Balmer lines, (3) the effect of a weak stellar wind, or (4) magnetic fields. From these points the iron line blanketing or the line broadening are the most likely reasons. Test calculations with D. Koesters LTE code showed that pressure ionization and quenching of atomic levels (Hummer & Mihalas 1988) is completely unimportant for the investigated Balmer lines ($H_\alpha$ to $H_\delta$).

Independent temperature estimates are possible from the ionization balance of iron and helium, and from the modeling of the ionization structure of the surrounding nebula. A more detailed discussion can be found in Napiwotzki (1992) and will be published in the Proceedings of the White Dwarf Workshop 1992 in Leicester.

We can conclude that the DA/DAO central stars are the evolutionary link between the hydrogen-rich central stars and the DA white dwarf sequence. It is likely that the temperature scale of hot white dwarfs is in error. Independent temperature estimates are important to clarify this problem.

*References*

Hummer, D.G., Mihalas, D., 1988, ApJ 277, 233
Napiwotzki, R., 1992, Proc. Kiel/CCP7 workshop on Atmospheres of Early Type Stars, eds. U. Heber, S. Jefferey, Springer, p. 310
Werner, K., 1986, A&A 161, 171

# INFRARED EMISSION-LINES AND THE STELLAR TEMPERATURE

SUELI M. VIEGAS and RUTH GRUENWALD

*Instituto Astronômico e Geofísico - USP, Av. Miguel Stefano, 4200, 04301-002 São Paulo, SP Brazil*

Observations of near infrared emission-lines are becoming available and can be a powerful tool to improve our knowledge on planetary nebulae properties. For wavelengths in the range 1 to 5 $\mu$m, the emission-lines correspond to atomic transitions of high ionized species of heavy elements. In particular, the [Si VI] 1.96$\mu$m and [Si VII] 2.48$\mu$m lines have already been detected (Ashley and Hyland, 1988).

The presence of such ions indicates a high stellar temperature ($T_*$), and intensity line ratios can provide a good stellar temperature indicator for objects for which other methods do not give reliable $T_*$ estimations. For example, the Zanstra method gives only a lower limit for stellar temperatures higher than 150000K; this method is also affected by the nebula optical depth. The Stoy method also fails at high stellar temperatures (Stasinska and Tylenda, 1986).

We analyze here the dependence of the line intensity ratio [Si VI] 1.96$\mu$m/[Si VII] 2.48$\mu$m with stellar temperature using the photoionization code Aangaba (Gruenwald and Viegas, 1992) and typical PN conditions ($n_H=10^2$-$10^6$ cm$^{-3}$; $T_*$=150000-400000K; $L_*$=1000-10000$L_\odot$; $[Z]$=Stasinska and Tylenda, 1986). The above line intensity ratio decreases with the gas density (7% from $n_H=10^2$ to $10^4$ cm$^{-3}$; 47% from $n_H=10^2$ to $10^6$ cm$^{-3}$) but depends very little on the stellar luminosity. This line intensity ratio is a good indicator of stellar temperatures between 150000 and 350000K.

Presently, there are only two PNe for which intensities of [SiVI] and [SiVII] infrared lines were measured. We apply the proposed method for stellar temperature determination for these two nebulae, considering stellar luminosities and gas densities given in the literature. Unfortunately, these two nebulae provide only extreme examples. For NGC 6302, only a lower limit for the stellar temperature can be obtained, since the observed intensity ratio points to the flat part of the curve intensity ratio vs. stellar temperature, indicating $T_* \geq 350000$K. This is consistent with values obtained with Stoy cases II and III (Pottasch, 1984). For NGC 6537, the observations provide only an upper limit for the [SiVII] line intensity. Therefore, only a lower limit for the line ratio and an upper limit for the stellar temperature (250000K) can be obtained. This result is also consistent with previous determinations with the Zanstra method (Pottasch, 1984). Furthermore, the observed intensity of these infrared lines relative to H$\beta$ can be reproduced with a Si abundance close to half solar, indicating that in these objects Si is not very depleted.

### References

Ashley, M.C.B. and Hyland, A.R. (1988), Ap.J. **331**, 532
Gruenwald, R.B. and Viegas, S.M. (1992), Ap.J.Suppl.Ser. **78**, 153
Pottasch, S. (1984), "Planetary Nebulae"
Stasinska, G. and Tylenda, R. (1986) A.&A. **155**,137

# ULTRAVIOLET AND OPTICAL SPECTRA OF CENTRAL STARS OF HALO PLANETARY NEBULAE

M. PEÑA and S. TORRES-PEIMBERT

Instituto de Astronomía, Universidad Nacional Autónoma de México, Apdo. Postal 70 264, 04510 Méx. D.F., México

and

M. T. RUIZ

Departamento de Astronomía, Universidad de Chile, Casilla 36-D, Santiago, Chile

ABSTRACT. Optical and UV spectrophotometric data are analyzed for the central stars of the known Population II planetary nebulae. From this, we derive visual magnitudes, spectral classification, color temperatures, luminosities and masses of the objects. It is found that all the stars show absorption type spectrum and most of them have normal H and He photospheric abundances, the only possible exceptions are M 2-29 and GJJC-1 which seem to be H-deficient stars.

The visual magnitudes, derived from UV stellar continuum, are in good agreement with magnitudes derived from photometric optical data, except for NGC 2242 for which we found a value V = 17.1 mag instead of the 15.02 reported by Maehara et al (1987), this implies a distance 2 kpc larger than that calculated by Torres-Peimbert et al. (1990). High signal-to-noise photometric observations of the central star are required to confirm these results.

The color temperatures derived for these objects range from less than 40,000 K to more than 80,000 K and, in general, they are consistent with effective temperatures calculated by other means. We found that the central star of NGC 4361 shows a UV continuum steeper than expected from the effective temperature determined from modelling of H and He absorption lines. From comparison with theoretical evolutionary tracks for nuclei of PN we found that the stellar masses of these objects spread over a na- rrow range from 0.55 to 0.58 M(⊙). The average mass is lower than the average masses derived for large samples of disk and bulge planetary nebulae.

References

Maehara, H., Okamura, S, Noguchi, T., He, X-T., Liu, J., Huang, Y.W. and Feng, X.C. (1987) A&A, 178, 221
Torres-Peimbert , S., Peimbert, M. and Peña, M. (1990) A&A, 233, 540

# A HIGH RESOLUTION FAR-UV SPECTRAL ATLAS OF CSPNs AND HOT WHITE DWARFS

R.W. TWEEDY
*Steward Observatory, University of Arizona, Tucson, USA*

A high-resolution IUE spectral atlas of central stars of planetary nebulae and hot white dwarfs has been produced (part of Tweedy, 1991, PhD thesis from the University of Leicester, UK), and examples from it are shown here. It has been sorted into an approximate evolutionary sequence, based on published spectroscopic analyses, from the cool 28,000K young central star He 2-138, through the hot objects like NGC 7293 and NGC 246 at 90,000K and 130,000K respectively, down to 40,000K DA white dwarfs like GD 2, which is the chosen cutoff for this selection. Copies of a revised version of this atlas, which will include more recent spectroscopic information and also white dwarfs down to 35,000K – to include the Si III object GD 394 – will be sent to anyone who requests one.

# THE WIND TEMPERATURE AND C/He AND O/He RATIOS OF THE WC10 CENTRAL STAR CPD–56°8032

M. J. BARLOW and P. J. STOREY

*Department of Physics & Astronomy, University College London,*
*Gower Street, London WC1E 6BT, U.K.*

We have analysed absolutely calibrated 3500–8900Å AAT spectrophotometry of the WC10 central star CPD–56°8032. After dereddening by $E(B-V) = 0.68$, the few line-free portions of the stellar continuum are well fitted by the 18000K blackbody which corresponds to the nebular H I Zanstra temperature. The Wolf-Rayet wind terminal velocity is sufficiently low (210–230 km s$^{-1}$) that it is possible to resolve virtually every fine structure component of every multiplet of C II, C III and O I listed by the NBS Multiplet Tables for this wavelength region.

Many of the observed C II recombination lines in the spectrum of CPD–56°8032 are due to the process of low temperature dielectronic recombination. They arise from the radiative decay, and subsequent cascade, of strongly autoionizing resonance states near the ionization limit of C+. The populations of such resonance states are determined by the balance between autoionization and radiative decay. Autoionization usually dominates, and the populations are then given by the Saha equation. The measured fluxes of such lines may therefore be used to infer the electron temperature. We have identified four such C II lines in the spectrum of CPD–56°8032; $\lambda\lambda 4621, 4961, 5113, 8798$. Two of the transitions ($\lambda 4621$ and $\lambda 5113$) are between states residing entirely in the continuum and should therefore be thought of as resonance features in the free-free emission associated with C2+, rather than as recombination lines. Our fit to these four dielectronic recombination lines of C II yields an emission measure of $9.0\pm0.1\times 10^{13}$ cm$^{-5}$ for C2+, along with a wind temperature of 12800K, which can be used for a recombination line abundance analysis of the other ions that are present in the wind. The stronger lines of He I (no lines of hydrogen are emitted by the wind) all show P Cygni profiles due to self-absorption, but the weak He I lines at 4713Å and 5047Å both show Gaussian profiles and are unblended. The dereddened fluxes measured for these lines have been used with high density recombination coefficients to derive a He+ emission measure (I/Q) equal to $5.82\pm0.3\times 10^{14}$ cm$^{-5}$. We have determined the abundance of O2+ in the wind from the O II recombination lines at $\lambda\lambda 4489.77, 4491.54$ and $\lambda\lambda 4890.93, 4906.98$ We derive an O2+ emission measure (I/Q) of $4.1\pm2.1\times 10^{13}$ cm$^{-5}$. C3+ was found not to make a significant contribution to the total abundance of carbon.

The C/He ratio of 0.15 by number is comparable to those found for Population I WC Wolf-Rayet stars, but the O/He ratio of 0.07 is significantly higher than is believed applicable for Population I WC stars. This may not be surprising in view of the completely different nuclear-burning histories of massive WC stars and PN WC stars. The relatively high oxygen abundance that is found implies that when CPD–56°8032 evolves to higher effective temperatures it will become an O VI emission line central star and, eventually, a low luminosity pulsating PG1159–035 star.

# MASS LOSS IN TWO LOW TEMPERATURE CENTRAL STARS OF PLANETARY NEBULAE

A. MODIGLIANI and M. PERINOTTO
*Dipartimento di Astronomia Univ. Firenze, L.go E. Fermi 5, I-50125 Firenze, Italy*

and

P. PATRIARCHI
*GNA/CNR c/o Osserv. Arcetri, L.go E. Fermi 5, I-50125 Firenze, Italy*

The high resolution IUE images SWP 16742 and SWP 42675 have been studied to derive the fast wind properties of the central stars of the two low temperature planetary nebulae IC 2149 and Tc 1. The first image is from the IUE archive, while the second one has been taken by us.

IC 2149 and Tc 1 exhibit clearly developed P Cygni profiles only in the lines of the ions: CIV, NIV, and, in Tc 1, also SiIV, consistent with a $T_{eff}$ of around 34,000 K. IC 2149 has a faint P Cygni profile also in the NV doublet at 1240 Å, which comes out in the spectrum with a very low S/N ratio. The OIV triplet, with its bluest (and strongest) component at 1339 Å, appears quite narrow in both stars, suggesting an essentially photospheric origin. The SiIV line appears in IC 2149 in absorption, with its blue edge shifted of about -150 km/s, from which can be inferred that it forms very close to the photospheric layers.

The observed P Cygni profiles have been matched with theoretical profiles calculated with the SEI method (Lamers, Cerruti-Sola, and Perinotto 1987, Ap. J. 314, 726). In this method the velocity and opacity law across the wind are parametrized.

Essentials of the SEI method are: a) the solution of the transfer equation in the line across the wind is performed exactly, under the Sobolev hypothesis for the calculation of the source function, and b) the effects of turbolence are considered via a Doppler widening.

From the theoretical best fitting parameters $\beta$ (velocity law), and $T,\alpha_1,\alpha_2$ (opacity law), the mass loss rate follows, provided we specify: the terminal velocity, the stellar radius, the fractional ionization of the ion, the corresponding elemental abundance, and, for excited lines, also the radiation temperature at the wavelenght of the line populating the lower level of the observed transition.

We obtain the following preliminary values. For IC 2149:

$\dot{M}q_i$ (CIV) > 2.0 $10^{-10}$ $M_\odot$ yr$^{-1}$;    $\dot{M}q_i$ (NIV) = 2.2 $10^{-8}$ $M_\odot$ yr$^{-1}$

Since the NV P Cygni profile is present, but very faint, we infer that the assumption of q(NIV) to be about 0.5 is correct to within a factor of 2. Then $\dot{M}$ (IC 2149) = 4. $10^{-8}$ $M_\odot$ yr$^{-1}$.

For Tc 1 we derive:

$\dot{M}q_i$ (CIV) > 1.9 $10^{-10}$ $M_\odot$ yr$^{-1}$;    $\dot{M}q_i$ (NIV) = 6.4 $10^{-7}$ $M_\odot$ yr$^{-1}$;    $\dot{M}q_i$ (SiIV) = 3.7 $10^{-9}$ $M_\odot$ yr$^{-1}$

The two last values are also low limits to the "true" $\dot{M}$ because $0 \leq q \leq 1$. From the same argument used for IC 2149, we infer a $\dot{M} = 10^{-6}$ $M_\odot$ yr$^{-1}$. This is the highest value of $\dot{M}$ so far obtained in central stars of planetary nebulae studied with the SEI method.

Its amount is such that the effects both on the evolution of the central star and on the interaction of the fast wind with the previously ejected material, are relevant.

# STARK BROADENING PARAMETERS OF C IV LINES FOR STELLAR PLASMA RESEARCH

M.S. DIMITRIJEVIC

*Astronomical Observatory, Volgina 7, 11050 Beograd, Yugoslavia*

and

S.SAHAL-BRECHOT

*Observatoire de Paris-Meudon, 92190 Meudon, France*

In order to complete available C IV broadening data needed for stellar plasma research, we have calculated Stark broadening parameters for 69 C IV multiplets of large principal quantum number. The results along with a discussion of the Stark broadening parameter regularities within spectral series will be published elsewhere (Dimitrijevic and Sahal-Brechot, 1992). As an example in Figs 1 and 2 the case of C IV $np^2P^0 - 9s^2S$ transitions, is presented. We can see that particularly for shifts the changes of Stark broadening parameters are relatively small, permitting the interpolation of new data or critical evaluation of mutual consistency of existing data.

## References

Dimitrijevic, M.S. , Sahal-Brechot,S. (1992) *A&AS*, in press.

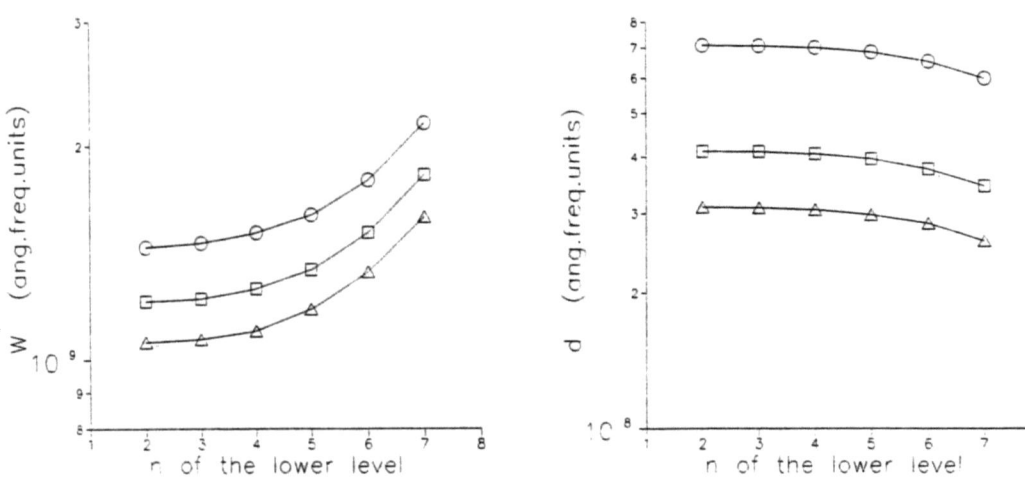

Fig. 1. Figure 1a, 1b. Stark full half widths (1a) and shift (1b) for the C IV $np^2P^0 - 9s^2S$ lines as a function of n for T=20,000K (circles); T=100,000K (squares) and 200,000K (triangles) at $N_e = 10^{13} cm^{-3}$

# E2 AND M1 TRANSITION PROBABILITIES IN IONS OF THE NITROGEN ISOELECTRONIC SEQUENCE CALCULATED USING MBPT

G. GAIGALAS, R. KISIELIUS, G. MERKELIS and M. VILKAS

*Institute of Theoretical Physics and Astronomy, Gostauto 12, 2600, Vilnius, Lithuania*

Forbidden electric quadrupole *(E2)* and magnetic dipole *(M1)* transitions are of extreme importance in astrophysics. Up to now the most extensive calculations for the nitrogen isoelectronic sequence have been done using the method proposed by C.J. Zeippen [1] or in MCHF approximation [2]. To account for electron correlations both these methods use a large list of configurations. We have chosen the stationary many-body perturbation theory (MBPT) [3] for the inclusion of the electron correlations. The calculations have been perfomed in the second order in the complete model space $1s^22s^22p^3 + 1s^22p^5$. Relativistic corrections have been accounted for in the Breit-Pauli approximation. In the Table we present probabilities for electric quadrupole *W(E2)* and magnetic dipole *W(M1)* transitions (in $s^{-1}$), wavelengths $\lambda$ (in A). The comparision of the results shows that our second order calculation data in the most cases are closer to term-energy corrected ones from [1].

| $1s^22s^22p^3$ ( $^4S_{3/2}$ - $^2D_{3/2}$ ) | | MBPT | | MCHF [2] | Recommended [1] |
|---|---|---|---|---|---|
| | | First order | Second order | | |
| O II | *W(E2)* | 4.30-5 | 2.11-5 | 1.88-5 | 2.36-5 |
| | *W(M1)* | 1.66-4 | 1.20-4 | 1.31-4 | 1.29-4 |
| | $\lambda$ | 3329.1 | 3821.1 | 3569.7 | 3727.1* |
| Ne IV | *W(E2)* | 4.05-4 | 2.55-4 | 2.24-4 | 2.46-4 |
| | *W(M1)* | 6.40-3 | 4.73-3 | 4.74-3 | 5.52-3 |
| | $\lambda$ | 2237.8 | 2451.5 | - | 2422.5* |
| S X | *W(E2)* | 2.44-? | 1.95 2 | 1.77-2 | 1.89-2 |
| | *W(M1)* | 1.73+1 | 1.43+1 | 1.40+1 | 1.50+1 |
| | $\lambda$ | 1160.0 | 1212.6 | 1186.7 | 1213.6* |

\* - $\lambda_{exp}$ from [1]

1. S.R. Becker, K. Butler, C.J. Zeippen, 1989, Astron. Astrophys. 221, 375
2. M. Godefroid, Ch. Froese-Fischer, 1984, J. Phys. B, 17, 681.
3. M.J. Vilkas, G. Gaigalas, G. Merkelis, 1991, Lithuanian J.Phys. 31, 84

# MBPT RESULTS FOR $\Delta n=0$ ELECTRIC DIPOLE TRANSITIONS

G. GAIGALAS, R. KISIELIUS, G. MERKELIS, Z. RUDZIKAS and M. VILKAS

*Institute of Theoretical Physics and Astronomy, Gostauto 12, 2600, Vilnius, Lithuania*

To identify spectra of Planetary Nebulae which usually have many atomic lines one needs very accurate theoretical atomic data.

Recently quasi-degenerate many-body perturbation theory (MBPT) for effective Hamiltonian proposed by I. Lindgren was adopted to calculate the energy levels for open-shell atoms [1]. We continue the applications of this method to the investigation of the electric dipole transitions. The computer programs to generate radial and spin-angular parts of matrix elements of effective Hamiltonian and transition operator have been worked out. The energy levels of the configurations considered were calculated in the complete model space. The first order relativistic corrections were taken into account in the Breit-Pauli approximation. The Table presents wavelengths $\lambda(A)$, absorption oscillator strenghts $f$ (in the length form) and ratios of line intensities $I(^1P-^1D)/I(^1P-^1S)$ for $1s^22s2p^5$ $^1P - 1s^22s^22p^4$ $^1D$ (a) and $1s^22s2p^5$ $^1P - 1s^22s^22p^4$ $^1S$ (b) transitions in various approximations.

| | Exp [2] | MBPT | | Recommended [3] | MCDHF [4] |
|---|---|---|---|---|---|
| | | First order | Second order | | |
| **Ne III** | | | | | |
| $\lambda_a(A)/\lambda_b(A)$ | 379/428 | 349/382 | 381/429 | 372/417 | 351/384 |
| $f_a/f_b$ | 0.194/0.078 | 0.368/0.146 | 0.337/0.129 | 0.271/0.110 | 0.377/0.150 |
| $I_a/I_b$ | 20.8 | 15.1 | 16.7 | 15.5 | 15.1 |
| **Mg V** | | | | | |
| $\lambda_a(A)/\lambda_b(A)$ | 277/312 | 261/288 | 278/313 | 274/308 | 262/289 |
| $f_a/f_b$ | | 0.289/0.120 | 0.273/0.108 | 0.227/0.098 | 0.294/0.122 |
| $I_a/I_b$ | | 14.7 | 15.8 | 14.6 | 14.6 |

Disagreement between experimental [3] and our second order values of oscillator strengths is expected to reduce in accounting for correlation corrections in matrix elements of electric dipole transition operator.

1. M.J. Vilkas, G. Gaigalas, G. Merkelis, 1991, Lithuanian J.Phys., **31**, 84
2. H.J. Flaig, K.-H. Schartner, E. Trabert, P.H. Heckmann, 1985, Physica Scripta, **31**, 225
3. K.L. Baluja, C.J. Zeippen, 1989, J. Phys. B, **21**, 15
4. K.T. Cheng, J.K. Kim, J.P. Desclaux, 1979, At. Data Nucl. Data Tables, **24**, 111

# III. HIGHLIGHTS ON THE NEBULAE

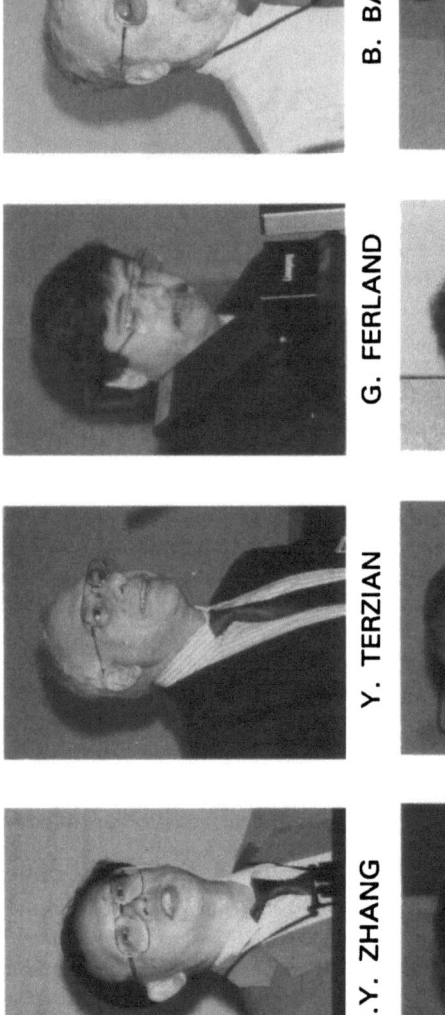

# ENERGY DISTRIBUTION OF PLANETARY NEBULAE (UV TO RADIO)

C.Y. ZHANG
*Department of Astronomy*
*The University of Texas at Austin*
*Austin, TX 78712*

**Abstract.** The past decade has seen significant progress in our understanding of spectral energy distribution of planetary nebulae over the entire wavelength range from UV to radio. In this review we show the detailed breakdown of the energy budget for a planetary nebula as a system of the three components, i.e., the central star, the gaseous nebula and the dust shell. This picture of the energy distribution is further discussed in the context of planetary nebula evolution.

## 1. Introduction

The major obstacle to fully comprehend the energetics in planetary nebulae (PN) had been the difficulty of obtaining the completely continuous spectrum, which has a comparable energy output in the infrared, ultraviolet, and optical wavelength ranges. Dramatic progress has been made in our understanding of spectral energy distribution (SED) of PN since the last decade. It largely results from our access to instruments covering a vastly broad wavelength range. The modern techniques and instruments, including the VLA, IRAS, through IUE to Einstein, EXOSAT and ROSAT, have allowed extension of the wavelength coverage from the conventional optical, to radio, infrared, UV, and even X-ray.

## 2. Components of a Planetary Nebula System

A PN system is made up of three major components: a central star, an ionized gas nebula, and a dust shell. These three major components can be further divided to include the ionized stellar wind from the star, a high temperature "bubble" formed by the interaction of this wind with the nebula and a neutral gas envelope if the nebula is ionization-bounded (Kwok 1982). The spectrum of the central star can be approximated by a blackbody of temperature 30,000 to 200,000 K. There can be departures from the Planck spectrum (especially in the UV) due to transfer effects in the stellar atmospheres. The ionized nebula is dominated by bound-free, free-free, and two-photon continuum emissions, and recombination and collisionally-excited line emissions. The dust envelope is responsible for thermal continuum emission at a dust temperature of about 100 K.

## 3. Energy Distribution of Planetary Nebulae

It is only recently that a significant number of PN have been observed over a wide spectral range from UV to radio for their energetics to be studied in a systematic manner. If a PN is ionization-bounded, the present wavelength coverage from the radio to 0.1 $\mu$m probably contains most of the energy emitted. For density-bounded nebulae, there could be flux missing in the far-ultraviolet.

## 3.1. PROGRESS IN OBSERVATIONS

UV observations have played an important role in our understanding of energy distribution of PN. It is crucial to observe them in this wavelength regime, where the energy distribution becomes more sensitive to the temperature of the central star, compared to that in the optical. The IUE observations of more than 200 PN have demonstrated that (1) the two-photon emission is one of the essential ingredients in nebular continuum (Pottasch 1984), and (2) the UV spectrum is vital for studying properties of stellar energy distribution (Kaler and Feibelman 1985, Heap and Augensen 1987, Bianchi et al. 1989).

In an attempt to determine accurate stellar magnitudes, narrow or intermediate band photometric data of PN have been accumulated, particularly through the work done by Webster (1988), Kaler (1976a, 1978), Kohoutek and Martin (1981; KM), and Shaw and Kaler (1985, 1989; SK). The above work was usually aimed at determination of the central star B and V magnitudes. For a review on the various techniques in this regard see Kaler (1989). SK have carried out narrow-band continuum and emission-line photometry for 145 southern hemisphere PN. They have obtained B and V magnitudes for 120 central stars after subtracting nebular continuum from the total flux recorded. The continuum fluxes at 3226, 3546, 4225, 5306, 6865, and 7901 Å were measured by KM using narrow-band photometry. This set of data are extremely useful in that the Balmer jump is clearly revealed in emission for their sample PN. The requirement, that any model to fit the SED of these PN must be able to fit the Balmer jump, gives very tight constraints on determinations of relative contributions from the nebula and the central star to the continuum in the wavelength range from UV to optical. It is particularly interesting to see the magnitudes of the central star and the global nebular continuum were derived simultaneously for 11 PN, based on the nebula-subtraction method using CCD imagery (Jacoby and Kaler 1989).

An unparalleled, extensive spectroscopic survey of about 1000 PN have been conducted by Acker and Stenholm (1987). This provides a unique means to study SED for a large sample of PN. Spectroscopic measurements of the continuum are free of contamination from line emission. In deriving stellar magnitudes, Tylenda et al. (1991a) have measured the continuum levels at 4 wavelength points. Bearing in mind that these authors were interested in stellar B and V magnitudes, they gave the ratios of stellar to total continuum at three wavelengths (4340, 4860, and 6560 Å) and the stellar flux at 4860 Å. In Tylenda et al. (1989), the stellar fluxes at both 4860 and 6560 Å were given. Since the aperture used was 4", the total continuum flux can be obtained for a nebula with a size less than 4". Kaler et al. (1990) presented spectrophotometry of 75 large PN and derived B and V magnitudes for 17 central stars of them. The total continuum including nebular and stellar parts is not available from their paper. Kaler et al. (1991) have obtained the continuum fluxes at 1300, 1500, 1750, 4000, 4310, and 5480 Å for PB6.

Near-infrared (near-ir) photometry has been obtained by several groups (Whitelock 1985; Kwok et al. 1986; van der Veen et al. 1989; Preite-Martinez and Persi 1989). Some PN are identified in a near-ir survey of PN candidates selected from their IRAS colors (Manchado et al. 1989; Garcia-Lario et al. 1990). Very recently,

Rudy et al. (1991a, b) have obtained the spectra of NGC 6572 and BD+303639 between 0.5 to 1.3 μm, and analyzed the continuum of them. Zhang and Kwok (1992) published the spectra of IC418 and IC5117 from 0.5 to 1.75 μm and modeled the continuum. These studies show that the near-ir is an important wavelength regime yet to be more extensively explored. Besides the rich emission lines, the stellar emission, bound-free, free-free emission, sometimes hot dust thermal emission, and possible dust features all contribute to the continuum in the near-ir.

IRAS has provided not only the most complete far-infrared (far-ir) photometric measurements (IRAS PSC), but also the very extensive far-ir spectroscopic data for PN (IRAS LRS). Volk and Cohen (1990) have extracted 170 spectra of PN from the LRS raw database. Zhang and Kwok (1990) have made a quantitative chemical analysis of solid material in 13 young PN, using the IRAS LRS spectra. They found that six of them show both oxygen- and carbon- rich dust features, and the silicate 18 μm feature is prominent in a number of young PN.

Strong line emission is one of the most prominent characteristics of PN, and a significant amount of energy is emitted in the atomic and even molecular lines. In the optical, the classical sources of these line data come from Kaler (1976b), Aller and Czyzak (1979, 1983), KM, Torres-Peimbert and Peimbert (1977) and others. Acker et al. (1989a) and SK provide the most recent results of line fluxes for a large number of PN. Line flux measurements are extended to the UV (Marionni and Harrington 1981; Boggess et al. 1981; Aller and Keyes 1981) and to far-ir (Pottasch et al. 1986; Zhang and Kwok 1990).

The absolute flux in the $H\beta$ line is one of the most important physical parameters for a study of PN, since all the line fluxes are scaled to $H\beta$. The most comprehensive measurements of the absolute $H beta$ flux is given by Acker et al. (1989a, b). This important quantity is now available for 880 PN owing to the most extensive spectroscopic survey done by the Acker's group.

Despite the impressive progress, the data of emission line fluxes cannot be complete, since different authors concentrated on limited wavelength ranges. In particular, in the near- and far-ir, emission lines have not been explored as thoroughly as in the optical.

Early in the 60's and 70's, single dish radio continuum measurements had established the spectral distribution of PN in the radio (for a review, see, Terzian 1991). Beginning from the 70's, high resolution mapping of PN with the VLA (Aaquist and Kwok 1990; Zijlstra et al. 1989; Ratag et al. 1990) have resulted in radio images, accurate flux densities, and angular sizes for 402 PN.

Kreysing et al. (these proceedings) reported the results from the ROSAT survey and found extended X-ray emission from six PN. The X-ray spectral distribution has ruled out the central as a source of the X-ray emission.

## 3.2. PLANETARY NEBULAE IN SYSTEMS AT KNOWN DISTANCES

Studies of PN in systems for which the distances can be determined are very important to our understanding of the intrinsic properties and evolution of PN. Knowledge of energy distribution of PN in the galactic bulge (GBPN) has significantly increased since the extensive spectroscopic data are now available for more than

300 GBPN (Acker et al. 1991; Tylenda et al. 1991b; Stasinska et al. 1991; Stasinska these proceedings). The B and V magnitudes of their central stars are obtained from the continuum in the spectra. This has led to determinations of the central star temperature and luminosity (Tylenda et al. 1991b). Progress has been also made in studies of PN in Magellanic Clouds (MCPN). Meatheringham and Dopita (1991) presented detailed spectroscopy in the optical for a sample of 41 MCPN and obtained central star magnitudes from the continuum. Other spectroscopic surveys of MCPN include those carried out by Monk et al. (1988) and Boroson and Leibert (1989).

One of the advantages to study the GBPN, and in particular, the MCPN, is that they are usually small so that the flux integrated over the entire nebula can be completely recorded. However, the disadvantage is that it is hard to measure them in the far-ir, where a significant fraction of the power is emitted. Contamination by foreground objects might be a problem for GBPN studies.

### 3.3. OBSERVED ENERGY DISTRIBUTION

Zhang and Kwok (1991) compiled observations of 66 compact PN in the whole wavelength range, and made a model fitting to the observed SED from 0.1 to about 100 $\mu$m (Fig. 1). Almost all nebulae show strong emissions over the three decades of wavelengths. The Balmer jump in emission can clearly seen (KM). To illustrate a wealth of emission lines, we plotted vertical lines at the line center wavelengths in Fig. 1. In contrast to the richness in lines in the optical, no lines are indicated in the near-ir, due to lack of the near-ir spectrum. The observed SED exhibits a double-peaked shape with a cut-off at about 0.1 $\mu$m, beyond which little is known.

## 4. Model of Energy Distribution

It is necessary to model the observed SED, in order to obtain a detailed breakdown of the energy coming from the different components of the system, because the contributions from a specific component cannot usually be directly measures in the observed spectrum.

### 4.1. MODEL COMPONENTS

The observed monochromatic flux is fitted by a following function:

$$F_\lambda = F_\lambda(star) + F_\lambda(gas) + F_\lambda(dust) \tag{1}$$

where $F_\lambda(star)$, $F_{lambda}(gas)$, and $F_{lambda}(dust)$ are the stellar photospheric continuum flux, the nebular gaseous continuum emission flux, and the nebular dust thermal emission flux, respectively. The nebular bound-free and free-free emission from HI, HeI and HeII, and hydrogen two-photon emission are taken into account in the effective gaseous continuum emission coefficient ($\gamma_{eff}$). Calculations of $\gamma_{eff}$ as a function of y1, y2, $T_e$ and $n_e$ are based on the formalism of Brown and Mathews (1970), with an extension to 1000 Å on the short wavelength side and to 20 $\mu$m on the long wavelength side. Four quantities, the angular radius of the star ($\theta_*$), the central star temperature ($T_*$), the dust optical depth, and the dust temperature ($T_d$), in this model are used as free parameters to be optimized.

## 4.2. RESULTS

The individual model components are shown along with the synthesized model continuum in Fig. 1. The stellar flux points in the optical wavelength range determined by Shaw and Kaler (1989) and Tylenda et al. (1989) are given for comparison with the best- fit model stellar components. They are in good agreement with the model in most cases. Extrapolation from optically thin radio continuum flux densities is shown as a dashed line in Fig. 1. Dotted lines represent the bound-free, free-free, and two-photon emission from the ionized gas. A Planck function with a single temperature can fit the IRAS points well, indicating that the emitting dust is located near the nebular shell and is not distributed over a large volume as in the case of HII regions.

## 4.3. PHYSICAL PARAMETERS

Four parameters are derived from the model fitting: the temperature ($T_d$) and optical depth ($\tau$) of the dust component, and the angular size ($\theta_*$) and temperature ($T_*$) of the central star.

The derived central star temperatures are in good agreement with the energy-balance temperature ($T_{EB}$) and the hydrogen Zanstra temperature [$T_Z(H)$]. The agreement with $T_Z(HeII)$ and the previously determined color temperature ($T_c$) is not particularly good. Three nebulae in the sample are in common with those of Mendez et al. (1988; MKHHG), who determined the effective temperatures spectroscopically. These are IC2448, He2-138, and M1-26, which have our $T_* = 80,000$, 33,100 and 33,100 K and the effective temperatures of $T_{eff} = 65,000, 27,000$, and 33,000 K from MKHHG respectively. The agreement between them is good in general, except for IC2448. It is not clear what causes this discrepancy (Zhang and Kwok 1991).

The total fluxes for the three model components can be obtained by integrating the individual component over wavelengths. These are the non-ionizing stellar flux ($F_*$), nebular flux due to bound-free, free-free, and two-photon emission ($F_n$), and the far-ir flux ($F_{IR}$). The total line flux ($F_l$) is a summation of all emission line fluxes. To obtain the total observed flux from all the components, one should correct for the part of energy that is absorbed from the stellar non-ionizing radiation and converted into the far-ir radiation by dust. This correction will be important when the infrared excess (IRE) is larger than unity and at the same time the central star temperature is lower than 40,000 K. The amount of correction also depends on the optical properties of dust grains. Assuming that the central star radiates like a blackbody and the dust absorption coefficient varies with wavelength as a power law with an index of -1, one can estimate the fraction of the energy absorbed in UV out of that absorbed in the entire wavelength range, y, which is a function of $T_*$. Only the amount of energy surpassing the $Ly\alpha$ flux, i.e., ($F_{IR} - F_{Ly\alpha}$), where $F_{IR}$ and $F_{Ly\alpha}$ are in units of $F_{H\beta}$, should be corrected for this effect, so that

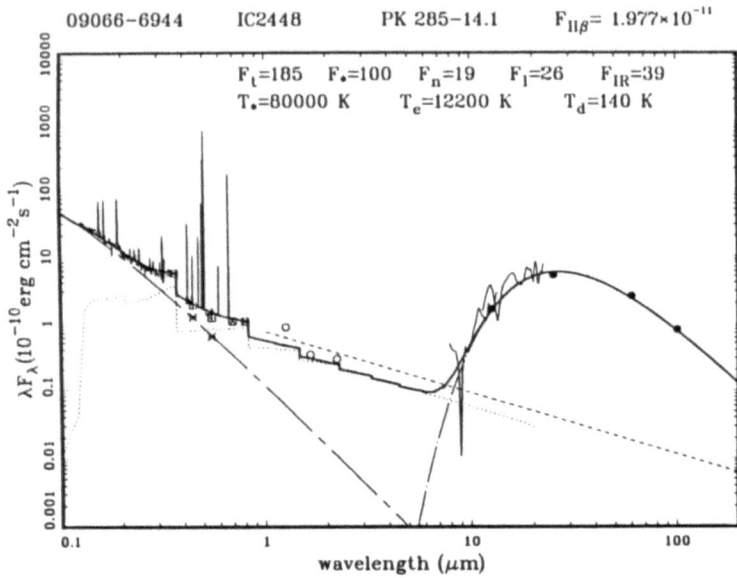

Fig. 1. The energy distribution of IC2448. Vertical lines illustrate emission lines. Model curves for the stellar, nebular, and dust components are shown as chain-dashed, dotted, and chain-dotted lines respectively. Thick solid lines show the synthesized energy distribution. Dashed lines indicate the level of free-free emission extrapolated from the optically thin radio continuum flux densities. Thin lines are the IUE and IRAS LRS spectra. Data points are from KM, SK, and Whitelock (1985)

the amount of energy absorbed by dust from the non-ionizing stellar radiation is now $F_c = (F_{IR} - F_{Ly\alpha}) \times (1 - y)$. Thus the total observed flux from a nebula is $F_t = (F_* - F_c) + F_n + F_l + F_{IR}$. It turns out that $F_c$ is less than about a few percent of $F_t$ for the majority sample nebulae, except for BD+303639 (30%) and M1-11 (20%).

We find that a significant fraction (38%) of the total observed flux is from the dust component. In the near-ir, almost all contribution to the observed fluxes are due to the nebular bound-free and free-free emission. In the visible, the star and nebula contribute approximately equally, depending on the central star temperature. In the UV, contribution from the central star often exceeds that of the nebular component.

## 4.4. OPTICAL THICKNESS

Supposing that the total flux integrating the Planck function at $T_*$ over the entire wavelength range (FBB) could represent the "true" total flux emitted by the central star, one can define an optical "thinness", $Q = (F_{BB} - F_t)/F_t$. It is found that the majority of the sample compact PN are optically thick to ionizing radiation. If $-0.4 < Q < 0.4$, the difference between $F_{BB}$ and $F_t$ can probably be attributed to the uncertainties in the fitting. A few PN with $Q > 0.6$ are likely to be optically thin. Seven PN have $Q < -0.6$, which is likely to be caused by (1) an excess in the far-UV of the stellar spectrum; or (2) the nebula being not spherically symmetric. For three nebulae in common with the sample of MKHHG we found that M1-26 and He2-138 are optically thick, while IC2448 is optically thin. Mendez et al. (1992) found that the fraction of stellar ionizing luminosity absorbed by the nebula for these three PN are 0.7, 0.96, and 0.23 respectively, which is consistent with our results.

While Mendez et al. (1992) concluded that the majority of their nebulae are optically thin, our sample seems to contain more optically thick ones. This is likely due to selection effects. We have selected compact nebulae, of which the size is sufficiently small to ensure that all the nebular emission is properly registered in the observations. MKHHG's sample contains the brightest central stars to allow for the analysis of absorption line profiles. It is conceivable that as a PN system evolves and a nebula expands with time, the nebula may evolve from optically thick to optically thin. It is, therefore, very probable that our sample contains mostly young PN.

For the 27 objects in our sample with $T_Z(H) < T_Z(HeII)$, only 5 have $Q > +0.5$, implying optical thinness to the ionizing radiation. The rest of them show mostly negative $Q$ values and $-0.15 < Q < 0.1$. These nebulae are unlikely to be far from optically thick stage. We suggest that for these nebulae, the Zanstra temperature discrepancy and Q nearly zero are indications that the photospheres of many of these central stars have an UV excess relative to a blackbody beyond the He+ ionization threshold due to a lower-than-solar helium abundance in the atmospheres (Henry and Shipman 1986; Clegg and Middlemass 1987; Husfeld et al. 1984).

## 4.5. RATIOS OF THE TOTAL FLUX OVER $H\beta$ AND OVER THE INFRARED FLUX

We find that the ratio of the total flux over the $H\beta$ flux has a median of 160, consistent with the earlier result of Gathier and Pottasch (1989) for a sample of more extended nebulae. However, the ratio has a wide range from 90 to 620. It varies with $T_*$ and the dust mean UV optical depth. The curves of $L_* / L_{H\beta}$ as a function of $T_*$ for an ionization-bounded nebula with $\tau_{UV} = 0.0, 0.5, 1.0, 1.5, and 2.0$ are plotted in Fig. 2 along with the data points.

Most nebulae with moderate $Q$ values which are encompassed by the theoretical curves are likely to be ionization-bounded and enriched with dust grains. The reason for many of them to be locate above the $\tau_{UV} = 0$ curve lies in the fact that they are young nebulae with strong dust emission. However, two nebulae with higher $Q$ values, IC2149 and IC2448, are probably density-bounded.

In contrast to the above nebulae with large positive $Q$, six nebulae with large negative $Q$ values are located well below the $\tau - UV = 0$ curve. It cannot be explained by any effects of dust grains or optical thickness of the nebulae. It appears that there may be an excess in the far-UV portions of the emergent flux from these central stars.

In a recent paper by Zhang and Kwok (1992) the age and core mass of the central

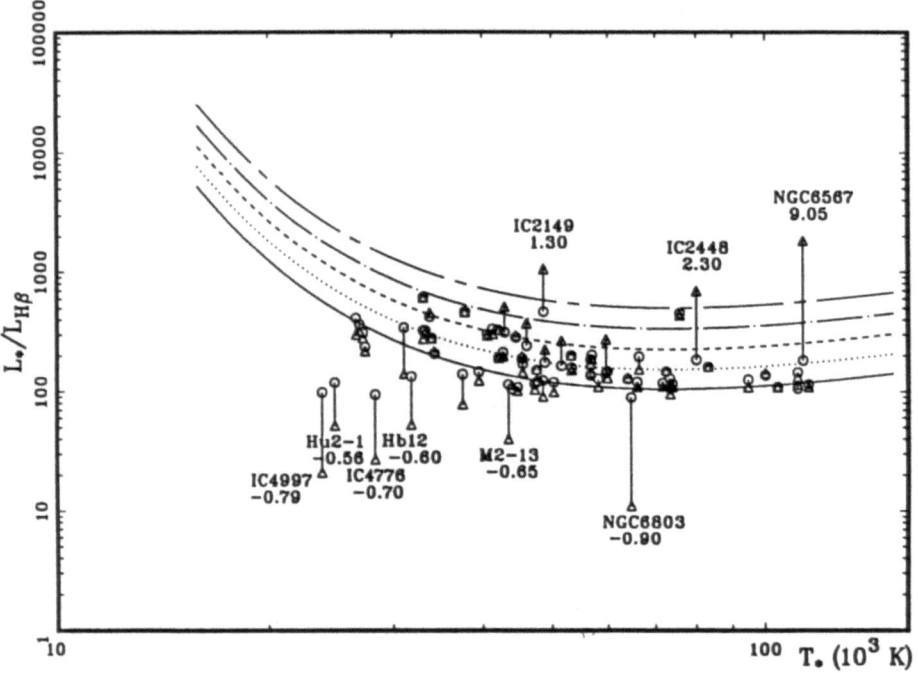

Fig. 2. $L_*$ / $L_{H\beta}$ as a function of $T_*$. Solid, dotted, dashed, chain-dotted, and chain-dashed curves are for mean UV optical depths of 0, 0.5, 1., 1.5, and 2 respectively. Open circles and triangles are data points of $F_t/F_{H\beta}$ and $F_{BB}/F_{H\beta}$ respectively. Vertical lines connect the pairs of data points belonging to the same objects.

star have been derived from the distance- independent parameters for 302 galactic PN. If the ratio of $F_{IR}/F_t$ is plotted against the central star age $(t_*)$, it appears that this ratio decreases with increasing age. This indicates that as a nebula ages, the dust tends to contribute less. This conclusion is confirmed by the theoretical modeling of the evolution of PN spectra (Volk 1992). It is conceivable that as the dust shell expands, the dust grains become cooler, and emit less far-ir radiation in the IRAS wavelength bands.

## 5. Conclusion

Significant progress has been made in our understanding of energy distribution of PN for a vastly wide wavelength coverage from UV to radio. It is expected further observations using the Space Telescope, ROSAT, ISO, and ground-based telescopes will fill many holes left in the current wavelength coverage. I would draw attention to near-ir surveys of the 2-MASS and DENIS projects. These deep surveys will provide us with unprecedented results and shed light on this less thoroughly explored wavelength range.

This work was supported by the NASA grant NAG 2-67.

# References

Aaquist, O.B., and Kwok, S. 1990, A&AS 84, 229
Acker, A., Koppen, J., Stenholm, B., and Jasniewicz, G. 1989a, A&AS 80, 201
Acker, A., Koppen, J., Stenholm, B., and Raytchev, B. 1991, A&AS 89, 237
Acker, A., and Stenholm, B. 1987, The Messenger 48, 16
Acker, A., Stenholm, B., and Tylenda, R. 1989b, A&AS 77, 487
Aller, L.H., and Czyzak, S.J. 1979, Ap&SS 62, 397
Aller, L.H., and Czyzak, S.J. 1983, ApJS 51, 211

Bianchi, L., Recillas, E., and Grewing, M. 1989, in IAU Symp. No. 131, p.307
Boroson, T.A., and Leibert, J. 1989, ApJ 339, 844
Boggess, A., Freibelman, W.A., and McCracken, C.W. 1981, NASA Publ. 2171, 663
Brown, R.L., and Mathews, W.G. 1970, ApJ 160, 939
Clegg, R.E.S., and Middlemass, D. 1987, MNRAS 228, 779
Gacia-Lario, P., Manchado, A., Pottasch, S.R., Suso, J., Olling, R. 1990, A&AS 82, 497
Gathier, R., and Pottasch, S.R. 1989, A&A 209, 369
Heap, S.R., and Augensen, H.J. 1987, ApJ 313, 268
Henry, R.B.C., and Shipman, H.L. 1986, ApJ 311, 774
Husfeld, D., Kudritzki, R.P., Simon, K.P., and Clegg, R.E.S. 1984, A&A 134, 139
IRAS Point Source Catalog, vers. 2, 1988, Joint IRAS Science Working Group, Washington DC: GPO
IRAS Science Team: 1986, IRAs Catalog and Atlases, Atlas of Low Resolution Spectra, A&AS 65, 607
Jacoby, G.H., and Kaler, J.B. 1989, AJ 98, 1662
Kaler, J.B. 1976a, ApJ 210, 113
Kaler, J.B. 1976b, ApJS 31, 517
Kaler, J.B. 1978, ApJ 226, 947
Kaler, J.B., and Feibelman W.A. 1985, ApJ 297, 724
Kaler, J.B. 1989, IAU Symp. No. 131, p.229
Kaler, J.B., Shaw, R.A., Feibelman, W.A., Imhoff, C.L. 1991, PASP 103, 67
Kaler, J.B., Shaw, R.A., Kwitter, K.B. 1990, ApJ 359, 392
Kohoutek, L., and Martin, W. 1981, A&AS 44, 325
Kwok, S. 1982, ApJ 258, 280
Kwok, S., Hrivnak, B.J., and Milone, E.F. 1986, ApJ 303, 451
Manchado, A., Pottasch, S.R., Garcia-Lario, P., Esteban, C., Mampaso, A. 1989, A&A 214, 139
Marionni, P.A., and Harrington, J.P. 1981, NASA Publ. 2171, 633
Meatheringham, S.J., and Dopita, M.A. 1991, ApJS 75, 407
Mendez, R.H., Kudritzki, R.P., Herrero, A., Husfeld, D., and Groth, H.G. 1988, A&A 190, 113
Mendez, R.H., Kudritzki, R.P., Herrero, A., 1992, A&A in press
Monk, D.J., Barlow, M.J., Clegg, R.E.S. 1988, MNRAS 234, 583
Pottasch, S.R., Baud, B., Beintema, D., Emerson, J., Habing, H.J., and Haris, S. 1984, A&A 138, 10
Pottasch, S.R., Preite-Martinez, A., Olnon, F.M., Mo, J.E., and Kingma, S. 1986, A&A 161, 363
Preite-Martinez, A., and Persi, P. 1989, A&A 218, 264
Ratag, M.A., Pottasch, S.R., Zijlstra, A.A., Menzies, J. 1990, A&A 233, 181
Rudy, R.J., Cohen, R.D., Rossano, G.S., Erwin, P., Puetter, R.C., Lynch, D.L.K. 1991a, ApJ 380, 151
Rudy, R.J., Rossano, G.S., Erwin, P., Puetter, R.C. 1991b, ApJ 368, 468
Shaw, R.J., and Kaler, J.B. 1985, ApJ 295, 537
Shaw, R.J., and Kaler, J.B. 1989, ApJS 69, 495
Stasinska, G., Tylenda, R., Acker, A., Stenholm, B. 1991, A&A 247,173
Terzian, Y. 1991, in Asteroids to Quasars, ed. P.M. Lugger (Cambridge Univ. Press), p.105
Torres-Peimbert, S., and Peimbert, M. 1977, Rev. Mex. Astron. Astrof. 2, 181
Tylenda, R., Acker, A., Gleizes, F., and Stenholm, B. 1989, A&AS 77, 39
Tylenda, R., Acker, A., Stenholm, B. Gleizes, F., and Raytchev, B. 1991a, A&AS 89, 77
Tylenda, R., Stasinska, G., Acker, A., and Stenholm, B. 1991b, A&A 246, 221
van der Veen, W.E.C.J., Habing, H.J., Geballe, T.R. 1989, A&A 226, 108

Volk, K. 1992, ApJS 80, 247
Volk, K., and Cohen, M. 1990, AJ 100, 485
Webster, B.L. 1988, MNRAS 230, 477
Whitelock, P.A. 1985, MNRAS 221, 63
Zhang, C.Y., and Kwok, S. 1990, A&A 237, 479
Zhang, C.Y., and Kwok, S. 1991, A&A 250, 179
Zhang, C.Y., and Kwok, S. 1992, ApJ 385, 255
Zhang, C.Y., and Kwok, S. 1992, A&A in press
Zijlstra, A.A., Pottasch, S.R., and Bignell, R.C. 1989, A&AS 79, 3292

# DISTANCES OF PLANETARY NEBULAE

YERVANT TERZIAN
Cornell University, NAIC, Ithaca, New York

## I. Introduction

One of the most fundamental physical parameter in astronomy is the distance to the objects we detect in the universe. For many classes of astronomical objects, accurate and proven methods have been developed to determine their distances. Such classes of objects include stars within ~100 pc from the sun, binary stellar systems, variable stars, stellar clusters, main sequence stars, and other galaxies. It has been, however, more difficult to develop satisfactory methods to determine accurate distances to the more than 1000 planetary nebulae that have been discovered in our galaxy.

During the 1977 IAU Symposium Meeting on Planetary Nebulae, which took place at Cornell University in Ithaca, New York, William Liller (Liller, 1978) reviewed the status of the distance scale of planetary nebulae and he stated: "At the Tatranská Lomnica meeting 10 years ago (1967), there seemed little hope that one day soon planetary nebulae distances would become reliable. That day is near if not here already". Fourteen years later in a recent paper, Lawrence H. Aller (Aller 1991) stated that "the 'bone in the throat' of the PN research is the determination of reliable distances for individual objects".

To be sure, much progress has been made during the last decade and this progress promises to continue. Recently several authors have tried various methods in deriving distances to planetary nebulae, however the amount of disagreement is still considerable.

It may be that statistical methods in deriving distances for many planetary nebulae are very unreliable because, as Julie H. Lutz (1989) clearly described five years ago, these objects are so diverse in their physical parameters such as, nebular masses, morphology and filling factors, and state of ionization that it is unlikely that any single method can be applied to all planetary nebulae. However, the comparison of distances of objects determined by various methods such as (a) binary stars and clusters; (b) HI 21 cm absorption; (c) visual extinction; (d) nebular expansion parallaxes, and (e) parameters of central stars, also show serious disagreements indicating large uncertainties inherent in these methods.

## II. Distance Comparisons

Table 1 shows a comprehensive comparison of distances derived or adopted by various authors for fourty planetary nebulae. This Table was formulated to include objects that Gathier (1987) had called "Standard Distances" derived from spectroscopic, expansion, reddening, and HI absorption methods. Most of these objects are the same ones used by Mallik and Peimbert (1988) as objects having distances independent of statistical arguments. The distances given by the above authors were assumed to be much better established than distances derived from statistical methods. The Table also shows Cudworth's (1974) distances from statistical parallaxes which were derived from proper motion measurements. The distances by Daub (1982) and Amnuel, et. al. (1984) were computed from various mass-radius and surface brightness-diameter relations and by using the measured radio fluxes at 5 GHz and 2.7 GHz; these distances are also indicated in Table 1.

Very recently Cahn et. al. (1992) have presented a list of recalibrated absolute $H\beta$ fluxes and have calculated Shklovsky distances according to the Daub (1982) scheme on the scale used earlier by Cahn and Kaler (1971). The recalibration was made from a set of objects with "most dependable known distances". Their extensive compilation includes distances to more than 600 planetary nebulae. Table 1 shows the distances from this work compared to the other studies.

More recently Zhang and Kwok (1992) have also derived distances to 142 galactic planetary nebulae. These distances are based on the stellar mass, surface gravity and luminosity inferred from the modeling of distance-independent parameters. The results of these distance determinations are also shown in Table 1 for comparison with the distances given by other authors.

The collected data in Table 1 give an indication of the degree of agreement or disagreement in the distance determinations and may also reflect on the accuracy of the various methods. In some cases the agreement seems reasonable and in other cases there are large differences. It is important to emphasize that Table 1 contains the planetary nebulae sample that has been claimed to have the most accurate distance determinations.

It seems instructive to compare a few distance determinations from Table 1. NGC 6572 has a range in the determination of its distance from 0.41 kpc (G87) to 3.3 kpc (ZK92), however, six of the seven entries have a distance of less than 0.9 kpc. The largest distance for NGC 7009 is 2.4 kpc (MP88) and the lowest is 0.58 kpc (G87) with five other intermediate values. Again the well known nebula NGC 3242 is shown with a large distance of 2.0 kpc (MP88) and a short distance of 0.50 kpc (G87), with five intermediate values. Even the distance to the Ring Nebula

TABLE 1. Planetary Nebulae Distances (kpc)

| Object | C74 | D82 | A84 | G87 | MP88 | CKS92 | ZK92 |
|---|---|---|---|---|---|---|---|
| NGC 40 | 1.8 | 1.1 | 0.70 | - | 1.0 | 1.2 | - |
| NGC 246 | 0.57 | 0.46 | 0.45 | 0.50 | - | 0.47 | - |
| NGC 1514 | 1.1 | 0.67 | 0.65 | 0.50 | - | 0.75 | - |
| NGC 1535 | 3.1 | 1.7 | 1.2 | - | 2.1 | 2.3 | 1.9 |
| NGC 2346 | - | 1.3 | 1.1 | 0.80 | 1.1 | 1.4 | - |
| NGC 2392 | 2.0 | 1.2 | 0.86 | - | 2.7 | 1.3 | - |
| NGC 2440 | - | 1.0 | 0.74 | 2.2 | 2.2 | 1.4 | - |
| NGC 2452 | - | 2.6 | 1.8 | 3.6 | 3.6 | 2.8 | 4.3 |
| NGC 2792 | - | 1.9 | - | 1.9 | 1.9 | 3.0 | 3.5 |
| NGC 2818 | - | 2.0 | 1.6 | - | 3.2 | 2.0 | - |
| NGC 2867 | - | 1.2 | 0.99 | - | 1.4 | 1.8 | 1.6 |
| NGC 3132 | - | 1.0 | 0.80 | 0.60 | 0.54 | 1.3 | - |
| NGC 3211 | - | 2.5 | 1.7 | 1.9 | 1.9 | 2.9 | 3.4 |
| NGC 3242 | 1.7 | 0.73 | 0.52 | 0.50 | 2.0 | 1.1 | 1.5 |
| NGC 3918 | - | 0.58 | 0.54 | 2.2 | 2.2 | 1.0 | 1.7 |
| NGC 5189 | - | 0.51 | 0.49 | 1.7 | - | 0.54 | - |
| NGC 5315 | - | 0.69 | 0.67 | 2.6 | 2.6 | 1.2 | 3.2 |
| NGC 6369 | - | 0.42 | 0.33 | 2.0 | 2.0 | 0.7 | 1.0 |
| NGC 6537 | - | 0.65 | 0.58 | 2.4 | 2.4 | 0.9 | - |
| NGC 6565 | - | 3.5 | 2.5 | 0.90 | - | 4.6 | 4.0 |
| NGC 6572 | 0.90 | 0.47 | 0.43 | 0.41 | 0.68 | 0.66 | 3.3 |
| NGC 6578 | - | 1.4 | 1.2 | 2.0 | 2.0 | 2.3 | 4.3 |
| NGC 6720 | 1.3 | 0.79 | 0.64 | 0.65 | - | 0.87 | 1.9 |
| NGC 6803 | - | 1.7 | 1.5 | 3.0 | 3.0 | 3.0 | - |
| NGC 6804 | 2.4 | 1.2 | 1.1 | - | 1.4 | 1.7 | - |
| NGC 6884 | - | 1.1 | 1.4 | 1.8 | 1.8 | 2.1 | - |
| NGC 6886 | - | 1.8 | 1.6 | 1.7 | 1.7 | 3.1 | - |
| NGC 6891 | 4.7 | 1.8 | 1.4 | - | 3.8 | 3.2 | 2.5 |
| NGC 7009 | 1.9 | 0.76 | 0.59 | 0.58 | 2.4 | 1.2 | 1.1 |
| NGC 7026 | 2.0 | 1.3 | 0.95 | 2.2 | 2.5 | 1.9 | 3.8 |
| NGC 7027 | 0.51 | 0.18 | 0.82 | 1.1 | 0.94 | 0.27 | - |
| NGC 7293 | 0.21 | 0.15 | 0.18 | - | 0.30 | 0.16 | - |
| NGC 7354 | - | 0.88 | 0.64 | 1.5 | 1.5 | 1.3 | 2.7 |
| NGC 7662 | 1.5 | 0.84 | 0.67 | 0.98 | - | 1.2 | 1.9 |
| IC 418 | 0.76 | 0.41 | 0.38 | - | 2.0 | 0.61 | 1.9 |
| IC 1747 | - | 0.19 | 1.2 | 2.5 | 2.5 | 2.9 | - |
| IC 2448 | - | 2.5 | - | - | 3.5 | 4.0 | 3.6 |
| He 2-108 | - | 4.1 | 2.5 | - | 8.3 | 4.3 | 4.0 |
| He 2-131 | - | 0.91 | 0.86 | 0.60 | 0.59 | 1.4 | 3.1 |
| He 2-138 | - | 2.2 | 1.9 | - | 5.0 | 3.6 | 2.3 |

C74 (Cudworth 1974); D82 (Daub 1982); A84 (Amnuel, et. al. 1984); G87 (Gathier 1987); MP88 (Mallik and Peimbert 1988); CKS92 (Cahn, et. al. 1992); ZK92 (Zhang and Kwok 1992).

NGC 6720 shows a discrepancy of a factor of three disagreement.

One of the brightest and most studied nebula is NGC 7027, its distance, from Table 1, ranges from 180 pc (D82) to 1100 pc (G87). The recent VLA expansion distance of this object derived by Masson (1986) is 940 pc which is compatible with the best general and thorough discussion on the distance of this object by Pottasch et. al. (1982), who report a distance range from 1 to 1.5 kpc, yet Cudworth's (1974) distance from statistical parallaxes is 510 pc. More recently Masson (1989) revised his expansion parallax distance for NGC 7027 to 880±150 pc., and Terzian, et. al. (1992) using the same method derive a preliminary distance of 1100±330 pc.

Generally, inspection of Table 1 is instructive, and depending on one or another argument, one could decide to ignore one or another entry to show that some (but not all) entries are in general agreement, however such arguments must be substantiated. It is possible to conclude that even with our best efforts, individual distances of planetary nebulae show disagreements of factors of two and three. Such large uncertainties severely limit our ability to study the physical parameters of these objects accurately and does not allow us to correctly assess the population of these objects in the galaxy.

Figure 1 gives a summary of the distances shown in Table 1 in the form of distance histograms. It is clear that there is disagreement in the distance scale. The so called 'long' distance is preferred by Zhang and Kwok (1992), Mallik and Peimbert (1988), and Cudworth (1974) followed closely by Gathier (1987) and Cahn et. al. (1992). The work by Zhang and Kwok represents the most extreme long distance scale where from Table 1 only one object has a distance less than 1 kpc. In contrast, the 'short' distance scale is represented by Daub (1982) and Amnuel et. al. (1984).

If we arbitrarily examine the number of planetary nebulae within 1 kpc from the 40 objects in Table 1, we find the following:

| | |
|---|---|
| Zhang and Kwok (1992) | 1 |
| Cudworth (1974) | 5 |
| Mallik and Peimbert (1988) | 7 |
| Gathier (1987) | 11 |
| Cahn et. al. (1992) | 12 |
| Daub 1982) | 20 |
| Amnuel et. al. (1984) | 23 |

The above comparison is not entirely exact because some authors do not report distances for a few objects, but it is sufficient to demonstrate the significant differences.

The example of the nebula He2-131 indicates the great uncertainty in the various distance methods that have been applied. Maciel (1985) summarized the

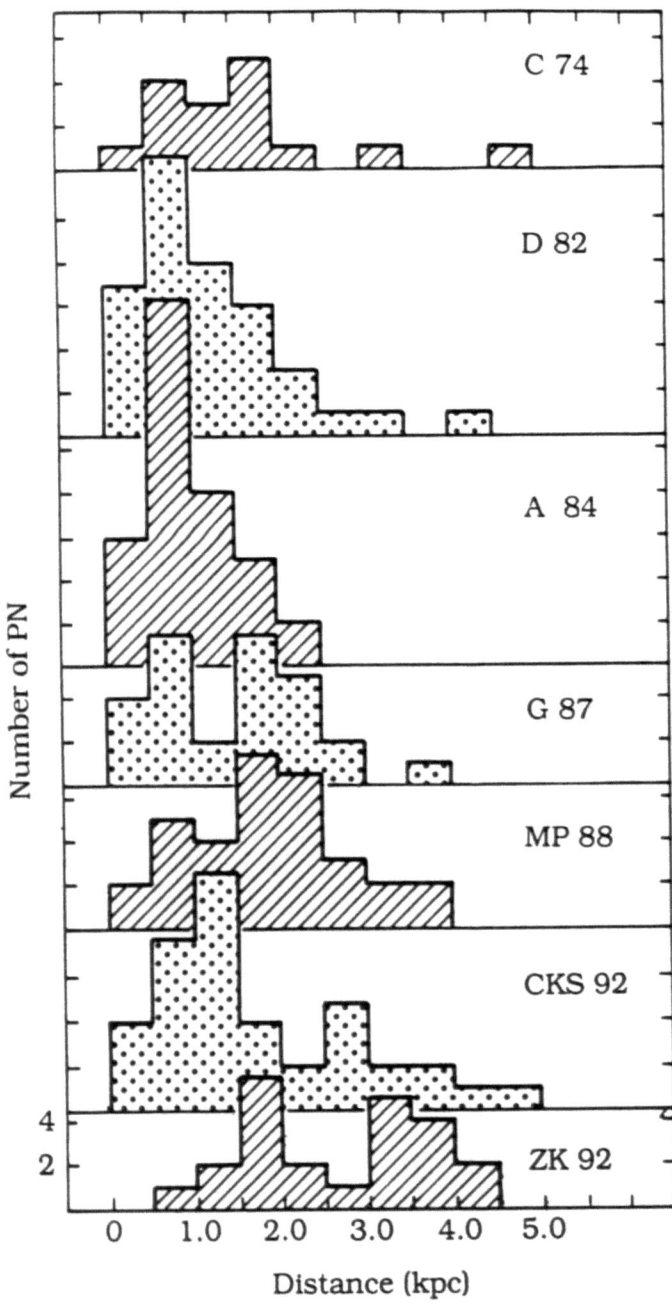

Figure 1. Histograms of Distance Scales

various derived distances of this object and Table 2 brings this data up to date. The reported range for the distance to He2-131 is from 0.6 to 4.5 kpc, and the data shows that the Shklovsky type distance methods prefer large distances and the extinction methods prefer shorter distances. However, Zhang and Kwok (1992) using stellar parameters derive a large distance of 3.1 kpc, and the Cahn, et. al. (1992) recalibrated Shklovsky distance is 1.4 kpc. We are still not converging with good understanding.

TABLE 2. He2-131 Derived Distances

| Reference | Method | Distance (kpc) |
| --- | --- | --- |
| Cahn and Kaler (1971) | Shklovsky (H$\beta$) | 3.3 |
| Cahn and Kaler (1971) | Shklovsky (H$\beta$) | 2.6 |
| Cahn and Kaler (1971) | Shklovsky (red) | 4.5 |
| Milne and Aller (1975) | Shklovsky (radio) | 4.0 |
| Milne and Aller (1975) | Effective absorption | 0.7 |
| Cahn (1976) | Shklovsky (H$\beta$) | 3.9 |
| Pottash (1980) | General extinction | 0.7 |
| Daub (1982) | Mass-radius relation | 0.9 |
| Maciel and Pottasch (1980) | Mass-radius relation | 1.6 |
| Amnuel, et. al. (1984) | Brightness-radius relation | 0.9 |
| Maciel (1985) | General extinction | 1.2 |
| Gathier (1987) | General extinction | 0.6 |
| Cahn, et. al. (1992) | Shklovsky (Recalib. H$\beta$) | 1.4 |
| Zhang and Kwok (1992) | Stellar Model | 3.1 |

### III. The Nearest Planetary Nebulae

Of special importance are the nearest planetary nebulae recently discussed by Ishida and Weinberger (1987). Most of the nearby objects are of low surface brightness, are highly evolved, and have angular sizes larger than two or three arc minutes. These are difficult objects to detect, however, during the last decade several such objects have been discovered and a comprehensive list has been given by Ishida and Weinberger. These authors indicae that there are 31 planetary nebulae within 500 pc from the sun. Five of these nebulae are included in Table 1 and show reasonable agreement with the distances given by Ishida and Weinberger. Here we present these nebulae in Table 3 with the addition of NGC 6572 and NGC 5189 which from our Table 1 have probable distances within 500 pc from the sun. The peculiar compact object Cn1-1 listed by Ishida and Weinberger has been omitted since its distance is uncertain and has an angular size <1". The objects NGC 3242, NGC 7027 and IC 1747 have probable distances >500 pc even though some authors indicate shorter distances in Table 1 and have not been included. Table 3 lists 32 objects including their positions and angular radii (d/2) mostly as summarized by Ishida and Weinberger.

TABLE 3. Planetary Nebulae Within 500 pc

| Object | R.A.    Decl. (1950) (h m)  (°  ′) | Distance (kpc) | d/2 (″) |
|---|---|---|---|
| S176 | 00 29.1 + 57 06 | 0.27 | 359 |
| NGC 246 | 00 44.5 − 12 09 | 0.47 | 125 |
| S188 | 01 27.4 + 58 07 | 0.22 | 270 |
| HFG1 | 02 59.4 + 64 44 | 0.37 | 450 |
| HW4 | 03 23.8 + 45 14 | 0.41 | 240 |
| NGC 1360 | 03 31.1 − 26 02 | 0.30 | 198 |
| IW1 | 03 45.4 + 49 51 | 0.33 | 390 |
| NGC 1514 | 04 06.1 + 30 39 | 0.40 | 64 |
| S216 | 04 37.3 + 46 35 | 0.04 | 3000 |
| A7 | 05 00.9 − 15 40 | 0.22 | 382 |
| IC418 | 05 25.2 − 12 44 | 0.48 | 6 |
| WDHS1 | 05 56.6 + 10 42 | 0.32 | 463 |
| PW1 | 06 15.4 + 55 38 | 0.24 | 600 |
| K2-2 | 06 49.8 + 10 02 | 0.48 | 207 |
| A21 | 07 26.2 + 13 21 | 0.27 | 319 |
| A29 | 08 38.1 − 20 44 | 0.41 | 201 |
| A31 | 08 51.5 + 09 05 | 0.24 | 485 |
| EGB6 | 09 50.3 + 13 59 | 0.35 | 359 |
| He2-77 | 12 06.4 − 62 59 | 0.33 | 11 |
| A35 | 12 50.9 − 22 36 | 0.36 | 400 |
| LT5 | 12 53.1 + 26 10 | 0.40 | 263 |
| NGC 5189* | 13 30.0 − 65 43 | 0.49 | 70 |
| A36 | 13 38.0 − 19 38 | 0.38 | 196 |
| NGC 6369 | 17 26.3 − 23 43 | 0.45 | 15 |
| NGC 6572** | 18 09.7 + 06 50 | 0.41 | 7 |
| S68 | 18 22.4 + 00 50 | 0.31 | 199 |
| A62 | 19 30.9 + 10 30 | 0.50 | 81 |
| NGC 6853 | 19 57.5 + 22 35 | 0.27 | 208 |
| A74 | 21 14.7 + 24 00 | 0.23 | 415 |
| IW2 | 22 12.0 + 65 40 | 0.26 | 449 |
| DHW5 | 22 18.4 + 70 41 | 0.40 | 264 |
| NGC 7293 | 22 26.9 − 21 06 | 0.16 | 402 |

Adapted from Ishida and Weinberger (1987).
* Distance from Amnuel (1984)
** Distance from Gathier (1987)

An inspection of Table 3 shows one large object S216 within 100 pc from the sun, and one object NGC 7293 in the range 100 to 200 pc. In the intervals 201 to 300, 301 to 400, and 401 to 500 pc there are 10, 11, and 9 objects respectively. Ishida and Weinberger have used distances from several recent studies and also Shklovsky distances. They conclude that the number of local objects is a lower limit and that the number of extended faint and old planetary nebulae in the galaxy must be very large. Using the available data for objects in Table 3, they derive a large space density of ~330 $kpc^{-3}$ and a birthrate of ~$8\times10^{-3}$ $kpc^{-3}$ $yr^{-1}$, which is a few times larger than the birthrate of white dwarfs. It is important to keep in mind that if the local sample, which is mostly composed of very extended faint and old nebulae is representative for the galaxy, then these objects must be very numerous. In fact Ishida and Weinberger deduce that the total number of planetary nebulae in the galaxy should be >$10^5$. These numbers could be much smaller if the distance scale of the local objects is increased significantly. It seems of great importance to study this sample more carefully.

Very recently a study of the central star of S216 was made by Tweedy and Napiwotzki (1992) where they report that Napiwotzki has estimated a distance to S216 of ~110 pc instead of 40 pc as indicated in Table 3. These authors derive an effective temperature for the central star of S216 of ~90,000K, and a surface gravity log g ~7.

## IV. Galactic Bulge Nebulae

The major problem in the study of the physical parameters of planetary nebulae and their central stars is the great uncertainty of their individual distances. However studies of these objects in other nearby galaxies whose distances are known by other methods can provide luminosity functions for the planetary nebulae which may be used to calibrate the sample in our own galaxy. Such studies will be discussed later and by other authors in this symposium. Here we mention the Galactic Bulge planetary nebulae population with the usual assumption that the distance to the Galactic Bulge is reasonably well known and that most of the planetary nebulae in the direction of the Galactic Bulge are part of it and hence at the same distance. This distance is about 7.8 kpc (Feast, 1987).

The most recent large scale study of planetary nebulae in the Galactic Bulge was performed by Stasińska, et. al. (1991) whose sample contains about 200 objects. These objects lie at a galactic latitude $|b^{II}|$ < 10° and a galactic longitude $|l^{II}|$ ± 10°; they have angular radii <20 arc sec; and their radio flux at λ6 cm is <100 mJy. Indeed Pottasch (1990) estimates that between 80 and 90% of the planetary nebulae in this region of the sky belong to the Galactic Bulge population. Stasińska, et. al. (1991) apply the Shklovsky distance method to these nebulae and determine their distance distribution. They use a nebular mass of 0.2 $M_\odot$ and a filling factor of 0.5, and conclude that their histogram of the distances of planetary nebulae is quite consistent with that of the Galactic Bulge. These authors discuss the possible distance uncertainties including the non-finite size of the Bulge, the absolute error of the Bulge distance, and the unknown variations in the extinction corrections. They conclude that most of the nebulae in the Bulge are density bounded, and hence are optically thin.

Following Stasińska's, et. al. (1991) work, Pottasch and Zijlstra (1992) analyzed a sample of planetary nebulae in the Bulge to judge the validity of the use of the Shklovsky distance method, and contrary to Stasińska, et. al. (1991), they conclude that this method does not give acceptable results.

Both Stasińska, et. al. (1991) and Pottasch and Zijlstra (1992) have used the same basic Shklovsky distance relation with a filling factor of 0.5. However, Pottasch and Zijlstra used only λ6 cm fluxes (since the Hβ flux is directly related to the radio flux), while Stasińska, et. al. used both. Both studies used $M_{ion} = 0.2$ $M_\odot$, but Pottasch and Zijlstra used nebular sizes determined only from radio interferometric observations (primarily Very Large Array results), and Stasińska, et. al. used primarily sizes from optical observations - this is the main difference in their analysis. Figure 2 shows the Shklovsky-distance histograms of the distance distributions from the two analyses. It is clear that there is substantial disagreement - Pottasch and Zijlstra show a median value of 11.5 kpc with a peak at 10.5 kpc and a possible peak at 16.5 kpc, while Stasińska, et. al. show that the maximum distribution peaks between 8 and 9 kpc with a median of 9.5 kpc.

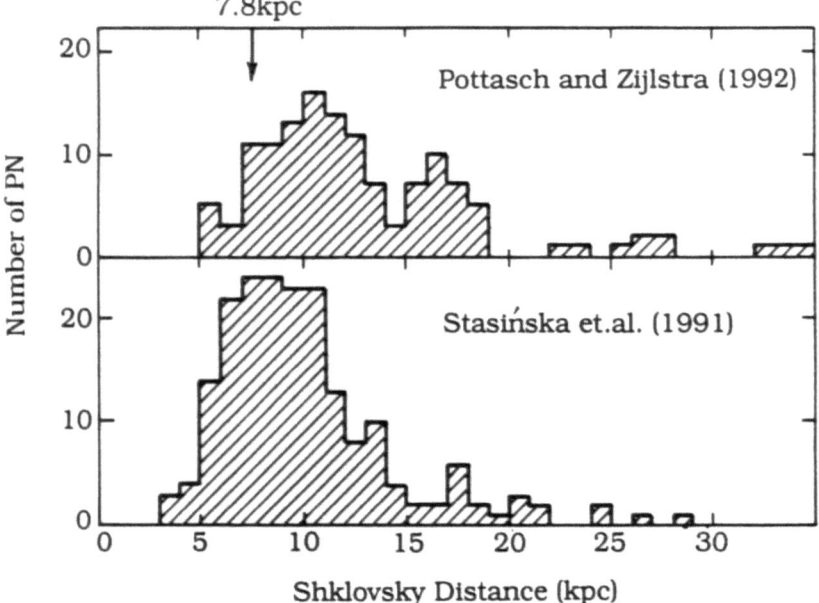

Figure 2. Galactic Bulge Distance Histograms

These results indicate that the Shklovsky distances are probably wrong and that the assumption of a constant ionized mass cannot be correct. It is more probable that a range of nebular masses exists. The data require a range from 0.01 to 0.3 $M_\odot$ to obtain agreement. A mean of 0.2 $M_\odot$ or larger will have the effect of overestimating the distances. A careful evaluation of the nebular radio/optical sizes should be performed to better assess the above conclusions.

## V. Expansion Parallaxes

One promising method for obtaining distances to planetary nebulae is to use their observed angular expansions together with the measured expansion velocities. Such a method has been used to determine distances to novae and the Crab Nebula. This method has been applied by Liller and Liller (1968) and a few other authors using optical images. Terzian (1980, 1987) suggested that this method can be applied by using radio images produced from VLA observations. Masson (1986) using VLA maps of NGC 7027 separated by only 2.8 years was able to derive a distance of 940 ± 200 pc, in good agreement with previous estimates for this object. In 1989, he refined this distance to 880±150 pc (Masson 1989). However, the same technique showed distances for the objects BD+30°3639 and NGC 6572 which were 2 to 4 times larger than previous estimates (Masson 1989). Terzian, et. al. (1992) have also obtained VLA observations of NGC 6210, NGC 6572, NGC 2392, NGC 3242, BD+30°3639 and NGC 7027 at two epochs separated by ~7 years and the results are being analyzed. Preliminary results show that the distance to NGC 7027 is 1100±330 pc, and that of BD+30°3639 is 3.3±1.1 kpc. A detailed analysis will be given elsewhere.

It is important to realize that the method of expansion parallaxes can provide important lower limits to the distances of the observed objects in the cases where no expansion is detected. Indeed Seaquist (1991) using VLA observations derived a lower limit of 5 kpc to the young planetary nebula Vy 2-2. Various authors have derived widely different distances to Vy 2-2. Acker (1978) reports a distance of 1.9 kpc from optical calibrations, Davis, et. al. (1979) adopt a kinematic distance of 20 kpc, Knapp and Morris (1985) find a distance of 9 kpc assuming a bolometric luminosity of $10^4$ $L_\odot$, but adopt 1 kpc due to the observed mass loss rates, and Clegg, et. al. (1989) give a distance of 2.5 kpc based on a calibration of planetary nebulae in the Magellanic Clouds. Clearly we do not know the distance to this object!

## VI. Extragalactic Planetary Nebulae

In recent years, significant progress has been reported in the identification of planetary nebulae in other galaxies other than the LMC, SMC, and M31 (e.g. Jacoby, et. al. 1990). Indeed the total number of identified planetary nebulae in other galaxies now rivals the number of known planetary nebulae in our galaxy. As an example, Jacoby, et. al. (1990) have catalogued 486 planetary nebulae in six early type galaxies in the core of the Virgo cluster, and Ford, et. al. (1989) reported on 665 planetary nebulae in other galaxies including those in the Leo Group of galaxies. It is ironic that since the distances of these galaxies are known to a reasonable accuracy by a variety of methods, then the distances of the planetary nebulae in these galaxies are also known with higher percentage accuracies than we know the distances to the ones in our own galaxy.

This wealth of new information has prompted Jacoby (1989) and his collaborators to use the planetary nebula luminosity functions to derive distances to other galaxies. The method used is to observe many planetary nebulae in a galaxy and to form their luminosity function, which is then compared to that observed in the calibrated galaxy M31. Using this method, Jacoby, et. al. (1990) derived a distance to the core of the Virgo cluster of 14.7±1.0 Mpc, and a Hubble constant between 81 and 94 km s$^{-1}$ Mpc$^{-1}$. However, Bottinelli, et. al. (1991) have argued that the six galaxies selected in the Virgo cluster were among the brightest and may not be representative of the whole Virgo population, hence they suggest that the derived distance to Virgo by Jacoby, et. al. (1990) is an underestimate, and indicate that the Hubble constant is more likely to be in the range 71 to 83 km s$^{-1}$ Mpc$^{-1}$.

Clearly the debate has begun, but the above already demonstrates significant advances in the study of planetary nebulae in other galaxies and points into a new powerful method in determining the distances of nearby galaxies.

## VII. Discussion and Conclusions

Accurate distances of planetary nebulae are crucial in order to study and understand their physical parameters. Due to lack of good distance determinations, in addition, several other important problems remain uncertain - such as the space density and total number of planetary nebulae in the galaxy which have implications on the galactic ultraviolet radiation from the central stars, the total processed mass returned to the interstellar medium, and in general to the chemical evolution of the galaxy.

Other important examples, where accurate distances are essential, include the study of the object M1-78 by Puche, et. al. (1988) which was classified as a planetary nebula, but the new large HI absorption distance determination now indicates that this object is more likely an HII region. Also the recent detection of helium-3 in the planetary nebula NGC 3242 (Rood, et. al. 1992) could provide information on the $^3$He/H abundance which may have cosmological implications. Unfortunately the uncertainty in the distance of NGC 3242 prevents accurate abundance determinations.

Due to the uncertain distance scale of planetary nebulae, their total number in the galaxy is not well known. Following Phillip (1989) the lowest estimate is ~2000 and the largest is 430,000, with most determinations in the range from 10000 to 30000. However, Ishida and Weinberger (1987) derive 140,000 planetary nebulae in the galaxy when they take into account the larger diffuse and old objects that they have been able to detect. Similarly their space density and birthrate have large uncertainties, and are in the range 30 to 100 kpc$^{-3}$, and 1 to $3 \times 10^{-3}$ PN kpc$^{-3}$ yr$^{-1}$, although larger values have been suggested.

The near future looks somewhat more promising for distance determinations of planetary nebulae. Pier, et. al. (1992) and Anguita, et. al. (1992) have already begun studies of trigonometric parallaxes for a few dozen nearby objects. It is estimated that CCD techniques can determine parallaxes with an error of the order of 0.001 arcsec in a time period of three years. Emphasis should also concentrate in finding planetary nebulae with binary central stars and deriving spectroscopic parallaxes.

The author wishes to thank J. B. Kaler, S. Kwok and C. Y Zhang for graciously providing unpublished data. This work was supported in part by the National Astronomy and Ionosphere Center which is operated by Cornell University under a cooperative agreement with the National Science Foundation.

## REFERENCES

Acker, A., 1978, Astron. Astrophys. Suppl., 33, 367.
Aller, L. H., 1991 Asteroids to Quasars, ed. P. M. Lugger, (Cambridge University Press), 85.
Amnuel, P. R., Guseinov, O. H., Novruzova, H. I., and Rustamov, Yu. S., 1984, Astrophys. Space Sc., 107, 19.
Anguita, C., Gutterrez-Moreno, A., and Moreno, H., 1992, This Symposium, Contr. Paper.
Bottinelli, L., Gouguenheim, L., Paturel, G., and Teerikorpi, P., 1991, Astron. Astrophys., 252, 550.
Cahn, J. H., and Kaler, J. B., 1971, Astrophys. J. Suppl., 22, 319.
Cahn, J. H., 1976, Astron. J., 81, 407.
Cahn, J. H., Kaler, J. B., and Stanghellini, L., 1992, Astron. Astrophys. Suppl., (in press).
Clegg, R. E. S., Hoare, M. G., and Walsh, J. R., 1989, Planetary Nebulae, ed. S. Torres-Peimbert, (Kluwer Academic Publ.), 443.
Cudworth, K. M., 1974, Astron. J., 79, 1384.
Daub, C. T., 1982, Astrophys. J., 260, 612.
Davis, L. E., Seaquist, E. R., and Purton, C. R., 1979, Astrophys. J., 230, 434.
Feast, M., 1987, The Galaxy, eds., G. Gilmore and B. Craswell, (D. Reidel Publ. Co.).
Ford, H. C., Ciardullo, R., Jacoby, G. H., and Hui, X., 1989, Planetary Nebulae, ed. S. Torres-Peimbert, (Kluwer Academic Publ.), 335.
Gathier, R., 1987, Astron. Astrophys. Suppl., 71, 245.
Ishida, K., and Weinberger, R., 1987, Astron. Astrophys., 178, 227.
Jacoby, G. H., 1989, Astrophys. J., 339, 39.

Jacoby, G. H., Ciardullo, R., and Ford, H. C., 1990, Astrophys. J., 356, 332.
Knapp, G. R., and Morris, M., 1985, Astrophys. J., 292, 640.
Liller, M. H., and Liller, W., 1968, Planetary Nebulae, eds. D. Osterbrock and C. R. O'Dell, (D. Reidel Publ. Co.), 38.
Liller, W., 1978, Planetary Nebulae, ed. Y. Terzian, (D. Reidel Publ. Co.), 35.
Lutz, J. H., 1989, Planetary Nebulae, ed. S. Torres-Peimbert, (Kluwer Academic Publ.), 65.
Maciel, W. J., and Pottasch. S. R., 1980, Astron. Astrophys., 88, 1.
Maciel, W. J., 1985, Rev. Mexicana Astron. Astrof., 10, 199.
Mallik, D. C. V., and Peimbert, M., 1988, Rev. Mexicana Astron. Astrof., 16, 111.
Masson, C. R., 1986, Astrophys. J., 302, L27.
Masson, C. R., 1989, Astrophy. J., 346, 243.
Milne, D. K. and Aller, L. H., 1975, Astron. Astrophys., 38, 183.
Phillips, J. P., 1989, Planetary Nebulae, ed. S. Torres-Peimbert (Kluwer Academic Publ.), 425.
Pier, J. R., Harris, H. C., Dahn, C. C. and Monet, D. G., 1992, This Symposium, Contr. Paper.
Pottasch, S. R., 1980, Astron. Astrophys., 89, 336.
Pottasch, S. R., Goss, W. M., Arnal, E. M., and Gathier, R., 1982, Astron. Astrophys., 106, 229.
Pottasch, S. R., 1990, Astron. Astrophys., 236, 231.
Pottasch, S. R., and Zijlstra, A. A., 1992, Astron. Astrophys., 256, 251.
Puche, D., et. al., 1988, Astron. Astrophys., 206, 89.
Rood, R. T., Bania, T. M., and Wilson, T. L., 1992, Nature, 355, 618.
Seaquist, E. R., 1991, Astron. J., 101, 2141.
Stasińska, G., Tylenda, R., Acker, A., and Stenholm, B., 1991, Astron. Astrophys., 247, 173.
Terzian, Y., 1980, Q. Jl. R. Astro. Soc., 21, 82.
Terzian, Y., 1987, Sky and Telescope, 73, 128.
Terzian, Y., Hajian, A., Bignell, C., and Phillips, J. A., 1992, (preprint).
Tweedy, R. W., and Napiwotzki, R., 1992, Monthly Notices R.A.S. (in press).
Zhang, C. Y., and Kwok, S., 1992, Astron. Astrophys. Suppl., (submitted).

Y. TERZIAN, S.R. POTTASCH

# ADVANCES IN NUMERICAL SIMULATIONS OF GASEOUS NEBULAE

GARY J. FERLAND

*Department of Physics & Astronomy, University of Kentucky, Lexington, KY 40506, USA*

and

*Cerro Tololo Inter-American Observatory, National Optical Astronomy Observatories[1], Casilla 603, La Serena, Chile*

## Abstract

I outline recent advances in numerical simulations of gaseous nebulae. These fall into three major areas; the Opacity Project and its extensions, the role of grains within the ionized gas, and the effects of mechanical heat on the nebula. These advances, together with improvements in stellar atmosphere calculations, should lead to a new generation of more realistic simulations of conditions in planetary nebulae and predictions of their emitted spectra.

## 1. Introduction

Planetary Nebulae are important both as a stellar phenomena and as a source of chemical enrichment of galaxy. Emission line analysis is the primary way one measures the chemical composition of the ejecta and the luminosity and temperature of the central star. With the continued development of large-scale codes designed to simulate both the central star and surrounding nebula, and ready access to cheap fast computers, numerical simulations of the conditions producing the observed spectrum are increasing being used as a tool to interpret spectroscopic observations. Some of these codes are summarized in the 1985 Meudon Meeting on Model Nebulae (Pequignot 1986), and in previous reviews in this series of meetings (Pequignot 1983 and Harrington 1989).

---

[1]The National Optical Astronomy Observatories are operated by the Association of Universities for Research in Astronomy, Inc., under cooperature agreement with the National Science Foundation.

In addition to their astronomical importance, planetary nebulae can be considered to be an especially clean laboratory for atomic molecular physics, and as an opportunity to test our ability to simulate conditions within an ionized plasma. We understand the geometry and the central source of ionizing photons, and the continuum emitted by the central star can be calculated from first principles, although this is in itself a very difficult problem. These are all assets we do not enjoy in extragalactic objects such as the giant HII regions or active nuclei. The ability to reproduce the spectrum of a PN is a necessary first step to validating simulations of far more poorly understood objects.

Although the boundary conditions of a planetary nebula simulation are well posed, there are several major complications which are now altering the standard picture, or answer, developed over the past thirty years. These are a) the results of the Opacity Project, the definitive photoabsorption-recombination data base, but one of almost overwhelming size and complexity, b) the effects of grains on the ionized gas, mainly in adding an additional source of photoelectric heating and opacity, and c) the role played by mechanical heating and ionization.

# 2. The Opacity Project

Numerical simulations of any non-equilibrium plasma require a wide range of collisional and radiative data. The major advance now occurring is in the photoionization-recombination data base, the result of the Opacity Project (hereafter OP; see Seaton 1987) and its extensions to recombination processes.

## 2.1. Photoabsorption Data

The original calculations of photoabsorption cross sections were either at only a few energies, or used approximations which resulted in an opacity which was smooth and could be well-fitted by simple power laws (see, for example, Table 2.7 of Osterbrock 1989). For many years the most complete photoionization data base was the central field calculations of Reilman and Manson (1979; RM). These showed some structure and had the great advantage of extending to medium X-ray energies. The OP has changed all this with complete data of high accuracy, but in a manner which itself presents two serious challenges, its size, and its complexity.

The OP includes all atoms and ions of astrophysically abundant elements, and includes the ground and excited states for n<10 and l<10. The predictions are the result of state-of-the-art R-matrix close coupling calculations, and the effects of auto-ionization resonances are explicitly included. An example of the type of data the OP routinely generates is shown in Figure 1, adopted from Nahar and Pradhan (1992a). The small-scale resonant structure shown there is characteristic of most OP data. Excited states can present even more complicated behavior; the so-called PEC (photoexcitation of core) resonances make excited state cross sections decidedly non-hydrogenic.

The first challenge of the OP is the shear volume of data. The full data base requires roughly a gigabyte of storage. Incorporating these data into existing radiative equilibrium codes is the second challenge. It is wrong to simply fit the OP data by a low order polynomial of some sort; the high-frequency resonant structure cannot be ignored. These contribute to the net photoionization rate when the gas is optically thin, and (as a corollary) the resonances will become optically thick in many circumstances. Each resonance must therefore be transferred if the depth-

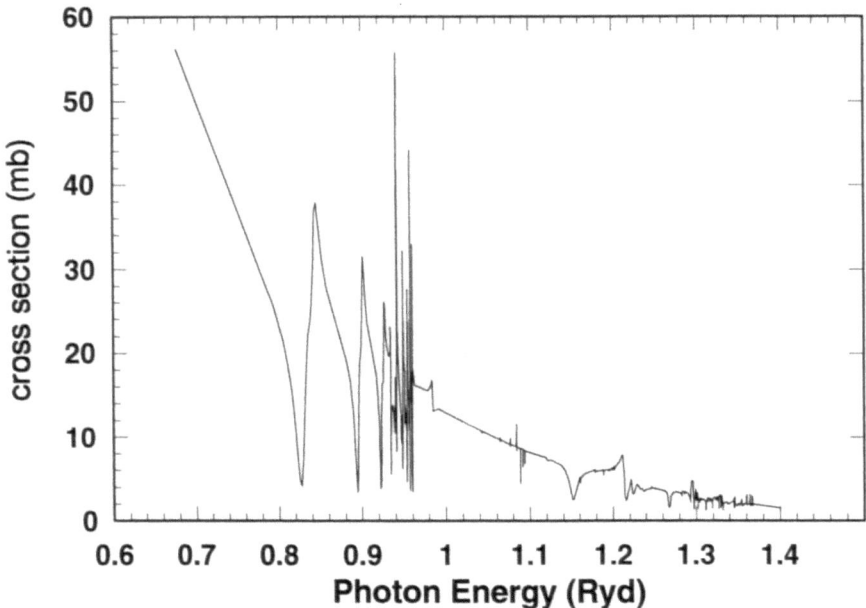

Figure 1, Photoabsorption cross section for the ground term of atomic Si.

dependent photoionization rate is to be computed. Moreover, if one of these resonances happens to coincide with a major ionizing emission line, such as the HeI or HeII Lyα lines, the photoionization rate of the heavy element could be changed by many orders of magnitude.

It will be necessary to develop new numerical methods to take advantage of the OP. The evaluation of photoioinization rates for a PN model, using this data in its raw form, frequency by frequency, and depth by depth, is beyond the capacity of even today's supercomputers. An obvious solution would be to adopt an opacity distribution function approach to both distill the OP data to a more manageable form and include the resonances in the photoionization rates. This will be the approach I will take, although this is going to take some time.

## 2.2. Recombination Rate Coefficients

Both radiative and dielectronic recombination rate coefficients can be obtained from OP photoionization cross sections by direct integration with the Milne relations. This effort is now underway (Nahar and Pradhan 1992b) but will be a massive undertaking and is far from complete. It would of course be inconsistent to combine OP photoabsorption data with other sources of recombination rates.

It is difficult to say what affects these changes in the photoionization - recombination database will produce in the computed spectrum. Predictions for lines from atoms of the third row will certainly change; currently there are no calculations of the dielectronic recombination rate coefficients for these elements at all. Furthermore, the OP cross sections sometimes disagree with the RM values by factors of several for these many-electron systems.

# 3. The Ubiquitous Grains

The physics of "classical" grains (large-mass particles with a time-steady temperature) is well understood (Spitzer 1948, Draine 1978; Draine and Salpeter 1979, Oliveira and Maciel 1986, Baldwin et al. 1991). Photoionization of grains establishes a floating potential, and the kinetic energy of the photoelectron heats ionized gas. In PNs grains are a source of photoelectric heating at least as important as helium, and in some cases they can be the dominant heating mechanism (Borkowski and

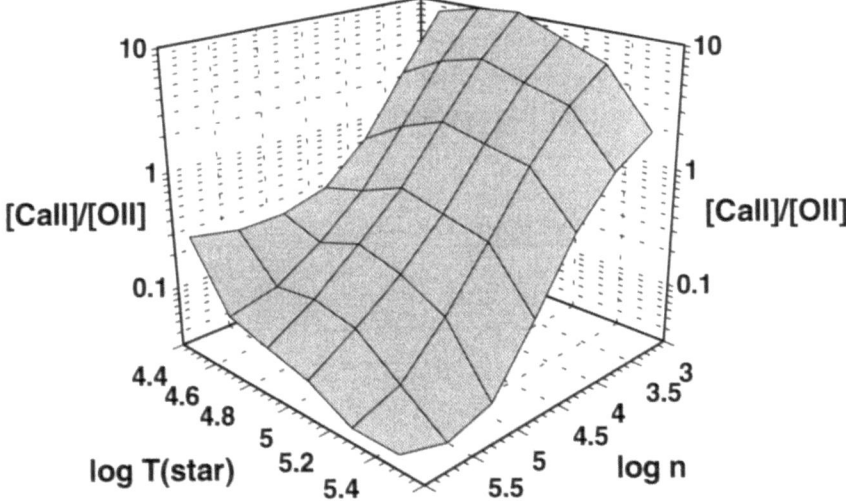

Figure 2. The [CaII]/[OII] ratio as a function of hydrogen density and stellar temperature.

Harrington 1991). No photoionization calculation can safely neglect grains in environments in which they are known to exist.

A simple argument can be used to test for extreme depletions due to the presence of grains (see also Shields 1975). Calcium is among the most highly depleted of the elements in ISM gas (approaching four orders of magnitude for dense gas, see Field 1974; Cowie and Sangila 1986), and thus offers a simple test for the existence of grains.

Figure 2 shows the results of including calcium with a solar abundance (Grevesse and Anders 1989) in a typical PN photoionization calculation (Ferland and Persson 1989; Kingdon and Ferland 1993). A spherical nebula with a small inner radius was assigned various densities. The central star was assumed to have a luminosity near the Eddington limit ($10^{38}$ erg s$^{-1}$) and assigned various blackbody temperatures. The Figure shows the predicted ratio of the intensity of the [CaII] doublet at λλ7291, 7325 to the nearby [OII] λ7325 multiplet. It shows that the unobserved [CaII] lines would be among the strongest lines in the near infrared spectrum *if calcium had a cosmic gas-phase abundance*. The conclusion is that Ca must be strongly depleted. As a corollary, it seems difficult to imagine circumstances in which any significant fraction of the grains within a nebula could be destroyed without liberating too large an amount of calcium (see also Harrington 1990). Several mechanisms could destroy grains over the lifetime of a

PN. If the calcium carriers are destroyed too, then [CaII] emission should "turn on" as the nebula ages and grains are destroyed. It is because of its extreme depletion that the search for [CaII] emission provides a fairly clean test for the destruction of grains.

# 4. Mechanical Heating

Photoionization (of both gas and dust) is the primary heating and ionization mechanism for planetary nebulae. Although it dominates the global energetics, this is not to say that there are not situations where other mechanisms come into play. The nuclei of planetary nebulae undergo rapid mass loss. The interaction of this fast wind with the slowly expanding PN envelope must occur, and acts to both heat and ionize the nebula. The effects will be above that provided by photoionizaiton.

Abell 30 provided one of the first examples of the possible influences of heating by a wind (Harrington and Feibelman 1984). Ultraviolet spectra revealed collisionally excited lines that were very strong relative to recombination lines. The ratio of these two is proportional to the mean ionizing photon energy through the Stoy (1933) ratio:

$$\frac{I(\text{strong collisional line})}{I(\text{recombination line})} \propto \frac{\text{cooling rate}}{\text{recombination rate}} \propto \frac{\text{heating rate}}{\text{photoionization rate}} \propto \langle h\nu - \text{IP} \rangle$$

The observation is that no reasonable choice of stellar continua can provide the amount of heat deposited per photoionization. Mechanical heating could provide this additional heating (although grain photoionization is a second possibility, see the work by Borkowski and Harrington in this volume).

The outer halo of NGC 7662 provides a second possible example of winds heating the nebula (Middlemass et al 1991). Observations of high electron temperatures and inferred anomalous heating could be explained with the interaction of the windy central star with the surrounding outer nebula.

A second line of evidence for mechanical heating may be provided by the Type 1 PN NGC 6302. Ashely and Hyland (1988) discovered the presence of significant emission from the high-ionization species $Si^{+5}$ and $Si^{+6}$. If the result of photoionization then the central star would need a temperature of roughly half a million degrees.

Lame and Ferland (1991; LF) pointed out that photoionization by a star this hot would produce very strong [OI] $\lambda 6300$ emission, the result of penetrating X-rays heating neutral gas. The argument is a general one, and could be used to set an upper limit to the temperature of the population of radiation-bounded nebulae. Figure 3, adopted from LF using their parameters but solar abundances, shows how the [OI]/H$\beta$ ratio increases with depth for photoionization by a very hot star. The geometry of NGC 6302 is exceedingly complex, well-shielded regions are evidenced by $H_2$ emission, and the reddening is large. However, if a continuum hard enough to produce the silicon emission also strikes the region with neutral hydrogen, then strong [OI] emission would result. LF argued that mechanical heating and ionization could produce the weak high-ionization lines, and would be consistent with the evidence of strong mass loss from the central star of this object.

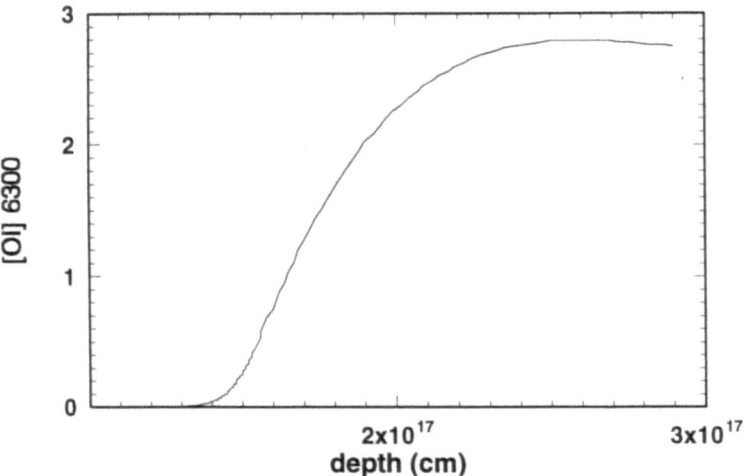

Figure 3. The intensity of [OI] 6300 relative to H$\beta$ is shown as a function of depth for a gas with solar abundances irradiated by a blackbody with a temperature of 450,000K.

## 5. Summary

Three major advances in the numerical simulations of ionized gas within Planetary Nebulae are now occurring. The predictions for even the simplest photoionization models will change once the results of the Opacity Project have been fully incorporated. Another advance is the inclusion of the thermal/opacity effects of grains within the ionized gas, which depletion patterns suggest are a general feature of PNs. Grain photoionization can sometimes be the dominant heating mechanism. Finally, mechanical heating due to stellar winds may account for weak or anomalous emission in some cases.

I thank the National Science Foundation for support, most recently through grant AST 90-19692 and by NASA through GO-2306. I acknowledge helpful conversations with Robin Clegg, Pat Harrington, Anil Pradhan, and Kevin Volk.

## 6. References

Ashley, M., and Hyland, A.R., 1988, Ap.J. 331, 532.
Baldwin, J., Ferland, G.J., Martin, P.G., Corbin, M., Cota, S., Peterson, B.M., and Slettebak, A., 1991, Ap.J. 374, 580.
Borkowski, K.J., and Harrington, J.P., 1991, Ap.J. 379, 168.
Cowie, L.L., and Sangila, A., 1986, Ann Rev Ast Ap 24, 499.
Draine, B.T., 1978, Ap.J. (Sup) 36, 595.
Draine, B.T., and Salpeter, E.E., 1979, Ap.J. 231, 77.
Field, G. B., 1974, Ap.J. 187, 453.
Grevesse, N., and Anders, E., 1989, *Cosmic Abundances of Matter*, AIP Conference Proceedings 183, Ed. C.J. Waddington, (New York: AIP).
Ferland, G.J. and Persson, S.E., 1989, Ap.J. 347, 656.
Harrington, J.P., 1990, *Planetary and Proto-Planetary Nebulae, from IRAS to ISO*, ed. P Martinez, p 277.
Harrington, J.P., 1989, *Planetary Nebulae*, IAU 131, ed. S. Torres-Peimbert, p 157.
Harrington, J.P., and Feibelman, W.P., 1984, Ap.J. 277, 716.
Kingdon, J., and Ferland, G.J., 1993, Ap.J. submitted.
Lame, N.J., and Ferland, G.J., 1991, Ap.J. 367, 208.
Middlemass, D., Clegg, R.E.S., Walsh, J.R., and Harrington, J.P., 1991, MNRAS 251, 284.
Nahar, S.N., and Pradhan, A.K., 1992a, Phys Rev in press.
Nahar, S.N., and Pradhan, A.K., 1992b, Ap.J. 397, 729.
Oliveira, S., and Maciel, W.J., 1986, Astro Space Sci 126, 211.
Osterbrock, D.E., 1989, *Astrophysics of Gaseous Nebulae and Active Galactic Nuclei*, University Science Press.
Pequignot, D., 1983, in *Planetary Nebulae*, IAU 103, D. Flower, ed., p 173.
Pequignot, D., 1986, *Workshop on Model Nebulae*, p 363, Pub. de l'Observatoire de Paris.
Reilman, R.F., and Manson, S.T., 1979, Ap.J. Sup. 40, 815.
Seaton, M.J., 1987, J.Phys. B 20, 6363.
Shields, G.A., 1975, Ap.J. 195, 475.
Spitzer, L., 1948, Ap.J. 107, 6.
Stoy, R.H., 1933, MNRAS 93, 588.

# EVOLUTION OF PN MORPHOLOGIES: CONCEPTS, MODELS AND OBSERVATIONS

BRUCE BALICK

*Astronomy Department, FM-20, University of Washington, Seattle WA 98195, U.S.A.*

ABSTRACT. The shapes of planetary nebulae (PNs) provide paleontological clues about the origin and evolution of the gas expelled in the late phases of stellar evolution. The morphological classes of planetaries and various structural components of the nebulae are interpreted as hydrodynamic interactions of episodes of relatively brief, axisymmetric wind-driven mass-loss events. Theoretical studies of the past five years are compared with extant data to show that astrophysical hydro models are achieving a very high level of success as explanations for the shapes of most PNs.

The most successful models are those for which the star is assumed to expel much or most of its mass in an equatorial wasteband. In stark contrast are dense ansae, dense and low-ionization knots of fast-moving gas, for which the mass distribution is decidedly polar. The origins of both the equatorial wastebands and polar knots remain decidedly enigmatic.

## 1. Introduction

PN morphologies have been classified in many different ways. Recent classification schemes seem to have in common that PN shapes span the extrema of circular nebulae (containing various concentric components shown schematically in fig. 1) and highly bipolar structures

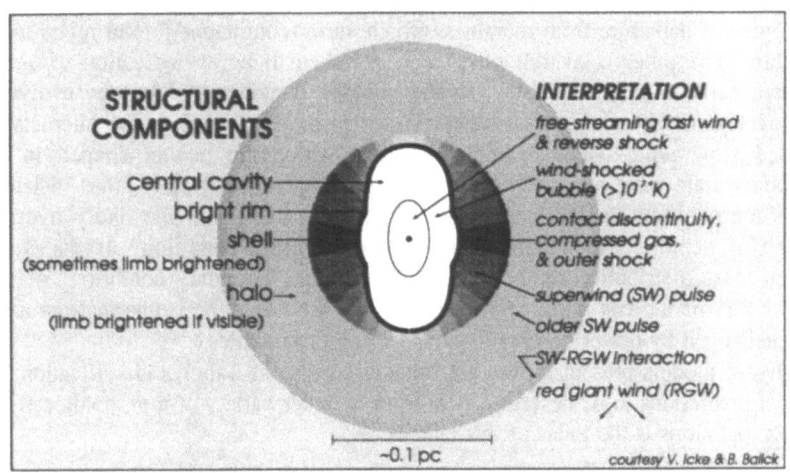

Fig. 1. Major PN components and their hydrodynamic interpretation

with prolate elliptical structures of various types in between (e.g. Balick [1]; Schwarz et al.* and Romano et al.*). The PNs in each morphological class have highly characteristic properties. There is strong evidence that much of the axisymmetry is imposed in the proto-PN phase (e.g. Sahai, et al.*, Bachiller et al.*, Shibata et al.*, Wolstencroft et al.*, [2]).

The organized shapes of PNs and their structural components contain information about the history of the nebula, especially the ways in which material is ejected from the rapidly evolving nucleus. These clues can be read if the processes that govern the evolution of the expelled gas can be understood.

The idea that PNs eject mass in episodes, each marked by different mass loss rates, velocities,and ejection patterns, has become soundly established in the past five years. Kwok, Purton, and their many collaborators introduced and established the concept of interacting winds into our field a decade ago. Their calculations assumed spherical symmetry.

Balick [1] argued that only a tiny fraction of PNs appear circular; most are far more interesting and needed to be investigated theoretically using two dimensional hydro. The seminal work in the realm of two-dimensional mass loss was that of Kahn and West [3], who assumed that early episodes of mass ejection expelled mass preferentially in an equatorial plane. They showed that a fast, symmetric wind ejected later that interacts with the older ejecta can produce elliptical or peanut nebular symmetry. Various groups (e.g. Soker and Livio; Balick, Icke, Frank, and Mellema [4-10] and Icke*, Mellema*, Igumenschev*, Zweigle et al.*, Pascoli*, Diesch and Grewing*, Sahu et al.*, and Frank's string talk) expanded upon this idea in some detail through extensive observations of kinematics and the development of both analytical and numerical hydrodynamic models containing many of the important physical processes that were not treated earlier. The highlights of these papers are reviewed here, and details can be found in specific papers cited later.

Because of their brightness and ease of observations, PNs are serving as a crucial testing ground of astrophysical hydrodynamics [8]. PNs have served physics faithfully in the same mode for theories of atomic structure and ionization and cooling processes.

## 2. Morphological Classes

We shall adopt Balick's qualitative two-parameter morphology classes here. One parameter is the degree of departure from roundness which varies continuously from round to elliptical to bipolar. The other is overall physical size, taken to be an indication of age, and so designated early, middle, and late. Round nebulae remain round as they evolve, but the degree of asymmetry increases in other classes as the fast stellar wind interacts with the structured slow wind. For example, nebulae which are peanut shaped in an early evolutionary state were assumed to develop into bipolars with open lobes in later stages. Balick's morphology classes were really simply guesses about the likely hydrodynamic evolution of nebulae which depend only on the "initial condition" of the equatorially enhanced mass distribution of the slow wind and the "boundary condition" of the speed and mass flux of the fast wind. (He assumed the nebulae to be isothermal, an assumption that is challenged by newer observational and theoretical studies.)

The hydro models presented here are broadly consistent Balick's classification, although the model predictions to be described here show a richer variety of morphologies. A review of the computations is the focus of this talk.

---

* This symbol denotes a contribution to the present conference.

## 3. Structural Components

### 3.1. HALOS

Halos are the largest and (by assumption) the oldest ejected material still visible. Halos are reviewed by Meaburn, Dyson, Frank, Balick, Chu and their many collaborators [12-17]; see also Kwitter et al.*, Lopez et al.*. Halo masses are comparable to those of the interior structure ($\approx 0.25$ $M_\odot$) [15]. Frank et al. [16] have argue that halos are characterized by ages of more than $10^4$ y. Frank and Balick* propose that they arise from episodic superwind ejections associated with thermal pulses of the central star.

Morphologically, halos are larger, fainter, and rounder than the structure interior to them [17] and are always limb brightened, suggesting significant ram pressure by an external confining medium. Some PNs, e.g. A 16, 30, 33, 34, and 39, appear to be halo-dominated with little or no interior structure. (A few halos are distorted by their motion relative to the local ISM [17-19].) No halo has been found to be associated with a bipolar PN, though this may be an artifact of ionization shadowing and the exact definition of "bipolar PN". Most halos expand at or below the sound speed (Bryce et al.* ,[12-14]) – the unusual halo of NGC 7662 is a likely exception [12] – and have slightly higher excitation temperatures [15].

The roundness of halos may indicate that early superwind ejections are less structured than later pulses. Alternately the roundness can evolve from hydrodynamic effects. Halos might eventually form as expanding bipolar lobes merge into one large bubble; NGC 2371-2 and 2440 may be examples.

As yet there are no detailed models of the growth of haloes. Proper calculations are difficult; the form, duration, and axisymmetries of the mass distribution of a superwind pulse (Vassiliadis and Wood*, [20]); the gas "softening" effects of uv photon heating; and radiative cooling mechanisms for neutral gas must all be included.

### 3.2. SHELLS

Shells are smaller, brighter, (presumably) younger than halos, and universally attached to interior bright rims. Occasionally they are also limb brightened, hinting at the presence of a partially confining medium (e.g. a halo). Frank et al. [16] showed that between the interior rims and exterior limbs shell emission measures (EM) decline linearly, and not $r^{-3}$ as might be expected from a steady wind. Shells show enhanced EM along their narrowest axis which is always orthogonal to the longest symmetry axis of the nebula, as illustrated in fig. 2. (See Bremer and Grewing*, Volk and Leahy*, and Soker et al. [21] for mapping techniques.)

Kahn, Balick and many others [1,3-11] assert that regions of enhanced EM arise from the higher-density, equatorial superwind wasteband. This forces PNs to grow preferentially along their poles. More precisely, an invisible fast wind blowing from the central star, characterized by a speed of $10^3$ km s$^{-1}$ and a mass loss rate $\approx 10^{-7}$ $M_\odot$, presently interacts with the shell (i.e. the latest pulse of superwind material) forming a strong shock and, with it, a region of snowplowed material which forms the bright rim inside the shell. Detailed hydrodynamic models of the evolution of such a system are form the Ph.D. theses of A. Frank and G. Mellema; encouraging recent results are being presented here (see Frank*, Mellema*, and the evolving hydro model video by Icke*), and others appear in the literature [4-11].

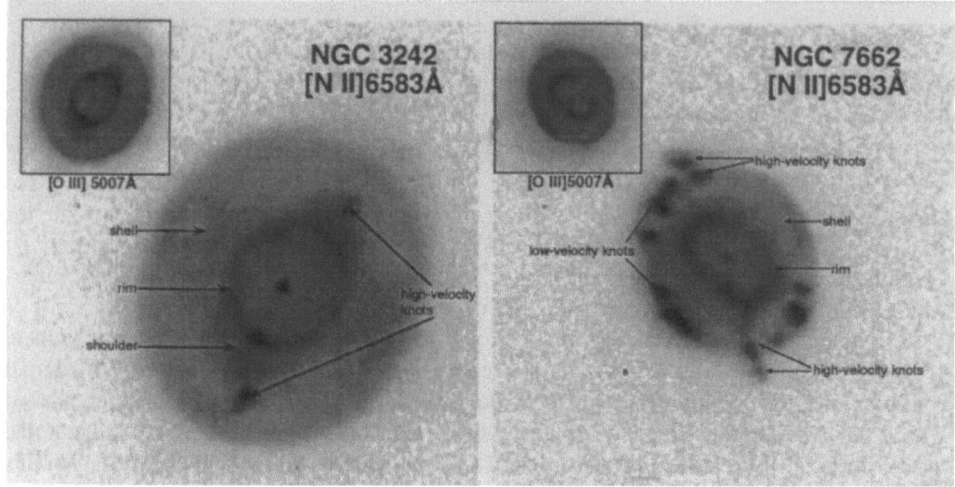

Fig. 2. Shells and Other Features of Elliptical PNs

Generally speaking, IR $H_2$ lines are distributed at the outer edges of shells where [O I] and [N II] lines are bright. For NGC 6720 and 6853 in which $H_2$ has been mapped, the molecules arise *inside* the ionized halo at the outer perimeter of the shell [e.g., 17].

3.3. CORES AND RIMS

Cores (i.e. central cavities) and the bright rims which surround them and define their geometry are ubiquitous amongst almost all PNs. The rims are generally described as having round, elliptical, peanut, and butterfly (or bipolar) shapes.

The interacting wind model accounts nicely for the morphological features of rims. A fast stellar wind enters and shocks material in the cavity to $10^{7-8}$ °K, creating a hot bubble at higher pressure than its surroundings (i.e. the latest pulse of superwind ejecta). The bubble expands supersonically into the denser, cooler gas outside of it (T ≈ $10^4$ °K; n ≈ $10^3$ cm$^{-3}$). In the process the interface region is shocked and compressed nearly isothermally, owing to the highly effective cooling processes available to the cooler gas. The compressed gas appears in projection as the rims. The hot bubble is invisible at optical-uv wavelengths.

The shape of the rims is governed by the initial distribution of the superwind ejecta which forms the shell. Since shells tend to have toroidal distributions, the hot bubble grows much faster towards the toroid's poles. Once the elliptical bubble "breaks out" along the poles the nebular shape becomes bipolar. Expansion velocities can be huge since the poorly confined hot gas tends to expand along the poles at its own sound speed of ≥ 50 km s$^{-1}$.

A snapshot of the early evolution of the hot bubble and its rim-like boundary is illustrated in fig. 3 under two sets of initial conditions. The models are from Mr. Frank's Ph.D. thesis and will be described by him in more detail in this conference. See also Icke *et al.* [8] for detailed models which neglect energy loss by radiation; such models could be applicable during the protoPN phase if radiative cooling is unimportant. Similar results were obtained by Soker and Soker and Livio [9,11] using a radiationless and numerically less precise "particle-in-cell" methodology.

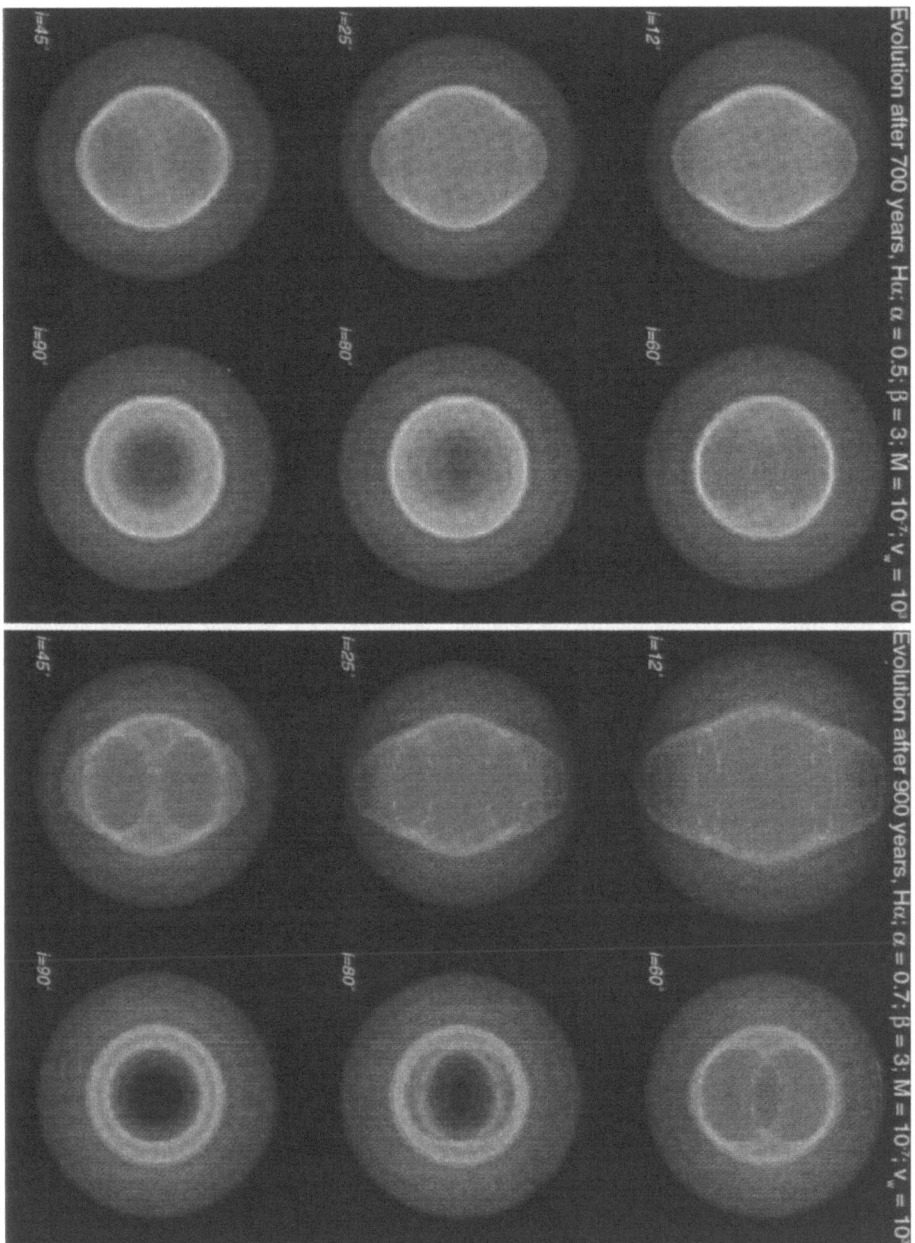

Fig. 3. Two "snapshots" of hydrodynamic models after tilting and projection onto the sky ($i$ = inclination angle).

## 4. Kinematics

Kinematics are, in a sense, the first derivative of the structure; they show how the structure evolves into the next era. More than this, kinematics are crucial tests of models. Model builders twiddle their parameters until they fit the morphologies. Until the models successfully *predict* the kinematics, they are little more than academic homework problems.

Space does not permit an adequate discussion of nebular kinematics. Over the past five years many observations at $\geq 10$ km s$^{-1}$ spectral resolution have appeared. Balick, Icke, O'Dell, and their many collaborators have mapped the two-dimensional kinematics of various PNs in H$\alpha$, [N II], and [O III] lines. See also Pismis et al.*, Diesch and Grewing*, and Sahu et al.*.

PNs have complex motions not well characterized by a single expansion velocity. NGC 2392, studied in greatest detail by O'Dell et al. [22] is an excellent case in point. The brightest inner nebulosity has velocities which span about 80 km s$^{-1}$. Outer, fainter nebulosity in the nebula's shell spans twice this range of velocities, and two small features appear at ±180 km s$^{-1}$ on opposite sides of the star. Most other nonbipolar PNs exhibit velocities which *decrease* at larger radii from the nucleus. As for the bipolars, the most detailed analysis is by Icke et al. [7] who find considerable nebula-to-nebula differences.

The interpretation of the kinematics is rich in ideas. Hydro models which are careful to permit conversion of kinetic (and thermal) energy into radiation are just now appearing for the first time. So far the results are sketchy. Nonetheless it is clear from one-dimensional calculations (some of which are to be presented by Frank) that the patterns of kinematics change qualitatively as shocks and ionization fronts propagate beyond various structural components of PNs. Generally the hot bubble expands into the shells and, so, accounts for a negative velocity gradient with radius. However, there are times when shocks propagate through the confining medium, and the velocity gradient tends towards zero.

Two-dimensional models (e.g. Frank*, Icke*, and Mellema*) show complex patterns of motions which are generally consistent with the various various data mentioned above.

## 5. Puzzles

Most all of the large morphological features of PNs. such as rims, shells, and halos, are qualitatively understandable if a theoretician is free to pick initial and boundary conditions for the models. Our ignorance hides in these choices: in what patterns do stars eject their mass prior to the PN stage of evolution (and *why??*); what are the time-dependent spectral and wind characteristics of the nucleus (and *why??*); etc.

The mechanisms of mass ejection are only poorly understood, and it is truly a puzzle how a star can eject material with quadrapole and higher moments in its distribution. Rotation, chaotic or nonradial pulsations, magnetic fields internal and external to the star, contact or close binaries; all of these mechanisms and others are all being explored. It is safe to say that no consensus is emerging, and that detailed images of protoPNs, such as the many results being reported here, are crucial to constrain our imaginations.

The wind characteristics of fast winds have been probed, primarily along the line of sight ending at the star. P-Cygni profiles have been observed in many ultraviolet lines and in many nuclei. Whether the winds are spherically symmetric, and how they evolve in time are open questions whose answers are likely to come eventually from theoretical models.

On the smaller scale, low-ionization knots of various distributions are common in PNs, as illustrated in fig. 2. Often the knots appear along the outer perimeter of shells (e.g. the low-velocity knots of NGC 7662 and 2392). Kahn and Breitschwerdt [23,24] argue that expansion of the PN in certain phases of its evolution can generate Rayleigh-Taylor instabilities and fragment the shell. They argue that the low-ionization knots are short-lived.

Pairs of knots along or near the nebular major axis, called "ansae", are not uncommon (fig. 2). The ansae are characterized by spectra very similar to Herbig-Haro objects in the strengths of their [S II], [O I], [N I], and Mg I line intensities and the weakness of all permitted lines, including H$\alpha$. These unusual line ratios suggest that the ansae are partially or completely collisionally excited by shocks with velocities of 50–80 km s$^{-1}$. Almost certainly ansae are plowing outward through the nebula along a symmetry axis. Many ansae are embedded in gas of much higher ionization. Curiously enough, Echelle observations show exactly these characteristic velocities for the ansae after reasonable correction for projection effects!

Ansae are always associated with elliptical PNs except for IC 2149 in which the ansae are the *only* bright nebular features. Some PNs, such as NGC 2440 and 5189, show several pairs of ansae along axes with different P.A.s. Other PNs and protoPNs exhibit ansae only as molecular or dusty features (e.g. Morris [25], Sahai *et al.*$^*$). Abell 30, a very old and large PN, has N-rich ansae as well. In other words, ansae are apparently ejected in symmetric pairs as early as the protoPN phase and throughout the subsequent lifetime of the PN. What are the ansae? How are they ejected? Only the central star knows for sure.

That H$_2$ is even detectable is somewhat surprising given that many host PNs are not ionization bounded. Even more enigmatic is why the H$_2$ in NGC 6720 and 6853 is found between their respective highly ionized halos and the interior shells.

Finally we note that although the existence of halos is no surprise, we have much to learn from them about the ways in which superwinds are ejected in the protoPN phase or even earlier (see reviews by [21,26]). Some halos develop comet-shaped knots, others are mottled, and yet others are smooth. Halos seem to provide fertile conditions in which to probe the formation of astrophysical instabilities, and discussed by Meaburn, Dyson, Bryce, and collaborators [13,14].

## 6. Summary

What have we learned since the 1987 IAU Symposium on PNs in Mexico City about the morphologies and evolution of PNs? The crucial importance of continuing wind shaping is firmly established. Most PN features can be well understood in terms of natural hydrodynamic process. Two-dimensional hydrodynamic models are now climbing onto very solid ground as the tools for understanding all of the gross morphological and kinematic features, and perhaps the fine-scale structure ($\leq 0.01$ pc) structure, of PNs. We understand that shocks are very important in the shaping and line excitations of PNs, particularly in the ansae. All of the morphological characteristics of PNs appear in protoPNs as well, so stellar mass loss and the attendant hydrodynamic processes that shape PNs exert their influence before some PNs become visible.

Yet we are sobered by the various puzzles listed in the previous section. And a detailed model is yet to be successfully computed for one very intensively studied PN, such as NGC 6751 [27], albeit with unlimited freedom to choose initial and boundary conditions!

## 7. Acknowledgements

It is a sincere pleasure to thank my close colleagues and friends, V. Icke, A. Frank, and G. Mellema, for their collaborations, driving enthusiasm, and warm hospitality. N. Soker, F. Kahn, and J. Dyson have engaged me in many insightful and thought-provoking conversations. Many other colleagues throughout the world who make the work reported here so stimulating and enjoyable.

Above all, I would like to thank the many authors of the 37 poster papers that are directly relevant to this talk, and to apologize to everyone whose work I could not or did not reference explicitly herein. These fabulous poster papers clearly define the state of the theoretical and observational art of our science at this time.

## 8. References

1 Balick, B. (1987) *Astron. J.* **94**, 671
2 Morris, M. and Reipurth, B. (1990) *Publ. Astron. Soc. Pacific* **102**, 446
3 Kahn, F. and West, K.A. (1985) *Mon. Not. Roy. Astron. Soc.* **212**, 837
4 Balick, B. Preston, H.L., and Icke V. (1987) *Astron. J.* **94**, 1641
5 Icke, V. (1991) *Astron. Astrophys.* **251**, 369
6 Icke, V., Balick, B., and Frank, A. (1992) *Astron. Astrophys.* **253**, 224
7 Icke, V., Preston, H.L., and Balick, B. (1989) *Astron. J.* **97**, 462
8 Icke, V., Mellema, G., Balick, B., Eulderink, F., and Frank, A. (1992) *Nature*, **355**, 524
9 Soker, N. (1989) *Astrophys. J.* **340**, 921
10 Soker, N. (1990) *Astron. J.* **99**, 1869
11 Soker, N. and Livio, M. (1989) *Astrophys. J.* **339**, 268
12 Chu, Y.-H. and Jacoby, G.H. (1989) in *IAU Symp. 131, Planetary Nebulae*, S. Torres-Peimbert (ed.) Kluwer Academic Publishers, Dordrecht, p. 198
13 Meaburn, J., Nicholson, R. Bryce, M., Dyson, J.E., and Walsh, J.R. (1992) *Mon. Not. Roy. Astron. Soc.* **252**, 535
14 Bruce, M., Meaburn, J., Walsh, J.R., and Clegg, R.A.S. (1992) *Mon. Not. Roy. Astron. Soc.* **254**, 477
15 Middlemass, D., Clegg, R.A.S., Walsh, J.R., and Harrington, J.P. (1991) *Mon. Not. Roy. Astron. Soc.* **251**, 284
16 Frank, A., Balick, B., and Riley, 1990, *Astron. J.* **100**, 1903
17 Balick, B., Gonzalez, G., Frank, A., Jacoby, G. (1992) *Astrophys. J.* **392**, 582
18 Borkowski, K.J., Sarazin, C.L., and Soker, N. (1990) *Astrophys. J.* **360**, 173
19 Soker, N., Borkowski, K.J., and Sarazin, C.L. (1991) *Astron. J.* **102**, 381
20 Wood, P.R. and Vassidiadis, E. (1992) *Late Evolution of Low-Mass Stars*, (to appear in *Highlights of Astronomy*, vol.. 9) Kluwer Academic Publishers, Dordrecht
21 Soker, N., Zucker, D.B., and Balick, B. (1992) subm. to *Astrophys. J.*
22 O'Dell, R., Weiner, L.D., and Chu, Y.-H. (1990) *Astrophys. J.* **362**, 226
23 Breitschwerdt, D. and Kahn, F.D. (1990) *Mon. Not. Roy. Astron. Soc.* **244**, 521
24 Kahn, F.D. and Breitschwerdt, D. (1990) *Mon. Not. Roy. Astron. Soc.* **242**, 505
25 Morris, M. (1987) *Publ. Astron. Soc. Pacific* **99**, 1115
26 Pascoli, G. (1992) *Publ. Astron. Soc. Pacific* **104**, 350
27 Chu, Y.-H., Manchado, A., Jacoby, G.H., and Kwitter, K.B. (1991) *Astrophys. J.* **376**, 150

# RING NEBULAE AROUND PN NUCLEI AND MASSIVE STARS

YOU-HUA CHU

*Astronomy Department, University of Illinois at Urbana-Champaign,*
*1002 W. Green Street, Urbana, IL 61801, USA*

**Abstract.** Planetary nebulae (PNe) and ring nebulae around massive stars are not just superficially similar in morphologies. For massive stars that evolve through red supergiant phase, the final fast wind would sweep up the slow red supergiant wind and form a bubble of stellar material, reminiscing the two-wind formation of a PN. Sometimes it can be really difficult to determine whether the central star of a ring nebula is a PN nucleus or a Pop I massive star. Parallel studies of PNe and ring nebulae around massive stars can greatly benefit each other.

## 1. Introduction

Stars lose mass as they evolve. The outflowing stellar mass interacts with the ambient medium and often forms one or more circumstellar shells. Since optically visible shells attract attention more easily, extensive studies exist only for shells whose central stars heat up and emit ionizing radiation at late evolutionary stages. For stars with initial masses of 8 $M_\odot$ or less, their circumstellar shells are recognized as "planetary nebulae". For stars with initial masses $\gg 10$ $M_\odot$, their shells are called "ring nebulae around massive stars" (Chu 1991). For stars with intermediate masses, the circumstellar shells may not be noticed until after the supernova explosion as in the case of SN1987A (Wampler *et al.* 1990).

"Planetary nebulae" and "ring nebulae around massive stars" share many similar physical properties and formation mechanisms. In fact, without knowing the distance it is often hard to distinguish whether the central star of a ring nebula is a massive star or a PN nucleus. In this paper I will discuss the similarities and differences between these two types of nebulae, examine the direct evidence of two-wind interaction for both nebulae, and plead the inclusion of ring nebulae in future PN conferences.

## 2. Mass Loss History and Shell Formation

The formation of a ring nebula, or a circumstellar shell, depends on the mass loss history of the central star. If the stellar evolution is known, the mass loss history of a star can be inferred from the observed mass loss of stars that represent earlier evolutionary stages. The formation and structure of circumstellar shells can then be hydrodynamically calculated.

### 2.1 Planetary Nebulae

The progenitor of a PN nucleus evolves off main sequence and starts to lose mass appreciably at red giant stage via slow wind then much more copiously at asymptotic giant branch (AGB) phase via "superwind". The superwind velocities range from a few to 80 km s$^{-1}$ with the majority being 10–25 km s$^{-1}$, and the mass loss rates range from a few $\times 10^{-8}$ to $\geq 10^{-4}$ $M_\odot$ yr$^{-1}$ with the majority being a few $\times 10^{-6}$ to $10^{-5}$ $M_\odot$ yr$^{-1}$ (Knapp 1989). The superwind tapers off at the post-AGB phase. As the star heats up and ionizes the circumstellar shell into a visible planetary nebula, it also turns on a fast stellar wind, of which the velocity may be 600 to 3500 km s$^{-1}$ and the mass loss rate may be $10^{-11}$ to $10^{-6}$ $M_\odot$ yr$^{-1}$ (Patriarchi & Perinotto 1991).

This fast wind inevitably will catch up and interact with the previous slow wind

(Kwok, Purton, & FitzGerald 1978). This wind-interaction mechanism was originally calculated for two spherically symmetric winds colliding inelastically (Kwok 1983). More sophisticated hydrodynamic calculation was later developed to include multiple and even non-spherical winds (Kahn 1989; Frank, Balick, & Riley 1990; Icke 1991). These models successfully explain many observed PN properties, especially the bipolar and butterfly PNe.

However, it ought be borne in mind that not every PN could be described by interacting fast wind and slow wind. 40% of PN nuclei do not show any evidence of fast stellar wind (Patriarchi & Perinotto 1991). Furthermore, many multiple shell PNe have density profiles and internal motions too complex to be explained by two- or three-wind models. For example, M2-2 has two shells with different expansion velocities and ellipticities and IC 3568 has an outer shell expanding faster than its inner shell (Chu 1989).

PN nuclei can lose mass more creatively than the aforementioned red giant wind, superwind, and fast stellar wind. Bipolar jets have been observed in NGC 2392 (Gieseking, Becker, & Solf 1985) and NGC 6543 (Miranda & Solf 1992). The spatiokinematic structure of the innermost, H-depleted nebulae in the born-again PNe A30 (Reay, Atherton, & Taylor 1983; Jacoby & Chu 1989) and A78 (Clegg et al. 1992) indicates a more violent ejection mechanism.

## 2.2 Ring Nebulae around Massive Stars

Massive stars lose mass via fast stellar wind even at the main sequence stage (Garmany et al. 1981). They evolve off main sequence toward later spectral types at roughly constant luminosity, then loop back toward the left of H-R diagram at turning points that depend on the initial stellar mass (Maeder 1983; Maeder & Maynet 1988). The most massive stars never reach red supergiant stage.

Stars with initial masses $\geq 50$ $M_\odot$ evolve into luminous blue variables (LBVs) near the turning points, eject H-rich envelopes and become Wolf-Rayet (WR) stars later (Humphreys 1991). LBVs are known to go through eruptions during which the mass loss rate reaches a maximum of a few $\times 10^{-4}$ $M_\odot$ yr$^{-1}$ and lasts $\sim 20$ yr (Lamers 1989; Humphreys 1989). WR stars have the most powerful stellar winds with typical wind terminal velocities of 2000–3000 km s$^{-1}$ and mass loss rates of $10^{-5}$–$10^{-4}$ $M_\odot$ yr$^{-1}$ (Barlow, Smith, & Willis 1981; Willis 1982; Abbott et al. 1986; Prinja, Barlow, & Howarth 1991).

Stars with initial masses <50 $M_\odot$ may evolve off main sequence and reach red supergiant phase (Maeder 1983), at which stage copious mass loss occurs at rates of a few $\times 10^{-5}$ $M_\odot$ yr$^{-1}$ with wind terminal velocities of 20–30 km s$^{-1}$ (Jura & Kleinmann 1990). As red supergiants evolve toward the left of H-R diagram, only stars with initial masses $\sim 40$ $M_\odot$ may lose enough H-rich surface material and become WR stars; the less massive ones evolve into blue superginats before exploding as supernovae.

Simplistically, a massive star starts to blow a bubble of interstellar material at the main sequence stage, ejects stellar material near the turning point of evolutionary track via LBV outbursts or red supergiant wind, and finally uses the hot fast stellar wind to sweep up the circumstellar material into a shell after it evolves into a WR star or a blue supergiant. In principle, an evolved massive star would be surrounded by multiple shells, e.g., NGC 6164-5 (Bruhweiler et al. 1981; Leitherer

& Chavarria-K. 1987). However, multiple shells are not usually observed because of the following reasons. The interstellar bubble blown at main sequence stage may be too tenuous, especially if the star belongs to an OB association inside a superbubble. The final bubble of stellar material may fade below detection limit in a few $\times 10^4$ yr (Miller & Chu 1993). Evolved stars may not be hot enough to ionize the circumstellar shells, if the spectral type is later than B0-1. The final stellar wind may not be powerful enough to sweep the circumstellar material into a shell that is dense enough to be easily observable and large enough to be easily resolvable from the stellar image, *e.g.*, the circumstellar nebula around P Cyg (Johnson *et al.* 1992).

The reality is inevitably more complex than this simple scenario. As Dyson (1992) reviewed in this symposium, the motion of stars, the time intervals between different episodes of mass loss, the instabilities in shells, *etc.* all influence the physical structures of the final ring nebulae. It is not surprising that no two ring nebulae look exactly alike (Chu, Treffers, & Kwitter 1983; Miller & Chu 1993). Massive stars, too, can lose mass bipolarly, *e.g.*, AG Car (Paresce & Nota 1989), HD 148937 (NGC 6164-5; Pismis 1974; Leitherer & Chavarria-K. 1987), and Sk -69°202 (SN1987A; Wampler *et al.* 1990).

### 3. Similarities and Differences between PNe and Ring Nebulae

Figure 1 demonstrates some obvious similarities and differences bewteen PNe and ring nebulae. NGC 6720 (the Ring Nebula) and NGC 7094 are galactic PNe, and DEM 39 and DEM 231 are ring nebulae around WR stars in the Large Magellanic Cloud (LMC). Note the striking similarity in morphology between NGC 6720 and DEM 231, and between NGC 7094 and DEM 39. However, this morphological similarity between LMC nebulae (50 kpc away) and galactic nebulae at a similar angular resolution also implies that WR rings are at least 50-100 times larger than PNe. Most ring nebulae around massive stars are a few pc in diameter; some may be as large as 100-200 pc across. It can also be seen that WR rings are often in gas-rich environments while PNe are not.

The known ring nebulae around massive stars have a variety of physical properties, representing different stages in the simple scenario described in §2. For example, NGC 2359 and NGC 3199 contain mostly interstellar material, thus representing the main sequence bubble; NGC 6888 and RCW58 contain mostly stellar material, thus representing the final bubble of stellar material; RCW104 conatins a mixture of stellar and interstellar material, thus representing an intermediate case in which the final stellar bubble has merged with the remnant main sequence interstellar bubble (Esteban *et al* 1992; Miller & Chu 1993). The largest ring nebulae, the 100-200 pc rings in the LMC (Chu & Lasker 1980), were probably remnant main sequence bubbles in tenuous medium or superbubbles blown by a group of massive stars.

PNe on the other hand contain almost exclusively stellar material. The only interstellar matter contamination in a PN occurs at the outer edge of red giant wind, which sweeps up the ambient interstellar matter and produces limb-brightening in the faint halo of a PN.

One type of ring nebula formation is physically similar to PN formation. For a massive star that evolves through red supergiant into WR or blue supergiant, its circumstellar material lost via red supergiant wind will be swept up by the final fast stellar wind into a shell. This is identical to the popular two-wind formation of PNe. Only recently are ring nebulae around massive stars modeled as interacting asymmetrical slow wind and spherical fast wind (NGC 6888: Garcia-Segura & Mac Low 1992; ring around SN1987A: Martin & Arnett 1992).

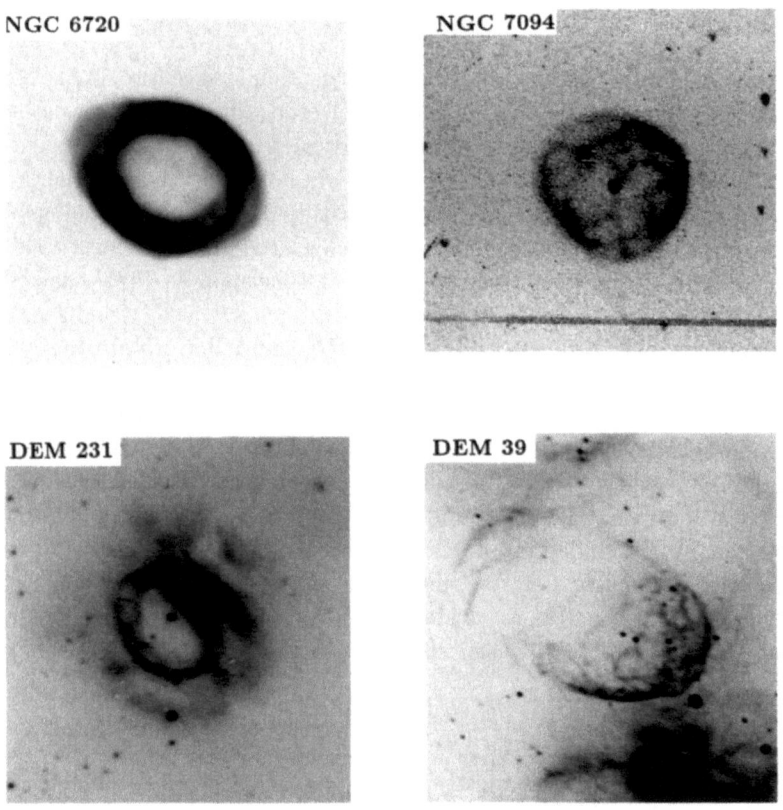

**Figure 1.** Images of two galactic PNe - NGC 6720 and NGC 7094, and two ring nebulae around WR stars in the Large Magellanic Cloud - DEM231 and DEM39. Note the morphological similarity between NGC 6720 and DEM231, and between NGC 7094 and DEM39.

## 4. Direct Evidence of Two-Wind Interaction

Around massive stars, the action of fast stellar winds is clearly manifested by the acceleration of ambient interstellar matter and the diffuse X-ray emission from bubbles. For example, the diffuse X-rays in NGC 6888 indicate the existence of $3-9\times 10^6$ K plasma (Bochkarev 1988), and the observed X-ray luminosity as well as the nebular morphology can be better modeled by fast stellar wind sweeping up the previous asymmetric slow wind (Garcia-Segura & Mac Low 1992), just like the conditions for elliptical PNe (Kahn 1983).

It is more difficult to observationally verify two-wind interaction in PNe, because the pre-shocked gas is already expanding outward and the fast stellar wind is weak. X-ray emission from PNe has been detected by EXOSAT and Einstein Observatory, but only point sources are found (Tarafdar & Apparao 1988; Apparao & Tarafdar 1989). Extended X-ray emission from PNe is first reported in this syposium by Kreysing *et al.* (1992) using ROSAT PSPC observations; among the 6 PNe reported, NGC 6543 and NGC 6853 provide the most convincing cases of extended emission. Future theoretical models need to explain the X-ray shell morphology of NGC 6543 and the very low plasma temperature implied by the X-ray spectrum of NGC 6853.

The shocked fast stellar wind is so hot that high ionization absorption lines should be present against the stellar spectra. Unfortunately, O VI $\lambda\lambda 1031.9, 1037.6$ lines are not accessible by IUE or HST, and the N V, C IV, and Si IV lines are dominated by the photoionized dense PN shells. Kaler, Feibelman, & Henrichs (1988) reported absorption components blue-shifted by up to 250 km s$^{-1}$ in high ionization lines in the UV spectrum of the nucleus of A78, and suggested it as direct evidence of fast stellar wind interacting with the PN shell. It ought to be noted that the H-depleted nebula in A78 has a very high expansion velocity (Clegg *et al.* 1992), and it is possible that the observed high-velocity absorption comes from the H-depleted nebula in front of the nucleus. High-velocity absorption in high ionization lines should be searched for in other PNe to verify its origin of interacting winds.

## 5. Real Confusion between PNe and Ring Nebulae

Many ring nebulae around massive stars were once cataloged as PNe, *e.g.*, AG Car (PK 289-0°1) and NGC 6164-5 (PK 336-0°1). There are also nebulae classified alternately between PNe and ring nebulae, *e.g.*, M1-67 and We21.

M1-67 was discovered by Minkowski (1946), and classified as a PN by Bertola (1964) because the radial velocity of the central star 209 BAC was very high ($\sim$200 km s$^{-1}$). However, no other PNe have WN8 nuclei, and the extinction of $A_V=4.1$ mag may require a distance larger than the 0.9 kpc adopted by Bertola (1964). Using the absolute magnitude calibration for Pop I WR stars (Smith 1973), Cohen & Barlow (1975) estimated a distance of 4.3 kpc, and proposed that M1-67 was a ring nebula around a Pop I WN8 star. Van der Hucht *et al.* (1985) compared the IR emission of M1-67 to that of the central region of RCW58 (an ejecta-type ring nebula around a Pop I WN8 star) and claimed that the color temperature of M1-67 was warmer than that of RCW58 but similar to those of PNe, hence concluded that M1-67 was a PN. Esteban *et al.* (1991) examined IRAS color temperature for more ring nebulae around massive stars and found that IRAS color temperature could not discriminate between PNe and ring nebulae around massive stars. Finally,

Crawford & Barlow (1991a) used the interstellar Na I absorption velocity profile and the galactic rotation to derive a distance of 4–5 kpc for 209 BAC, making M1-67 a ring nebula around a Pop I WN8 star again.

We21 (Weaver 1974) is a less well-known WN8 star in a small ring nebula (Duerbeck & Reipurth 1990). It was assumed to be a PN at discovery; however, the velocity profile of the interstellar Na I absorption suggested a distance of 11.5 kpc (Crawford & Barlow 1991b). This large distance requires We21 to be a luminous Pop I WN8 star.

At present the massive, Pop I WN8 nature of We21 and 209 BAC is favored. Nevertheless it is possible that unexpected new evidence would change the nebular nature back to PN.

## 6. Epilogue

According to the original definition, "planetary nebulae" should have included all ring nebulae around massive stars, too. The segregation of PNe and ring nebulae is, to some extent, artificially made by stellar astronomers. These two types of nebulae represent circumstellar shells produced by mass loss from low-mass and high-mass stars, respectively. The apparent gap at the intermediate stellar mass is owing to the invisibility of circumstellar shells around evolved B supergiants. From the nebular point of view, PNe and ring nebulae would be best studied comparatively. Future conferences about PNe may consider including ring nebulae around massive stars, especially the ones consisting of mostly stellar material.

Finally, I would like to point out the fact that circumstellar shells have been systematically searched only around galactic and Magellanic Cloud WR stars and galactic Of stars. More effort should be devoted to surveys of circumstelalr shells around other evolved massive stars.

## References

Abbott, D.C., Bieging, J.H., Churchwell, E., & Torres, A.V. 1986, ApJ, **303**, 239
Apparao, K.M.V. & Tarafdar, S.P. 1989, ApJ, **344**, 826
Barlow, M.J., Smith, L.J., & Willis, A.J. 1981, MNRAS, **196**, 101
Bertola, F. 1964, PASP, **76**, 241
Bochkarev, N.G. 1988, Nature, **332**, 518
Bruhweiler, F.C., Gull, T.R., Henize, K.G., & Cannon, R.D. 1981, ApJ, **251**, 126
Chu, Y.-H. 1991, in IAU Symp. 143, *Wolf-Rayet Stars and Interrelations with Other Massive Stars in Galaxies*, eds. K.A. van der Hucht & B. Hidayat, (Dordrecht: Kluwer), p. 349
Chu, Y.-H. 1989, in IAU Symp. 131, *Planetary Nebulae*, ed. S. Torres-Peimbert, (Dordrecht: Kluwer), p.105
Chu, Y.-H. & Lasker, B.M. 1980, PASP, **92**, 730
Chu, Y.-H., Treffers, R.R., & Kwitter, K.B. 1983, ApJS, **53**, 937
Clegg, R.E.S., Devaney, M.N., Doel, A.P., Dunlop, C.N., Major, J.V., Myers, R.M., & Sharples, R.M. 1992, in this volume
Cohen, M. & Barlow, M.J. 1975, Astrophys. Letters, **16**, 165
Crawford, I.A. & Barlow, M.J. 1991a, A&A, **249**, 518
Crawford, I.A. & Barlow, M.J. 1991b, A&A, **251**, L39
Duerbeck, H.W. & Reipurth, B. 1990, A&A, **231**, L11
Esteban, C., Vílchez, J.M., Smith, L.J., & Manchado, A. 1991, A&A, **244**, 205
Esteban, C., Vílchez, J.M., Smith, L.J., & Clegg, R.E.S. 1992, A&A, **259**, 629
Frank, A., Balick, B., & Riley, J. 1990, AJ, **100**, 1903
Garcia-Segura, G. & Mac Low, M.-M. 1992, in ASP conference *"Massive Stars: Their Lives in the Interstellar Medium"*, in press

Garmany, C.D., Olson, G.C., Conti, P.S., & Van Steenberg, M.E. 1981, ApJ, **250**, 660
Gieseking, F., Becker, I., & Solf, J. 1985, ApJL, **295**, L17
Humphreys, R.M. 1989, in IAU Colloquium 113, *Physics of Luminous Blue Variables*, eds. K. Davidson, A.F.J. Moffat, & H.J.G.L.M. Lamers, (Dordrecht: Kluwer), p. 303
Humphreys, R.M. 1991, in IAU Symp. 143, *Wolf-Rayet Stars and Interrelations with Other Massive Stars in Galaxies*, eds. K.A. van der Hucht & B. Hidayat, (Dordrecht: Kluwer), p. 485
Icke, V. 1991, A&A, **251**, 369
Jacoby, G.H. & Chu, Y.-H. 1989, in IAU Symp. 131, *Planetary Nebulae*, ed. S. Torres-Peimbert, (Dordrecht: Kluwer), p. 183
Johnson, D.R.H., Barlow, M.J., Drew, J.E., & Brinks, E. 1992, MNRAS, **255**, 261
Jura, M. & Kleinmann, S.G. 1990, ApJS, **73**, 769
Kahn, F.D. 1989, in IAU Symp. 131, *Planetary Nebulae*, ed. S. Torres-Peimbert, (Dordrecht: Kluwer), p. 411
Kaler, J.B., Feibelman, W.A., & Henrichs, H.F. 1988, ApJ, **324**, 528.
Knapp, G.R. 1989, in IAU Symp. 131, *Planetary Nebulae*, ed. S. Torres-Peimbert, (Dordrecht: Kluwer), p. 381
Kreysing, H.C., Diesch, C., Zweigle, J., Satubert, R., & Grewing,M. 1992, A&A, in press; this volume
Kwok, S. 1983, in IAU Symp. 103, *Planetary Nebulae*, ed. D.R. Flower, (Dordrecht: Reidel), p. 293.
Kwok, S., Purton, C.R., & FitzGerald, P.M. 1978, ApJL, **219**, L125
Lamers, H.J.G.L.M. 1989, in IAU Colloquium 113, *Physics of Luminous Blue Variables*, eds. K. Davidson, A.F.J. Moffat, & H.J.G.L.M. Lamers, (Dordrecht: Kluwer), p. 135
Leitherer, C. & Chavarría-K., C. 1987, A&A, **175**, 208
Maeder, A. 1983, A&A, **120**, 113
Maeder, A. & Maynet, G. 1988, A&AS, **76**, 411
Martin, C. & Arnett, D. 1982, in ASP conference *"Massive Stars: Their Lives in the Interstellar Medium"*, eds. J. Cassinelli & E. Churchwell, in press
Miller, G.J. & Chu, Y.-H. 1993, ApJS, in press
Minkowski, R. 1946, PASP, **58**, 305
Miranda, L.F. & Solf, J. 1992, A&, **260**, 397
Paresce, F. & Nota, A. 1989, ApJL, **341**, L83
Patriarchi, P. & Perinotto, M. 1991, A&A, **91**, 325
Pismis, P. 1974, Rev. Mex. Astron. Astrof., **1**, 45
Prinja, R.K., Barlow, M.J., & Howarth, I.D. 1990, ApJ, **361**, 607
Reay, N.K., Atherton, P.D., & Taylor, K. 1983, MNRAS, **203**, 1079
Smith, L.F. 1973, in IAU Symp. 49, *Wolf-Rayet and High Temperature Stars*, eds. M.K.V. Bappu & J. Sahade, (Dordrecht: Reidel), p. 15
Tarafdar, S.P. & Apparao, K.M.V. 1988, ApJ, **327**, 342
van der Hucht, K.A., Jurriens, T.A., Olnon, F.N., The, P.S., Wesselius, P.R., & Williams, P.M. 1985, A&A, **145**, L16
Wampler, E.J., Wang, L., Baade, D., Banse, K., D'Odorico, S., Gouiffes, C., & Tarenghi, M. 1990, ApJL, **362**, L13
Weaver, W.B. 1974, ApJ, **189**, 263
Willis, A.J. 1982, MNRAS, **198**, 897

# THE NEUTRAL ENVELOPES OF PLANETARY NEBULAE: MOLECULES AND H I

P.J. HUGGINS

*Physics Department, New York University, 4 Washington Place, New York, NY 10003, U.S.A.*

ABSTRACT. This paper summarizes recent developments in the study of planetary nebulae using observations of molecular lines and the 21 cm line of H I. The observations reveal that many planetary nebulae are surrounded by envelopes of neutral gas, whose mass often exceeds that of the ionized nebulae. They also provide valuable information on the physical and chemical properties of the envelopes, their structure, and kinematics. The neutral envelopes firmly link the formation of planetary nebulae with the mass loss by AGB stars, and can play an important role in the subsequent evolution of the nebulae.

## 1. Introduction

Rapid progress has been made during the last few years in the study of the neutral envelopes of planetary nebulae (PNe). The envelopes of $\approx 70$ PNe have now been detected in the lines of one or more molecular species or H I. They are found in compact PNe (e.g., BD+30°3639 and IC 4997) as well as much more evolved PNe (e.g., the Dumbbell and Helix nebulae), and sufficient data have now been obtained that some general conclusions can be made about the properties of the envelopes and their role in the formation and evolution of the ionized nebulae.

The development of this field confirms earlier ideas, based on indirect evidence, that many, if not most PNe must be surrounded by envelopes of neutral gas. The concept of ionization bounded nebulae has a long history in the study of PNe, and the mass-radius relation, first discussed by Pottasch (1980), strongly suggested the presence of envelopes of neutral gas which become ionized as the nebulae expand. The typical ionized mass of an evolved PN is about 0.3 $M_\odot$; on the basis of the mass-radius relation, one might expect roughly this amount of neutral matter around a compact PN, and in some objects this turns out to be the case. This amount of cool, neutral matter at representative distances of 0.5–10 kpc, is, in fact, quite difficult to detect in H I, $H_2$, or in a trace species, except under special circumstances. Accordingly, the development of this field has, to a large extent, paralleled the capabilities of the available instrumentation.

The data on the neutral envelopes of PNe have grown dramatically since the earlier reviews of this field by Black (1983) and Rodriguez (1989). The volume of recent data means that the present paper is, necessarily, highly condensed, and a more extended discussion of some of the issues raised here is given by Huggins (1992).

Table 1. Main Probes of the Neutral Envelopes of PNe

| Species | Wavelength | Detections | Comments | References |
|---|---|---|---|---|
| H I | 21 cm | 10 | mainly in absorption | 1 |
| $H_2$ | 2.1 $\mu$m | 33 | thermal? E/k$\approx$6700 K | 2 |
| CO | 1.3, 2.6 mm | 45 | thermal E/k=6,16 K | 3 |
| OH | 18 cm | 13 | maser | 4 |
| HCN, etc. | mm | 5 | thermal | 5 |
| OI etc. | opt./ir | 5 | NaI/OI,CII (ir PDR lines) | 6 |

1. Taylor et al. (1990) Gussie (1992) 2. Webster et al. (1988) Zuckerman & Gatley (1988) 3. Huggins & Healy (1989) Huggins et al. (1992a) 4. Zijlstra et al. (1989) 5. Cox et al. (1992) 6. Dinerstein (1991)

## 2. Observations

The principal atomic and molecular species which have been used to study the neutral envelopes of PNe are listed in Table 1, together with the approximate number of PNe detected in each. The envelopes which have been mapped in one or more lines are listed in Table 2. The references in the tables are to recent or summary papers from which other contributions can be found, and are the main references for the discussion which follows. The typical angular resolution of the maps is 3″–30″, and the observations of the millimeter and radio lines (but not the infrared lines) usually provide high velocity resolution ($\approx$ 1 km s$^{-1}$, or better) with which the kinematics can be studied.

The information obtained from the various lines of the neutral species is, to a large extent, complementary. The millimeter lines of CO are the most useful general probes of the molecular gas, and have been the most widely observed; the IRAM 30-m telescope has proved to be particularly effective for these studies. The infrared $H_2$ lines have been detected in nearly as many PNe as CO, but their interpretation is still not entirely clear; the emission arises from high lying energy levels which may be excited in shocks, or by ultraviolet radiation or collisions at high temperatures in the photodissociation regions (PDRs) formed by the central star. The relative importance of these processes in most PNe has not yet been sorted out (e.g., Dinerstein 1991, Huggins 1992, Tielens 1992). There is, however, a good correlation between the PNe which have been detected in CO and $H_2$. The data on other molecules and atoms is less extensive, but is nevertheless important for outlining a more complete picture.

Certain general features of the neutral envelopes are readily apparent from the observations. Among the most important is the continuity between AGB envelopes and PNe implied by the CO and OH observations. For example, OH maser emission can be seen in O-rich AGB envelopes, proto-PNe, and young PNe; and in Vy2-2 and OH0.9+1.3, which have been mapped at high resolution (Table 2), the OH surface brightness and the shell sizes (0.6–3×10$^{16}$ cm) are characteristic of AGB stars. The relatively small dimensions of the maser shells means that they will, of course, be rapidly destroyed by the developing nebulae. H I is seen in young PNe, but rarely in AGB envelopes, so is likely to be produced as a result of the evolution of the central star. Perhaps the most surprising feature is the survival of substantial amounts of molecular gas in the envelopes of some very evolved PNe (e.g. the Dumbbell, and the Helix). Specific aspects of the envelopes are discussed in the sections that follow.

Table 2. Maps of the Neutral Envelopes of PNe

| Species | PNe | Refs |
|---|---|---|
| H I | BD+30°3639, IC 418, NGC 6302 | 2,2,1 |
| $H_2$ | CRL618, He2-111, He2-114, Hf48, IC 4406, Mz1 | 1,3,3,3,1,3 |
|  | NGC 2346, NGC 2818, NGC 2899, NGC 3132 | 4,3,3,1 |
|  | NGC 4071, NGC 5189, NGC 6445, NGC 6720 | 3,3,3,4&5 |
|  | NGC 6772, NGC 6781, NGC 6853, NGC 7027 | 3,4,4,1 |
| CO | CRL618, IC 4406, IRAS 21282+5050, M1-7, M2-51 | 6,7&9,6,6&8,8 |
|  | M4-9, NGC 2346, NGC 3132, NGC 6072, NGC 6563 | 8,10,7,9,9 |
|  | NGC 6720, NGC 6772, NGC 6781, NGC 7027 | 10,10,10,12 |
|  | NGC 7293, VV47 | 11&8,10 |
| OH | M1-92, NGC 6302, OH0.9+1.3, Vy2-2 | 15,14,13,13 |
| HCN etc. | CRL618 (HCN), NGC 7027 (HCO+) | 16,17 |

1. References prior to 1988 can be traced through the review by Rodriguez (1989) 2. Taylor et al. (1989, 1990) 3. Webster et al. (1988) 4. Zuckerman & Gatley ('1988) Zuckerman et al. (1990) 5. Greenhouse et al. (1988) 6. Shibata et al. (1992) 7. Sahai et al. (1990, 1991) 8. Forveille & Huggins (1991) 9. Cox et al. (1991) 10. Bachiller et al. (1989a, 1989b, 1992) 11. Healy & Huggins (1990) 12. Bieging et al. (1991) 13. Shepherd et al. (1990) 14. Payne et al. (1988) 15. Seaquist et al. (1991) 16. Neri et al. (1992) 17. Likkel (1992)   See also Clegg et al. and Sahai et al. in these proceedings

## 3. The Physical and Chemical Properties of the Envelopes

3.1 The Molecular Gas

The structure of the molecular component of the neutral envelopes is well documented in the CO and $H_2$ maps referenced in Table 2. In general, the large scale structure of the envelopes shows a good deal of symmetry, and is dominated to a greater or lesser extent by the degree to which the matter is concentrated in a preferred plane, or equator. In some compact PNe (e.g., NGC 7027, or CRL 618), the CO envelope is fairly complete, and is considerably more extended than the ionized nebula, consistent with the idea that the central star has very rapidly evolved from the mass losing AGB phase, and that the young nebula has formed with the outer envelope still intact. In more evolved PNe, the gas is primarily found in toroidal structures around the waist of the nebula, in thin shells, or in isolated condensations. Among the best examples are: NGC 6720 and NGC 6781, where the molecular gas is in thin, ellipsoidal shells; the bow-tie PN NGC 2346, where the gas lies around the waist and to some extent along the walls of the bow tie; and VV 47, where the gas is found within the bright, symmetric, optical lobes. The optical morphology of the PNe with neutral envelopes clearly reflects the underlying structure of the envelopes.

The available maps also show that the molecular gas in the envelopes is not homogenous, but is clumped on the smallest size scales observed. This is the case in essentially all of the PNe which have been well resolved in CO, ranging from the extended envelope of the compact PN NGC 7027 to the thin shells of the much more evolved Helix nebula; in these two cases the clump scale size is $\lesssim 0.02$-0.01 pc. An extreme form of clumping is seen in the numerous cometary globules inside the ionized cavity of the Helix, where the optical condensations are $\approx 1''$ in size; recent observations show that these, too, have cores of molecular gas (Huggins et al. 1992b).

The physical conditions in the molecular gas have been studied in several PNe using the ratios of the rotational lines of CO (e.g. Bachiller et al. 1992). The excitation temperatures are typically $\approx 20$ K, and the densities are rather high ($\gtrsim 3\times 10^3$ cm$^{-3}$, and possibly much higher). The infrared $H_2$ lines are not excited under these conditions, but require high temperature regions due to shocks or PDRs. In line with this, the mass of excited $H_2$ is invariably found to be small ($\lesssim 1\times 10^{-3}$ $M_\odot$) and is much less than the total mass of molecular gas in the envelopes. Under the conditions found in the bulk of the molecular gas, the CO(2-1) line provides a useful estimate of the total mass, whose main uncertainty (apart from the distance) is the CO/H ratio (Huggins & Healy 1989); using the C or O abundance from the ionized gas then gives a good lower limit to the total mass. With this approach, the molecular envelopes detected in CO have masses in the range $1-1\times 10^{-3}$ $M_\odot$; more detailed estimates are available in a few cases, e.g., Jaminet et al. (1991) find the mass of molecular gas is 1.4 $M_\odot$ in NGC 7027. Upper limits on nearby PNe without CO emission are typically $\lesssim 1\times 10^{-2}$ $M_\odot$. There is, therefore, a very large range in the mass of molecular gas in PNe; even in some old PNe (e.g. NGC 6781 and the Helix) the mass of molecular gas is roughly comparable to the mass of ionized gas. The large variation found is attributable to a combination of population and evolutionary effects (§4).

## 3.2 The Atomic Gas

There is considerably less data on the properties of the atomic component of the neutral envelopes. For nebulae which are optically thick, it is expected that the radiation field of the central star longward of the Lyman limit will establish a warm, mainly atomic PDR at the periphery of the ionized gas. Evidence for this comes from the study of the C II and O I infrared fine structure lines in NGC 7027 by Ellis and Werner (1985; see also Tielens 1992). Although the lines are not spatially resolved, the line ratios indicate a gas temperature of 500 K, and a density of $7\times 10^4$ cm$^{-3}$, which are consistent with the idea that the lines originate in the PDR. The mass of this region (inferred from the C/H ratio) is 0.3 $M_\odot$, which is roughly that expected if the PDR extends for $\approx 1$ visual magnitude of extinction into the neutral shell. Further work has been reported by Dinerstein et al. (1991), who find roughly similar physical conditions in 4 additional PNe.

From direct observations of the H I gas using the 21 cm line, the location of the H I is known in only three cases, all young PNe (Table 2). In NGC 6302 the HI is found in the dark lane around the waist of the nebula, and in BD+30°3639 the H I is concentrated near the H II region, and may lie within the molecular shell (Bachiller et al. 1991). In these two cases the masses of H I and $H_2$ are very roughly comparable, and it is likely that the H I is photo-produced from $H_2$ in PDRs between the molecular and ionized gas. In IC 418 the situation is rather different. The H I emission is very extended (0.4 pc) compared to the ionized nebula and the mass of the extended H I envelope (0.4 $M_\odot$) is very much larger than the mass of the molecular component (based on CO); thus most of the extended remnant wind is atomic. H I is also seen in several additional PNe in absorption against the continuum (N(H I) is typically a few$\times 10^{20}$ cm$^{-2}$), but others like NGC 7027 are not detected. Most of those detected have little or no molecular emission, so they, like IC 418, may belong to a population of PNe with relatively thin envelopes. For the PNe detected in H I, lower column densities are found in the larger PNe (Taylor et al. 1990), as might be expected from the expansion of the envelopes. Further work, however, is needed before a coherent picture of the atomic gas is developed.

## 3.3 Kinematics

The kinematics of the neutral envelopes are important clues to the dynamical evolution of PNe. Although the typical expansion velocity of the ionized gas in a PN is $\approx$ 20–25 km s$^{-1}$, and that of an AGB envelope is $\approx$ 14 km s$^{-1}$, much higher velocity molecular winds appear to be a fairly common feature of the intervening proto-PN phase. The highest velocity molecular wind is found in CRL 618, which has, in fact, already developed into a very young PN. It shows a normal, molecular, AGB envelope expanding at 19 km s$^{-1}$, and a high velocity, bipolar, molecular wind expanding at $\approx$ 200 km s$^{-1}$(e.g., Neri et al. 1992). A number of other young PNe also show evidence for a high velocity molecular component, although are less spectacular than CRL 618. Some other young PNe with low velocity molecular envelopes show high velocity atomic jets (e.g., M1-16, Schwarz 1992). The origin of these high velocity phenomena are not well understood. They are clearly able to modify the structure of the envelopes, but their effects are not yet well documented.

In all the more evolved PNe with molecular emission or H I absorption, the typical expansion velocities of the bulk of the neutral gas are moderate, $\approx$ 15–30 km s$^{-1}$, similar to typical PN expansion velocities. In a few cases where higher velocity wings are seen, mapping in CO shows that this can be attributed to wind velocities which increase towards the poles of a mainly equatorial distribution of matter. The best example of this is the Ring type PN, NGC 6781, where high quality CO mapping is well accounted for by a thin ellipsoidal shell with an expansion velocity which increases with latitude (Bachiller et al. 1992). In the most extended PNe where the individual condensations can be observed (e.g., the Helix and VV47), the CO linewidths are very small ($\approx$ 1 km s$^{-1}$; Forveille & Huggins 1991).

The kinematic relation between the neutral and ionized gas is an important diagnostic of the shell development. In several PNe which have been well observed in CO, the expansion velocity of the neutral envelope appears to be larger than that of the ionized gas, an extreme example being BD+30°3639, where the CO expansion velocity is 52 km s$^{-1}$(Bachiller et al. 1991), and those of [O III], H$\alpha$, and [N II] are 23–27 km s$^{-1}$. Thus the general trend for lines of lower ionization species to have larger expansion velocities, which is seen in the ionized nebulae, extends to some of the neutral shells as well. Unfortunately, few of the PNe which have been mapped in detail in CO have also been mapped in one or more of the ionic lines so that they can be compared in detail. One interesting exception is the Helix, where the CO around the ring structure observed by Healy & Huggins (1990) appears to be kinematically part of a more complete shell of atomic gas seen in [N II] (Walsh & Meaburn 1987). Further optical studies of the PNe whose neutral envelopes have been mapped in detail (Table 2) would be important for developing a clearer understanding of their dynamics.

## 3.4 Chemistry

An additional aspect of the envelopes in which recent progress has been made is the chemistry of the neutral gas. A rich chemistry takes place in the thick envelopes of AGB stars, and an increasing number of molecular species are also found in the much harsher environments of PNe. 14 species are detected in the very young PN CRL 618 (e.g., Bujarrabal et al. 1988) where the envelope still retains some of the AGB characteristics; fewer species are seen in more evolved objects, but five (CO, HCN, HNC, CN, and HCO+) have now been detected in NGC 6072, and IC 4406 (Cox et al. 1992), where the central stars are on cooling tracks ($T \approx 100,000$ K, $L \approx 100 - 400$ $L_\odot$) and the envelopes have developed well beyond the AGB phase.

Certain systematic trends in the abundances in the envelopes are already apparent from the data in hand, the most striking being a dramatic increase in the relative abundances of HCO+ and CN as the PNe form (Cox et al. 1992). Interferometer maps also show that the HCO+ in NGC 7027 is found in a thick torus around the periphery of the ionized gas (Likkel 1992, Deguchi et al. 1992). A range of different processes need to be considered in understanding the unusual abundances found in the envelopes. The shielding of dissociating UV radiation is an overriding consideration, and the strong clumping of the gas mentioned in §3.1 clearly plays a role, although the effects of UV radiation (and possibly x-rays) from the central star can also initiate the synthesis of some species under appropriate conditions. High temperature processing through shocks is also likely to be important, especially in the proto-PN and early PN phases. A clear picture has not yet emerged (Cox et al. 1992, Howe et al. 1992), in part because we do not understand in detail the physical conditions in the gas. As in AGB envelopes and the ISM, there is a strong probability that the chemistry in PNe will eventually prove to be invaluable in understanding other aspects of the envelopes.

## 4. Evolution

On the basis of the observational results described above it is possible to outline some aspects of the evolution of the envelopes of PNe. This is more easily done for the molecular gas component, since the data are more extensive. From the molecular line surveys it is clear that the possession of a substantial molecular envelope is related to the population type of the PN. The $H_2$ emission (Webster et al. 1988, Zuckerman & Gatley 1988) and CO emission (Huggins & Healy 1989) is correlated with galactic latitude, bipolarity, and PN type, with the implication that higher mass AGB stars are those that form PNe with substantial molecular envelopes. Several factors probably contribute to this. The higher mass cores move more rapidly across the H-R diagram and form compact PNe while the envelope is still optically thick to dissociating radiation; their mass loss rates on the AGB are also probably larger, so the envelopes are thicker; and their tendency to bipolarity and filamentary structure provide additional shielding of the molecular gas.

It is also possible to study quantitatively the evolution of the molecular envelopes once the PNe have formed. Huggins & Healy (1989) have shown that for PNe detected in CO, the ratio $M_m/M_i$ (where $M_m$ is the mass of the molecular gas, and $M_i$ is the mass of ionized gas) decreases systematically with increasing nebular size. This relation is shown in Fig. 1, and has recently been confirmed and extended (Huggins et al. 1992). The interpretation of this diagram is as follows. When the central stars first cross the H-R diagram and form compact PNe, the system is still largely molecular ($M_m/M_i \gg 1$). As the gas expands, the mass of ionized gas increases at the expense of the neutral envelope. The masses of ionized and molecular gas are roughly comparable by the time the PN has a radius of $\approx 0.1$ pc, and only a residual amount of molecular gas remains when the PN is several tenths of a parsec in size. The observed structure of the molecular envelopes at different stages of evolution (§3.1) strongly supports this evolutionary scenario.

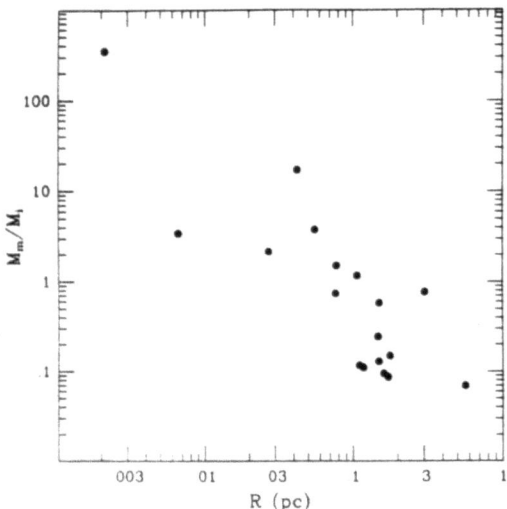

Fig. 1. The molecular/ionized mass ratio versus nebular radius (Huggins & Healy 1989)

Other nearby PNe which are not seen in CO or $H_2$ emission probably lose their molecular envelopes early in crossing the H-R diagram. Some of these are seen in atomic lines, but the systematics of their evolution remain to be sorted out.

## 5. Concluding Remarks

The main conclusions that can be made from the results described here can be summarized as follows. i) The presence of neutral envelopes around many PNe firmly links PN formation with mass loss by AGB stars and proto-PNe. ii) The neutral envelopes provide interesting laboratories in which to study highly unusual physical conditions and a unique chemistry. iii) The envelopes are found to play a key role in the evolution of a broad range of PNe, including many of the most familiar objects, for a large part of their observable lifetimes. In view of the fact that this field is still at a relatively early stage of development, further work is likely to prove very worthwhile.

**Acknowledgements.** It is a pleasure to acknowledge collaborative programs with Drs. R. Bachiller, P. Cox, T. Forveille, and A. Omont which form the basis for part of this paper. This work was supported in part by the National Science Foundation.

## References

Bachiller, R., Planesas, P., Martin-Pintado, J., Bujarrabal, V., Tafalla, M., 1989a, A&A, 210, 366

Bachiller, R., Bujarrabal, V., Martin-Pintado, J., Gomez-Gonzalez, J., 1989b, A&A, 218, 252

Bachiller, R., Huggins, P.J., Cox, P., Forveille, T., 1991, A&A, 247, 525
Bachiller, R., Huggins, P.J., Cox, P., Forveille, T., 1992, A&A, in press
Bujarrabal, V., Gomez-Gonzalez, J., Bachiller, R., Martin-Pintado, J., 1988, A&A, 204, 242
Bieging, J.H., Wilner, D., Thronson, H.A., 1991, ApJ, 379, 271
Black, J.H., 1983, in: Planetary Nebulae, ed. D.R. Flower, Reidel, Dordrecht, p. 91
Cox, P., Huggins, P.J., Bachiller, R., Forveille, T., 1991, A&A, 250, 533
Cox, P., Omont, A., Huggins, P.J., Bachiller, R., Forveille, T., 1992, A&A, in press
Deguchi, S., et al., 1992, ApJ, 392, 597
Dinerstein, H.L., 1991, PASP, 103, 861
Dinerstein, H.L., Haas, M.R., Werner, M.W., 1991, BAAS, 23, 915
Ellis, H.B., Werner, M.W., 1985, in: Mass Loss From Red Giants, eds. M. Morris & B. Zuckerman, Reidel, Dordrecht, p. 309
Forveille, T., Huggins, P.J., 1991, A&A, 248, 599
Greenhouse, M.A., Hayward, T.L., Thronson, H.A., 1988, ApJ, 325, 604
Gussie, G., 1992, private communicaton
Healy, A.P., Huggins, P.J., 1990, AJ, 100, 511
Howe, D.A., Millar, T.J., Williams, D.A., 1992, MNRAS, 255, 217
Huggins, P.J., 1992, in: Mass Loss on the AGB and Beyond, ed. H. Schwarz, ESO, in press
Huggins, P.J., Healy, A.P., 1989, ApJ, 346, 201
Huggins, P.J., Bachiller, R., Cox, P., Forveille, T., 1992a, in preparation
Huggins, P.J., Bachiller, R., Cox, P., Forveille, T., 1992b, ApJL, submitted
Jaminet, P.A., et al., 1991, ApJ, 380, 461
Likkel, L., 1992, ApJL, submitted
Neri, R., et al., 1992, these proceedings
Payne, H.E., Phillips, J.A., Terzian, Y., 1988, ApJ, 326, 368
Pottasch, S.R., 1980, A&A, 89, 336
Rodriguez, L.F., 1989, in: Planetary Nebulae, ed. S. Torres-Peimbert, Kluwer, Dordrecht, p. 129
Sahai, R., Wootten, A., Clegg, R.E.S., 1990b, A&A, 234, L1
Sahai, R., Wootten, A., Schwarz, H.E., Clegg, R.E.S., 1991, A&A, 251, 560
Schwarz, H.E., 1992, A&A, in press
Seaquist, E.R., Plume, R., Davis, L.E., 1991, ApJ, 367, 200
Sheperd, M.C., Cohen, R.J., Gaylard, M.J., West, M.E., 1990, Nature, 344, 522
Shibata, K.M., et al., 1992, these proceedings
Taylor, A.R., Gussie, G.T., Goss, W.M., 1989, ApJ, 340, 932
Taylor, A.R., Gussie, G.T., Pottasch, S.R., 1990, ApJ, 351, 515
Tielens, A.G.G.M., 1992, these proceedings
Walsh, J.R., Meaburn, J., 1987, MNRAS, 224, 885
Webster, B.L., Payne, P.W., Storey, J.W.V., Dopita, M.A., 1988, MNRAS, 235, 533
Zijlstra, A.A., et al., A&A, 217, 157
Zuckerman, B., Gatley, I., 1988, ApJ, 324, 501
Zuckerman, B., Kastner, J.H., Balick, B., Gatley, I., 1990, ApJ, 356, L59

# PHOTODISSOCIATION REGIONS AND PLANETARY NEBULAE

A.G.G.M. TIELENS

*MS 245-3, NASA Ames Research Center, Moffett Field, CA 94035-1000, USA*


ABSTRACT: FUV photons (<13.6eV) from the central star create a region of warm ($\approx 1000$K) atomic and molecular gas around Planetary Nebulae (PN). This paper reviews theoretical and observational characteristics of such regions, commonly called photodissociation regions or PDRs. PDRs around PN differ in some aspects from those in other galactic objects and this is briefly discussed with an emphasis on time dependent effects. It is concluded that, in evolved PN ($t_{exp} > 10^3$ yr), molecules will only survive inside dense clumps ($>10^6$ cm$^{-3}$). $H_2$ emission from such dense gas will show a thermal spectrum in the low v states. Finally, the physical conditions in the PDR associated with NGC 7027 are compared to those in other galactic and extragalactic PDRs

## 1. Introduction

Low mass stars lose a considerable fraction of their mass in the form of a dusty molecular wind on the asymptotic giant branch, possibly starting as a gentle breeze ($\dot{M} \approx 10^{-6}$ M$_\odot$/yr; v=10 km/s) which later on turns into a superwind ($\dot{M} \approx 10^{-4}$ M$_\odot$/yr; v=15 km/s). During the rapid evolution of the central star in the transition phase and the early parts of the planetary nebula stage, increasingly higher energy UV photons will illuminate this circumstellar shell and carve out an ionized and an atomic zone in the molecular gas (Pottasch 1980). The ionized gas gives rise to the visibly dominant planetary nebulae itself while the atomic and molecular gas region, commonly called a PhotoDissociation Region (PDR), only lights up in the infrared and submillimeter through atomic finestructure lines, molecular rotational/vibrational lines and the dust continuum.

Over the last two decades, numerous infrared and submillimeter studies have been made of the characteristics of AGB outflows (cf., Habing elsewhere in this volume). By comparison, observations of atomic and molecular gas surrounding planetary nebulae are more limited (cf., Huggins elsewhere in this volume). Molecular observations have largely been limited to the CO 1-0 and 2-1 lines and the $H_2$ 1-0 S(1) line (Huggins and Healy 1989; Zuckerman and Gatley 1988). The predominantly neutral atomic gas in the PDR has been observed in the 21cm HI line (Taylor et al. 1990) and the [CII] 158μm and the [OI]

63 and 146μm finestructure lines (Dinerstein 1991). These observations reveal that the mass of atomic and molecular material surrounding planetary nebulae can be substantial. For example, for the young compact PN NGC 7027, a conservative lower limit, based upon optically thin CO emission, places the molecular gas mass at $\approx 0.2$ $M_\odot$ (Huggins and Healy 1989), including various realistic correction factors would increase this to $\approx 1.4 M_\odot$ (Jaminet et al. 1991). Submillimeter dust continuum emission observations also yield $\approx 1 M_\odot$ (Sopka et al. 1985). In contrast, the ionized gas mass is only $\approx 0.08$ $M_\odot$, comparable to the atomic gas associated with the CII zone ($\approx 0.1 M_\odot$).

Clearly, atomic and molecular gas is an important component of PNe and further studies of their properties are crucial for a proper understanding of the dynamical and morphological evolution of PNe. This paper reviews our theoretical and observational understanding of PDRs. Since little work has been done on PDRs surrounding PNe, the emphasis will be on PDRs associated with HII regions and reflection nebulae (§2). In §3 some of the differences of PDRs in PNe will be pointed out. The origin of the observed $H_2$ emission - shocks versus fluorescence - will be assessed in §4. Physical conditions in the NGC 7027 PDR are contrasted in §5 with those in galactic and extragalactic PDRs. Finally, in §6, the importance of clumps in PNe is reemphasized.

## 2. PDR Models

Photodissociation regions (PDRs) are regions where FUV (<13.6eV) photons dominate the heating and/or chemical composition of the gas. PDRs are associated with HII regions, reflection nebulae, bright rim clouds, galactic nuclei, AGNs, and of course PNe. Indeed, penetrating FUV photons from the average interstellar radiation field dominate the chemical composition and heating of molecular clouds for $A_V$ <4. Thus, besides HI, most of the molecular mass is in PDRs as well. Detailed theoretical models have been developed for the interaction of FUV photons with atomic and molecular gas and dust (Tielens and Hollenbach 1985a hereafter TH; Wolfire et al. 1989; Sternberg and Dalgarno 1989; Burton et al. 1990; Hollenbach et al. 1991). Figure 1 shows the results of a detailed calculation for the physical structure of the PDR in Orion (density $n_0 = 2.3 \times 10^5$ cm$^{-3}$; incident UV field $G_0 = 10^5$ times the average interstellar radiation field, $1.6 \times 10^{-3}$ erg cm$^{-2}$ s$^{-1}$; Habing 1968). Basically, the penetrating FUV photons dissociate and ionize the molecular gas to neutral and singly ionized atoms (ie., H$^o$, C$^+$, O$^o$, Si$^+$, ...). Deeper in, due to dust attenuation ($A_V$=2), hydrogen transforms into $H_2$. Similarly, around $A_V$=4, the carbon balance shifts from C$^+$ to C$^o$ and CO (Fig. 1). Except for the fraction locked up in CO, most of the oxygen stays in atomic form until very deep in the cloud ($A_V$>10) ion-molecule chemistry produces $O_2$. Besides dominating the chemical balance, FUV photons also control the energy balance. FUV photons can ionize dust grains and PAH molecules and the excess kinetic energy carried away by the photoelectrons will heat the gas. Typically, about 0.5% of the flux absorbed by the dust is converted into gas heating this way. The remainder ($\approx$99.5%) heats the dust and is reradiated as IR continuum. Even when the UV is attenuated ($A_V$>4), penetrating red and infrared photons still keep the dust warm and gas-grain collisions dominate the gas heating. The gas cools through the emission of atomic fine structure lines (ie., [OI] 63μm & [CII]158μm) at the surface (Av<4 mag) and molecular rotational lines (ie., CO) deeper in (Fig. 1). At the surface, the gas is generally much hotter than the dust while deeper in the gas temperature drops to somewhat below the

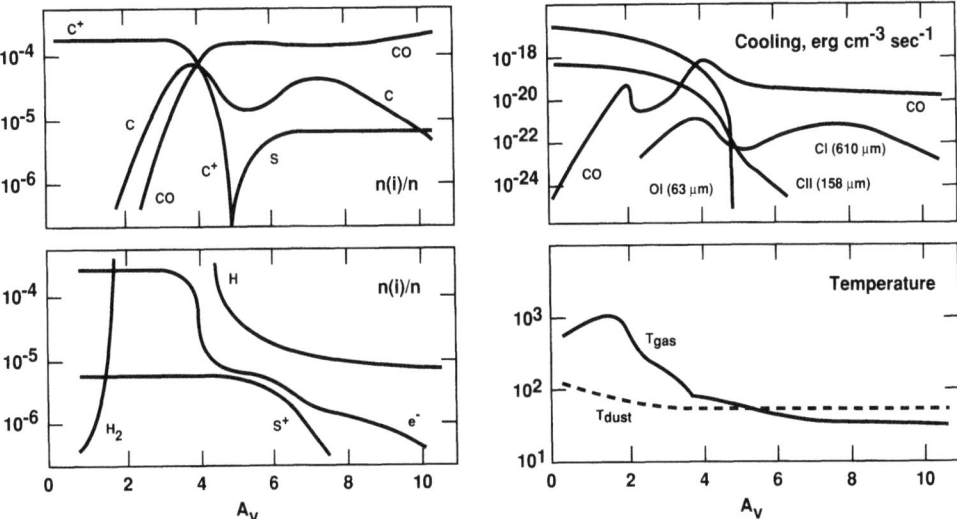

Figure 1: Structure of the PDR in Orion ($n_o=2.3\times10^5$ cm$^{-3}$, $G_o=10^5$; Tielens and Hollenbach 1985). Left: Abundances relative to total hydrogen as a function of visual extinction into the PDR. Right: Cooling in various lines and gas and dust temperature in the PDR.

dust temperature (Fig. 1).

The chemical structure of a PDR is largely dependent on the penetration of the UV photons and thus on the dust extinction. For species whose photodissociation is dominated by line absorption, self-shielding can be important as well. However, for bright PDRs, this requires high densities. Thus, when $n_o/G_o > 50$ cm$^{-3}$, $H_2$ self-shielding becomes important and the $H^o/H_2$ transition is pulled to the cloud surface. Typically, hydrogen goes then molecular around $A_v\approx0.3$. The upwardly revised CO photodissociation rates make CO self-shielding of little importance in PDRs (van Dishoeck and Black 1988). However, the increased $H_2$ abundance near the surface when $H_2$ self shielding is important leads to rapid CO formation and as a result appreciable amounts of CO are present near the surface as well (Burton et al. 1990). Further at these high densities and temperatures, collisional deexcitation of UV-pumped vibrationally excited $H_2$ is an important heating source for the gas. High density, bright PDRs have therefore a thin ($A_v\approx0.03$) very warm ($T\approx2000K$), partly molecular, surface layer.

In summary, PDR models predict a large column density - $N_o\approx8\times10^{21}$ cm$^{-2}$ ($A_v\approx4$) - of warm, atomic gas ($\approx500$ K) at the surface and a somewhat cooler ($\approx50$ K) column of molecular gas. The PDR will be bright in the far-IR dust continuum and PAH emission features at 3.3, 6.2, 7.7, and 11.3μm. Particularly, these near-IR emission features traces the penetration of the UV photons. The atomic surface layer ($A_v<4$ mag) will emit copiously in the [OI], [CII], [SiII], and [CI] atomic finestructure lines. The fluorescing near-IR rovibrational lines of $H_2$ will originate around $A_v=2$ mag. Rotational CO emission arises from the cooler molecular gas deeper in ($A_v>4$ mag). Finally, the very warm surface layer in high density PDRs will be bright in high level CO (ie., 17-16, 14-13) and, reflecting the importance of collisional deexcitation, $H_2$ emission in the low v states from this gas will show a thermal near-IR spectrum. High v states will still be fluorescing.

## 3. PDRs in Planetary Nebulae

The physics and chemistry of PDRs in PNe is in some aspects different from that in most other regions. First, in contrast to other regions, PNe may show strong time dependent effects associated with the evolution of the central star (ie., increase in $T_{eff}$). Second, the dynamical evolution of the nebula may result in weak shocks driven into the atomic and molecular gas. The expected shock velocities range from 5-15km/s during the ionization phase to almost 30km/s in the nebular expansion phase (Marten and Schönberner 1991). Shocks driven into high density clumps will be much weaker. Third, the elemental abundance ratio can be quite different from that in galactic sources (ie., N/O, C/O). Fourth, planetary nuclei may emit appreciable fraction of their energy in the form of soft X-rays, which can penetrate through the ionized volume and heat and ionize the atomic and molecular gas. In that respect PDRs in PNe may resemble those around AGNs. Finally, cosmic rays, which play a major role in interstellar chemistry, are tied to the galactic magnetic field. With a typical gyration radius of $10^{11}$ cm, cosmic rays do not penetrate into the bubble blown by the red giant and thus they play no role in the chemistry of AGB or PN ejecta. Space does not permit a thorough discussion of all of these effects and here we will concentrate on the time dependent effects.

First, consider $H_2$ reformation: H atoms will chemically adsorb on graphitic and silicate surfaces with binding energies of $\approx 2eV$ (Tielens and Allamandola 1987). Even at dust temperatures of 1000K, the grain surface will be completely covered by chemisorbed H. Since dust temperatures in PNe are generally in excess of 50K, physisorbed H will evaporate before reacting with another H. $H_2$ formation occurs then through direct reaction of impinging H atoms with chemisorbed H. Taking the activation barrier of this reaction into account, the efficiency of this reaction is $\varepsilon \approx 0.3 \exp(-1000/T)$ of the H-collision rate, with T the gas temperature (Tielens and Allamandola 1987). For T=500K, appropriate for a PDR, $\varepsilon \approx 0.04$. The timescale to reach $H/H_2$ equilibrium is then $\approx 7 \times 10^9/n_0$ yr (Hill and Hollenbach 1978). Clearly, $H_2$ will only reform inside dense clumps (>$10^6$ cm$^{-3}$).

Now, consider $H_2$ photodissociation: The photodissociation timescale is a strong function of the effective temperature of the star. Spaans et al. (1992) have calculated photodissociation rates for constant luminosity (6000$L_\odot$) star at a distance of $5 \times 10^{16}$ cm ($t_{exp}=10^3$ yr), corresponding to $G_0=5 \times 10^5$ (Fig. 2). For a 6000K star, the $H_2$ photodissociation lifetime, about $10^3$ times that in the average ISM, is much longer than the expansion timescale. However, for stars earlier than A5, $H_2$ photodissociation becomes important and for a typical young PN with $T_{eff}$>25000K an $H_2$ molecule survives less than a year. Thus, at that point, an $H_2$ dissociation wave will propagate rapidly into the molecular gas. This wave will stop when dust shielding has increased the $H_2$ lifetime (ie., $A_v \approx 4$). However, for a typical spherically expanding superwind, the circumstellar dust shell becomes transparent after only 600 yr. Alternatively, in clumps, $H_2$ self shielding could be important if $H_2$ reforms rapidly ($n_0 > 10^6$ cm$^{-3}$) and this will produce a regular PDR with an $H^0$-$H_2$ transition around $A_v \approx 0.3$-2, depending on $n_0/G_0$ (§2). In conclusion, in evolved PNe ($t_{exp} > 10^3$ yr), $H_2$ will only be present inside dense clumps.

A similar discussion can be held for CO. Various channels contribute to CO reformation (ie., O+CH, $CH_2$, $CH_3^+$). The reformation timescale is governed by the radiative association of $C^+$ and $H_2$ and is $\approx 10^8/n_0$ yr. However, as these chemical routes necessitate $H_2$, efficient CO reformation also occurs only inside dense clumps. As for $H_2$, CO will survive photodestruction through the transition stage to the PN phase (Fig. 2). CO

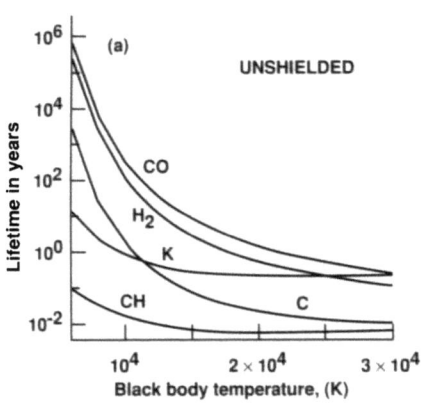 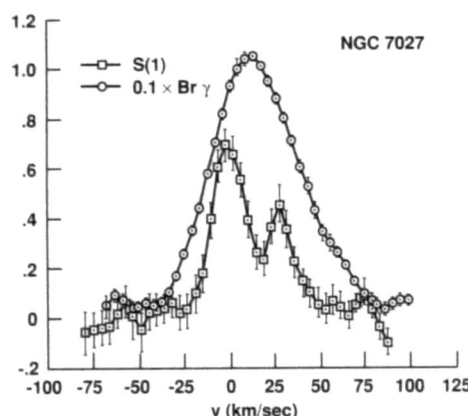

Figure 2: Photo-ionization and dissociation timescales as a function of the effective temperature of the stellar radiation field (Spaans et al. 1992).

Figure 3: The observed profile of the $H_2$ 1-0 S(1) line compared to the Br$\gamma$ line in NGC 7027 (Geballe 1990).

self shielding is generally of little importance in PDRs but, in the PN phase, dust will raise the lifetime to longer than $10^3$ yr for $A_v>2.5$. Thus, CO ejected by the red giant can survive into the PNe phase as long as the clump/envelope is thick enough. However, as for $H_2$, in evolved PNe, CO will only be present in high density clumps.

Many molecules can be dissociated by rather mild UV photons compared to CO and $H_2$ and as a result have short lifetimes even for a 6000K star (Figure 2). Again, in the transition phase, such molecules in the AGB superwind can survive due to dust shielding ($A_v>4$ or more). Their presence in evolved PNe, however, implies high density clumps. In such dense clumps an active chemistry will take place (Cox et al. 1992). HCN formation, for example, might be initiated through reactions of N with various small hydrocarbon radicals (CH, $CH_2$, $CH_3^+$). Likewise, $HCO^+$ might result from reactions of O with these same species.

## 4  $H_2$ emission: shocks versus UV fluorescence

The near-IR spectrum of $H_2$ can be readily classified as dominated by either thermal collisions or UV-pumped IR fluorescence. The 1-0 S(1)/2-1 S(1) ratio, for example, is characteristically different (10 versus 1.8). In the past, a thermal ratio has generally been taken to imply that the emission is due to shock heated gas and most observations of $H_2$ emission from PNe has been interpreted along these lines (Zuckerman and Gatley 1988). In recent years, it has however been realized that high density PDRs will have characteristics very similar to low velocity shocks; ie., a thin layer of warm molecular gas (§2; Sternberg and Dalgarno 1989; Burton et al. 1990). Thus, they will also give rise to a thermal $H_2$ NIR spectrum. In fact, recent studies of NGC 7027, show that its thermal $H_2$ emission occurs from high density PDRs rather than from shocks. First, morphologically, the $H_2$ emission, as well as the 3.3μm PAH emission, is seen to bridge the gap between the ionized gas and the CO emission as expected for a PDR (Graham et al. 1992). In this, NGC 7027 resembles well known galactic PDRs such as the Orion Bar (Sellgren et al. 1990). Second, assuming a preshock density of $10^6$ cm$^{-3}$ (cf., §5), the observed H$_2$ intensity ($\approx 10^{-3}$ erg/(cm$^2$ s sr) after dereddening) implies a shock velocity of >10 km/s and >20 km/s for a J or C shock, respectively (Burton et al. 1992). However, the H$_2$ line profile reveals a shell expanding at $\approx 15$ km/s (Fig 3; Geballe 1990). This velocity is very similar to the CO expansion velocity and less than that of HII. Certainly, there is no evidence for shocked gas at 25-35km/s with respect to the stellar systematic velocity required to explain the observed intensity (Graham et al. 1992). Indeed, in general, H$_2$ molecules will only survive inside dense clumps (§3 & 6). Thus, as long as T>1000K ($G_0>10^3$), the H$_2$ spectrum should be thermal. Conversely, a fluorescing H$_2$ spectrum is only expected from the H$_2$ photodissociation wave "racing" through the interclump gas (ie., for young PN).

## 5   Physical conditions in PDRs around PNe

Infrared and submillimeter observations can be used in a simple procedure to determine average physical conditions in PDRs. In particular, the ratios of [OI] 63/146 μm and [OI] 63/[CII] 158μm are sensitive to density and temperature. Over much of the relevant parameter space for warm, dense PDRs ($10^2<n_0<10^6$ cm$^{-3}$; 100<T<1000K), curves of constant ratio of these emission lines are nearly perpendicular in the $n_0$-T plane and thus these ratios give a good indication of the physical conditions in the emission zones (cf., Genzel et al. 1989). Alternatively, rather than using the [OI] 63μm line, which may well be optically thick, one might combine the [OI] 146/[CII] 158μm with the observed heating efficiency. The latter can be determined by comparing the flux in the dominant cooling lines ([OI]+[CII]) with the total IR dust continuum. Then, adopting a dust model, this efficiency can be translated into a relation between $n_0$ and T since photoelectric heating efficiency depends on the grain charge which itself depends on these parameters (cf., TH). Although, at first sight, any relation based upon dust physics might be suspect, it should be emphasized that this efficiency relation is semi-empirically founded on a direct observational study of the heating of the ISM (Pottasch et al. 1979). Fortunately, both approaches give very similar results. Moreover, they agree well with other determinations of the physical conditions in PDRs, including such direct ones as the temperature determination from the rotational lines of H$_2$ (Parmar et al. 1992). As a result, such studies have now been applied to a large number of galactic and extragalactic sources (Wolfire et al. 1990; Burton et al. 1990).

Table 1 summarizes the results obtained this way for the PDR in NGC 7027 and compares them to those for the Orion Bar, the galactic center and M82. We conclude from this comparison that the physical conditions in these PDRs are all very similar, with average densities of $\approx 10^5$ cm$^{-3}$ and temperatures of $\approx 500$K. However, the observations also imply the existence of higher density clumps ($10^6$-$10^7$ cm$^{-3}$) with a small volume filling fraction (Burton et al. 1990). We suggest that, in NGC 7027, this dense gas is part of the highly clumped circumstellar torus (diameter$\approx 6$"), obvious in Brα, H$_2$ 1-0 S(1), and 3.3μm feature (Graham et al. 1992). With an expected scale size of $\approx 5\times 10^{15}$ cm (0.3"), the PDR will form an unresolved layer outside the ionized gas in this torus. The expected HI column density for such high density gas is small ($10^{20}$ cm$^{-2}$; cf., §2). Coupled with its high temperature ($\approx 2000$), this may explain the non-detection of HI absorption in NGC

Table 1: Physical conditions in PDRs derived from atomic finestructure lines.

| Object | NGC 7027 | Orion Bar | Sgr A | M82 | notes |
|---|---|---|---|---|---|
| n (cm$^{-3}$) | 7. (4) | 1. (5) | 1. (5) | 5. (4) | 1 |
| T (K) | 400 | 500 | 750 | 400 | 2 |
| $G_0$ | 1. (4) | 4. (4) | 1. (5) | 1. (4) | 3 |
| $M_a/M_m$ | 0.1 | 0.04 | 0.04 | 0.1 | 4 |
| $n_c$ | 1. (6) | 1. (7) | 1. (7) | - | 5 |
| $f_v$ | 0.3 | 0.1 | 0.05 | - | 6 |

notes: 1) Average density. 2) Average Temperature. 3) Incident UV field in units of the average interstellar radiation field (1.6x10$^{-3}$ erg cm$^{-2}$ s$^{-1}$; Habing 1968). 4) Ratio of atomic to molecular gas mass. 5) Clump density. 6) Clump volume filling factor. Ref. Wolfire et al. (1990) & Burton et al. 1990).

7027 (Huggins 1992). Most of the material in these clumps (N≈2x10$^{22}$ cm$^{-2}$) is in molecular form. The lower density PDR gas evident from the atomic finestructure lines is likely associated with the low surface brightness (fluorescing) H$_2$ and PAH emission extending about 3" further out from the ionized gas (Graham et al. 1992). The density (10$^5$ cm$^{-3}$) derived from the observed scale size of the emission is in good agreement with that derived from the FIR lines (Table 1). The absence of HI absorption of this gas (N$_H$≈4x10$^{21}$ cm$^{-2}$) against the HII region suggests that it is located in two lobes, above and below the ionized torus.

## 6. Clumps

A clumped distribution of molecular emission seems to be an inherent property of all PNe (Huggins 1992; Cox et al. 1992). As discussed in §3, high density clumps are a necessity for molecular survival and/or an active chemistry in PN ejecta. These clumps may reflect instabilities during the superwind phase when the torus was created or result from Rayleigh-Taylor instabilities during the PN phase when the hot bubble pressed upon the superwind (cf., Balick elsewhere in this volume). The pressure in the NGC 7027 neutral clumps, ≈10$^9$ cm$^{-3}$ K, is comparable to the pressure of the ionized gas (n$_e$=6x10$^4$ cm$^{-3}$, T$_e$=14,500 K; Roelfsema et al. 1991). During the ionization phase of the PN, the pressure on the clumps will be constant (Marten and Schönberner 1991). Since the incident FUV field (≤13.6eV) will decrease due to geometrical dilution, the temperature in the neutral gas will slowly drop and the density increase in this phase. During the expansion phase (t$_{exp}$>3000 yr), the pressure on the clump will decrease and the clump density will start to decrease.

As for neutral globules in HII regions, the lifetime of these clumps will be determined by erosion due to photo-ionization (Huggins et al. 1992). The stellar ionizing photon flux, $N_*$, will be attenuated by the ionized layer on the clump surface. The flux of photons, $N$, arriving at the ionization front of a clump of size R$_c$ at a distance r from the star is then approximately given by $N \approx \{ [3N_*/4\pi r^2] C^2/\alpha R_c \}^{0.5}$ with $\alpha$ the recombination coefficient and C the sound speed in the ionized gas (Spitzer 1978). Adopting a hot black body, $N_* \approx 6.3 \times 10^{52}/T_{eff}$ photons s$^{-1}$ for L=6000L$_\odot$ and $N \approx 2 \times 10^{11}$ [10$^5$K/T$_{eff}$]$^{0.5}$ [5x10$^{16}$/r(cm)] photons s$^{-1}$. Thus, the column density eroded by photoionization is then given by $N(H) = \int N \, dt$, which is only weakly dependent on time for a clump coasting at a constant velocity; ie., $N(H) \approx 7 \times 10^{21} [10^5 K/T_{eff}]^{0.5} [15/v(km/s)]$ $\ln[t_{exp}/t_0]$ cm$^{-2}$ where $t_0$ is the time at which the ionizing flux is turned on ($\approx 500$ yr). Thus, clumps with column densities of at least $2 \times 10^{22}$ cm$^{-2}$ will survive throughout the PN phase. Such clumps are optically thick in the UV and thus molecules will survive. Moreover, adopting a typical clump scale size, $R_c \approx r_i/3 \approx 10^{15} - 10^{16}$ cm with $r_i$ the shell size at clump formation, the density is high enough to drive an active chemistry. In conclusion, clumps that survive during the PNe phase are by necessity molecular.

ACKNOWLEDGEMENTS: This review benefitted greatly from stimulating discussion with and preprints made available by Pierre Cox, James Graham, David Hollenbach, and Pat Huggins. Theoretical studies of photodissociation regions and their observational characteristics at NASA Ames are supported through NASA grant 188-41-53.

## References

Burton, M.G., Hollenbach, D.J., & Tielens, A.G.G.M., 1990, ApJ, 365, 620.
Burton, M.G., Hollenbach, D.J., & Tielens, A.G.G.M., 1992, ApJ, Nov, in press.
Cox, P., Omont, A., Huggins, P.J., Bachiller, R., and Forveille, T., AA, in press.
Dinerstein, H.L., 1991, PASP, 103, 861.
Geballe, T.R., 1990, in *Molecular Astrophysics*, ed. T. Hartquist, (Cambridge Univ Press, Cambridge), p.345.
Genzel, R., Harris, A.I. & Stutski, J., 1999, in *IR Spectroscopy and Astronomy*, ed. Kaldeich, B.H., ESA, SP-290, p115.
Graham, J.R., et al., 1992, ApJ, in press.
Habing, H.J., 1968, Bull. Astr. Inst. Neth., 19, 421.
Hill, J.K., & Hollenbach, D.J., 1978, ApJ, 225, 390.
Hollenbach, D.J., Takahashi, T., & Tielens, A.G.G.M, 1991, ApJ, 377, 192.
Huggins, P.J., and A.P. Healy, 1989, ApJ, 346, 201.
Huggins, P.J., 1992, in *Mass Loss on the AGB and Beyond*, ESO, in press.
Huggins, P.J., Bachiller, R., Cox, P., and Forveille, T., 1992, ApJL, Sept, in press.
Jaminet, P.A., et al., ApJ, 380, 461.
Marten, H. & Schönberner, D., 1991, A&A, 248, 590.
Parmar, P.S., Lacy, J. & Achtermann, J.,1992, ApJL, 372, L25.
Pottasch, S.R., 1980, A&A, 89,336.
Roelfsema, P.R., Goss, W.M., Pottasch, S.R., and Zijlstra, A., 1991, A&A, 251, 611.
Sopka, R.J., et al, 1985, ApJ, 294, 242.
Spaans, M., van Dishoeck, E.F., & Tielens, A.G.G.M., 1992 in preparation.
Spitzer, L., 1978, *Physical Processes in the Interstellar Medium*, (Wiley and Sons, NY).
Sternberg, A. & Dalgarno, A., 1989, ApJ, 338, 197.
Taylor, A.R., Gussie, G.T. & Pottasch, S.R., 1990, Ap.J., 351, 515.
Tielens, A.G.G.M. & Hollenbach, D.J., 1985, ApJ, 291, 722 (TH).
Tielens, A.G.G.M. and Allamandola, L.J., 1987, in *Interstellar Processes*, eds. D. Hollenbach and H.Thronson, (Reidel, Dordrecht), p.397.
van Dishoeck, E.F. & Black, J.H., 1988, ApJ, 334, 771.
Wolfire, M.G., Tielens, A.G.G.M, & Hollenbach, D.J., 1990, ApJ, 358, 116.
Zuckerman, B. & Gatley, I., 1988, Ap.J, 324, 501.

# DUST IN PLANETARY NEBULAE AND IN POST-AGB OBJECTS

M. J. BARLOW

*Department of Physics & Astronomy, University College London,*
*Gower Street, London WC1E 6BT, U.K.*

**Abstract.** Developments in the field since the review of Roche (1989) are discussed. Section 1 describes the results from detailed models of the dust emission from planetary nebulae and summarises the derived dust-to-gas mass ratios. The properties of the remarkable gas-poor planetary nebula IRAS 18333–2357, located in the globular cluster M22, are discussed in Section 2. The use of thermal infrared imaging to map the distribution of different dust components is reviewed in Section 3, and two recently applied optical probes of nebular dust properties are described: emission line spectropolarimetry, and spectroscopy of the Extended Red Emission (ERE). The final section describes recent results from infrared spectroscopy. The spectra of carbon-rich post-AGB objects have provided valuable new insights into the nature and evolution of the materials responsible for the UIR bands, while unusual oxide grains and crystalline ice emission have been detected around oxygen-rich post-AGB objects. The spectra of a small number of objects show evidence for the simultaneous presence of both oxygen-rich and carbon-rich material.

## 1. Dust to Gas Mass Ratios in Planetary Nebulae

Harrington, Monk & Clegg (1988) modelled the dust emission from the carbon-rich nebula NGC 3918 using graphite and amorphous carbon grain models with power-law size distributions of the type discussed by Mathis, Rumpl & Nordsieck (1977; MRN). Because graphite grains have a steeper infrared emissivity law and are less efficient emitters, they will be hotter than amorphous carbon grains of the same size. To fit the overall energy distribution, Harrington et al. had to have a lower limit of $0.04\mu m$ for the radius of graphite grains (compared to $0.005\mu m$ for the standard MRN insterstellar grain mixture), since grains smaller than this produced too much flux at $10\mu m$ and shorter IR wavelengths. On the other hand, their amorphous carbon model required a lower grain radius cut-off of $0.0005\mu m$ in order to produce enough $10\mu m$ flux. Both grain models had maximum grain radii of $0.25\mu m$, similar to the MRN insterstellar mixture. The graphite and amorphous carbon grain models for NGC 3918 implied overall dust-to-gas mass ratios of $8\times10^{-4}$ and $3\times10^{-4}$, respectively. Harrington et al. noted that if the observations had been modelled assuming single-temperature blackbody emission (as is sometimes found in the literature), a misleadingly high dust-to-gas mass ratio of $2\times10^{-3}$ would have been derived.

Hoare, Roche & Clegg (1992) have modelled the overall dust emission from the C-rich PN NGC 7027 and BD+30°3639, again using both graphite and amorphous carbon models but this time also utilizing their own JCMT sub-mm and mm-wavelength photometry as well as the *IRAS* and shorter wavelength infrared data. For both PN they found that their graphite grain models were unable to produce enough sub-mm flux but that an acceptable fit could be obtained with amorphous carbon models. Figure 1 shows their fit to the IR and sub-mm energy distribution of NGC 7027. The dotted line shows the predicted emission from dust in the ionized zone – the remaining emission comes from dust in the surrounding neutral zone.

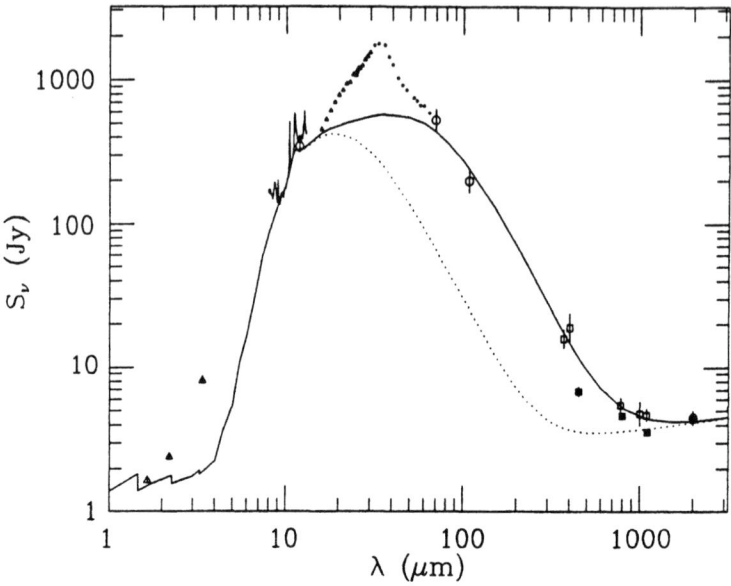

Fig. 1. The observed infrared energy distribution of NGC 7027 is shown, along with an amorphous carbon grain model fit (solid line). The dotted line shows the predicted emission from the dust in the ionized zone alone. From Hoare, Roche & Clegg (1992).

They estimated a dust-to-gas mass ratio of $7 \times 10^{-4}$ in the ionized zone, based on the observed attenuation of the C IV 1550Å resonantly scattered line (use was made of the Middlemass (1990) photoionization model of NGC 7027, which predicted the flux emitted in this line before dust attenuation). For the neutral zone around NGC 7027, Hoare et al. estimated a dust-to-gas mass ratio of $1.5 \times 10^{-3}$, not significantly different from the ratio derived for the ionized zone in view of the uncertainties associated with the CO-based neutral mass estimate. An amorphous carbon dust visual optical depth of 0.4 was estimated for the neutral zone.

Inspection of Figure 1 shows that there is a very prominent feature peaking at 30–35$\mu$m which is not fitted by the model (Hoare et al. used unpublished *KAO* spectrophotometry by Moseley, Glaccum & Silverberg). This 30$\mu$m feature had previously been observed in the spectra of carbon stars and IC 418 (Forrest et al. 1981; Moseley & Silverberg 1986) and an identification with solid MgS has been discussed by Goebel & Moseley (1985). However, in the case of IC 418 about 25% of the total IR luminosity is emitted in the 30$\mu$m band (the feature is similarly prominent in NGC 7027; Fig. 1). It would seem surprising if material composed of relatively low abundance elements such as Mg and S could emit such a large fraction of the total flux. Further, Middlemass (1988) found magnesium to be undepleted in the nebular gas phase in IC 418. Given the high abundance of carbon in all the objects so far found to emit the 30$\mu$m feature, a carbonaceous origin for the feature would seem more likely. There is also an excess over the model predictions in the 2–5$\mu$m region of the spectrum of NGC 7027 (Figure 1). This phenomenon has been noted in other PN too and temperature spiking in small grains has been suggested as the origin for this short-wavelength excess.

TABLE I
Dust-to-Gas Mass Ratios in the ionized zones of C-rich PN

| Name | C/O | $n_e$ (cm$^{-3}$) | $M_{dust}/M_{gas}$ | Reference |
|---|---|---|---|---|
| NGC 7027 | 2.3 | 50000 | $7 \times 10^{-4}$ | Hoare et al. (1992) |
| BD+30°3639 | 2.0 | 15600 | $7 \times 10^{-4}$ | Hoare et al. (1992) |
| IC 418 | 1.3 | 9000 | $6 \times 10^{-4}$ | Hoare (1990) |
| NGC 3918 | 1.6 | 6000 | $3 \times 10^{-4}$ | Harrington et al. (1990) |
| NGC 7662 | 1.7 | 2500 | $3 \times 10^{-4}$ | Hoare (1990) |

Table I collects together all the dust-to-gas mass ratios that have been derived to date for the ionized zones of carbon-rich PN. They were all derived from detailed photoionization plus dust modelling, using amorphous carbon grains with MRN size distributions. The mean nebular electron densities for the five PN span a range of a factor of twenty (50000 cm$^{-3}$ down to 2500 cm$^{-3}$, yet the derived dust-to-gas mass ratios decrease by no more than a factor of two over the same range. There does not therefore seem to be evidence for significant destruction of dust after incorporation into the ionized region of PN. The highest PN dust-to-gas mass ratios in Table I are about a factor of ten lower than the ratio of $7 \times 10-3$ that is believed to prevail in the general interstellar medium (MRN).

## 2. IRAS 18333–2357, a Remarkable Planetary Nebula in M22

The infrared source IRAS 18333–2357 near the centre of the globular cluster M22 (D=3 kpc) was identified as a planetary nebula by Gillett et al. (1989). The nebula is remarkable in (a) being bow-shaped, with an off-centre exciting star (V=14.3), apparently as a result of the cluster's motion through the interstellar medium; and (b) in showing no detectable hydrogen or helium emission lines, its spectrum being dominated by [O III] and [Ne III] lines. In this latter respect it shows similarities to the inner nebulosities of A30 and A78. From ionized gas mass estimates based on the forbidden lines, and from dust mass estimates based on silicate or amorphous carbon optical constants, Gillett et al. derived a dust-to-gas mass ratio of approximately unity for the nebula, implying equal quantities of heavy elements in the gas and in the grains.

Borkowski & Harrington (1991) have carried out detailed modelling of the gas and dust emission from IRAS 18333–2357. The 50000 K central star was found to have a luminosity of 3680 L$_\odot$, with 600 L$_\odot$ being absorbed and re-emitted in the infrared by the dust. Because of the severe deficiency of hydrogen and helium in the nebula, they found that they were unable to account for the observed optical forbidden line strengths based on the normal nebular heating mechanism of photoelectron ejection from H and He during photoionization. Instead they were able to show that photoelectron ejection from the high relative abundance dust grains can heat the nebular gas to temperatures of $\sim 10^4$ K and account for the observed optical forbidden lines. The C/O ratio in the nebula is not well-constrained and Borkowski & Harrington opted for amorphous carbon grains on the grounds that a low metallicity giant star could not produce a high abundance of silicate grains, whereas a high mass in carbon grains would be feasible following helium-burning and the 3rd dredge-up. Graphite and vitreous carbon have well-characterized photoelectron yield curves and Borkowski and Harrington obtained a good fit to the nebular observations using these yields. This is of significance for interstellar medium studies as a whole because photoelectron ejection from grains by UV photons (first proposed by Watson 1972) is believed to be the dominant heating mechanism for diffuse interstellar clouds – IRAS 18333–2357 has provided direct proof that this heating mechanism operates as hypothesized.

## 3. The Spatial Distribution of Dust

As reviewed by Roche (1989), near-IR array observations have for some time made it possible to image the distribution of emission in the $3.3\mu m$ UIR ('unidentified infrared') band in PN and other objects. The advent of mid-infrared (10 and $20\mu m$) arrays has made it possible to directly probe the thermal emission from the main grain components, by imaging in bandpasses centred on the continuum and on the main dust features (e.g. the silicate, silicon carbide and UIR bands). These maps can be compared with the distribution of ionized gas mapped using mid-IR lines such as [Ne II] or [S IV], or with radio maps of the nebular free-free emission. The mid-infrared continuum emission originates from the warmer dust immersed in the ionized regions – imaging at far-IR wavelengths would be necessary to reveal the cooler dust continuum associated with neutral material surrounding a PN. In the case of planetary nebula precursors (post-AGB objects) which are not yet ionized, infrared imaging can provide the sole means of determining the size and morphology of the extended material.

Hora et al. (1990) have presented high resolution (0.78 arcsec pixel$^{-1}$ 16×16 pixel images of NGC 6572 and BD+30°3639 in five mid-IR passbands. NGC 6572 is known to emit a strong 10.5-12.5$\mu m$ SiC emission feature. Hora et al. found that their 11.2$\mu m$ image (which would include 11.25$\mu m$ UIR band emission, if present) showed a stronger and more extended eastern lobe than did their 12.36$\mu m$ image (which should probe just the SiC emission). They concluded that 11.25$\mu m$ UIR band emission was responsible for the difference between the two images. Mid-infrared imaging capabilities are developing very rapidly, as evidenced by the number of posters presenting new results at this meeting. Mid-IR images of planetary nebulae are presented by Hora et al. (Poster III-75), Meixner et al. (Poster III-76) and Deutsch et al. (Poster III-125), while mid-IR imaging data on post-AGB objects and on AGB stars are presented by Meixner et al. (Poster III-124) and by Skinner et al. (Poster III-102), respectively.

It is also possible to map dust spatial distributions at optical wavelengths. Only a very few PN have a high enough circumnebular dust optical depth for reflection nebulae to be detected or extinction variations to be mapped (e.g. NGC 7027: Middlemass, Clegg & Walsh 1989; NGC 6302: Ashley 1990) but a number of other optical methods are available for mapping dust distributions in PN. Walsh & Clegg (Poster III-98) have measured polarized [O III] and H$\alpha$ line profiles from six PN at high spectral resolution. At the nebular centres they found the polarization to be very low or undetectable at the line peaks but to increase to $\sim$5% in the line wings, with stronger polarization in the red wing. The light scattered from the neutral halo around NGC 7027 was found to be polarized by up to 35%. Walsh and Clegg interpreted their results in terms of single-scattering 'moving mirrors' dust models. The emission seen at line centre is predominantly unscattered and so the polarization is low. Since the atoms and dust grains in an expanding nebula are red-shifted with respect to each other, photons will be predominantly scattered in the red wings of lines, where the intrinsic emission is lower, so the degree of polarization will be higher.

Furton & Witt (1990) have detected and mapped the spatial distribution of extended red emission (ERE) around NGC 7027, using long-slit spectrophotometry. This emission, which was first discovered around the C-rich post-AGB object HD 44179 (the Red Rectangle), manifests itself as a very broad luminescence band extending from 6000Å to longer than 8000Å. The ERE from the Red Rectangle shows a highly structured spectrum (Schmidt, Cohen & Margon 1980) and the objects in whose spectrum it has been detected all exhibit the 3.3$\mu$m and longer wavelength UIR bands. The ERE is therefore believed to be arise from hydrocarbons, with hydrogenated amorphous carbon (HAC) being one candidate (Duley 1985). Furton & Witt (1990) carefully allowed for nebular continuum emission processes in deriving the spectral shape and strength of the ERE in NGC 7027. They found that the strength of the ERE peaked just beyond the ionized boundary of NGC 7027 and was perhaps more extended than the 3.3$\mu$m UIR band emission. Furton & Witt (1992) have observed twenty more PN in the same manner, positively identifying ERE in the spectra of seven of them, all of which had C/O ratios in excess of unity. However, they noted several PN in whose spectra UIR band emission had been detected but for which they only obtained ERE upper limits (J900, NGC 3242 and NGC 6543).

## 4. Infrared Spectroscopy of Planetary Nebulae and Post-AGB Objects

One of the most important developments since the last meeting has been the identification of a significant number of sources from the *IRAS* database with post-AGB objects (e.g. Manchado et al. 1989, van der Veen, Habing & Geballe 1989, Hrivnak, Kwok & Volk 1989). These are objects whose optical counterparts typically have F or G-type supergiant spectra and are believed to be transiting from the cool AGB phase to the hot planetary nebula phase, the timescale for which should be at most a few thousand years. Since they are not yet hot enough to ionize the material around them, they were only recognized as potential post-AGB objects by virtue of the large infrared dust excesses detected by the *IRAS* survey. These objects are of great interest since the dust around them is situated in a completely different physical environment from that encountered in either the AGB or PN phases and so spectroscopy can potentially reveal new aspects of the dust properties. Unlike AGB stars, which have a fairly continuous energy distribution resulting from the attenuation of their photospheric distributions by hot dust which re-emits in the near-IR and longer wavelengths, these post-AGB objects are characterized by 'double-peaked' energy distributions (Hrivnak et al. 1989). This is because the dust created during the high mass loss rate AGB phase has moved out and cooled to fairly low temperatures (100–200 K) whose emission peaks at 10–20$\mu$m, leaving a dip at near-IR wavelengths because of the relative lack of hotter dust particles. It is convenient to discuss separately the carbon-rich (C/O>1) and oxygen-rich (C/O<1) objects because of the completely different infrared spectra that they present.

### 4.1. CARBON-RICH OBJECTS

Roche (1989) has presented the statistics for PN of the incidence of the various mid-IR dust features (silicates, silicon carbide, UIR bands) versus the nebular C/O ratio, which showed that silicates are associated with nebulae having C/O<1, whereas SiC emission occurs in nebulae with C/O$\geq$1. Strong UIR band emission is in general associated with PN having C/O>2 (Barlow 1983). Cohen et al. (1989) have shown that there is a very good correlation between the nebular C/O ratio and the fraction of the total infrared flux appearing in the 7.7$\mu$m band (a feature attributed to C-C stretches in aromatic hydrocarbons; Allamandola et al. 1989).

Kwok, Volk & Hrivnak (1989) discovered a strong broad 21$\mu$m emission band in the *IRAS* LRS spectra of four carbon-rich post-AGB stars, the first features to be found in the 20$\mu$m spectral region of objects exhibiting the mid-infrared UIR bands, and emitted by the coolest objects yet found with UIR bands (spectral types F3I to G5I). Buss et al. (1990) showed that the UV photon fluxes from these F-G stars are insufficient to pump the observed infrared bands, implying that they must be excited by optical photons. Buss et al. obtained 5–13$\mu$m spectra of two of Kwok et al.'s sources and found 6.2$\mu$m, 6.9$\mu$m and 8.0$\mu$m emission features, but with the 6.9$\mu$m feature much stronger relative to the 6.2$\mu$m feature than in hot UV-emitting objects such as NGC 7027. In addition, the strong 7.7$\mu$m feature attributed to aromatic C-C stretching was shifted longwards to 8.0$\mu$m. While both the 6.2$\mu$m and 7.7$\mu$m features are normally attributed to aromatic C-C stretch modes, the 6.9$\mu$m feature was attributed by Buss et al. to CH deformation modes in $CH_2$ and $CH_3$ groups cross-linking aromatic domains. They also found evidence for strong underlying plateau emission in the 6–9$\mu$m and 10–13$\mu$m regions and associated these with clusters of PAH molecules containing up to 500 carbon atoms. They suggested that such clusters could more easily survive in the less harsh radiation environment around F and G type stars compared to PN. As well as the plateau emission, Buss et al. proposed that these clusters could produce the 21$\mu$m emission feature by means of aromatic out-of-plane ring bending modes. However, Allamandola et al. (1989) predicted such modes to be significantly weaker than the modes responsible for the shorter wavelength features, whereas the reverse appears to be the case for the 21µm band sources. Sourisseau, Coddens & Papoular (1992) have suggested that the 21µm band is due to N-C-N and O-C-N deformation modes of amides such as urea.

Justtanont, Barlow & Skinner (Poster IV-122) have obtained 10 and 20µm spectra at a resolving power of R=60–80 for three of the Kwok et al. 21µm band sources. They resolve up to six separate peaks in the 7.5–13.3µm region. The 11.2–11.3µm feature normally attributed to out-of-plane C-H bending modes is shifted to 11.55µm and the strong 12µm 'plateau' seen in the lower resolution spectrum of Buss et al. is found to contain three other peaks, at 10.7µm, 12.2µm and 12.7µm, the former two not having been clearly observed anywhere else before. The overall 10.5-13.5µm plateau observed in UIR band sources has been attributed by Allamandola et al. to bending modes of (with increasing wavelength) nonadjacent, doubly-adjacent and triply-adjacent peripheral H atoms on the surfaces of polycyclic aromatic hydrocarbon (PAH) clusters. Justtanont, Barlow & Skinner found the 21µm band in the spectra of IRAS 04296+3429 and IRAS 22272+5435 to be be resolved into regularly spaced peaks at 20.45µm, 21.05µm, 21.7µm and 22.4µm, with an energy spacing of about 14 cm$^{-1}$ between successive peaks. Omont (1993) has presented *KAO* spectra which show that two of the 21µm band sources have even stronger 30µm emission bands.

Roche, Aitken & Smith (1991) published 7.7–13.2µm spectra at R=110 of a number of PN and post-AGB objects that exhibit UIR-band spectra. They found the asymmetric profile of the 11.25µm band to vary significantly from object to object, with a few objects exhibiting a subsidiary peak at 11.05µm. In their R=400 spectra of three PN and one post-AGB object, Witteborn et al. (1989) found the '11.3µm' feature to peak at 11.22µm. Tokunaga et al. (1988) presented R=1500 spectra of NGC 7027 and the Red Rectangle, which showed the Red Rectangle 3.29µm band to be significantly the narrower and to peak at a slightly longer wavelength. Geballe et al. (1989) found the Red Rectangle to have a relatively weak and broad subsidiary 3.4µm feature at the stellar position but that this feature increased in prominence at 5"N from the central star, where it showed a 3.4–3.6µm band structure similar to that observed in UIR band sources such as NGC 7027 (e.g. Nagata et al. 1988). The 3.29µm band is normally attributed to aromatic C-H v=1–0 stretch vibrations. However, the subsidiary bands in the 3.4–3.6µm region have been alternatively identified with (a) C-H vibrations of aliphatic sidegroups attached to PAH structures (de Muizon et al. 1986); or (b) to higher vibrational aromatic C-H stretch transitions (e.g. v=2–1, v=3–2 'hot bands') which are shifted from the v=1–0 transition by anharmonic effects (Allamandola et al. 1989). The discovery that the 3.4–3.6µm 'sidebands' are *stronger* than the 3.29µm band in the spectrum of the G-supergiant post-AGB source IRAS 05341+0852 (Geballe & van der Veen 1990), as well as in the spectra of two of the G-supergiant post-AGB objects which have the 21µm band (Geballe et al. 1992), rules out 'hot bands' as the origin of the side-bands in these sources. For these sources at least, interpretation (a) of the 3.4–3.6μm bands as due to aliphatic sidegroups attached to aromatic clusters appears to be the most viable hypothesis at present. This also accords with the interpretation above that the strong 6.9μm feature and the 10.5–13.5μm plateau in the spectra of the 21μm band sources indicate relatively large clusters with attached sidegroups. The conditions around the cooler post-AGB objects seem conducive to the survival of these structures, which are probably rapidly destroyed in the harsh environment encountered on entering the PN phase. Magazzu & Strazzulla (Poster III-92) have shown that the 3.4μm feature in the spectrum of the planetary nebula Hb 5 is also unlikely to be a hot band, since although F(3.4)/F(3.3) is quite high, there is no v=2–0 feature at 1.67μm corresponding to the sum of the energies of the 3.3μm and 3.4μm features. The 3.4μm emission band therefore appears to be due to aliphatic C-H stretch vibrations. Lequeux & Jourdain de Muizon (1990) found a broad 3.4μm *absorption* band (with $\tau=0.09$) in the spectrum of the C-rich post-AGB object GL 618 and attributed it to aliphatic C-H stretch vibrations of the kind invoked to explain similar absorption features arising in interstellar sightlines, such as that to the Galactic Centre.

4.2. OXYGEN-RICH OBJECTS

Hrivnak et al. (1989) have shown that the thermal infrared portion of the double-peaked energy distributions of many oxygen-rich post-AGB objects can be adequately fitted by cool classical silicate emission. However, the energy distributions of a few O-rich post-AGB objects are not so easily fitted. Justtanont et al. (1992) have presented 10 and 20μm spectra of HD 161796 (F3Ib) and HD 179821=GL 2343 (G5Ia). There is a broad emission feature between 10 and 12μm in the spectra of both objects, which can be fitted very well by radiation-disordered olivine. The mid-IR energy distribution of both objects shows a 13–16μm rise that is too steep to be fitted by any combination of cool continuum plus silicate and Justtanont et al. (1992) argued that the rise could be explained by iron oxides of the type invoked by Cox (1990) to explain a similar steep rise seen in the *IRAS* LRS spectra of a number of compact H II regions.

Forveille et al. (1987) noticed that the source IRAS 09371+1212 has unique IRAS colours, showing a huge peak in the 60μm band. They found the source to be associated with CO emission, and with an optical reflection nebula showing an M4 spectrum. They suggested that it was a post-AGB object with the peak in the 60μm band being due to emission in the 40–70μm bands of water ice and christened it the 'Frosty Leo Nebula'. Their suggestion of the presence of ice was soon confirmed by Rouan et al. (1988), who found a deep ice absorption band at 3.1μm in its spectrum. Omont et al. (1990) obtained a 35–70μm *KAO* spectrum of IRAS 09371+1212, which confirmed that it had prominent ice emission features at 44μm and 62μm. The presence of the 62μm feature implied that the ice had to be crystalline rather than amorphous and their modelling yielded a best-fit grain temperature of 47 K. The 44μm ice band was also detected in emission in the spectra of two other cool O-rich post-AGB objects, OH 127.8+0.0 and OH 231.8+4.2.

There are a number of PN with spectral characteristics which indicate the simultaneous presence of C-rich *and* O-rich material. IRAS 07027-7934 was identified by Menzies & Wolstencroft (1990) with a young planetary nebula having a cool carbon-rich WC central star and they noted that its LRS spectrum showed the UIR bands. While this is indicative of a C-rich nebula, Zijlstra et al. (1991) found that it also had a strong OH maser, normally indicative of O-rich material. The Type I bipolar planetary nebula NGC 6302 has a spectrum which shows the 8.7$\mu$m and 11.3$\mu$m UIR bands (Roche & Aitken 1986) but additionally has a weak OH maser (Payne, Phillips & Terzian 1988). NGC 6302 also has a prominent 18-19$\mu$m silicate emission feature (Barlow, Skinner & Justtanont, unpublished).

A possible interpretation of the presence of materials indicative of both O-rich and C-rich environments is as follows. The C+N abundance derived for NGC 6302 by Aller et al. (1981) is the same as the mean for the non-Type I ('standard') planetary nebulae in their Table 4 and its O abundance is also the same as the mean value found for standard PN. In standard PN, carbon is greatly enhanced by the third dredge-up, which brings up primary carbon from the helium-burning shell, while in NGC 6302 it is nitrogen which is greatly enhanced. Since the first and second dredge-ups merely recycle CNO via the CN and CNO cycles (Iben & Renzini 1983), it is clear that the third dredge-up must have occurred in NGC 6302 too, followed by envelope-burning (a process whereby dredged-up carbon is converted to nitrogen by the CN cycle at the bottom of the hydrogen envelope of sufficiently massive AGB stars; Iben & Renzini 1983). Carbon dredge-ups, which can take the C/O ratio to above unity, are nearly instantaneous but it takes more time for envelope burning to convert sufficient C to N to drop the surface C/O ratio to below unity. Therefore between successive carbon dredge-ups there can be a C-rich envelope phase followed by an O-rich envelope phase, with carbon-rich grains and oxygen-rich grains being alternately synthesised in the AGB star's outflow. Both types of grains can then be observed later in the object's evolution.

## References

Allamandola, L.J., Tielens, A.G.G.M. and Barker, J.R.: 1989, *ApJS*, **71**, 733
Aller, L.H., Ross, J.E., O'Mara, B.J. and Keyes, C.D.: 1981, *MNRAS*, **197**, 95
Ashley, M.C.B., 1990, *Proc. Astr. Soc. Austral.*, **8**, 360
Barlow, M.J.: 1983, in Proc. IAU Symp. No. 103, *Planetary Nebulae*, ed. D.R. Flower, Reidel, 105
Borkowski, K.J. and Harrington, J.P.: 1991, *ApJ*, **379**, 168
Buss, R.H. et al.: 1990, *ApJ*, L23
Cohen M., et al.: 1989, *ApJ*, **341**, 246
Cox, P.: 1990, *A&A*, **236**, L29
de Muizon, M., Geballe, T.R., d'Hendecourt, L.B. and Baas, F.: 1986, *ApJ*, **306**, L105
Duley, W.W.: 1985, *MNRAS*, **215**, 259
Forveille, T., Morris, M., Omont, A. and Likkel, L.: 1987, *A&A*, **176**, L13
Forrest, W.J., Houck, J.R. and McCarthy, J.F.: 1981, *ApJ*, **248**, 195
Furton, D.G. and Witt, A.N.: 1990, *ApJ*, **364**, L45
Furton, D.G. and Witt, A.N.: 1992, *ApJ*, **386**, 587
Geballe, T.R., Tielens, A.G.G.M., Allamandola, L.J., Moorhouse, A. and Brand, P.W.J.L.: 1989, *ApJ*, **341**, 278
Geballe, T.R., Tielens, A.G.G.M., Kwok, S. and Hrivnak, B.J.: 1992, *ApJ*, **387**, L89

Geballe, T.R. and van der Veen, W.E.C.J.: 1990, *A&A*, **235**, L9
Gillett, F.C., et al.: 1989, *ApJ*, **338**, 862
Goebel, J.H. and Moseley, S.H.: 1985, *ApJ*, **290**, L35
Harrington, J.P., Monk, D.J. and Clegg, R.E.S.: 1988, *MNRAS*, **231**, 577
Hoare, M.G.: 1990, *MNRAS*, **244**, 193, 1990
Hoare, M.G., Roche, P.F. and Clegg, R.E.S.: 1992, *MNRAS*, **258**, 257
Hora, J.L., Deutsch, L.K., Hoffman, W.F. and Fazio, G.G.: 1990, *ApJ*, **353**, 549
Hrivnak, B.J., Kwok, S. and Volk, K.M.: 1989, *ApJ*, **346**, 265
Iben, I. and Renzini, A.: 1983, *Ann. Rev. Astr. Astrophys.*, **21**, 271
Justtanont, K., Barlow, M.J., Skinner, C.J. and Tielens, A.G.G.M.: 1992, *ApJ*, **392**, L75
Kwok, S., Volk, K.M. and Hrivnak, B.J.: 1989, *ApJ*, **345**, L51
Lequeux, J. and Jourdain de Muizon, M.: 1990, *A&A*, **240**, L19
Manchado, A., Pottasch, S.R., Garcia-Lario, P., Esteban, C. and Mampaso, A.: 1989, *A&A*, **214**, 139
Mathis, J.S., Rumpl, W. and Nordsieck, K.H.: 1977, *ApJ*, **217**, 425 (MRN)
Menzies, J.W. and Wolstencroft, R.D.: 1990, *MNRAS*, **247**, 177
Middlemass, D.: 1988, *MNRAS*, **231**, 1025
Middlemass, D.: 1990, *MNRAS*, **244**, 294
Middlemass, D., Clegg, R.E.S. and Walsh, J.R.: 1989, *MNRAS*, **239**, 5P
Moseley, S.H. and Silverberg, R.F.: 1986, in *Interrelationships among Circumstellar, Interstellar and Interplanetary Dust*, ed. J.A. Nuth and R.E. Stencel, NASA CP-2403, A18
Nagata, T., et al.: 1988, *ApJ*, **326**, 157
Omont, A.: 1993, in *Astronomical Infrared Spectroscopy: Future Observational Directions*, ed. S. Kwok, ASP Conf. Ser., in press
Omont, A., et al.: 1990, *ApJ*, **355**, L27
Payne, H.E., Phillips, J.A. and Terzian, Y.: 1988, *ApJ*, **326**, 368
Roche, P.F. 1989, in Proc. IAU Symp. No. 131, *Planetary Nebulae*, ed. S. Torres-Peimbert, Kluwer Academic Publishers, 117
Roche, P.F. and Aitken, D.K.: 1986, *MNRAS*, **221**, 63
Roche, P.F. Aitken, D.K. and Smith C.H.: 1991, *MNRAS*, **252**, 282
Rouan, D., Omont, A., Lacombe, F. and Forveille, T.: 1988, *A&A*, **189**, L6
Schmidt, G.D., Cohen, M. and Margon, B.: 1980, *ApJ*, **239**, L133
Sourisseau, C., Coddens, G. and Papoular, R.: 1992, *A&A*, **254**, L1
Tokunaga, A.T., et al.: 1988, *ApJ*, **328**, 709
van der Veen, W.E.C.J., Habing, H.J. and Geballe, T.R.: 1989, *A&A*, **226**, 108
Watson, W.D.: 1972, *ApJ*, **176**, 103
Witteborn, F.C. et al.: 1989, *ApJ*, **341**, 270
Zijlstra, A.A., Gaylard, M.J., te Lintel Hekkert, P., Menzies, J., Nyman, L.-A. and Schwarz, H.E.: 1991, *A&A*, **243**, L9

# ON THE DISTANCES TO GALACTIC PLANETARY NEBULAE

C.Y. ZHANG
*Department of Astronomy, The University of Texas at Austin, Austin, TX 78712*

ABSTRACT. The distances are determined for 142 Galactic planetary nebulae, using two methods. Method A is based on the stellar mass and surface gravity, while Method B makes use of the stellar luminosity. These properties of the central star are inferred from the modeling of the distance-independent parameters (Zhang and Kwok 1992). The results from the two methods used in this paper are consistent with each other. A distance of 10.8 kpc is found for K648 using Method A of this work. This is in excellent agreement with the distance of 10 kpc of its hosting globular cluster M15. The distances obtained in this work are in good agreement with the distances of Mendez et al. (1992), based on the recently developed non-LTE model atmospheres including spherical extension and stellar winds. The agreement of our distances with the individually determined distances of Gathier et al. (1986a, 1986b), using the extinction-distance method and the HI 21 cm absorption method, is worse than that found between ours and Mendez et al.'s. The distances obtained by Gathier et al. are likely often to be underestimated. The various statistical distance scales, except for Cudworth's (1974) distance scale, show no agreement with the distances determined in this work.

CYZ was supported by the NASA grant NAG 2-67.

## References

Cudworth, K.M. (1974) AJ 79, 1384
Gathier, R., Pottasch, S.R., and Pel, J.W. (1986a) A&A 157, 171
Gathier, R., Pottasch, S.R., and Goss W.M. (1986b) A&A 157, 191
Mendez, R.H., Kudritzki, R.P., Herrero, A. (1992) A&A in press
Zhang, C.Y. and Kwok, S. (1992) A&A, submitted

# A COMPARISON OF NEBULAR DISTANCE SCALES

M. SAMLAND and J. KÖPPEN
*Institut für Theoretische Physik und Sternwarte, D-2300 Kiel, Germany*

A. ACKER
*Observatoire de Strasbourg, 11, rue de l'Université, 67000 Strasbourg, France*

and

B. STENHOLM
*Lund Observatory, Box 43, S-221 00 Lund, Sweden*

Determination of the positions of central stars of planetary nebulae in the HR-diagram requires the knowledge of nebular distances. For almost all nebulae, these can only be given in terms of statistical scales. These scales have in common that they assume all nebulae to have the same structure (e.g. constant density) and that a unique ionized mass-radius relation exists. If the mass-radius relation is given by $M_{ion} = M_0 \cdot (R/R_0)^\eta$, the distance $d(\mathrm{pc})$ of planetary nebulae can be expressed as a function the de-reddened H$\beta$-flux (erg cm$^{-2}$s$^{-1}$) and the angular radius $\Theta$(arcsec):

$$d = \left(\frac{5.5891 \cdot 10^{-6}}{\epsilon\, F_0(H\beta)}\right)^{\frac{1}{5-2\eta}} \left(\frac{M_0}{R_0^\eta}\right)^{\frac{2}{5-2\eta}} \left(\frac{\Theta}{206265}\right)^{\frac{2\eta-3}{5-2\eta}}$$

$M_0$ and $R_0$ are in solar masses and pc ($T_e = 10000$ K, He/H $= 0.1$). The parameter $\eta$ characterizes the distance scale: e.g. Shklovsky (1956) $\eta = 0$, Maciel & Pottasch (1980) $\eta = 1$, Pottasch (1984) $\eta = 3/2$, Daub (1982) $\eta = 5/3$, and Kwok (1985) $\eta = 9/4$.

For 284 planetary nebulae we take de-reddened H$\beta$-fluxes from the ESO-Strasbourg spectrophotometric survey (Acker et al. 1989) and the angular diameters from Schneider et al. (1983) and compute the distances with the different methods. Using Ambartzumyan temperatures, we determine the positions of the central stars in the HR-diagram. We find that, for $\eta > 2$, the distances show a very large scatter, but for all distance scales with $\eta < 2$, the positions agree well with theoretical evolutionary tracks for post-AGB-stars (Schönberner 1983). We note that with $\eta = 3/2$ the distance of a nebula is only a function of the H$\beta$-flux which depends on the temperature of the central star. Therefore all central stars are arranged on a crescent-shaped line in the HR-diagram.

## References

Acker A., Köppen K., Stenholm B., Jasniewicz. 1989 A&AS 80, 201
Daub C. T. 1982, ApJ 260, 612
Kwok S. 1985, ApJ 290, 568
Maciel W. J., Pottasch S. R. 1980, A&A 88, 1
Pottasch S. R. 1984, *Planetary Nebulae*, Reidel, Dordrecht
Schneider S. E., Terzian Y., Purgathofer A., Perinotto M., 1983, ApJS 52, 399
Schönberner D.: 1983, ApJ 272, 708
Shklovsky I. S.: 1956, *Sov. Astron. J.* 33, 108

# TRIGONOMETRIC PARALLAXES OF PLANETARY NEBULAE

J. R. PIER, H. C. HARRIS, C. C. DAHN and D. G. MONET

*U.S. Naval Observatory Flagstaff Station, U.S.A.*

**Abstract.** Parallaxes are presented for nine Planetary Nebulae central stars.

Nine planetary nebulae fields are now routinely observed on the U.S. Naval Observatory CCD trigonometric parallax program (Monet *et al*, 1992). Due to the small parallaxes of the PN central stars, the corrections from relative to absolute parallax are crucial. To aid in this correction, spectroscopic and/or photometric parallaxes were obtained for all of the reference frame stars. Reasonable solutions are beginning to emerge for three of the central stars, while the remaining fields will require many more frames before a parallax solution can be given with any comfortable degree of certainty.

Results are presented in the Table below. The PN field identification is given in the first column, column 2 lists the number of CCD frames used in the solution, the third column gives the epoch range covered by the frames employed and the fourth column lists the absolute parallax and mean error in the parallax (in milliarcseconds) for the central star. The absolute V magnitude in the fifth column is based upon the absolute parallax and CCD photometry. The comment appearing in the last column is a qualitative assessment of the stability of the solution based upon the repeatability of the results as various parameters (removing certain frames and/or reference stars) were changed from solution to solution.

Trigonometric Parallaxes of Planetary Nebulae Central Stars

| PN | #Frames | Epoch | $\Pi \pm$ m.e. (mas) | $M_V$ | Comment |
|---|---|---|---|---|---|
| A 21 | 35 | 89.2–92.1 | 2.9 ± 0.6 | 8.3 | Fragile |
| A 24 | 12 | 88.0–92.1 | 4.0 ± 1.6 | — | Insufficient Data |
| A 29 | 18 | 89.2–92.3 | -1.0 ± 2.2 | — | Insufficient Data |
| A 31 | 18 | 88.1–92.3 | 4.9 ± 1.7 | 9.0 | Preliminary |
| A 74 | 48 | 88.5–92.5 | 1.2 ± 0.8 | 7.: | Preliminary |
| NGC 6720 | 142 | 88.4–92.5 | 2.0 ± 0.6 | 7.3 | Fragile |
| NGC 6853 | 116 | 88.5–92.5 | 2.8 ± 0.5 | 6.3 | Stable |
| PW 1 | 43 | 87.9–92.1 | 3.1 ± 0.7 | 8.0 | Stable |
| S 216 | 57 | 89.7–92.1 | 8.3 ± 1.4 | 7.3 | Stable |

## Reference

Monet, D.G., Dahn, C.C., Vrba, F.J., Harris, H.C., Pier, J.R., Luginbuhl, C.B. & Ables, H.D.: 1992, *Astron.J.* **103**, 638.

# INFRARED EXCESS (IRE) AS AN INDICATOR OF PN DISTANCE

GEORGE JACOBY

*Kitt Peak National Observatory, National Optical Astronomy Observatories,*
*P.O. Box 26732, Tucson, Arizona 85726, USA*

The VLA observations presented by Pottasch et al. (1988, A&A 205, 248) and Ratag et al. (1990, A&A 233, 181) demonstrate that $IRE$ (Pottasch et al. 1984, A&A 138, 10) correlates very well with $F(6\text{ cm})$ for the Galactic center PN, especially when $IRE$ is large The correlation (see figure) can be approximated by:

$$IRE = 0.0294 + \frac{64.78}{F(6\text{ cm})} + \frac{321}{F(6\text{ cm})^2} + \frac{195.96}{F(6\text{ cm})^3}$$

Since $IRE$ is a ratio of observed fluxes, it is distance independent, and we can predict $F(6\text{ cm})$, which is distance dependent, from the $IRE - F(6\text{ cm})$ relation calibrated at the GC. Comparing the predicted value for $F(6\text{ cm})$ with the observed value yields a distance to any target PN. The relation is tightest among PN having high $IRE$, and is most applicable to young, dense, dusty PN; it is not a good method to derive PN distances in general.

This method yields distances with accuracies of 15% for $60 < IRE < 1500$, 30% for $10 < IRE < 60$, and is unreliable for most PN which have $IRE < 10$. Distances for a selection of high $IRE$ objects agree adequately (factor of 2) with other techniques, except for the Shklovsky distances (assuming $M_n = 0.2$, and $\epsilon = 0.5$). Given that these objects are young, the generally large Shklovsky distances are consistent with an excessively large assumed value for the ionized mass for the stage of evolution. Inverting the Shklovsky procedure yields typical ionized masses of 0.02 to 0.10 $M_\odot$. The dashed arrow in the figure indicates the direction and magnitude a point will move due to a factor of 2 error in the IR data.

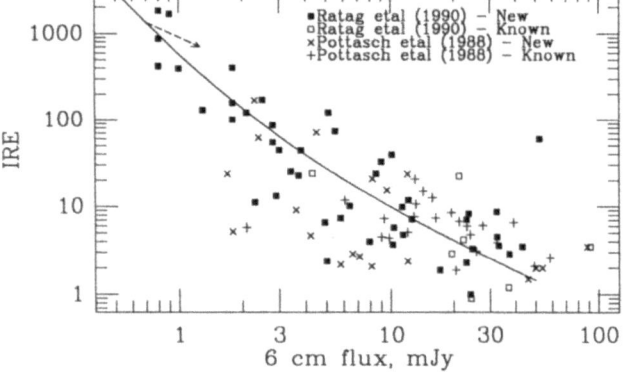

# INTERSTELLAR EXTINCTION OF PLANETARY NEBULAE

G. STASIŃSKA[1], R. TYLENDA[1,2,3], A. ACKER[2], B. STENHOLM[4]
[1] *DAEC, Observatoire de Paris-Meudon, France*
[2] *Observatoire de Strasbourg, France*
[3] *Copernicus Astronomical Center, Torun, Poland*
[4] *Lund Observatory, Sweden*

We have compared the extinction of planetary nebulae (PN) as derived from the Balmer decrement and from the ratio of radio to Hβ fluxes.

From a compilation of observational data we have selected PN for which the 6 cm (5 GHz) radio flux has been measured with the VLA and for which reliable measurements of the Hβ flux and the Hα/Hβ ratio are available. After a careful discussion we have rejected objects having a 6 cm flux below 10 mJy or a 6 cm brightness temperature below 10 K, as their radio flux measurements are presumably less accurate. In this way we have obtained a sample of ~130 PN for which we have derived the extinction constants both from the Hα/Hβ ratio ($C_{opt}$) and from the radio to Hβ flux ratio ($C_{rad}$). The results are compared in the figure below which shows that $C_{opt}$ is systematically larger than $C_{rad}$. The orthogonal regression line (dashed in the figure) is:
$C_{opt} = (1.19\pm0.03)C_{rad} + (-0.01\pm0.03)$.

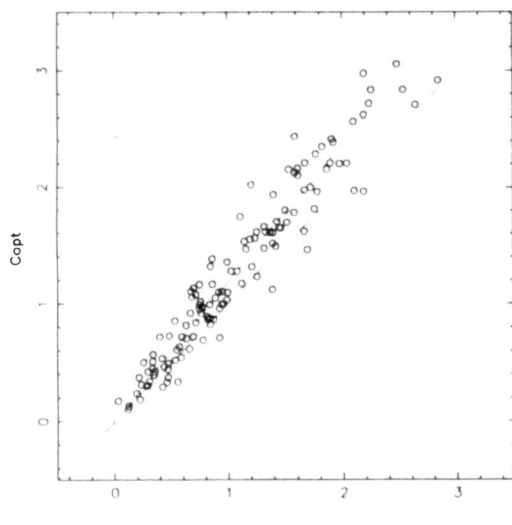

This result is opposite to the well known effect observed in extragalactic HII regions which is generally explained by the presence of large amounts of local dust (inside and/or surrounding the HII region). We argue that the effect we have found for the PN cannot be entirely attributed to systematic observational errors although the problem of nonlinearity of some optical detectors needs to be clarified. Circumnebular dust, although undoubtfully present in some - especially young - PN, cannot be responsible for the bulk of the extinction observed for most of the PN. Instead, we suggest that, for most of distant PN, the total to selective extinction, $R_V$, is significantly lower than 3, the value usually adopted for the standard interstellar extinction law.

A detailed report on the subject is in press in Astronomy and Astrophysics.

# INTERSTELLAR REDDENING TOWARDS S188, HW4 AND We1-6

## W. SAURER[*,**]

*Institut für Astronomie, Leopold-Franzens-Universität, Technikerstr. 25, A-6020 Innsbruck, Austria*

**Abstract.** Photoelectric UBV photometry of stars was carried out in the angular vicinity (radius = 15') of 3 planetary nebulae. An extinction distance relation was constructed for the line of sight towards each planetary nebula. This relation was verified by published data on nearby stars and star clusters. The distances derived are 850 ± 300 pc for PK 128 −04 1 (S 188, Simeis 22), 400 ± 200 pc for PK 149 −09 1 (HW 4), and (less reliable) 800 pc ± 400 pc for PK 224 +01 1 (We 1-6). For the central star of PK 224 +01 1 our observations gave V = 15.76 ± 0.03 mag, (B−V) = −0.08 ± 0.03 mag and (U−B) = −0.87 ± 0.03 mag.

**Key words:** planetary nebulae: individual, PK 128 −04 1, PK 149 −09 1, PK 224 +01 1

The determination of individual distances of planetary nebulae (PN) remains still a serious problem. As a consequence, the number of distances known with uncertainties smaller than 40% is limited to a few dozen.

One method to derive reliable distances to individual PN is the "extinction distance method" or "reddening distance method".

To demonstrate the influence of the patchy interstellar obscuration, we have carried out star counts on the Palomar Observatory Sky Prints (POSS) in several fields (area = 5.4 arcmin$^2$), symmetrically arranged within 1° around the PNs. The result is that the patchiness of the interstellar medium does not play a predominant role for the regions around PK 128 −04 1 and PK 149 −09 1, whereas this might not be true for PK 224 +01 1.

To obtain an accurate extinction distance relation a two-dimensional classification of all measured stars is required. As a matter of fact, this is not uniquely possible by applying three-colour photometry only. We have tried to overcome our ignorance of the luminosity class (LC) and spectral type in the following way. From the two colour diagram we derived E(B−V) and the corresponding value of the distance for the LC IAB, III and V using the relation E(U−B)/E(B−V) = 0.72 + 0.05 E(B−V) and the intrinsic colours and visual absolute magnitudes given by Schmidt-Kaler (1982, Landolt-Börnstein). The aim of this method is to find a most probable functional relationship between reddening and distance. The extinction-distance relations of two PN studied are shown below.

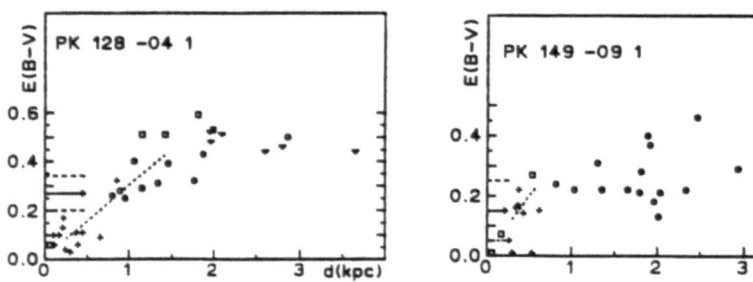

[*] Visiting Astronomer at the Centro Astronomico Hispano-Aleman, Calar Alto, operated by the Max-Planck-Inst. für Astr., Heidelberg, jointly with the Spanish Nat. Comm. for Astronomy.
[**] This work was supported financially by the local government (Tiroler Landesregierung).

# THE SHKLOVSKY PARADOX

DAVID BUCKLEY

Department of Physics, East Stroudsburg University of Pennsylvania

Department of Physics & Astronomy, Univ. of Massachusetts, Amherst

and

STEPHEN E. SCHNEIDER and DAVID VAN BLERKOM

Department of Physics & Astronomy, Univ. of Massachusetts, Amherst

ABSTRACT. Shklovsky estimated distances to planetary nebulae (PNs) based on an assumed constant ionized mass and the relationship between flux and radius under the assumption of a constant density, fully ionized shell. He found that a mass of $\sim 0.2$ $M_\odot$ yielded the best results. Estimates of the ionized masses of PNs with independently determined distances also rarely exceed a few tenths of a solar mass. This is surprising since many PNs are thought to derive from high mass progenitors (up to 8 $M_\odot$). Recent optical work (Plait and Soker, 1990) and our own computer simulations show that this simple mass estimation method may severely underestimate the total ionized mass. This is because of a halo of lower density ionized material which often contributes only a small fraction of the PN luminosity even though it may contain many times the mass of the dense inner shell. The precipitous drop in surface brightness (both optical and radio) beyond the inner part of the ionized shell also lead to underestimates of the PN's actual ionized radius. Since the evolution of PNs is driven by the expansion of the nebular shell coupled with the evolution of the nucleus (PNN), we ran several simulations using a simple momentum conserving two-wind model as well as employing density profiles derived by more sophisticated energy conserving models, with a wide range of wind parameters and using two different models of PNN evolution. From our simulations (assuming a 4 $M_\odot$ progenitor) we derive an apparent "Shklovsky Mass" – defined as the ionized mass that would be derived from the observationally determined fluxes and radii of our model PNs. While the total ionized masses and Strömgren radii of the model PNs varied widely depending on the PNN and wind parameters, the derived Shklovsky mass consistently remained below one solar mass. This result is almost independent of the total ionized mass or the mass of the progenitor envelope and is fairly insensitive to the wind parameters chosen as input to the models. The observed spread of masses (based on an error analysis of the work of Gathier, 1987) is similar to the mass dispersion in our models for PNs of moderate age. This may explain why the Shklovsky distance method has been found to agree well with kinematic distances (Schneider and Terzian, 1983) even though the fundamental assumptions may be inappropriate for the nebula as a whole.

## References

Gathier, R. (1987) Astr. Astrophys. Suppl., **71**, 245.
Plait, P. and Soker, N. (1990) Astron. J., **99**, 1883.
Schneider, S. and Terzian, Y. (1983) Astrophys. J., **274**, L61.

# REDDENING DISTANCES FOR PLANETARY NEBULAE FROM BROAD BAND BVIc CCD IMAGING

D.L. POLLACCO and G. RAMSAY

*Department of Physics and Astronomy, North Haugh , St. Andrews, Fife KY16 9SS, Scotland*

If the zero age main sequence is expressed in the (V–I) versus (B–V)–(V–I) plane the reddening lines are found to lie at a great enough angle to allow reasonably accurate spectral type classification for stars later than $\sim$ F5. Earlier spectral types can also be identified but with lower accuracy. Comparison with the Q method of UBV photometry and with spectra of some of the program stars shows that the BVIc technique produces reliable results. As late–type stars constitute the most numerous spectral types and are plentiful in all galactic plane directions BVIc reddening distances can be derived close to the desired direction (although to smaller distances than techniques that utilize early type stars). The applicability of the technique is further enhanced by the use of CCDs which generally have a spectral response well suited for BVIc imaging observations. Using the new technique the distance to the PN NGC2440 was found to be $(3100 \pm 320)$pc.

# THE IONIZATION STRUCTURE OF PLANETARY NEBULAE

T. BARKER

*Department of Physics and Astronomy, Wheaton College, Norton, MA 02766, USA*

Ground-based and satellite spectrophotometric observations of emission-line intensities over the spectral range 1400-7200 Å have been made in 5 or more positions in a total of 9 planetary nebulae.

Since 1980, I have measured emission-line intensities over the spectral range 1400-7200 Å in 5 or more positions in a total of 9 planetary nebulae (NGC 6720, NGC 7009, NGC 6853, NGC 3242, NGC 7662, NGC 6826, NGC 1535, NGC2392, and NGC 2440). The goals of these studies include: (1) to observe elements in more stages of ionization than is possible from optical spectra alone; this provides a check on optical ionization correction procedures, which are still needed for nebulae which are too faint to observe with the IUE satellite; (2) to get particularly accurate total abundances for each nebula by averaging measurements made in different nebular locations; such differences can be sensitive tests of theories of mixing of processed material in the planetary progenitors; (3) to test theoretical ionization models. In most nebulae, electron temperatures determined from different indicators such as $H^+$, $O^+$, $O^{++}$, $S^+$, and $S^{++}$, and $Ne^{3+}$ agree fairly closely, but for others, the more highly-ionized species indicate significantly higher temperatures. For all planetaries studied so far, optically-measured abundances agree well with UV determinations, with the exception of $C^{++}$, where the optical measurement is systematically higher. Since this discrepancy is systematically greater closer to the central star, I believe that the $\lambda 4267$ C II line intensity is not being interpreted correctly, either because it is blended with another line, or because the excitation mechanism for the $\lambda 4267$ line is not fully understood. Standard ionization- correction equations give consistent results for the different positions which are in excellent agreement with abundances calculated using ultraviolet lines, and there is no evidence for abundance gradients in any of the nebulae. The elemental abundances do differ slightly between nebulae, implying differences both in the composition of the gas that formed the progenitor stars and differences in the amount of mixing of processed material in the stars before envelope ejection. Average logarithmic abundances (relative to H = 12.00) are: He = 11.00, O = 8.74, N = 8.12, Ne = 8.12, C = 8.67, Ar = 6.27, and S = 6.82.

# CCD PHOTOMETRY OF NGC 2453

D.C.V. MALLIK, RAM SAGER and A.K. PATI
*Indian Institute of Astrophysics, Bangalore 560034, India*

We have undertaken a detailed photometric study of the open cluster NGC 2453 in Puppis with a view to determining accurately its distance modulus and to investigate its possible association with the PN NGC 2452. We have observed the core region of the cluster in V and I photometric bands using a Thomson-CSF TH 7882 CCD chip with the format 384×576 pixels, at the f/13 cassegrain focus of the 102-cm telescope at Vainu Bappu Observatory, Kavalur, India. Each image frame covers a field of $2'.3 \times 3'.4$ of the sky. Bias frames and flat field exposures on the twilight sky were obtained as per standard practice.

The data were initially processed using STARLINK package and the magnitude estimation of a star has been done using DAOPHOT. The figure below shows a V, V-I plot of the core region of the cluster. A total of 275 stars have been plotted in the diagram. We find a well populated main sequence down to $V \geq 20^m.0$. This is the first time the unevolved main sequence of the cluster has been observed. We have adopted a $E(B - V) = 0^m.47$ from earlier work and used $A_V = 3.06 * E(B - V)$ to correct for interstellar extinction. ZAMS from Walker (1985, MNRAS **213**, 889) was used to fit the cluster CMD after applying a color correction $E(V - I) = 1.25E(B - V)$. A true distance modulus of $12^m.9 \pm 0^m.25$ has been obtained which puts the cluster at a distance of $3.80 \pm 0.47 kpc$. The PN NGC 2452 has a directly determined extinction distance of $3.57 \pm 0.47 kpc$ (Gathier 1984, Ph.D. thesis, Groningen).

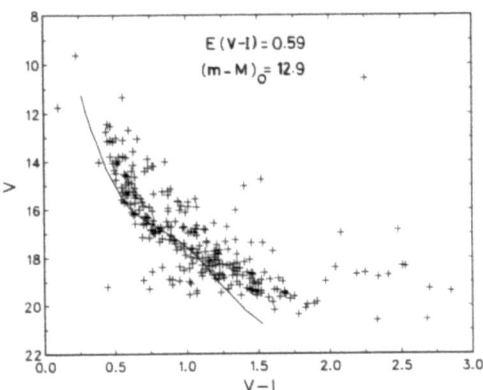

# ELECTRON DENSITY AND NITROGEN ABUNDANCE FROM FIR LINES

R.H. RUBIN, S.W.J. COLGAN, E.F. ERICKSON, M.R. HAAS, S.D. LORD and J.P. SIMPSON
*NASA Ames Research Center, M.S. 245 - 6, Moffett Field, CA 94035, USA*

In order to study the physical properties of nebulae and determine their elemental abundances, it is important to observe lines from many different ionic species. Such studies have been enhanced in recent years with the measurement of lines in the far-infrared (FIR). The [N III]57$\mu$m line provides a way to assess the $N^{++}$ abundance – which is not readily done from any other spectral region. Recent detection of the [N II]122 and 205$\mu$m lines provides a new way to assess both the electron density in the $N^+$ region and the total N abundance in an object. When there are few observations to warrant a detailed modeling approach, it may be necessary to use another approach which has been referred to as a semi-empirical method (hereafter SEM) (*e.g.* Aller 1984). We delineate a SEM scheme for doing this and apply it to observations for the H II region G333.6−0.2.
Electron Densities: There are several FIR diagnostics for $N_e$. These include the ratio of [O III]52/88$\mu$m, [S III]19/33$\mu$m, and [N II]122/205$\mu$m – which provide $N_e$[O III], $N_e$[S III], and $N_e$[N II] respectively. Each acts as an excellent discriminant of $N_e$ that is insensitive to $T_e$. The [N II] lines provide a new tool to examine low density, lower ionization gas. For conditions prevalent in most nebulae, these lines should readily show the effect of collisional deexcitation because their low critical densities ($N_{crit}$ ∼40 and 260 cm$^{-3}$) are much lower than for most lines used in nebular analyses.
The $N^{++}/N^+$ ratio and Stellar Effective Temperatures $T_{eff}$: We provide a way to estimate $N^{++}/N^+$ based on the measurement of [N III]57 and [N II]122, 205$\mu$m lines. We assume a 2-component model, with $N^+$ and $N^{++}$ zones, each having uniform density – using $N_e$[N II] and $N_e$[N III] (obtained from $N_e$[O III] and/or $N_e$[S III]). These yield the requisite volume emissivities and hence the $N^{++}/N^+$ ratio. Once $N^{++}/N^+$ is found, we may estimate $T_{eff}$ by using curves relating these two, generated from H II region models (Rubin 1985). Because there is a fairly tight correlation (with a much smaller dependence on total number of ionizing photons/s and density), $T_{eff}$ may be inferred for an ionization bounded H II region as long as integrated fluxes have been measured. This technique has the advantage that it does not depend on prior knowledge of an abundance ratio.
The Total N Abundance − (N/H): With $N^{++}/N^+$, we have, for the first time using FIR spectroscopy alone, the capability to assess the two dominant ionization states from the same element in H II regions or low ionization PN (with negligible $N^{+3}$ and $N^0$). To obtain N/H, an extinction corrected flux must be available for the same location/beam as used for the FIR fluxes for a generic hydrogen recombination line (or appropriate continuum surrogate). This derivation of N/H requires neither that all of the flux in the lines be measured nor that the object be ionization bounded.
Application and Conclusions: The first discrete source in which both FIR [N II]122, 205$\mu$m lines have been measured is G333.6−0.2 − a luminous, obscured southern H II region, which has sufficiently low ionization that a substantial fraction of N is expected to be $N^+$. ¿From measurements made with the Kuiper Airborne Observatory in a 45″ beam with the facility cryogenic grating spectrometer, we obtain a preliminary log $N_e$[N II] (cm$^{-3}$) ∼2.5. This is significantly smaller than the value of log $N_e$[O III] and log $N_e$[S III] ∼3.6. This difference may be due to the lower density in the $N^+$ zone as well as to a biasing effect that causes lower $N_e$ values to be deduced for $N_e$[N II] (Rubin 1989). For G333.6, $N^{++}/N^+$ = 13. This large value cannot be used to infer $T_{eff}$ (except as an upper limit), because the 45″ beam does not encompass all the [N II] flux. The analysis for N/H is still in progress; however, we estimate that N/H increases merely by ∼7% due to the contribution of $N^+$/H to a 45″ beam centered on the source.

## References
Aller, L.H. 1984, *Physics of Thermal Gaseous Nebulae* (Dordrecht: Reidel).
Rubin, R.H. 1985, *ApJS*, **57**, 349.
Rubin, R.H. 1989, *ApJS*, **69**, 897.

# FILLING FACTORS AND IONIZED MASSES OF PLANETARY NEBULAE

FRANCESCA R. BOFFI

*Dept. of Physics and Astronomy, University of Oklahoma, 440 West Brooks, Norman, OK 73019, USA*

and

LETIZIA STANGHELLINI

*Osservatorio Astronomico di Bologna, via Zamboni 33, I-40126 Bologna, Italy*

We present a study on the filling factors and ionized masses of four sets of galactic and extra-galactic planetary nebulae (PNe) at known distances. The calculation of filling factors and ionized masses has been pursued as to get a deeper insight on the evolution of this class of objects. We used a galactic set of PNe, another set of nebulae that are averagely located near the galactic center, and two sets of nebulae in the Magellanic Clouds. As input data, we used the electron densities derived from the forbidden line intensity ratios, the $H\beta$ nebular fluxes, the distances of galactic PNe derived from extinction, and the distances of galactic center PNe and of extra-galactic PNe derived from galaxy (or galactic region) memberships. All these quantities, plus the input angular radii, have been selected among the most recent measurements available in the literature. We obtained several interesting results. (1) The calculated filling factors are on average much smaller than what is usually assumed, independently for each set. (2) The ionized masses are all in good agreement with the theoretical predictions, with the possible exception of the Galactic Bulge PNe. (3) Both filling factors and ionized masses cover a wide range of values as it is shown in the cumulative histograms below (filled circles=local PNe; open circles=galactic center PNe; filled squares=LMC PNe; open squares=SMC PNe).

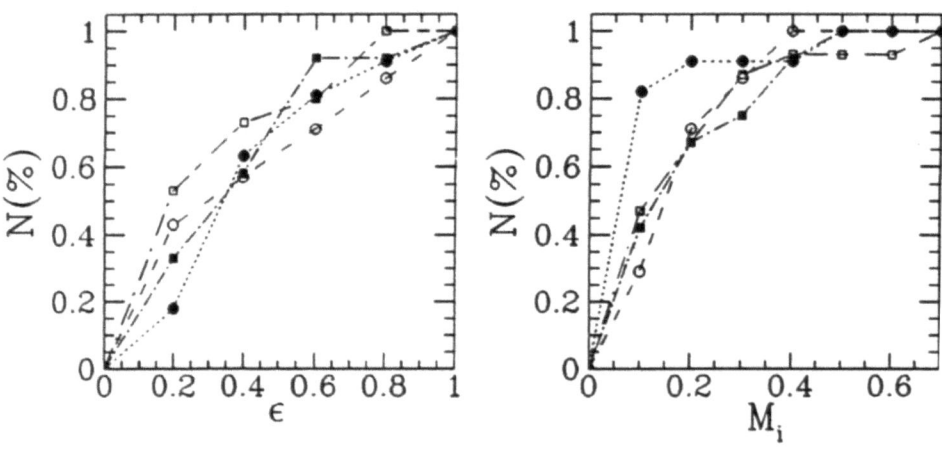

# MEAN ELECTRON DENSITIES, DISTANCES AND FILLING FACTORS FOR GALACTIC PLANETARY NEBULAE

ROBIN L. KINGSBURGH and M.J. BARLOW

*Dept. of Physics & Astronomy, U.C.L., Gower Street, London WC1E 6BT, U.K.*

ABSTRACT: Mean electron densities are presented for over 100 planetary nebulae (PN). Distances to the majority of these PN are then derived, based on calibrations from observations of Magellanic Cloud PN. Absolute radii and filling factors have also been determined. A trend is seen in that for larger radii, smaller filling factors are found, however we show that such a trend results from uncertainties in the observed angular diameters.

We present [O II] 3726, 3729 Å doublet ratios and electron densities for ∼120 galactic planetary nebulae. For ∼100 of the objects, the doublet ratios represent integrations over the whole of the nebula which were obtained via trailed spectrograph exposures. Additionally, we present integrated [S II] 6716, 6731 Å doublet ratios and electron densities for 63 southern galactic PN. We find that the densities derived from the integrated [O II] and [S II] doublet ratios are in excellent agreement with each other.

Distances have been derived for the majority of the nebulae, using calibrations recently derived from Magellanic Cloud PN. For PN which are optically thin in the hydrogen Lyman continuum, we have derived distances using a variant of the Shklovsky method (constant ionized hydrogen mass) which uses the mean [O II] electron density and the measured radio flux and which does not require knowledge of the filling factor or nebular angular radius. For PN which are optically thick in the Lyman continuum, the constant H$\beta$ flux method was used to derive distances. The typical [O II] density at the transition point between an optically thick and thin nebula is 4400 cm$^{-3}$. Since the optically thin and thick methods both overestimate the distance when applied to inappropriate nebulae, the smaller of the two distance estimates is adopted for each nebula.

An extensive comparison is made between the distances derived here and previously published distances and distance scales. It is shown that the present distances, based on Magellanic Cloud calibrations, give consistency with independent distance estimates. They also yield much greater self-consistency between central star masses derived from luminosity versus T$_{eff}$ comparisons on the one hand, and from absolute magnitude versus evolutionary age comparisons on the other hand.

For the PN in our sample, rms electron densities, filling factors and absolute radii have also been derived. The derived filling factors are found to decrease with increasing absolute angular radius, but we argue that this effect can be attributed to the effects of measurement uncertainties in the adopted angular radii.

This work is fully presented in Kingsburgh & Barlow (1992, MNRAS 257,317) and Kingsburgh & English (1992, MNRAS, in press).

# THE IONIZATION AND THERMAL STRUCTURE OF NGC 2392 AND NGC 3242

## XIAO-WEI LIU and JOHN DANZIGER
*European Southern Observatory, Garching bei München, FRG*

In Liu and Danziger(1992b) we present $T_e$ derived from the Balmer discontinuities for 14 objects by long slit spectrophotometry. It is found that on the average the $T_e$ derived in this way are systematically lower than those derived the forbidden line ratios, indicating the presence of large temperature fluctuations(Peimbert 1967, 1971). The spectra obtained have been analyzed point-by-point along the slit for NGC 2392 and NGC 3242, for which $t^2 = 0.13$ and 0.058, respectively, as measured on the summed spectra integrated along the slits.

For each pixel along the slit, we derive the $N_e$ from [OII] $I(\lambda 3729)/I(\lambda 3726)$ and [ArIV] $I(\lambda 4711)/I(\lambda 4740)$, the $T_e$ from the [OIII] forbidden line ratio and from the observed Balmer discontinuity, $N(He^+)/H^+$ and $N(He^{++})/H^+$ respectively from HeI$\lambda$4471 and HeII$\lambda$4686. In the case of NGC 2392, $T_e$ derived the [NII] lines is also obtained. Uniform reddening has been assumed in both NGC 3242 and NGC 2392 for which we have adopted $c = 0.12$ and $c = 0.22$ respectively.

1. NGC 3242    The [OIII] temperature is quite uniform along the slit. In contrast, the Balmer temperature show large amplitude variations. Within the bright shell, where both the HI$\beta$ and HeII$\lambda$4686 are strongest, $T_e(Bac)$ are close to $T_e(OIII_{na})$, whereas both inside and outside the bright shell $T_e(Bac)$ drops rapidly below $T_e(OIII_{na})$ by as much as 3000 K. It seems such large variations of the Balmer temperature is responsible for the large value of $t^2=0.058$ found for this object.

2. NGC 2392    It has rather complicated structure both in ionization and thermal structure. The most striking feature is that its ionization degree increases outward. The ionization degree and [OIII], [NII] temperatures are lowest within the central bright disk. In contrast here the Balmer temperatures are highest and comparable to that of [OIII]. The densities in this central disk seems quite uniform, with $N_e$ about 2200 cm$^{-3}$. Outside the central disk, the $N_e$ decreases rapidly to values below 600 cm$^{-3}$, then flattens and decreases to nearly zero, although with some fluctuations. The intensity of HeII$\lambda$4686 line charges almost exactly oppositely as the $N_e$ does. At the meantime, the [OIII] and [NII] temperatures increase rapidly whereas the Balmer temperature decreases. At some positions [OIII] temperature reaches values as high as 18000 K, about 4000 K higher than those within the central disk. The Balmer temperatures at these positions are only about 10000 K. Throughout the selected slit, the [NII] temperatures are systematically lower than that of [OIII]. Both of them show a similar variation pattern.

## References

Liu, X. W. and Danziger, J., 1992a, submitted.
Liu, X. W. and Danziger, J., 1992b, to be submitted.
Peimbert, M., 1967, *Ap.J.*, **150**, 825.
Peimbert, M., 1971, *Bol. Observ. de Tonantzintla Y Tacubaya*, **6**, 29.

# SELF-CONSISTENT PHOTOIONIZATION MODELS OF PLANETARY NEBULAE LUMINESCENCE

V.V. GOLOVATY

*Astronomical Observatory of Lviv University, 290005 Lomonosov str. 8, Lviv, Ukraine*

and

YU. F. MALKOV

*Crimean Astrophysical Observatory, 334413 Nauchny, Crimea, Ukraine*

The modern self-consistent photoionization model of planetary nebula luminescence is described. All of the processes which play an important role in the ionization and thermal equilibrium of the nebular gas are taken into consideration. The diffuse ionizing radiation is taken into account completely. The construction of the model is carried out for the radial distribution of gas density in the nebular envelope which is consistent with isophotal map of the nebula. The application of the model is illustrated on the example of the planetary nebulae BD+30°3639 and NGC 7293. For each nebula, the intensities of the emission lines of ten basic chemical elements in the UV, optical and IR spectral ranges are calculated and matched with observational data. Both the chemical composition of the nebular gas and the continuum of the central star at the wavelengths $\leq 912$Å are determined during the process of model calculation. It is shown that the continuum of the central star at $\leq 912$Å does not correspond to the blackbody spectrum but agrees with the spectrum of the corresponding non-LTE model atmosphere. The radial distributions of electron density, electron temperature and other parameters in the nebular envelopes are found. The optical thickness of the nebulae in the Lyman continuum is derived.

# TEMPERATURE FLUCTUATIONS IN PN

RUTH GRUENWALD and SUELI M. VIEGAS
*Instituto Astronômico e Geofísico - USP, Av. Miguel Stefano, 4200, 04301-002 São Paulo,SP Brazil*

For planetary nebulae, empirical abundances can be obtained from the observed emission-lines as long as the electron density, the electron temperature, and the ionization corrections factor are determined. However, due to temperature fluctuations in the emitting gas, the evaluation of the temperature from the observational data is strongly dependent on the method used. The temperature fluctuation is usually characterized by the mean square temperature fluctuation, $t^2$ (Peimbert and Costero, 1969 - PC).

Theoretical $t^2$ values have been discussed in detail for H II regions (Gruenwald and Viegas, 1992 - GV). These results show that $t^2$ decreases with the gas density. The stellar temperature is also an important parameter, but the $t^2$ dependence is not monotonic. Although planetary nebulae are denser, the stellar temperature can be higher than that of the H II region ionizing star. The temperature fluctuation could then still be important.

Theoretical $t^2$ values are obtained for typical PN conditions ($n_H = 10^2$-$10^6$cm$^{-3}$; $T_* = 30000$-$300000$K; $L_* = 300$-$20000 L_\odot$; $[Z]=$ Stasinska and Tylenda,1986) using the photoionization code Aangaba (GV). The main conclusions are the following: a) $t^2$ generally increases with stellar temperature, and can be high even at high densities; b) $t^2$ is higher for S ions; c) $t^2(H^+)$ is not negligible, and increases very much with stellar temperature and luminosity, and also with the gas density; d) the effect of density is also important for $N^+$ ions; e) $t^2(O^{++})$ is small, but $t^2(Ne^{++})$ can be important at high stellar temperatures.

Since temperature fluctuations are more important for planetary nebulae with high stellar temperatures, the effect of $t^2$ on ionic abundance determinations in such nebulae is analyzed. The abundances of $N^+$, $O^+$, $O^{++}$, $Ne^{++}$, $S^+$, and $S^{++}$ ions relative to $H^+$ are obtained for 47 high central stellar temperature PNe ($T_* \geq 10^5$K). Two cases are considered: a)no temperature fluctuation, considering $T_{[NII]}$ for the region where low ionization lines are formed, and $T_{[OIII]}$ where high ionization lines and $H\beta$ are produced; b)the line temperature (PC) is obtained for each line from $T_{[NII]}$ (low ionization lines) or $T_{[OIII]}$ (high ionization lines and $H\beta$) and $t^2$ for the corresponding ion (from the models) . Data for $T_*$, $L_*$, $n_e$, $T_{[NII]}$, and $T_{[OIII]}$ for each nebulae are obtained from the literature. The calculations show that the ionic relative abundances are higher in case $b$, but, due to the combined effect of $T_*$, $L_*$, and gas density, $t^2$ of the analyzed objects are not high. Consequently, the differences in the ionic abundance determinations by the two methods are less than 20%, except for $S^+$ ions, which deviation can reach 48%.

## References

Gruenwald, R.B. and Viegas, S.M. (1992), Ap.J.Suppl.Ser. 78, 153
Peimbert, M. and Costero, R. (1969), Bol.Obs.Ton.Tac. 6,29
Stasinska, G. and Tylenda, R. (1986) A.&A. 155,137

# RADIATION CHARGE EXCHANGE AND RADIATION ION-ATOM RECOMBINATION AS A SOURCE OF CONTINUAL E-M RADIATION FROM ASTROPHYSICAL PLASMA

A.A. MIHAJLOV
*Institute of Physics, P.O.Box 57, 11001 Beograd, Yugoslavia*

M.S. DIMITRIJEVIC
*Astronomical Observatory, Volgina 7, 11050 Beograd, Yugoslavia*

and

A.M. ERMOLAEV
*Departement of Physics, University of Durham, Science Laboratories, Durham DH1 3LE, UK*

We show that for the study of emission from weakly ionized low temperature hydrogen plasmas, the processes $A^+ + A \to A^+ + A + \hbar\omega$ and $A^+ + A \to A_2^+ + \hbar\omega$ (where A denotes a neutral atom in ground state and, $A^+$ and $A_2^+$ atomic and molecular ions) must be treaded as a source of continual electro-magnetic radiation from low temperature plasma. Both reaction channels are treated separately and the corresponding total and separate spectral intensities are determined for hydrogen plasma at $T < 10000K$. The obtained results have been also compared with the corresponding spectral intensities for electron-ion bremsstrahlung and electron-ion photorecombination.

Our results (which will be published in Mihajlov et al.,1992) show that in the case of low temperature plasma one must particularly be carefull concerning the continuous EM-radiation spectrum nature. Namely, at typical values of electron and atom component ratio in hydrogen plasma, investigated ion-atom radiation processes might completly determine the character of spontaneous EM-radiation spectrum. We expect similar results in the case of helium plasma. If this fact is not taken into account, serious errors in plasma diagnostic might follow. The important astrophysical cases of interest are hydrogen clouds, circumstellar hydrogen shells and e.g. solar photosphere and chromosphere.

### References

Mihajlov, A.A., Dimitrijevic, M.S., and Ermolaev, A.M. (1992), in press.

# OBSERVATIONS OF THE BOWEN FLUORESCENCE MECHANISM AND CHARGE TRANSFER IN PLANETARY NEBULAE I.

### XIAO-WEI LIU and JOHN DANZIGER
*European Southern Observatory, Garching bei München, FRG*

Deep, medium resolution, long-slit spectrophotometric data have been taken for a number of high-excitation planetary nebulae, covering a wavelength range from 3100 Å to 7200 Å with some selected regions observed at higher resolution. For about half the objects, the whole optical region has been observed, from 3100 Å to 11600 Å. Accurate flux calibration is achieved over this whole wavelength range. These data allow a detailed quantitative study of the Bowen fluorescence mechanism and the charge transfer reaction $O^{3+}(^2P_{3/2,1/2}) + H^0 - O^{2+}(^{2s'_a+1}L'_{a\ j'_a}) + H^+$ in planetary nebulae, the primary goal of this program. In this paper measurements of $O^{2+}$ Bowen fluorescence and charge transfer lines are presented. We show that $LS$ coupling fails for the $O^{2+}$ permitted transitions studied in this program and intermediate coupling may be a better assumption. Efficiencies of the Bowen fluorescence mechanism are derived for 15 objects, and a wide range of possible values is apparent. There is a remarkable linear positive correlation between the Bowen efficiency and the fractional abundance of oxygen in the ionization stage of $O^{2+}$ and the fractional abundance of helium in the form of $He^+$. Evidence that the Bowen efficiency is anticorrelated with the electron temperature, as first noted by Likkel and Aller, is established. The Bowen efficiency drops substantially when the nebular expansion velocity $2V_{exp}(O^{2+}) > 55$ km/sec. For lower expansion velocities there is no detectable correlation between these two quantities. There are no observable differences in Bowen efficiency among objects of different morphological type nor between objects excited by stars of different spectral types as suggested by Likkel and Aller.

The OIII$\lambda$3261, $\lambda$3265, $\lambda$3267 lines are found to originate from radiative and dielectronic recombination. $O^{3+}$ abundances are derived for some objects using these lines.

# BOWEN RESONANCE FLUORESCENCE LINES OF OIII IN PLANETARY NEBULAE

C. R. O'DELL and C. O. MILLER
*RICE UNIVERSITY*
*P. O. Box 1892*
*Houston, Texas, USA*

**Abstract.** New and published line intensities are used to test the role of the pumping of the Bowen Lines of O III by absorption of the O1 and O3 lines and population by Charge Exchange. Although general agreement is obtained, noticeable deviations from theory exist.

We have obtained uniquely complete observations down to the ozone cutoff of the atmospheric window of the Bowen Resonance Fluorescence Lines of O III in the moderate ionization planetary nebula NGC 6210 and the high-ionization objects NGC 7027 and 7662 (O'Dell & Miller 1992). These data are added to an additional eight well measured high-ionization nebulae (Likkel & Aller 1986). Comparison of the observed lines with those predicted theoretically for the dominant O1 excitation shows general agreement, although some disagreements are present. We also measured lines arising from pumping by the secondary O3 line, which occurs farther away from the center of the He II Ly$\alpha$ line, and find good agreement with 20 year old theoretical predictions (Harrington 1972) for NGC 7027 and 7662 but unexpectedly high pumping rates in NGC 6210. The one unblended line resulting only from O3 pumping may, however, be underestimated due to atmospheric ozone absorption. Examination of the role of Charge Exchange in explaining the excess of radiation from certain levels indicates that the process is important but does not quantitatively agree with theoretically predicted rates (Sternberg *et al.* 1988, Gargaud *et al.* 1989) for transitions out of the $^3S_1$ state. If the O3 pumping has been underestimated, the $^3S_1$ discrepancy disappears. The high-ionization nebulae are all very similar and stand in contrast with NGC 6210, which shows unexpectedly high rates of pumping by O3 and population by Charge Exchange. These differences are probably due to different distribution within the nebula of the pumping line (He II Ly$\alpha$), O III, and the key ingredient in Charge Exchange (H$^0$).

### References

Gargaud, I., McCarroll, R., & Opradolce, L.: 1989, *A&A* **208**, 251
Harrington, J. P.: 1972, *ApJ* **176**, 127
Likkel, L. & Aller, L. H.: 1986, *ApJ* **301**, 825
O'Dell, C. R. & Miller, C. O.: 1992, *ApJ* **390**, 219
Sternberg, A., Dalgarno, A., & Roueff, E.: 1988, *Comm. Ap.* **13**, 29

# DIELECTRONIC RECOMBINATION IN THE GASEOUS NEBULAE AS A COOLING PROCESS

## A.F. KHOLTYGIN

*S.-Petersbourg Univ. Astron. Obs., 198904, St.Petergof, Russia*

The dielectronic recombination (DR) is of importance at 'low' (nebular) temperature [1]. This process leads to cooling the electron gas in nebulae. The cooling rate by recombination of ion $X^{+n}$ is

$$L_{dr}(T_e) = \sum_j \frac{4\pi^{3/2} a_0^3}{(kT_e/Ry)^{3/2}} \frac{g_j W_j^a}{g^+} exp\left(-\frac{\Delta E_j}{kT_e}\right) \frac{W_j^r}{W_j^a + W_j^r} \Delta E_j.$$

Here $W_j^r$ and $W_j^a$ are respectively the radiation and autoionization probabilities for the autoionization state $j$ of ion $X^{+n}$, $\Delta E_j$ is the energy of this state, $g_j$ and $g^+$ are respectively the statistical weights of the state $j$ and the ground state of the ion $X^{+n+1}$. We have calculated $L_{dr}(T_e)$ for all ions of C. It is shown that the process of DR cooling is important only for nebulae with extraordinarily high abundances ($\{C^{+i}/H^+\} > 0.01$) of these ions.

## References

[1]. Nussbaumer H., Storey P.J. (1984) 'Dielectronic recombination at low temperature.II: Recombination coefficients for lines at C, N, O', Astronomy and Astrophysics, 56, 293-312.

# HIGH DISPERSION SPECTRA OF BRIGHT PLANETARY NEBULAE

SIEK HYUNG and LAWRENCE H. ALLER
*University of California, 8979 Math Science bldg., 405 Hilgard Ave.*
*Los Angeles, California 90024-1562*

The Hamilton Echelle Spectrograph at the coudé focus of the Shane 3 $m$ telescope at Lick Observatory permits us to obtain high resolution spectra of bright planetary nebulae over the spectral range from 3500Å to 10300Å. Not only is it possible to separate pairs such as $\lambda$ 5198, 5200 [NI], 4860 HeII and 4861 HI, but one may secure the profiles of Doppler broadened lines. It appears to be possible to separate the HI and HeI contributions of the 3889 line in Hu 1-2. The nebulae most suitable for observation are those of high surface brightness. NGC 7027 was the first nebula intensively studied with this equipment (Keyes et al. 1990); subsequently we have observed NGC 2440, 6537, 6543, 6567, 6572, 6741, 7009, 7662, IC 351, 418, 2149, 2165, 4634, 4997, 5217, Hb 12, and Hu 1-2. In NGC 7009 we measured line fluxes at the ends of major and minor axes. Small compact objects were centered on the slit; other were observed in the bright ring. These planetary nebulae cover a range in excitation level, chemical composition, and evolutionary status.

A comparison of Hamilton Echelle data with those obtained with an image tube scanner reveals that echelle data are often an order of magnitude more accurate. Part of this superiority arises from the high spectral resolution of the echelle, which not only enables one to separate lines in blends but also permits the position of the continuum to be estimated more confidently; other benefits arise from from the linearity of the CCD detector. High spectral resolution can increase the number of lines for plasma diagnostic uses and it yields data on internal motions as noted above. Improved quality and quantity of nebular line measurements are essential to assess theoretical $A$-values and collision strengths since no laboratorial checks are possible. The high spectral resolution permits as to separate lines of stellar and nebular origin. Stellar lines are often diffuse and their profiles invariably differ from the Doppler broadened nebular emissions.

An objective of this investigation is to provide data for improved abundance determinations and theoretical nebular models. For a number of these nebulae such as NGC 7009 and IC 4997 we observed many permitted lines of CII, NII, OII, SiII etc. that are excited by recombination and cascade. Iron is observed in various ionization stages in several objects. Some of the weaker lines may arise from shock excitation and assist us in the construction improved sophisticated nebular models in which evolutionary effects are taken into account.

# IMAGING SPECTROPHOTOMETRY OF THE RING NEBULA

NANCY JO LAME and RICHARD W. POGGE
*The Ohio State University*
*Columbus OH 43210, USA*

We present new results from a program of emission-line imaging spectrophotometry of planetary nebulae using the Ohio State University Imaging Fabry-Perot Spectrograph (IFPS). High-quality emission-line maps of the important diagnostic lines [NII]$\lambda\lambda$5755,6583, [SII]$\lambda\lambda$6717,6731, [OI]$\lambda$6300, [OIII]$\lambda$5007, H$\alpha$, and H$\beta$ have been obtained. Maps of the ionization structure ([S II]/H$\alpha$, [N II]/H$\alpha$, [O III]/H$\beta$, and [O I]/[O III]), temperature in the N$^+$ region, density in the S$^+$ region, and Balmer decrement across the nebula are presented. These show considerable variation in ionization state, temperature and density. This detailed information will provide powerful constraints on photoionization models for the Ring Nebula.

# PLANETARY NEBULAE WITH THE STRONG [NII] EMISSION LINES

L. N. KONDRATJEVA

*Astrophysical Institute of Kazakh Academy of Sciences, Kamenskoje Plato,
480068 Alma-Ata, Kazakhstan*

**Abstract.** The sample of planetaries with the strong [NII] emission lines are analysed. All available parameters are discussed. The enrichments of all elements and the tendency to rather low surface brightness of ne bulae in H$\beta$ flux are remarked. The deficiency of hydrogen in envelopes is proposed as the possible reason of observational spectra.

## 1. INTRODUCTION

The spectra of planetary nebulae with the strongest [NII] emission lines can't be represented with traditional photoionization models because the computed intensities are lower than observational values. The only way out is to change the abundances in order to come to an agreement with observations. The spectra in which the strongest [NII] lines coexist with HeII emission are of special interest. Usually such objects are considered to be massive ionization- bounded nebulae of the Population I type. The aim of our research was to analyse the physical parameters of such object in order to single out the features peculiar just to them.

## 2. RESULTS

We have compiled a list of about 300 planetaries studied in the wide wavelength region and then the objects with I([NII]$\lambda$6548+$\lambda$6583)/ I(H$\alpha$ )$\geq$ 1.0 and I(HeII$\lambda$4686)/ I(H$\beta$ )>0.2. To avoid the N abundance anomaly we have choosen only those objects with [NII]/[SII] ratio close to the average value of about 10. Our sample consists of 20 planetaries. They have quite normal $T_e$ and $N_e$ and $T_{eff}$ >60000K. All available abundances of HE, N, NE, O, S, Ar are significantly higher than the average values adopted for planetaries. The derived surface brightnesses are rather low indicating the small nebular masses or the low stellar luminosities.

## 3. DISCUSSION

All observational and derived features may be explained if we suppose the defiency of hydrogen in selected envelopes. If so the nebular masses must be reduced as a factor of 2-4 in dependence on the hydrogen deficit degree. The stellar luminosities must be low enough to adjust the small nebular masses with the large optical depths. Thus the central stars may be expected to take place in the lower part of H-R diagram for nuclei, somewhere close to white dwarfs, and our sample is appeared to be rather old planetaries. The extreme cases of hydrogen deficit are observed in A30 and A78, their secondary envelopes are almost devoid of hydrogen. Probably the nebulae with selected spectra are the results of the second throw out as well.

## PROBABLE TYPE I PLANETARY NEBULAE

H. MORENO, A. GUTIERREZ-MORENO and G. CORTES

*Departamento de Astronomía, Universidad de Chile, Casilla 36-D, Santiago, Chile*

ABSTRACT: During a program of observations of planetary nebulae (PN) made at CTIO in 1982 using the 1.5 m telescope equipped with a Cassegrain spectrograph and a Vidicon detector, and repeated in 1991 using the same configuration with a CCD detector, we found four PN for which $H\alpha <$ [N II] $\lambda 6583$: He2-109, He2-145, He2-152, and He2-163. The analysis of the data was made using the computing facilities of Cerro Tololo Computing Center at La Serena, and of the Centro de Procesamiento Digital de Imágenes of the Departamento de Astronomía de la Universidad de Chile.

The general conclusions concerning these four PN are as follows:

1) He2-109 is not a Type I PN, but it is an interesting object due to its large O abundance [log $N(O)/N(H) = 9.16$] which is rather unusual. Shaw and Kaler (1989) have observed this PN and in their results they give an observed value for [O III] $\lambda 4959$ that is 17% higher than ours; this seems to confirm our large O abundance.

2) He2-145 has characteristics which are close to those of a Type I PN [the abundance by number $y = 0.115$, log $N(N)/N(H) = 8.51$, log $N(O)/N(H) = 8.55$] but it does not quite satisfy the necessary conditions, being the value of y slightly smaller than is needed.

3) He2-152 is clearly a Type I PN, with a very high He contents ($y = 0.29$); it does not show any bipolar or filamentary structure and looks nearly circular.

4) He2-163 is a special object, since it shows variability in some of its lines. In three of the five spectra obtained in 1982 [N II] $\lambda 6583$ appears slightly stronger than $H\alpha$, while in the other two it appears slightly fainter. There are clear differences in some other emission lines, for example in the [O III] and [O II] lines: $\lambda 3727$ decreases in strength from 1982 to 1991, while the contrary is true for $\lambda 5007$.

### References

Shaw, R.A., and Kaler, J.B. 1989, APJS, **69**, 495.

# EXTENDED X-RAY EMISSION FROM PLANETARY NEBULAE

H.C. KREYSING, C. DIESCH, J. ZWEIGLE, R. STAUBERT and M. GREWING
*Astronomisches Institut der Universität Tübingen, Waldhäuserstr. 64, D-7400 Tübingen, Germany*

We present first results from the ROSAT All Sky Survey on X-ray emission of planetary nebulae (PNe). For the first time extended X-ray emission from PNe was detected. This is the case for NGC 6543, NGC 6853, A 12, NGC 4361 (and LoTr 5). X-ray emission compatible with a point source was detected from BD+30°3639, however, the spectral distribution of the X-ray photons is leading to temperatures beyond $2 \cdot 10^6$ K. Thus in all cases, with the possible exception of LoTr 5, the central star of the PNe can be excluded as the main source of the observed X-ray emission. X-ray images and ROSAT spectra for all detected PNe are presented. The best observed PN in X-ray emission is NGC 6543. Due to the close vicinity to the north ecliptic pole, this object was regularly observed, every 90 minutes during the whole half year of the ROSAT All Sky Survey, resulting in 41 ksec of integration time. In addition NGC 6543 was observed in a 50 ksec pointed observation to the north ecliptic pole, taken in June 1990 during the calibration phase (Kreysing et al. 1992). A comparison of the semi-ring-like distribution of the X-ray emission of NGC 6543 with optical CCD-images shows, that most of the X-ray emission seems to originate from the boundary region between the nebula and the halo. Neither the central star nor the hot wind from the central star wind is the main source of the X-ray emission, as proposed by the interacting stellar wind model (Kwok 1982). An alternative model employing a possible coronal heating mechanism has been discussed by Kreysing (1992); accoustic waves, travelling outward from the nebula, encounter a sudden density decline at the boundary to the halo. As a consequence the waves degenerate into shock waves, dissipating their energy in a thin region of only some $10^{15}$ cm into the ambient medium.

|  | d [kpc] | [cts/sec] | $T_x$ [K] | $N_H$ [cm$^{-3}$] | $L_x$ [erg/sec]* |
|---|---|---|---|---|---|
| BD+30°3639 | 0.2-2.8 | 0.18±0.03 | $2.5 \cdot 10^6$ | $1.4 \cdot 10^{21}$ | $4.3 \cdot 10^{32}$-$8.5 \cdot 10^{34}$ |
| NGC 6543 | 1.2 | $(8.1±0.4) \cdot 10^{-3}$ | $1.7 \cdot 10^6$ | $6.9 \cdot 10^{20}$ | $(1.7±0.1) \cdot 10^{32}$ |
| A 12 | 2.4 | 0.10±0.02 | $3.3 \cdot 10^6$ | - | $6.9 \cdot 10^{32}$-$2.3 \cdot 10^{34}$ |
| NGC 4361 | 0.86 | 0.05±0.02 | $<2.0 \cdot 10^5$ | $4.9 \cdot 10^{20}$ | $(2.0±0.8) \cdot 10^{34}$ |
| LoTr 5 | 0.4 | 0.04±0.01 | $<2.0 \cdot 10^5$ | $<10^{18}$ | $(1.3±0.3) \cdot 10^{31}$ |

* the X-ray luminosity is given in range 0.1-2.4 keV

## References

Kreysing H.C. et al. 1992, As.& Ap., accepted.
Kreysing H.C, 1992, submitted, Ph.D. thesis, University of Tübingen
Kwok S. 1982, ApJ. 258, 280

# SPATIAL VARIATIONS IN UV-OPTICAL LINES ACROSS THE RING NEBULA

R.J. DUFOUR
*Rice University, Houston, TX, USA*

and

R. QUIGLEY
*Western Washington University, Bellingham, WA, USA*

IUE spectra of the Ring Nebula (M57=NGC 6720) were taken 16-18 May 1991 using the large aperture (10 X 20 arc sec oval) at low dispersion. SWP and LWP spectra of seven locations were acquired at 10 arc sec intervals extending along a PA = 124° line and passing through the central star of the nebula. This direction was such that the long axis of the large aperture was aligned and overlapped, thus enabling the line profiles to be "spliced" together. This provided continuous spatial variation curves at approximately 1 arc sec spatial resolution over the entire diameter (80 arc sec) of the main body of the Ring Nebula.

Spatial variation curves for the lines of CIV 1549Å, HeII 1640Å, CIII] 1909Å, CII] 2326Å, and [NeIV] 2424Å are presented and compared with corresponding curves for prominent optical lines extracted from longslit spectra taken at McDonald Observatory. Overall, we find the variations of the UV lines consistent with that expected from a basic model of the ionization and density structure of the nebula. Some implications of these regarding calculations of C/N/O/Ne abundances in PN are discussed.

Acquisition of the IUE observations was supported by NASA grant NAG5-262 and our collaboration made possible by the NASA JOVE Program.

# COLLIMATING DISCS AND BIPOLAR FLOWS IN SH 2-71

L. CUESTA and J. P. PHILLIPS

*Instituto de Astrofísica de Canarias, E-38200 La Laguna, Tenerife, Spain*

Sh 2-71 appears to represent a diffuse, ellipsoidal nebulosity extending over a range $\Delta\alpha \times \Delta\delta = 1.7 \times 3$ arcmin$^2$. Early spectroscopy by Glushkov et al (1975), and Chopinet and Lortet-Zuckerman (1976) suggested the presence of a high excitation central star, and Kaler (1983) has more recently determined Zanstra temperatures $T_z(\text{HeII}) > 7.7 \cdot 10^4$ K. The observed central star clearly constitutes a much cooler companion, and luminosity variations in this source have been attributed to binary eclipse.

We have recently undertaken a programme of high and low resolution spectrophotometry and NIR photometry of this source using the 2.5 m Isaac Newton Telescope (Observatorio del Roque de Las Muchachos), and 1.5 m Carlos Sanchez Telescope (Observatorio del Teide). As a consequence, we determine that:

(i) There is evidence for H$\alpha$ velocities extending over a range $\Delta V \sim 1026$ km sec$^{-1}$, with the principal emission peaks separated by $\Delta V \sim 230$ km sec$^{-1}$.

(ii) The kinematics of the exterior shell appear to be unusual; line velocities appear to shift rapidly over the nucleus, and mapping of the source reveals a distinct bilobal pattern, with high velocities to the N and W, and lower velocities to the SE.

The kinematics of the exterior shell are interpreted in terms of a tilted disk outflow, in which material is ejected with initial velocity $\sim 19.6$ km sec$^{-1}$, and subsequently decelerates towards the outer disk limits. Analysis of disk velocities and emission measures suggests overall mass-loss rates $\sim 10^{-6}$ M$_\odot$ yr$^{-1}$, although these values may have been higher in the past, whilst disk densities are likely to be low, and of order $\leq 10^2$ cm$^{-3}$. The disk inclination is estimated to be $\geq 50°$ to the line of sight.

It is proposed that this disk is also responsible for collimating the central high velocity wind into a bipolar flow, and we estimate associated core emission measures $1.6 \cdot 10^6$ cm$^{-6}$ pc, together with densities $n_e > 2.1 \cdot 10^4$ cm$^{-3}$. Mass loss rates appear to be dM/dt $\sim 2.0 \cdot 10-9$ M$_\odot$ yr$^{-1}$.

Finally, it is suggested that the cooler companion may be filling its Roche lobe, and mass loss from this component is responsible for creating both the bipolar flow, and collimating disk. The earlier spectral type B8V attributed to this star is almost certainly wrong, unless derived distances to this source are grossly in error. We propose, rather, that nuclear fluxes be attributed to a blend of A7V and hot central star continua; a combination which enables reasonable fits to the optical continuum, and would result in binary eclipse amplitudes $\Delta m_{pg} \simeq 0.75$ mags comparable to the luminosity variations found by Kohoutek (1979).

## M4-18: THE LOW EXCITATION PN AROUND A WC11 STAR

R.SURENDIRANATH and N.KAMESWARA RAO
*Indian Institute of Astrophysics, Bangalore 560034, India*

A photoionization model is presented for M4−18. The model includes the presence of dust mixed with gas. The model shows sulphur and nitrogen are underabundant (relative to solar) and partially succeeds in explaining the observed IR excess.

M 4−18 has been known to be a low excitation nebula of type WC11 (van der Hucht et al. 1981, Goodrich & Dahari 1985). All the nebulae of this WC11 group show spectra dominated by numerous stellar C II and C III lines. A large IR excess is typical of these objects. We have developed a model code (Surendiranath 1992) appropriate for a low excitation nebula having gas and dust mixed together. Combining our new CCD spectra taken at Kavalur, with published observations from UV to radio wavelengths, a photoionization model was made for M 4−18. Abundances of all elements except sulphur and nitrogen are found to be normal (relative to solar). The mean $T_e$ and $N_e$ are 7150 K and 6400 cm$^{-3}$ as per the model. The assumed presence of amorphous carbon grains having a power law distribution of sizes in the range of $0.04 \leq a \leq 0.55$ $\mu m$, explains the 12 and 25 $\mu m$ IRAS bandfluxes well while failing in the 60 and 100 $\mu m$ bands as well as in the range 1 to 10 $\mu m$. The model requires a hotter central star and this is interpreted in terms of a "born again AGB star" that has ejected a nebula for a second time.

### References

Goodrich, R.W., and Dahari, O. (1985) Astrophys. J. **289**, 342.
Surendiranath, R. (1992) Ph.D. Thesis, Bangalore University.
van der Hucht et al. (1981) Space Sci. Rev. **28**, 227.

# EXTENDED NEBULAE AROUND WC11 STARS: IRAS 17514-1555

D.L. POLLACCO

*Department of Physics and Astronomy, North Haugh, St. Andrews,*
*Fife KY16 9SS, Scotland*

Imaging and long-slit spectroscopy of the nebula surrounding the WC11 star IRAS 17514-1555 are presented that allow, for the first time, the nebula to be studied without contamination from the underlying WC11 star. These data show that the nebula is photoionized and does not appear to have an obvious scattering component. The nebula consists of two shells: a low density outer region (diameter 18 arcsec) where $n_e \sim 100 \text{cm}^{-3}$ and an unresolved inner nebula (diameter < 1.5 arcsec) where $n_e > 10^4 \text{cm}^{-3}$. The abundances in the nebula are similar to those found in Type I PN (He and N enriched). Simple models are constructed which demonstrate that the nebula is probably in equilibrium with the central star ($T_* \sim 30{,}000$ K). The nebular and stellar properties are compared to those of other late-type WC stars and the evolutionary status of IRAS 17514-1555 discussed.

# KINEMATICS OF THE PLANETARY NEBULA Hb 5.
## A Progress Report.

P. PIŞMIŞ[1], M. MANTEIGA[2], A. MAMPASO[3], G. CRUZ-GONZALEZ[1]
[1] *Instituto de Astronomía, UNAM. Apdo. Postal 70-264, 04510 México*
[2] *LAEFF Villafranca. Apartado 50727, 28080-Madrid, Spain*
[3] *Instituto de Astrofísica de Canarias, 38200-La Laguna, Tenerife, Spain*

The planetary nebula Hubble 5 shows a very striking bi-symmetry; a 180° rotation around its center would bring the object into fair coincidence. Similar to this nebula is NGC 6537 and, to a lesser degree, NGC 6302. Jet outflows are reported from this "post-main sequence" nebula over a range of 260 km s$^{-1}$ (Phillips and Mampaso, 1988). Otherwise not much has appeared on this object in the literature.

We have taken CCD direct images of Hb 5 in several spectral bands and in the Hα line with the 2.1m reflector of the Observatory at San Pedro Mártir, México. Long slit spectra were obtained in May of 1992 with the IDS spectrograph of the 2.5m Isaac Newton Telescope at the Observatory Roque de los Muchachos, Spain. The 6 slits passing through the center of the object are along position angles 5, 43, 67, 88, 107 and 335 degrees, respectively. The latter direction is the symmetry axis of the nebula.

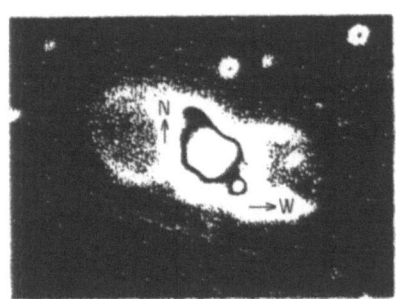 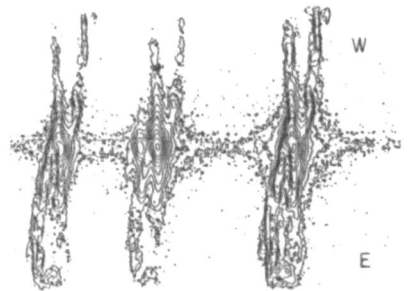

The left hand figure is an Hα CCD image of Hb 5 while the right hand one is a section of a typical spectrum at PA 67°; it shows Hα flanked by the [NII] lines. A wealth of information is contained in our material. Here we call attention to some properties leaving to a future publication a detailed treatment of our data.

A few relevant results: 1) The spectral lines and their splitting affords evidence for the expanding hollow structure of the lobes. 2) At about 4 arcsec from the center of Hb 5 the bright "wing" of the NE lobe is receding while its SW bright couterpart is approaching the observer. The difference in their projected velocity is roughly 100 km s$^{-1}$. 3) It appears that the direction defined by the lobes is inclined to the plane of the sky. 4) Based on the markedly unequal intensity of the edges of the lobes and the bi-symmetry of this phenomenon, we may state that such structure is not solely produced by interaction with the ambient matter, as commonly assumed; but most likely the morphology at the source of ejection is responsible for it. The wind causing the outflow is therefore not isotropic; rather it emanates from a bipolar source on the central star. Magnetic phenomena are expected to be at work in such a picture.

# A SYMMETRIC JET-LIKE STRUCTURE IN THE PLANETARY NEBULA FG 1

J.A. LÓPEZ and M. TAPIA

*Instituto de Astronomía, UNAM, Apartado Postal 877, Ensenada, Mexico*

and

M. ROTH

*Las Campanas Observatory, Casilla 601, La Serena, Chile*

Fg 1 (He 2-66) is a southern planetary nebula that presents an elliptical shape. Deep CCD imaging and long-slit spectroscopy have been obtained at Las Campanas Observatory of this object. The images were obtained in the light of H$\alpha$+[NII], [SII] $\lambda\lambda$6716,6731, HeII $\lambda$4686 and in the broad-band $R_{gunn}$ filter. The spectra were obtained oriented N-S, P.A. 90°, and P.A. 85°. The instrumental combination yields a spectral resolution of $\sim$ 2 Å FWHM, covering a spectral range $\simeq \lambda\lambda$ 6290 – 6805 Å.

The discovery of a symmetric jet-like strukture, consisting of two strings of ionized nebular knots, highly reminiscent of Herbig-Haro objects is reported. The strings are bent in opposite directions and span 2 arcmin to either side of the PN. The main body of Fg 1 consists of an orthogonal system of elliptical structures. The major axis size for the brighter of these structures is $\simeq$ 38 arcsec. The common axes of the strings pass right through the central star and coincides with the minor axis of the bright elliptical structure. The spectra from the opposite innermost knots, intersected by the slit, show opposite sign velocities, indicating an expansion radial velocity $V_{exp}$ 43 km s$^{-1}$ and line ratios typical of collisionally excited gas.

The pairs of opposite knots that make up the strings are symmetric and nearly equidistant with respect to the planetary nebula nucleus (PNN). These knots are interpreted as multiple ansae that have been formed and blown away in episodic events by symmetric, collimated flows probably produced by a pressing source. It is estimated that the outermost ansae system was produced between 3.42 10$^4$ and 1.34 10$^4$ years ago, when Fg 1 was in proto-planetary stage.

The details of this work will be published in *Astronomy and Astrophysics*.

# MONOCHROMATIC CCD IMAGES OF THREE PLANETARY NEBULAE

WALTER A. FEIBELMAN

*NASA-Goddard Space Flight Center, Lab. for Astronomy and Solar Physics, Greenbelt*

We present monochromatic false-color images of three planetary nebulae obtained by means of a CCD camera at the f/13.5 focus of the 0.91m reflector of the Cerro Telolo Interamerican Observatory in August, 1985. The resultant image scale is 0.363 arcsec square per pixel. The objects, IC 1297, NGC 7009, and M2-9 were imaged through narrow band filters in the V band, [O III] $\lambda 5007$, and H$\alpha$ $\lambda 6563$, respectively. The V band filter corresponds to the standeard Johnson V magnitude. The $\lambda 5007$ filter has a FWHM of 14 Å. The H$\alpha$ filter has a FWHM of 73 Å, thus including contributions from the [N II] $\lambda\lambda 6548, 6584$ lines. IC 1297 shows a double shell structure with a pronounced bright rim at the extreme edge of the faint outer envelope, suggesting that it is density bounded. NGC 7009 (the "Saturn" nebula), in addition to the well-known double shell and extended W-W ansae, also shows a faint large, circular outer halo which strongly suggests that NGC 7009 is a member of the rare class of triple-shell nebulae. The bipolar object M2-9 (the "Butterfly" nebula) shows a bright, non-stellar central core from which the wings extend in a nearly N-S direction. Condensations in the wings are seen particularly well for the [O III] $\lambda 5007$ images. The CCD images were obtained by B. Schaefer at CTIO and were processed at the Goddard Image Processing Facility.

# HIGH-RESOLUTION CCD IMAGERY OF NGC 6537 AND NGC 7027

S.R. HEAP
*Laboratory for Astronomy and Solar Physics*
*Goddard Space Flight Center, Greenbelt MD 20771 U.S.A.*

Using Hα and Hβ images obtained at the CFHT, we derived the spatially varying reddening within NGC 6537 and NGC 7027. We then corrected the observed Hβ picture for dust extinction to obtain pictures more in accord with radio maps. The left panels below show NGC 6537; the right panels, NGC 7027. The top panels show the observed Hβ picture, and the bottom panels, the Hβ picture corrected for spatially varying extinction. Note that the bright optical knot in NGC 7027 corresponds to a hole in the extinction.

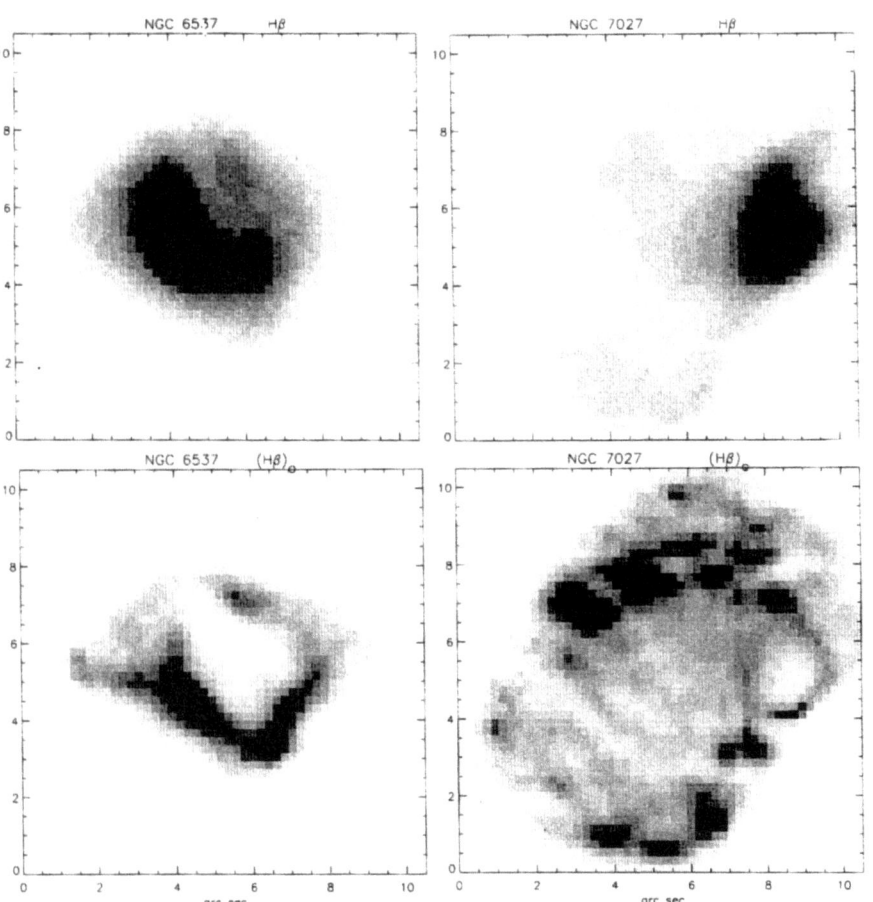

## OPTICAL IMAGERY OF NGC6302

JOAQUÍN BOHIGAS

*Instituto de Astronomía, U.N.A.M., Apartado Postal 877, Ensenada,*
*22830 Baja California, Mexico*

Images with narrow band filters centred at the most significant emission lines, such as H$\beta$, H$\alpha$, [SII] 6717 and [SII] 6731, were taken to explore the physical conditions of NGC 6302. Observations were secured on May 1991 with the 2.0m telescope of the Observatorio Astronómico Nacional at San Pedro Mártir, B.C., Mexico, the f/7.5 secondary, and a 384 × 576 CCD chip. The most significant results are:

1. Excitation mechanism. The image of the H$\alpha$/[SII] lines ratio reveals that photoionization and shocks are equally important throughout most of the object. Photoionization is dominant in the central region and a few isolated patches. Elongated structures pointing towards the central region can be identified with a shock impinging the wall of a bipolar cavity.
2. Extinction. The image of the H$\alpha$/H$\beta$ lines ratio shows that extinction changes over all distance scales. C(H$\beta$) is nearly equal to 1.7 in the central region. It falls off very rapidly away from it, and at a distance of 10 arcsec it is approximately equal to 1.3. In the mean, H$\alpha$/H$\beta$ is 1.25 times larger in the western lobe than in the eastern.
3. Excitation. Different lines ratios, such as [SII] 6724/[SIII] 9069, [OIII] 5007/[OI] 6300 and HeI 5876/H$\beta$, indicate that excitation is larger in the eastern lobe. For instance, HeI 5876/H$\beta \simeq 0.15$ in the eastern lobe and $\simeq 0.13$ in the western region.
4. Electron density was calculated from the [SII] 6717 and [SII] 6731 images. The density structure is very complex. Within the central region the highest density is found at both sides of the dark lane. No clear correlation exists between density and shock excited regions. Wave-like patterns can be recognized in the density image. These can be attributed to instabilities, sporadic ejections events and/or multiple shocks.

In conclusion, physical parameters change over all distance scales. This should be acknowledged when calculating chemical abundances.

# A TWODIMENSIONAL IONISATION MODEL OF NGC 2440

M. BÄSSGEN (1), C. DIESCH (1) and M. GREWING (1,2)

(1) Astronomisches Institut der Universität
Waldhäuserstraße 64, D-7400 Tübingen
(2) Institute for Radioastronomy at Millimeter Wavelengths (IRAM)
300,rue de la Piscine,Domaine Universitaire,F-38406 St.Martin d'Hères, France

**Abstract.** NGC 2440 is one of the most prominent butterfly- shaped nebula in the sky. It has a very rich spectrum up to high ionic species. Its central star is handled as one of the hottest central stars. We present a 2-dimensional cylindrically symmetric ionisation model of this nebula. Under the assumption that the main nebula axis lies perpendicular to the line of sight and achieving the inverse Abel-transformation, the density distribution could be calculated using a monochromatic $H\alpha$-CCD-image. The gridsize of the model was chosen to 0.38 arcsec (=pixel size of the image). Due to the fact that in a cylindrically symmetric model only a half plane has to be calculated, the total grid has a size of (-100..100, 0..100). The procedures determining the ionisation state is principally the same we used for our 3D-model of NGC 3132 (Bässgen et al 1990). If the ionic densities and the electron temperatures in all volume elements are known, emission line strengths and artificial monochromatic images can be calculated. These can be compared with "real" monochromatic images.
The physical parameters which yielded the best agreement with observations are given in the following table.

| | |
|---|---|
| Distance of the object | 500 pc |
| Teff of Central Star | 125,000 K |
| Luminosity of Central Star | 260 $L_\odot$ |

That means that the object is much closer than previous estimations suggested. We also could not confirm proposed central star temperatures of more than 200,000 or 300,000 K. In a paper which is in preparation we will present our results in a more detailed manner.

### References

Bässgen M., Diesch, C., Grewing, M. 1990, Astron. Astrophys. 237, 201

# EXTENDED STRUCTURES IN THE PLANETARY NEBULAE HE2-111 AND HE2-119

J.A. LÓPEZ and M. TAPIA

*Instituto de Astronomía, UNAM, Apartado Postal 877, Ensenada, Mexico*

and

M. ROTH

*Las Campanas Observatory, Casilla 601, La Serena, Chile*

He 2-119 is a bright, elliptical, planetary nebula of relatively large size (117 arcsec along the major axis and 74 arcsec along the minor one). Digital, unsharp masking CCD imaging of this object is presented, revealing a filamentary, nearly bipolar inner structure. In addition, the discovery of an extended, faint halo in this object is reported. The halo has a diameter of 208 arcsec and has a nearly circular form. He 2-119 thus becomes a new member of the group of planetary nebulae with halos. Its general characteristics are discussed.

As for He 2-111 (the mandrill nebula) this is a planetary nebula best known for its core appearence, however, its structure is that of an extremely large bipolar (lobe size $\sim$ 5.4 arcmin), probably produced by a Nova-like event. Deep CCD imaging is presented, showing most of its entire extent. As for the case of He 2-119, digital, unsharp masking image processing techniques are used to display the bright core and the extended, faint lobes at levels that show the internal filamentary structure in both cases. A number of noteworthy morphological characteristics are pointed out in this most peculiar bipolar nebula.

The details of this work will be published in *Rev. Mexicana Astron. Astrofis.*

# CCD IMAGING OF PLANETARY NEBULA HALOS

K.B. KWITTER
*Astronomy Dept., Williams College, Williamstown, MA 01267 USA*

Y.-H. CHU
*Astronomy Dept., University of Illinois, Urbana, IL 61821 USA*

and

R.A. DOWNES
*Space Telescope Science Institute, Baltimore, MD 21218 USA*

We have obtained deep CCD images of 14 PN to search for and examine faint halos. These images were obtained with H$\beta$, [OIII], H$\alpha$, and [NII] narrowband interference filters on the Burrell Schmidt at Kitt Peak. Table 1 summarizes the observations. We have found interesting features around NGC 1360, NGC 3587 (The Owl), and NGC 6853 (The Dumbbell), and we have obtained new images of the very faint outer halo of NGC 7293 (The Helix). Two CCDs were used: TI6, an 800x800 chip, with 1.45"/pix and a 19.3 arcmin square field., and ST2K, a 2048x2048 chip, of which only 1200x1200 pixels were read out, yielding 2.07"/pix and a 40 arcmin square field.

TABLE 1. Planetary Nebulae Observed

| Nebula | Chip | Filters | Nebula | Chip | Filters |
|---|---|---|---|---|---|
| NGC 40 | TI6 | H$\beta$, [OIII], H$\alpha$, [NII] | NGC 6309 | ST2K | [OIII], H$\alpha$, [NII] |
| NGC 246 | ST2K | H$\alpha$ | NGC 6543 | ST2K | [OIII], H$\alpha$, [NII] |
| NGC 650-1 | ST2K | H$\alpha$, [OIII] | NGC 6853 | TI6 | H$\beta$, [OIII], H$\alpha$, [NII] |
| NGC 1360 | ST2K | [OIII], H$\alpha$, [NII] | NGC 6720 | ST2K | [OIII], H$\alpha$, [NII] |
| NGC 1514 | TI6 | H$\beta$, [OIII], H$\alpha$, [NII] | NGC 7293 | ST2K | [OIII], H$\alpha$, [NII] |
| NGC 2022 | TI6 | H$\beta$, [OIII], H$\alpha$, [NII] | NGC 7354 | TI6 | [OIII] |
| NGC 3587 | ST2K | [OIII], H$\alpha$, [NII] | NGC 7662 | TI6/ST2K | [OIII]/H$\beta$, H$\alpha$,[NII] |

Figure 1 is NGC 1360 in [OIII]; the lobes projecting to the northeast and southwest are also visible in H$\alpha$ and [NII]. Figure 2 is NGC 3587 in [OIII]; the off-center, round halo seen here is also visible, though not nearly as prominent, in H$\alpha$ and [NII]. Figure 3 is NGC 6853 in [OIII]; the extended, almost bipolar halo is also evident, though clumpier, in H$\alpha$ and [NII] (Kwitter et al. 1991). Figure 4 is NGC 7293 in H$\alpha$; the scalloped projections to the east and the diffuse clump to the northeast were first noticed by Malin (1982). The clump is also bright in [NII]. More detailed analysis of these halos is in progress (Kwitter et al. 1992).

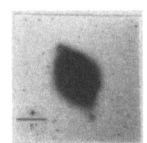
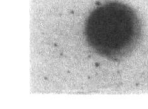
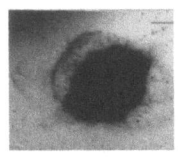
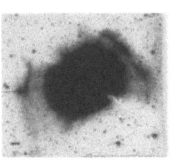

Fig. 1 NGC1360-[OIII]   Fig. 2 NGC3587-[OIII]   Fig. 3 NGC 7293-H$\alpha$   Fig. 4 NGC6853-[OIII]

## References

Kwitter, K.B., Downes, R.A., and Chu., Y.-H. 1991, "Schmidt CCD Images of the Dumbbell Nebula (NGC 6853)," Bulletin of the A.A.S., 23, 914.Kwitter, K.B., Chu., Y.-H., and Downes, R.A. 1992, "A Study of Planetary Nebula Halos," in prepara tion.

Malin, D.F. 1982, "A Look at Some Unstable Stars," Sky and Telescope, 63, 22.

# NEAR- AND MID-INFRARED IMAGING OF THE PN IC 418

J. L. HORA
*Institute for Astronomy, 2680 Woodlawn Dr., Honolulu, HI 96822, USA*

L. K. DEUTSCH
*NASA/Ames Research Center, M/S 245-6, Moffett Field, CA 94035, USA*

W. F. HOFFMANN
*Steward Observatory, University of Arizona, Tucson, AZ 85721, USA*

G. G. FAZIO
*Smithsonian Astrophys. Obs., 60 Garden Street, Cambridge, MA 02138, USA*

and

K. SHIVANANDAN
*Center for Advanced Space Sensing, NRL, Washington, DC 20375, USA*

**Abstract.** We present high-resolution near- and mid-infrared images of the planetary nebula IC 418 at 1.2 ($J$), 1.6 ($H$), 2.2 ($K$), 9.8, and 11.7 $\mu$m. The near-IR images were obtained with a 64x64 pixel Hg:Cd:Te array camera, and the mid-IR images were obtained using the new 20x64 pixel Mid-Infrared Array Camera (MIRAC). The size of IC 418 in the near-IR is seen to vary with wavelength, being largest at $K$ and smallest at $J$. Differences in the morphology of the nebula are seen between the $J$ and $K$ images. There is excess near-IR emission in the center of the nebula, after subtracting out the emission from the central star. Faint halo emission is detected at $H$ and $K$, extending to a total diameter of approximately 40″. The 9.8 $\mu$m emission is distributed nonsymmetrically, with the peak in the NE lobe.

The images of IC 418 at 2.2 and 9.8 $\mu$m are shown in Figure 1. Images of the nebula at the other wavelengths are presented elsewhere (Hora 1991). The brightest nebular emission in both images is in the NE lobe, at roughly the same position angle. However, there is no 9.8 $\mu$m SW emission lobe, and no emission peak from the region near the central star. The FWHM size of the 2.2 $\mu$m image is ~1″ larger than the 9.8 $\mu$m image. No emission from the halo region is detected at 9.8 $\mu$m.

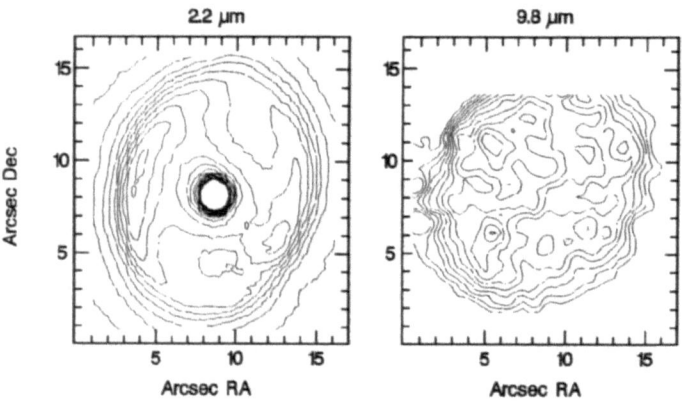

Figure 1. Contour images of IC 418 at 2.2 and 9.8 $\mu$m.

## References

Hora, J. L. 1991, Ph. D. Dissertation, University of Arizona

# MID-IR (8-13μm) IMAGES OF PLANETARY NEBULAE

M. MEIXNER, J.F. ARENS and J.G. JERNIGAN

*Dept. of Astr. and Space Sci. Lab., U.C. Berkeley, Berkeley, CA 94720 USA*

and

J.R. BALL and C.J. SKINNER

*LLNL, IGPP, L-413, P.O. Box 808, Livermore, CA 94550 USA*

We present mid-IR (8-13μm) images of dust in six young planetary nebulae: IC 418, IRAS 21282+5050, NGC 6790, M4-18, M2-9 and IC 5117. The images were taken at UKIRT and the IRTF with the Berkeley mid-IR camera which was developed at the Space Sci. Lab. in UC Berkeley and is supported by IGPP and LEA, LLNL. In IC418, M2-9 and IRAS 21282, the spatial distributions of dust and of ionized gas are measurably different. In M4-18, IC5117 and NGC 6790 the spatial distributions are similar, so apparently the dust and the gas are well mixed. Spatial resolution is the key to discerning the differences in the gas and dust morphologies, and in 2 of the 3 cases where the spatial distributions appear similar, higher resolution may reveal differences. For example, we can discern that the [NeII] peaks further out than the SiC in IC 418 because it is so close and thus large in angular size compared to our $\sim 0.''8$ resolution. In IRAS 21282+5050, we have compared the spatial distribution of the different features attributed to Polycyclic Aromatic Hydrocarbons to the dust continuum (Fig.1). Our data shows that while the 8.5 and 11.3 μm emission peaks outside the 10μm dust continuum, the 12.5μm emission is spatially coincident with it. Hence, not all PAH emission peaks outside of the dust continuum. The plateau emission, sampled by our 12.5μm band, arises from smaller PAHs than the size that causes the 11.3μm PAH feature. Thus, one interpretation of our observations is that the grain size distribution changes with radius: smaller PAHs in the center, increasing size with radius.

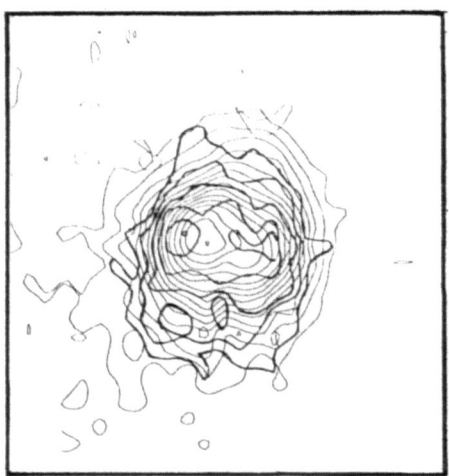

Fig. 1. IRAS 21282+5050: 8.5μm (heavy line) overlaid on 10μm

# IMAGING OF MAGELLANIC CLOUD PLANETARY NEBULAE WITH THE HUBBLE SPACE TELESCOPE

M. A. DOPITA [1], S. J. MEATHERINGHAM [1], P. R. WOOD [1]
H. C. FORD [2], R. C. BOHLIN [2]
T. P. STECHER [3], S. MARAN [3] and J. P. HARRINGTON [4]

[1] *Mt. Stromlo and Siding Springs Observatories, Institute of Advanced Studies, The Australian National University,* [2] *The Space Telescope Science Institute,* [3] *NASA Goddard Space Flight Center,* [4] *The University of Maryland.*

We have obtained Hubble Space Telescope (HST) Planetary Camera (PC) images of a number of Magellanic Cloud planetary nebulae. The objects, except for SMP 83 were observed as part of the Cycle I GO program. The observations were made in the [O III] $\lambda 5007$Å line. The object SMP 83, was observed as part of the GTO program, and in this case observations were also made in the H$\alpha$ line using the F650N filter. In order to characterise the point spread function, a star was placed at the same point on the chip as the PN. This allowed us to determine the diameters of barely resolved PN in an accurate manner, by convolving the PSF with a function until it matched the appearance of the PN image. The results are given in Table 1.

**Table 1:** Dimensions and Dynamical Ages of the LMC PN inferred from HST Images.

| Object | Size (arc sec.) | Radius (cm) | Dynamical Age (yr) |
|--------|-----------------|-------------|---------------------|
| SMP 02 | 0.25x0.25 | 9.32e+16 | 2980 |
| SMP 08 | 0.06x0.07 | 2.43e+16 | 310 |
| SMP 20 | 0.21x0.43 | 1.61e+17 | 1969 |
| SMP 35 | 0.64x0.86 | 2.80e+17 | 2150 |
| SMP 47 | 0.13x0.21 | 7.84e+16 | 320 |
| SMP 72 | 1.72x2.02 | 6.97e+17 | - |
| SMP 76 | 0.13x0.13 | 5.24e+16 | 570 |
| SMP 83 | 0.70x1.46 | 4.04e+17 | 1544 |
| SMP 85 | 0.07x0.07 | 2.60e+16 | 1070 |
| SMP 87 | 0.79x0.82 | 3.02e+17 | 2560 |
| SMP 96 | 0.17x0.52 | 1.94e+17 | 1007 |

All the low excitation PN (E.C < 4) are compact objects, barely resolved with HST. These have [O III] diameters which are much smaller than was expected on the basis of photoionisation models, even when the fact that the $O^{++}$ zone occupies only a fraction of the ionised volume has been taken into account. This result can therefore be taken to mean that the $O^{++}$ zone is denser than the model. A second result is that the dynamical ages of these PN, as shown in Table 1, are much shorter than the evolutionary timescale of the central stars. This situation can only be maintained if there exists a dense, slowly expanding core of un-ionised gas in these PN, perhaps associated with shell ejection in the last helium flash. This gas would have to have an expansion velocity of only ~1-3 km.s$^{-1}$ in order to have remained compact to the present day.

The results described in this paper are based on observations with the NASA/ESA *Hubble Space Telescope*, obtained at the Space Telescope Science Institute, which is operated by the Association of Universities for Research in Astronomy, Inc., under NASA contract NAS5-26555.

# HUBBLE SPACE TELESCOPE IMAGES OF FOUR MAGELLANIC CLOUD PLANETARY NEBULAE

M. J. BARLOW

*Department of Physics & Astronomy, University College London, Gower Street, London WC1E 6BT, U.K.*

and

J. C. BLADES, S. OSMER & THE FAINT OBJECT CAMERA I.D.T.

*Space Telescope Science Institute, 3700 San Martin Drive, Baltimore, MD 21218, USA*

**Abstract.** Using the Faint Object Camera on-board the Hubble Space Telescope, we have obtained images of four planetary nebulae in the Magellanic Clouds, namely N2 and N5 in the SMC and N66 and N201 in the LMC. Each nebula was imaged through narrow-band filters isolating [O III] $\lambda 5007$ and H$\beta$, for a nominal exposure time of 1000 seconds in each filter. The f/96 optical chain of the FOC was used, yielding 512×512 0.022 arcsec square pixels. Considerable detail is evident on the raw images and after deconvolution using the Richardson-Lucy algorithm, structures as small as 0.06 arcsec are easily discernible. Figure 1 shows NS and EW intensity cross-cuts through the deconvolved [O III] $\lambda 5007$ images of SMC N2 and SMC N5. SMC N2 is a slightly ellleptical ringlike nebula, with its greatest elongation in the EW direction. The peak-to-peak dimensions of the ring are $0.21 \times 0.26$ arcsec$^2$. SMC N5 has a circular ring shape, with the [O III] $\lambda 5007$ image showing a clearly defined, nearly uniform structure, apart from a bright patch at the northern edge. The peak-to-peak diameter of the ring is 0.26 arcsec and the ring itself is significantly narrower than that of SMC N2, with a width as small as 0.06 arcsec (FWHM) in some places. LMC N201 is very compact, with a FWHM of 0.21 arcsec in the H$\beta$ image. The Type I PN LMC N66 is a multi-polar nebula, with the brightest part having an extent of about 2 arcsec. Its structure is extremely complex, with several bright knots and faint loops visible outside the two bright lobes. A full description of our results can be found in Blades et al., *ApJ*, **398**, L41-44.

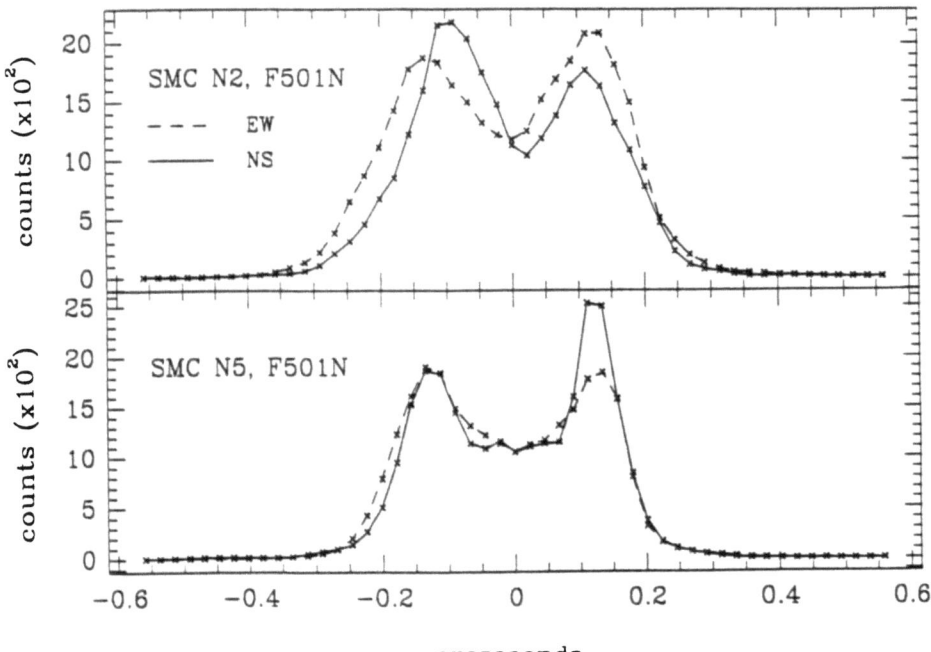

# Hα MORPHOLOGICAL CLASSIFICATION OF PLANETARY NEBULAE

HUGO E. SCHWARZ and ROMANO L.M. CORRADI
*ESO, Casilla 19001, Santiago 19, Chile*

and

LETIZIA STANGHELLINI
*Osservatorio Astronomico di Bologna, via Zamboni 15, Bologna, Italia*

Using the data from the catalogue of PNe images of Schwarz et al. (1992), we have classified 255 nebulae into 5 morphological groups. The images were taken through a filter centred on the Hα line and included the [NII]6583 line. Most of the observations were made with the 3.5m ESO NTT during commissioning time.

The main groups are called: stellar (st), irregular (i), elliptical (e), bipolar (b), and point-symmetric (p). The ellipticals are subdivided into: simple (e), multiple event (em), and inner structure (es) types. The bipolars into: simple (b), and multiple event (bm). The point-symmetric objects into: simple (p), and multiple events (p-m).

Preliminary simple statistical tests show that there are significant differences between the physical properties of the various groups. When comparing the HR diagram of the whole sample, with that of sub-groups, especially the bipolar nebulae show markedly different behaviour. Their mass and luminosity distributions deviate strongly from those of the whole sample, as well as from those of other sub-groups. Details can be found in the poster of Stanghellini et al., presented here.

## References

Schwarz, H.E., Corradi, R.L.M., and Melnick, J. 1992 AASS, in the press

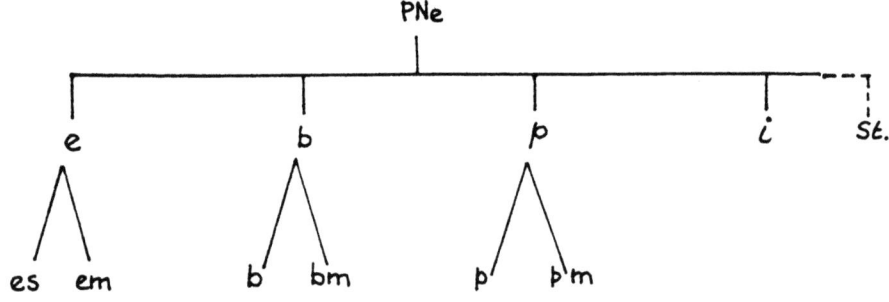

# A PC-BASED QUICKLOOK-PROGRAM FOR PN IMAGES

M. BÄSSGEN and M. BREMER
*Astronomisches Institut der Universität*
*Waldhäuserstraße 64, D-7400 Tübingen*

**Abstract.** We developed a computer program, running on a PC which allows a quicklook of monochromatic PN-images. Until now the data base consists of about 300 images obtained by B. Balick, M. Grewing, L. Bianchi and ourselves. This data base can easily be enhanced. Four monochromatic images (Hα, NII, OIII and HeII, if available) are shown simultaneously. It is also possible to show up to six objects in one emission line simultaneously to compare special features. Different look-up-tables are possible. The program does not compete with reduction packages, its only intention is to do some 'quick morphological studies' in a comfortable and easy way. In the present version the program needs a PC equipped with a mouse, a TSENG based VGA graphics card, 640 K (EMS rec.) and about 15 MB free disk space (for the images).

## POINT SYMMETRY IN PLANETARY NEBULAE

ROMANO L.M. CORRADI and HUGO E. SCHWARZ
*ESO, Casilla 19001, Santiago 19, Chile*

and

LETIZIA STANGHELLINI
*Osservatorio Astronomico di Bologna, via Zamboni 15, Bologna, Italia*

We present narrow band images and spectra illustrating the morphological and kinematical characteristics of a new class of planetary nebulae. It consists of all those objects in the catalogue by Schwarz et al. (1992) whose morphology shows no other symmetries than point-reflection about the central source. This peculiar shape is often determined by pairs of ansae or blobs separated from the central body of the nebula. The kinematical data show both low (e.g. IC 4634) and high (He 2-186; $V_{exp} > 135$ km s$^{-1}$) velocity outflows.

The peculiar property of the pairs of blobs is that: a) the radially opposed blobs in a given pair are red, respectively blue shifted, and b) the different pairs can have reversed shifts. This means that red and blue shifted material is seen on the same side of the central nebula.

We explain these properties as being the result of multiple mass ejections from interacting binary systems composed of a hot star, accreting matter from a mass-losing primary, and forming an accretion disk. The observed blobs then are the result of discrete mass ejection events. The relative blue/redshifted patterns are caused by precession of the disk driving the mass outflow from the system. The observed morphology and kinematics are then explained by projection of the precession cone onto the plane of the sky.

### References

Schwarz, H.E., Corradi, R.L.M., and Melnick, J. 1992 AASS, in the press

# SPECTROPHOTOMETRY AND MULTICOLOUR IMAGERY OF THE PLANETARY NEBULA AROUND THE P CYGNI STAR AG CARINAE

C. ROSSI and A. ALTAMORE
*Ist. Astronomico, Univ. La Sapienza, Via Lancisi 29, 00161 Roma, Italy*
R.D.D. COSTA, A. DAMINELI NETO and J.A. DE FREITAS PACHECO
*Inst. Astron. e Geofís., Caixa Postal 9638, 01065 São Paulo SP, Brasil*
and
A. CASSATELLA, A.R. MARENZI, P. PERSI, V.F. POLCARO and R. VIOTTI
*Ist. Astrofisica Spaziale, CNR, Via Fermi 21, 00044 Frascati RM, Italy*

Some of the high luminosity stars in our Galaxy are surrounded by planetary-like nebulae formed by material ejected from the central star. The most interesting case is that of the ring nebula PK 289-0°1 around the P Cygni star AG Car. Long slit spectroscopy shows that nitrogen is overabundant and oxygen underabundant in the nebula. The H$\alpha$/[NII] ratio is lower in the nebula with respect to the surrounding H II region, possibly as a result of the N overabundance in the stellar wind. The emission line peak separation confirms a model of a distorted spherical shell expanding at 66 km s$^{-1}$. The scattered star's spectrum is observable near the star, suggesting the presence of circumstellar dust grains. A nebular mass of at least 2.7 M$_\odot$ is derived. While the nebula in the H$\alpha$ imagery reveals the ring-like shape with many structures, in the blue it is much fainter and smoother. No nebular emission was detected in the JHK bands, suggesting a low dust temperature.

# AN ITERATIVE METHOD FOR THE RECONSTRUCTION OF TWO-DIMENSIONAL DENSITY DISTRIBUTIONS

M. BREMER

*Astronomisches Institut der Universität, Waldhäuserstraße 64, D-7400 Tübingen*

M. GREWING

*Astronomisches Institut der Universität, Waldhäuserstraße 64, D-7400 Tübingen*

and

*Institute for Radioastronomy at Millimeter Wavelengths (IRAM), 300, rue de la Piscine, Domaine Universitaire, F-38406 St .Martin d'Heres, France*

**Abstract.** As hydrodynamic models of planetary nebulae advance from 1D to 2D calculations, it becomes desirable to make the same step for reconstruction techniques, which aim at deriving from the 2D intensity distributions of PNe images their 3D structure. A basic step is the determination of a symmetry axis and its orientation in space which can be described by two angles, one measured in the plane of the sky, and one measured with respect to the tangential plane. While the first one can be determined from the images, e.g. by applying the criterion of maximum normalized correlation between the object halves, which we found to yield the best results, the second angle is treated as a free parameter (see below). The steps of the iterative reconstruction algorithm are the following (with image(x,y) = input image):

1. Fill a 3D kartesian density grid(xyz) (random or continuous), observing the constraint
$\int grid(xyz)dz = image(x,y)$, where z is along the line of sight.
2. Transform to cylindrical symmetry system (r,z',$\varphi$) and read out density(r,z') averaging over $\varphi$.
3. Return density(r,z') into grid(xyz) and normalize the $\int grid(xyz)dz$ to match the input image.
4. Return to (2.) as long as two subsequent density(r,z') estimations differ by more than a specified limit.

This algorithm can be applied to CCD images without previous smoothing and cannot produce negative densities like the inverted Abel Transform (IAT). However, the results are not unique. A special status for the observer's point of view (condensations followed by low density regions are allowed in the line of sight) can be included or suppressed by the careful choice of weighting functions $g(\varphi)$ at $\varphi$-averaging (for example, $g(\varphi) = |sin\varphi|^m$ : m=0 special, increasing m less special). An entirely different, noniterative approach using the IAT and deconvolving the tilt was constructed and tested with less success. This algorithm turned out to be extremely instable (oscillations, negative densities) and was therefore given up. Using the iterative algorithm with monochromatic CCD H$\alpha$ images as input, we present density distributions for several objects with different morphologies under different assumed tilt angles. Unfortunately, several of them are 'plausible' under hydrodynamical considerations, i.e. the uniqueness problem remains to be solved.

# RING NEBULAE AROUND POPULATION I WR STARS: IS THEIR ORIGIN SIMILAR TO THE PNe?

TATIANA A. LOZINSKAYA

*Sternberg Astronomical Institute, Moscow Lomonosov University, Russia*

MICHAEL A. DOPITA

*Mt. Stromlo and Siding Spring Observatories, The Australian National University, Australia*

and

YOU-HUA CHU

*Department of Astronomy, University of Illinois at Urbana-Champaign*

Interference filter CCD images have been obtained in H$\alpha$ and [O III] $\lambda$5007 for all population I WR stars in the LMC and SMC using the 2.4 m telescope at Siding Spring Observatory ANU. We have found new ring nebulae in the LMC around Br#13, 25, 31 and 56. Ring nebulae around Br 31 and Br 13 most probably represent stellar ejecta. Nebula around Br 56 is a filamentary wind-blown bubble type.

Our new deep survey of WR stars in the LMC confirms the statistics of galactic WR ring nebulae. Only about 16% of WR stars in the LMC are found to be associated with small-size shells of the ejecta and wind-blown types.

Why do we not see wind-blown ring nebula around every WR star? The most reasonable explanation is that a stellar shell is ejected at a velocity of about 100 km/s and a fast stellar wind of the central WR star sweeps up the ejected circumstellar material.

Statistics of WR-rings agree with the suggestion that all WR stars eject circumstellar shells, if taken into account is the detectability of ejected and swept up interstellar material in the LMC. Therefore we adopt a "two-wind model" for the origin of the small-size ring nebulae around population I WR stars. In terms of the model WR - ring nebulae represent a large-scale version of a PN. Most "big" shells and supershells in the LMC which contain a WR star could actually be created by common winds and SNe of the parental OB association.

# PHOTODISSOCIATION REGIONS IN PLANETARY NEBULAE

V. ESCALANTE

*Instituto de Astronomía, UNAM, Ap. Postal 70-264, México, DF 04510, México*

and

A. GÓNGORA-T.

*Instituto de Física, UNAM, Ap. Postal 20-364, México, DF 01000, México*

**ABSTRACT.** The photodissociation rate of $H_2$ molecules by UV photons from PN central stars is generally several orders of magnitude larger than the rate produced by the average interstellar field (Sternberg 1988, Escalante et al. 1991). Thus, in neutral envelopes of PN's, $H_2$ molecules are destroyed quickly, and a photodissociation region forms around ionization bounded nebulae. Observations of $H_2$ in PN's reveal that not all the hydrogen is photodissociated, and it has been suggested that this is due to the existence of dense disks around the ionized region (Zuckerman and Gatley 1988).

In order to test different geometries and to examine the possibility that the atomic hydrogen observed in association with some PN's has been produced by radiation from the central star, we have used a photodissociation model in a gas with uniform temperature, and with formation rates of hydrogen molecules on grain surfaces typical of the interstellar medium. The model predicts the HI column density and mass, the 21 cm flux density and the brightness temperature profile. We have considered spherical envelopes with variable density ($n \propto r^{-2}$) and disk structures like the ones observed in bipolar PN's with variable and constant density. We used expansion velocities measured by Taylor et al. (1990).

The results show that radiation of the central star can photodissociate the entire envelope in the cases with a spherical envelope and low column density ($N(H) \lesssim 5 \times 10^{20}$ cm$^{-2}$), and produce a region of extended low intensity, optically-thin 21 cm emission around the ionized zone. In the case of a molecular disk the central star may not photodissociate all the gas. The observed intensity of the 21 cm emission in NGC 6302, which is believed to have such a disk (Gómez et al. 1989), suggests that if the photodissociated region is formed in a disk-like envelope, the width of the disk must be considerably larger ($\gg 0.01$ pc) than the typical dimensions of the ionized region. The mass of atomic H in spherical envelopes with variable density can be $\sim 0.1 M_\odot$, but it depends strongly on model parameters. For NGC 6302 we predict a mass of atomic H of $0.05 M_\odot$ in good agreement with the value obtained by Gómez et al. (1989).

### References

Escalante, V., Sternberg, A., and Dalgarno, A., 1991, *Ap. J.*, **375**, 630.
Gómez, Y., Moran, J.M., Rodríguez, L.F., and Garay, G., 1989, *Ap. J.*, **354**, 862.
Sternberg, A., 1988, *Ap. J.*, **332**, 400.
Taylor, A.R., Gussie, G.T., and Pottasch, S.R., 1990, *Ap. J.*, **351**, 515.
Zuckerman, B., and Gatley, I., 1988, *Ap. J.*, **324**, 501.

# A MODEL OF THE CHEMISTRY IN THE NEUTRAL SHELL OF A PLANETARY NEBULA

S. N. GOULDSWORTHY and D. R. FLOWER
*Physics Department, University of Durham Durham, England*

A model of the neutral region of a planetary nebula has been constructed, building on an existing program (Abgrall et al. 1992, Astr. Astrophys. 253, 525). It incorporates a large set of equations governing the formation and destruction of molecular species and also covers photo-dissociation/ionization reactions and cosmic ray interactions. The radiation field impinging on the nebula is modelled as a $10^5$ K diluted Planck spectrum, truncated below 91.2 nm, augmented by spectral emission lines from the ionized region (data from G. Stasinska, private communication) and the hydrogen (2s→1s) two-photon continuum. The chemical species involved in the reactions are composed of seven elements - H, He, C, O, N, S and Fe - with H and He dominating the elemental abundances. The model considers a chemical environment which is carbon-rich (i.e. C/O > 1).

The transfer of radiation in the ultraviolet absorption lines of $H_2$ and CO is solved in order to accurately determine photodissociation rates for these two important molecular species. For other species the photorates decrease exponentially with optical depth owing to dust absorption and are further reduced through geometrical dilution of the radiation field. Where possible, photorates are calculated more accurately by direct numerical integration of the product of the local photon field strength and the cross-section of the process. This method obviously requires knowledge of the cross-section as a function of wavelength in the relevant wavelength range.

Dust grains cause an exponential decrease in the field intensity which is wavelength dependent. The grains are assumed to be a mixture of silicate and graphite and the wavelength dependence of their absorption coefficient was taken from the literature (Mathis et al. 1983, Astr. Astrophys. 128, 212).

The results of the model show the expected trends in the abundances of major molecular species. Molecular hydrogen and CO rapidly form as the dust optical depth $\tau_v$ increases, owing to the efficacy of self-shielding. Recent observations have detected other molecules within planetary nebulae e.g. HCN, HNC and $HCO^+$ in NGC 2346 (Bachiller et al. 1989, Astr. Astrophys. 210, 366); CN, HCN, HNC and $HCO^+$ in NGC 7027 (Cox et al. 1992, Astr. Astrophys. in press). The chemical modelling includes these species and many other molecules that may exist in the PN environment, e.g. $NH_3$, CS, OH, $C_2H$, $C_3H_2$, $HeH^+$. In its present form, the model predicts a column density of $HCO^+$ of about $10^{12}$ cm$^{-2}$ at $\tau_v = 5$ compared with a corresponding value of about $10^{21}$ cm$^{-2}$ for H. For HCN and HNC column densities comparable to $HCO^+$ are obtained, and for CN a larger value of about $10^{15}$ cm$^{-2}$ is computed.

# CO LINE EMISSION IN PLANETARY AND PROTO-PLANETARY NEBULAE: THE MOLECULAR ENVELOPE

G. SILVESTRO AND I. PORRO

*Istituto di Fisica Generale dell'Università, via P. Giuria 1, I-10125, Torino, Italy*

Planetary nebulae (PNe) are believed to form when the envelope of gas and dust, ejected at the end of the AGB phase of stellar evolution, is heated by the hot stellar nucleus, left after mass loss due to the "superwind" is terminated. The presence of residual molecular gas in PNe should therefore be important for understanding the formation and subsequent evolution of the nebulae. Observations (Bachiller et al. 1988) of CO line radiation in proto-planetary nebulae (PPNe), the transition phase from AGB stars to PNe, allowed to set very low upper limits on the molecular gas mass and to predict almost complete dissociation of the molecular envelope at the end of the PPN stage. On the other hand, observations (Huggins & Healy 1989; Healy & Huggins 1990) of CO lines detected intense signals from several PNe, including some evolved sources.

The two apparently contradictory sets of data are critically discussed. We point out the uncertainties in the results for PPNe, due to the small number of surveyed sources and to the lack of well defined identification criteria for this class of objects. New methods are suggested, which could allow to increase the number of identified PPNe and thus obtain more reliable estimates for the molecular mass in their envelopes.

## References

Bachiller, R., Gomez-Gonzales, J., Bujarrabal, V. & Martin-Pintado, J. (1988) Astron. Astrophys. 196, L5
Healy, A.P. & Huggins, P.J. (1990) Astron. J. 100, 511
Huggins, P.J. & Healy, A.P. (1989) Astrophys. J. 346, 201

# CO INTERFEROMETRIC MAPS OF CIT 6 AND CRL 618

M. MEIXNER and W.J. WELCH
*Dept. of Astronomy, U.C. Berkeley, Berkeley, CA 94720 USA*

We present CO (1-0) interferometer maps of carbon star, CIT 6, and carbon rich proto-planetary nebula, CRL618. Both objects were mapped with the Berekeley-Illinois-Maryland millimeter array (BIMA). The resulting resolution is $\sim 4''$ ; and structure larger than $\sim 1'$ is completely filtered out. Two velocity resolutions were employed; 0.4 km s$^{-1}$ and 0.8 km s$^{-1}$ permitting a velocity coverage of 52 and 104 km s$^{-1}$. We find evidence for non-spherical expanding envelopes in both objects. The interferometer detected all the single dish CO flux in CRL618 (840 Jy km s$^{-1}$), and 70% of the CO flux in CIT 6 (or 1030 Jy km s$^{-1}$). For CIT 6, we obtained single dish maps at the NRAO 12m to measure the zero spacing flux distribution. We measure an envelope diameter of 90'' ($2.6 \times 10^{17}$ at 190 pc) for CIT 6 and 15'' ($4.5 \times 10^{17}$ at 2 kpc) for CRL 618 and expansion velocities of 17 km s$^{-1}$ for both. The small scale structure seen by the interferometer in CIT 6 has a core and an elongation to the SW. We interpret this elongation as the beginning of an equatorial density enhancement and predict that that CIT 6 will evolve into a bipolar planetary nebula. In CRL 618, the elongated structure seen in CO is perpendicular to the optical reflection nebula, which has an east-west orientation. Hence the Asymptotic Giant Branch (AGB) wind appears as an expanding waist band around this bipolar nebula giving credence to the idea that equatorial density enhancements in AGB winds produce bipolar planetary nebulae (Balick 1987).

We measure a continuum flux of 1.6 Jy in CRL 618, confirming previous observations (Martin-Pintado et al. 1988). This continuum arises from the free-free emission in the central HII region which is unresolved in our observations. No continuum is detected in CIT6.

CRL 618 has been observed to have a high velocity molecular outflow (Cernicharo et al. 1989). We have marginally detected this outflow which is unresolved in our observations; and are working on improving the continuum subtraction in order to solidify this detection.

## References

Balick, B.: 1987, *AJ* **94**, 671
Cernicharo, J., Guelin, M., Martin-Pintado, J., Penalver, J., and Mauersberger, R.: 1988, *AA* **222**, L1
Martin-Pintado, J., Bujarrabal, V., Bachiller, R., Gomez-Gonzalez, J. and Planesas, P.: 1988, *AA* **197**, L15

# THE SPATIO-KINEMATIC STRUCTURE OF THE CO ENVELOPES IN EVOLVED PLANETARY NEBULAE

R. BACHILLER[1], P.J. HUGGINS[2], P. COX[3], T. FORVEILLE[4]

[1] *Centro Astronomico de Yebes (Spain)*, [2] *New York University (USA)*, [3] *Observatoire de Marseille (France)*, [4] *Observatoire de Grenoble (France)*

We report high angular resolution mapping of the CO ($J=2\rightarrow1$ and $1\rightarrow0$) lines in three evolved planetary nebulae (PNe): NGC 6781, NGC 6772, and VV 47. The CO $2\rightarrow1$ observations of the ring-like nebula NGC 6781 provide the most detailed map to date of the kinematic structure of a PN envelope. The data are well explained with a model consisting of a thin, clumpy, ellipsoidal shell, which is open at the ends and is expanding with a velocity proportional to distance from the star. The molecular shell of NGC 6772 appears similar, but the gas is more confined to an equatorial ring and is much more incomplete. The molecular gas in VV 47 is in two clumpy lobes, which are likely to be the only surviving molecular condensations from an earlier, more extended equatorial distribution of the same kind. The average CO excitation temperature of these PNe is found to be >23 K from the CO $2\rightarrow1/1\rightarrow0$ line ratio, and the mass of molecular gas is estimated to be 0.1, 0.02, and 0.002 $M\odot$ in NGC 6781, NGC 6772, and VV 47, respectively. It appears that the ring-like PNe are formed from the dissociation and ionization of neutral ellipsoidal shells; destruction of the envelope begins with the rapid ionization of the least dense polar caps, and continues until the densest molecular material at the nebular waist is fully ionized.

# HIGH RESOLUTION OBSERVATIONS OF CO IN PNe

K.M. SHIBATA and S. DEGUCHI
*Nobeyama Radio Observatory, National Astronomical Observatory, Japan*

T. KASUGA
*Department of Instrument and Control Engineering, Hosei University, Japan*

S. TAMURA
*Astronomical Institute, Tohoku University, Japan*

N. HIRANO
*Laboratory of Astronomy and Geophysics, Hitotsubashi University, Japan*

and

O. KAMEYA
*Mizusawa Astrogeodynamics Observatory, Japan*

**Abstract.** We have made aperture synthesis observations of $^{12}$CO(J=1-0) emission IRAS 21282+5050, CRL 618 and M 1-7 using the Nobeyama Millimeter Array (NMA). We observed with 3 or 2 configurations and obtained an angular resolution of 3".2 x 3.1, 3".6 x 3".5 and 4".3 x 3".8.

In CRL 618, the CO emission has a size of about 15" x 12" and concentrates to the center of the continuum emission. CO emission extends from the NW to the SE at the lower contours. At the higher contours it elongates to the EW and has a weak extension to the S at the western part of the emission. In the velocity channel maps, the emissions have only one peak and change complexly their shape at each velocity. No systematic trend with velocity can be seen in these maps. As a first approximation the CO gas in CRL 618 has a spherical structure in its base. It is considered that the ionized cavity in the molecular gas cannot be resolved because of its small size. The total flux density of the continuum emission at 115 GHz is 1.48 Jy. We may consider that all the 1.48 Jy come from a f-f emission of the ionized gas. Our value is consistent with the cylindrical isothermal ionized gas model with a power-law density distribution (Martín-Pintado et al. 1988).

The integrated CO map in M 1-7 shows that the CO emission elongates toward a p.a. = 60° (perpendicular to the optical major axis) and has a size of about 10" x 20". Two emission peaks are shown in each side of the nebular center. Pos.-vel. diagrams clearly show the toroidal structure around the minor axis of the optical image. The CO structure is very similar to that of NGC 2346 (Bachiller et al. 1989). Obviously, M 1-7 is in an earlier evolutionary stage than NGC 2346 and its ionized gas will develop faster toward the pole of the CO toroid than toward the equatorial direction and then become a bipolar nebula like NGC 2346.

For IRAS 21282, we conclude that CO gas forms an expanding toroid whose axis lies along the EW direction and is normal to the line of sight. Both from the ratio of the size of CO gas to that of ionized gas and from the intensity ratio of [OIII] to H$\beta$, we conclude that IRAS 21282+5050 is in an earlier evolutionary stage than NGC 7027 and 2346. Detailed discussions are given in Shibata et al. (1989).

### References

Bachiller et al., 1989, *Astron. Ap.*, **210**, 366.
Martín-Pintado et al., 1988, *Astron. Ap.*, **197**, L15.
Shibata et al., *Ap. J. (Letters)*, **345**, L55.

# 1.6-1.75 AND 3.1-3.75 μm SPECTRUM OF Hb5

## A. MAGAZZÙ and G. STRAZZULLA

*Osservatorio Astrofisico and Istituto di Astronomia, Citta' Universitaria, I-95125 Catania, Italy*

We have recently obtained high resolution IR spectra ($\lambda/\Delta\lambda \approx 1000$) of the planetary nebula Hb 5, over the range 1.6–1.75 μm and 3.1-3.75 μm (Magazzú & Strazzulla 1992) at the European Southern Observatory (ESO). Emission bands have been detected the most prominent being at 3.3, usually attributed to PAH molecules, and 3.4 μm. For the predicted (de Muizon et al. 1986) first armonic (at 1.67 μm) of the 3.3 μm band we get an upper limit at least 45 times weaker than the 3.4 μm feature. This, together with the high ratio of fluxes F3.4/F3.3, challenges the hypothesis that the 3.4 μm band is the first hot band. We suggest that the first hot band may be identified with a small feature at 3.46 μm. In this case the first armonic would fall at 1.686 μm where we detect a small feature.

From the measured flux ratios we evaluated (considering the absorption of a 9 eV photon) the peak temperature of the emitting species ($T_{\text{peak}}$) and the total number of atoms per molecule ($N_t$) (Magazzú & Strazzulla, 1992). In order to have an idea of the possible size distribution, we also calculated, using data from the literature (Aitken & Roche 1982; 1984) the ratio F(11.3)/F(3.3). We showed that the colour temperature (~1420 K) obtained from the ratio $F(1.686)/F(3.3)$ is larger than that obtained from $F(11.3)/F(3.3)$ (~1050 K) and, accordingly, species emitting at 3.3 μm are smaller in size than those emitting at 11.3 μm.

In Conclusion:

-The 3.4 μm band in Hb 5 cannot be identified with the first hot band of anharmonically emitting PAHs.

-An interesting possibility is that the first hot band may be identified with a small feature at 3.46 μm. We have evidenced for the first time a size distribution of PAHs, the smaller being hotter and responsible for emission at lower wavelength.

-The overall spectrum observed in the 3 μm region seems difficult to be accounted for only by PAH molecules.

## References

Aitken D.K., Roche, P.F., 1982, MNRAS 200, 217
Aitken D.K., Roche, P.F., 1984, MNRAS 208, 751
de Muizon M., Geballe T.R., d'Hendecourt L.B., Baas F., 1986, ApJ 306, L105
Magazzù A., Strazzulla G., 1992, A&A in press

# CHEMISTRY IN THE MOLECULAR ENVELOPE OF NGC 7027

P. COX[1], R. BACHILLER[2], P.J. HUGGINS[3], A. OMONT[4], S. GUILLOTEAU[5]

[1] Observatoire de Marseille, 2, pl. Leverrier, 13248 Marseille, France

[2] Centro Astronomico de Yebes, Guadalajara, Spain

[3] Department of Physics, New York University, NY 10003, USA

[4] I.A.P., 98b, bd. Arago, 75014 Paris, France

[5] IRAM, Grenoble, France

We report here a systematic study of the molecular content in the neutral envelope of NGC 7027. The measurements were done in September 1991 with the 30-meter IRAM telescope on Pico Veleta (Spain). The frequencies and transitions covered are given in Table 1. Long integrations were performed in each transitions toward the central position of NGC 7027, and for the strongest lines we obtained detailed maps. The results of the observation are given in Table 1 (in units of main beam temperature).

The main conclusions of this systematic search and mapping of molecules in the neutral envelope of NGC 7027 are:

1) the basic features of the chemistry in PNe described in Cox et al. (1992) - and references therein - also characterize the on-going chemistry in the envelope of NGC 7027 which is likely dominated by photodissociation, shocks, and ion-molecule reactions.

2) the detection of $N_2H^+$(1-0) is the first detection of this ion in any stellar envelope.

3) From the higher and lower transitions of HCN, $HCO^+$, and CN, we estimate that the emitting regions have typical densities of a few $10^5$ cm$^{-3}$.

A detailed analysis of the data will appear elsewhere.

References

Cox, P., Omont., A., Huggins, P.J., Bachiller, R., Forveille, T.: 1992, Astron. Astrophys. in press

Table 1: Observational results towards the central position of NGC 7027

| Molecule | Line | Frequency (MHz) | $T_{mb}$ (peak) (K) | Area (K km/s) | r.m.s (mK) | Comments |
|---|---|---|---|---|---|---|
| HCN | 1-0 | 88631.6 | 0.29 | 9.6 | 48 | mapped |
| HCN | 3-2 | 265880 | 0.58 | 10.4 | 76 | |
| HNC | 1-0 | 90663.6 | 0.014 | 0.58 | 6.1 | |
| $HCO^+$ | 1-0 | 89188.6 | 1.12 | 21.6 | 16 | mapped |
| $HCO^+$ | 3-2 | 267557 | 2.10 | 55.1 | 220 | |
| CN $^a$ | 1-0 | 113488 - 113520 | 0.48 | 19.8 | 18 | mapped |
| | | 113123 - 113191 | 0.15 | 12.8 | 18 | mapped |
| CN $^a$ | 2-1 | 226849 - 226913 | 0.78 | 31.5 | 30 | mapped |
| | | 226615 - 226697 | 0.37 | 21.2 | 36 | mapped |
| $N_2H^+$ | 1-0 | 93174.0 | 0.06 | 1.57 | 10 | |
| $C_2H$ | 1-0 | 87316.9 | | $<0.5^b$ | 20 | |
| $C_2H$ | 3-2 | 262004.2 | | $<2.0^b$ | 80 | |
| $C_3N$ | 13-12 | 128622.1 | | $<0.25^b$ | 10 | |

$^a$ Sum of the hyperfine lines in high and low frequency components, respectively

$^b$ The upper limits on the area are derived from the noise level of the spectra (1 r.m.s) times $V_{exp}$, where $V_{exp}$ is 25 km s$^{-1}$

# AXISYMMETRIC DUST-SHELLS IN PLANETARY NEBULAE

W. HOPFENSITZ (1) and M. GREWING (1,2)

*(1)Astronomisches Institut der Universität*
*Waldhäuserstraße 64, D-7400 Tübingen*
*(2)Institute for Radioastronomy at Millimeter Wavelengths (IRAM)*
*300,rue de la Piscine,Domaine Universitaire,F-38406 St.Martin d'Hères, France*

Aspherical mass loss is likely to be a common feature in the late stages of stellar evolution when the star arrives at the asymptotic giant branch. In the transition phase from the red giant to a proto-planetary nebula (PPN) the star developes an extended circumstellar dust-shell which is in many cases ellipsoid or disk-shaped. An algorithm for treating the radiative transfer problem in dense axisymetric dust shells is presented. The program has been used to determine the temperature distribution in disk-shaped dust envelopes in PNe. The implications of both composition and the spatial structure of the dust shell has been investigated. We confined the viewing angle to the equatorial plane to maximize the effects of the disk geometry. Our computations show that the observed morphology is mainly affected by the geometry of the envelope whereas the integrated spectrum is mainly determined by the characteristics of the grains themselves. The observed morphology in the optical and near infrared wavelength region is determined by scattered starlight in the thin polar region where the optical depth is low both to the the observer and to the central star whereas scattered light in the equatorial plane is blocked by the optical thick disk. This results in the familiar hourglass shaped morphology perpendicular to the orientation of the disk. Towards longer wavelengths the disk becomes more transparent thus the bipolar shape becomes less prominent. At wavelength longer than 5 $\mu$m the disk appears as a compact round source produced by scattered light at the inner rim of the disk. In the Mid and Far Infrared thermal emission of dust is the dominant emission process. Observational evidence for dust disks in PNe can be found in several wavelength regions. For example M 2-9 has been observed by Aspin (1984) in the optical region with a CCD Imaging Polarimeter. The polarisation maps show a high degree of polarisation in the lobes whereas emission from the equatorial plane is little or not polarized. In the NIR the change of the appearance from the bipolar to the compact round morphology predicted by our models is confirmed by the presented images in the $J$ (1.2 $\mu$m), $H$(1.6 $\mu$m) and $K$(2.2 $\mu$m) filters obtained with the ESO InfraRed Array Camera. This confirms the general picture of M 2-9 as a proto- planetary nebula being surrounded by an extended, optically thick dust disk.

## References

Aspin C., 1984 Astron. Astrophys. Vol. 134 p. 333

# MOLECULAR-LINE OBSERVATIONS OF THE REMNANT AGB ENVELOPES AROUND PLANETARY NEBULAE

R. SAHAI

*Chalmers University of Technology, 41296 Gothenburg, Sweden*

and

A. WOOTTEN

*National Radio Astronomy Observatory, Charlottesville, VA 22903, USA*

and

R.E.S. CLEGG

*Royal Greenwich Observatory, Madingley Road, Cambridge CB3 0EZ, UK*

ABSTRACT: We present recent results from a "search and mapping" program of molecular line emission (mainly CO) from remnant AGB envelopes around planetary nebulae (PNe), using the SEST (La Silla, Chile). New detections in CO J=2-1 include NGC2899 (0.02K), NGC6369 (0.14K) & NGC7009 (0.08K). In many of the detected PNe, notably NGC3132, IC4406, NGC6302, M1-16, and CPD-56°8032, the molecular envelopes contain 2 kinematically distinct outflows. Mapping of the strongest of these shows (1) that the fast (e.g. $V_{exp} \gtrsim 40$-60 km s$^{-1}$ in NGC3132, IC4406) outflows have bipolar spatial structure, and (2) there exists an equatorial density enhancement in the slower, more massive [$\dot M$ ($M_\odot$yr$^{-1}$)>5 10$^{-6}$(NGC3132), >2 10$^{-5}$(IC4406)] outflows, which presumably collimates the fast outflow (e.g. Sahai et al. 1990, A & A, 234, L1; Sahai et al. 1992, A & A, 251, 560). The fractional CO abundance in the envelope, f(CO), is probably rather low (<10$^{-4}$), as a result of photodissociation by the stellar and interstellar UV radiation [e.g. f(CO)$\lesssim$10$^{-5}$ in IC4406]. HCN, HCO$^+$, and $^{13}$CO have also been detected in several PNe, and sensitive upper limits set on CS, C$^{18}$O, & C$^{17}$O (in M1-16), and SO (in NGC3132). Some results are tabulated below, and calculations to estimate the molecular masses, mass-loss rates and molecular abundances are in progress.

| Mol./Line | Data | $V_c^a$ | $\Delta V^b$ | $\Delta V_{hi}^c$ | $T_{mb}(K)$ | Comments |
|---|---|---|---|---|---|---|
| M: Map, S: Spectrum | | | M1-16 | | | |
| CO J=2-1 | M | 50.5 | 44 | 60 | 1.8 | HVW$^d$, B$^e$ (NW-SE)$^f$, opt. thick |
| CO J=1-0 | S | 50.5 | 43 | 60 | 0.43 | HVW, opt. thick |
| $^{13}$CO J=1-0 | S | 50.2 | 38 | -- | 0.14 | [$^{12}$C]/[$^{13}$C]<10, $\dot M$>10$^{-4}$ $M_\odot$yr$^{-1}$? |
| HCN J=1-0 | S | ≈50 | ≈43 | 160? | 0.025 | Very high-vel(-20 to 140) emission? |
| HCO$^+$ J=1-0 | S | ≈50 | ≈43 | ≈70 | 0.043 | |
| N$_2$H$^+$ J=1-0 | S | ≈47 | ≈31 | -- | 0.03 | Double-peaked line?(marginal det.) |
| | | | CPD-56°8032 | | | |
| CO J=2-1 | M | -57 | 50 | 186 | 0.58 | HVW, B (NWW-SEE), opt. thick |
| CO J=1-0 | S | -57 | 50 | 120 | 0.13 | HVW |
| $^{13}$CO J=1-0 | S | ≈-57 | ≈50 | -- | 0.04 | [$^{12}$C]/[$^{13}$C]<10, $\dot M$>10$^{-4}$ $M_\odot$yr$^{-1}$? |
| HCO$^+$ J=1-0 | S | ≈-57 | ≈50 | -- | ≈0.035 | |
| | | | NGC3132 | | | |
| CO J=2-1 | M | -25 | 30 | 70 | 0.35 | HVW, B (≈NW-SE), opt. thick |
| CO J=1-0 | S | -25 | 30 | 70 | 0.24 | HVW, opt. thick |
| $^{13}$CO J=1-0 | S | ≈-25 | ≈18 | -- | 0.035 | |
| HCO$^+$ J=1-0 | S | ≈-23 | ≈55 | -- | 0.045 | Broad profile |
| | | | NGC6302 | | | |
| CO J=2-1 | M | -36 | 60 | 110 | 0.9 | HVW, B Expanding Lobes (≈E-W) |
| CO J=1-0 | M | ≈-30 | ≈55 | ≈100 | ≈0.25 | (contamination from I.S. emission) |
| HCO$^+$ J=1-0 | S | ≈-36 | ≈30 | -- | 0.025 | |
| HCN J=1-0 | S | ≈-40 | ≈40 | -- | 0.025 | |

a- VLSR (km s$^{-1}$); b & c- Full-widths at zero intensity (km s$^{-1}$) of main & high-velocity components in central spectrum; d- High-vel. wings in spectrum; e (f)- Bipolar (axis)

# NGC 7027: NEW 7.8 - 20.0 μm ARRAY CAMERA AND $H\alpha/H\beta$ CCD IMAGE ANALYSIS OF DUST, PAH AND IONIZED GAS DISTRIBUTION

D. Y. GEZARI, M. D. THORNLEY, S. R. HEAP, S. N. SHORE, F. VAROSI, S. J. MEATHERINGHAM and S. P. MARAN

*NASA/GODDARD*

Infrared array camera images of the planetary nebula NGC7027 at seven wavelenghts between 7.8 and 20μm taken at the IRTF telescope are compared to $H\alpha$ and $H\beta$ images obtained at the CFHT, and published 2-, 6-, and 20-cm VLA maps, all made with better than ~ 1 arcsec resolution. The mid-infrared images have also been deconvolved using the Maximum Residual Likelihood algorithm to show structural details on the ~ 0.3 arcsec scale. The visible, infrared and radio images are aligned with ±0.2 arcsec astrometric precision using common features and visible stars. Image algebra can then be readily performed. The images are analysed to determine the spatial distribution of polycyclic aromatic hydrocarbon (PAH) grains and ionized gas relative to the continuum dust emission. The results show the extent of the mid-infrared dust continuum, the PAH emission features at 8.65 and 11.25μm, and would reveal any contribution from 9.8 μm "silicate" emission. The optically thicker 20-cm emission more closely resembles the extinction-corrected $H\alpha$ distribution. The large grains giving rise to the visible extinction (derived from the $H\alpha/H\beta$ ratio) are distribued differently from the mid-infrared and ionized gaz, indicating the location of the outer cold dust. The 11.6μm image could contain both PAH and continuum grain emission, and we have attempted to separate the two components. It appears that the overall emission distributions are quite similar for all the infrared wavelengths, so the various dust constituents are generally uniformly mixed. However there are indications that the 11.6μm PAH emission distribution is patchy and not identical to the continuum dust, although not distinctly different in radial distance from the center. The ionized gas distribution peaks inside that of the dust continuum. The visible/infrared/radio spatial analysis, and color indices calculated from the infrared images, are used to develop a detailed new morphological model of NGC7027 based on this comprehensive multi-wavelength data set.

# POLARIZED LINE PROFILES IN PLANETARY NEBULAE

J.R.WALSH[1] and R.E.S.CLEGG[2]

[1] Space Telescope Coordinating Facility, ESO, Karl-Schwarzchild-Strasse 2
D-8046, Garching bei München, Germany
[2] Royal Greenwich Observatory, Madingley Road, Cambridge, CB3 0HZ, United Kingdom

There is much direct and indirect evidence for the presence of dust in Planetary Nebulae (PN): variations in extinction across the face of the nebulae; IR emission with strong 25 and 60$\mu$m fluxes; broad near-IR emission lines of Silicate, SiC and PAH grains; optically thick lines, such as [C IV]1550Å, have lower strength on account of the increased path length due to dust scattering; a centro-symmetric pattern of polarization vectors in a few PN (Leroy et al., A&A, **160**, 171, 1986). An observational programme has begun to study the polarization profiles of bright emission lines in PN arising from dust scattering within the nebulae. The first results are discussed.

Six PN were observed with the 4.2m William Herschel Telescope and the I-SIS spectrograph. A Wave Plate Polarimeter with a half-wave plate located above the spectrograph slit was employed to measure linear polarization at high spectral resolution. Of the six PN observed, five (NGC 650, NGC 1535, NGC 2346, NGC 7027 and IC418) showed observable polarization in the emission lines and only one, NGC 2392, showed none detectable. In the core regions (generally the bright centre of the nebulae) the polarization is low or zero, accountable by interstellar polarization between the PN and the Earth. In 3 out of the 5 PN with detectable polarization, it is larger in the red wing of the emission line than in the blue wing. The offset regions observed in NGC 7027 (over the neutral and molecular halo) have very large polarization, upto 35% at [O III]5007Å.

A PN has strong line emission so it is expected that this would mask any polarization. However dust mixed with the gas scatters photons and since the relative velocity between two atoms in an expanding nebula is equivalent to a redshift, scattering occurs predominantly in the red wing of the line. Since the emission is lower in the line wings, dilution of the polarized emission is less than in the line centre and the characteristic polarization profile can be seen. Some PN also show increased polarization in the blue wing indicating either that some of the dust is not well mixed with the ionized gas or does not simply partake in the expansion with the gas. The high polarization values in the outer halo of NGC 7027 can be explained in an identical way, however there is almost no local unpolarized emission to dilute the polarized flux and in addition scattering angles close to 90° can occur.

A polarization model was constructed for NGC 7027 using a 3-D form of the nebula for the emitting gas and scattering by graphite dust. Apertures placed on the model nebula at positions corresponding to those observed could show both the effect of increased polarization in the red wing for line profiles at the centre of the nebula and the strong polarization profiles at offset positions over the neutral halo. Thus modelling the polarization profiles can provide useful information on the scattering properties of the grains (complementing the emission properties from IR data) and the extent and velocity field of the dust within PN.

# MILLIMETRE AND SUBMILLIMETRE CONTINUUM OBSERVATIONS OF PLANETARY NEBULAE

M.G. HOARE and P.F. ROCHE

*Astrophysics, Dept of Physics, Oxford University, Nuclear Physics Building, Keble Rd., Oxford OX1 3RH, UK.*

and

R.E.S. CLEGG

*Royal Greenwich Observatory, Madingley Rd, Cambridge CB3 0EZ, UK.*

Continuum emission has been detected at millimetre and submillimetre wavelengths from a total of thirteen planetary nebulae, all but one for the first time. The observations were obtained with the liquid $^3$He-cooled UKT14 bolometer on the JCMT and UKIRT telescopes on Mauna Kea, Hawaii.

We find that Type I and young, compact planetaries both have significant dust emission at 450 and 800 microns, while at 1100 and 2000 microns the free-free emission from the nebular ionised gas dominates. More evolved PNs do not usually show any 'cool-dust' excess above the observed free-free level at 800 microns. We have also spatially resolved the cool dust in the neutral regions surrounding NGC 7027 and BD+30.3639 for the first time. In NGC 7027, the 450 micron emission appears to correlate spatially with the optical extinction in front of the nebula and with observed CO emission.

We have combined detailed photo-ionisation models of the ionised zones in NGC 7027 and BD+30.3639 with detailed dust models which allow for a distribution of particle sizes in both the ionised and neutral zones. We match our present data plus ground-based and IRAS fluxes, between 1 and 1100 microns, using amorphous carbon dust. (Graphite grains are not able to provide enough submillimetre flux). In the ionised regions of these two young, compact nebulae we deduce that the dust-to-gas ratios are no higher than about $7 \times 10^{-4}$ (a value an order of magnitude lower than that for the diffuse interstellar medium). This value for compact PNs is similar to the ratio in large, older PNs, and so from the present sample we have no evidence so far that the dust content of PNs drops as the objects evolve.

The dust-to-gas ratio in the neutral region around NGC 7027 is found to be about $1.5 \times 10^{-3}$, a result that implies that grains are not significantly destroyed in the ionized region of this nebula.

# IV. PLANETARY NEBULAE CONNECTION: EVOLUTION FROM THE AGB

# THE THIRD DREDGE-UP: STATUS AND PROBLEMS

J.C. LATTANZIO

*Dept. of Mathematics, Monash University, Clayton, Vic., 3168 Australia*

and

*Institute of Astronomy, Madingley Rd., Cambridge, CB3 0HA, England*

ABSTRACT. In this paper I review our understanding of the third dredge-up phenomenon and the nucleosynthesis which occurs during thermally-pulsing AGB evolution.

## 1. Introduction

Asymptotic Giant Branch (AGB) stars are among the most complex and most interesting of all stars. Yet this complexity makes understanding them very difficult because they exhibit the very phenomena which stellar physicists find the hardest to model, such as convection and mass loss. Still, much progress has been made in understanding these stars (Iben and Renzini 1983, Lattanzio 1989a). In this paper we are concerned primarily with their nucleosynthesis and the periodic "dredge-up", or penetration of the convective envelope into regions of nuclear processing. This results in mixing to the surface of the nuclear burning products. Note that dredge-up plays a crucial role, linking theory to observations, *i.e.* the interior to the photosphere. As such, one may expect that trying to determine characteristics of dredge-up would be restricted by errors and uncertainties in both theory and observations. This is indeed true, especially since the extent and occurrence of dredge-up itself is very sensitive to changes in modelling parameters etc. Yet we will see that much can still be done.

## 2. Anatomy of a Shell Flash

Figure 1 shows the four phases of a shell flash, or thermal pulse, as defined by Iben (1981a). Cross-hatched regions denote convection, and wavy lines indicate radiative energy transport. During the "on" phase the high ($\sim 10^8 L_\odot$) luminosity of the helium shell drives a convective zone, hereafter called the Inter-Shell Convective Zone (ISCZ), which consists of roughly 23% $^{12}C$ and 75% $^4He$. In the "power-down" phase the ISCZ disappears as the He-shell decreases in its output. The legacy of the ISCZ is the enhanced $^{12}C$ abundance in the region between the H and He shells. Finally, during the "dredge-up" phase, the inward advance of the convective envelope eventually penetrates the erstwhile ISCZ and mixes freshly produced $^{12}C$ to the surface. The convection then recedes, and the cycle repeats every $\sim 10^4$ years or so (depending on the core mass), after a period of quiescent H burning.

Figure 2 shows two consecutive thermal pulses, and defines some parameters. The mass of material processed by the H shell between pulses is $\Delta M_H$, and the amount dredged to the surface is $\Delta M_{dredge}$. Two very important parameters are the mass of the hydrogen exhausted core, $M_H$, and the dredge-up parameter, $\lambda = \Delta M_{dredge}/\Delta M_H$. The results of many

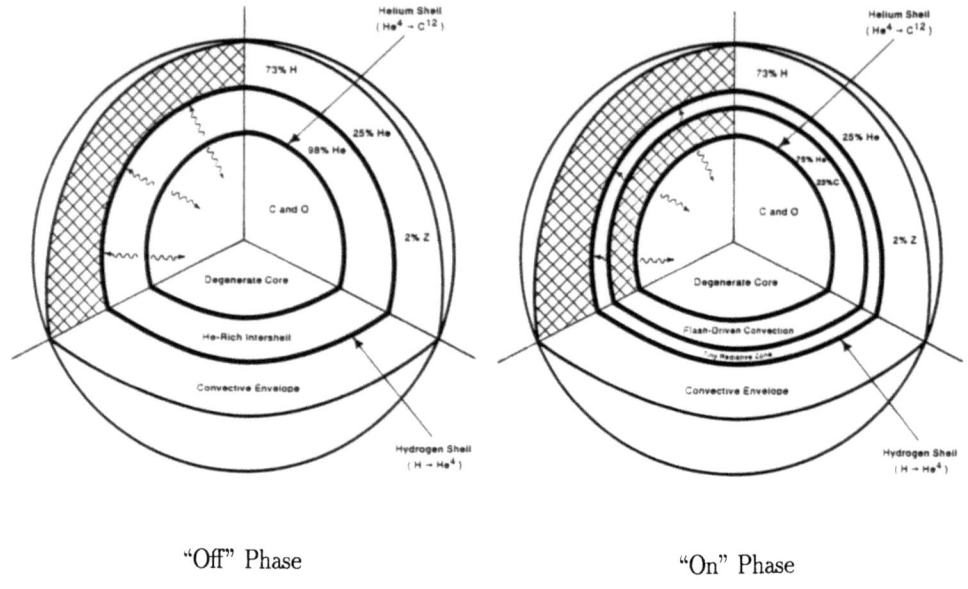

"Off" Phase  "On" Phase

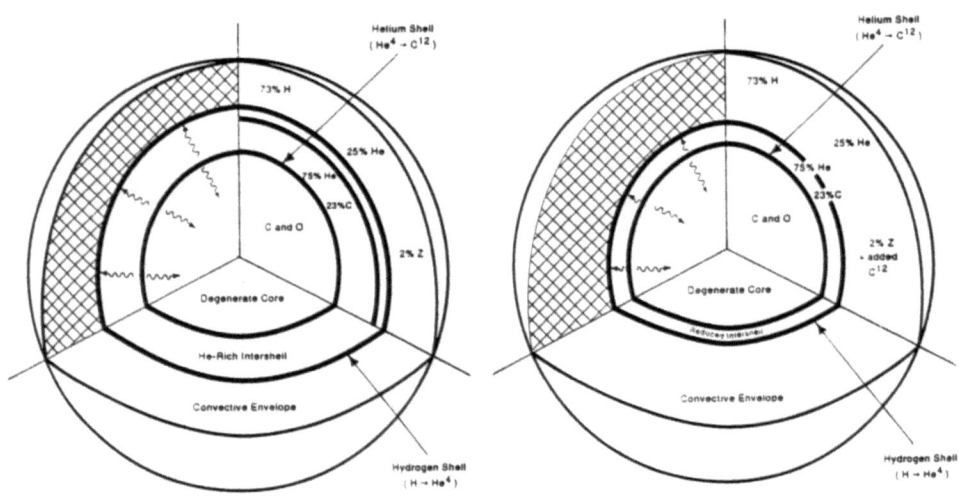

Figure 1: The four phases of a thermal pulse.

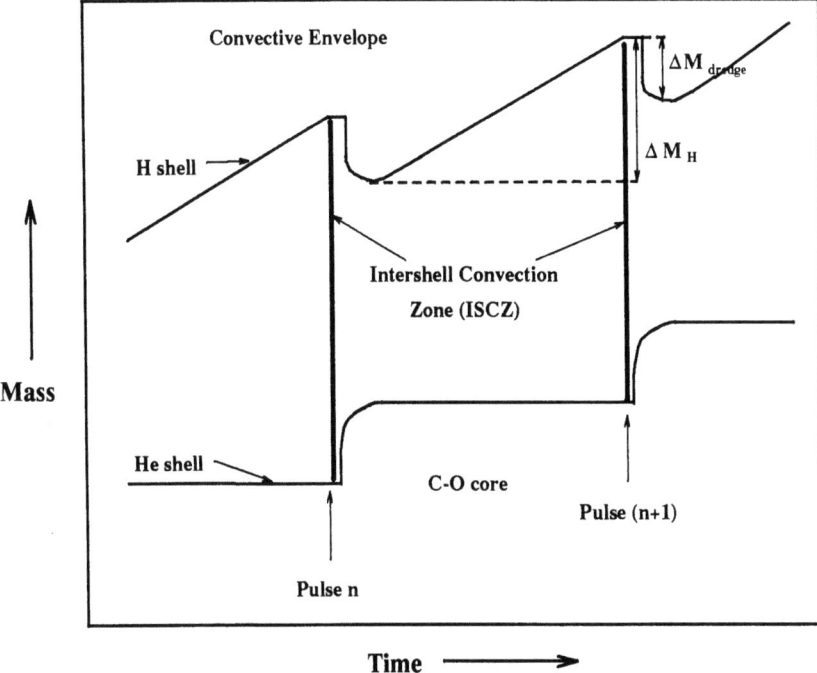

Figure 2. Schematic diagram showing the time evolution of important mass boundaries in two consecutive thermal pulses.

calculations (*e.g.* Wood 1981, Iben and Renzini 1983, Lattanzio 1986, 1989b, 1991) show that dredge-up becomes easier when $M_H$ increases, $M_e = M - M_H$ increases, $Z$ decreases, $Y$ decreases, or $\alpha = \ell/H_P$ = mixing-length parameter increases.

## 3. Making $^{12}$C in Shell Flashes

The simplest nucleosynthesis which occurs in thermally pulsing AGB stars is the production of $^{12}$C, which was outlined in the previous section. Unfortunately, a direct study of this phenomenon is very expensive in computer time. But results of those calculations which have been performed show that most of the AGB evolution can be parametrized in terms of the core-mass. For example, the peak quiescent luminosity reached before a pulse isgiven by $L = aM_H + b$, where $a$ and $b$ are constants (see Paczynski 1971). Using such expressions we can simulate detailed evolutionary calculations for far more cases than are possible by direct methods. Such calculations are called "synthetic" evolutionary calculations, and were used by Iben (1981b) and Renzini and Voli (1981) to show that theory produces too many high luminosity C-stars and not enough of low luminosity (when compared to the observed luminosity functions in the Magellanic Clouds). The problem with these calculations was that in 1981 there were no models available for the range of 1–3 $M_\odot$, which is appropriate for most of the Cloud stars. Thus the parametrized evolutionary models were based on calculations then available ($M \leq M_\odot$ and $M \geq 3M_\odot$). Now the most important parameter is the core mass at the first thermal pulse, and this is a sensitive function of initial mass (and metallicity) in this mass range (see Lattanzio 1986, Boothroyd and Sackmann 1988d). This limits the quantitative accuracy of these early calculations.

With new detailed evolutionary calculations available in the appropriate mass range (Lattanzio 1986,1989b; Lattanzio and Malaney 1989; Boothroyd and Sackmann 1988a,b,c,d) the synthetic calculations were repeated by Bryan et al. (1990) with updated parametrizations. They were able to match the observed luminosity functions of AGB stars, as well as the O-rich to C-rich number ratio, by adopting a new mass loss formula. The dredge-up was taken as constant, with $\lambda \simeq 0.25$, and assumed to occur at each pulse, whereas the models experience quite a few pulses before dredge-up begins. Indeed, there is a minimum core mass for dredge-up, which depends on the total (initial) mass and the composition. Part of the agreement found by Bryan et al. between models of C stars and observations must, unfortunately, be due to this assumption, which is *not* consistent with the models. (Note that this only affects when stars become C-stars, and not the initial-final mass relation, etc., which is also discussed by Bryan et al.)

There are also three further refinements which should be included in synthetic models. The recent evolutionary calculations revealed some dependence on metallicity, which has so far been neglected. Also, as pointed out frequently by Iben, there is actually a sizeable variation in luminosity during the pulse cycle, with a significant fraction of the interpulse period spent at a luminosity which is lower than that given by the core-mass–luminosity relation. Finally, the variation of physical parameters with $M_H$ is much more rapid during the first "few" pulses ($\sim 10$) than it is in the later, so-called "full amplitude" pulses. All of these effects will alter the predicted AGB star luminosity distributions if included in the synthetic calculations. Indeed, at this meeting, a poster was presented by Groenewegen and de Jong which included these effects (see elsewhere in this volume).

The calculations of Groenewegen and de Jong (1992) assumed $\lambda = $ constant, and that dredge-up begins when $M_H$ exceeds $M_H^{min}$. Using values for these parameters from recent calculations they are unable to match the observed C-star luminosity functions in the Large Magellanic Cloud (LMC) or the Galaxy. A good fit is found for $\lambda = 3/4$ and $M_H^{min} = 0.58 M_\odot$, in contrast to the theoretically determined values of $\lambda \simeq 1/3$ and $M_H^{min} = 0.62$–$0.63 M_\odot$ (Lattanzio 1989b, Lattanzio and Malaney 1989). So we are forced to a conclusion very similar to that of Iben (1981b): dredge-up occurs at smaller core masses than indicated by models, and at higher (average) efficiency (at least at these small core masses: note that Vassiliadis and Wood (1992, at this meeting) find $\lambda$ approaching the maximum possible value of unity in their calculations of a $5 M_\odot$ model). In this context it remains to be seen what will be the effects of the recent OPAL opacities (Rogers and Iglesias 1992) on dredge-up calculations. Likewise, the dredge-up parameter is actually an unknown (!) function of total mass, core mass, envelope mass as well as composition. The determination of this dependence is a laborious task, but one which we need to complete.

Finally a small warning. As pointed out by Bryan et al. (1990), the mass loss formula used in synthetic calculations has a significant effect on the results, through the termination of AGB evolution. This is another uncertainty in all the synthetic models calculated so far.

## 4. Making s-Process Elements in Shell Flashes

The unambiguous observational evidence for a correlation between C/O and mean abundances of s-process elements in AGB stars (*e.g.* Smith and Lambert 1990a) is direct evidence for the association of s-process enhancements with shell-flashes and dredge-up, since these are clearly responsible for the increase in C/O. There are two suspected neutron sources for this s-processing: $^{22}Ne(\alpha,n)^{25}Mg$ and $^{13}C(\alpha,n)^{16}O$. We will deal with these separately below.

## 4.1 $^{22}$NE AS THE NEUTRON SOURCE

**4.1.1 *Getting the* $^{22}$*Ne.*** During H shell burning the CN(O) cycle converts virtually all of the $^{12}$C and $^{16}$O present into $^{14}$N. Later, during He burning, we find the sequence of alpha captures $^{14}$N$(\alpha,\gamma)^{18}$F$(\beta^+\nu)^{18}$O$(\alpha,\gamma)^{22}$Ne converts all $^{14}$N to $^{22}$Ne.

**4.1.2 *Getting the neutrons.*** At the next thermal pulse the hot He burning can process the $^{14}$N (from H burning) into $^{22}$Ne, and if the temperature at the base of the ISCZ exceeds about $300\times10^6$ K then we get $^{22}$Ne$(\alpha,n)^{25}$Mg releasing neutrons.

**4.1.3 *Problems.*** The high temperature required for this reaction limits its applicability to stars of large core masses ($M_H \gtrsim 0.9 M_\odot$) and (hence) total masses ($M \gtrsim 7 M_\odot$). But the majority of LMC (and Galactic) stars are thought to have masses in the range of 1–3 $M_\odot$, based on many different lines of reasoning (e.g. kinematics, spatial distibution, pulsation). Also, no enhancements of $^{25}$Mg/$^{24}$Mg have been observed (Lambert 1989, 1991) which would be expected if $^{22}$Ne were the neutron source in these stars (but see §6 below).

## 4.2 $^{13}$C AS THE NEUTRON SOURCE

**4.2.1 *Getting the* $^{13}$*C.*** Sackmann (1980) suggested that the large expansion following a strong pulse could propel the erstwhile ISCZ, rich in $^{12}$C, to temperatures low enough for recombination to begin. This, she suggested, could be an important source of extra opacity in the models. Calculations by Iben and Renzini (1982a,b) found that this extra opacity led to the formation of a semiconvective zone at the base of the convective envelope during what would be the dredge-up phase. The main effect of this semiconvection is to mix a small quantity of H downward into the C-enriched regions. When the H shell is re-ignited these protons are then captured by the $^{12}$C to form $^{13}$C. (An important feature of this mechanism is that there is not enough H to continue the CN cycle through to the formation of $^{14}$N.)

**4.2.2 *Getting the neutrons.*** During the next thermal pulse the ISCZ engulfs the layer of $^{13}$C, mixing it with He, and initiating $^{13}$C$(\alpha,n)^{16}$O. This reaction is quite efficient, requiring temperatures no higher than $\sim 150\times 10^6$ K, which are easily found in the ISCZ of models of low mass stars.

**4.2.3 *Problems.*** The largest uncertainties in the above scenario are the details of the semiconvection. Different authors find different amounts (*e.g.* Boothroyd and Sackmann 1988d, Hollowell 1988, Hollowell and Iben 1989) or none at all (Lattanzio 1991). This is possibly due to the use of different opacities (and interpolation schemes *etc.*). Also, it seems that this mechanism only works for low $Z$ and small envelope masses (Iben and Renzini 1984). This may be a problem, since $s$-process enhancements are clearly seen in stars of solar composition.

**4.2.4 *A Further Complication.*** Bazan and Lattanzio (1992) have drawn attention to the fact that all extant evolutionary calculations have ignored the energy released by the $^{13}$C$(\alpha,n)^{16}$O reactions, as well as the associated $s$-processing. By including this, in an approximate manner, they found some significant changes in the evolution. Firstly, the peak neutron density achieved in the ISCZ increases, but seems to be limited by the fact that the base of the ISCZ moves outward in mass, which decreases the maximum temperature in the convective zone. Without this, the peak neutron density could create problems for certain $s$-elements. But a further effect of this change in the shape of the convective boundary is the near quenching of the ISCZ ! This raises the possibility of mini-pulses of convection within a single thermal pulse. This could have profound effects on the nucleosynthesis, and will be discussed further in §7.5

## 5. Making $^{19}$F in Shell Flashes

Recent observations by Jorissen et al. (1992) have shown that F/O correlates with C/O in AGB stars. Again, as was the case with s-process elements, this implicates thermal pulses and dredge-up in the production of $^{19}$F. Forestini et al. (1992) investigated many possible reaction chains and sites for the $^{19}$F production, and concluded that the most likely is $^{14}$N$(\alpha, \gamma)^{18}$F$(\beta^+\nu)^{18}$O$(p, \alpha)^{15}$N$(\alpha, \gamma)^{19}$F occurring in the He shell. The catch here is that protons are needed for the p-captures on $^{18}$O. Possible sources of protons are $^{14}$N(n,p)$^{14}$C or $^{26}$Al(n,p)$^{26}$Mg, both of which require neutrons. And the neutron source ? It "must" be $^{13}$C again. The CN cycle does not seem to produce enough $^{13}$C for this purpose, so again we are forced to conclude that some extra $^{13}$C is present (see §4). But the problem here is getting enough $^{19}$F without over-producing $^{18}$O. The models seem to require more $^{13}$C than the semiconvective scheme, discussed earlier, is capable of producing (by about a factor of 5). In any case, this requires much more H be mixed down to produce the $^{13}$C, but without producing $^{14}$N (which is the next step in the CN cycle when there is abundant H). There are clearly quantitative problems with explaining the observed $^{19}$F abudances, but the possibility of using this to determine the distribution of $^{13}$C is exciting.

## 6. Hot Bottom Burning

A final complication is hot bottom burning, where the base of the convective envelope is hot enough for nucleosynthesis to occur. This has been invoked to explain the lack of high luminosity C-stars by assuming that the $^{12}$C dredged to the surface is then cycled to $^{14}$N by CN reactions at the base of the envelope. Hot bottom burning has been suggested as the mechanism resposible for the production of Li in the Li-rich LMC AGB stars discovered by Smith and Lambert (1989, 1990b), which are indeed bright AGB stars which are O-rich rather than C-rich. Recent calculations by Sackmann and Boothroyd (1992) showed that these stars can indeed be explained by this mechanism. Note that temperatures of 50–100 $\times 10^6$K are often found (Blocker and Schonberner 1991, Forestini et al. 1992, Lattanzio 1992). This is hot enough for $^{25}$Mg$(p, \gamma)^{26}$Al, with two important consequences. Firstly, it would destroy any $^{25}$Mg produced by the $^{22}$Ne neutron source (see §4.1), and secondly this could be the source of $^{26}$Al enhancements seen in SiC grains (see the recent meeting "Nuclei in the Cosmos", held in Karlsruhe 1992).

In a broader context, however, hot bottom burning must be seen as an unfortunate occurrence. It is already difficult using interior models, linked by dredge-up, to predict the composition of AGB star photospheres. To this we must now add the possibility of nuclear processing of the envelope, which consists of the initial envelope, and the products of earlier nuclear burning in the interior. A very complicated problem in nucleosynthesis and mixing.

## 7. Summary

Just as parametrized models of $^{12}$C production can be used to determine the dredge-up parameter, a *simultaneous* fit of $^{12}$C, s-process elements and $^{19}$F production would be a powerful constraint on dredge-up. Unfortunately we are nowhere near advanced enough in our understanding of the s-processing or $^{19}$F production to make such calculations possible. Instead, we must look at each separately.

## 7.1 $^{12}$C PRODUCTION

We are getting closer to reality, but we still need more efficient dredge-up, beginning at lower core masses.

## 7.2 s-ELEMENT PRODUCTION

It seems that $^{22}$Ne can be the neutron source only in massive stars. Lower masses ($M \lesssim 7M_\odot$) require the $^{13}$C source, which has many quantitative problems. Specifically, it is yet to be found to operate in stars near solar metallicity, or with significant envelope masses.

## 7.3 $^{19}$F PRODUCTION

Clearly this is associated with dredge-up and shell flashes. Can we use this to determine the distribution and amount of $^{13}$C present ? How can we avoid over-producing $^{18}$O ? Are the $^{18}$O rich planetary nebulae also rich in $^{19}$F ?

## 7.4 HOT BOTTOM BURNING

We need models of hot bottom burning if we are to understand how the observed photospheric composition is related to the composition of the dredged-up material.

## 7.5 FANTASY ?

I discussed earlier the exploratory calculations of Bazan and Lattanzio (1992) which showed that previously ignored energy sources associated with $^{13}$C ingestion and s-processing could lead to changes in the evolution of the boundaries of the ISCZ. Although sensitive to the details of the $^{13}$C ingestion, the possibility arose of mini pulses of convection within a single thermal pulse. It seems that the shape of the ISCZ may be critically dependent on the uncertain distribution of $^{13}$C in the model. It is possible to imagine a distribution which could cause the ISCZ to split into two regions, an outer one which produces the $^{19}$F, and an inner zone for the s-processing. By separating the sites of this nucleosynthesis, it may be possible to satisfy the observational constraints on the neutron density and the $^{18}$O abundance. Perhaps some overlap is required between these zones, say at the start (or finish ?) of a thermal pulse. It is also quite possible that some details of convection, not reproduced by the mixing-length theory, may be required to solve all of the above problems.

### Acknowledgements

It is a pleasure to thank the many people who try to help me understand AGB stars, specifically A. Boothroyd, M. Busso, R. Gallino, D. Lambert, R. Malaney, V. Smith and P. Wood.

### References

Bazan, G., and Lattanzio, J. C., 1992, *Astrophys. J.*, submitted.
Blocker, T., and Schonberner, D., 1991, *Astron. Astrophys.*, **244**, L43.

Boothroyd, A. I., and Sackmann, I.-J., 1988a, *Astrophys. J.*, **328**, 632.
Boothroyd, A. I., and Sackmann, I.-J., 1988b, *Astrophys. J.*, **328**, 641.
Boothroyd, A. I., and Sackmann, I.-J., 1988c, *Astrophys. J.*, **328**, 653.
Boothroyd, A. I., and Sackmann, I.-J., 1988d, *Astrophys. J.*, **328**, 671.
Bryan, G. L., Volk, K., and Kwok, S., 1990, *Astrophys. J.*, **365**, 301.
Forestini, M., Goriely, S., Jorissen, A., and Arnould, M., 1992, preprint
Forestini, M., Paulus, G., and Arnould, M., 1991, *Astron. Astrophys.*, **252**, 597.
Groenewegen, M. A. T., and de Jong, T., 1992, preprint (see also this volume).
Hollowell, D., 1988, Pd. D. Thesis, University of Illinois
Hollowell, D., and Iben, I., Jr., 1989, *Astrophys. J.*, **340**, 966.
Iben. I., Jr., 1981a, in *Physical Processes in Red Giants*, Eds. I. Iben Jr. and A. Renzini, p3.
Iben, I., Jr., 1981b, *Astrophys. J.*, **246**, 278.
Iben, I., Jr., and Renzini, A., 1982a, *Astrophys. J. Lett.*, **259**, L79.
Iben, I., Jr., and Renzini, A., 1982b, *Astrophys. J. Lett.*, **263**, L188.
Iben, I., Jr., and Renzini, A., 1983, *Ann. Rev. Astr. Astrophys.*, **21**, 271.
Iben, I., Jr., and Renzini, A., 1984, *Phys. Rept*, **105**, 329.
Jorissen, A. A., Smith, V. V., and Lambert, D. L., 1992, preprint
Lambert, D. L., 1989, in *Evolution of Peculiar Red Giants*, Eds. H. R. Johnson and B. Zuckerman, Cambridge (C.U.P.), p101.
Lambert, D. L., 1991, in *Evolution of Stars: The Photospheric Abundance Connection*, Eds. Eds. G. Michaud and A. Tutukov, p299.
Lattanzio, J. C., 1986, *Astrophys. J.*, **311**, 708.
Lattanzio, J. C., 1989, in *Evolution of Peculiar Red Giants*, Eds. H. R. Johnson and B. Zuckerman, Cambridge (C.U.P.), p161.
Lattanzio, J. C., 1989b, *Astrophys. J. Lett.*, **344**, L25.
Lattanzio, J. C., 1991, *Astrophys. J. Suppl. Ser.*, **76**, 215.
Lattanzio, J. C., 1992, *Proc. Astron. Soc. Aust.*, in press.
Lattanzio, J. C., and Malaney, R. A., 1989, *Astrophys. J.*, **347**, 989.
Paczynski, B., 1971, *Acta Astron.*, **21**, 417.
Renzini, A., and Voli, M., 1981, *Astron. Astrophys.*, **94**, 175.
Rogers, F. J., and Iglesias, C. A., 1992, *Astrophys. J. Suppl. Ser.*, in press.
Sackmann, I.-J., 1980, *Astrophys. J. Lett.*, **241**, L37.
Sackmann, I.-J., and Boothroyd A. I., 1992, *Astrophys. J. Lett.*, **392**, L71.
Smith, V. V., and Lambert, D. L., 1989, *Astrophys. J. Lett.*, **345**, L75.
Smith, V. V., and Lambert, D. L., 1990a, *Astrophys. J. Suppl. Ser.*, **72**, 387.
Smith, V. V., and Lambert, D. L., 1990b, *Astrophys. J. Lett.*, **361**, L69.
Vassiliadis, E., and Wood, P. R., 1992, in this volume.
Wood, P. R., 1981, in *Physical Processes in Red Giants*, Eds. I. Iben Jr. and A. Renzini, p135.

# CARBON- AND OXYGEN-RICH PROGENITORS OF PLANETARY NEBULAE

H.J. HABING and J.A.D.L. BLOMMAERT
*Sterrewacht Leiden, P.O. Box 9513, 2300 RA Leiden, The Netherlands*

## 1. Introduction

This review concerns stars on the Asymptotic Giant Branch that are about to develop into planetary nebulae (=PNe). They are still termed "stars", and properly so, and yet they have already, in statu nascendi, the structure of a PN. There are two different kinds: oxygen-rich stars (spectral class: M) and carbon- rich stars (spectral class: C).

There is, we estimate, a consensus that AGB stars are immediate progenitors of PNe. The frontier of present day research is thus beyond the status of the AGB stars and concerns the next generation of questions: How do AGB stars make the transition? What mechanism inside the stars is the main cause? However, to have a firm beginning it seems useful to summarise in Section 2 the main arguments for our confidence in the AGB stars as progenitors of PNe.

For an introduction into this topic we strongly recommend the proceedings of a meeting in 1989 in Montpellier (France): "From Mira to Planetary Nebula: Which path for stellar evolution?" (Mennessier and Omont, 1990)

## 2. AGB Stars: The Progenitor Stars of Planetary Nebulae

In 1956 Shklovskii proposed that the core of a red giant might become first the central star of a planetary nebula and later a white dwarf. Several arguments argue convincingly for Shklovskii's proposal: **1.** Calculations of the evolution before and on the Asymptotic Giant Branch (AGB) produce stellar models with a remarkably sharp distinction between a small core of very dense, degenerate matter (very similar to a white dwarf) and a huge envelope with a very low density: core and envelope contain (at least early in the AGB phase) comparable amounts of matter, and yet the core radius is of the order of $10^{-4}$ of that of the envelope and thus the density in the core is of the order of $10^{12}$ times that in the envelope. Qualitatively the situation is very similar to that in a PN. **2.** Quantitatively there is also agreement: According to the model calculations the star will, for most of its time, burn quietly hydrogen into helium at a luminosity that is a linear function of the mass of the core (the "Paczynski relation"). From the observed luminosities of AGB stars in our Galaxy ($3000 \lesssim L_* \lesssim 20,000 L_\odot$) one concludes that the masses of the AGB cores range between 0.51 and 0.83 $M_\odot$. This range is very similar to that of white dwarfs and of the central stars of PNe (Weidemann, 1990). **3.** An important category of AGB stars are the OH/IR stars. These stars have a galactic distribution that is remarkably similar to that of the PNe: not only is there close similarity between the two distributions in the plane of the sky (the l-b diagram) but also in the characteristic distributions of longitude versus radial velocity (the l-V diagram).

One concludes that PNe and OH/IR stars both belong to the "old disk" population and thus both come from the same main sequence stars.

Originally Shklovskii's proposal posed one major difficulty: A planetary nebula contains less ionized gas than the original envelope- a conclusion that is confirmed again at this conference (original envelope mass between 0.4 and several time $M_\odot$; ionized mass between 0.01 and 0.2 $M_\odot$). Yet this does not longer pose a serious objection against Shklovskii's proposal because: 4. In young open clusters with turn-off masses of a few times $M_\odot$ white dwarfs have been detected of mass less than 1.0 $M_\odot$ (Weidemann, 1990). This informs us that the envelope is largely ejected, but it does not tell us why and how this happens. 5. AGB stars have been discovered that expell their outer layers at a rate of the order of 1 $M_\odot$ in $10^4$ to $10^5$ year: infrared stars; OH/IR stars and C-stars like IRC+10216. The discovery that high mass-loss rates exist is necessary, but not sufficient evidence; at present we have no generally accepted proof that this high mass-loss phase lasts sufficiently long, and yet there seems to be a consensus that it does; the point will be discussed further down. 6. Haloes of neutral material have been found around the ionized gas of PNe, and those may contain the envelope mass (see especially the work by Balick et al., 1992, and this symposium).

## 3. Samples of AGB Stars

### 3.1. IN THE MAGELLANIC CLOUDS

The LMC and the SMC offer the possibility to study stars all at the same distance and statistical properties of complete samples may be discovered. A disadvantage is their distance: m-M=18.47 for the LMC and 18.6 for the SMC. Over the last ten years much research has been done on AGB stars in the Magellanic Clouds. We list some developments.

1. The study of field stars in the LMC (Reid and Mould, 1984) and in the SMC (Reid and Mould, 1990). The authors conclude that the observations contradict predictions from earlier model calculations (e.g. Renzini and Voli, 1981). The Paczynski relation (mentioned above) predicts $M_{bol} = -7.1$ if the core mass equals the Chandrasekhar limit. One expects quite a few stars near that limit, but in fact there are few stars with $M_{bol} < -6.0$. Reid and Mould suggest that in these early model calculations the importance of mass loss has been underestimated.

2. The study of AGB stars in clusters. The Magellanic Clouds contain many stellar clusters, and the ages of these clusters spread very nicely over the range from 10 Myr to 10 Gyr - unlike the situation in our Galaxy where the clusters are either very young (< 1 Gyr) or very old (15 Gyr). Because AGB stars are of intermediate $M_{ms}$ and intermediate age, they can be found in most Magellanic Cloud clusters. A recent paper, also a summary, of much work done on individual cluster stars is by Frogel et al. (1990); they discuss 39 clusters and about 400 AGB stars. The work shows clearly that younger clusters produce brighter AGB stars. In the LMC carbon stars occur in clusters with an age between 0.1 Gyr and a few Gyr; in the SMC carbon stars occur even in the oldest clusters. For the LMC one concludes that stars with $M_{ms}$ below 3 to 5 $M_\odot$ become carbon stars when they

reach a certain, high luminosity during their climb along the AGB.

3. Recent deep searches for LPV's in the LMC have been made by Reid et al. (1988) and by Hughes (1989). See also the discussion by Hughes and Wood (1990). More than 1000 LPV's were found, half of them SRa and half Mira variables. The distribution of the pulsation periods shows one strong peak at about 200 days, a sharp downfall for shorter periods and a more shallow fall-off for longer ones. This peaked distribution strongly resembles that for Mira variables in the solar neighbourhood except that for the local Miras the peak is shifted to somewhat longer periods. A specific problem in understanding the results is the total number found: on various grounds Hughes and Wood argue that about 15 times more Mira's should be expected. This, they argue, points to a fundamental lack of understanding; perhaps an AGB star is an LPV star for less than 10% of its time. A point not noted by Hughes and Wood, is that the luminosity distribution of their LPV sample is very similar to that of the field AGB stars found by Reid and Mould (1984).

4. The studies mentioned sofar all started with photographic studies, usually in the I band. We know however that in our Galaxy the AGB stars sometimes eject matter (especially true for the LPV's) and that they are then very reddened. To find them you have to search in the infrared, and preferentially beyond 3 $\mu m$. In the Magellanic Clouds, the IRAS survey is of little help: Even heavily obscured stars were faint to IRAS, and in addition the poor angular resolution (several arcminutes) leads to confusion. Nevertheless several studies have been undertaken: Reid et al. (1990) (followed by Reid, 1991) and Wood et al. (1992). Stars with strong circumstellar shells have been found, and in 5 of them Wood et al. detected the 1612 MHz OH maser; these stars are the brightest AGB stars, with $-6 \lesssim M_{bol} \lesssim -7$. It is clear that many more highly reddened objects must exist in the LMC and have not yet been detected (see the next review paper by Whitelock).

These studies of AGB stars in the Magellanic Clouds allow us to get good statistics, i.e. a reliable luminosity distribution, and an insight of what AGB stars become carbon-rich. The scarcity of AGB stars brighter than $M_{bol} = -6$ suggests that the final stages of AGB stars are not determined by nuclear processes around the core but rather by mass loss processes near the surface of the stars: AGB stars bleed to death, before they can explode.

3.2. IN THE GALAXY

AGB stars in our Galaxy are obviously much closer by and can be studied in more detail. Yet, the study of the statistical properties of samples of AGB stars is hindered by the difficulty of determining individual distances. This difficulty plays less a role if one studies stars at about the same distance, e.g. at the galactic centre or in the bulge. Here the problem is of course interstellar extinction. Luckily there are a number of windows; for a large piece of work by a team of several astronomers see a recent paper (Terndrup et al. 1991). Here we will discuss briefly a very recent study by ourselves.

The so-called Palomar-Groningen Field #3 (l= $0°$, b= $-10°$) -an area of roughly $6.5° \times 6.5°$ for which the extinction is small and well known- has been searched for long period variables and for RR Lyrae stars on B- and R-plates by Plaut (1970) (recently Wesselink (1987) upgraded the results somewhat). These samples can be complemented with a sample of AGB stars with thick circumstellar shells by using the IRAS Point Source Catalogue. This is the subject of a thesis by one of us (J.B.) (to be completed in the autumn of this year; related, earlier studies are: Whitelock et al. 1991 and van der Veen and Habing, 1990). Blommaert concludes that the samples of optically detected Mira variables and the IRAS sample agree so closely in their average luminosity and in the distribution of their luminosities, that both samples must originate from the same parent population of main sequence stars. Because the stars in the IRAS sample have significant longer pulsation periods (on average 450 days against on average 250 days for the optical sample) they must be further evolved: their envelope mass has decreased. Their evolution has been dominated by mass loss, not by core growth.

## 4. Mass Loss - a New Evolutionary Scenario

4.1. OBSERVATIONS AND MODELS OF CIRCUMSTELLAR SHELLS.

Following the publication of the IRAS Point Source Catalog, and stimulated by vastly improving facilities for millimeter and near-infrared observations a large number of measurements of circumstellar emission have been made over the last 10 years. An excellent compilation of maser observations of more than 3000 stars (with close to 1000 succesful detections) is the one by Benson et al. (1990), and a similarly impressive compilation of 1069 CO line observations of 384 circumstellar shell stars by Loup et al. is in press.

Once again, the distinction between continuum and line observations is useful. Continuum observations, from a few $\mu$m to mm wavelengths measure the solid particle distribution, molecular lines (sometimes the 21-cm line) measure the gas; the spectral line also gives the outflow velocity, an item of great importance in the interpretation. To start with the continuum observations: if you want to read up on the topic then for silicate grain models start with Justtanont and Tielens (1992) and for shells containing carboneous grains with Griffin (1990) or Chan and Kwok (1990). These papers show that the modelling has become a well established technique, and that the observed continuum over the full range from $1\mu$m to several mm is explained by basically a simple, spherically symmetric model. Once the expansion velocity, $V_{exp}$ has been derived from spectral line measurements, a fair measurement of the mass loss rate of the dust can be obtained; Justtanont and Tielens compare several different ways to make such estimates (see also the review by Van der Veen and Olofsson in Mennessier and Omont, 1990).

Molecular line emission concerns maser transitions and thermal transitions. We will not dicuss here the maser transitions, although much interesting work on this topic has been done; a recent textbook (Elitzur, 1992) and the proceedings of a recent symposium (Astrophysical Masers (Washington), eds. Glegg and Nedoluha) will have to help you out. Thermal line emission is always observed at millimeter and submillimeter wavelengths. Over the last 5 years or so these measurements have become easier, more sensitive and more reliable. The analysis of the observations is again usually done with spherically symmetric models of continuous outflow.

Such models are increasingly complex, but incorporate also more (and still reliable) physics- for two recent papers see Sahai (1990) and Kastner (1992).

## 4.2. THE IRAS 2-COLOUR DIAGRAM.

A useful tool to study the properties of circumstellar shell sources is the IRAS 2-colour diagram: a diagram of the $12\mu m/25\mu m$ versus the $25\mu m/60\mu m$ flux density ratio. Van der Veen and Habing (1988) defined areas in this diagram that contain a significant concentration of different types of objects (C-rich stars; Mira's; OH/IR stars; PNe; etc.). These definitions have been found useful. An important addendum to their discussion is a recent paper by Omont et al. (1992) on the locations of carbon stars. Two parameters determine the spectrum and thus the position of an object in the 2-colour diagram: the optical depth of the dust, $\tau$, and $r_0$, the inner radius where the dust forms. In the 2-colour diagram a well defined curve indicates the position of a star that would gradually increase its mass loss rate and thus the optical depth in its shell (Rowan-Robinson and Harris, 1983; Bedijn, 1987). Actually there are two curves: one for carbon rich dust, the other for silicate dust.

If the mass loss remains constant for more than a few thousand years the object will occupy a definite position on these curves in the 2-colour diagram. If the mass loss stops, the existing shell will expand, the inner radius becomes larger and the object will describe a "loop" through the diagram (Willems and de Jong, 1986). These loops have been invoked to explain why so many carbon stars appear to have normal stellar colours between 12 and $25\mu m$, but a large excess at $60\mu m$; these stars suffered a high mass loss rate, but some 10,000 yr ago the mass loss rate stopped on a short time scale: as a witness of past events the star has still a "detached, fossil shell" around it. Egan and Leung (1991) have confirmed this conclusion by an accurate analysis of the $60\mu m/100\mu m$ IRAS colours of carbon stars.

Recently Zijlstra et al. have analysed the IRAS colours of large samples of known M-giant stars and of known carbon stars. They show (a) that all carbon stars are loosing mass heavily, or have finished such a phase at most $10^5$ yr ago and (b) that a small, but significant fraction of the M-giants have also old shells around them. Earlier Olofsson et al. (1990) had found detached shells around carbon stars from millimeter line measurements, and they concluded that carbon stars are capable of interrupting the mass-loss process, even when the rate is high, and then later start again (a phenomenon they tried to link with the thermal pulses); Zijlstra et al. conclude that also M-giants on the AGB show this interruptive mass-loss behaviour.

## 4.3. THEORY OF MASS-LOSS.

Understanding the evolution of the AGB stars implies understanding mass loss. Yet major questions like: "What causes the star to eject matter?"; "How will the mass loss rate develop in time?"; "Is the envelope lost in a smooth, steady proces, or is the proces intermittent and sometimes interrupted?" have today been answered only tentatively. Nevertheless- even a tentative answer is much more than none at all, so we have seen considerable progress. For a good discussion of the problem we refer to the review by Hearn in Mennessier and Omont, 1990.

A promising modelling development is by Bowen and Willson- see their most recent paper in 1991. Mass loss is a two stage proces: because of the stellar pulsations the atmosphere of the star is very extended; this brings enough matter to such a height that solid particles begin to condense; these particles are being driven out by light pressure and carry the gas along. The mass loss rate is essentially determined by the density of the gas at the point where the condensations take place. The resulting outflow velocity depends on how much momentum is transfered by the stellar light. The mass loss rate increases until the envelope gets so thin that the photosphere begins to contract and the star becomes a protoplanetary nebula. Most of the considerations of Bowen and Willson had already been formulated semi-intuitively by Bedijn (1988).

Bowen and Willson attach little significance to the thermal pulses- in their view these are only a temporary interruption. Wood and Vassiliadis (1992) however think that the thermal pulses play a major role: each thermal pulse initiates a period of stellar oscillations, the star becomes a Mira variable for a short time and, because of the pulsations, it loses mass at a high rate. Following every next thermal pulse more mass is lost until finally the matador pulse appears and the bull is being slaughtered: the star becomes a planetary nebula, the feast can begin and the steak can be fried.

Weighing Bowen and Willson's model against that by Wood, it seems that the first has more hydrodynamic modelling to it, but that the observations prefer the second model: mass loss may well be intermittent and not continuous. Apart from the arguments presented by Zijlstra et al. and by Olofsson et al. there are also the following considerations: (1). In the LMC the luminosity distribution of the LPV's agrees closely with that of the non- variable AGB M-giants. This might suggest that the Mira phase appears at one or several intermediate moments, and not as the final stage of the M-giant evolution. (2). Interrupted mass loss may also explain why there are so few Mira's in the LMC and in the solar neighbourhood in comparison to the number of PNe (see Wood's review in Menessier and Omont, 1990).

## 5. Carbon Rich stars: a difficult group

The first carbon stars were recognised in optical spectra before the turn of the century. More recent are the discoveries of dust-enshrouded carbon stars like IRC+10216. For a recent compilation see Groenewegen et al. (1992) or the discussion by Omont et al.(1992). A difficult question is that of the luminosity distribution; in the Galaxy we do not have the means to obtain a direct determination of distances. Work on carbon stars in the Magellanic Clouds (Frogel et al., 1990) indicates that the distribution of luminosities there is narrow (standard deviation less than 0.50 magnitude), but in the LMC the distribution shifts systematically with the age of the cluster in which the stars have been observed and a comparison between various environments (see e.g. Lequeux in Mennessier and Omont, 1990) reveals large differences between metal-poor galaxies, like the SMC and a metal-rich environement, like the galactic bulge. There is, we think considerable uncertainty about the luminosities of the galactic carbon stars.

The origin of the carbon-richness was felt to have been understood with the discovery by Iben (see his review of 1991) that in the interior of AGB stars, after a thermal pulse newly produced carbon may be added to the (hydrogen) envelope; a M-giant then turns into a C-star when the carbon abundance exceeds that of oxygen. This theoretical result appears to be robust, but there remain major problems with detailed model predictions of when and what stars turn into C-stars (Lattanzio, this symposium). The carbon stars detected in the Magellanic Clouds are all rather bright with an average lumonosity around 8000 $L_\odot$, or $M_{bol} = -5$. But in the bulge of our Galaxy carbon stars of much lower luminosity ($M_{bol} >$-2.5) have been found (see Westerlund et al., 1991). These stars may never have been AGB stars and their carbon richness probably has another cause than a dredge-up after a thermal pulse. More enigmatic carbon stars are the two with thick circumstellar shells found at high galactic latitude (Cutri et al., 1989). Are these true halo objects, or are they escaped members of the galactic disk population? A few planetary nebulae in the halo, most of them carbon-rich, may pose the same riddle (Clegg et al., 1987).

Progress in this field will be difficult to obtain in view of the inability of the observers to provide reliable luminosity functions of carbon stars inside our own Galaxy, and the difficulties that the modellers have in calculating the details of the dredge-up.

## 6. Last Remarks

That PNe originate from AGB stars appears to be an hypothesis with many solid arguments in favour of it and no solid arguments against it. Yet, we know that the cause, the duration and the time behaviour of mass loss is incompletely understood and we realize that this process is essential for the evolution of AGB stars. About the identification of the bright M-giants and LPV's as AGB stars there is no uncertainty: solid spectroscopic evidence exists of dredge- up phenomena and thus of the existence of thermal pulses. Thus the origin of PNe is largely understood, be it not in detail.

Yet, this success should not close our eyes for the short comings in the present theory: notably on the question of why and how a red giant begins to loose mass when it becomes pulsational unstable; and that while the AGB may provide most of the progenitors, there still may be progenitors of a different breed: the carbon stars in the galactic centre and in the galactic halo seem to tell us that. Common envelope evolution in double stars (read the beautiful review by Iben, 1991) is perhaps part of an alternative way of evolution.

Of the later stages of stellar evolution we may have seen much of the light, but we have not yet seen it all.

## References

-Balick, B., Gonzalez, G., Frank, A., Jacoby, G. 1992, Astrophys. J. 392, 582.
-Bedijn, P.J. 1987, Astron. Astrophys. 186, 136.
-Bedijn, P.J. 1988, Astron. Astrophys. 205, 105.
-Benson, P.J., Little-Marenin, I., Woods, T.C., Attridge, J.M., Blais, K.A., Rudolph, D.B.,

Rubiera, M.E., Keefe, H.L. 1990, Astrophys. J. Suppl. Ser. 74, 911.
-Blanco, V.M. 1988, Astron. J. 95, 1400.
-Bowen, G.H., Willson, L.A. 1991, Astrophys. J. Lett. 375, L315.
-Chan, S.J., Kwok, S. 1988, Astrophys. J. 334, 362.
-Clegg, R.E.S., Peimbert, M., Torres-Peimbert, S. 1987, Month. Not. R.A.S. 224, 761.
-Cutri, R.M., Low, F.J., Kleinmann, S.G., Olszewski, E.W., Willner, S.P., Campbell, B., Gillett, F.C. 1989, Astron. J. 97, 866.
-Egan, M.P., Leung, C.M. 1991, Astrophys. J. 383, 314.
-Elitzur, M. 1992, Astronomical Masers (Kluwer Publ., Dordrecht).
-Frogel, J.A., Mould, J., Blanco, V.M. 1990, Astrophys. J. 352, 96.
-Griffin, I.P. 1990, Mon. Not. R.A.S. 247, 591.
-Groenewegen, M.A.T., de Jong, T., van der Bliek, N.S., Slijkhuis, S., Willems, F.J.: 1992 Astron. Astrophys. 253, 150.
-Hughes, S.M.G. 1989, Astron. J. 97, 1634.
-Hughes, S.M.G., Wood, P.R., 1990, Astron. J. 99, 784.
-Iben, I. ApJ Suppl. Ser. 1991 76, 55.
-Justtanont, K., Tielens, A.G.G.M. 1992, Astrophys. J. 389, 400.
-Kastner, J.H. 1992 Astrophys. J. (december 10 issue)
-Loup, C., Forveille, T., Omont, A., Paul, J.F. Astron. Astrophys. Suppl. Ser. 1992 (in press)
-Mennessier, M.O., Omont, A. 1990 (editors) From Mira's to Planetary Nebulae: Which path for stellar evolution? Editions Frontiéres (Gif sur Yvette).
-Olofsson, H., Carlström, U., Eriksson, K., Gustafsson, B., Willson, L. 1990, Astron. Astrophys. Lett. 230, L13.
-Omont, A., Loup, C., Forveille, T., te Lintel-Hekkert, P., Habing, H., Sivagnanam, P. 1992 Astron. Astrophys. (in press)
-Plaut, L. 1970, Astron. Astrophys. 8, 341.
-Reid, N. 1991, Astrophys. J. 382, 143.
-Reid, N., Tinney, C., Mould, J. 1990, Astrophys. J. 348, 98.
-Reid, N., Glass, I., Catchpole, R.M., 1988, Mon. Not. R.A.S. 232, 53.
-Reid, N., Mould, J. 1984, Astrophys. J. 284, 98.
-Reid, N., Mould, J. 1990, Astrophys. J. 360, 490.
-Renzini, A., Voli, M. 1981, Astron. Astrophys. 94, 175.
-Rowan-Robinson, M., Harris, S. 1983, Mon. Not. R.A.S. 202, 767.
-Sahai, R. 1990, Astrophys. J. 362, 652.
-Shklovskii, I.S. 1956, Astron. Zh. 53, 315.
-Terndrup, D.M., Frogel, J.A., Whitford, A.E.: 1991, Astrophys. J. 378, 742
-van der Veen, W.E.C.J., Habing, H.J. 1988, Astron. Astrophys. 194, 125.
-van der Veen, W.E.C.J., Habing, H.J. 1990, Astron. Astrophys. 231, 404.
-Weidemann, V. 1990, Ann. Rev. Astron. Astrophys. 28, 103.
-Wesselink, T. 1987, thesis Katholieke Universiteit van Nijmegen (the Netherlands).
-Westerlund, B.E., Lequeux, J., Azzopardi, M., Rebeirot, E., 1991, Astron. Astrophys. 244, 367.
-Whitelock, P., Feast, M., Catchpole, R. 1991, Mon. Not. R.A.S. 248, 276.
-Willems, F.J., de Jong, T. 1986, Astron. Astrophys. 196, 173.
-Wood, P.R., Vassiliadis, E. 1992, in Highlights of Astronomy 9, tbd.
-Wood, P., Whiteoak, J., Hughes, S., Bessell, M., Gardner, F., Hyland, A. (preprint)
-Zijlstra, A.A., Loup, C., Waters, L.B.F.M., de Jong, T. 1992 Astron. Astrophys. Lett. (submitted)

# PLANETARY NEBULAE FROM MIRAS?

PATRICIA A. WHITELOCK

*South African Astronomical Observatory, P.O. Box 9, Observatory 7935, South Africa*

and

MICHAEL W. FEAST

*Institute of Astronomy, Cambridge University, Madingley Road, Cambridge CB3 0HA, England*

ABSTRACT. The properties of large amplitude AGB variables (Miras) are reviewed with particular emphasis on their relevance for evolution into PNe. We concentrate on recent observations in the LMC and Galactic Bulge. The available evidence is consistent with the idea that most Miras evolve into PNe and that most old disk stars go through the Mira and PNe phases. However, there are some suggestions that the most metal rich stars in the Galactic Bulge go through neither phase or do so very rapidly. The lifetime of a typical Bulge Mira must be more than $10^5$ yrs and Miras outnumber PNe in the Bulge by about 10 to 1.

## 1. Introduction

This review deals with stars near the top of the AGB and their possible evolution into planetary nebulae. The category of highly evolved AGB stars encompasses the Miras, including those found in globular clusters, the optically invisible OH/IR stars and their carbon-rich analogues. These are collectively referred to here as *Mira Variables*. Observationally we see such stars as long period (P>100 day), large amplitude ($\Delta M_{bol} \gtrsim 0.5$ mag) pulsators. If the Mira has an optical counterpart then it will have a late spectral-type (Me, Se or Ce); otherwise it will be a strong IR source and possibly an OH maser. We know that these stars must be very close to the top of the AGB and we know that they have high mass-loss rates, typical values being between $10^{-7}$ and $10^{-4}$ $M_\odot$ $yr^{-1}$. Kinematic and other evidence indicates a general increase of initial mass with pulsation period from the ~100 day Miras at one end of the sequence to the ~2000 day OH/IR stars at the other.

There has been a lot of recent activity in the field of AGB evolution particularly following the results from the IRAS satellite. This review covers only a few highlights that seem particularly relevant to the relationship with PNe. In particular we concentrate on Miras in two widely different environments: the LMC and the Galactic Bulge. A more detailed review of late-type variables in the Bulge can be found in Whitelock (1992).

## 2. The LMC

The recent results of very extensive LMC surveys of Miras have been published by, Reid et al. (1988), Feast et al. (1989), Hughes (1989) and Hughes & Wood (1990). In addition to these major surveys some thick-shelled sources have been discovered in the LMC as a result of investigations of IRAS sources by Reid et al. (1990), Wood et al. (1991, 1992), and Reid (1991, 1992). Of particular interest are the OH sources for which Wood et al. have measured periods; the first periods for extragalactic OH stars. They are in excess of 1000 days and it is interesting that these OH/IR stars fit rather well on an extrapolation of the Mira period-luminosity (PL) relation derived for Miras with periods less than 420 days (Feast et al. 1989). Given the extreme difficulty in determining the distances to OH/IR stars in the Galaxy it will be very interesting to see more observations of these and other LMC masers.

The Magellanic Clouds contain both oxygen-rich (M & S type) and carbon-rich (C type) Miras as discussed in the above references. Of particular interest here is that the oxygen-rich Miras obey a period-luminosity-colour (PLC) relation (Feast et al. 1989). The existence of a PLC relation implies that the instability strip in the HR diagram has a finite width. Therefore some change in period of an individual Mira might be expected as it evolves through the strip. The observations however suggest rather little evolution in log P at least among short period stars (see Feast & Whitelock 1987, Whitelock 1990). A proper comparison of the PLC relation with theory requires the calculation of accurate evolutionary tracks incorporating the effects of mass loss. Although such detailed models are not available, Wood (1991) and Feast (1992) have discussed the implications of the PLC relation. Wood has suggested that the PLC relation exists because the effective temperature of the giant branch at a given luminosity varies with metal abundance. Feast has pointed out that the tracks derived by Wood imply that the short period Miras, at least, have already evolved away from the AGB due to the depletion of their envelopes by mass loss. In other words, Miras may already be post-AGB stars.

## 3. The Bulge

We (Glass, Whitelock, Feast & Catchpole in preparation) are in the process of analyzing multi-phase observations of the Mira variables in the Sgr I Baade window ($l = 1°.4, b = -2°.6$). The analysis is still in a preliminary stage but suggests that these Bulge Miras diverge somewhat from the LMC Mira PL and period-colour (PC) relations. Figure 1 shows the PC relation for Miras in the Bulge and the LMC. A few Bulge stars with extreme colours, $(J-K)_o \gtrsim 4$, have been omitted from this plot. Such extreme colours are found for IRAS Miras and are the product of circumstellar reddening in stars with thick dust shells (Whitelock et al. 1991). It is clear from Fig 1 that the longer period Bulge sources have distinctly redder colours than would be predicted from the LMC PC relation.

The PL relation is difficult to derive directly in the Bulge because of the finite distance spread in the line of sight. However if the LMC PL relation is compared with the measured luminosity, distance moduli can be derived for individual Miras. The distribution of distance moduli can then be compared with that predicted from a model of the Bulge and the distance to the Galactic Centre ($R_o$) derived. This approach results in $R_o$=9.1 kpc if the

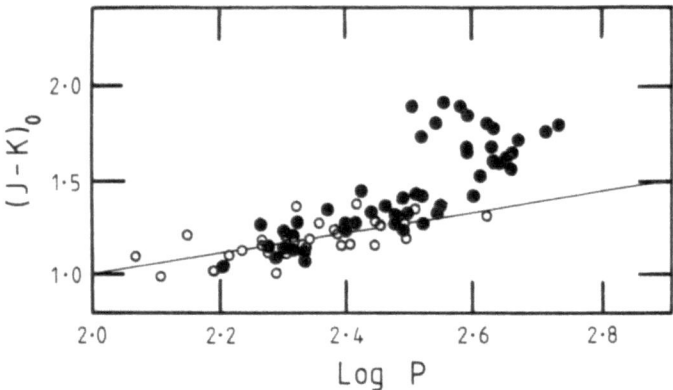

Figure 1. The PC relation for Miras in the LMC (open circles) and the Bulge (closed circles). The solid line is a least-squares fit to the LMC data.

bolometric PL relation is used, or $R_o$=8.6 kpc if the K(2.2$\mu$m) PL relation is used (omitting the few sources with optically thick shells). That is, the PL relations do not give consistent results and they both imply a rather large distance to the centre (Feast 1987 gives $R_o$=7.8±0.8 kpc). In other words the stars are fainter than the LMC PL relations predict. If instead of using the LMC PL relation to calculate the distances we use the LMC PLC relation, then the derived distance is $R_o$=8.2 kpc from both the K and the bolometric PLC relations. Figure 2 illustrates the model fit for distances derived from the bolometric PLC relation. This result confirms the existence of the PLC relation and implies that the Bulge Miras, or at least the longer period ones, occupy a slightly different part of the instability strip from the LMC Miras. Wood (1990) has suggested on theoretical grounds that we might expect metal rich Miras to behave in this way.

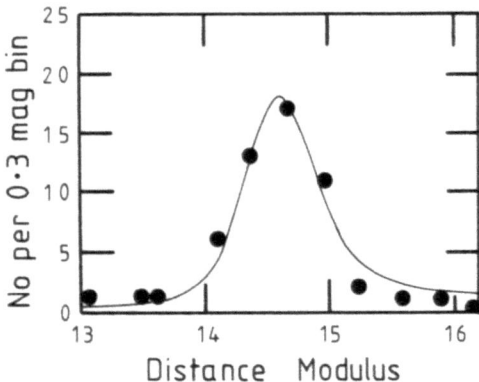

Figure 2. The distribution of distances for Miras in the Sgr I window. The observations are compared to a model with $R_o$=8.2 kpc, a volume density distribution $\rho \propto a^{-2}$ and $c/a = 0.6$. The standard error of a single distance modulus is assumed to be 0.2 mag.

Whitelock et al. (1991) did a survey of Miras with high mass-loss rates in the Galactic Bulge. They colour selected IRAS sources from two strips across the Bulge between $b = +7°$ and $+8°$ and $b = -7°$ and $-8°$ with $345° < l < 15°$. They monitored these for several years in the near-infrared finding 113 Miras and deriving periods for 104 of them. Bolometric luminosities were then derived from the PL relation and apparent luminosities were calculated by integrating the flux from the IRAS and near-infrared photometry. Hence distances to individual Miras were derived. Note that the PLC relation cannot be used to derive the luminosities of and hence distances to these stars because use of the PLC relation requires the photospheric colour which is unknown for these thick shelled sources. An estimate of the error introduced by use of the PL relation rather than the PLC relation may be made from the fact that $R_o = 8.6$ kpc is derived from use of the PL for these sources in comparison to the currently favoured value of $R_o = 7.8$ kpc. The effect of this will be noted below where appropriate.

Whitelock et al. (1991) used a slightly modified version of the expression derived by Jura (1987) to estimate the mass-loss rates ($\dot{M}$) for these Miras. The circles in Fig 3 show the derived mass-loss rates as a function of pulsation period and amplitude. The use of the PL relation to derive the distances will result in an overestimate of the mass-loss rate by about 20%. Sources with periods over 600 days are shown as filled circles on the period/$\dot{M}$ diagram; sources with amplitudes over 1.8 mag are similarly indicated on the $\Delta K/\dot{M}$ diagram. It is clear that at a given period there is a range of mass-loss rates and that the stars with the largest mass-loss rates have the largest pulsation amplitudes. This illustrates the now well known and presumably causal connection between pulsation and mass loss (see also Wood's paper in these proceedings).

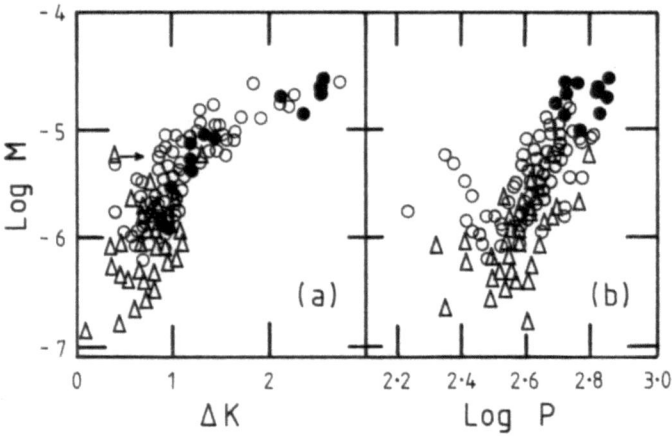

Figure 3. Mass-loss rates ($\dot{M}$) as a function of (a) pulsation amplitude and (b) period. Circles are from Whitelock et al. (1991) and triangles are from the extension of the survey as described in the text.

The Bulge survey was recently extended in the 3.8 square degree area where the $-7°$ to $-8°$ strip overlaps the Groningen field number 3, near to $l = 0°$. The intention was to identify all of the Miras in this one area. JHKL photometry was obtained for all of the 12μm IRAS sources in this region other than the 7 already examined as part of the original survey. From the 89 stars examined 47 have the colours of Miras and periods were determined for 40 of these. The periods of these 40 covered roughly the same range as those with the extreme colours but with a somewhat lower average period. From blue plates Plaut (1971) and Wesselink (1987) found a total of 45 Miras in the overlap region, most of which have periods less than 300 days. Only 4 of these are in common with the 47 in the IRAS detected sample, suggesting that there are other Miras that were detected neither by IRAS nor on the blue plates and that the survey is therefore not complete.

Mass-loss rates were calculated for the Miras with IRAS fluxes in the same way as those described above and are plotted as triangles in Fig 3. For those with only upper limits for the 25μm flux a value of 0.56 times the 12 μm flux was used to estimate $\dot{M}$, this being the average value for those which did have 25μm fluxes. It may be that the mass-loss rates for these stars are overestimated with respect to the red group in that the expansion velocity of 15 km s$^{-1}$ which is assumed for all the sources is probably an overestimate for those with thinner dust shells. Nevertheless these Miras extend the sequence shown by the red group; they exhibit lower mass-loss rates and lower pulsation amplitudes over the same period range.

There were a total of 95 Miras in the 3.8 square degrees at $b = -7°.5$ providing a lower limit for the total number of Miras per unit area at this latitude. An estimate is also available for the total number of Miras in two other inner Bulge fields, Sgr I and NGC 6522, from the work of Lloyd Evans (1976) and Glass (1986). Both of these are probably less than complete but are unlikely to be more so than the $-7°.5$ field. Figure 4 is modified from Feast et al. (1992) and shows the distribution of Miras from the three Bulge fields as a function of galactic latitude compared to that of late-M stars (from Blanco 1988). The Mira fall off is more gradual than that of the M stars; the surface density of Miras changes as $R^{-2}$ (where R is distance from the galactic centre) in contrast to a dependency of $R^{-3.2}$ (corresponding to a volume density of $R^{-4.2}$) found by Blanco and Terndrup (1989) for stars later than M5.

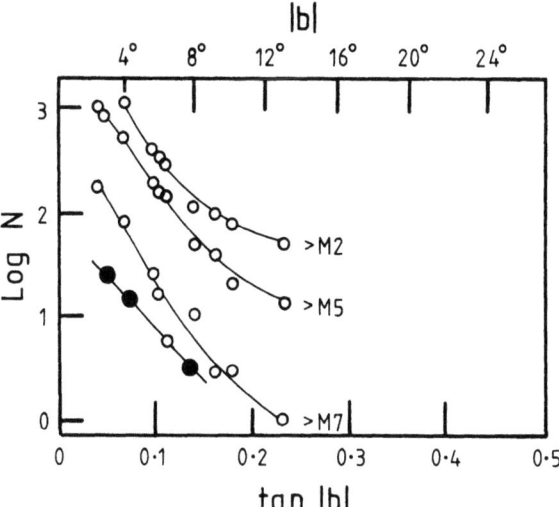

Figure 4. The galactic latitude dependence of surface number density for late-M stars (open circles) and Miras (closed circles).

This result must be regarded as tentative until deep surveys of Miras in a number of fields in the Bulge are obtained. The M stars whether they are on the giant branch or asymptotic giant branch are generally thought to be the progenitors of the Miras so this difference in distribution is rather surprising. It implies that either the most metal-rich M stars do not become Miras (perhaps they become AGB-manqué stars, the existence of which has been postulated to explain the uv-excess measure for extreme metal-rich ellipticals, e.g. Renzini & Greggio (1990)) or that the Mira descendants of metal-rich M giants may exist but have lifetimes much shorter than their globular cluster counterparts. An alternative possibility is that the late M-giants evolve predominantly into long-period Miras and that these have a much steeper density gradient than do short period ones.

## 4. Mira Lifetimes

As stated above, the $-7°.5$ field contains 95 Miras in 3.8 square degrees, 7 of which are high mass-loss objects observed in the first survey. The total number of Miras we would expect in the two strips covered by the original survey which contains 113 high mass-loss Miras, is therefore $(113/7) \times 95 = 1534$. Approximately 13% of these will be foreground objects reducing the number to 1300. In order to estimate the total number of Miras in the Bulge it is necessary to assume a model for the Mira distribution. The model fit to Fig 2 suggests a volume density distribution $\rho \propto R^{-2}$, while Fig 4 which shows that the surface density distribution goes as $R^{-2}$ would imply $\rho \propto R^{-3}$. This difference is probably the result of the bar-like structure of the Bulge (Whitelock & Catchpole 1992 and references therein) which should be explicitly modelled to determine the number of Miras. In fact our current knowledge is inadequate to do this effectively and so for the purpose of making an estimate we assume an ellipsoidal Bulge with a volume density distribution $\rho \propto R^{-2}$. The total number of Miras within a radius of 2 kpc from the centre is then *at least* $2 \times 10^4$. The effect of increasing the exponent for the volume density distribution will be to increase the estimate of the total number of Miras in the Bulge.

It is then possible to use the fuel consumption theorem from Renzini & Buzzoni (1986) to estimate the lifetime of a typical Bulge Mira. This expression:

$$t_j = N_j/B(t)L_T$$

relates the duration of a specific evolutionary phase, $t_j$, to the number of objects in that phase, $N_j$, via the total luminosity of the population, $L_T$, and the stellar death rate per unit luminosity, $B(t)$. A lifetime of at least $10^5$ yrs is derived. This can be compared with other estimates. Renzini and Greggio (1990) using the same expression find $2.5 \times 10^5$ yrs for the Globular Cluster Miras. The cluster Miras are of short period ($< 300$ days). Jura and Kleinmann (1992) have estimated a lifetime of $2 \times 10^5$ for Miras with intermediate periods (300-400 days). The Miras in the Bulge cover a very large period range from 100 to perhaps 800 days, but the bulk of them will have periods less than 400 days.

The numbers of Miras can be compared with the numbers of PNe. Pottasch (1992) has estimated that there are at least 1000 PNe within $10°$ of the centre. An estimate for Miras within the same radius is $1.4 \times 10^4$ giving a ratio of Miras to PNe of 14. The only Bulge field surveyed in depth for both PNe and Miras is the Baade window near NGC 6522, which contains 114 Miras per square degree and 9 PNe per square degree (from Shaw &

Wirth (1985) and Acker, private communication) giving a ratio of 13. Pottasch estimates the observable lifetimes of Bulge PNe as 8000 yrs compared to our estimate of $> 10^5$ yrs for Miras; this gives a ratio of >12. All these are in remarkably good agreement given the uncertainties and are consistent with all Bulge PNe evolving from Miras even if not all M giants become Miras.

The view that all or most Miras evolve into PNe is also consistent with other evidence. PNe evolve from stars with a range of initial masses, but the bulk of PNe appear to have evolved from low mass stars. This is true for PNe both in the Galactic Bulge (where there is an absence of young stars) and in the extended solar neighbourhood where the kinematics of PNe shows that the bulk of the objects must belong to an old disk population (radial velocity dispersion $\sim$40 km s$^{-1}$ and asymmetric drift $\sim$20 km s$^{-1}$, cf Oort (1965), Pottasch (1983)). In view of this it has been puzzling that estimates of the scale heights of PNe in the Galaxy have given relatively small values, 100–150 pc, compared with the scale height of the old disc population ($\sim$300–350 pc, cf Freeman 1987). However, the scale height estimates depend critically on the very uncertain distances to the PNe. Recently Zijlstra & Pottasch (1991) have estimated a scale height of 250±50 pc for PNe by a method which seems likely to be less dependent on the adopted distance scale. This result coupled with the discussion at this meeting by Peimbert (his table relating birthrate to scale height) would suggest that all old disk stars pass through a PN phase and that the PN birthrate per unit luminosity ($\dot{\xi}$) in the galactic disk is $\sim 20 \times 10^{-12} yr^{-1} L_\odot^{-1}$ consistent with the value found in other galaxies by Peimbert and with the value of the stellar death rate predicted by Renzini & Buzzoni (1986). The decrease in $\dot{\xi}$ discussed by Peimbert for galaxies brighter or redder than our own suggests that these galaxies contain a significant population which does not evolve through the PN phase. One possibility would be that they contain an old super-metal-rich population and that these stars do not become PNe (or that super-metal-rich PNe have short lifetimes). This would be consistent with the results of Ratag et al. (1992) and of Clegg (these proceedings) which show that the PNe in the Galactic Bulge evolve predominantly from the relatively metal poor part of the Bulge population.

It would evidently be important to study the relative numbers and distributions of PNe, Miras and giant stars in the Galactic Bulge as a function of metallicity (which in the case of Miras seems related to period). Such a study would place our understanding of the relation of Miras to PNe on a much firmer footing.

## Acknowledgements

We are grateful to Ian Glass and Robin Catchpole for permission to quote results from work in preparation. MWF acknowledges financial support from the Royal Society.

## References

Blanco, V. M. (1988) *AJ*, **95**, 1400.
Blanco, V. M. & Terndrup, D. M. (1989) *AJ*, **98**, 843.
Feast, M. W. (1987) In: *The Galaxy*, p1, Eds. G. Gilmore & B. Carswell, Reidel, Dordrecht.
Feast, M. W. (1992) In: *Highlights of Astronomy*, in press, Ed. J. Bergeron, Kluwer, Dordrecht.

Feast, M. W. & Whitelock, P. A. (1987) In: *Late Stages of Stellar Evolution*, p33, Eds. S. Kwok & S. R. Pottasch, Reidel, Dordrecht.
Feast, M. W., Glass, I. S., Whitelock, P. A. & Catchpole, R. M. (1989) *MNRAS*, **241**, 375.
Feast, M. W., Whitelock, P. A. & Sharples, R. (1992) In: *The Stellar Populations of Galaxies*, IAU Sym **149**, p77, Eds. B. Barbuy & A. Renzini, Kluwer, Dordrecht.
Freeman, K. C. (1987) *Ann Rev A & A*, **25**, 603.
Glass, I. S. (1986) *MNRAS*, **221**, 879.
Hughes, S. (1989) *AJ*, **97**, 1634.
Hughes, S. & Wood, P. R. (1990) *AJ*, **99**, 784.
Jura, M. (1987) *ApJ*, **313**, 743.
Jura, M. & Kleinmann, S. G. (1992) *ApJ Sup*, **79**, 105.
Lloyd Evans, T. (1976) *MNRAS*, **174**, 169.
Oort, J. H. (1965) In: *Galactic Structure*, p455, Eds. A. Blaauw & M. Schmidt, Univ of Chicago Press.
Plaut, L. (1971) *A & A Sup*, **4**, 75.
Pottasch, S. R. (1992) preprint.
Pottasch, S. R. (1984) *Planetary Nebulae*, Reidel, Dordrecht.
Ratag, M. A., Pottasch, S. R., Dennefeld, M., Menzies, J. W. (1992) *A & A*, **255**, 255.
Renzini, A. & Buzzoni, A. (1986) In: *Spectral Evolution of Galaxies*, p195, Eds. C. Chiosi & A. Renzini, Reidel, Dordrecht.
Renzini, A. & Greggio, L., (1990) In: *Bulges of Galaxies*, p47, Eds. B. J. Jarvis & D. M. Terndrup, ESO Conf. & Workshop Proc. 35.
Reid, N., Glass, I. S. & Catchpole, R. M. (1988) *MNRAS*, **232**, 53.
Reid, N., Tinney, C. & Mould, J. (1990) *ApJ*, **348**, 98.
Reid, N. (1991) *ApJ*, **382**, 143.
Reid, N. (1992) these proceedings.
Shaw, R. A. & Wirth, A. (1985) *PASP*, **97**, 1071.
Wesselink, Th. (1987) Ph D Thesis, Nijmegen.
Whitelock, P. A. (1990) In: *Confrontation Between Stellar Pulsation and evolution*, ASP conf ser **11**, p365, Eds. C. Cacciari & G. Clementini.
Whitelock, P. A. (1992) In: *Galactic Bulges*, IAU sym **153**, in press, Kluwer, Dordrecht.
Whitelock, P. A. & Catchpole, R. M. (1992) In: *The Center, Bulge and Disk of the Milky Way*, in press, Ed. L. Blitz, Kluwer, Dordrecht.
Whitelock, P. A., Feast, M. W. & Catchpole, R. M. (1991) *MNRAS*, **248**, 276.
Wood, P. R. (1990) In: *From Miras to Planetary Nebulae*, p67, Eds. M. O. Mennessier & A. Omont, Editions Frontiers.
Wood, P. R., Moore, G. K. G. & Hughes, S. M. G. (1991) In: *The Magellanic Clouds*, IAU Sym **148**, p259. Eds. R. Haynes & D. Milne, Kluwer, Dordrecht.
Wood, P. R., Whiteoak, J. B., Hughes, S. M. G., Bessell, M. S., Gardner, F. F. & Hyland A. R. (1992) preprint.
Zijlstra, A. A. & Pottasch, S. R. (1991) *A & A*, **243**, 478.

# EVOLUTION FROM THE AGB: VARIABILITY

DIMITAR D. SASSELOV

*Harvard-Smithsonian Center for Astrophysics, Cambridge, MA 02138, USA*

ABSTRACT. Luminous low-mass stars of intermediate $T_{\text{eff}}$ – candidates for post-AGB transition objects to PNNi, are all expected to be pulsationally variable. Different modes and their combinations may be involved, leading to a variety of observed behaviour.

## 1. Introduction

All stars *on* and *close* to the asymptotic giants branch (AGB) are intrinsic variable stars. This variability involves dynamical time scales (months) and is generally caused by large scale motions of the stellar envelope, which will be referred to as *pulsations* for simplicity.

The pulsations of a stellar envelope can (and appear to) influence strongly the stellar AGB and post-AGB evolution. Such pulsations cannot influence the nuclear burning regions of a star, yet they affect most of the volume of its interior – the extended, non-adiabatic, envelope. Thus pulsation is related to or, in fact, drives mass loss and may determine the onset and scale of post-AGB evolution (Schönberner & Blöcker 1992; Wood, this volume).

Intrinsic stellar variability is also a very powerful diagnostic tool for stellar interiors and evolution. This is especially true for post-AGB stars which evolve on comparatively very short time scales.

## 2. Classes of Variable Stars

The region of the HR diagram of the AGB and to immediately hotter $T_{\text{eff}}$ is populated with a large variety of different variable stars. However, their differences are often superficial or misleading, because many classes of variable stars defined earlier in this century were based mainly on light curves criteria alone.

There are four classes of variables that are relevant to our discussion of post-AGB objects: RV Tauri, UU Herculis, SRd, and R CrB stars. The variability of central stars of planetary nebulae is outside the scope of our discussion.

### 2.1. *RV Tauri stars*

This is a small class of pulsating variable stars, apparently luminous, and with a characteristic light curve – alternating deep and shallow minima. A handful of stars with similar light curves are known in globular clusters; most field members have peculiar infrared excesses (Jura 1986); several exhibit decreasing pulsation periods (Percy et al. 1991), and as a consequence the RV Tauri stars are usually associated with low-mass post-AGB stellar variability. It should be remembered however, that the RV Tauri phenomenon refers only to brightness variability in F–K giants, while the RV Tauri light-curve behaviour has not been proven to arise in low-mass stars alone (see review by Wahlgren 1992).

There are no known RV Tau stars in binaries. There are no established Magellanic Clouds members either. As a result their luminosities are poorly known.

## 2.2. UU Herculis stars

This is an even smaller class of pulsating variable stars, apparently luminous, at high galactic latitudes, and low-amplitude Cepheid-like (but unstable) light curves. Their variability, though peculiar (Fernie 1992), is not a straightforward discriminant. They are of class F to early G, and span in metalicity from deficient to solar. Many of them were found to have substantial IR excess and were associated with transit objects between the AGB and PNNi. Despite the largely disparate characteristics of these high-latitude objects, it is noteworthy that there are much more (known and suspected) spectroscopic binaries among the UU Her stars, than for any of the other three classes here (Waelkens, 1992). However, parameters for any of these binaries are yet to be determined. No Magellanic Clouds members are known.

## 2.3. R CrB stars

The R CrB stars have strong features as a class (occasional deep minima), and seem less prone to selection effects. However, their semi-regular pulsational variability falls within the general definition of the SRd class. The R CrB stars are few; none are known in binaries, and just a couple are found in the Magellanic Cloud. They occupy the same region of the HR diagram as the above classes, but their evolutionary status is not clear (see Whitney, Soker, and Clayton 1991).

## 2.4. SRd stars

The yellow semi-regular stars (SRd) class has a very loose definition: it contains almost all yellow giants and supergiants which are neither Cepheids nor RV Tauri stars. This is a disparate mixture of Population I and II stars, as well as other sparsely observed variables (like RV Tauri or UU Herculis) hastily assigned to it.

In addition to the classified stars, there are also others that are loosely associated with some of the above classes and our topic, like FG Sge and $\epsilon$ Aur (F).

## 3. Models

On the HR diagram these four classes of variable stars and the post-AGB transition objects, are expected to coexist with evolved massive Population I stars. Most notable of these – the classical Cepheids, define a narrow instability strip in that region. Similar temperatures and luminosities, but different masses, imply a very different internal structure, hence pulsation characteristics.

A comparison between a low-mass post-AGB star at spectral class F, and a Cepheid of the same $T_{\text{eff}}$ and $L$ (but 10 times more massive) is instructive. On one hand both stars have very similar structure with respect to a radial mass coordinate. Hence, the driving mechanisms of their oscillations are similar. On the other hand, the two stars are very different with respect to a radial depth coordinate: the low-mass star contains most of its mass in the tiny hot core (of less than 0.01 $R_{ph}$), which is surrounded by an extended envelope with virtually zero density gradient. As a result, any pulsations excited in such a low-mass supergiant are going to be strongly non-adiabatic and non-linear. For example, a 0.8 $M_\odot$, 20000 $L_\odot$ star with $T_{eff}$ = 5500 K will have a thermal time-scale $t_{th} \approx 4$ days in its envelope. However, the envelope dynamical time-scale will be much greater, $t_{dyn} \approx 40$ days, leading to strong non-linear effects (see theoretical models by Fadeyev (1984), Saitou, Takeuti, and Tanaka (1989), Aikawa (1991) and others).

One of the most reliable facts established by the different theoretical models to-date, is that luminous low-mass stars of intermediate $T_{eff}$ are unstable to different modes of pulsation through a wide range of temperatures and luminosities. This is in contrast to the massive stars in that region of the HR diagram, in which pulsation is strongly confined in parameter space. In the luminous low-mass stars radial modes different from the fundamental are often dominant, and this may explain the observed preference towards lower amplitudes of pulsation (see Aikawa 1992). As can be expected, the pulsational behaviour of luminous low-mass stars is very unstable and complex.

In general the current theoretical understanding of the pulsations of luminous low-mass stars remains very poor. One important challenge is the treatment of convection (Saitou 1992). The effects of radiative transfer have been shown to be significant, as well (Zalewski 1992).

So, we might expect a post-AGB transition object to look like an intermediate temperature supergiant, pulsating with a period of about 1-2 months, and more often with low amplitude (below 0.2 mag), and almost certainly in a semi-regular fashion. While no specific form of the light curve can be ascribed to the post-AGB objects, there is no good understanding of the peculiar light curves observed among likely candidates (Zsoldos 1990; Fernie 1992). The models point to the essential similarity between SRd and RV Tau light curves (Takeuti 1992; Aikawa 1992). However, such results are more strongly affected by the problems mentioned above, as well as by the treatment of the stellar atmosphere, and should be used with caution.

## 4. Pulsation and Evolution

The evolution of stars from the AGB to the white dwarf phase (towards higher temperatures and smaller radii) is very fast by normal stellar standards. A temperature increase at a mean rate of several Kelvins per year is expected (Schönberner & Blöcker 1992). Such a rapid increase should lead to detectable color and period changes on a time scale of decades. No completely convincing example has been observed yet, but there are several very hopeful cases.

Potentially, the pulsation period of a star is a more sensitive indicator of evolutionary change than is color. This is especially true for classes of variables, like RV Tauri and UU Herculis, with stable periods (despite light curve instabilities). Generally, the pulsation period, $P \sim T_{eff}^{-n}$, where $n$ has values between 3.0 and 3.7, depending on the model's density distribution and the mode involved. The temperature, $T_{eff}$, increases with time at a constant luminosity, $L$, as determined by the evolutionary calculations, $e.g.$ Schönberner & Blöcker (1992). The stellar luminosity, $L$, is a parameter. Thus obtained, the rate of change of the pulsation period is a robust result of general validity: for radial or non-radial, non-adiabatic, and non-linear oscillations (see Bruggen and Smeyers 1987).

The expected rates of period change are between 0.1 and 2 days per decade; the pulsation period decreases. Such period changes are indeed observed in five RV Tauri stars with extensive photometric data, which may confirm that at least some RV Tauri stars are post-AGB objects.

## 5. Conclusions

Stellar pulsation on the AGB and after the AGB is intimately related to the driving, the amount, and the spacial distribution of mass loss. Thus pulsation may determine the speed of post-AGB evolution, as well as play a role in shaping the planetary nebula (*e.g.* the inner axisymmetric regions of NGC 6826 and NGC 6543, see Soker and Harpaz 1992).

Stellar pulsation on the AGB and after the AGB is the rule, rather than the exception. This is a theoretical expectation for single stars, but is also true for binaries. The probability that most AGB and post-AGB stars have companions, often in a common envelope, is very high (see Livio, this volume). The stellar oscillations excited even by a Jupiter-like companion may be significant and cause low-amplitude semi-regular variability (Soker 1992). This adds additional variety to the types of oscillation modes that can be exhibited by post-AGB objects and suggests more caution when using traditional variable star classes. Light curve characteristics may not have a direct relation to evolutionary states, as exemplified by RV Tauri and SRd stars, as well as SRa and SRb stars (see Kerschbaum and Hron, this volume).

## 6. References

Aikawa, T. 1991, ApJ, 374, 700.
Aikawa, T. 1992, MNRAS, in press.
Bruggen, P., and Smeyers, P. 1987, A&A, 186, 170.
Fadeyev, Yu.A. 1984, Ap&SS, 100, 329.
Fernie, J. D. 1992, in Luminous High-Latitude Stars, ed. D. Sasselov, ASP Conf. Series, in press.
Jura M. 1986, ApJ, 309, 732.
Percy, J.R., Sasselov, D.D., Alfred, A., & Scott, G. 1991, ApJ, 375, 691.
Saitou, M. 1992, in Non-Linear Phenomenae in Stellar Variability, IAU Colloq. 134, eds. M.Takeuti and J.R.Buchler, Ap&SS, in press.
Saitou, M., Takeuti, M., and Tanaka, Y. 1989, PASJ, 41, 297.
Soker, N. 1992, ApJ, 386, 190.
Soker, N., and Harpaz, A. 1992, PASP, in press.
Schönberner, D., and Blöcker, T. 1992, in Luminous High-Latitude Stars, ed. D. Sasselov, ASP Conf. Series, in press.
Takeuti, M. 1992, *ibid.*
Waelkens, C. 1992, *ibid.*
Wahlgren, G.M. 1992, *ibid.*
Whitney, B.A., Soker, N., and Clayton, G. C. 1991, AJ, 102, 284.
Zalewski, J. 1992, PASJ, 44, 27.
Zsoldos E. 1990, Ap&SS, 165, 111.

# PROTO-PLANETARY NEBULAE

SUN KWOK

*Department of Physics and Astronomy, University of Calgary, Calgary, Alberta, Canada T2N 1N4*

ABSTRACT. The progress in the search for proto-planetary nebulae (PPN) in the last 5 years is reviewed. An observational definition of PPN is developed and a list of current PPN candidates is given. The optical, infrared, and radio properties of PPN are summarized and compared with theoretical models.

## 1. Introduction

Proto-planetary nebulae (PPN) are defined as transition objects between the end of the asymptotic giant branch (AGB) and the planetary nebulae (PN) phases (Kwok 1987). Over 1000 PN were detected in the *IRAS* survey. According to the evolutionary models of Schönberner (1983), the fraction of time a star spends in the PPN phase is ~10% of the entire PN lifetime. If this is the case, we can expect ~100 PPN candidates in the *IRAS* Point Source Catalog (PSC). PPN candidates can be identified by two ways: either by searching for stars in existing optical catalogs with appropriate *IRAS* colors, or to search for the optical counterparts of low-temperature *IRAS* sources. The PPN candidates discovered to date are the result of the combination of these two strategies.

## 2. Search for Proto-Planetary Nebulae

### 2.1. PPN CANDIDATES ASSOCIATED WITH KNOWN OPTICAL STARS

The first attempt to search the *IRAS* Point Source Catalog for associations with G-type stars was by Odenwald (1986). Excess emission was found in ~4% of the 150 G supergiants detected by *IRAS*. Several of the supergiants identified are RV Tauri variables, which have long been known to possess infrared excesses at 10 µm (Gehrz 1972). The evolutionary status of RV Tauri stars as post-AGB objects of low mass has been discussed by Jura (1986). A systematic identification of *IRAS* sources with objects with existing optical spectra has been carried out by Bidelman. Bidelman's extensive list contains several stars with intermediate spectral types (F-K) with large infrared excesses which can be candidates for PPN (Bidelman 1985).

It has been known for some time that a number of high-latitude F supergiants can be better explained as old, halo objects observed in the post-AGB phase (Bond, Carney, and Grauer 1984). Among the properties exhibited by various members of this class (but not by all) are high space velocity, low metal abundance (Bond and Luck 1987), circumstellar dust (Parthasarathy and Pottasch 1986), and molecular envelopes (Likkel *et al.* 1987). The supergiant spectral classification just reflects the low gravity of these stars, but not their intrinsic luminosity. The prototype of this class of 89 Her, which is a MK spectral standard of type F2 Ia. 89 Her is located at a galactic latitude of 23°, which will put it at an unreasonably large distance above the plane if it has a luminosity of a Population I supergiant. The strongest support for its post-AGB status is the infrared excess (Gillett et al. 1970), which is likely to originate from the remnant of the CSE of its AGB progenitor.

Two other classes of variable stars are also likely to be post-AGB stars. RV Tauri stars are characterized by light curves showing alternate deep and shallow minima, with periods of 50-150 days, and have spectral type F, G, or K (Preston et al. 1963). The circumstellar envelopes of RV Tauri stars can also be seen in OH (Bujarrabal et al. 1988, Alcolea and Bujarrabal 1991) and CO. The class of UU Her stars is designated by Sasselov (1984) to represent small-amplitude variable stars of periods 40-100 days and located at high galactic latitudes. Infrared excesses similar to 89 Her have been found in RV Tauri (Gehrz 1972), but not in UU Her (Gehrz and Woolf 1970). A comprehensive study of 25 high galactic latitude supergiants with infrared excesses was made by Trams (1991).

2.2. COOL IRAS SOURCES AS PPN CANDIDATES

The search for PPN candidates by ground-based observations of *IRAS* sources satisfying certain color criteria has been carried out by several groups (Kwok 1987, van der Veen 1988, Slijkhuis 1992). Basically, candidates are selected from a certain region of the *IRAS* color-color diagram. The optical counterparts of these candidates are identified with a ground-based telescope or by positional coincidence on sky survey plates. The spectral energy distribution can be determined by visible and infrared photometry. The nature of the sources can be further defined by optical, infrared, and millimetre spectroscopy and optical/infrared imaging.

This strategy of PPN search depends critically on the correct identification of the *IRAS* sources. Many cool *IRAS* sources are evolved AGB stars which suffer from large circumstellar extinction and will not have an optically-bright counterpart (Kwok, Hrivnak, and Boreiko 1987). Identification by positional coincidence will easily lead to the wrong identification of the *IRAS* source, in particular in the Galactic plane where the field is crowded. Even when the source is identified by searching in the K band, there are still possibilities of confusion because most red stars are strong near-infrared emitters. The only sure way of obtaining the correct identification is by searching around the *IRAS* position at 10 or 20 µm and compare the observed fluxes with the *IRAS* 12 and 25 µm PSC fluxes. It is our experience that while the *IRAS* positions are generally good, there are cases where the actual source is located more than one arc min away from the PSC position.

A typical energy distribution of PPN is shown in Fig. 1. The most notable feature of a PPN is the "double-peaked" spectral distribution. The low temperature component corresponds to the remnant of the AGB envelope and the high temperature component corresponds to the reddened photosphere of the central star.

2.3. CONFUSION OF PPN WITH OTHER OBJECTS

Since there is a proliferation of the use of the term PPN in the literature, it is essential to sort out which ones are real. The common causes of confusion are with: (i). PN; (ii). AGB stars; (iii). massive stars; (iv). HII regions, and (v). wrong association with the *IRAS* sources. Since PN have large far infrared excesses and often have near-infrared excesses due to nebular emission (Zhang, this volume), they are easily mistaken as PPN. Their nature can, however, be conclusively determined by optical spectroscopy or radio imaging. The difference between AGB stars and PPN is more subtle: if the two components are clearly separated, it is likely to be a PPN. AGB stars are also likely to have large-amplitude variability. As for massive stars in transition from the red to the blue, it is very difficult to separate from PPN

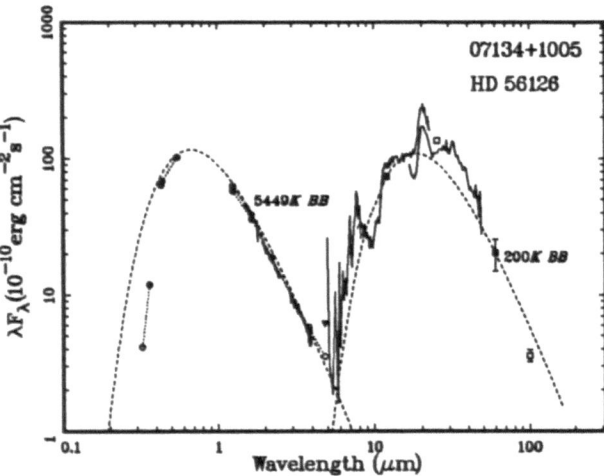

Figure 1. Spectral energy distribution of the PPN candidate 07134+1005. The solid line between 7 and 23 μm is the IRAS LRS. The H, K, and L spectra are from Kwok et al. (1990), the 5-8 μm spectrum is from Buss et al. (1990), and the 30-60 μm spectrum is from Omont et al. (1992).

without some independent distance estimates. For example, IRC+10420 (F8 Ia) shares many common properties with PPN but is widely believed to be a massive star.

We have developed the following observational definition for a PPN:

i.  A PPN candidate should show clear evidence of the remnant of the AGB envelope. These include: (a) large infrared excess with color temperatures between 150-300 K; and (b) molecular emission (CO or OH) showing expansion velocity of 5-30 km s$^{-1}$ typical of AGB winds.

ii. There should be evidence that the circumstellar envelope is detached from the photosphere and is not the result of an ongoing mass loss process.

iii. If the central star is bright enough for its spectral type to be determined, it should be of spectral types B-K with luminosity class I.

iv. There should not be large-amplitude photometric variability as the result of pulsation of a massive ($>10^{-3}$ $M_\odot$) hydrogen envelope above the core.

We have tabulated in Table 1 26 objects that we consider to be good candidates for PPN.

## 3. Optical Properties of PPN

A number of PPN candidates show emission activities in Hα. The Hα profiles range from P Cygni, inverse P-Cygni, to shell (Waters, this volume). Velocity monitoring of a number of PPN candidates have found several binary candidates (Hrivnak & Woodsworth, this volume). $C_2$ and $C_3$ molecules have been detected in a number of PPN, suggesting that these are highly carbon-rich objects (Hrivnak, this volume).

Table 1
Proto-Planetary Nebulae Candidates

| IRAS | other names | Sp. Type | l | b | Infrared | Mol. Line | Ref |
|---|---|---|---|---|---|---|---|
| 04296+3429 | - | G0 Ia | 166.2 | -9.1 | 21 μm | CO | a |
| 05113+1347 | - | G8 Ia | 188.8 | -9.1 | 21 μm | - | b |
| 05381+1012 | - | G | 195.5 | -10.6 | - | - | |
| 06530-0213 | - | F0 I | 215.4 | -0.1 | - | CO | c |
| 07134+1005 | HD 56126 | F5 I | 206.7 | +10.0 | 21 μm | CO | a,d |
| 10215-5916 | DM-583221 | G5: I | 285.1 | -1.9 | silicate | - | d,h |
| 12175-5338 | SAO 239853 | A9 Iab | 298.3 | +8.7 | - | - | d |
| 17150-3224 | - | G2 I | 353.8 | +3.0 | - | OH,CO | c,g,h |
| 17436+5003 | HD 161796 | F3 Ib | 77.1 | +30.9 | - | OH,CO | d,e |
| 17441-2411 | - | - | 4.2 | +2.2 | - | CO | c,h |
| 18025-3906 | - | G2 I | 353.3 | -8.7 | - | OH | c |
| 18095+2704 | - | F3 Ib | 53.8 | +20.2 | silicate | OH | f,g |
| 19114+0002 | HD 179821 | G5 Ia | 36.6 | -5.0 | silicate | OH,CO | d,g |
| 19454+2920 | - | - | 65.2 | +2.1 | featureless | CO | i |
| 19475+3119 | 31°3797 | F3 Ia | 67.1 | +2.7 | - | CO | |
| 19477+2401 | - | - | 60.8 | -0.9 | - | OH | i |
| 19480+2504 | - | - | 61.8 | -0.6 | featureless | CO | |
| 19500-1709 | HD 187885 | F3 I | 24.0 | -21.0 | - | CO | d,g |
| 20000+3239 | - | G8 Ia | 69.7 | +1.2 | 21 μm | CO | b |
| 20004+2955 | V 1027 Cyg | G7 Iab | 67.4 | -0.4 | silicate | - | d,h |
| 20028+3910 | - | - | 75.5 | +4.2 | featureless | CO | h |
| 22223+4327 | DO 41288 | G0 Ia | 96.7 | -11.5 | 21 μm | CO | |
| 22272+5435 | HD 235858 | G5 Ia | 103.3 | -2.5 | 21 μm | CO | a,g,k |
| 22574+6609 | - | - | 112.0 | +6.0 | 21 μm | CO | j |
| 23304+6147 | - | G0 Ia | 113.9 | +0.6 | 21 μm | CO | a |
| 23321+6545 | - | - | 115.2 | +4.3 | - | OH,CO | h |

(a). Kwok, Volk, and Hrivnak 1989, (b). Kwok et al. 1992, (c). Slijkhuis 1992, (d). Hrivnak, Kwok, and Volk 1989, (e). Parthasarathy and Pottasch 1986, (f). Hrivnak, Kwok, and Volk 1988, (g). van der Veen 1988, (h). Volk and Kwok 1989, (i). Kwok, Hrivnak, and Boreiko 1987, (j). Hrivnak and Kwok 1991, (k). Pottasch and Parthasarathy 1988.

## 4. Circumstellar Dust and Gas Features

### 4.1. OXYGEN-RICH PPN: SILICATES.

Since the 10 and 18 μm features of silicates are commonly observed in oxygen-rich AGB stars, it is expected that these features will also be observable in oxygen-rich PPN. It has been noted that the 10 μm feature in PPN will be much less prominent than in AGB stars because of the decline in dust temperature and the shift of the spectral peak to longer wavelengths (Kwok 1980). This effect is quantitatively confirmed by the detached-shell radiative-transfer model of Volk and Kwok (1989) who find that 10 μm feature not only weakens but also broadens. A search of the *IRAS* LRS has yielded a number of sources with the predicted shape at 10 μm (Volk and Kwok 1989). A number of these objects are later confirmed to be PPN (e.g., 18095+2704, 10215-5916, and 20004+2955). The existence of oxygen-rich PPN confirms the evolutionary connection between OH/IR stars and PN.

### 4.2. CARBON-RICH PPN

The dominant circumstellar dust features in carbon-rich AGB stars are SiC and (in extreme carbon stars) graphite. The strength of the 11.3 μm SiC feature is weaker than the 10 μm silicate feature, and graphite is featureless in the mid-infrared. These facts have led us to assume that carbon-rich PPN will not have strong identifying features in the 10-20 μm range. The discovery of strong emission feature at 21 μm in four PPN therefore came as a surprise (Kwok, Volk, and Hrivnak 1989). The *IRAS* LRS of these four sources show a prominent feature at 21 μm with an almost flat (in $\lambda F_\lambda$) continuum between 12 and 18 μm. The 21 μm feature has been confirmed by both airborne (Omont et al. 1992) and ground-based observations (Barlow, this volume).

A number of suggestions have been made for the origin of this feature. Cox (1990) found similar features in a number of HII regions, and suggests that the feature is due to $Fe_2O_3$ or $Fe_3O_4$. Sourisseau et al. (1992) suggest that the 21 μm feature arises from a mixture of coal, SiC, and urea. The strength of the 21 μm feature suggests that it originates from abundant atomic species. The fact that it is associated with carbon-rich objects suggests that carbon may be a major constituent.

In addition to the well-known 3.3 μm PAH emission feature that are commonly observed in PN and HII regions, there exist emission features in the 3.4-3.5 μm region which, although present in PN, are found to be strongest in PPN (Geballe and van der Veen 1990, Geballe et al. 1992). Several identifications have been proposed for the carriers of the 3.4-3.6 μm emission features. Barker, Allamandola, & Tielens (1987) and Allamandola et al. (1989) have suggested that the 3.40 μm and 3.51 features are hot bands of the fundamental C-H vibrational stretch at 3.29 μm in PAHs. de Muizon et al. (1986) and Jourdain de Muizon, d'Hendecourt, & Geballe (1990) propose that these and other weak features are the fundamental C-H stretches of sidegroups of PAHs. They also have suggested that overtones and combinations of C-C vibrations may be responsible for the underlying plateau.

Airborne observations of two of these 21 μm sources (07134+1005 and 22272+5435) in the 5-8 μm region showed that the 6.9 and 8 μm are also present in these PPNs (Buss et al. 1990). These features closely resemble the unidentified emission bands observed in HII regions and PN. While the 3.3 and 6.2 μm features observed in HII regions and PN are thought to be fluorescently excited by UV photons, it cannot be the case in these PPN because of their late spectral types. If these bands in PPN are excited by visible photons, then the molecules responsible must be larger (>100

C atoms) than interstellar PAH molecules (Buss et al. 1990). Based on the strength correlation between the 3.4-3.5 μm features and the 6.9 μm feature in 22272+5435, Geballe et al. suggest that the former is due to the stretching mode of $CH_2$ and $CH_3$ groups. Because of the different temperatures of the central stars in PN and PPN, the relative strengths of the 3.4-3.5 μm to the 3.3 μm features are due the different amount of UV and visible photons available.

### 4.3. ATOMIC AND MOLECULAR LINES

The photospheric spectrum of PPN are similar to those expected of intermediate spectral type supergiants. The infrared spectrum of F-type PPN is dominated by hydrogen absorption lines, and up to ten members of the Brackett series have been detected in one object. For PPN with spectral types later than G, the photospheric spectrum is dominated by vibrational absorption bands of CO (v=2-0 up to v=6-4). Most interestingly, CO have been observed to be in *emission* in several PPN (Hrivnak, Kwok, & Geballe, this volume).

## 5. Molecular Rotational Emissions

### 5.1. OH MASER EMISSION

While oxygen-rich AGB stars exhibit maser emissions in OH, $H_2O$, and SiO, not all of these emissions are expected to persist through the PPN stage. Theoretical calculations by Sun and Kwok (1987) suggest that only stars with high mass loss rates will have their 1612 MHz emission detectable through the PPN stage. An evolutionary scenario on how the strengths of these maser lines will vary after the star left the AGB is outlined by Lewis (1989). For example, the PPN 18095+2704 shows strong OH main (1665/1667 MHz) lines than the 1612 MHz line, which is usually stronger in AGB stars (Lewis, Eder, and Terzian 1990). A OH and $H_2O$ survey of cool *IRAS* sources by Likkel (1989) have detected several PPN candidates (17436+5003, 19114+0002, 19477+2401, 23321+6545), all in the main lines.

On particular interest is the OH 1667 MHz main-line emission maser emission from 11385-5517 (HD 101584). The blue- and red-shifted components are well separated in bipolar lobes on opposite sides of the stellar position (te Lintel Hekkert et al. 1992). The measured expansion velocity increases from 9 km $s^{-1}$ at the inner edge of each lobe to 40 km $s^{-1}$ at the outer edge. It is likely that the OH emission occurs along the polar axis of an equatorial disk of circumstellar dust.

### 5.2. CO THERMAL EMISSION

Since the lower rotational transitions of the CO molecule have been detected in the circumstellar envelopes of over 200 AGB stars (Knapp and Morris 1985), it is expected that CO emission should still be detectable in PPN after the shell has detached from the photosphere. Indeed, the early candidates of PPN (e.g., AFGL 618, AFGL 2688) all show strong CO emission. CO observations of PPN have been performed by many groups, and the results are reviewed by Huggins (this volume).

The mass loss rates responsible for the creation of the remnant AGB envelope in PPN are generally estimated by applying the formula of Knapp and Morris (1985). Since the energy distributions of the PPN are well determined, the momentum flux ($\dot{M}V$) represented by the derived mass loss rates can be compared to the radiative momentum implied by total observed fluxes

($4\pi D^2 F/c$). The ratio of these two momentum fluxes is generally referred to as $\beta$ (Knapp 1986). While the value of $\beta$ for AGB stars are ~1, the corresponding values are higher for PPN and highest for PN (Likkel 1989; Hrivnak and Kwok 1991).

## 6. Morphology

Since two of the first discovered PPN candidates (AFGL 618 and AFGL 2688) show bipolar morphology, there is a great deal of interest in exploring the relationship between PPN and bipolar nebulae. Hrivnak and Kwok (1991) found that many PPN have similar spectra in the mid-infrared but have vastly different optical brightness. They interpret this discrepancy as the result of PPN with non-spherically-symmetric envelopes being viewed at different orientations. The optical bright ones are likely to be pole-on systems while optically-faint ones are edge-on. Scattered light escaping from the polar directions in edge-on systems will form a bipolar nebula in the visible. PPN candidates found to show bipolar morphology include 17150-3224 and 17441-2411 (Slijkhuis 1992; Langill et al., this volume).

## 7. Theoretical Models

Nebular evolution of PPN have been calculated by Volk (1992) and Szczerba (this volume). In these models, the central star evolution models of Schönberner (1983) and a variety of mass loss rates at the end of the AGB are used to calculate the dust emission spectrum at different times beyond the AGB. In the case of Volk (1992), the gas emission spectrum is calculated using the ionization model *CLOUDY* (Ferland, this volume). Tracks on the *IRAS* colour-colour diagram is also calculated by convolving the *IRAS* bandpasses with the model spectra. In order to be in agreement with the observations, Volk (1992) found that the Schönberner (1983) tracks have to be speeded up and the PPN phase does not exceed ~800 yr.

## 8. Conclusions

Significant progress has been made since the last IAU Symposium on Planetary Nebulae in the identification and observation of transition objects between the AGB and PN phases. Many PPN (especially the carbon-rich ones) show unique infrared properties. The detection of bipolar morphology in several PPN candidates suggests that the mass loss process in the last stages of the AGB is not entirely spherically symmetric. Most importantly, the study of PPN will give us the important clues to identify the formation mechanism of PN.

I wish to thank my collaborators Bruce Hrivnak, Kevin Volk, and C.Y. Zhang for helpful comments. This work is supported by an operating grant from the Natural Sciences and Engineering Research Council of Canada.

## References

Alcolea, J., & Bujarrabal, V. 1991, *A&A*, **245**, 499
Allamandola, L.J., Tielens, A.G.G.M., & Barker, J.R. 1989, *ApJS*, **71**, 733
Barker, J.R., Allamandola, L.J., & Tielens, A.G.G.M. 1987, *APJ*, **315**, L61
Bidelman, W.P. 1985, *BAAS*, **17**, No. 4, 841
Bond, H.E., Carney, B.W., & Grauer, A.D. 1984, *PASP*, **96**, 176

Bond, H.E., & Luck, R.E. 1987, in *The 2nd Conf. on Faint Blue Stars*, ed. A.G.D. Philips, D.S. Hayes, & J.W. Leibert (Schenectady:L.Davis), p. 527
Bujarrabal, V., Bachiller, R., Alcolea, J., & Martin-Pintado, J. 1988, *A&A*, **206**, L17
Buss, R.H. et al. 1990, *ApJ*, **365**, L23
Cox, P. 1990, *A&A*, **236**, L29
de Muizon, M., Geballe, T.R., d'Hendecourt, LB., & Baas, F. 1986, *ApJ*, **306**, L105
Geballe, T.R., & van der Veen, W.E.C.J. 1990, *A&A*, **235**, L9
Geballe, T.R., Tielens, A.G.G.M., Kwok, S., & Hrivnak, B.J. 1992, *ApJ*, **387**, L89
Gehrz, R.D. 1972, *ApJ*, **178**, 715
Gehrz, R.D., & Woolf, N.J. 1970, *ApJ*, **161**, L213
Gillett, F.C., Hyland, A.R., & Stein, W.A. 1970, *ApJ*, **162**, L21
Hrivnak, B.J., Kwok, S. 1991, *ApJ*, **368**, 564
Hrivnak, B.J., Kwok, S., & Volk, K. 1988, *ApJ*, **331**, 382
Hrivnak, B.J., Kwok, S., & Volk, K. 1989, *ApJ*, **346**, 265
Jourdain de Muizon, M., d'Hendecourt, & Geballe, T.R. 1990, in *Infrared Spectroscopy in Astronomy*, ed. B.H. Kaldeich, ESA SP-290, p. 177
Jura, M. 1986, *ApJ*, **309**, 732
Knapp, G.R. 1986, *ApJ*, **311**, 731
Knapp, G.R., & Morris, M. 1985, *ApJ*, **292**, 640
Kwok, S. 1980, *ApJ*, **236**, 592
Kwok, S. 1987, in *Late Stages of Stellar Evolution*, eds. S. Kwok & S.R. Pottasch (Reidel:Dordrecht), p. 321
Kwok, S., Hrivnak, B.J., & Boreiko, R.T. 1987, *ApJ*, **321**, 975
Kwok, S., Volk, K., & Hrivnak, B.J. 1989, *ApJ*, **345**, L51
Kwok, S., Hrivnak, B.J., & Geballe, T.R. 1990, *ApJ*, **360**, L23
Lewis, B.M. 1989, *ApJ*, **338**, 234
Lewis, B.M., Eder, J., & Terzian, Y. 1990, *ApJ*, **362**, 634
Likkel, L. 1989, *ApJ*, **344**, 350
Likkel, L., Omont, A., Morris, M., & Forveille, 1987, *A&A*, **173**, L11
Odenwald, S.F. 1986, *ApJ*, **307**, 711
Omont et al., 1992, in preparation
Parthasarathy, R., & Pottasch S.R. 1986, *A&A*, **154**, L16.
Pottasch, S.R., & Parthasarathy, M. 1988, *A&A*, **192**, 182
Preston, G.W., Krzeminski, W., Smak, J., & Williams, J.A. 1963, *ApJ*, **137**, 401
Sasselov, D.D. 1984, *Ap. Sp. Sc.*, **102**, 161
Schönberner, D. 1983, *ApJ*, **272**, 708
Slijkhuis, S. 1992, Ph.D. thesis, University of Amsterdam
Sourisseau, C., Coddens, G., & Papoular, R. 1992, *A&A*, **254**, L1
Sun, J., & Kwok, S. 1987, *A&A*, **185**, 258
te Lintel Hekkert, P., Chapman, J.M., & Zijlstra, A.A. 1992, *ApJ*, in press
Trams, N.R. 1991, Ph.D. thesis, Rijksuniversiteit Ulrecht
van der Veen, W.E.C.J. 1988, Ph.D. thesis, University of Leiden
Volk, K. 1992, *ApJS*, **80**, 347
Volk, K., & Kwok S. 1989, *ApJ*, **342**, 345

# POST-AGB CANDIDATES

L.B.F.M. WATERS

*Astronomical Institute Anton Pannekoek, University of Amsterdam, Kruislaan 403, 1098 SJ Amsterdam*

*SRON Laboratory for Space Research Groningen, P.O. Box 800, 9700 AV Groningen*

and

K.C. SAHU

*Kapteyn Astronomical Institute, P.O. Box 800, 9700 AV Groningen*

**Abstract.** A review is given of the properties of post-AGB stars. Methods to find candidate post-AGB stars are briefly discussed, along with some results and selection effects involved. Their photospheric parameters ($T_{eff}$, log g, abundances), energy distributions and binarity are also discussed. Evidence for mass loss, both past and present, is reviewed.

**Key words:** post-AGB stars – mass loss – evolution – binarity

## 1. Introduction

The transition of low and intermediate mass stars from the Asymptotic Giant Branch (AGB) into a Planetary Nebula (PN) is one of the shortest phases in the evolution of such stars, but nevertheless is of importance for a better understanding of the formation mechanism of PNe and of the way the star leaves the AGB. Observations play a dominant role in attempts to improve our knowledge, because the theory of stellar evolution cannot predict the course of events from first principles. This review describes our current understanding of the physics of transition objects: several methods to find post-AGB stars, their photospheric properties and the circumstellar environment. The binary nature of some well-known post-AGB stars is discussed and the effect of binarity on the geometry of the circumstellar material, and on the photospheric abundance patterns. For other recent reviews on this subject we refer to Waters et al. (1992) and Waelkens & Waters (1992).

## 2. Selection of post-AGB candidates

Three methods are used in the literature to find post-AGB stars: searches of the IRAS colour-colour diagram, correlations of stellar catalogues with the IRAS database, and the study of luminous stars far from the galactic plane. All three methods have severe selection effects and will result in finding a limited subset of post-AGB stars.

The most widely used method is based on the change of the far-IR colours of AGB stars as they move off the AGB towards higher effective temperature. The remnant of the AGB wind will cool and become transparent at near-IR and optical wavelengths, while the IRAS colours of the star become redder. It was noticed by Pottasch et al. (1988), and also by van der Veen et al. (1989), that such a relation does indeed exist in the IRAS color-color diagram. In this diagram, the red stars are found mostly around the black body curve of 500 K while the PN are found to have an effective temperature of 100 to 200 K. The OH/IR stars, thought to be

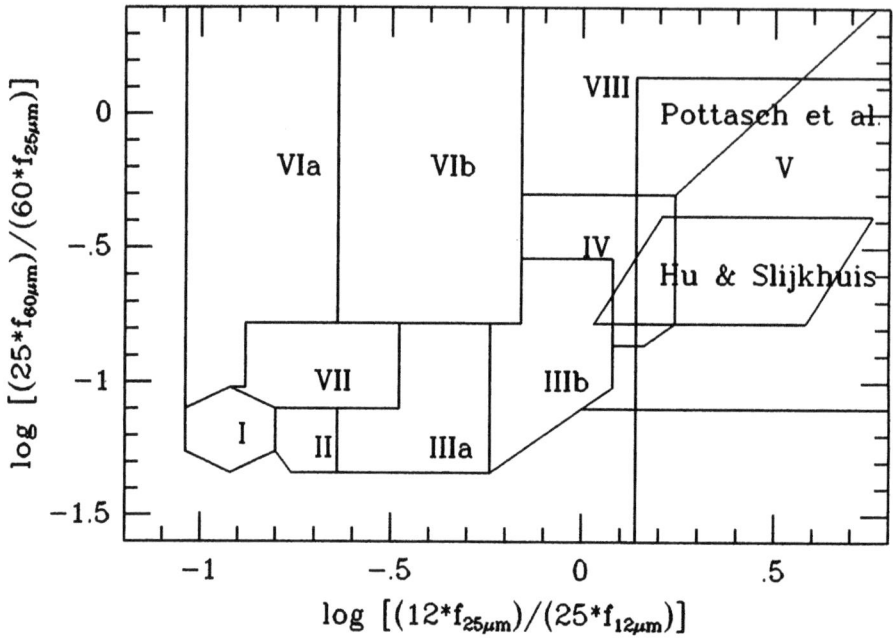

Fig. 1. The IRAS colour-colour diagram with the different regions corresponding to various classes of objects (van der Veen and Habing, 1989). The colour criteria used by Hu and Slijkhuis is shown by a trapezium and the rather relaxed criteria used by Pottasch et al. is shown by a bigger rectangle.

in the transition phase between the two stages, fall between the two temperature ranges as can be seen from Fig. 1. This roughly corresponds to the region V of van der Veen et al. (1989). So this method of looking at a fixed region of the color-color diagram has been employed by various groups, the details of which are briefly described below.

Hu and Slijkhuis (1991, 1992) have aimed at the follow-up multi-frequency study of a complete sample of post-AGB candidates from the IRAS catalog, in order to determine the nature and properties of these sources. To keep the size of the sample reasonable, they have however used rather stringent criteria, which is represented by the rectangular box shown in Fig. 1. A further criterion used in their sample is that the 12 micron flux is larger than 2 Jy, which makes the sample volume limited. About 30% of the known PNe lie in their colour-colour box. Out of the 98 sources that satisfy these criteria, 42 are southern sources. 23 out of these have no optical counterparts in the ESO survey films and the rest have been taken up for further study in the optical region by obtaining spectra and images. 17 are identified as bright stars which have spectral types F, G and M.

The approach of Pottasch and co-workers (Pottasch et al. 1988; Manchado et al. 1989 a,b; Sahu et al. 1992; Garcia-Lario, 1991) has been to use rather relaxed colour criteria to select post-AGB and PN candidates from the IRAS catalog. The criteria used by them, the details of which are given by Pottasch et al. (1988), is given by: $f_{12\mu m}/f_{25\mu m} < 0.35$ and $f_{25\mu m}/f_{60\mu m} > 0.3$ (shown by the bigger box in Fig. 1), and good IRAS flux quality at 12 and 25 microns. (The flux quality at 60 micron need not be high since lower quality flux in IRAS catalog only means an upper limit, and hence the criteria still hold). The aim of this has been not only to pick up post-AGB stars but also new PN's, although the method more efficiently picks up the dusty and hence younger PN. This criterion gives a sample of ~1000 sources, 500 of which are known PNe. The rest have been taken up for further optical, near-IR and radio continuum study.

Out of the ~200 sources in the sample of Pottasch that are observed in the optical and near-IR region so far, about 70 are OH/IR stars 35 of which are non-variable. About 35 are new or possible PN, and about a similar number are probable post-AGB stars many of which show H$\alpha$ line emission. The rest are still unknown, the nature of which are currently under investigation.

The objects observed so far have some selection effects. The ones that can be more easily seen on the Palomar plates seem to have been preferentially taken up for optical study. For a small fraction of them, particularly in a crowded field close to the galactic plane, the object observed may not be the correct IRAS counterpart although the objects which show post-AGB nature are almost certainly the correct counterparts. In this project, the objects that do not show any post-AGB nature are being re-observed by taking images in H$\alpha$ and other bands. These images will be analysed to check whether there is any object in the IRAS uncertainty ellipse that shows post-AGB characteristics. For example, if there is some object in the IRAS box which shows enhanced H$\alpha$ emission, that may be the correct counterpart, the spectrum will then be taken to confirm the nature.

A specially interesting example of this sample is IRAS 06562-0337 in terms of its fast evolution (Garcia-Lario et al. 1992). A spectrum of this object taken in 1987 shows mainly the hydrogen lines in emission (Fig. 2). A spectrum obtained about 2 years later shows many more emission lines including the forbidden lines [OIII] $\lambda\lambda$ 4959, 5007 Å, characteristic of a high density (as derived from the line ratios), young planetary nebula. The increase in intensity of the He I $\lambda$ 5876 Å line during this period, and the change in the H$\beta$ to continuum ratio both indicate an increase of stellar temperature. The object also shows strong CO emission in millimetric wavelengths. A spectrum obtained in 1992 however shows that the forbidden lines have all disappeared. Since the spectrum obtained in 1988 shows evidence of mass loss in the form of a P-Cygni profile, the authors have interpreted this as the final mass loss episode which precedes the formation of a planetary nebula. If this interpretation is correct, the object should soon show the forbidden lines in emission again, and monitoring this object may provide an opportunity to follow the evolution to a PN.

Similar work by others (e.g. Kwok et al. 1987; Parthasarathy & Pottasch 1986; Hrivnak et al. 1988, 1989; van der Veen et al. 1989; Kwok 1992) also shows that many post-AGB stars can be found in the IRAS colour-colour diagram in the region between those of the reddest AGB stars and the locus of the PNe. These searches yield objects that are bright in the IR, i.e. have large amounts of circumstellar dust:

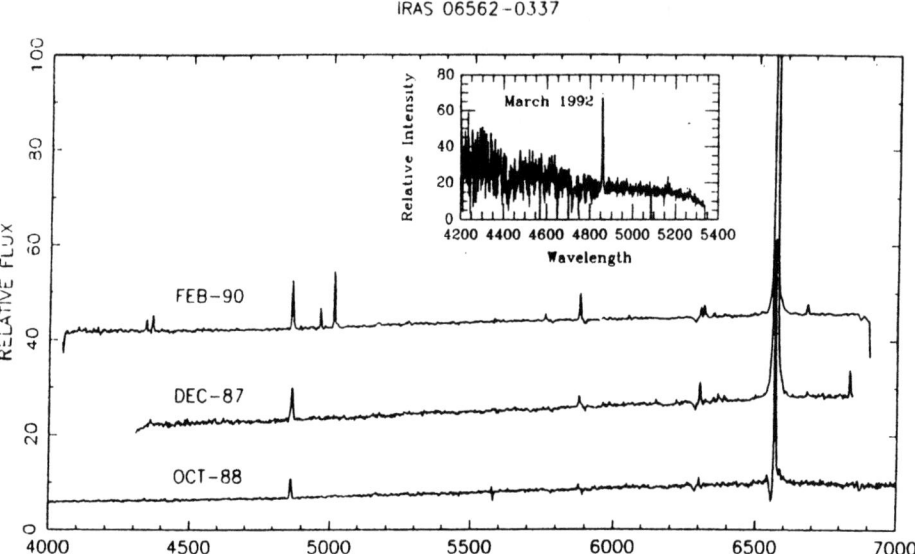

Fig. 2. The optical spectra of IRAS 06562-0337 obtained at different times (Garcia-Lario et al. 1992). Notice the onset of forbidden line emissions in 1990, and the significant increase of the He I $\lambda$ 5976 Å line, indicating the increase in the intensity of stellar temperature.

$L_{IR} \gg L_{opt}$. Therefore preferentially the more massive objects will be found.

The cross-correlation of optical catalogues with the IRAS database also has been successful (see e.g. Lamers et al. 1986; Waelkens et al. 1987; Parthasaraty & Pottasch 1988; Trams et al. 1990; Waelkens et al. 1990; Stencel & Backman 1991; Oudmaijer et al. 1993). No restriction on the IRAS colours is used, and so confusion with other kinds of stars with circumstellar dust is a problem. The study of Oudmaijer et al. (1993) of the correlation between the SAO catalogue and the IRAS PSC not only resulted in the finding of new post-AGB stars, but also yielded some pre-main-sequence objects as well as Vega-type stars. Detailed study of individual objects is necessary to establish their nature. The use of the SAO catalogue introduces a bias towards objects that have small amounts of dust or that have non-spherically symmetric dust envelopes. In most cases $L_{IR} \approx L_{opt}$.

The third approach is used by e.g. de Boer and co-workers and Dufton and collaborators. They study the properties of faint blue stars in the halo of the Galaxy by means of abundance analyses, and find objects with Fe deficiencies and CNO enhancements (Conlon et al. 1991; de Boer et al. 1992). In one case a low-excitation PN was found around one of the halo B stars (Conlon et al. 1992), which is also an IRAS source. No other star was found in the IRAS PSC (Conlon, private communication), so $L_{IR} \ll L_{opt}$.

It is interesting to note that most searches for post-AGB stars have yielded stars with spectral type F or G, irrespective of the method used (apart from the halo B stars). Two different effects may play a role in preferentially finding F and G type objects: (1) for low mass stars, the evolution to higher $T_{eff}$ is so slow that by the time they reach spectral type A or B, the dust shell has already cooled to flux levels below the IRAS detection limit. (2) for more massive objects, the evolution from G to F roughly takes 1000 years (e.g. Schönberner 1983), but after that $T_{eff}$ increases much more rapidly (Oudmaijer et al. 1992).

Since detailed optical studies of faint post-AGB stars have not (yet) been carried out, most of our knowledge of their properties stems from studies of objects that are optically bright. As mentioned above, the bright post-AGB stars may not be a representative sample because of the bias towards low amounts of circumstellar dust.

## 3. Radial velocity variations: pulsation and binarity

For bright post-AGB stars accurate radial velocities (RV) can be measured and thus also RV variations. Such studies have revealed that the UU Her stars show RV variations that can be explained in terms of pulsations (see e.g. Sasselov, these proceedings). Typical amplitudes are a few km/s and pulsation periods are of the order of 60 to 80 days. A systematic study of RV variations for a large sample of post-AGB candidate stars found in the IRAS PSC is being carried out by Waelkens and Mayor (1992) using CORAVEL. Preliminary results of this study indicate that most stars show evidence for RV variations. A significant fraction however shows RV variations with an amplitude and period that cannot be explained by pulsation, but must be due to orbital motions. Most orbital periods found are of the order of 100 to several hundred days, but longer orbital periods cannot be excluded at present due to the limited time coverage. Examples of binary post-AGB stars are HR 4049 (Waelkens et al. 1991), 89 Her (Arrelano-Ferro 1984; Waters et al. 1993) and HD 46703 (Waelkens & Mayor 1992).

The binary nature of post-AGB stars has interesting implications for their evolution: the orbital parameters found in e.g. 89 Her and HR 4049 suggest that these stars filled their Roche lobe on the AGB, and would have experienced a common envelope phase. It is not clear how these systems could have survived such a phase, especially in view of the (significant) non-zero eccentricities that are found: a common envelope phase tends to rapidly circularize the orbit.

## 4. Binarity and hot dust

The IR energy distribution of post-AGB stars not only shows evidence for cool dust (T ≈ 100-200 K) but in some stars hot dust (T ≈ 500-1300 K) also is found (Trams et al. 1989). The cool dust usually is explained in terms of a detached dust shell expelled while the star was on the AGB, but the hot dust must be very close to the star and have been ejected very recently. Trams et al. (1989) interpreted this as strong evidence for significant post-AGB mass loss. More detailed study of the stars with hot dust shells shows that they are binaries (Waters et al. 1991), which suggests that the presence of a companion triggers the presence of hot dust. In HR 4049 there is evidence that the hot dust is in a disc around the system (Waelkens et al. 1991); this model is also applicable to 89 Her (Waters et al. 1993).

TABLE I
Basic properties of some stars with depleted atmospheres

| HD | name | Sp. t. | $T_{eff}$ | log g | dust | [Fe/H] | [S/Fe] | log(C/O) |
|---|---|---|---|---|---|---|---|---|
| 89353 | HR4049 | F I | 7500 | 1.0 | yes | -4.8 | +4.4 | -0.23 |
| 52961 | | G : | 6500 | 1.0 | yes | -4.8 | +3.7 | -0.25 |
| 44179 | Red Rectangle | F I | 7500 | 1.0 | yes | -3.5 | +3.3 | -0.0 8 |
| | BD+39 4926 | F I | 7500 | 1.0 | no | -2.9 | +3.0 | -0.60 |
| 46703 | | F I | 6000 | 0.4 | yes | -1.6 | +1.2 | -0.24 |

Note to Table 1: data taken from Waelkens et al. 1992 and references therein

If the hot dust is in orbit around the system rather than in an outflow, it may be the remnant of a disc formed while the star was still on the AGB. In this case the temperature of the hot dust is not a measure of the time that has elapsed since it was ejected, but merely reflects the distance between the star and the inner radius of the circum-system disc.

The presence of hot dust which is 'trapped' in a binary system is not restricted to post-AGB stars. The C-rich AGB stars with O-rich dust shells may also be binaries in which the O-rich dust was ejected by the AGB star when it was still an M star. The dust either is in a circum-system disc or in a disc around the (unseen) companion (Morris 1989). Other examples are the massive binary 3 Puppis, and the interacting binary $v$ Sgr, showing that the mechanism that creates such discs also operates in massive Pop. I stars.

## 5. Chemical abundances

Several post-AGB stars have been the subject of detailed abundance determinations (for a recent review see Bond 1991). These studies have shown that there is a wide range in metallicities, from [Fe/H] $\approx$ -5.0 to 0.0! The stars with extremely low metallicities (e.g. HR 4049, HD 52961) show remarkable abundance patterns that cannot be explained by nucleosynthesis: their C,N,O, and S abundance is solar within an order of magnitude, but many metals (Fe, Cr, Ti, Ca, Mg) are strongly underabundant (e.g. van Winckel et al. 1992). Mathis & Lamers (1992) have suggested that the low metal abundance is not primordial, but is the result of selective depletion of certain metals due to the formation of dust grains. The gas and dust are separated in the circumstellar environment, for which Mathis & Lamers suggest either the dusty AGB wind, or an accretion disc in the case of a binary system. Clean gas is accreted on to the photosphere, while the dust grains containing the metals are pushed out by stellar radiation pressure. Waters et al. (1992) point out that 4 out of 5 stars with extreme Fe abundance are in a (wide) binary system, and propose that the clean gas was accreted from a circum-system disc rather than fallen back from the AGB wind.

## 6. Hα emission

Almost all post-AGB candidates show some degree of emission in Hα. This (variable) emission usually has the shape of a shell profile, i.e. a Violet and Red emission peak combined with a deep central absorption. Some stars show a P Cygni profile (HD 101584; SAO173329; see Waters et al. 1992), while HR 4049 shows alternating P Cygni and shell profiles (Waelkens et al. 1991). The study of Tamura (1992) has shown that in HD 56126 the Hα shell profile changes shape and strength on timescales of minutes. In the stars that show P Cygni profiles an estimate can be made of the outflow velocity, which in most cases is several hundred km/s, of the order of the surface escape velocity (assuming a mass of 0.6 $M_\odot$).

The shell profiles have several possible explanations: a disc around the star, infall of material, or pulsation. In most cases pulsation is a likely explanation. The RV variations indicate that many stars indeed show evidence for low-amplitude pulsations. Such pulsations will cause shock waves in the outer atmosphere that will ionize part of the gas, resulting in Hα emission. The shell profiles in post-AGB stars look remarkably similar to the Hα emission seen in RV Tau variables (Lèbre & Gillet 1991, 1992), although the emission is much stronger in the latter. The difference in line strength may be due to the difference in pulsation amplitude, which can be several tens of km/s in RV Tau stars, but is usually less than 10 km/s in the F supergiants.

## References

Arrelano-Ferro, A.: 1984, PASP 96, 641
Bond, H.E.: 1991, in 'Evolution of stars: the photospheric abundance connection', IAU Symp. 145, Eds. G. Michaud & A. Tutukov, p. 341
Conlon, E.S., Dufton, P.L., Keenan, F.P., McCausland, R.J.H.: 1991, MNRAS 248, 820
Conlon, E.S., 1992, Contributed paper (this volume)
de Boer, K.S., Moehler, S., Theissen, A.: 1992, in proc. of a workshop on post-AGB stars, Leuven, October 1991, ed. C. Waelkens, Leuven University Press (in press)
Garcia-Lario, P., Manchado, A., Pottasch, S.R., Suso, J., Ollig, R.: 1989, in in "From Mira's to Planetary Nebulae: Which Path for Stellar Evolution?", eds. M.O. Mennessier and A. Omont, Editions Frontieres, page 474
Garcia-Lario, P., Manchado, A., Sahu, K.C., Pottasch, S.R.: 1992, A&A Lett. (in press)
Garcia-Lario, P.: 1991, Ph. D. Thesis, University of La Laguna
Hrivnak, B.J., Kwok, S., Volk, K.M.: 1988, ApJ 331, 832
Hrivnak, B.J., Kwok, S., Volk, K.M.: 1989, ApJ 346, 265
Hu, J.Y., Slijkhuis, S., de Jong, T., 1992
Kwok, S., Hrivnak, B.J., Boreiko, R.T.: 1987, ApJ 312, 303
Kwok, S., 1992, Review paper, (this volume)
Lamers, H.J.G.L.M., Waters, L.B.F.M., Garmany, C.D., Perez, M.R., Waelkens, C.: 1986, A&A 154, L20
Lèbre, A., Gillet, D.: 1991, A&A 246, 490
Lèbre, A., Gillet, D.: 1992, A&A 255, 221
Manchado, A., Garcia-Lario, P., Pottasch, S.R.: 1989a, A&A 218, 267
Manchado, A., Garcia-Lario, P., Pottasch, S.R.: 1989b, A&AS 156, 57
Mathis, J.S., Lamers, H.J.G.L.M.: 1992, A&A 259, L39
Morris, M.; 1989, in "From Mira's to Planetary Nebulae: Which Path for Stellar Evolution?", eds. M.O. Mennessier and A. Omont, Editions Frontieres, page 520
Oudmaijer, R.D., van der Veen, W.E.C.J., Waters, L.B.F.M., Trams, N.R., Waelkens, C., Engelsman, E.: 1993, A&AS (in press)
Parthasarathy, M., Pottasch, S.R.: 1986, A&A 154, L16

Parthasarathy, M., Pottasch, S.R.: 1988, A&A 203, 117
Pottasch, S.R., Parthasarathy, M.: 1988, A&A 192, 182
Pottasch, S.R., Bignell, C., Olling, R., Zijlstra, A.A.: 1988, A&A 205, 248
Quin, D.A., Lamers, H.J.G.L.M.: 1992, A&A (in press)
Sahu, K.C., Pottasch, S.R., Van De Steene, G., Manchado, A, Garcia-Lario, P.: 1993 (in press)
Schönberner, D.: 1983, ApJ 272, 708
Slijkhuis, S., Hu, J.Y., de Jong, T.: 1991, A&A 248, 547
Slijkhuis, S.: 1992, Thesis, University of Amsterdam
Stencel, R.E., Backman, D.E.: 1991, ApJS 75, 905
Tamura, S.: 1992, in 'Luminous High Latitude Stars', ed. D.D. Sasselov, ASP Conference Series (in press)
Trams, N.R., Waters, L.B.F.M., Waelkens, C., Lamers, H.J.G.L.M., van der Veen, W.E.C.J.: 1989, A&A 218, L1
Trams, N.R., Waters, L.B.F.M., Lamers, H.J.G.L.M., Waelkens, C., Geballe, T.R., Thé, P.S.: 1991, A&AS 87, 361
Trams, N.R., Waters, L.B.F.M., Waelkens, C., Lamers, H.J.G.L.M.: 1992, A&A (submitted)
van der Veen, W.E.C.J., Habing, H.J., Geballe, T.R.: 1989, A&A 226, 108
Waelkens, C., Waters, L.B.F.M., Cassatella, A., Le Bertre, T., Lamers, H.J.G.L.M.: 1987, A&A 181, L5
Waelkens, C., van Winckel, H., Bogaert, E., Trams, N.R.: 1991a, A&A 251, 495
Waelkens, C., Lamers, H.J.G.L.M., Waters, L.B.F.M., Rufener, F., Trams, N.R., Le Bertre, T., Ferlet, R., Vidal-Madjar, A.: 1991b, A&A 242, 433
Waelkens, C., Mayor, M.: 1992, in 'Luminous High Latitude Stars', ed. D.D. Sasselov, ASP Conference Series (in press)
Waelkens, C., Waters, L.B.F.M.: 1992, in 'Luminous High Latitude Stars', ed. D.D. Sasselov, ASP Conference Series (in press)
Waters, L.B.F.M., Trams, N.R., Waelkens, C.: 1991, in "The Infrared Spectral Region of Stars", eds. C. Jaschek and Y. Andrillat, Cambridge University Press, p. 40
Waters, L.B.F.M., Waelkens, C., Trams, N.R.: 1992, A&A 262, L37
Waters, L.B.F.M., Waelkens, C., Mayor, M., Trams, N.R.: 1993, A&A (in press)

# PLANETARY NEBULAE WITH BINARY NUCLEI

MARIO LIVIO

*Space Telescope Science Institute, 3700 San Martin Drive, Baltimore, MD 21218, USA*

and

*Dept. of Physics, Technion, Haifa 32000, Israel*

## ABSTRACT

Planetary nebulae with close binary nuclei are reviewed. It is shown that these systems can be used as a source of information for the physics of the common envelope phase in the evolution of binary systems. Mechanisms for the production of bipolar planetary nebulae are examined and it is concluded that presently the action of binary companions to the central stars appears to provide the most promising mechanism. Other systems (e.g. novae, supernovae) in which similar processes may be operating are discussed.

## 1. INTRODUCTION

The existence of planetary nebulae (PNe) with close binary nuclei can be regarded as the most direct evidence for the occurrence of common envelope (CE) phases in the evolution of binary systems (e.g. Paczynski 1976; Ostriker 1975; Livio, Salzman and Shaviv 1979 and see Bond, Ciardullo and Meakes 1992 for a recent review). The occurrence of a CE phase in classical nova systems will be discussed separately in section 3.

A compilation of all the PNe with confirmed *close* binary nuclei is given in Table 1 (adapted from Bond and Livio 1990). Three more objects which surely contain binary (or possibly triple) nuclei are given in Table 2. In the latter case, however, the observed periodic photometric variations almost certainly represent the rotation periods of the cool stars (Bond, Ciardullo and Meakes, private communication) and thus, the orbital periods remain unknown. It has been suggested that LoTr 5 is a triple system (Jasniewicz et al. 1987; Malasan et al. 1991), with the internal binary having a period of about 2 days or 1.75 days (Jasniewicz and Acker, this volume, Malasan et al. 1991). This conclusion however,

Table 1. Planetary Nebulae with Close Binary Nuclei (adapted from Bond and Livio 1990)

| Planetary Nebula | Central Star | Orbital Period (days) |
|---|---|---|
| A41 | MT Ser | 0.113 |
| DS1 | KV Vel = LSS 2018 | 0.357 |
| A63 | UU Sge | 0.465 |
| A46 | V477 Lyr | 0.472 |
| HFG 1 | V664 Cas | 0.582 |
| K1-2 | VW Pyx | 0.676 |
| A65 |  | 1.00 |
| HtTR 4 |  | 1.71 |
| Sp 1 |  | 2.91 |
| NGC 2346 | V651 Mon | 15.99 |

Table 2. Planetary Nebulae with Binary Central Stars Which Show Photometric Variability Which Probably Represents the Rotation Period of the Cool Star (from Bond, Ciardullo and Meakes, private communication)

| Planetary Nebula | Central Star | Period of Optical Variability (days) |
|---|---|---|
| A35 | LW Hya = BD$-22°3467$ | 0.76 |
| LoTr 1 |  | 3.3 or 6.6 |
| LoTr 5 | IN Com = HD 112313 | 5.9 |

should be re-examined, because of the fact that the rotation period of the cool star (with a period of about 5.9 days) coupled with amplitude variations that are probably due to star spots, may generate apparent false periodicities. The suggestion that Sh2-71 has a period of $P = 68\overset{d}{.}06$ (Jurcsik, this volume) is presently not confirmed by observations of Bond (private communication). More observations on this object are needed. Two other suggestions for nebulae with binary central stars need to be mentioned. The central star of M1-67 (which is suggested to be a PN, van der Hucht et al. 1985), QR Sge ≡ 209 BAC, is a binary with a period of 2.358 days (Moffat et al. 1982). However, the nebula in this case was only identified in the infrared and the central star is a Wolf-Rayet object. Thus, this object is probably more similar to rings around massive stars (see Chu, this volume) than to ordinary PNe. The central star of PK 331-5°1 (PC 11, HD 149427) was also found to be a binary (Parthasarathy, Pottasch and Clavel, this volume). However, this system was also identified as a D-type symbiotic star.

A few new observations of known binary central stars should be mentioned. These include observations of UU Sge (Pollacco and Bell 1992; Malasan and Yamasaki 1992 and Walton, Walsh and Pottasch, see this volume), HFG 1 (Malasan and Yamasaki 1992), A 46 and A 65 (Walsh, Walton and Pottasch, this volume) and x-ray observations of LoTr 5 (Kreysing et al. this volume).

The main thing to note about *all* the objects in Table 1 (and possibly in Table 2) is that *they had to undergo CE evolution*! Indeed, all binaries containing at least one compact component with orbital periods satisfying $P_{orb} \lesssim$ days, which are not located in very dense clusters (and therefore are probably not capture products), had to pass through a CE phase in order to be formed.

## 2. A BRIEF DESCRIPTION OF THE PHYSICS OF THE CE PHASE

Common envelope is considered to be an essential phase in the formation of such objects as cataclysmic variables, binary pulsars and at least some x-ray binaries (e.g., Eggleton and Tout 1989; Tutukov, Yungelson and Iben 1992). One of the scenarios leading to Type Ia supernovae involves the merger of two white dwarfs and requires one (e.g., Webbink 1984) or two (e.g., Iben and Tutukov 1984) CE phases.

The occurrence of a CE phase is supposed to be often the result of *a dynamical mass transfer event*. This, in turn, can be the consequence of the following two things happening together:

(1) Mass being transferred from the more massive to the less massive component (it is easy to show that this leads, if mass and angular momentum are conserved, to a reduction in the binary separation).
(2) The mass losing star has a deep convective envelope (as in the AGB phase; in this case the star tends to expand when losing mass).

These two conditions lead to a situation in which the mass losing star cannot contract as rapidly as its Roche lobe and an unstable mass transfer ensues (e.g. Paczynski and Sienkiewicz 1972; Edwards and Pringle 1987). The high mass transfer rate (which can exceed the Eddington limit) overwhelms the secondary star, and drives it out of thermal equilibrium. The secondary starts expanding and the system rapidly evolves into a CE configuration (e.g. Yungelson 1973; Webbink 1977; Prialnik and Livio 1985).

The main effect of the CE is to reduce the binary separation by the drag which the two components experience, while the gravitational energy can be deposited (in some cases) into the ejection of the common envelope. Thus, we immediately realize that the final outcome of the CE phase is determined mainly by the efficiency with which orbital energy is deposited into mass ejection. This efficiency is conveniently expressed by the parameter (Tutukov and Yungelson 1979; Livio and Soker 1988)

$$\alpha_{CE} \equiv \frac{\Delta E_{bind}}{\Delta E_{orb}} \quad (1)$$

where $\Delta E_{bind}$ represents the binding energy of the ejected material and $\Delta E_{orb}$ is the total change in the binary's orbital energy. A value of $\alpha_{CE} = 1$ means that every ejected mass element receives (from orbital energy) exactly the energy it needs for escape.

One of the most important tasks of the study of CE evolution is to determine the value of $\alpha_{CE}$, both using observations and theoretically.

At least two processes can act to reduce the value of $\alpha_{CE}$:

(1) Efficient energy transport
(2) Non-spherical effects.

The first effect is important in situations in which the timescale for energy transport is short compared to the orbital decay timescale. In such cases, the energy that is released by the decay of the orbit is transported to the surface and radiated away without causing almost any mass motions (e.g. Meyer and Meyer-Hofmeister 1979). Such situations can arise when the secondary is of a very low mass (e.g. a brown dwarf, Livio and Soker 1984), or in the very outer layers of CE configurations resembling AGB stars. In both of these cases, the drag luminosity represents only a small perturbation to the star's own luminosity.

The second effect ((2) above) which can reduce the value of $\alpha_{CE}$ is related to the fact that multi-dimensional numerical hydrodynamic calculations revealed that *mass is ejected preferentially in the orbital plane* (e.g. Taam and Bodenheimer 1989, 1991; Livio and Soker 1988 and see Fig. 1). This has the effect that only a part of the envelope mass, is given more energy than it needs to escape, thus reducing the overall efficiency. In the hydrodynamical calculations, values of $\alpha_{CE}$ in the range 0.3–0.6 were obtained.

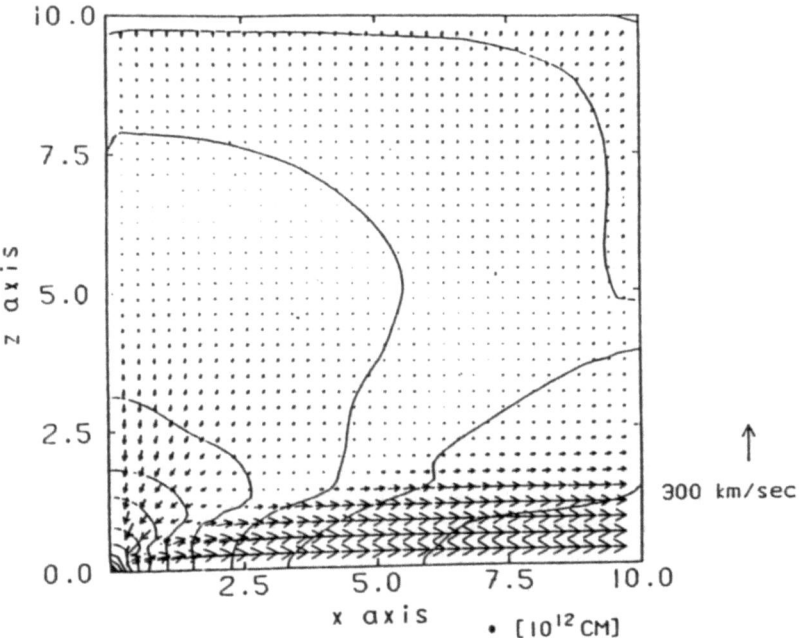

Figure 1. The velocity field and density contours in a common envelope calculation. The z axis is perpendicular to the binary's orbital plane. From Taam and Bodenheimer (1989).

In principle, one could consider other energy sources which can contribute to envelope ejection during the CE phase, while retaining the formal definition of $\alpha_{CE}$ as in eq. (1). If such additional sources indeed participate in the ejection of the envelope, then this could result in values of $\alpha_{CE}$ that are even larger than unity.

Processes during the CE phase that can (in principle) increase the value of $\alpha_{CE}$ are:

(1) Spin-up of the envelope and the dynamo generation of magnetic fields (see section 4).
(2) The excitation of non-radial modes (e.g. Soker 1992a, b, and see section 4).
(3) The tapping of the recombination energy in the hydrogen and helium ionization zones (originally suggested for PN ejection by single stars, Paczynski 1967).
(4) The injection of new fuel (e.g. hydrogen into the helium burning shell) by circulations which develop in the CE (e.g. Taam and Bodenheimer 1989).

Until now, none of the above processes was included in CE calculations, thus their actual effect on the value of $\alpha_{CE}$ is still unknown.

In order to obtain an idea of the reduction in the separation that can be expected for a typical cataclysmic binary progenitor, we can crudely estimate $\Delta E_{bind}$ and $\Delta E_{orb}$ in eq. (1) to obtain

$$\frac{GM_1^2}{a_i} \simeq \alpha_{CE} \frac{GM_{1R}M_2}{a_f}, \qquad (2)$$

where $M_1$, $M_2$ are the initial masses, $a_i$ and $a_f$ are the initial and final separations (respectively) and $M_{1R}$ is the remnant mass of the primary, following the CE phase. This leads for typical parameters to a final separation of

$$a_f \simeq 2.4 R_\odot \left(\frac{\alpha_{CE}}{0.4}\right) \left(\frac{M_{1R}}{M_\odot}\right) \left(\frac{M_2}{0.3 M_\odot}\right) \left(\frac{M_1}{5 M_\odot}\right)^{-2} \left(\frac{a_i}{500 R_\odot}\right). \qquad (3)$$

## 3. PLANETARY NEBULAE WITH BINARY NUCLEI AND COMMON ENVELOPE PHYSICS

Because of the fact that the existence of PNe with close binary nuclei can be regarded presently as the closest to direct evidence for the occurrence of a CE phase, these objects provide us with the best source of information on the physics of the CE. PNe with binary central stars can be used in this respect in at least three ways:

(i) Using assumptions about the distribution of binary systems in primary masses, in initial separations and initial mass ratios, together with some assumptions about CE evolution, it is possible to calculate the expected distribution in the masses of the two components of the binary central stars, in their orbital periods, etc. This approach was used by de Kool (1990) and by Yungelson and Tutukov (this volume). In principle, the obtained distributions can be used (by comparison with the observed objects) to place constraints on $\alpha_{CE}$. However, the small number of objects that is presently known is not sufficient to enable us to place meaningful constraints yet.

(ii) A second method (Iben and Tutukov 1989) used similar assumptions as in (i) above, in combination with attempts to reconstruct the evolutionary path of V651 Mon, LSS 2018, MT Ser and UU Sge (Table 1), in order to obtain a semi-empirical determination of $\alpha_{CE}$. Unfortunately, uncertainties in the binary parameters of the systems (as well as in the theoretical evolutionary paths) do not allow yet firm conclusions to be drawn. This is, however, a potentially very promising approach.

(iii) Observations of the morphology of the PNe with binary nuclei can provide tests for both the CE phase and the "interacting winds" (Kwok 1982; Kahn 1982) model for the shaping of the nebulae.

As explained in section 2, the ejection of the CE occurs mainly in the orbital plane (within $\sim 12°$). This produces a "density contrast" between the equatorial (orbital plane) and polar directions. In the interacting winds model, the fast and dilute wind that is emitted by the hot, exposed nucleus, runs into the slowly moving material, shocks it and snowplows into it. Balick (1987, and see this volume) proposed a simple morphological scheme in which PNe are divided into three basic types: round, elliptical and butterfly. The key parameter in this scheme is the density contrast between the equatorial and polar directions. A moderate contrast results in an elliptical morphology while a high contrast in a "butterfly." The pioneering, exploratory hydrodynamic calculations of this interacting winds scenario performed by Soker and Livio (1989) and the much more accurate and refined recent calculations by Icke *et al.* (1992), Frank *et al.* (this volume) and Mellema *et al.* (1991), have shown that it is indeed possible to reproduce many of the observed morphology characteristics (see also Dyson, this volume). Bond and Livio (1990) presented (or provided reference to) imagery of all the PNe with close binary nuclei and showed that the observed morphologies are consistent with the expectations based on the density contrast generated by CE evolution (e.g. Fig. 2).

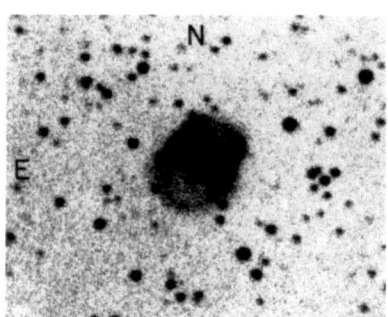
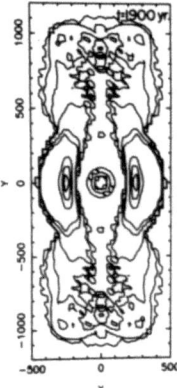

Figure 2. Left panel: direct photograph of Abell 41 (from Grauer and Bond 1983). Right panel: density contours obtained from interacting winds, when a density contrast exists between the equatorial (horizontal) and polar directions (from Soker and Livio 1989).

It is interesting to note that the image of NGC 2346 shows a striking similarity to that of supernova 1987A (Fig. 3, see also Panagia et al. 1991). The possibility of a bipolar morphology for SN 1987A was anticipated by Soker and Livio (1989). It is therefore not very surprising that in a recent review, Podsiadlowski (1992) concludes that it is most likely that the progenitor of SN 1987A had a binary companion.

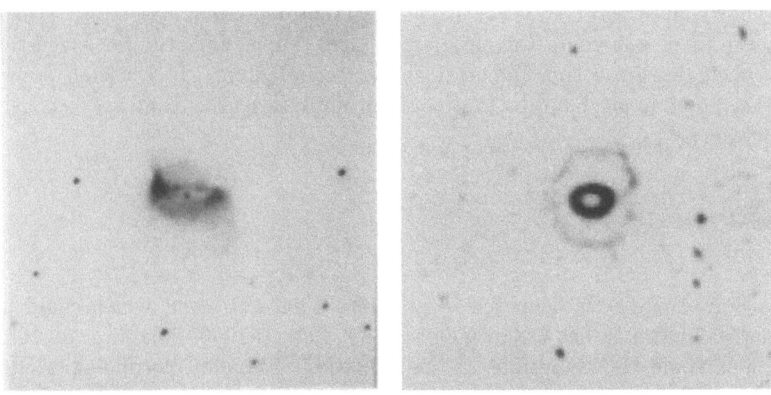

Figure 3. Left panel: H$\alpha$ CCD image of NGC 2346 (from Bond and Livio 1990). Right panel: deconvolved nebular image of SN 1987A in [OIII] $\lambda$5007 (from Wampler et al. 1990).

Another class of objects (known to be binaries) that experience a short CE phase are classical novae. We note the following properties:

(1) Classical nova systems are binaries with orbital periods of the order of a few hours, implying separations of the order of a solar radius.
(2) Classical novae at maximum are characterized by luminosities of the order of the Eddington luminosity, which for typical white dwarf masses imply a luminosity of order $L \sim (2-5) \times 10^4 \; L_\odot$.
(3) The effective temperatures of novae at maximum are of order $T_{eff} \lesssim 10^4$ K.

Properties (2) and (3) above, result in photospheric radii in excess of $4 \times 10^{12}$ cm, much larger than the orbital separation. Therefore, *the secondary star is necessarily engulfed in the expanding nova envelope*. An examination of nova light curves reveals that some novae remain in this CE phase for periods of months. It should be noted, however, that the mass of the nova envelope is very low, typically of order $\Delta M_{env} \sim 10^{-4} \; M_\odot$. Thus, the density contrast that can be generated by mass ejection that is induced by the engulfed binary companion is never extreme. On the basis of our discussion of the morphologies that can be expected from an interacting winds scenario, one could therefore expect some nova shells to exhibit an elliptical (prolate) morphology. This is indeed the case for some novae (e.g. DQ Her, Barden and Wade 1988; Wade 1990).

## 4. MECHANISMS THAT CAN PRODUCE A DENSITY CONTRAST

In section 2 I described numerical calculations which demonstrated (admittedly, under-simplifying assumptions) that binary companions can generate the density contrast in the slow wind that is required for bipolar or "butterfly" morphologies. I also presented evidence showing that in the cases in which close binary companions are known to exist, the morphology that is observed is consistent with the predictions based on CE evolution. Using Occam's razor one could therefore argue that this is sufficient for suggesting that bipolar morphologies are a consequence of binary companions. It is useful, however, to examine other possible mechanisms that could produce a density contrast (see also Soker and Harpaz 1992).

### 4.1. MAGNETIC FIELDS

Dynamo generated magnetic fields are in principle a natural agent which could produce axial symmetry. Indeed, it has been suggested (e.g. Pascoli 1990; Pascoli et al. 1992) that a magnetic field could be responsible for the formation of bipolar morphologies. However, two things should be noted. In a recent work, Tout and Pringle (1992) estimated the strength of the azimuthal magnetic field obtained in fully convective stars. Using their formalism one obtains for an AGB star (with $M = 1\ M_\odot$, $R = 400\ R_\odot$, $L = 7000\ L_\odot$) rotating at 10% of its break-up angular velocity, an azimuthal field of $B_\phi \sim 80$ G, which is more than an order of magnitude lower than the field required in the model of Pascoli (1990). Furthermore, Pascoli et al. (1992) actually *assumed* the ejection of an equatorial torus with a large scale azimuthal field, rather than showing that such a configuration (which is incidentally unstable) is indeed obtained.

Thus, it is probably fair to conclude at this point that while magnetic fields may play a role in the production of the observed axial symmetry, this has not been demonstrated yet. In fact, a considerable spin-up of the AGB star is probably necessary for dynamo generated magnetic fields to play an important role. Amusingly enough, such a spin-up could be provided by a secondary companion (see below). This may mean that a binary companion is required for magnetic effects to be important.

### 4.2. NON-RADIAL p-MODES

It has been suggested that towards the end of the AGB phase, non-radial oscillations of the star become more important, potentially leading to axisymmetric mass loss (Soker and Harpaz 1992). This is supposed to happen as a consequence of the fact that the transition region (between the inner envelope, where the oscillations are quasi-adiabatic and the outer envelope, where they are highly non-adiabatic) for the non-radial modes moves inwards, towards the hydrogen ionization zone, thus leading possibly to an increased driving. Soker and Harpaz (1992) did conclude, however, that in order for the non-radial modes to lead to a substantial equatorial (or polar) mass ejection, a relatively rapid rotation (which could be induced by a binary companion) is needed.

## 4.3. EFFECTIVENESS OF BINARY COMPANION

In order to appreciate the effectiveness with which a binary companion can produce axially symmetric effects in the CE phase it is sufficient to examine the following few numbers. A companion of 0.1 $M_\odot$ that is spiralling inside a configuration corresponding to an AGB star of mass 0.94 $M_\odot$ (core mass of 0.59 $M_\odot$) and radius 350 $R_\odot$, at a distance of 10 $R_\odot$ from the center, can produce a drag luminosity that is as high as $10^5$ $L_\odot$. Since this luminosity is about 16 times larger than the local Eddington luminosity (for a given AGB star model), significant mass motion in the orbital plane can ensue. The angular momentum that such a companion can deposit into an AGB star's envelope (by spiralling-in from the stellar radius) is sufficient (in principle) to cause the entire convective envelope to rotate at the break-up angular velocity. Furthermore, even with an orbital period as long as 10000 yrs, a massive companion can cause deviations from spherical symmetry in the wind from the AGB star at the 10-20% level (see also Soker 1991).

Finally, it has been recently suggested by Bjorkman and Cassinelli (1992), that dense equatorial disks can form from radiatively driven winds of rapidly rotating stars. They have shown that the wind is concentrated to the equatorial plane (by conservation of angular momentum and orbital dynamics) if the rotation velocity of the star is at least about half the break-up speed. The wind shocks in the plane and the ram pressure confines the equatorial material. Again, for this mechanism (suggested for B stars) to be operative in AGB configurations, a very significant spin-up of the star has to occur. The most natural way to produce such a spin-up is via a binary companion.

## 5. INCIDENCE OF RELATIVELY CLOSE BINARY CENTRAL STARS

Out of 108 planetary nebula central stars that have been searched for binarity, 10-14 objects (Tables 1 and 2) proved to be close binary nuclei (Drummond 1980; Drilling 1985; Bond and Grauer 1987; Bond and Ciardullo (ongoing)). This is a slightly lower fraction than the one expected on the basis of the following finding. The search for multiplicity in 164 primaries with spectral types F7-G9 (Duquennoy and Mayor 1991) found that 61% of them have companions (with masses larger than about 0.1 $M_\odot$) and about 28% of those have separations that are smaller than 1000 $R_\odot$. Thus, about 17% of all of these primaries can interact through a CE evolution phase.

I should note that even very low mass companions (e.g. brown dwarfs) can produce significant deviations from spherical symmetry, although more massive secondaries are probably required in order to produce a very pronounced bipolar morphology.

## 6. SUGGESTIONS AND DISCUSSION

On the basis of the discussion in the previous sections the following suggestions (I think it may be premature to call them "conclusions") emerge:

(1) The central stars of all the PNe that show a clear axial symmetry may have (or had) binary companions. In this respect it is important to note that in the calculations of Yungelson and Tutukov (this volume), mergers of the central star with the companion (especially low mass companions) occur at a rate of 0.12/yr (the total rate of CE

events is 0.5/yr). Thus, even PNe that presently appear to have single central stars may either have very low mass companions or may have had binary companions in the past (mergers in fact have been suggested to occur for EGB 5 and PHL 932, Mendez et al. 1988a, b).

(2) Clear "bipolar" or "butterfly" morphologies probably result from relatively massive secondaries (what is really important is the ratio $M_{secondary}/M_{envelope}$) and/or common envelope phases that occur when the primary is in a relatively less evolved AGB configuration (this produces a higher density contrast). Zuckerman and Gatley (1988) found that "butterfly" PNe are more concentrated to the galactic plane than PNe in general. An examination of their statistics and the more recent statistics obtained by Stanghellini (private communication and this volume), reveals only a rather marginal effect. However, such a finding (if confirmed) is consistent with the above discussion. Massive secondaries (needed for "butterfly" morphologies, see above) tend to be associated (in the statistical sense) with relatively more massive primaries. It could therefore be expected that such systems will be more concentrated to the plane.

A very important aspect of bipolarity is the occurrence of "jets" in some objects. The most spectacular examples of such jets were found by Schwarz and co-workers (see e.g. NGC 6309, Corradi and Schwarz, this volume), but a few were known previously (see e.g. K1-2, Bond and Livio 1990; Lutz and Lame 1989).

The important thing to note here is that although the mechanism for jet formation (e.g. in AGNs and young stellar objects) is not entirely clear yet, there seems to be a general concensus that it requires the presence of an accretion disk (e.g. Pringle 1992). The presence of such disks has been advocated by Morris (1987) for some time. In fact, the observations of Corradi and Schwarz and those of K1-2 are consistent with the presence of a precessing jet (incidentally, the similarity with some AGN jets is quite striking).

(3) Classical novae provide an excellent opportunity to study CE physics.

(4) The carbon star V Hya, an AGB star which is rapidly rotating and which shows evidence for a bipolar structure in its circumstellar envelope, may represent the exciting possibility of a common envelope phase in progress (Kahane et al. 1992).

(5) It is extremely important to attempt to determine observationally the masses of both components for all binary central stars. This will provide invaluable information for the testing of CE theory.

**Acknowledgements:** I am grateful to Howard Bond, Icko Iben and Noam Soker for very helpful discussions. I am truly obliged to Sarah Stevens-Rayburn for her help with the references. This work has been supported in part by the Israel Academy of Sciences at the Technion and by the Director's Research Fund at Space Telescope Science Institute.

# References

Balick, B. (1987), *A.J.*, **94**, 671.
Barden, S. C. and Wade, R. A. (1988) in *A.S.P. Conference Series 3, Fiber Optics in Astronomy*, ed. S. C. Barden (San Francisco: Astron. Soc. Pacific), p. 113.
Bjorkman, J. E. and Cassinelli, J. P. (1992) in *A.S.P. Conference Series 22, Nonisotropic and Variable Outflows from Stars*, ed. L. Drissen et al. (San Francisco: Astron. Soc. Pacific), p. 88.
Bond, H. E. and Livio, M. (1990) *Ap.J.*, **355**, 568.
Bond, H. E. and Grauer, A. D. (1987) in *IAU Colloq. 95, The Second Conference on Faint Blue Stars*, eds. A. G. D. Philip et al. (Schenectady: L. Davis Press), p. 221.
Bond, H. E., Ciardullo, R. and Meakes, M. G. (1992) in *IAU Symp. 151, Evolutionary Processes in Interacting Binary Stars*, ed. Y. Kondo et al. (Dordrecht: Kluwer Academic Publishers), p. 577.
Drilling, J. S. (1985) *Ap. J.*, **294**, L107.
Drummond, J. (1980) Ph.D. Thesis, New Mexico State University.
Duquennoy, A. and Mayor, M. (1991) *Astr. Ap.*, **248**, 485.
Edwards, D. and Pringle, J. E. (1987) *M.N.R.A.S.*, **229**, 383.
Eggleton, P. P. and Tout, C. A. (1989) *Space Sci. Rev.*, **50**, 165.
Grauer, A. D. and Bond, H. E. 1983, *Ap. J.*, **271**, 259.
van der Hucht, K. A. Jurriens, T. A., Olnon, F. M., The, P. S. Wesselius, P. R. and Williams, P. M. (1985) *Astr. Ap.*, **145**, L13.
Iben, I. Jr. and Tutukov, A. V. (1984) *Ap. J. Suppl.*, **54**, 335.
Iben, I. Jr. and Tutukov, A. V. (1989) in *IAU Symp. 131, Planetary Nebulae*, ed. S. Torres-Peimbert (Dordrecht: Reidel), p. 505.
Icke, V., Balick, B. and Frank, A. (1992) *Astr. Ap.*, **253**, 224.
Jasniewicz, G., Duquennoy, A. and Acker, A. (1987) *Astr. Ap.*, **180**, 145.
Kahane, C., Audinos, P., Barnbaum, C. and Morris, M. (1992) preprint.
Kahn, F. D. (1982) in *IAU Symp. 103, Planetary Nebulae*, ed. D. R. Flower (Dordrecht: Reidel), p. 305.
de Kool, M. (1990) *Ap. J.*, **358**, 189.
Kwok, S. (1982) *Ap. J.*, **258**, 280.
Livio, M., Salzman, J. and Shaviv, G. (1979) *M.N.R.A.S.*, **188**, 1.
Livio, M. and Soker, N. (1988) *Ap. J.*, **329**, 764.
Livio, M. and Soker, N. (1984) *M.N.R.A.S.*, **208**, 763.
Lutz, J. H. and Lame, N. J. (1989) in IAU Symp. 131, *Planetary Nebulae*, ed. S. Torres-Peimbert (Dordrecht: Reidel), p. 462.
Malasan, H. L., Yamasaki, A. and Kondo, M. (1991) *A. J.*, **101**, 2131.
Malasan, H. L. and Yamasaki, A. (1992) preprint.
Mellema, G., Eulderink, F. and Icke, V. (1991) *Astr. Ap.*, **252**, 718.
Mendez, R. H., Kudritzki, R. P., Herrero, A., Husfeld, D. and Groth, H. G. (1988a) *Astr. Ap.*, **190**, 113.
Mendez, R. H., Groth, H. G., Hasfeld, D., Kudritzki, R. P. and Herrero, A. (1988b) *Astr. Ap.*, **197**, L25.
Meyer, F. and Meyer-Hofmeister, E. (1979) *Astr. Ap.*, **78**, 167.

Moffat, A. F. J., Lamontagne, R. and Seggewiss, W. (1982) *Astr. Ap.*, **114**, 135.
Morris, M. (1987) *PASP*, **99**, 1115.
Ostriker, J. P. (1975) talk presented at IAU Symp. 73, *The Structure and Evolution of Close Binary Systems*.
Paczynski, B. (1976) in IAU Symp. 73, *The Structure and Evolution of Close Binary Systems*, eds. P. Eggleton, S. Mitton and J. Whelan (Dordrecht: Reidel), p. 75.
Paczynski, B. and Sienkiewicz, R. (1972) *Acta Astron.*, **22**, 73.
Paczynski, B. (1967) paper presented at IAU General Assembly, Comiss. 35.
Panagia, N., Gilmozzi, R., Macchetto, F., Adorf, H.-M., and Kirshner, R. P. 1991, *Ap. J.*, **380**, L23.
Pascoli, G. (1990) in From Miras to Planetary Nebulae: Which Path for Stellar Evolution?" eds. M. D. Mennessier and A. Omont (Editions Frontieres), p. 136.
Pascoli, G., Leclerco, J. and Poulain, B. (1992) *PASP*, **104**, 182.
Podsiadlowski, P. (1992) *P.A.S.P.*, in press.
Pollacco, D. L. and Bell, S. A. (1992) *M.N.R.A.S.*, submitted.
Prialnik, D. and Livio, M. (1985) *M.N.R.A.S.*, **216**, 37.
Pringle. J. E. (1992) in *Astrophysical Jets*, eds. D. Burgarella, M. Livio and C. O'Dea (Cambridge: Cambridge University Press), in press.
Soker, N. (1991) *Ap. J.*, **367**, 593.
Soker, N. (1992a) *Ap. J.*, **386**, 190.
Soker, N. (1992b) *Ap. J.*, **389**, 628.
Soker, N. and Livio, M. 1989, *Ap. J.*, **339**, 268.
Soker, N. and Harpaz, A. (1992) *P.A.S.P.*, in press.
Taam, R. E. and Bodenheimer, P. (1989) *Ap. J.*, **337**, 849.
Taam, R. E. and Bodenheimer, P. (1991) *Ap. J.*, **373**, 246.
Tout, C. A. and Pringle, J. E. (1992) *M.N.R.A.S.*, **256**, 269.
Tutukov, A. V. and Yungelson, L. R. (1979) in IAU Symp. 83, *Mass Loss and the Evolution of O-Type Stars*, eds. P. S. Conti and C. W. H. de Loore (Dordrecht: Reidel), p. 415.
Tutukov, A. V., Yungelson, L. R. and Iben, I. Jr. (1992) *Ap. J.*, **386**, 197.
Wade, R. A. (1990) in *Physics of Classical Novae*, IAU Colloq. 122, eds. A. Cassatella and R. Viotti (Berlin: Springer-Verlag), p. 179.
Wampler, E. J., Wang, L., Baade, D., Banse, K., D'Odorico, S., Goniffes, C. and Tarenghi, M. 1990, *Ap. J.*, **362**, L13.
Webbink, R. F. (1977) *Ap. J.*, **211**, 486.
Webbink, R. F. (1984) *Ap. J.*, **277**, 355.
Yungelson, L. R. (1973) *Nauch Informatsii*, **27**, 93.
Zuckerman, B. and Gatley, I. (1988) *Ap. J.*, **324**, 501.

# THERMAL PULSES AND PLANETARY NEBULA EJECTION

P.R. WOOD and E. VASSILIADIS
*Mount Stromlo and Siding Spring Observatories, Private Bag, Weston Creek P.O., Canberra, ACT 2611, Australia*

ABSTRACT. Thermal pulses in AGB stars cause large luminosity variations at the stellar surface. The role of these luminosity variations in the production of planetary nebulae is discussed. Results of theoretical evolution calculations which include mass loss modulated by thermal pulses are presented.

## 1. Thermal pulses and AGB evolution

The hydrogen and helium burning shells of AGB stars do not burn smoothly. For most of the time, the hydrogen burning shell is active while the helium burning shell is essentially dormant. Periodically, at intervals of typically $5 \times 10^4$ - $10^5$ years for low mass stars, the helium shell ignites violently and burns up the helium produced by the hydrogen shell since the last phase of helium shell activity. Then, when the inter-shell helium supply is exhausted, helium shell burning dies out and hydrogen burning recommences. These periodic bursts of activity by the helium shell, first discovered by Schwarzschild and Harm (1965), are called *helium shell flashes* or *thermal pulses*. In the present context, their most important effect is to modulate the luminosity appearing at the surface of the star: a typical example of thermal pulse behaviour is shown in Figure 1. Two important features

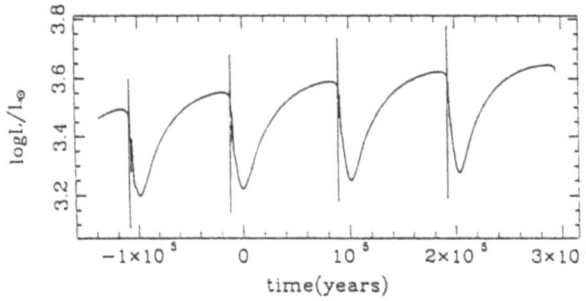

Figure 1. The time dependence of the luminosity in a thermally pulsing AGB star with a core mass $M_{core}$ = 0.58 $M_\odot$ and total initial mass 1.0 $M_\odot$.

should be noted in these light curves. Firstly, there is a brief spike of luminosity which lasts ~500 years (Wood and Zarro 1981; Boothroyd and Sackmann 1988) and which is a direct result of the escape of energy released by the initial burst of helium burning. In low mass stars, the peak of this luminosity spike is typically 50% higher than the maximum luminosity during hydrogen burning. Secondly, there is a slow variation of luminosity by a factor of ~2 throughout the whole shell flash cycle: the initial decline to luminosity minimum coincides with the phase of helium shell burning while the recovery to luminosity maximum corresponds to the phase of hydrogen burning. Note that the long-term increase in average luminosity resulting from the increase in the core mass is dominated by the luminosity variations resulting from shell flashes.

## 2. Ejection mechanisms

Mechanisms invoked for the ejection of planetary nebulae (PNe) from single stars generally involve some instability of the envelope, and theoretical studies indicate that this instability is greater the greater the luminosity of the star. It therefore seems that thermal pulses and the surface luminosity variations that they produce will influence planetary nebula ejection.

One of the earliest mechanisms suggested for ejecting PNe was the dynamical instability of the envelope (Roxburgh 1967; Paczynski and Ziolkowski 1968). Roxburgh showed that as a star evolved up the giant branch, it would eventually become luminous enough that the fundamental pulsation mode eigenvalue $\sigma^2$ would become negative, giving rise to an unstable, exponentially growing expansion of the envelope. Paczynski and Ziolkowski (1968) also showed that in AGB stars the total envelope energy could indeed be positive so that once the envelope started to expand in the unstable mode it would have enough energy to escape to infinity. Many subsequent non-linear studies of this possibility (ex. Smith and Rose 1972; Wood 1974; Tuchman, Sack and Barkat 1979) indicated that AGB stars did not eject their envelopes via dynamical instabilities but instead underwent violent, irregular pulsation in the fundamental mode and suffered large amounts of mass loss. These results led Wood and Cahn (1977) to suggest that a the switch in pulsation mode of Mira variables from first overtone to fundamental mode that led to planetary nebula ejection by this process. However, recent observational and theoretical studies of Mira variables suggest that these stars are already steadily pulsating in the fundamental mode (Willson 1982; Wood 1990), so that both the non-linear pulsation studies exhibiting violent pulsation and the mode switch suggestion of Wood and Cahn (1977) must be incorrect (but see Tuchman 1991). It now seems that mass loss in AGB stars is produced by the joint action of steady pulsation and radiation pressure on grains (Wood 1979; Bowen 1988).

One other mass loss mechanism that may apply for more massive AGB stars is radiation pressure ejection of the envelope (Wood and Faulkner 1986). In this mechanism, if the luminosity at a helium shell flash exceeds the Eddington luminosity $L_{Ed} = 4\pi cGM_{core}/\kappa$ at the edge of the core (where the opacity is mainly due to electron scattering), then no hydrostatic solution for the envelope exists and the likely result is envelope ejection. Wood and Faulkner (1986) show that the condition $L > L_{Ed}$ occurs at the peak of a luminosity spike in AGB stars with core mass $\gtrsim 0.9$ $M_\odot$ but with envelope mass $\lesssim 1.5$ $M_\odot$. Since core masses $\gtrsim 0.9$ $M_\odot$ are only likely to develop in AGB stars with mass $\gtrsim 5$ $M_\odot$, and the initial envelope mass will then be ~4 $M_\odot$, the radiation pressure ejection mechanism could only work to remove the last ~1.5 $M_\odot$ of envelope material. It should also be noted that in the ~500 year duration of the luminosity spike, the total momentum of radiation emitted from the central core would be insufficient to eject this much envelope material (M. Morris, private communication). However, the total internal energy of these very luminous envelopes is often positive (Paczynski and Ziolkowski 1968), so that radiation pressure combined with gas pressure forces may be sufficient to eject a large amount of material.

## 3. When does planetary nebula ejection occur

The possibility that the luminosity spikes produced by helium shell flashes cause planetary nebula ejection is now examined. The two Mira variables R Hya and R Aql are currently in the luminosity spike phase of AGB evolution. These two variables have undergone period changes over the last century or two that are so rapid that they can only be understood if the stars are currently undergoing shell flashes (Wood and Zarro 1981). Both variables have had their mass loss rates determined by measurements of the circumstellar CO microwave emission (Wannier and Sahai 1986). A plot of the mass loss rates of Mira variables against pulsation period shows a clear increase of mass loss rate with period (Wood 1986; Schild 1989; Wood 1990). R Aql has a mass loss rate significantly in excess of the value given by the trend line while R Hya has a mass loss rate less that expected for a typical Mira variable of its period (Wood 1990). In both cases, the total amount of mass that will be lost at current mass loss rates over the ~500 year duration of the luminosity spike (~$5 \times 10^{-4}$ $M_\odot$ for R Aql and ~$5 \times 10^{-5}$ $M_\odot$ for R Hya) is small compared to a typical planetary nebula mass. These results do not seem to support the suggestion that shell flash induced luminosity spikes lead to planetary nebula ejection. Basically, if we accept that dust-enshrouded AGB stars such as OH/IR stars are currently in the stage of ejecting planetary nebulae, and if we adopt typical mass loss rates for these objects in the solar vicinity of ~$10^{-5}$ $M_\odot$ yr$^{-1}$ (Knapp 1987), then in 500 years (the duration of the luminosity spike) only ~0.005 $M_\odot$ will be ejected, this mass being much smaller that a typical planetary nebula mass. Clearly, mass loss must occur over a significantly longer time interval, and planetary nebula mass loss must therefore occur mainly during the inter-flash, hydrogen burning evolution.

## 4. Low mass AGB evolution with empirical mass loss rates

In order to explore the process of planetary nebula ejection theoretically, Vassiliadis and Wood (1992a) have evolved stars in the mass range $0.89 \leq M/M_\odot \leq 5$ from the main sequence to the white dwarf stage. The most important feature of these calculations is the mass loss formula, which is an empirical one based on measurements of the mass loss rates

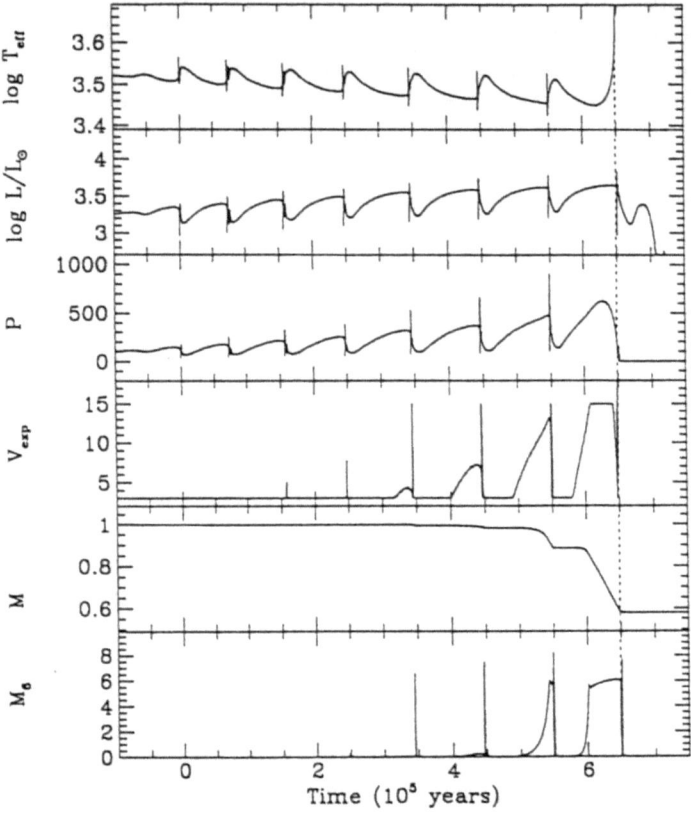

Figure 2. The time dependence of effective temperature $T_{eff}$, luminosity L, pulsation period P (days), stellar wind expansion velocity ($V_{exp}$), mass M ($M_\odot$) and mass loss rate $\dot{M}_6 (10^{-6} M_\odot \text{ yr}^{-1})$ in a thermally pulsing AGB star of initial mass 1.0 $M_\odot$ and abundance (Y,Z) = (0.25,0.008). The vertical dotted line represents the end of the AGB phase and a change to a lower mass loss rate.

of Mira variables, OH/IR stars and other dust-enshrouded red giant stars. It is now believed that stellar pulsation is a crucial factor in the production of mass loss from AGB stars (Castor 1981; Hearn 1990) and plots of mass loss rate against pulsation period in these stars shows a clear exponential increase in $\dot{M}$ with P up to a limiting value which corresponds to the radiation pressure driven limit $\dot{M} = L/cv_\infty$ (Wood 1986; Schild 1989; Wood 1990). The mass loss formulae given in Wood (1990) were used in the calculations to be described here.

Some properties of a 1 $M_\odot$ AGB star evolving with mass loss are shown plotted against time in Figure 2; the zero of time coincides with the first major thermal pulse. The luminosity shows the characteristic spikes and the slow increase during hydrogen burning. The most important feature of these calculations is the behaviour of the mass loss rate. In all low mass (M ≲ 2.5 $M_\odot$) stars, the mass loss rate is negligible except during the last few shell flash cycles, and even then Ṁ is significant only during the high luminosity phases of hydrogen burning, and during the brief luminosity spike. However, the plot of total mass M against time shows that only a negligible amount of mass is lost during the luminosity spike: nearly all mass loss occurs during quiescent hydrogen burning.

The reason for the mass loss behaviour shown in Figure 1 is the very rapid increase in mass loss rate with pulsation period, which varies roughly as $R^2/M \propto L/MT_{eff}^4$ (Wood 1990). During the latter part of the hydrogen burning phase of the shell flash cycle, L is high and $T_{eff}$ is low so the period (and mass loss rate) is correspondingly large. Once the stellar mass starts to decrease, the reduced total stellar mass acts to further increase P and Ṁ. In the present calculations, the mass loss rate was not allowed to increase beyond the radiation pressure limit, which seems to coincide to within a factor of ~2 with the mass loss rates in very high mass loss rate AGB stars in the Galaxy (Knapp 1986), in the LMC (Wood et al 1992) and at the Galactic Centre (Whitelock et al 1991). This high mass loss phase can be equated to the *superwind* of Renzini (1981).

The fact that the superwind mass loss rate turns on and off indicates that there should be AGB stars with hollow circumstellar shells, or even multiple shells. If the central star has completely dissipated its envelope and has become hot enough to excite the surrounding circumstellar material, then the resulting object could appear as a multiple shell planetary nebula. The observational evidence for such situations is described in section 6. (It is worth noting at this point that if the maximum mass loss rate were set to *twice* $L/cv_\infty$, consistent with errors in the measured mass loss rates and with the possibility of extra momentum being provided by scattering of photons within the circumstellar envelope, then complete envelope loss may occur in a single shell flash cycle for stars of $M \sim M_\odot$.)

As seen in Figure 2, the mass of an AGB star decreases significantly only during the last few shell flash cycles. This has important implications for stellar evolution calculations that seek to explain the dredge-up of carbon and s-process elements on the AGB. The dredge-up efficiency on the AGB is significantly enhanced if the envelope mass is large (Wood 1981). If most envelope mass is lost only during the last few flash cycles then, for the majority of AGB evolution, the full envelope mass is available to enhance the dredge-up process. However, if an enhanced Reimers (1975) mass loss formula is used on the AGB, mass loss occurs at lower luminosities than with the formula used above, and the likelihood of dredge-up occurring in model calculations will be reduced.

## 5. Intermediate mass AGB evolution

The behaviour of AGB stars of mass 5 $M_\odot$ (Figure 3) is quite different from the behaviour at low masses. In particular, the deep envelope convection in massive AGB stars rapidly extinguishes helium burning at the shell flash so that there is very little luminosity variation throughout the shell flash cycle. The mass loss increases steadily to superwind values over $\sim 10^5$ years and then stays at that value until the envelope is completely dissipated. Once the envelope mass is reduced below $\sim 1.5\ M_\odot$, the deep envelope convection recedes and the surface luminosity variations seen in low mass AGB stars appear.

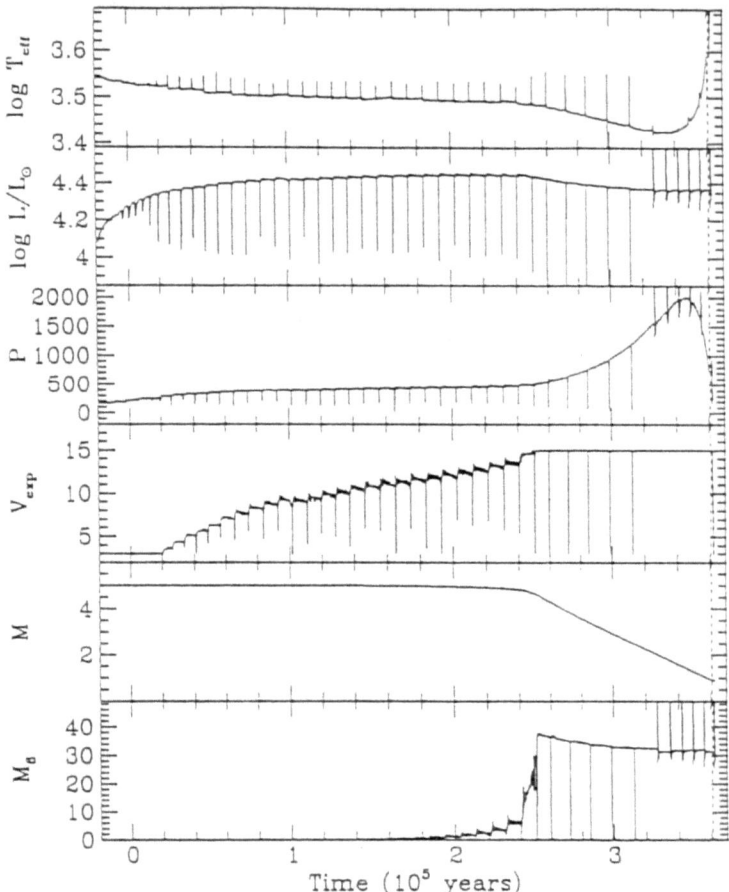

Figure 3. The time variability of a thermally pulsing AGB star of initial mass 5.0 $M_\odot$ and abundance $(Y,Z) = (0.25, 0.008)$. Symbols have the same meaning as in Figure 2.

## 6. Observational evidence for mass loss modulated by thermal pulses

Given the modulation of the mass loss rate by thermal pulses shown in Figure 2, we would expect at the very least that there should exist AGB stars in which a superwind mass loss episode had been turned off by a helium shell flash. Such stars would be surrounded by hollow circumstellar shells but would otherwise appear as normal M-, S- or C-type AGB stars.

Many stars with hollow circumstellar have now been identified using results from IRAS. Willems and de Jong (1988) found that detached dust shells were a common feature of carbon stars and they suggested that thermal pulses were the cause of these hollow shells. More recently, Zijlstra *et al* (1992) have used IRAS fluxes to find similar detached shells around oxygen-rich AGB stars.

More direct evidence for hollow shells around AGB stars has been found by Olofsson *et al* (1990, 1992). In these studies, the circumstellar shell has been detected by its CO microwave emission: mapping across the shell with the telescope beam clearly indicates the existence of a hollow shell. With the telescope beam centred on the star S Sct, Olofsson *et al* (1992) also found a weak central feature in the CO emission line profile that they interpreted as due to a weak, low velocity wind presently blowing from the central star. However, time-dependent hydrodynamic models of the winds around thermally pulsing AGB stars (Wood and Vassiliadis 1991; Vassiliadis and Wood 1992b) show that matter falls back onto the star during the low luminosity, post-flash evolution. The central CO emission peak found in S Sct could therefore be due to infalling material rather than a weak expanding wind.

Given the clear evidence for thermal pulse modulation of the mass loss shells around AGB stars, we would expect planetary nebula shells to show similar evidence for thermal pulse modulation. Multiple shell planetary nebulae are a possible result of the on-off-on existence of superwind mass loss on the AGB. Some multiple shell PN that have times between successive ejections which are consistent with the inter-pulse timescale are noted in these proceedings by Frank, Balick and van der Veen (1992).

**References**

Blöcker, T. and Schönberner, D. 1990, A&AL, 240, L11.
Boothroyd, A.I. and Sackman, I.-J. 1988, ApJ, 328, 632
Bowen, G.H. 1988, ApJ, 329, 299
Castor, J.I. 1981, in *Physical Processes in Red Giants*, eds. I. Iben and A. Renzini (Dordrecht: Reidel), p. 285
Frank, A., Balick, B. and van der Veen, W. 1992, in IAU Symposium 155 *Planetary Nebulae*, eds. R. Weinberger and A. Acker (Kluwer: Dordrecht), (these proceedings)
Hearn, A.G. 1990, in *From Miras to Planetary Nebulae: Which Path for Stellar Evolution*, eds. M.O. Mennessier and A. Omont (Editions Frontières), p.121
Iben, I. 1984, ApJ, 277, 333.
Knapp, G.R. 1986, ApJ, 311, 731
Knapp, G.R. 1987, in *Late Stages of Stellar Evolution*, eds. S. Kwok and S.R. Pottasch (Reidel: Dordrecht), p.103
Olofsson, H., Carlström, Eriksson, K., Gustafsson, B. and Willson, L.A. 1990, A&A, 230, L13
Olofsson, H., Carlström, Eriksson, K. and Gustafsson, B. 1992, A&A, 253, L17
Paczynski, B. 1971, Acta Astronomica, 21, 417.
Paczynski, B. and Ziolkowski, J. 1968, Acta Astronomica, 18, 255.
Reimers, D. 1975, in *Problems in Stellar Atmospheres and Envelopes*, eds. B. Baschek, W.H. Kegel and G. Traving (Berlin: Springer), p.229
Roxburgh, I.W. 1967, Nature, 215, 838
Schild, H. 1989, MNRAS, 240, 63

Schwarszchild, M. and Härm, R. 1965, ApJ, 142, 855
Smith, R.L. and Rose, W.K. 1972, ApJ, 176, 395
Tuchman, Y. 1991, ApJ, 383, 779
Tuchman, Y., Sack, N. and Barkat, Z. 1979, ApJ, 234, 217
Vassiliadis, E. and Wood, P.R. 1992a, ApJ, to be submitted
Vassiliadis, E. and Wood, P.R. 1992b, Proc. Astr. Soc. Australia, in press
Wannier, P.G. and Sahai, R. 1986, ApJ, 311, 335
Whitelock, P., Feast, M. and Catchpole, R., 1991, MNRAS, 248, 276
Willems, F.J. and de Jong, T. 1988, ApJL, 309, L39
Willson, L.A. 1982, in *Pulsations in Classical and Cataclysmic Variable Stars*, eds. J.P. Cox and C.J. Hansen (Boulder: Joint Institute for Laboratory Astrophysics), p.269
Wood, P.R. 1974, ApJ, 190, 609
Wood, P.R. 1979, ApJ, 227, 220
Wood, P.R. 1981, in *Physical Processes in Red Giants*, eds. I. Iben and A. Renzini (Dordrecht: Reidel), p. 135
Wood, P.R. 1986, in *Stellar Pulsations*, eds. A.N. Cox, W.M. Sparkes and S.G. Starrfield (Berlin: Springer-Verlag), p.250
Wood, P.R. 1990, in *From Miras to Planetary Nebulae: Which Path for Stellar Evolution*, eds. M.O. Mennessier and A. Omont (Editions Frontières), p.67
Wood, P.R. and Cahn, J.H. 1977, ApJ, 211, 499
Wood, P.R. and Faulkner, D.J. 1986, ApJ, 307, 659.
Wood, P.R. and Vassiliadis, E. 1991, Highlights of Astronomy, 9, in press
Wood, P.R. and Zarro, D.M. 1981, ApJ, 247, 247
Wood, P.R, Whiteoak, J.B., Hughes, S.M.G., Besell, M.S., Gardner, F.F. and Hyland, A.R. 1992, ApJ, 397, in press
Zijlstra, A.A., Loup, C., Waters, L.B.F.M. and de Jong, T. 1992, A&A, submitted

# BASIC PROBLEMS OF PLANETARY NEBULA GAS DYNAMICS

J.E. DYSON

*Department of Astronomy, University of Manchester, Manchester M13 9PL, U.K.*

## Abstract

Planetary Nebulae result from the interaction of fast winds and radiation fields with slow ejecta produced in the late stages of evolution of moderately low mass stars. We review the basic principles of the interaction and describe some of the modifications needed to produce realistic models for comparison with observational data. The present status of numerical modelling is discussed and some important problems outlined.

## 1 Introduction

Planetary nebulae (PNe) provide an exciting opportunity for the study of high speed gas dynamics under conditions far removed from terrestrial. The interactions taking place in them are of importance both intrinsically and for other areas of astrophysics. These interactions involve a rich variety of physical phenomena whose elucidation can come only from combined observational and theoretical studies.

Most theoretical investigations of PNe predicate the 'Interacting Winds' model of Kwok et al (1978). A fast ($u_F \geq 2000$ km s$^{-1}$) wind from the PNe nucleus interacts with slower ($u_s \approx 10$ km s$^{-1}$) ejecta from previous red giant (RG) and asymptotic giant branch (AGB) phases of the nucleus' evolution. The ejecta are swept up into a shell by the fast wind and photoionized by the radiation field of the nucleus. In general terms this model is strikingly successful; the global properties of PNe (e.g. sizes and expansion velocities) are well reproduced. However, the wide range of PNe morphologies and properties and great complexities revealed in the detailed study of specific objects, demand considerable elaboration of this basic picture. In order to construct more realistic theoretical models, a number of areas demand clarification. Firstly, conditions in the slow ejecta need specification. Molecular line observations of AGB and post-AGB objects (e.g.Olofsson 1992) show that the slow ejected envelopes possess complex density and velocity structure. Secondly, the behaviour of the sources of momentum, energy and mass input which interact with the slow ejecta must be known. Finally, it is necessary to include all relevant physical processes such as hydrodynamic mixing at interfaces. Given the often large uncertainties and parameter ranges in these areas, it is hardly surprising that PNe are so varied in form.

The comparison of models with particular objects is the only way forward, but is complicated by the effects of evolution. Even the simplest models demonstrate appreciable differences in structure and evolution when parameter values are varied within perfectly plausible ranges. Studies of the—in many ways analogous—object RCW 58 (a Wolf-Rayet nebula), show that unless comprehensive observational data is available, totally inappropriate (though superficially applicable) models may be used (e.g. Smith et al 1984, 1988; Hartquist et al 1986; Arthur et al 1992).

The emphasis in this review is on the physical consequences as different modifications to the simple models are made and on the relevance of the effects produced to observational data. The results of detailed numerical studies are discussed where appropriate.

## 2 Basic Models with Spherical Symmetry

The simplest model assumes constant mass ejection rates and velocities in the various stages of mass loss and a constant Lyman continuum output rate from the nuclear star. With characteristic parameters, most of the flow momentum initially resides in the AGB (or superwind) ejecta. The impact of the fast wind on the slow ejecta produces the familiar 'two-shock' flow pattern. Provided $u_F$ is great enough ($u_F \gtrsim 1000$ km s$^{-1}$) and the shocked fast wind cools only by adiabatic expansion, a shell of compressed slow ejecta is driven outwards by the shocked wind pressure at constant velocity $V_s$, typically a few times $u_s$. The total momentum content in the flow is increased by a factor of a few. The shell is initially optically thick in the Lyman continuum and becomes optically thin at times in the range $10^{(3-5)}$ yr, i.e. spanning characteristic evolutionary time scales.

Significant effects can occur if even relatively minor variations are introduced into the model. For example, if there is a time delay between the end of the slow wind and the onset of the fast wind, the swept-up shell may be initially optically thin and later become optically thick. A low emission measure 'halo' of unshocked slow wind can form which persists for a few thousand years after the shell becomes optically thick (Kahn 1989).

Evolutionary models of the nuclear stars show that the winds and radiation fields gradually power up to their full terminal velocities and luminosities (Schönberner 1983; Pauldrach et al 1988). Kahn & Breitschwerdt (1990) and Breitschwerdt & Kahn (1990) allowed for this in a semi-analytic treatment of the early stages of PNe evolution, assuming a constant wind momentum output rate. The wind speed is initially low enough for the shocked wind to cool radiatively; the shell is driven by direct momentum transfer. The changeover to a pressure driven shell occurs at time $t_*$ ($\approx 10^3$ yr) when $u_F \approx 150$ km s$^{-1}$. Complex events can be contained within this simple scenario. Ionization of the swept up shell becomes dynamically significant at a time $t_i$ (generally $\leq t_*$). At times $t < t_i$; $t_*$, the shocked fast and slow winds are cool. For times $t_* > t > t_i$ the inner part of the shell is ionized, its increased pressure drives the stellar wind shock back towards the star, the outer shock accelerates and the neutral-ionized interface in the shell can become Rayleigh-Taylor (R-T) unstable. At times $t > t_*$, the now finite pressure of hot shocked fast wind drives a second shell into ionized previously shocked cool ejecta. This shell can trap the Lyman continuum and the previously ionized ejecta can recombine. The resultant pressure drop can lead to acceleration of the second shell and another bout of R-T instability producing cool small ($\ell \sim 10^{15}$ cm) fragments.

The role of R-T instability is important in this model and is an example of a physical process requiring careful investigation. Hartquist & Dyson (1987) argued that shell acceleration alone is not sufficient to cause shell fragmentation and that shells might reseal themselves by the ablation of gas off fragments as the hot gas flows round them; some additional mechanism (e.g. thermal instability) may be necessary for genuine fragmentation. Observational support for this view comes from data on filaments in the Vela Supernova (Meaburn et al 1988), and it remains an open question.

Purely radial variations in the density of the slow wind may be reasonably conjectured as a result of time dependent ejection, and may have important consequences. Molecular hydrogen line features with velocities exceeding 250 km s$^{-1}$ are observed towards the protoplanetary nebula CRL 618 (Burton & Geballe 1986). Serious problems exist for shock accelerated models; molecules are dissociated even in magnetically moderated shocks of velocity $\geq$ 50 km s$^{-1}$. The ram pressure acceleration of coherent clumps is more magic than physics. Hartquist & Dyson (1987) have suggested that the re-acceleration of shells whose velocity has dropped for a time below 50 km s$^{-1}$, by the presence of suitably steep density gradients in the slow ejecta, may provide a way round these problems.

## 3 Effects of Non-Uniform Slow Winds

Because of the very high sound speed in the shocked fast wind, deviations of shell shapes from spherical symmetry are induced only by non-uniformity in the slow wind density distribution. Balick (1987) qualitatively demonstrated that many of the global features of PNe morphology (e.g. bipolarity) can be explained if there exists a density contrast between the polar and equatorial regions of the slow wind. There are several possibilities for producing such non-uniformity. Probably the most plausible involves the presence of a binary system (Morris 1987; Soker 1990; Soker and Livio 1990; Bond & Livio 1990; Soker 1992a). However there may have to be mechanisms involving single stars. Soker & Harpaz (1992a) have, for example, suggested that non-radial oscillation modes can lead to axi-symmetric mass loss towards the end of the AGB phase, producing greater asymmetry in the core regions of PNe than in the haloes. However, Soker & Harpaz (1992b) note that a companion is probably still necessary to excite non-radial modes.

Many general models have been calculated which include such variations, ranging from relatively simple quasi-analytic approaches (e.g. Kahn & West 1985; Ghanbari 1989; Henney & Dyson 1992) to detailed numerical modelling (e.g. Soker 1989; Soker & Livio 1989; Mellema et al 1991; Icke et al 1992). The quasi-analytic approaches have the merit of greater generality and many relevant physical processes can be included at least on an approximate level. Detailed numerical models face formidable difficulties (e.g. Icke et al 1992) due to the extreme variations in conditions in different regions of the flow. These models show many intriguing features, such as cocoon and jet-like features (Mellema et al (1991), but, as pointed out by these authors, the assumption of axial symmetry encourages caution in the interpretation of them. Three-dimensional calculations are essential.

Generally, these models have supported Balick's (1987) proposals, but to compare models with actual objects, the detailed ionization structure must be solved along with hydrodynamics. Recent calculations by Mellema and by Frank (this meeting) have made considerable progress in this aspect. Both morphology and internal kinematics are essential to discriminate between models. Chu (1989), for example, has noted examples of PNe which have a core-halo morphology where the less bright outer halo appears to expand as fast or even faster than the inner core. This is hard to explain on the simple interacting wind model. Finally, all such calculations invoke smooth distributions of slow ejecta and, as discussed below, flows in clumpy media behave very differently.

## 4 Flows in Clumpy Media

PNe appear clumpy on a variety of scales in both ionized and neutral gas. Ionized features have scale sizes from $\ell \sim 10^{15}$ cm (e.g. in NGC 7293) to $\ell \sim 10^{16-17}$ cm (e.g. ansae). A substantial fraction of the total PNe mass can reside in molecular bearing material which is fragmented down to scales limited only by instrumental resolution (Bachiller et al 1990; Forveille & Huggins 1991; Huggins 1992). There are two extreme possibilities for the origin of these fragments; either they are primordial or generated during the interaction processes taking place. We concentrate on the former possibility here.

In general, the clumps would be embedded in an interclump medium. Their importance is four-fold. Firstly, gas flow around clumps ablates material into the flows and the subsequent mass injection can affect flow physics, chemistry and dynamics (Hartquist et al 1986; Arthur et al 1992). Secondly, the clumps can determine where the incident wind is shocked. If the clumps are not too numerous, the shock position is essentially determined by the interclump medium and it moves outwards in a similar manner to the case of smoothly distributed gas; however, mass loading behind the shock can profoundly affect the post-shock flow (Hartquist et al 1986; Arthur et al 1992). If, on the other hand, the clumps are very numerous, the interclump material plays little role in determining the position of the wind shock. The wind is decelerated primarily in bow shocks around the clumps. If the clumps last long enough, the positions at which the wind is shocked either remain spatially fixed or move outwards with more or less the clump velocities. This latter situation may be relevant to NGC 7293 where many small clumps of primordial origin could be present (Dyson et al 1989). Thirdly, the coexistence of the clumps and global flows results in interface (or boundary layer) formation. Very important effects such as the enhancement of radiative losses, can occur there. The study of such interfaces is a major area of interest in a wide variety of astrophysical environments (Hartquist & Dyson 1988, 1992). Finally, structures intermediate in scale size between the boundary layers and the global flows may produce observable features in PNe (Dyson et al 1992–Section 5).

Many qualitatively different flows can be produced in the shocked wind region as a result of mass loading. The situation where there are relatively few mass loading centres has been discussed in the context of RCW 58 (Smith et al 1988). As noted above, however, the structure of NGC 7293 suggests the presence of many small distributed clumps. It is likely then that there is very effective enhanced radiative cooling in the shocked stellar wind gas and it may percolate ballistically through the clumps. If the rate of mass ablation per unit volume into the shocked flow is constant and ablated mass dominates the flow it is straightforward to show (Dyson & Hartquist 1992) that the gas flow velocity $v \propto r^{-3}$ and density $\rho \propto r^4$, where $r$ is the distance from the star. This flow matches at a contact discontinuity to interclump material swept up by the outwards facing shock (possibly after going through another shock). The appearance of such a flow depends crucially on where the radiated energy originates. If the entire bubble radiates as a result of photoionization by the stellar radiation field, the emission will be concentrated in a sheet like structure near the contact discontinuity if it largely originates in the interclump flow; on the other hand it will be more uniformly distributed (though clumpy on a small scale) if it originates at clump flow interfaces. The mixing of cold clump and hot shocked wind gas may lead to enhanced soft X-ray emission from intermediate temperature ($\sim 10^6$ K) gas (cf. Kreysing et al, these Proceedings). Whatever the details, it is clear that clumps can significantly affect the structure and dynamics of PNe.

## 5 Specific Structural Features of PNe

### 5.1 Haloes

The bright cores of many PNe are surrounded by extended envelopes with radii extending up to 0.9 pc in extreme cases (Chu et al 1987; Chu 1989; Frank et al 1990; Balick et al 1992). The envelope may give information on the ejection history in the slow wind phase (Frank et al 1990; Balick et al 1992), and provide constraints on the models for the formation of the structure (Soker et al 1992). Balick et al (1992) divide the envelopes into shells which often have roughly linearly outwards decreasing brightness distributions and haloes which are of low brightness and always limb brightened. Limb brightening can be the result of halo confinement by some external medium of adequate pressure (Frank et al 1990). This could either be RG ejecta (Balick et al 1992) or the local interstellar medium (Borkowski et al 1990; Soker et al 1991). Differentiation between the two possibilities is clearest for cases where the bright halo rims have a bow shock morphology, suggesting the motion of the star relative to an ISM (Soker et al 1991). However, none of the calculations so far made has included dynamical effects associated with photoionization. Generally haloes are less elliptical than the inner cores. This may simply be a hydrodynamic effect since flows expanding with velocities about equal to their internal sound speed tend to even out their pressure (or—for isothermal flows—density) gradients (Balick et al 1992). Alternatively, the later stages of slow wind ejection may be much more non- spherical (e.g. Plait & Soker 1990; Soker & Harpaz 1992a).

The kinematic structures of haloes are varied. Chu (1989) has noted cases where the halo appears to expand faster than the core. She also noted others where the simple dynamical timescale is greater for the core than the halo. Wang (1992) has suggested that the interaction of a fast wind with a constant velocity but time varying asymmetric slow wind can lead to fast expanding lobes with an apparent lifetime less than the slower moving waist regions. Another possibility (Dyson 1992; Dyson & Hartquist 1992) is that the cores are extremely clumpy and the stellar wind is so mass loaded that it exists transsonically from the core. Isothermal transsonic flows can accelerate to mach numbers of two or three over a reasonable range of halo to core radius.

As a specific example, NGC 6720 poses particular problems (Balick et al 1992). Molecular emission from $H_2$ occurs from shells apparently within the ionized haloes. The molecular gas expands more rapidly than the surrounding ionized gas, thus ruling out inclined bipolar flows. Balick et al (1992) suggest two possibilities. Firstly, the core is clumpy and the exterior halo is photoionized by UV photons which leak through between clumps. In this case the halo could not be mass loaded wind whose velocity should not drop much below that of the core clumps. Alternatively, if the more recently ejected core is optically thick, the halo could be in the process of recombining (cf. Section 2). In either case, the core gas must be ejected faster than the halo gas.

Mass loaded flows exitting from clumpy cores can have interesting consequences if they interact with pre-existing halo material. Three PNe with extended haloes have halo electron temperatures (from [OIII]) up to 50% higher than the core temperatures (Middlemass et al 1990). Dyson (1992) has suggested that these are produced as a moderate (2–3) mach number mass loaded flow shocks against condensations in a halo. Forbidden line emission is enhanced by the temperature increase, but because of the bow shock geometry, the line widths would be much lower than the pre-shock flow velocities. Meaburn et al (1991) have shown that the [OIII] lines over a bright condensation in the halo of NGC 6543 indeed have low line widths.

## 5.2 Ansae

Some elliptical PNe have bright knots or 'ansae' along the major axis on either side of the nucleus. They appear to contain neutral material since low ionization lines and molecular emission is associated with them. They can be inside or outside the main body of nebular emission; usually they are moving outwards relative to the bulk of the nebula. It is tempting to associate them with the tips of fast moving lobes in bipolar flows. However, numerical calculations (e.g. Mellema et al 1991) do not produce suitable structures. A jet-like origin (Soker 1990) where the ansae are produced at the working surface of a jet (cf. HH objects) has many attractions. Lopez et al (this meeting) have shown a particularly striking jet-like structure in FG 1 (He2-66).The collimation of jets from a rotating star has been discussed by Soker (1992b). However, it depends critically on the very badly understood distribution of angular momentum within the star. Jet-like activity is known to occur in symbiotic stars (e.g. R Aquarii). Links between symbiotic stars and at least some PNe are becoming apparent (Lutz et al 1989). It may be that some ansae or jet-like features in PNe are formed by whatever (badly understood) processes are operative in R Aq or in the young stars associated with HH objects.

## 5.3 Small Scale Structure

The origin and morphology of small scale clumps in PNe are of great interest. Mechanical processes such as R–T instability (Section 2) may provide a way of introducing them into PNe. Alternatively they may be primordial and reflect the presence of very clumpy mass ejection (Olafsson 1992; Dyson et al 1989). NGC 7293 possesses many small 'cometary' globules at the inner edge of the ionized region. The globule parameters inferred by Meaburn et al (1992) from dust absorption are in excellent agreement with those required for the globules to be of primordial origin (Dyson et al 1989). The globules have long tails with widths about equal to the globule head diameters. Such a 'wind-swept' appearance argues for some type of stellar wind interaction. This cannot be due to the interaction of a supersonic wind with a globule (Dyson et al 1992). The central star of NGC 7293 has evolved too far to have a fast wind, and the interaction of supersonic streams with embedded gas sources produces very broad tails (Dyson et al 1992). A more likely possibility is that the globule is immersed in subsonic hot shocked wind gas which has been 'bottled up' following the fast wind phase. Small pressure differentials produced as the hot gas leaks out of the core are responsible for the confined long tails (Dyson et al 1992).

# 6 Conclusions

Considerable progress has been made in understanding the general gas dynamical processes occurring in PNe. However, many challenging areas remain. The development of 3-D numerical codes which incorporate radiative heating and cooling is a prime candidate. There are significant processes (such as the importance of mixing) which are badly understood and their incorporation into phenomenological models should be a priority. The great importance of the structure of the slow ejecta to all models is abundantly clear and the increasing wealth of data relating to the envelopes of AGB and post-AGB stars is invaluable. It is clear that even at the simplest level, there can be no such thing as a single evolutionary path for all PNe. Ultimately the confrontation of models of specific objects with the observational data on them must be the way forward.

Myf Bryce, Franz Kahn and Alex Raga made helpful comments on the manuscript.

## References

Arthur, S. J., Dyson, J. E. & Hartquist, T. W. (1992) Mon. Not. R. astr. Soc. (submitted).
Bachiller, R., Bujarrabal, V., Martin-Pintado, J., Planesas, R. & Gómez-Gonzalez, J. (1990) Astrophys. Space Sci. **171**, 195.
Balick, B. (1987) Astron. J. **94**, 671.
Balick, B., Gonzalez, G., Frank, A. & Jacoby, G. (1992) Astrophys. J. (in press).
Bond, H. & Livio, M. (1990) Astrophys. J. **355**, 568.
Borkowski, K. J., Sarazin, C. L. & Soker, N. (1990) Astrophys. J. **360**, 173.
Breitschwerdt, D. & Kahn, F. D. (1990) Mon. Not. R. astr. Soc. **244**, 521.
Burton, M. G. & Geballe, T. R. (1986) Mon. Not. R. astr. Soc. **223**, 13p.
Chu, Y.-H. (1989) In S. Torres-Peimbert (ed.), *Planetary Nebulae*, IAU Symposium 131, Kluwer Academic Publishers, Dordrecht, p.105.
Chu, Y.-H., Jacoby, G. H. & Arendt, R. (1987) Astrophys. J. Suppl. **64**, 529.
Dyson, J. E. (1992) Mon. Not. R. astr. Soc. **255**, 460.
Dyson, J. E. & Hartquist, T. W. (1992) Astro. Lett. & Comm. **28**, 301.
Dyson, J. E., Hartquist, T. W. & Biro, S. (1992) Mon. Not. R. astr. Soc. (submitted).
Dyson, J. E., Hartquist, T. W., Pettini, M. & Smith, L. J. (1989) Mon. Not. R. astr. Soc. **241**, 625.
Forveille, T. & Huggins, P. J. (1991) Astron. Astrophys. **248**, 599.
Frank, A., Balick, B. & Riley, J. (1990) Astron. J. **100**, 1903.
Ghanbari, J. (1989) Ph.D. Thesis, University of Manchester.
Hartquist, T. W. & Dyson, J. E. (1987) Mon. Not. R. astr. Soc. **228**, 957.
Hartquist, T. W. & Dyson, J. E. (1992) Quart. J. R. astr. Soc. (in press).
Hartquist, T. W., Dyson, J. E., Pettini, M. & Smith, L. J. (1986) Mon. Not. R. astr. Soc. **221**, 715.
Henney, W. J. & Dyson, J. E. (1992) Astron. Astrophys. (in press).
Huggins, P. J. (1992) In H. E. Schwarz (ed.), *Mass Loss on the AGB and Beyond*, Proc. 2nd ESO/CITO Workshop (in press).
Icke, V. (1991) Astron. Astrophys. **251**, 369.
Icke, V., Balick, B. & Frank, A. (1992) Astron. Astrophys. **253**, 224.

Kahn, F. D. (1989) In S. Torres-Peimbert (ed.), *Planetary Nebulae*, IAU Symposium 131, Kluwer Academic Publishers, Dordrecht, p.411.
Kahn, F. D. & Breitschwerdt, D. (1990) Mon. Not. R. astr. Soc. **242**, 505.
Kahn, F. D. & West, K. A. (1985) Mon. Not. R. astr. Soc. **212**, 837.
Kwok, S., Purton, C. R. & Fitzgerald, P. M. (1978) Astrophys. J. **219**, L125.
Lutz, J. H., Kaler, J. B., Shaw, R. A., Schwarz, H. E. & Aspin, C. (1989) Publ. Astr. Soc. Pacific **101**, 966.
Meaburn, J., Hartquist, T. W. & Dyson, J. E. (1988) Mon. Not. R. astr. Soc. **230**, 243.
Meaburn, J., Nicholson, R., Bryce, M., Dyson, J. E. & Walsh, J. R. (1991) Mon. Not. R. astr. Soc. **252**, 535.
Meaburn, J., Walsh, J. R., Clegg, R. E. S., Walton, N. A., Taylor, D. & Berry, D. S. (1992) Mon. Not. R. astr. Soc. **255**, 177.
Mellema, G., Euderink, F. & Icke, V. (1991) Astron. Astrophys. **252**, 718.
Middlemass, D., Clegg, R. E. S., Walsh, J. R. & Harrington, J. P. (1990) In M. O. Mennessier & A. Omont (eds.), *From Miras to Planetary Nebulae: Which Path for Stellar Evolution?*, Editions Frontières, Gif-sur-Yvette, p.420.
Morris, M. (1987) Publ. Astr. Soc. Pacific **99**, 1115.
Olofsson, H. (1992) In H. E. Schwarz (ed.), *Mass Loss on the AGB and Beyond*, Proc. 2nd ESO/CITO Workshop (in press).
Pauldrach, A., Puls, J., Kudritzki, R. P., Méndez, R. H. & Heap, S. R. (1988) Astron. Astrophys. **207**, 123.
Plait, P. & Soker, N. (1990) Astron. J. **99**, 1883.
Schwarz, H. E. (1992) In H. E. Schwarz (ed.), *Mass Loss on the AGB and Beyond*, Proc. 2nd ESO/CITO Workshop (in press).
Schönberner, D. (1983) Astrophys. J. **272**, 708.
Smith, L. J., Pettini, M., Dyson, J. E. & Hartquist, T. W. (1984) Mon. Not. R. astr. Soc. **211**, 679.
Smith, L. J., Pettini, M., Dyson, J. E. & Hartquist, T. W. (1988) Mon. Not. R. astr. Soc. **234**, 625.
Soker, N. (1989) Astrophys. J. **340**, 927.
Soker, N. (1990) Astron. J. **99**, 1869.
Soker, N. (1992a) In H. E. Schwarz (ed.), *Mass Loss on the AGB and Beyond*, Proc. 2nd ESO/CITO Workshop (in press).
Soker, N. (1992b) Astrophys. J. (in press).
Soker, N., Borkowski, K. J. & Sarazin, C. L. (1991) Astron. J. **102**, 1381.
Soker, N. & Harpaz, A. (1992a) Astrophys. J. (in press).
Soker, N. & Harpaz, A. (1992b) (preprint).
Soker, N. & Livio, M. (1989) Astrophys. J. **339**, 268.
Soker, N., Zucker, D. B. & Balick, B. (1992) Astrophys. J. (submitted).
Wang, L.-F. (1992) (in preparation).

# INTERACTION OF PLANETARY NEBULAE WITH THE INTERSTELLAR MEDIUM

K.J. BORKOWSKI
*Department of Astronomy, University of Maryland, College Park, MD 20742, USA*

ABSTRACT. Interaction of planetary nebulae (PNe) with the interstellar medium (ISM) is quite common, in accord with our understanding of the large-scale structure of the ISM. The characteristic signature of the interaction is a dipole asymmetry resulting from a distortion and compression of the nebula in the direction of stellar motion. Studies of interacting PNe provide us with information on the structure of the ISM, and on the evolution of old PNe and PN halos.

## 1. Introduction

PNe initially expand freely because their densities exceed the ISM density by orders of magnitude. However, once the PN density drops below a critical value because of the expansion, an interaction takes place between the PN and the ISM.

The strength of the interaction depends on the ISM density and pressure distribution within the layer occupied by PNe. Because most PNe are located outside of the thin H I layer of dense gas in the Galactic plane, the distribution of gas at large distances from the plane is needed. The presence of an extended, 1.5 – 2 kpc thick layer of ionized gas has been recently established (Reynolds 1991a; Nordgren, Cordes, & Terzian 1992). Reynolds (1991b) found an average electron density $n_e = 0.08$ cm$^{-3}$ in this layer. There is also neutral hydrogen at large distances from the Galactic plane (Dickey & Lockman 1990), with similar (or higher) densities. The third phase of the ISM is composed of the hot ($\sim 10^6$ K) ionized gas with a low density of $\sim 0.001$ cm$^{-3}$. Within the Galactic plane, average densities might be several times higher than $n_0 \sim 0.1$ cm$^{-3}$, typical of the warm neutral and ionized gas above the plane.

Another parameter which strongly influences the interaction is the relative velocity $v_*$ between the PN and the ISM, usually dominated by the stellar motion. The ratio of the ISM ram pressure to its thermal pressure is equal to $[(v_* + v_{exp})/c]^2$ in the leading (upstream) nebular section, where $v_{exp}$ denotes the nebular expansion velocity and $c$ is the isothermal sound speed in the ISM. In the warm medium $c = 10$ km s$^{-1}$, and with the typical values of $v_* = 60$ km s$^{-1}$ and $v_{exp}=20$ km s$^{-1}$, the ISM ram pressure exceeds its thermal pressure by a factor of 60. The ISM ram pressure is thus usually expected to dominate the interaction process, distorting nebulae in the direction of the stellar motion.

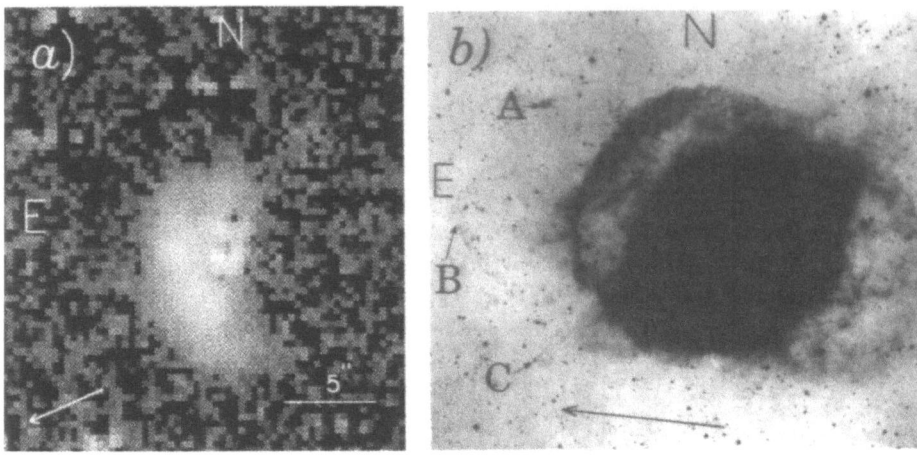

Figure 1: (a) The continuum-subtracted [O III] λ5007 image of the M22 PN. Arrow indicates the direction of the transverse cluster motion. A characteristic dipole asymmetry is typical for interacting PNe, with the nebula compressed and distorted in the direction of stellar motion. (b) The Helix nebula in the red light (courtesy of David Malin), showing faint outlying features. Arrow indicates the proper motion direction of the central star.

## 2. Observational Evidence

As suggested by theoretical considerations, a good observational diagnostic for the PN-ISM interaction is a distortion of the nebula in the direction of the stellar motion. This results in a characteristic dipole asymmetry, with the brighter, ram-pressure compressed nebular section located in the leading (upstream) direction. This morphology is clearly seen in interacting PNe PHL932 (Arp & Scargle 1967), A35 (Jacoby 1981), and S216 (Reynolds 1985). An examination of morphologies of old, nearby PNe revealed a number of nebulae with this dipole asymmetry (Borkowski, Sarazin, & Soker 1990). The implication is that the PN-ISM interaction should be quite common. This might account for low expansion velocities of old PNe, particularly those located in the Galactic plane (Gieseking, Hippelein, & Weinberger 1986; Hippelein & Weinberger 1990). I will now highlight the basic features of the interaction by discussing the most interesting individual cases.

An asymmetric morphology of A35 (Jacoby 1981), with an emission enhancement visible ahead of the fast-moving (125 km s$^{-1}$) star, is characteristic of the PN-ISM interaction. However, a bow-shock structure visible in the [O III] 5007 Å emission line is the most striking nebular feature. Its presence implies that the nebular expansion in the direction of the stellar motion was halted, and the nebular gas is now accelerated past the central star by the ram pressure of the ISM (Jacoby 1981). A35 is in its final evolutionary stage, with the nebula being stripped away from its central star. The nebular r.m.s. electron density is $\sim 10$ cm$^{-3}$ and should be approximately equal to $(v_*/c)^2 n_0 = 150 n_0$. We obtain $n_0 \sim 0.07$ cm$^{-3}$, in good agreement with expectations. The bow shock forms where the ram pressure of the accelerated PN gas is balanced by the stellar wind ram pressure. A similar structure is seen in the PN 623+71, a nebula surrounding a cataclysmic binary (Krautter,

Klaas, & Radons 1987; Hollis et al. 1992). Because the central star of A35 is also a close binary, the enhanced stellar winds in close binary systems may play a crucial role in the bow shock formation.

The most spectacular example of a fast-moving, interacting nebula is provided by a hydrogen-poor PN in the globular cluster M22. Figure 1a (Borkowski, Tsvetanov, & Harrington 1992) demonstrates that because of the high cluster velocity of 200 km s$^{-1}$ (Cudworth 1986), the PN has been very strongly distorted. The estimates of the ambient ISM density are somewhat dependent on modeling, but are again consistent with $n_0 \sim 0.1$ cm$^{-3}$. The ISM ram pressure will strip the M22 PN from the cluster within several $\times 10^5$ yr. Stripping of the M22 PN by the ambient ISM is the first direct evidence for removal of gas from globular clusters.

A different picture of the interaction emerges by looking at the well-known Helix nebula (NGC 7293) in Figure 1b. The motion of the central star (Cudworth, private communication) coincides with the direction determined by the conical filament B and the cometary object A discovered by Malin (1982). Kwitter, Chu, & Downes (these proceedings) demonstrated through narrow-band imaging that excitation conditions in these features differ appreciably from those in the nebula proper. These features are most probably sections of the red giant wind (or superwind) which have been compressed in the east by the ram pressure of the ISM. During the course of the PN evolution, a bipolar outflow (Walsh & Meaburn 1987) expanded in the distorted wind (whose density is higher in the direction of nebular motion), resulting in a strong asymmetry between the two lobes of the outflow. What we are seeing here is an interaction of a PN halo (or an outer shell) with the ISM, which in the Helix nebula has been further modified by the PN dynamics. The halo interaction could have begun long before the PN formation – for example, a ram-pressure-distorted red giant wind was detected around Mira (Bowers & Knapp 1988). A systematic investigation of PN halos should reveal more cases like NGC 7293, and some might have been already detected (e. g., NGC 6853 – Borkowski et al. 1990; NGC 6751 - Chu et al. 1991; IC 4593 - Balick et al. 1992; Zucker & Soker 1992; NGC 3587 - Kwitter et al., these proceedings). Even for spherically symmetric halos, Frank et al. (1990) found that their morphologies require a confining medium with the pressure of $10^3$ K cm$^{-3}$, equal to the thermal pressure of the warm ionized ISM. Therefore, the PN-ISM interaction should influence PN halos.

Among large asymmetric nebulae there are two relatively bright nebulae, A21 (YM29) and S188, located in the Galactic plane. They appear to be moving through dense ($n_0 \geq$ 1 cm$^{-3}$) ISM, which led to an extreme asymmetry, stellar displacements away from the nebular center, and to the creation of a small H II region behind A21 (Weinberger 1989). Recent proper motion measurements by Pier et al. (these proceedings) firmly establish that the extreme asymmetry of A21 is indeed caused by the interaction with the ISM, because the ram-pressure compressed nebular section is located in the direction of the stellar motion. (The same conclusion follows from Pier et al. measurements of the proper motion of the central star of another asymmetric PN A31.) These results demonstrate that proper motion measurements are very useful in studying the interaction of PNe with the ISM.

## 3. Summary

Theoretical work on interacting PNe has been limited. The interaction of a moving PN has been considered in the framework of a thin-shell approximation (Smith 1976; Isaacman

1979). Soker et al. (1991) performed numerical calculations and refined the analytic theory, finding satisfactory agreement with the thin-shell approximation. The thin-shell approximation breaks down for fast-moving nebulae such as the M22 PN, where the cooling time scale for the shocked ISM is longer than the flow time scale. In this case, Soker et al. (1991) found that the Rayleigh-Taylor instability is likely to fragment the nebular shell into knots, a result which has been confirmed by observations (Fig. 1a).

Available observational data on the interacting nebulae reveal varied and complex dynamical processes which are not explained by the current theory. There is ample evidence for instabilities: linear streaks in A35 and A31, a conical filament and knots with tails in NGC 7293 (Fig. 1b), and strongly filamentary nebulae A21 and S188. More theoretical work is also needed to understand the structure of bow shocks in A35 and the PN 623+71, the interaction of ionization-bounded nebulae such as S216 (Reynolds 1985), and the time-dependent evolution of miniature H II regions such as seen behind A21.

The current developments have already led to interesting results, which should encourage further studies of interacting PNe. The evolution of old nebulae and of PN halos is commonly affected and may even be governed by the interaction. The interaction of PNe with the ISM is a useful probe of the ISM. With a sufficient effort, it should be possible to determine the filling fraction of the hot ISM phase where the interaction is much weaker than in the warm ISM. For nearby PNe, proper motions of central stars would be very helpful in this respect. Significant improvements are now possible over the last systematic study by Cudworth (1974), as shown by Pier et al. at this conference.

## References

Arp, H., & Scargle, J. D.: 1967, *ApJ* **150**, 707
Balick, B., Gonzalez, G., & Frank, A.: 1992, *ApJ* **392**, 582
Borkowski, K. J., Sarazin, C. L., & Soker, N.: 1990, *ApJ* **360**, 173
Borkowski, K. J., Tsvetanov, Z., & Harrington, J. P.: 1992, *ApJ*, in press
Bowers, P. F., & Knapp, G. R.: 1988, *ApJ* **332**, 299
Chu, Y.-H., Manchado, A., Jacoby, G. H., & Kwitter, K. B.: 1991, *ApJ* **376**, 150
Cudworth, K. M.: 1974, *AJ* **79**, 1384
Cudworth, K. M.: 1986, *AJ* **92**, 348
Dickey, J. M., & Lockman, F. J.: 1990, *ARA&A* **28**, 215
Frank, A., Balick, B., & Riley, J.: 1990, *AJ* **100**, 1903
Gieseking, F., Hippelein, H., & Weinberger, R.: 1986, *A&A* **156**, 101
Hippelein, H., & Weinberger, R.: 1990, *A&A* **232**, 129
Hollis, J. M., Oliversen, R. J., Wagner, R. M., & Feibelman, W. A.,: 1992, *ApJ* **393**, 217
Isaacman, R.: 1979, *A&A* **77**, 327
Jacoby, G. H.: 1981, *ApJ* **244**, 903
Krautter, J., Klaas, U., & Radons, G.: 1987, *A&A* **181**, 373
Malin, D. F.: 1982, *Sky and Tel.* **63**, 22
Nordgren, T., Cordes, J., & Terzian, Y.: 1992, *AJ*, in press
Reynolds, R. J.: 1985, *ApJ* **288**, 622
Reynolds, R. J.: 1991a, in IAU Symposium No. 144, The Interstellar Disk/Halo Connection in Galaxies, ed(s)., H. Bloemen, Dordrecht: Kluwer, 67
Reynolds, R. J.: 1991b, *ApJ* **372**, L17
Smith, H.: 1976, *MNRAS* **175**, 419
Soker, N., Borkowski, K. J., & Sarazin, C. L.: 1991, *AJ* **102**, 1381
Walsh, J. R., & Meaburn, J.: 1987, *MNRAS* **224**, 885
Weinberger, R.: 1989, in Reviews in Modern Astronomy 2, ed(s)., G. Klare, Springler-Verlag: Berlin, 167
Zucker, D. B., & Soker, N.: 1992, *ApJ*, submitted

# SPINDLES, SPHERES AND A FEW JETS: THE RADIATION GASDYNAMICS OF PLANETARY NEBULAE

ADAM FRANK

*Astronomy Department, FM-20, University of Washington, Seattle WA 98195, U.S.A.*

ABSTRACT. We present the results of accurate numerical radiation-gasdynamic models for spherical (1-d) and aspherical (2-d) planetary nebulae (PNe). The models confirm and surpass previous efforts by recovering the detailed morphologies, kinematics, and ionization structure of real PNe. Our 1-d simulations include the evolution of the central star and a superwind/AGB wind interaction. The importance of the forward wind-blown shock and time-dependant ionization in redistributing mass in the nebulae is demonstrated. Our 2-d simulations, the first of their kind, confirm that aspherical PNe can develop through the interaction of a spherical fast wind with an aspherical superwind. We demonstrate that the full range of nebular morphologies can be recovered by considering the combination of wind/wind interactions, the ionization state of the gas and the projection angle on the sky

## 1. Introduction

The last decade has seen a number of 1 and 2-d interacting wind PNe simulations of various degrees of sophistication. These models have been successful in demonstrating the the role of interacting stellar winds in the formation of PNe (see Marten and Schoenberner (1991) and Soker and Livio (1989) for the best examples of the work done so far). Previously published models have, however, suffered from a less than accurate treatment of the radiation or the gasdynamics or both. In an effort to produce models which can be compared, in detail, with observations of the morphology, kinematics, and ionization structure of real PNe we have constructed a new radiation-gasdynamic code which simulates the evolution of interacting wind PNe. Our code uses an accurate gasdynamic numerical method. We can simulate the interaction of the winds without appeal to artificial boundary conditions and resolve all the important nonlinear waves. In addition we have included the majority of important time-dependant radiation transfer and microphysical processes in a self-consistent way allowing us to model the details of the nebular ionization, heating and cooling. A natural consequence of this treatment is the production of detailed numerical "observations" of the model for comparison with what is seen on the sky.

## 2. Spherically symmetric models

Our models begin with a '3-wind' interaction similar to that used by Marteen and Schoenberner (1991). A fast wind projects into a superwind which is already expanding

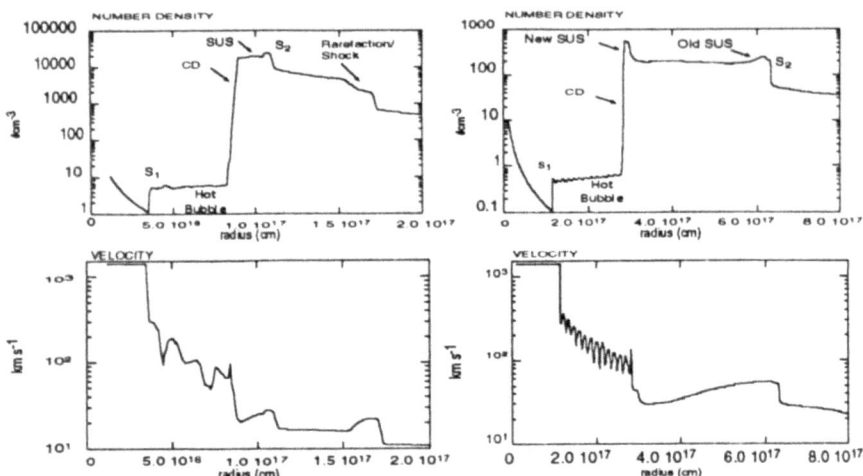

Fig 1. Density and Velocity during two phases of model PNe evolution. The frames on the left are taken from the Hydrogen Ionization Phase. The frames on the right are taken from the Shock Break-Out Phase.

into an AGB wind. The interaction of the winds occurs in the presence of a central star evolving along a .644 $M_\odot$ Schoenberner track.

We find the evolution of the model is determined by the ionization state of the gas and the evolution of the principle gasdynamic waves. Fig 1. shows the density and velocity in what we have called the Hydrogen Ionization (HI) and Shock Break-Out (SBO) evolutionary phases (in general the model evolution can be broken up into 4 distinct phases, Frank (1992)). Examination of the density in the HI phase demonstrates the codes' ability to resolve the classical wind interaction features: an inner shock ($S_1$), hot bubble, contact discontinuity (CD), swept-up shell (SUS), and an outer or forward shock ($S_2$). We also observe a rarefaction and shock wave occurring at the superwind/AGB wind boundary driven by an isothermal (due to ionization) pressure gradient. The velocity shows these same features as well as numerical noise in the hot bubble region. Eventually the forward shock will reach and accelerate through the superwind/AGB wind boundary. When this occurs the dynamics of the model changes appreciably and we enter the SBO phase. The density and pressure in the swept-up gas decrease in response to the acceleration allowing the hot bubble to drive a new shock into the superwind. A two shell morphology, clearly apparent in fig 1, develops and is maintained until the end of the simulations at 8000 y.

The observation of the SBO phase in model PNe is new and demonstrates the importance of resolving the energy conserving hot bubble, the isothermal forward shock and the swept up shell. The forward shock and SUS are crucial to modeling the morphological and kinematic evolution of PNe. Fig 2 shows the H$\alpha$ intensity profiles for two real PNe and for the model HI and SBO phases. The model is successful in recovering the familiar double peak profile as well as the linear shells, crowns and sharp edges investigated in Frank, Balick, and Riley (1990). Synthetic kinematical images demonstrate that the model successfully

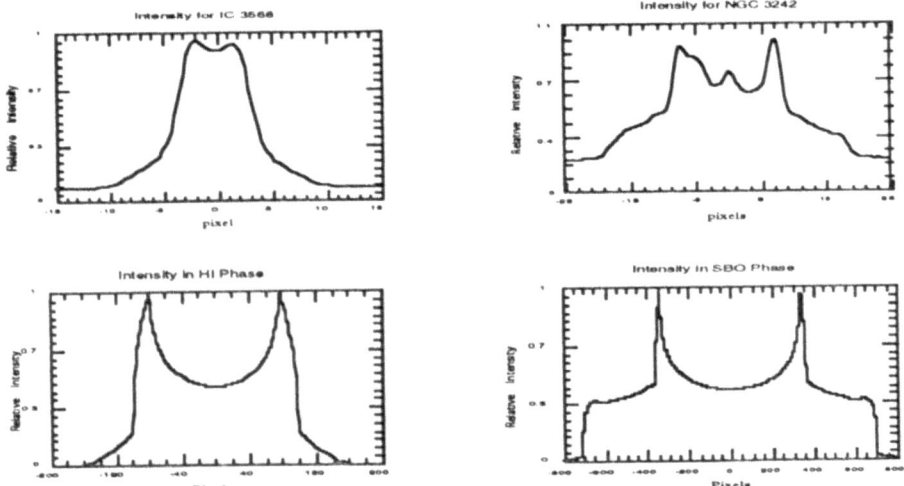

Fig 2. Hα intensity profiles for two real PNe and during two phases of model evolution. Note that the model intensity profiles have not been smoothed to account for seeing.

predicts the velocity patterns in real PNe. With the SBO phase we can embrace the full range of attached multiple shell PNe kinematical classifications enumerated by Chu et al (1989). Finally, comparison of model intensity profiles for emission lines of various ions (HeII, NII, OIII) with data show that the code is doing a good job of recreating the microphysical state of the nebular gas. In particular the models predict HeII/HeIII ionization fronts trapped in SUS for a significant fraction of the evolution of the nebulae. Many PNe show these kinds of fronts (Balick 1987).

## 3. Aspherical models

Beginning with the assumption of an aspherical superwind Mellema, Eulerink, and Icke (1991) and Icke, Balick, and Frank (1992) have performed accurate simulations of the adiabatic interaction of stellar winds in 2-d. Their results confirm that elliptical and bipolar PNe can be embraced by this generalized interacting winds scenario. They have also observed new, gasdynamical flow patterns of great potential significance to the study of PNe, these include flow focusing via an elliptical inner shock and an effective mechanism for the production of jets (Icke *et al* 1992).

In order to explore more realistic scenarios and to make more detailed contact with observations we have used a 2-d version of our radiation-gasdynamic code and calculated a number of PNe models (Frank 1992, Frank *et al* 1992). Our results demonstrate that the generalized interacting wind scenario, when combined with the effects of radiation losses and projection effects, can account for the full range of PNe morphologies. As with real PNe, the morphologies which develop in our models range from mildly elliptical, through

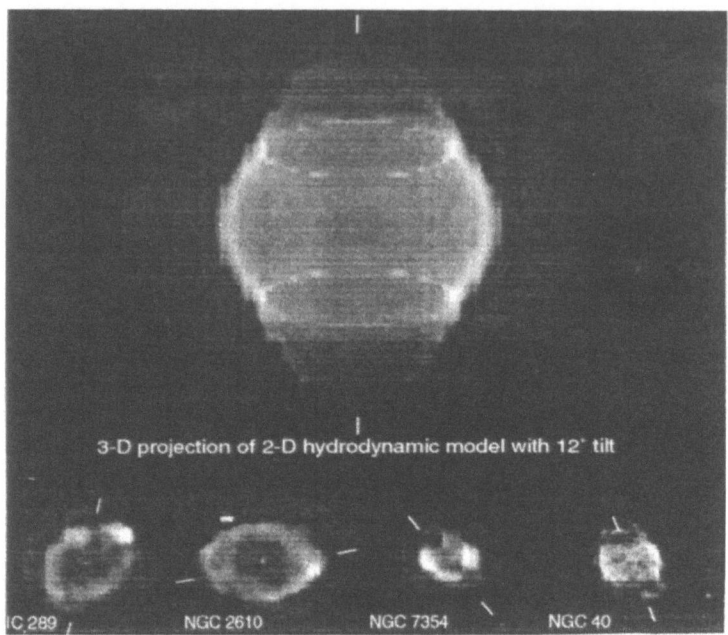

Fig 3. Hα intensity map for 2-d model inclined inclined 12° from the plane of the sky. Also shown are Hα intensity maps for 4 PNe with similar morphologies.

spindle-like bipolars, to true peanut shaped bipolars. Fig 3 shows an Hα intensity map for a 900 year old bipolar model inclined 12° from the plane of the sky. Below the model we show a number of real PNe with similar morphologies. We are in the process of calculating kinematic images of these models for more detailed comparisons with the data. The next step is to perform highly structure comparisons of the models with individual PNe.

**4. References**
Balick, B., (1987), " The Shapes and Shaping of PNe", *A.J.*, **94**, 671
Chu, Y.H., (1989), "Multiple Shell PNe", *Proceedings of IAU Symposium No. 131*, ed Torres-Peimbert,S., (Kluwer, Dordrecht)
Frank, A., (1992), "Radiation-Gasdynamics of PNe", Thesis, University of Washington
Frank, A., Balick, B. and Riley, J., (1990), "Stellar Wind Paleontology", *A.J.*, **100**, 1903
Frank, A, Balick, B, Icke, I., and Mellema, G, (1992), "Radiation-Gasdynamics of Aspherical PNe", in preparation
Icke, I., Mellema, G., Balick, B., Eulerink, F., and Frank, A., (1992), "Collimation of Astrophysical Jets by Interial Confinement", *Nature*, **355**
Icke, I., Balick, B., and Frank, A., (1992), "The Hydrodynamics of Aspherical PNe II",*A.A.*, **253**, 224
Mellema, G., Eulerink, F., and Icke, I.,(1992), "Hydrodynamical Models of Aspherical PNe", *A.A.*, **252**, 718
Marten, H., and Schonberner, D., (1991), "On the Dynamical Evolution of PNe", *A&A*, **248**, 590
Soker, N., and Livio, N., (1989), "Interacting Winds and the Shaping of PNe", *Ap.J.*, **339**, 268

# DYNAMICAL STRUCTURES OF PLANETARY NEBULAE - MODELS AGAINST OBSERVATIONS

H. MARTEN[1], K. GĘSICKI[2] AND R. SZCZERBA[2]

[1] *Institut für Theoretische Physik und Sternwarte der Universität Kiel*
*Olshausenstr. 40, D-2300 Kiel, Germany*
[2] *Copernicus Astronomical Center, Laboratory of Astrophysics*
*ul. Chopina 12/18, 87-100 Toruń, Poland*

**Abstract.** Hydrodynamical calculations of Planetary Nebulae (PNe) over 25000 yrs of evolution which include timedependent effects of ionization as well as variable central stars winds and parameters consistent with stellar model calculations show a great variety of velocity and density structures which strongly deviate from the often assumed "homogeneous shells, expanding with constant velocities" (Marten & Schönberner, 1991, hereafter *MS*). By means of a static photoionization code we calculate the surface brightness in the 10 most prominent nebular emission lines for density structures of the model sequence "VS" as given by *MS*. The obtained radial emissivities are used together with the velocity and temperature structure of the ionized gas to calculate (thermally broadened) line profiles in order to derive "measured" expansion velocities. We compare the theoretical surface brightnesses with observations and demonstrate some difficulties in the interpretation of nebular expansion velocities, expansion distances and ages.

## 1. Detailed Model Analysis

The first nebular phase is characterized by completely ionization bounded models. Their radial structure is dominated by an ionization front which drives a density front into the neutral material. A typical velocity structure of a model during this "Ionization phase" is shown in Fig. 1a. The pressure of newly ionized matter accelerates the outer material near the ionization front and decelerates the inner nebular parts. However, an absolute expansion inwards is prohibited by the pressure of a hot bubble. Fig. 1b presents the corresponding normalized line profiles in [OII] $\lambda$ 3729 and [OIII] $\lambda$ 5007, calculated for an aperture with a diameter of 20% of the inner nebular rim which is placed at the PN center, so that we obtain double-peaked profiles. The peak separation which is often used as a measure of twice the expansion velocity is also given in Fig. 1b. It is obvious that lines from different ionization stages of the same element can be used to read off a velocity structure during this early phase of evolution, because these nebulae show clearly separated ionization zones: The emission in [OIII] mainly comes from the inner, low-density/low-velocity regions, while [OII] is generated in the higher-density, faster material farther outside (for the density structures of ionization bounded models see Fig. 6 of *MS*). Consequently, these PNe are also somewhat smaller in [OIII] than in [OII]. Furthermore, it is worth to notice that the measured peak separation can only give a *mean* velocity within the main emission region of the respective line, but at no time the minimum (5 km/sec) or maximum (25 km/sec) matter velocity.

As long as the star evolves with a constant luminosity, the growing energy input into the hot bubble dominates over its expansion cooling and causes an effective nebular compression. During this second, "Compression phase", the growing bubble pressure creates an inner region of high density while the ram pressure of the slow moving AGB-wind prevents the nebula from expanding too fast (Fig. 2a).

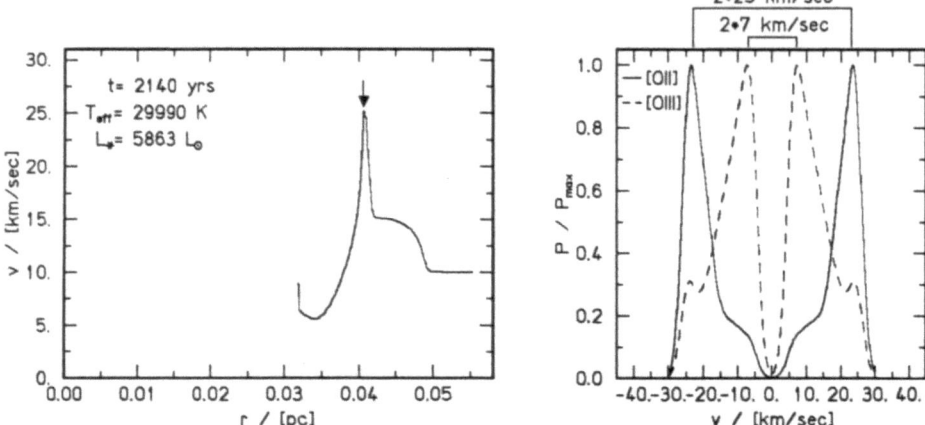

Fig. 1. Left panel (1a): Typical velocity structure of a nebular model according to *MS* during the "Ionization phase". The ionization front is marked by an arrow, model parameters are given in the upper left corner. Right panel (1b): Normalized line profiles in [OII] $\lambda$ 3729 and [OIII] $\lambda$ 5007 for the same model as measured through the nebular center (see text). The line intensities relative to $H_\beta = 100$ are 16.8 for [OII] and 11.1 for [OIII], respectively.

Fig. 2b presents the corresponding normalized surface brightnesses in the doublett [OII] $\lambda\lambda$ 3727,29 and in [OIII] $\lambda$ 5007. Models during the "Compression phase" show a typical, often observed two-shell structure (Balick, 1987; Frank et al., 1990; Balick, 1992; Frank, 1992). The observed inner bright shell can be identified with the high-density, compressed nebular material, while the faint outer shell (with a sharp outer rim) is generated by the outer, low density nebular matter expanding into slow moving wind material. The image in [OII] shows a limb-brightening which has been called a "crown" by Franck et al. (1990). Fig. 2b demonstrates that the question of whether we observe a "crown" in an emission line or not is not only determined by the density but also by the detailed ionization structure of the nebula.

When the stellar luminosity drops, the mechanical wind power decreases by about three orders of magnitude within a short time. The work done by adiabatic expansion then dominates the energetics of the hot bubble, rapidly decreasing its pressure. Consequently, the velocity of the inner PN material decreases and the nebula enters into its "Late phase" where the relative shell thickness becomes larger again. The surface brightnesses in different nebular lines show more and more "centrally filled" objects which seem to have no central hole, like the (very) old planetaries PW 1, A 16 or A 30 .

## 2. Expansion velocities, Ages and Distances

*MS* found that $R_N$, the (outer) radius at 10% of the maximum surface brightness in $H_\beta$, is a rather good definition for the outer nebular rim. Therefore, we define the "nebular expansion velocity", $v(R_N)$, to be the radius change of the $H_\beta$-image per unit time: $v(R_N) = dR_N/dt$. This velocity is shown in Fig. 3, together with the maximum matter velocity, $v_{max}$, as well as the velocity at the maximum density,

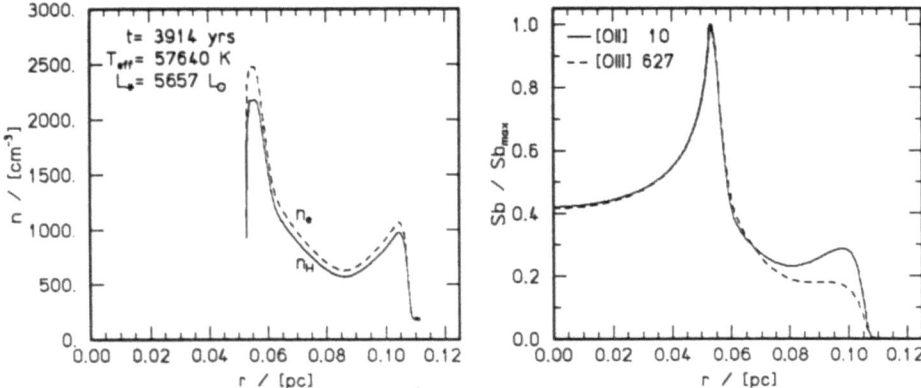

Fig. 2. Left panel (2a): Density structure of a nebular model according to MS during the "Compression phase". The model parameters are given in the upper left corner. Right panel (2b): Normalized surface brightness in [OII] $\lambda\lambda$ 3727,29 and [OIII] $\lambda$ 5007 for the same model. Total line intensities (relative to $H_\beta = 100$) are given in the upper left corner.

$v(\rho_{max})$, in the resp. shells of the whole model sequence. The conspicuous discrepancy between $v_{max}$ and $v(\rho_{max})$ around $t_{evol} = 3500$ yrs arises from the fact that our models during the early "Compression phase" show the largest velocities near the outer rim, where the densities are smallest. The contribution of these regions to the emission lines is so small that $v_{max}$ can neither be determined from the HWHM nor from the peak separation of the line profiles. As the growing bubble pressure further compresses the nebula and the velocity gradients within the shell become less steep, the situation becomes better again. However, Fig. 3 also shows that the maximum measurable velocity in most cases is significantly smaller than the above defined nebular expansion velocity, since $v(R_N)$ represents the velocity of a *shock*! As long as the nebula is optically thick, the ionization front is trapped in a density front and the shock velocity is similar to the shown matter velocities. When the nebula becomes optically thin, the outer rim expands into pre-ionized AGB matter and the corresponding shock becomes faster than the post-shock matter. On the other hand, our models also warn us against a generalization of this result, since short evolutionary phases where $v(R_N)$ becomes even smaller than $v_{max}$ cannot be ruled out in general. This may especially be the case when the maximum matter velocity is determined by a very high pressure of the hot bubble while at the same time $v(R_N)$ is determined by a relatively small ratio of the nebular to the AGB-wind density. In our sequence, $v(R_N)$ becomes comparable to $v_{max}$ soon after the pressure of the hot bubble has decreased ($t_{evol} \approx 10000$ yrs) and it takes a further few 1000 yrs until the fast and dense matter from the inner rim has reached the outer rim, again increasing the shock velocity $v(R_N)$.

In some cases, the angular expansion of the nebular image per year together with a velocity derived from line profiles is used to derive expansion distances (Terzian, 1992). In our model sequence, this would lead to a systematic underestimation of the distance and an overestimation of the expansion age, $t_{exp}$, during most of the

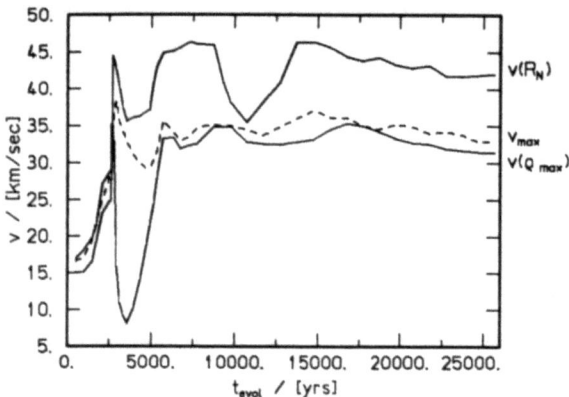

Fig. 3. Expansion velocity of the $H_\beta$-image, $v(R_N)$, maximum matter velocity, $v_{max}$, and matter velocity at the maximum density, $v(\rho_{max})$, within the PNe shells as a function of the model age.

nebular evolution, since the measurable matter velocities are too small compared to the expansion velocities. For example, for our final model at $t_{evol} = 25000$ yrs ($R_N \approx 1$ pc), we obtain $t_{exp} \approx 30000$ yrs from $v_{max} = 33$ km/sec.

## 3. Conclusions

The evolution of planetaries is a result of the competition between the pressure of a bubble of hot, shocked stellar wind gas, the pressure of the ionized nebular region and the ram pressure of the slowly moving AGB matter. During the whole lifetime of a PN, its structure is altered several times, so that it is impossible to derive a mass-loss history only from the present, "observed" radial density and velocity distribution. Dynamical calculations seem to be the only possibility to learn something about the past and future of a today observed object. For example, from these calculations it turns out that two shell planetaries can be explained *without* the need of a double mass-loss event. The analysis of dynamical PN calculations furthermore showed that expansion velocities and expansion distances might be systematically underestimated, while expansion ages of old planetaries are likely to be overestimated, apart from the difficulty that the observed nebular velocities are not constant, but are correlated with the stellar temperature (Heap, 1992). This is confirmed by dynamical calculations as well.

Balick, B., 1987, AJ 94, 671
Balick, B., 1992, these proceedings
Frank, A., 1992, these proceedings
Frank, A., Balick, B., Riley, J., 1990, AJ 100, 1903
Heap, S., 1992, these proceedings
Marten, H., Schönberner, D., 1991, A&A 248, 590
Terzian, Y., 1992, these proceedings

# MASS-LOSING AGB STARS IN THE MAGELLANIC CLOUDS

NEILL REID

*California Institute of Technology*

Asymptotic giant branch stars are the immediate precursors to the planetary nebula stage of stellar evolution. It is clear that the latter stages of a stars life on the AGB are accompanied by either continuous or episodic mass-loss, with the final convulsion being the ejection of the envelope (the future planetary shell), the gradual exposure of the bare CO core and the rapid horizontal evolution to the blue in the H-R diagram. Thus, the structure of the planetary nebula luminosity function, particularly at the higher luminosities (although this phase is extremely rapid), is intimately tied to the luminosity function of the AGB.

The exact point of termination of the AGB phase of evolution, however, remains a matter of some debate, with the Magellanic Clouds providing the debating hall. Clearly some AGB stars achieve luminosities close to the canonical theoretical limit of $M_{bol} = -7.1$ (Wood, Bessell & Fox (1983)), but, as is well known, the numbers fall short of the predictions. Several mechanisms have been proposed to account for this deficit, with the current favourites either being the intervention of high mass-loss rates terminating evolution at $M_{bol} \sim -5.5$ to -6, or the presence of efficient envelope convection in the more massive stars leading to more rapid evolution beyond $M_{bol} = -7.1$ (Blocker & Schonberner, 1991). (Note, however, that no AGB stars have been observed at such high luminosities (P.R. Wood et al., preprint)). We have used IRAS observations to search for dusty AGB stars, discovering ten optically-invisible 'cocoon' stars (Reid, 1991) in a 15-square degree region of the northern LMC. We have extended this survey to cover the entire LMC, the SMC and six dwarf systems. However, although we identify sixteen new candidates in the LMC (and none in any of the other galaxies), none have striking IR properties - all are consistent with $M_{bol} \sim -5.5$ AGB stars undergoing mass loss at rates of $\sim 10^{-6} M_\odot yr^{-1}$. We have, however, potential identifications of a further 20 LMC planetary nebulae. Further observations are required to confirm the identification and characteristics of these objects.

This research was supported partially by a NASA IRAS ADP grant

## References

Blocker, T. & Schonberner, D. 1991, *Astr. Astrophys.*, **244**, L43
Reid, I.N., 1991, *Astrophys. J.*, **382**, 143
Wood, P.R., Bessell, M.S. & Fox, M.W., 1983, *Astrophys. J.*, **272**, 99

# 10μm IMAGES OF AGB STARS & SUPERGIANTS

C. J. SKINNER, G. HAWKINS and M. M. MEIXNER

*Institute of Geophysics & Planetary Physics, L-413, Lawrence Livermore National Laboratory, P.O.Box 808, Livermore, CA94551-9900, U.S.A.*

and

J. G. JERNIGAN and J. F. ARENS

*Space Sciences Laboratory, University of California, Berkeley.*

We present mid-IR and far-IR images of a variety of AGB stars and red supergiants. The mid-IR images were all taken with the Berkeley/Livermore mid-IR array camera, which employs a 10×64 pixel Hughes photoconductor. All the images reported here were taken using a 10% bandpass CVF, at various wavelengths in the 10μm atmospheric window. These were supplemented by *IRAS* images, some constructed from survey scans, others made as Additional Observations during the pointed phase of the *IRAS* mission. We have so far observed 11 such sources with our mid-IR camera, and report here that only two of them (R Aql and V Hya) appear to be unresolved.

The red supergiants α Ori and μ Cep were both pointlike at 8.2μm and 8.5μm, but extended at longer wavelengths due to emission in the silicate bands (which peak around 9.7μm and 18μm ). Neither source is spherically symetric. α Ori shows general extension, but this is accentuated to the NE at wavelengths of 9.7μm , 12.5μm , 60μm and 100μm . μ Cep shows a remarkable elongation in the E-W direction, both in our mid-IR images at wavelengths of 9.7μm , 11.3μm and 12.5μm , and in KAO observations we have made at 50μm and 100μm . On the other hand, NML Cyg, another red supergiant, which has an optically thick dust shell (unlike the previous two sources), appears to exhibit symmetry in its dustshell, which is elliptical at 9.7μm , the major axis being roughly in the EW direction.

The oxygen-rich Mira R Cas is marginally extended in the ESE direction in our images at 10.0μm , and pointlike at 8.5μm . The *IRAS* AOs reveal an extraordinarily elongated image, with the elongation in the ESE direction, almost aligned with our mid-IR images.

The carbon star IRC+10216 shows similar asymmetries at wavelengths of 10.0μm 60μm and 100μm . However, the asymmetries are in different directions at different wavelengths. The more extreme C-star AFGL3068 is also extended, but somewhat smaller; in this case we find no evidence for anything other than spehrical symmetry.

Thus, in the majority of our sources, mass-loss appears to occur in a preferred direction, which does not alter on the timescale reflected in the 100μm observations (a few millenia). IRC+10216 also appears to lose mass preferentially in certain directions, but the direction appears to vary on a timescale of a few hundred years. We have no explanation for why mass should be lost preferentially in a singular direction. Neither do these observations help to clarify the appearance of bipolar symmetry in some PN.

# MID-IR SPECTRA OF AGB AND POST-AGB STARS

C. J. SKINNER

*Institute of Geophysics & Planetary Physics, L-413, Lawrence Livermore National Laboratory, P.O.Box 808, Livermore, CA94551-9900, U.S.A.*

and

M. J. BARLOW, K. JUSTTANONT and R. J. SYLVESTER

*Dept. of Physics & Astronomy, University College London, Gower Street, London WC1E 6BT, Great Britain*

We have generated models of complete nova energy distributions, including the effects of the nova dust shell emission and absorption. In order to account for the dust, we require optical constants throughout the UV and EUV. Draine & Lee (1984) is the only source of such data for silicates. However, their 'astronomical silicate' has a strong resonance in the UV which causes circumnova dust to be heated far above its sublimation temperature in our models of Nova Her 1991. We have therefore recalculated the silicate optical constants, using the Kramers-Kronig relations, including a $\lambda^{-1.0}$ far-IR dependence and improved 9.7$\mu$m feature, and decreasing the UV and EUV efficiencies of the grains. With these new silicate constants we can reproduce the observed behaviour of the dust shell in Nova Her 1991. Since our optical constants are consistent with the behaviour of Nova Her 1991, an EUV source, they should also be appropriate for the treatment of PN.

The 10$\mu$m silicate feature in Nova Her is most unusual. Instead of peaking near 9.7$\mu$m, as in most O-rich AGB stars, the feature is very broad with a flat top, extending from $\sim$ 9.7$\mu$m to 11.5$\mu$m. We observe such behaviour in silicate features in several PPN. In the case of AFGL2343 and HD161796 we ascribed the broad, rather flat silicate profiles to their cool dust shells (Justannont et al., 1992). A third post-AGB source, OH53.8+20.2, demonstrates this behaviour particularly strongly. Contrarily, a fourth post-AGB star of the same type, V1027 Cyg, shows a remarkably strong 9.7$\mu$m silicate feature which strongly resembles that seen in AGB stars, typified by the strong feature in the well known M-Supergiant $\mu$ Cep. The three post-AGB sources with unusual silicate profiles, AFGL2343, HD161796 and OH53.8+20.2, all have rather cool dust shells, indicating that these stars left the AGB long ago. On the other hand V1027 Cyg has a much hotter energy distribution, indicating that it left the AGB only recently. We suggest that the unusual silicate profiles have arisen because the warm central stars have annealed the dust in the time since these stars left the AGB. Similarly, in Nova Her 1991 the dust forms in and is exposed continuously to an exceptionally harsh radiation field.

## References

Draine, B. T., Lee, H-M.: 1984, 'Optical Properties of Interstellar Graphite and Silicate Grains', *Ap. J.* **285**, 89

Justtanont, K., Barlow, M. J., Skinner, C. J., Tielens, A. G. G. M.: 1992, 'The Nature of Dust around the Post-asymptotic Giant Branch Objects HD161796 and HD179821', *Ap. J.* **392**, L75

# COMPARATIVE ANALYSIS MIRAS/PPN

D. BARTHES, M.O. MENNESSIER and F. GLEIZES

*GRAAL, CNRS and Univ. Montpellier II, F 34095 Montpellier Cedex 5, France*

and

A. LÈBRE

*DASGAL, Obs. Paris-Meudon, F 92195 Meudon Principal Cedex, France*

The spectrum of R Sct, a star firstly classified as an RV Tau one and located in the highest part of the instability strip, indicates a Na ejected shell at about 1000 $R_*$ with an expansion velocity close to 50 km/s (Lèbre and Gillet, 1991). This confirms the post AGB character of this star as already mentionned by Jura (1986). So it is interesting to compare the R Sct with some "classical" RV Tauri (AC Her and U Mon) and some "exceptional" miras (R Cen, U CMi and R Nor) with light curves exhibiting a double maximum and with spectral types earlier than those of the other miras with similar periods.

Such a comparison using the power spectra of their light curves gives the following results:

- R Sct and the three peculiar miras are pulsating on the fundamental mode with resonance with the first overtone, instead of the classical miras. This is indicated by the agreement between the power spectra and the linear nonadiabatic models of Tuchman et al.(1992)
 $(0.5 < M_*/M_\odot < 0.8; 10^3 < L_*/L_\odot < 10^4)$.
- The RV Tau stars AC Her and U Mon are pulsating on the first overtone with resonance with the fundamental mode.

This is a possible indication that R Cen, R Nor and U CMi are leaving the Mira stage after a mode switching from the first overtone to the fundamental mode. R Sct is confirmed as a post-AGB star and probably a PPN.

## References

Jura M., 1986, Ap.J., 309, 732
Lèbre A., Gillet D., 1991 A&A, 251, 549
Tuchman Y., Lèbre A., Mennessier M.O., Yarri A., A&A, to be submitted

# MECHANISMS FOR RADIO CONTINUUM EMISSION OF LONG-PERIOD VARIABLE STARS

G. M. RUDNITSKIJ

*Sternberg Astronomical Institute Moscow State University Moscow V-234, 119899 Russia*

The radio continuum emission of long-period variables (LPVs) in the continuum is considered. For some LPVs (e.g., R Aql and V Hya) weak emission, observed at centimeter wavelengths, can be explained by thermal free-free radiation of the ionized gas behind the shock front in the stellar atmosphere. The ionized layer behind the shock remains optically thick at the wavelengths as short as 1 cm; the brightness temperature found for R Aql at 2 cm by Drake et al. (1987, AJ, 94, 1280), $T_b=18000$ K, is consistent with the expected temperature behind the shock.

For R Aql, at least two nonthermal radio continuum events (with flux densities up to 250 mJy) were recorded - in 1973 by Woodsworth and Hughes (1973, Nature, Phys. Sci., 246, 111) and in 1982 by Estalella et al. (1983, A&Ap, 124, 309) I suggest as an explanation for these events synchrotron radiation by electrons accelerated at the shock front. The conditions for the shock acceleration in an LPV's atmosphere are discussed.

Amplification of radio continuum emission (both thermal and nonthermal) by masering molecules (OH, $H_2O$) in the circumstellar envelope may account also for the light-curve-correlated variability of circumstellar masers.

# HISTORY OF THE LIGHT CURVES AND MOLECULAR MASER EMISSION OF THE MIRAS U ORI AND R LEO

I. L. ANDRONOV and L. S. KUDASHKINA

*Odessa Astronomical Observatory Shevchenko Park Odessa, 270014 Ukraine*

and

G. M. RUDNITSKIJ

*Sternberg Astronomical Institute Moscow State University Moscow V-234, 119899 Russia*

We have collected all the available data on light curves, OH, $H_2O$ and SiO maser observations for a sample of Mira-type variables. We consider in detail the data on two stars, U Ori and R Leo. There is a net correlation between optical and radio line variations for all the three molecular species in these stars. More pronounced maser flares seem to follow brighter-than- average visual maxima of the stars. We discuss also the drastic changes in the type of the OH maser radio emission which happened in these stars some years ago. Implications for the mechanisms of maser pumping and the evolutionary status of these stars (probably undergoing the helium flash) are discussed.

# ON THE POSSIBLE RELATIONSHIP BETWEEN THE PHOTOMETRIC PARAMETERS OF THE AGB STARS AND THEIR EVOLUTIONARY STATUS

L.S. KUDASHKINA and I.L. ANDRONOV
*Department of Astronomy, Odessa State University,*
*T.G.Shevchenko Park, Odessa 270014 Ukraine*

The statistical study of the Long-Period Variables, provided by various authors, show, that there is a relationship between the photometric characteristics and the physical properties. For example, the asymmetry of the light curve may be related with the strength of the shock wave. The presence of the maser-fed line emission characterizes the definite evolutionary stage of the Long-Period Variable.

In this work, we present the results on the Mira - type stars with the periods $P < 600$ days, which may be subdivided into 4 groups according to the value of P:

1) $P < 200$ days,
2) $200 < P < 300$ days,
3) $300 < P < 400$ days,
4) $P > 400$ days.

The stars from the groups 2 and 3 have the most stable light curves with the significant asymmetry, but the stars from the group 3 are often observed as the maser sources.

The following photometric characteristics are studied: the amplitude, the asymmetry, the variability of the maximum brightness, the period and its stability, the presence and the duration of the hump at the ascending branch of the light curve, as well as the spectral class. For the group of the Mira-type stars, the light curves published by the AAVSO members were digitized, and the best fit period and the corresponding Fourier coefficients were computed and converted into the amplitudes and phase differences. The optimum number of the harmonics varies in the range from 1 to 7 for the different stars.

One of the aims is to choose the criterium, which characterizes the evolutionary stage of the Long-Period Variable, and is based on the observational characteristics (photometry and polarimetry in the optical and IR bands, spectral properties, maser emission). The various correlations between the photometric parameters are studied for the individual stars (cycle-to-cycle variations), as well as for the groups.

# SPLINE FITS: MODELLING THE OBSERVATIONS

I.L. ANDRONOV
*Department of Astronomy, Odessa State University,
T.G.Shevchenko Park, Odessa 270014 Ukraine*

The main results on the elaboration of the algorithms and programs for the search of the 'hidden periodicities' in the 'unevenly spaced' data, are discussed. They are mainly based on the application of the cubic spline-functions practically for all purposed of the variable star research - period determination, approximation of the phase curve, search for the possible period changes and/or the shape of the phase curve, detection of the periodic components in the case of the multi-harmonic oscillations, restoration of the smoothed function by removing the 'apparat function', numerical integration etc. Contrary to the commonly used splines with the arguments of the basic points coinciding with that of the real observations, we use the smoothing by the spline with the number of the basic points m, which is smaller than the number of the observations n. The main expressions for the corresponding fitting curves and their statistical properties were published by Andronov (1988, 1989), as well as the FORTRAN programs for the fits and the integration (Andronov, 1986, 1987, respectively).
For the evaluation of the smoothed curves at the small computers, the method of the 'Running parabolae' is proposed, the corresponding programs for which are significantly smaller as compared with that applying the cubic spline - functions (Andronov, 1990). The statistical properties of the periodograms obtained for the moments of the 'characteristic times' only are discussed by Andronov (1988, 1991).

## References

Andronov I.L.: 1985, Preprint Ukr.Inst.Sci.-Tech.Res. (UkrNIINTI, Kiev) **131**, 38 pp.
Andronov I.L.: 1986, Preprint UkrNIINTI **358**, 64pp.
Andronov I.L.: 1987, Publ.Astr.Inst.Czechoslovakia **70**, 161
Andronov I.L.: 1988, Astron. Nachr. **309**, 121
Andronow I.L., 1989, Die Sterne **65**, 20
Andronov I.L.: 1990, Cinem.& Phys.Celest.Bodies (Kiev) **6**, 87
Andronov I.L.: 1991, Cinem.& Phys.Celest.Bodies (Kiev) **7**, 78

# NEAR IR-PHOTOMETRY OF SEMIREGULAR VARIABLES

F. KERSCHBAUM and J. HRON

*Institut für Astronomie, Univ. Wien, Türkenschanzstr. 17, A-1180 Wien, Austria*

During the past decade much effort has been spent also on observation and theoretical modelling of AGB stars. However, the Semiregulars have been almost neglected. Kerschbaum & Hron (1992) used GCVS- and IRAS-data to define samples of SRa and SRb stars. These were compared with optical Miras with respect to their pulsational properties, effective temperatures, mass loss rates, luminosities, scale heights and galactic volume densities.

260 stars of their sample have now new NIR-photometry. In Fig. 1 temperature and chemistry-dependent J-K colour index is ploted versus pulsational period. It confirms the results obtained from V-[12$\mu$m] in Kerschbaum & Hron (1992). By fitting combinations of two blackbodies to visual, the new JHKLM and IRAS-data estimations of typical temperatures as well as luminosities of photospheres and dustshells were made. The two main results are that the short period O-rich SRb's have relatively high photospheric temperatures and that the relative luminosities of the two fitted blackbodies correlete well with the IRAS-LRS subclasses.

## References

Feast, M.W. et al., 1982, MNRAS 201, 439

Kerschbaum, F., Hron, J., 1992, A&A, in press

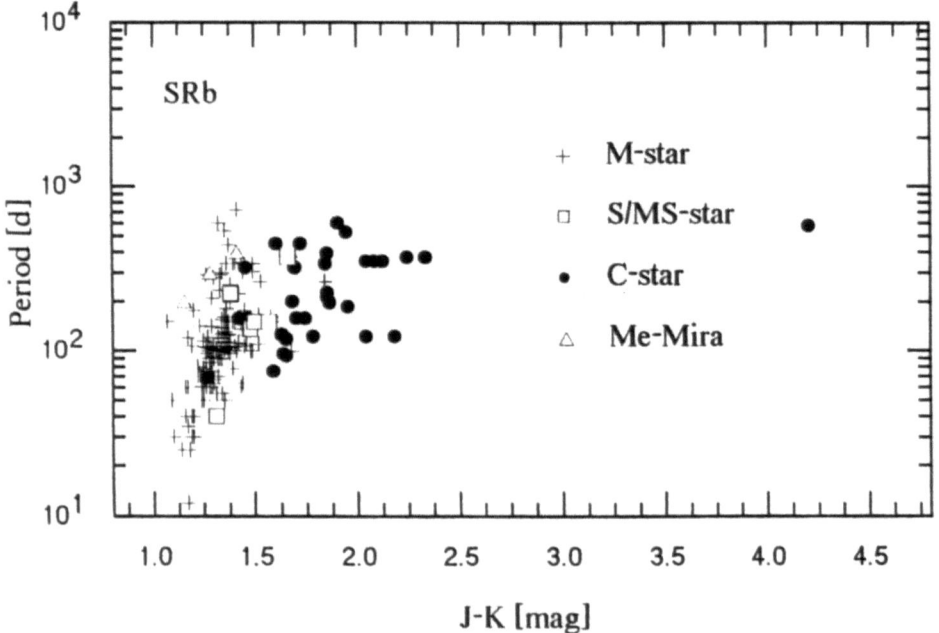

Fig. 1. Distribution of period as function of J-K Colour the SRb's. J-K colours of Me-Miras at periods of 200, 300 and 400 days are indicated (Feast et al. 1982).

# SPACE DISTRIBUTION OF SHORT PERIOD MIRA VARIABLES

J. HRON

*Institut für Astronomie der Universität Wien,*
*Türkenschanzstraße 17, A-1180 Wien, Austria*

The group of short period Mira variables (periods less than 200 days) is thought to be a mixture of metal poor stars and disk stars. We have compiled JHKL photometry for 159 stars in this period range, the great majority of the data coming from new observations. The photometry is used to separate metal poor from disk stars, following Hron (1991, A&A 252, 583), and to define volume limited subsamples. Distances were derived from the $K$-magnitudes and the period/$K$-luminosity relationship.

The mean distances from the galactic plane for the two metallicity groups confirm our photometric metallicity criteria (Fig. 1). We present first results on the scale heights and number densities of the metal poor and disk stars. These parameters were determined by two different methods which give slightly, but systematically, different values. The disk stars have a scale height comparable to that of Miras with longer periods. The scale height of the metal poor stars is about 1 kpc. Thus both the metallicity and scale height of this group of stars are typical for stars in the galactic thick disk, although the galactic kinematics apparently are different from thick disk stars. The value of the scale height and its independence from galactocentric distance strongly favour a flattened space distribution, rather than a spherical one.

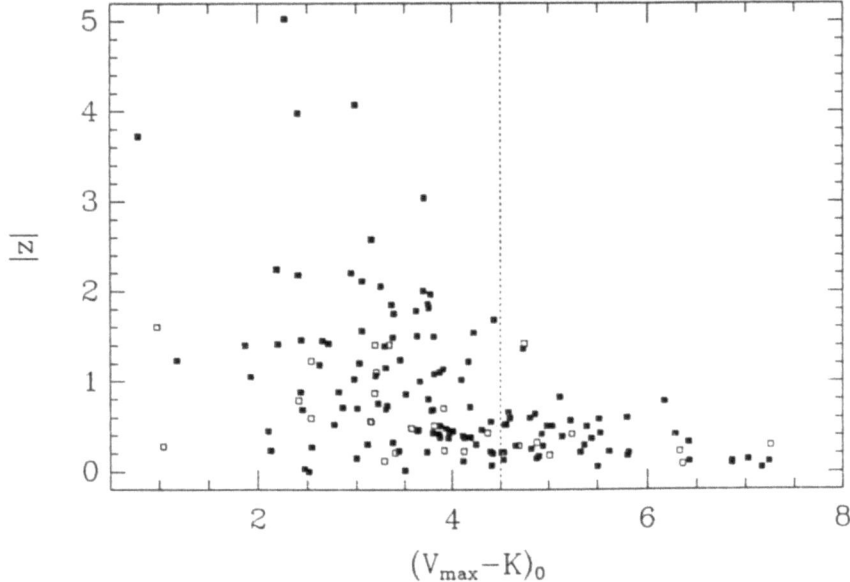

Fig. 1. $(V_{max} - K)_0$ versus distance from the galactic plane. Open squares are stars with $P < 140^d$.

# THE GALACTIC DISTRIBUTION OF O-RICH AGB STARS

B.W. JIANG and J.Y. HU
*Beijing Observatory, Academia Sinica, Beijing, 100080, China*

In 1976 Georgelin et al.(1976) combined optical and radio observational data of HII regions to study the structure of our Galaxy. The distance of each HII region can be obtained from its radio continuum flux caused by free-free emission and the flux of the Balmer lines with high accuracy. The results clearly show three arms and are used to draw the spiral structure of our Galaxy. It is commonly accepted now.

Habing(1988) pointed out that about 4/5 of all AGB stars are located in the galactic disk. There are some advantage using AGB stars as a tool of studying the galactic structure: 1. They are powerful infrared sources and mainly radiated in the infrared region. 2. The interstellar extinction in IRAS wavelengths can be neglected. and 3. IRAS completed a whole sky survey, and it is possible to select a complete sample from IRAS PSC.

A sample of AGB stars with O-rich circumstellar envelopes is selected from the IRAS PSC in the regions of IIIa, IIIb and IV defined by van der Veen and Habing(1988). The distance for each object is obtained assuming that the luminosities of O-rich AGB stars are identical. The spatial distribution of the sample shows a similar spiral structure presented by Georgelin et al.(1976). This confirms that the O-rich AGB stars are a disk population and that they are predominantly located in the arms. Fitting with the spiral structure of Georgeling et al. we derive that the average luminosity of the objects in the sample is $8300L_o$. It is larger than $4000L_o$ derived by Habing(1988). We consider this is result of selective effect. According to the dynamics of the Galaxy, only AGB stars with $M_0$ larger than $2-3M_o$ are found in the spiral arms. They are more luminous than others. The average luminosity, we derived in present work, is based on the O-rich AGB stars located in the arms.

Georgelin, Y.M., Georgelin, Y.P., Sivan, J.P.,1976,A&A,49,57.
Habing, H.J.,1988, A&A,200,40.
van der Veen,W.E.C.J., Habing, H.J.,1988,A&A, 194,125.

# ICE MANTLE FORMATION IN THE ENVELOPES OF OH/IR STARS

S.B. CHARNLEY[†]
Space Science Division, NASA Ames Research Center, California, USA
and
R.G. SMITH
Dept. of Physics, University of New South Wales, Campbell, Australia

ABSTRACT. We have computed ice column densities for a sample of O-rich late-type stars.

Water ice is observed in the outflows from several late-type stars (e.g. Smith et al. 1988) and it is of some observational and theoretical importance to understand how common ice mantles may be and how their formation depends upon the physical conditions in the envelope, such as the radial distributions of density and temperature. We have developed a simple theory of the gas-grain interaction in circumstellar shells (Jura & Morris 1985; Charnley et al. 1992) and used it to compute ice column densities ($N_{ice}$) and mantle thicknesses (d) for a sample of well-studied OH/IR stars (Herman & Habing 1985). The models are most sensitive to the dust and gas mass loss rates, as well as to the inner radius of the envelope, $R_i$, which is poorly determined. Representative results are presented in the Table below where $R_i$ was estimated according to the prescription of Herman et al. (1986).

| OBJECT | $R_i/R_*$ | $R_i$ (cm) | $R_{ice}$ (cm) | $N_{ice}$ (cm$^{-2}$) | d ($\mu$m) |
|---|---|---|---|---|---|
| 12.3-0.2 | 7.2 | 5.8(+14) | 6.3(+16) | 1.4(+17) | 1.4(-02) |
| 12.8-1.9 | 9.6 | 2.4(+14) | 2.3(+16) | 2.5(+15) | 2.8(-03) |
| 13.1+5.0 | 8.4 | 4.3(+14) | 4.1(+16) | 3.0(+16) | 1.2(-02) |
| 16.1-0.3 | 15.0 | 5.7(+13) | 4.3(+15) | 2.5(+16) | 1.7(-03) |
| 17.7-2.0 | 10.8 | 1.3(+14) | 9.8(+15) | 1.6(+17) | 9.0(-03) |
| 20.2-0.1 | 10.0 | 2.5(+14) | 2.4(+16) | 2.4(+16) | 5.2(-03) |
| 21.5+0.5 | 5.9 | 1.7(+15) | 1.6(+17) | 8.6(+16) | 2.6(-02) |
| 26.4-1.9 | 11.1 | 1.1(+14) | 1.0(+16) | 1.0(+16) | 5.9(-03) |
| 26.5+0.6 | 8.9 | 2.5(+14) | 2.4(+16) | 3.9(+16) | 9.3(-03) |
| 30.1-0.7 | 10.0 | 2.1(+14) | 1.6(+16) | 5.2(+16) | 4.8(-03) |
| 30.7+0.4 | 6.8 | 7.5(+14) | 7.1(+16) | 4.3(+16) | 1.4(-02) |
| 32.0-0.5 | 6.6 | 6.3(+14) | 6.0(+16) | 5.5(+16) | 9.3(-03) |
| 32.8-0.3 | 6.1 | 1.1(+15) | 1.0(+17) | 9.0(+16) | 1.9(-02) |

## References

Charnley, S.B., Smith, R.G. & Brown, P.D. 1992, in preparation.
Herman, J. and Habing, H.J. 1985. Physics Reports, Vol. 124, No. 4, pp 255-314.
Herman, J., Burger, J.H. and Penninx, W.H. 1986, Astr.Ap., 167, 247.
Jura, M. and Morris, M. 1985, Ap.J., 292, 487.
Smith, R.G., Sellgren, K. and Tokunaga, A.T. 1988, Ap.J., 334, 209.

[†] NAS/NRC Resident Research Associate

# FIRST RESULTS OF A NEAR IR MONITORING PROGRAM OF OH/IR STARS

P. GARCÍA-LARIO

*LAEFF - Estación de Villafranca del Castillo. Apdo. 50727. E-28080 Madrid (Spain)*

D. ENGELS

*Hamburger Sternwarte. Gojenbersweg 112. D-2050 Hamburg 80 (Germany)*

and

A. MANCHADO

*Instituto de Astrofísica de Canarias. E-38200, La Laguna, Tenerife (Spain)*

ABSTRACT. We present the first results of a long-term monitoring program of observations in the near infrared of a selected sample of OH/IR stars included in the IRAS Point Source Catalogue. The observations have been made using the 1.5m Sánchez Magro Telescope (SMT) at Izaña (Tenerife, Spain) since the beginning of 1991 and are still in progress. They are being complemented with observations made using the 1m ESO photometric telescope (La Silla, Chile). The sample includes 30 OH/IR stars with a variety of infrared and OH maser luminosities, expansion velocities, LRS classes and position in the IRAS two-colour diagram.

During the first phase of these observations we have identified for the first time their near infrared counterparts. In some cases we have obtained a first estimation of the amplitude $\Delta m$ and period $P$ of their photometric variations. As an example, we present the light curves of IRAS 16037+4218 ($\Delta m = 0.75$ mag ; P = 369 days) and IRAS 06297+4045 ($\Delta m = 1.29$ mag ; P = 697 days). Our data confirm the strong variability of these objects, which in some cases reaches more than 2 magnitudes. The near infrared colours observed in the $J - H$ vs $H - K$ colour-colour diagram can be explained as dust emission corresponding to temperatures ranging from 700 K to more than 1500 K, some of the bluest objects showing similar colours to those presented by Mira variables. More evolved objects are usually associated to the reddest colours while there seems to exist a clear correlation between the near infrared colours and the phase, the reddest colours appearing associated to the minimum brightness.

It is our intention to study the photometric properties observed in the near infrared and the correlation of the amplitudes and periods found with their characteristics derived from IRAS data (colour, variability index,...) and from OH maser measurements (expansion velocity, luminosity,...). A substantial progress will be made when more data will be available. This will help us to better understand the short transition phase after the end of the AGB phase which precedes to the formation of a new planetary nebula (PN), poorly known because of the lack of observational data.

# PLANETARY AND PROTO-PLANETARY NEBULAE IN THE IRAS TWO-COLOUR DIAGRAM

P. GARCÍA-LARIO

LAEFF - Estación de Villafranca del Castillo. Apdo. 50727. E-28080 Madrid (Spain)

A. MANCHADO

Instituto de Astrofísica de Canarias. E-38200, La Laguna, Tenerife (Spain)

and

S.R. POTTASCH

Kapteyn Laboratorium, Postbus 800. NL-9700, AV Groningen (The Netherlands)

ABSTRACT. We present a study of the distribution of post-AGB stars and planetary nebulae (PNe) in the IRAS two-colour diagram. From the analysis of the distribution of spectral types of post-AGB stars in this diagram we conclude that the position at which a star leaves the infrared sequence of colours observed in AGB stars depends only on the initial mass of the progenitor star. The evolution after the end of the AGB phase takes place, in a first approximation, at a constant value of $[25] - [60]$.

- Low mass stars leave this "infrared main sequence" ($IRMS$) at a low value of $[25] - [60]$, evolving slowly to the right in the IRAS two-colour diagram towards hotter spectral types. When the light of the central star is seen again after the phase of total obscuration, the spectral type is still K or M.

- High mass stars leave the $IRMS$ at a higher value of $[25] - [60]$, evolving faster towards hotter spectral types. When the central star is seen again it shows an A or B spectral type.

The position of the RV-Tauri stars in the IRAS two-colour diagram is consistent with their possible nature as low mass post-AGB stars, while many of the so called "BQ[ ] stars" are probably high mass progenitors of type I PNe. The presence of hot dust, detected in the near infrared, and H$\alpha$, sometimes with strong P-Cygni profiles, in many of these post-AGB stars confirms that the mass loss does not completely stop after the end of the AGB phase.

Similar results are found when we study the distribution of evolved planetary nebulae in this diagram. From the study of the galactic distribution, excitation class and type of PNe we conclude that low mass PNe are located in the lower part of the IRAS two-colour diagram while high mass PNe can only be found in the upper part of it. We suggest that bipolarity in PNe is a phenomenon that could be related to the post-AGB activity in high mass stars. The results here presented are consistent with the observational results found in OH/IR stars and confirms the new evolutionary interpretation of the sequence of infrared colours observed by IRAS.

# A NEW EVOLUTIONARY INTERPRETATION OF THE IRAS TWO-COLOUR DIAGRAM

P. GARCÍA-LARIO

LAEFF - Estación de Villafranca del Castillo. Apdo. 50727. E-28080 Madrid (Spain)

A. MANCHADO

Instituto de Astrofísica de Canarias. E-38200, La Laguna, Tenerife (Spain)

and

S.R. POTTASCH

Kapteyn Laboratorium, Postbus 800. NL-9700, AV Groningen (The Netherlands)

ABSTRACT. A new evolutionary interpretation of the sequence of colours observed in the IRAS two-colour diagram by AGB and post-AGB stars is given, which is capable of explaining the observational properties of both kind of objects. It is useful to define a parameter $\lambda$ to define the position of a given star in this "infrared main sequence" ($IRMS$). Adopting

$$[12] - [25] = \frac{1}{1.096} Ln\lambda \qquad [25] - [60] = -2.42 + 0.72\lambda$$

and from the analysis of the expansion velocities, mass loss rates and luminosities observed in a selected sample of non-variable OH/IR stars with no optical counterpart in the Galactic bulge as a function of $\lambda$, we conclude that the position in the IRAS two-colour diagram at which a star leaves the $IRMS$ ($\lambda_{max}$) only depends on the initial mass $M_i$ of the progenitor star, so that only massive objects can reach the upper end of this sequence. The relation found is:

$$\lambda_{max} = 3.2 \, log \, \frac{M_i}{M_\odot} + 2.45$$

Expansion velocities increase with the initial mass while every point in the $IRMS$ is found to be associated to a certain value of the mass loss rate. This model also predicts the evolution with time of the mass loss rate during the AGB as a function of the initial mass of the progenitor star, and confirms that most known planetary nebulae are the result of the evolution of considerably massive stars (between 2-3 solar masses) which means that the contribution of processed material to the interstellar medium is considerably higher than what theoretical models predict. Type I PNe are the result of the evolution of $3 - 5$ $M_\odot$ progenitors while progenitors with $M_i \leq 1.2$ $M_\odot$ probably do not give PNe. The model is also in agreement with the narrow distribution of core masses found in central stars of PNe and white dwarfs and with the usual expansion velocities found in OH/IR stars.

# A SYSTEMATIC STUDY OF IRAS SELECTED PROTO-PLANETARY NEBULA CANDIDATES

J.Y. HU and B.W. JIANG
*Beijing Observatory, Academia Sinica, Beijing, 100080, China*
and
T. DE JONG and S. SLIJKHUIS
*Astronomical Institute, University of Amsterdam, The Netherlands*

A complete sample of PPN candidates was selected mainly based on the IRAS colors(Hu et al.,1990). For total 62 unknown objects in the sample we have made 1. ground-based infrared astrometry; 2. optical identifications; 3. near infrared and optical photometry; 4. optical spectroscopic observations; 5. radio molecular line observations(OH maser and CO thermal lines); and 6. check the optical variations from historical plates. Some additional observations such as high resolution spectraoscopic, optical/NIR CCD imaging observations for particularly interested objects were also carried out.

We conclude that the nature of most objects in the sample is PPN on the following grounds.

1. 20 out of 21 objects with sufficiently high S/N in their spectra for a reliable luminosity classification, appear indeed to be supergiant-like(but IRAS16552-3050 is a giant).

2. The average luminosities are 3000Lo for 10 confirmed carbon-rich objects and 7300Lo for 15 confirmed oxygen-rich objects. They are comparable with AGB stars and fainter than real supergiants. Note the luminosities were derived based on the kinematic distances.

3. CO lines were detected from 15 objects and show profile with a width of 20- 40km/sec. This is the typical of evolved stars(but IRAS16279-4759 shows narrow profile from molecular cloud). In all 16 OH maser sources the 1612MHz transition is dominate. This is also the typical of evolved stars with dense circumstellar envelope.

4. The scale height of sample is $Z_0=0.27$kpc(based on the kinematic distances) which is comparable with planetary nebulae of 0.3kpc.

5. We have checked the optical variation of 24 southern objects which are with optical counterparts, all are nonvariable(but IRAS14122-5947).

Hu J.Y., de Jong T., Slijkhuis S.,1990, 'From Miras to Planetary Nebulae', 487-489.

# FI LYR: A CANDIDATE BINARY SYSTEM CONSISTING OF CARBON-RICH AND OXYGEN-RICH COMPANIONS

J.J. WANG, J.Y. HU and X. ZHOU

*Beijing Observatory, Academia Sinica, Beijing, 100080, China*

FI Lyr=IRAS18401+2854 is listed in the GCVS as type of SRb, the spectral type is M. In the IRAS LRS Catalog it is classified as 41, namely having the SiC feature at 11.3$\mu$m from a carbon-rich dust shell. We have obtained optical and near infrared spectra in the same period, but they show different chemical natures. We suggest that FI Lyr is a good candidate of a binary system consisting of carbon-rich and oxygen-rich companions.

The near infrared spectrum was obtained on April 2, 1992 using the 1.26m telescope at Beijing Observatory equipped with CVF spectrometer and InSb detector. The coverage of near infrared spectrum is from 2.9-4.0$\mu$m, in which clearly shows the feature of $C_2H_2$ and HCN at 3.1$\mu$m commonly observed in the carbon stars. The optical spectrum was obtained on March 10, 1992 using the 1.93m telescope equipped CARELEC at Haute-Provence Observatory. A linear CCD was used as the detector. It is a typical late M-type spectrum. We found in the IRAS LRS spectrum of FI Lyr an emission feature from 9 to 12$\mu$m and centred at 10.7$\mu$m. It is longer than silicate feature at 9.7$\mu$m of oxygen-rich objects and shorter than SiC feature at 11.3$\mu$m existed in carbon stars. Additionally there is a hump around 18$\mu$m, resemble with silicate emission from oxygen-rich object.

Skinner et al.(1990) observed the near infrared spectra of a group of M-type stars with SiC feature and carbon stars with silicate feature and found that the near infrared chemical nature is always identical with the optical. But this is not true in the case of FI Lyr. We consider FI Lyr to be a binary system consisting an M-type giant star and a carbon star.

Skinner,C.J., et al., 1990,MNRAS,243,78.

# NEAR-INFRARED SPECTROSCOPY OF PROTO-PLANETARY NEBULAE

B.J. HRIVNAK

Dept. of Physics & Astronomy, Valparaiso University, Valparaiso, IN 46383, USA

S. KWOK

Dept. of Physics & Astronomy, University of Calgary, Calgary, AB T2N 1N4, CANADA

and

T.R. GEBALLE

Joint Astronomy Center, 665 Komohana St., Hilo, HI 96720, USA

Sixteen candidates for proto-planetary nebulae have been observed with low-resolution infrared spectroscopy in the H and K bands, and 6 in the L band, using the United Kingdom Infrared Telescope. In the H band, the objects show hydrogen Brackett lines in absorption. In the K band, absorption bands (del v=2) of CO were observed, and in three cases the CO bands are in emission. The CO spectrum of IRAS 22272+5435 was found to change from emission to absorption over a three-month interval. This CO emission can be interpreted as an indication of some recent episodes of mass loss in these objects. Four of the objects were found to possess an emission feature at 3.3 um, usually associated with PAHs, and two of these show an unusually strong 3.4 um emission feature (Geballe, Tielens, Kwok, & Hrivnak 1992, ApJ, 387, L89).

# HIGH RESOLUTION OPTICAL IMAGING OF
# PROTO–PLANETARY NEBULAE

P.P. LANGILL and SUN KWOK
*Dept. of Physics and Astronomy, University of Calgary,*
*Calgary, Alberta, Canada. T2N 1N4*

and

B.J. HRIVNAK
*Dept. of Physics, Valparaiso University, Valparaiso IN, USA. 46383*

Several Proto–Planetary Nebulae (PPN) candidates have been observed at the Canada France Hawaii Telescope with the High Resolution Camera (HRCam) (M$^c$Clure et al. 1989, PASP, 101, 1156). Candidates were selected based on their IRAS [12]–[25] and [25]–[60] colors and their lack of variability. With HRCams' small angular pixel size (0.13"/pixel) and active optics we achieved an average angular resolution of 0.65" in V and I waveband images.

We have determined standard V and I magnitudes for these PPN some of which had no previous optical counterpart. These measurements have allowed us to map the flux distributions of the PPN from the optical to the infrared. Measured distributions have been modeled by a detached dust shell radiative transfer code (Volk, K. and Kwok, S. 1989, Ap.J. 324, 345) which helps to determine parameters of the dust shell and where the PPN is in its evolution between the asymptotic giant branch phase and the planetary nebulae phase.

By observing in the optical we are able to study the light scattered from the dust in the PPN envelopes. With the high resoution achieved we are able to probe their morphologies. Comparing the PPN images to field star images we have determined that several PPN have extended dust shells. A few have also been found to be bipolar. The study of the morphology of PPN dust shells gives invaluable clues as to the initial conditions which lead to the large variety of PN morphologies (Balick, B. 1987, Ap.J. 94(3), 671).

# ON THE EVOLUTION OF PROTO-PLANETARY NEBULAE

J.Y. HU and B.W. JIANG

*Beijing Observatory, Academia Sinica, Beijing, 100080, China*

and

S. SLIJKHUIS

*Astronomical Institute, University of Amsterdam, The Netherlands*

The ages of PNNe in our sample (Hu et al., 1990) were derived from the inner radius of the dust shell using the radiative transfer model fitting with SED and expansion velocities of envelopes obtained from radio molecular lines observations (if without radio data a $V_{exp}$ of 15 km/sec was assumed). The objects are classified into the following groups:

1. invisible;
2. M-type;
3. F (or G)-type; and
4. F (or G)-type with $H_\alpha$ emission.

The average ages of different groups are 97, 171, 247 and 477 years respectively. We considered that it is an evolutionary sequence.

The luminosities of the objects in the sample in which CO or OH maser lines were detected were derived using kinematic distances and total flux densities. The luminosities are relative to the IRAS color both for C-rich and O-rich objects. More luminous objects have a larger log(S60/S25). When an AGB star evolves into the PPN phase, it leaves the AGB evolution track and turns to the right on the IRAS color-color diagram (logS25/S12 vis logS60/S25). The model calculation (Slijkhuis et al., 1992) shows that in the early stages the objects evolve to the right almost parallel with the X-axis. Based on the relations of luminosity and logS60/S25 we considered that the turn-off points from the AGB evolution track on the IRAS color-color diagram to the PPN phase are relative to the luminosities i.e. the initial mass. Massive stars turn off in the upper part of AGB evolution track and low mass stars turn off in the lower part.

## References

Hu J.Y., de Jong T., Slijkhuis S., 1990, 'From Miras to Planetary Nebulae', 487–489.
Slijkhuis S., 1992, Ph.D. thesis, Chapter 6, Univ. of Amsterdam

# THE MOLECULAR FEATURES IN THE OPTICAL SPECTRA OF THE PROTO-PLANETARY NEBULAE

J.Y. HU

*Beijing Observatory, Academia Sinica, Beijing, 100080, China*

An AGB star evolves into the proto-planetary nebula(PPN) phase, when the stellar surface temperature is hotter than 5000K(Schonberner, 1983) . But the remnant of AGB dust/gas envelope is still thick in the early stages of PPN evolution. The observed optical spectrum of PPN is a spectrum of central star(T* hotter than 5000K, the spectral type earlier than G5) overlaped with molecular features formed by molecules in the envelope. The features are changing during the envelope turns to optical thin. Based on our systematic observations of IRAS selected PPN candidates(Hu et al. 1990) we found that molecular features in the spectra evolve as following:

1. When object just exposes from dust shell, the column density of molecules in the envelope is very high. The molecular absorption bands are dominant in the spectra, which are resemble to the M supergiant. There are 6 objects from our 62 PPN candidates sample in this stage.

2. When column density of molecules decreases, the spectra mainly show F-(or G-)type supergiant features from photosphere and partly also from envelope (especially the neutral atomic lines). But we still can detect the molecular bands such as $C_2$, CN in the spectra of carbon-rich objects(3 objects in our sample, IRAS23304+6147, 20000+3239 and 19477+2401), TiO of oxygen-rich object(one object, IRAS20406+2953), and ZrO and LaO of S-type transit object(one object, IRAS19454+2920).

3. The evelope becomes optical thin and molecules absorb photons from central star then radiate by fluorescence process. In the spectrum apear the molecular emission bands resemble to the cometary spectra. A well known object in which spectrum $C_2$ Swan bands in emission is CRL 2688(Crampton et al. 1975). We found another object, IRAS14429-4539, in the sample showing similar features.

Crampton, D., et al. 1975, ApJ, 198, L135.
Hu, J. Y.,de Jong, T., Slijkhuis, S.,1990, in 'From Miras to Planetary Nebulae', p487.
Schönberner, D., 1983, ApJ, 272, 708.

# OPTICAL SPECTROSCOPY OF SIX CARBON-RICH PROTO-PLANETARY NEBULAE

B.J. HRIVNAK

*Dept. of Physics & Astronomy, Valparaiso University, Valparaiso, IN 46383, USA*

We are engaged in a program of optical spectroscopy of proto-planetary nebulae (PPN). The objects were initially selected from the IRAS database on the basis of their strong infrared excesses, indicating dust temperature of 150 – 250 K. Spectra have been obtrained at medium resolution for the purpose of obtaining the spectral type, luminosity class, and to search for chemical peculiarities resulting from the post–AGB nature of the objects. We found our PPN candidates to typically display F-G supergiant spectra.

In this paper, we discuss in particular 6 PPN which show molecular carbon, IRAS 04296+3429, 05113+1347, 20000+3239, 22223+4327, 22272+5435 and 23304 +6147. From our spectra, we find the following results:
1. all 6 show $C_2$ $\lambda4735, 4717$ (and $\lambda5165$ when spectra cover this wavelength);
2. 4 of 6 also show $C_3$, which had previously been reported only in AFGL 2688 and late N–type carbon stars;
3. all 6 show strong BaII $\lambda4554$ and strong SrII and YII (s–process elements).

Those carbon and s-process enhancements are in accord with the general expectations of the effect of thermal pulses and mass loss on the AGB, and support the post–AGB nature of the objects.

These 6 carbon–rich PPN have several other properties in common, some of which are quite rare.
1. 5 are among the small group of 9 known IRAS sources which display an unidentified emission feature at $21\mu m$
2. 3 of them have been observed spectroscopically in the $3\mu m$ region, and all 3 show the $3.3\mu m$ emission feature.
3. 2 of these 3 are among the few sources with an unusually strong $3.4\mu m$ feature, comparable in strength to the $3.3\mu m$ feature.
4. Molecular line observations detect CO and HCN in these objects.

These and other common properties are summarized in the table. We suggest that the carbon richness and low stellear temperatures (T=4500–6000 K) are important parameters in relating these properties.

# NEAR–INFRARED IMAGING OF PROTO-PLANETARY NEBULAE

R.E.S. CLEGG

*Royal Greenwich Observatory, Madingley Rd, Cambridge CB3 0EZ, UK.*

and

N. A. WALTON and M.J. BARLOW

*Dept of Physics & Astronomy, University College London, Gower St, London WC1E 6BT, UK.*

It is not really known how low and intermediate mass stars eject mass to form PNs. We present preliminary results from a programme of near–IR imaging, in which we study a sequence of objects, from extreme AGB stars through proto-planetaries to young, compact PNs. We aim to study the sequence of morphologies, to see where the onset of bipolar shaping occurs, and to use the IR molecular hydrogen lines to map neutral regions around ionized nebulae.

The observations are from the 4.2m William Herschel Telescope at La Palma, with the near–IR camera 'FAST' developed by Prof. Genzel's group at MPE, Garching. A CVF and a Fabry–Perot together provide a 'tunable filter' with resolving power of 900. Observations were taken in H I Brackett $\gamma$, the $H_2$ v=1-0 and 2-1 S(1) lines, He II $n$=6-4, [Si VI] and [Si VII] lines, all in the 'K' band window.

We present preliminary results for four objects. For NGC 7027, the $H_2$ v=1-0 S(1) map shows a remarkable set of 4 'loops' outside the bright ionised ring. (The ring itself is completely absent from the $H_2$ maps). This $H_2$ emission is mostly collisionally–excited, and it traces warm, dense portions of the neutral shell. The hydrodynamical model predictions of Mellema et al. (A & Ap., 1992) for two–wind, bipolar structures with exit shocks may explain the 4 loops. A further map in the $v$=2-1 line will trace the relative distributions of collisionally–excited and UV–excited $H_2$.

CRL 2688 shows two symmetric pairs of $H_2$ emission knots, and our new deep maps reveal loops of emission joining the pairs. Comparison of the $v$=1-0 and 2-1 S(1) line maps indicates a gas kinetic temperature of about 1200 K. This must refer to shocked gas, since there is no UV source in this object.

For BD+30°3639 our $H_2$ line map shows a 15 × 10" elliptical 'neutral' nebula around a compact, 7" diameter spherical ionized nebula. The $H_2$ emission is however extremely non–uniform. Since the $H_2$ line is probably excited by UV photons, it is suggested that dust clouds between the central star and the outer parts of the large elliptical shell provide greatly–varying extinction in the UV.

Our images of the compact PN Vy 2-2 to have extended emission in the $H_2$ $v$=1-0 S(1) line, at PA about 30 deg E of N. This corresponds to an extension seen in the 4.9 GHz VLA map of Seaquist & Davis (1983), and suggests that the emission may be shock–excited along two polar flows, as is the case for the two main lobes in CRL 2688.

# MID-INFRARED SPECTROSCOPY OF FOUR 21µm EMISSION BAND SOURCES

K. JUSTTANONT and M. J. BARLOW
*Dept. of Physics & Astronomy, University College London, Gower Street, London WC1E 6BT, U.K.*

and

C. J. SKINNER
*Institute of Geophysics & Planetary Physics, L-413, Lawrence Livermore National Laboratory, P.O.Box 808, Livermore, CA94551-9900, U.S.A.*

We report 10 and 20µm spectroscopic observations of four C-rich post-AGB objects which exhibit the unidentified emission feature at 21µm. The observations were carried out in October 1990 and May 1991 using CGS3 on UKIRT. The spectral resolutions were 70 for the wavelength range of 7.4–13.3µm and 80 for the region between 15.4–24.1µm. Three of the sources reported here are from the list of Kwok, Volk & Hrivnak (1989), i.e., IRAS 04296+3429; IRAS 07134+1005 and IRAS 22272+5435. Figure 1 shows the full spectrum of IRAS 04296+3429 and IRAS 22272+5435. The 10µm spectra of these objects exhibit UIR bands whose peaks all fall longwards of the usual peak wavelengths associated with such features. This may be related to the fact that they are the lowest excitation objects so far found to exhibit UIR emission bands. We also found narrow emission features superimposed on the long wavelength wing of the 21µm emission bands of IRAS 04296+3429 and IRAS 22272+5435. The fourth object we observed, SAO 163075, was found to also exhibit a (weak) 21µm emission feature. However, there is no PAH features in the 10µm region, apart from the plateau at 12µm.

Reference

Kwok, S., Volk, K.M., Hrivnak, B.J., 1989, ApJ, 345, L51

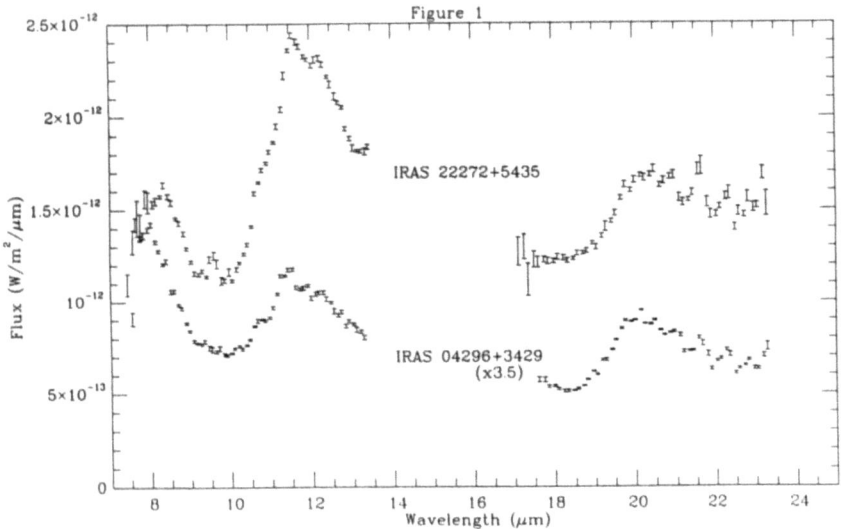

# UKIRT CGS3 OBSERVATIONS OF NEW IRAS 21 MICRON SOURCES

SUN KWOK
*Dept. of Physics & Astronomy, University of Calgary, Calgary, Canada*
BRUCE J. HRIVNAK
*Dept. of Physics & Astronomy, Valparaiso Univerity, Indiana, U.S.A.*
TOM R. GEBALLE
*Joint Astronomy Center, Hilo, Hawaii, U.S.A.*
and
PHILLIP L. LANGILL
*Dept. of Physics & Astronomy, University of Calgary, Calgary, Canada*


An unidentified emission feature at 21 μm has been detected in the IRAS Low Resolution Spectra (LRS) of 5 IRAS sources (Kwok, Volk, and Hrivnak 1989, Hrivnak and Kwok 1991). The sources are generally found to be F and G supergiants with cool, detached dust shells. We have searched for additional 21 μm sources in the LRS database and have obtained ground-based UKIRT spectra at 10 and 20 μm in an attempt to confirm the LRS feature.

The LRS spectra of 05113+1347, 20000+3239, and 22223+4327 show a peak around 21 μm, a flat plateau between 12 and 18 μm, and a drop between 7 and 11 μm which are the characteristics of 21 μm sources such as 07134+1005. UKIRT CGS3 spectra of these 3 sources plus 05341+0852 have been obtained with spectral resolution of CGS3 is ~52 at 10 μm and ~72 at 20 μm. The spectra of the 05341+0852 is different from the others in that the 10 μm band is higher than the 20 μm band. We are therefore less certain about the 21 μm feature in this object.

Ground-based visible, near- and mid-infrared photometry have been obtained for the 21 μm sources. The energy distribution of the sources show the "double-peaked" distribution characteristic of proto-planetary nebulae (see Kwok, this volume).

All the 21 μm sources have been found to show carbon-rich photosphere with $C_2$ and/or $C_3$ features (see Hrivnak IV:120). Unidentified 3.4-3.5 μm features are also observed in addition to the 3.3 μm PAH feature (Geballe et al. 1992). The strength of the 21 μm feature implies that it originates from an abundant element. The carbon-rich nature of the sources suggests that the carbon atom may be a major constituent of the molecule/grain responsible for the 21 μm feature.

## References

Geballe, T.R., Tielens, A.G.G.M., Kwok, S., & Hrivnak, B.J. 1992, *ApJ*, **387**, L89
Hrivnak, B.J., & Kwok, S. 1991, *ApJ*, **368**, 564
Kwok, S., Volk, K., Hrivnak, B.J. 1989, *ApJ*, **345**, L51

# MID-IR (8-13μm) IMAGES OF PROTO-PLANETARY NEBULAE

M. MEIXNER, J.F. ARENS and J.G. JERNIGAN

*Dept. of Astr. and Space Sci. Lab., U.C. Berkeley, Berkeley, CA 94720 USA*

and

C.J. SKINNER and G. HAWKINS

*LLNL, IGPP, L-413, P.O. Box 808, Livermore, CA 94550 USA*

We present mid-IR (8-13μm) images of dust in seven proto-planetary nebulae (PPN), GL2343, HD 161796, 89 Her, OH 0739-1435, CRL2688, IRAS 22272+5435, and CRL618. The images were taken at UKIRT and the IRTF with the Berkeley mid-IR camera which was developed at the Space Sci. Lab. in UC Berkeley and is supported by IGPP and LEA, LLNL. The results presented here are part of an on-going mid-IR imaging project to study the morphological development of a star as it evolves from the Asymptotic Giant Branch (AGB) to the planetary nebula (PN) stage. In particular, we aim to establish when non-spherical symmetry which is evident in so many PN arises. Four of the objects are oxygen rich. Of these, GL2343 (Fig. 1) and HD161796 are found to have spherical dust shells in the mid-IR with diameters of 6."7 and 3", respectively. OH 0739 is marginally resolved, but has a hint of elongation that is aligned with the bipolar nebula evident in the near-IR. 89 Her is unresolved at 1."1 resolution. 89 Her, HD 161796 and GL2343, all high latitude supergiants, appear to form an evolutionary sequence as evidenced by the size of their circumstellar dust shells. Three of the objects are carbon rich. CRL2688 and IRAS 22272 +5435 are found to have elongated structures which suggest a bipolar morphology. CRL 618 is unresolved at 1."7, but is known to be bipolar from optical studies. Hence, with the exception of OH 0739, a known binary, the oxygen rich PPN have spherical dust shells while the carbon rich have bipolar dust shells. Our small number of observations do not provide solid statistics; however, this trend suggests that nebula bipolarity is linked to carbon-rich chemistry.

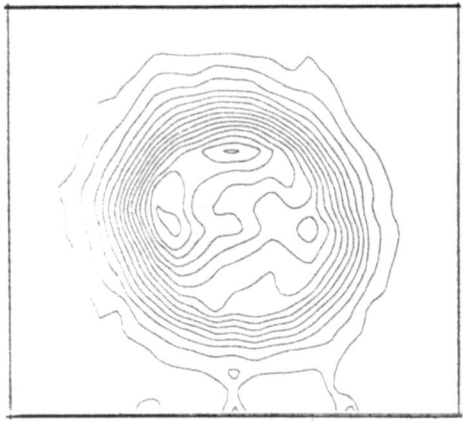

Fig. 1. GL 2343 at 12.5μm

# THE MORPHOLOGY OF MID-INFRARED UIR FEATURE EMISSION IN THE PPN M 2-9 AND IRAS 21282+5050

L. K. DEUTSCH
NASA/Ames Research Center, M/S 245-6, Moffett Field, CA 94035, USA

J. L. HORA
Institute for Astronomy, 2680 Woodlawn Dr., Honolulu, HI 96822, USA

W. F. HOFFMANN
Steward Observatory, University of Arizona, Tucson, AZ 85721, USA

G. G. FAZIO
Smithsonian Astrophys. Obs., 60 Garden Street, Cambridge, MA 02138, USA

and

K. SHIVANANDAN
Center for Advanced Space Sensing, NRL, Washington, D.C. 20375, USA

**Abstract.** The carbon-rich PPN M 2-9 and IRAS 21282+5050 are known to exhibit UIR feature emission at 7.7, 8.6, and 11.22 $\mu$m. The two nebulae have been imaged in the mid-IR with the UA/SAO/NRL Mid-Infrared Array Camera (MIRAC) in bandpasses which include UIR feature emission (8.8 and 11.22 $\mu$m). Near-infrared images of the nebulae have also been taken with the NICMOS 3 Hg:Cd:Te array camera at $J$, $H$, and $K$ for M 2-9 and at $K$ for IRAS 21282+5050.

**M 2-9.** The mid-IR emission from M 2-9 comes from the nebula's optically bright center. At 8.8 and 9.8 $\mu$m, the images are slightly extended roughly perpendicular to the major axis of the nebula. The image at 8.8 $\mu$m is larger, suggesting that the UIR feature carriers are spatially distinct and more extended than the mid-IR continuum emission. This is consistent with observations of other more evolved PN such as BD +30°3639 and NGC 6572 (Hora et al. 1990, Deutsch 1990). The nebular images at 10.0, 11.22, and 11.7 $\mu$m are not resolved; further mid-IR observations of M 2-9 are planned.

The near-IR images of M 2-9 presented show the changing characteristics of the nebula with increasing wavelength. At $J$, bright knots of emission are clearly seen predominantly along the inner eastern edges and at the ends of each lobe. At $H$, the knots are somewhat fainter, while the band across the northern lobe has become more dominant. At $K$, the knots at the ends of the lobes are barely visible, while the northern band is quite distinct.

**IRAS 21282+5050.** In the images presented at 8.8 and 11.22 $\mu$m, the nebula appears similar in size and morphology at the two wavelengths, showing round extension and a roughly E-W elongation with a FWHM of 3.6" at a P.A. of $\sim$70° E of N. The nebular image at $K$ with the central star subtracted shows two small symmetric lobes near the center of a more extended, round emission structure. A line drawn through the center of the lobes lies at a P.A. of $\sim$70° E of N, the same as in the mid-IR images. The emission at $K$ is less extended than in the mid-IR.

### References

Deutsch, L.K. 1990, *Ph.D. Thesis*, Harvard University
Hora, J.L., Deutsch, L.K., Hoffmann, W.F., and Fazio, G.G. 1990, *Ap.J.*, 353, 549

# VISUAL EXTINCTION AND PHYSICAL CONDITIONS IN THE BIPOLAR NEBULA M2–9

A. RIERA

Departament de Física i Enginyeria nuclear, Escola Universitària de Vilanova i la Geltrú, Universitat Politécnica de Catalunya, Barcelona (Spain)

and

Departament d'Astronomia i Meteorologia, Universitat de Barcelona, Av. Diagonal 647, E-08028 Barcelona (Spain)

ABSTRACT. A detailed analysis of the visual extinction and physical conditions of the extreme bipolar nebula M2-9 has been made from optical spectra acquired with the Intermediate Dispersion Spectrograph of the 2.5m Isaac Newton Telescope (Observatorio del Roque de los Muchachos, La Palma) in combination with a 235mm camera and the IPCS detector covering the spectral range 3500–7500 Å with spectral resolution of 4 Å.

Visual extinction values through the core and the lobes have been determined from HI Balmer serie lines including self-absorption from the $2^2s$ metastable level. An H$\alpha$ optical depth of 17 in the core and visual extinction values of 4.5 and 2.4 mag in the core and the lobes, respectively (lower than values previously estimated assuming case B recombination) are estimated.

Visual extinctions have also been calculated using the method developed by Allen (1979) which includes transauroral/nebular ratios of [SII] and [OII]. Values of $A_v$ derived through the core are slightly lower than values obtained from HI Balmer lines, as expected if forbidden emission lines originate in an outer region. $A_v$ values calculated through the lobes from both methods are in agreement within the errors.

A study of the physical conditions within the nebula using dereddened fluxes of forbidden lines of $O^+$, $O^{++}$, $N^+$ and $S^+$ reveals a high stratification with a two-tiered structure in which the auroral lines come from a high–density region ($n_e \sim 10^5 \to 10^7$ cm$^{-3}$) where nebular lines are suppressed while the nebular lines come from a low–density zone ($n_e \sim 10^3 \to 10^4$ cm$^{-3}$).

Both characteristics reported for M2–9: density stratification within the nebulosity and self–absorbed HI Balmer emission lines are common to symbiotic stars and some young PN. Analyses of the spectrum of M2-9 over a large spectral range (from centimeter to UV wavelengths) shows that its core shares characteristics with symbiotic stars and young PNe. However it is not well fitted by either object and could have recently evolved from one to the other as has been proposed for other bipolar nebulae such as He2-104 and Mz3.

## References

Allen, D.A., 1979, MNRAS 186, 1P.

# COMPLEX MOTIONS IN PLANETARY NEBULAE

VINCENT ICKE

*Sterrewacht Leiden, Postbus 9513, 2300 RA Leiden, The Netherlands*

## 1. The Hydrodynamic Model

I have made an extensive series of numerical simulations of aspherical PNs. This interacting-winds model consists of a point source of fast tenuous gas embedded in a flattened cloud of dense slow gas which is two-dimensional and cylindrically symmetric. I used a hydrocode specially designed to handle the extremely large gradients between the winds to second order accuracy. The outer shock shapes correspond very well to my analytic predictions. This shock may form cusps which compress the gas to form two rings on opposite sides of the equatorial plane.

## 2. The Reverse Shock

The central source becomes surrounded by a standoff shock in the fast wind. The oblateness of the slow wind causes this shock to become prolate or barrel-shaped. This causes shock focusing of the fast flow.

## 3. The Contact Discontinuity

A cool tongue-like deformation extends from the contact discontinuity, forming a chimney along the symmetry axis. Turbulent hot gas streams back from the head of the outflow. These two effects exert a confining pressure, leading to the formation of extremely well collimated jets in the fast gas.

## 4. Some Observable Consequences

The inner shock and the contact discontinuity show a very complex interplay of hydrodynamic effects. This produces a fascinating richness of internal motions and density distributions. Some of these can be readily identified with known PN features such as "helix" rings. Others are presented as predictions.

Balick, B., Preston, H.L., Icke, V.: 1987 *Astron. J.* **94**, 1641
Icke, V.: 1987 *Astron. Astrophys.* **202**, 177
Icke, V.: 1991 *Astron. Astrophys.* **251**, 369
Icke, V., Balick, B., Frank, A.: 1991 *Astron. Astrophys.* **253**, 224
Icke, V., Mellema, G., Balick, B., Eulderink, F., & Frank, A.: 1992 *Nature* **355**, 524.
Icke, V., Preston, H.L., Balick, B.: 1989 *Astron. J.* **97**, 462
Mellema, G., Eulderink, F., Icke, V.: 1992 *Astron. Astrophys.* **252**, 718

# THE NATURE OF THE HIGH VELOCITY FLOW IN CRL 618

R.NERI, M.GUÉLIN, S.GUILLOTEAU and R.LUCAS

*I.R.A.M, Domaine Universitaire, 300 rue de la Piscine, F - 38406 St-Martin-d'Hères*

and

S.GARCIA-BURILLO and J.CERNICHARO

*C.A.Y. - I.G.N., S - 19080 Guadalajara*

Using the IRAM interferometer, we have mapped with a $2\rlap.{''}4 \times 3\rlap.{''}4$ resolution the $J = 1 \rightarrow 0$ HCN line emission in the proto-planetary nebula CRL 618. Our maps resolve the $200\,\mathrm{km s}^{-1}$ molecular outflow (Cernicharo et al. 1989), as well as the slowly expanding circumstellar envelope (Bujarrabal et al. 1988), allowing a very precise positioning ($\leq 0\rlap.{''}1$) of these components with respect to the central HII region. 70% of the HCN envelope emission comes from a very compact, spherically symmetric core of size $\simeq 3\rlap.{''}2$. The core surrounds the high velocity gas which appears localized in a number of small 'clumps' ($\leq 0\rlap.{''}5$) – see figure. The large range of velocities observed in the 'clumps' suggests that we are not observing a decelerating molecular flow, but the impacts of a bipolar outflow on the slowly moving core, close to the HII region. The collision of a neutral gas outflow with high density regions (the 'clumps') results in the generation of dissociative shock-waves pushing and tearing the inner surface of the envelope. CRL 618 appears to have reached the stage where the stellar winds begin to disrupt and to scrape through the massive envelope, shortly before it evolves towards a Planetary Nebula.

The molecules have certainly been destroyed by the fast shock before the gas was accelerated to $200\,\mathrm{km s}^{-1}$. Expansion and cooling of the accelerated gas in the post-shock region allow their reformation (see the insert). We infer from the observed geometry and velocities that HCN reformation occurs in $\leq 50$ y. The small size of the HCN high velocity sources implies large excitation temperatures ($200\,\mathrm{K} \leq T_{ex} \leq 1000\,\mathrm{K}$) and a large HCN abundance $x(\mathrm{HCN})/x(\mathrm{CO}) \geq 0.1$. This is the first time that the HCN abundance and formation time scale have been measured behind a fast shock. These results will be presented by Neri et al. 1992.

**Figure**: Morphology of CRL 618's core, as derived from the HCN data. The figure width is $\simeq 4''$.

### References

Bujarrabal et al.: 1988, 'PPN: the case of CRL 618', *A&A*, **204**, 242

Cernicharo et al.: 1989, *A&A*, **222**, L1

Neri et al.: 1992, 'CRL618: The nature of the $200\,\mathrm{km s}^{-1}$ outflow', *A&A*, in press

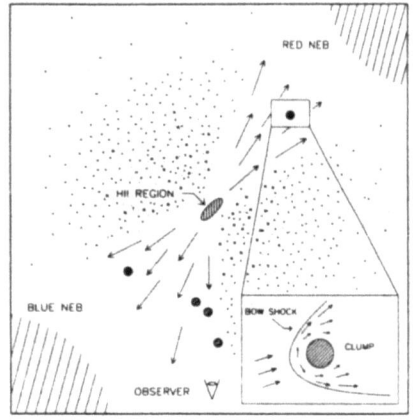

# HIGH VELOCITY OUTFLOWS IN IRAS 17423−1755

A. RIERA

Departament d'Astronomia i Meteorologia, Universitat de Barcelona, Av. Diagonal 647, E-08028 Barcelona (Spain) and Universitat Politcnica de Catalunya

P. GARCíA-LARIO

LAEFF - Estación de Villafranca del Castillo. Apdo. 50727. E-28080 Madrid (Spain)

A. MANCHADO

Instituto de Astrofísica de Canarias. E-38200, La Laguna, Tenerife (Spain)

and

S.R. POTTASCH

Kapteyn Laboratorium, Postbus 800. NL-9700, AV Groningen (The Netherlands)

ABSTRACT. IRAS 17423−1755 is a new transition object between the post-AGB phase and the planetary nebula (PN) stage included in an observational program of IRAS sources with infrared colours similar to those of PNe. We have taken B, V, R, I, Z, H$\alpha$ and [OIII] CCD images at the 1m JKT (Roque de los Muchachos, La Palma), where a clearly marked bipolar structure with a total extension of about 10" can be seen, and long slit (low and high resolution) optical spectra at the 4.2m WHT, 2.5m INT (RM, La Palma) and ESO 1.5m (La Silla) telescopes.

The spectrum of the core corresponds to a B star with strong emission lines of HI Balmer and Paschen series, FeII, [FeII], OI, CaII, and [CaII], sharing characteristics with BQ[ ] stars, which are stars in the post-AGB phase, while on the other hand, proto-PNe usually show a similar morphology (He2−123, M2−9 or M1−16).

The study of the OI $\lambda$ 8446 Å emission reveals that it is due to starlight excitation and/or Ly$\beta$ fluorescence. The detection of strong CaII triplet lines implies the existence of a predominantly neutral HI region (i.e. a cool circumstellar disk) with densities from $10^9$ to $10^{15}$ cm$^{-3}$. A rich [FeII] emission line spectrum is observed, which is probably excited by collisions in a gas with $n_e \sim 10^5 - 10^6$ cm$^{-3}$.

The most remarkable kinematic features are the widening of the assymetric H$\alpha$ profile at the core, the tilt in velocity of the lobes and the high–velocity feature (jet) connecting the core and the redshifted lobe. Line profiles observed in the lobes are characterized by large radial velocities ($\gtrsim$ 425 km s$^{-1}$), extraordinary line widths and double–peaked profiles, indicating that the emission arises from the cooling region behind a bow–shock, with a shock velocity of $\sim$ 260 km s$^{-1}$. The jet decelerates as it moves away from the central source, its velocity decreasing from 920 km s$^{-1}$ to 750 km s$^{-1}$. From the high values of the [NII] (6548 + 6584)/H$\alpha$ ratios ($\sim$ 4) measured in the lobes and in the jet, we deduce that the outflowing material is nitrogen enriched gas of stellar origin.

Theoretical models predict the formation of bow–shocks at the end of highly collimated outflows, when a high velocity wind with an axial jet interacts with the surrounding material. It is argued that assymetric mass loss in the post-AGB phase could be the responsible of the formation of bipolar structures, as an alternative to the binary hypothesis.

# ROTATION-PULSATION COUPLING IN THE BIPOLAR PREPLANETARY NEBULA, V HYA

MARK MORRIS and CECILIA BARNBAUM

*Department of Astronomy The University of California at Los Angeles Los Angeles, CA 90024 USA*

High-resolution optical spectra were taken of the carbon star V Hydrae at 10 different epochs spanning two stellar periods using the Hamilton Echelle Spectrograph at Lick Observatory. Velocities were determined at each epoch by performing a cross-correlation analysis against the spectra of standard stars with previously-determined velocities. The velocities of individual atomic absorption and emission lines, and their variations with phase, were also determined. The rising and falling of the photosphere is clearly in evidence, with an amplitude about the mean of $\pm 5$ km s$^{-1}$ and a phase consistent with the expectation that maximum light occurs at minimum radius. In addition, the spectra were subjected to a rotational broadening analysis, in which we determined V sin(i), where V is the presumed equatorial rotation velocity, by minimizing the differences between the spectra of V Hya and a number of artificially broadened comparison stars showing no evidence for broadening. Mechanisms other than stellar rotation, including both turbulence and opacity, were found to be unlikely contributors to the broadening. The very unusual rotation velocity (for a red giant) was found to vary between about 10 and 18 km s$^{-1}$, with an average of $\approx 14$ km s$^{-1}$. The relative phase of the rotation velocity curve is consistent with the hypothesis that the pulsation leads to a periodically varying moment of inertia. The assumption of angular momentum conservation, coupled with the radial velocity and V sin(i) curves, permits a deduction of the equilibrium stellar radius: The similar magnitudes of the rotation and pulsation velocities imply that they are dynamically coupled, leading to a latitude-dependent circulation which probably affects the mass loss. This interaction should also have a strong effect on the convective structure of the envelope. The angular momentum implied for the V Hya atmosphere leads us to the conclusion that it is a common-envelope binary, and that it is a precursor of a binary nucleus in a bipolar planetary nebula.

# INFLUENCE OF THE STELLAR WINDS ON THE POST-AGB EVOLUTION.

R. SZCZERBA

*N. Copernicus Astronomical Center, Chopina 12/18, 87-100 Toruń, POLAND*

Modelling of evolution of dusty shells around post-AGB star of 0.598 $M_\odot$ mass has been performed by means of the Yorke's hydrodynamical code. In this code dust and gas are treated as two separate components coupled by friction, and the spectral appearance of the model is computed by solving the angle and frequency dependent continuum radiation transfer problem. Calculations were carried out for dust composed of silicate grains. The emergent spectrum of the model has been convolved with the IRAS photometric band profiles to construct theoretical tracks on the IRAS two-color diagram.

The dynamical calculations have been started with the ejection of 0.2 $M_\odot$ during 2000 years. The "superwind" cessation has been coupled to the pulsation period ($P_o$ = 100 or 75 days) of the star. A number of evolutionary sequences, beginning from the end of "superwind" till the nebula ionization, have been calculated assuming different stellar winds after departure from the AGB (see Górny - this volume). Adopting the initial dust-to-gas mass ratio of $7.5 \cdot 10^{-3}$ (during the whole evolution) and the initial velocity of 5 km/s (during the "superwind") at the dust condensation radius ($T_d$ = 1000 K), terminal velocity of the main shell close to 15 km/s has been obtained.

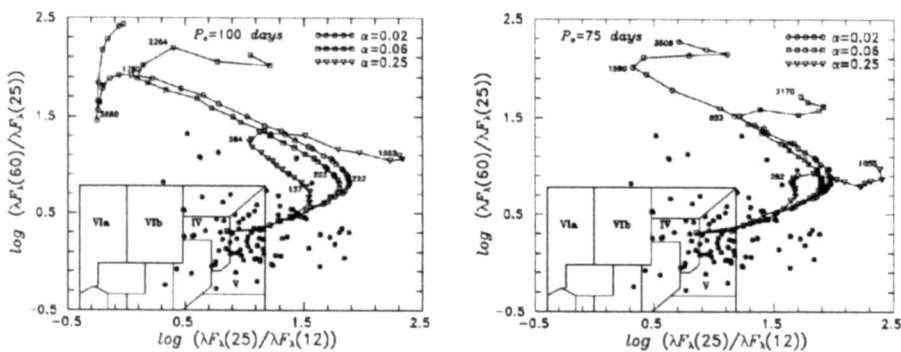

**Figure:** *The influence of the moment of "superwind" cessation and the stellar wind on the track positions in the IRAS diagram.*

From the Figure it can be seen that evolutionary behaviour of the calculated models strongly depends on the post-AGB mass loss rates as well as on the exact moment of "superwind" cessation. Both this factors greatly influence the stellar evolutionary timescales which in consequence determine the spectral behaviour of the models. Comparison between the calculated sequences and the observational data should allow to put constraints on the post-AGB evolution scenario.

# AXIALLY SYMMETRIC DYNAMICS OF PNe

LIFAN WANG

*Dept. of Astronomy, The University of Manchester*

and

*Center for Astrophysics, The University of Science and Technology of China*

Following Kahn & West (1985), we investigate the formation of PNe in the slow and fast wind interaction scheme by assuming the slow wind axially symmetric. We have further assumed that the mass loss rate for the slow wind is not steady. It is found that the final morphology of the nebula depends not only upon the initial degrees of the seed asymmetry in the slow wind, but also upon the time variations of the mass loss rate. As an example, we show in some detail the case where the central star first blows an axially symmetric slow wind during its red giant stage, this wind is followed by a superwind while the star is on the AGB, these slow wind is later overtaken and shaped by the fast wind during the post-AGB branch. It is found that a small initial asymmetry can be amplified and reproduces the various observed morphologies of the PNe.

# THE CHEMICAL COMPOSITION OF POST AGB STARS

M. PARTHASARATHY

*Indian Institute of Astrophysics, Bangalore 560034, India*

P.GARCIA LARIO

*Laboratorio de Astrofisica Espacial Y Fisica Fundamental, Estacion de Villafranca del Castillo, Apartado de Correos 50727, E-28080 Madrid, Spain*

and

S.R. POTTASCH

*Kapteyn Astronomical Institute, Postbus 800, Nl-9700 AV Groningen, The Netherlands*

The F-type supergiant HD 56126 (F5I) is an IRAS source with detached cold dust shell with characteristics similar to the dust shells around planetary nebulae. From an analysis of high resolution and high signal to noise ratio spectra metal and CNO abundances have been determined. It is found that in HD 56126 [C/H] = $-0.01$, [N/H]=+0.17, [O/H] = $-0.02$, [S/H]=+0.01 and [Fe/H]< $-1.0$. The C/N/O abundance ratios and CNO abundances relative to Fe of HD 56126 and related post AGB stars suggest that they have on their photospheres the material processed by triple alpha, CN and ON cycles.

From the high resolution spectra we derive the radial velocity of HD 56126 to be +105 km/sec, which suggest that HD 56126 is a high velocity star belonging to the old disk.

The abundance ratios [C/Fe], [N/Fe], [O/Fe] and [S/Fe] in HD 56126 and related post AGB stars are large and cannot be explained by nucleosynthesis and or mixing. These large ratios are due to the depletions of Fe but not of CNO and S elements. The depletion of refractory elements and the lack of depletions in CNO and S in the atmosphere of these post AGB stars is similar to that observed in the interstellar medium. Since most of these stars have circumstellar dust shells, the depleted refractory elements appear to be locked up in dust grains. The results suggest that in the recent past (possibly during the OH-IR stars stage) the outer atmospheres of these star expanded and cooled to the limit of the condensation temperature of refractory elements. Formation of cores of dust grains very close to the stars and the resulting dust driven mass loss may be able to explain the observed abundance peculiarities in these post AGB stars.

Some of the post AGB stars such as BD +39 4926, HR 4049, HD 52961 and HD44179 are extremely deficient in Fe and most likely in other refractory elements. However their CNO, S and Zn abundances do not share this extreme deficiency. Why some of these stars show extreme deficiency of refractory elements while others do not suggests that some of them had very dense cool envelopes where dust formation was very efficient which has depleted the refractory elements by a large fraction. Or these stars have gone through several cycles of envelope expansion, cooling to the limit of condensation temperature of refractory elements and dust formation close to the star. There seems a process operating in these stars which has driven out only the dust but not the gaz making the photospheres with depleted refractory elements.

# UBVRI POLARIZATION MEASUREMENTS OF POST AGB STARS

M. PARTHASARATHY and S.K. JAIN
*Indian Institute of Astrophysics, Bangalore - 560034, India*

UBVRI polarization measurements of 25 post AGB stars with circumstellar dust shells (CDSs) were made. Most of them show A, F, G, K supergiant type spectra. IRAS data of these stars show evidence for the presence cool detached circumstellar dust envelopes. Some of these stars also have warm dust shells. Many show significant polarization. The observed polarization in UBVRI is most likely due to scattering of the central star radiation by CDSs. Some of these stars show variation in polarization and position angle. The polarization data suggest that several may have aspherical or bipolar CDSs.

From an analysis of the IRAS data a new class of stars were detected. They have CDSs with far infrared colours and flux distributions similar to the dust shells of PNe. Most of them show A, F, G and K supergiant-like spectra in the optical. Parthasarathy and Pottasch (1986), and Lamers et al. (1986) first detected these shells around high galactic latitude A - F supergiants and interpreted them as the result of severe mass loss experienced during their AGB stage. These stars are considered to be in the post AGB stage, evolving from the tip of the AGB to the left in the H - R diagram into the region of PNe.

In order to study the CDSs we made polarization measurements in UBVRI with a polarimeter on the 1-m telescope of the Indian Institute of Astrophysics, Kavalur, India. Some of the post AGB stars found show significant intrinsic polarization. Several of these stars are at high galactic latitudes; thus the expected interstellar polarization is much less than the observed values. In addition, the wavelength dependence of polarization of light of these stars is found to different from that of interstellar polarization. Also the position angle for interstellar polarization is independant of wavelength. For some of the stars we find significant change in position angle. Hence, the observed polarizations are intrinsic to the stars.

The detection of linear intrinsic polarization suggests the nonspherical distribution of circumstellar matter. Since these stars have CDSs, this indicates that the dust shells are responsible for the observed polarization. The observed polarization of light is due to the scattering of the radiation of the central star in non-spherical CDSs. Here, stellar light gets polarized by being scattered off the dust grains; an asymmetry in dust grain distribution around the star would then produce a net observable polarization.

# HIGH-RESOLUTION RADIAL VELOCITY AND Hα STUDY OF PROTO-PLANETARY NEBULAE

B.J. HRIVNAK

*Dept. of Physics and Astronomy, Valparaiso University, Valparaiso, IN 46383 USA*

and

A.W. WOODSWORTH

*Dominion Astrophysical Observatory, 5071 W. Saanich Rd., Victoria, BC V8X 4M6 CANADA*

We are engaged in a program to monitor radial velocity variability in proto-planetary nebulae (PPN). Observations are being made with the radial velocity spectrometer at the DAO, with a precision of $\pm 0.5$ km s$^{-1}$. Radial velocity variability can arise from binary motion and/or pulsation in these post-AGB stars. The demonstration of a binary nature for some of these objects can have important implications for the understanding of their physical properties, and for the shaping of their circumstellar shells.

We have found the following results to date:
1. all 9 PPN studied show radial velocity variability with peak–to–peak amplitudes ranging from 5 to 13 km s$^{-1}$;
2. 4 objects clearly display periodic radial velocity variations, 67134+1005, 18095+2704, 22223+4327, 22272+5435, with preliminary periods ranging from 95 to 340 days.

These observations are continuing.

We have also made high resolution (0.1–0.2 Å) spectroscopic studies to investigate Hα emission in these same objects. It can be produced by shock waves which drive mass loss, or by other activity in the athmospheres: The PPN all display spectra of F and G supergiants. They have each each been observed between 1 and 7 times, with the following results to date:
1. Hα emission is common in the F stars, as it is seen in 3 of 5;
2. in the G stars 1 in 4 shows emission above the continuum;
3. P Cygni, reverse P Cygni and shell profiles are seen;
4. Hα emission varies in strength in the objects.

# Hα PROFILES OF SELECTED CANDIDATES FOR PROTO-PLANETARY NEBULAE

S. TAMURA

*Astronomical Institute, Tôhoku University, Sendai 980, Japan*

The high-dispersion Hα line profiles of the candidate stars for proto-planetary nebulae(PPNs) are presented. Samples include mostly F supergiant stars. Volk and Kwok(1989) summarized four classes of stars which had been described in various literature as PPNs. They selected nearly one hundred new candidates of PPNs due to IRAS data. Our goal is to make clear the mass-losing phenomena and to establish the adequate models for the explaination of observed data from such stars. F supergiant stars also attract our attention to the spectroscopic variability in connection to their extended, rarefied and therefore complicated atmosphere. Observations were made at the Okayama Astrophysical Observatory with the 188-cm reflector and its coudé spectrograph. We used the intensified Reticon as the detector which has one dimensional array of 1024 pixels. The size of one pixel is 25 $\mu$m and corresponds to 0.13 Å and 5.9 $km/sec$ at Hα. Our stellar samples are HD 46703, HD 56126, HD 112374, R CrB, HD 161796, 89 Her, IRAS 18095+2704, AFGL 2343, HD 187885, and UU Her. We obtained the Hα profiles and found a couple of absorption lines like the FeI$\lambda$6569, FeII$\lambda$6516, and CI$\lambda$6587 among them. An express report and a short discussion about a part of these samples was already presented by Tamura and Takeuti(1991). Subsequent reports appeared and will appear in the articles of Tamura, Takeuti, and Zalewski(1992); Tamura and Takeuti (1992). We show various kinds of Hα profiles which consist of absorption and emission components and show clear evidence of activity in the extended atmosphere. We classify them into (i)P Cyg type profile, *e.g.* 89 Her, (ii)inverse P Cyg type profile, *e.g.* HD 187885, (iii)profiles with the central reversal at the bottom of absorption, *e.g.* HD 46703, (iv)profile with shoulder emission components, *e.g.* HD 56126, (v) asymmetric absorption profile, *e.g.* AFGL 2343. Concerning HD 161796 the Hα profile seems to have only an absorption component. It should be attributed to time variations, because the profile observed by Luck *et al.*(1990) is similar to (iv). In addition to a wide variety of the Hα profiles we have noticed their time variations ranging over different time spans, month, day, even in several ten minutes. Arellano Ferro(1985) noticed the change of Hα profiles of luminous yellow supergiants in various time scales too. We obtained radial velocities and compared with the expected values by the galactic rotation. We can recognize our sample stars are far from the galactic kinematics of disk members.

## References

Arellano Ferro, A., 1985, *Rev. Mexicana A. A.*, **11**, 113
Luck, R. E., Bond, H. E., and Lambert, D. L., 1990, *Ap.J.*, **357**, 188
Tamura, S. and Takeuti, M., 1991, *IAU Inform. Bull. Var. Stars*, 3561
Tamura, S. and Takeuti, M., 1992, *Proc. Internatinal workshop on luminous high-latitude stars*
Tamura, S., Takeuti, M., and Zalewski, J., 1992, *Sci. Rep. Tôhoku Univ. 8th Ser.*, **12**, 145
Volk, K. M. and Kwok, S., 1989, *Ap.J.*, **342**, 345

# LSIV -12° 111 – A NEWLY EMERGING HALO PLANETARY NEBULA

E.S. CONLON, P.L. DUFTON, F.P. KEENAN and R.J.H. McCAUSLAND

*Department of Pure and Applied Physics, The Queen's University of Belfast, Belfast BT7 1NN, Northern Ireland*

**Abstract.** We report on multi-wavelength observations of a young halo planetary nebula, LSIV −12° 111. This object was previously classified as an emission–line young B–type star but a model atmosphere abundance analysis of high resolution optical spectra revealed it to be an evolved object, probably in the post-asymptotic giant branch (post-AGB) evolutionary phase. The presence of an infrared excess and low excitation nebular emission lines implies that the central star may have just started to photoionize the remnant (AGB) circumstellar material. Here we discuss the nebular and dust properties of LSIV −12° 111 and re-determine some metal abundances for the central star. These results are used to constrain the evolutionary status of this unique halo planetary nebula.

# A VERY RAPID-EVOLVING YOUNG PLANETARY NEBULA

A. MANCHADO

*Instituto de Astrofísica de Canarias, E-38200 La Laguna, Tenerife, Spain*

P. GARCÍA-LARIO

*Instituto de Astrofísica de Canarias, E-38200 La Laguna, Tenerife, Spain*

*Laboratorio de Astrofísica Espacial y Física Fundamental, Apartado de Correos 50727, E-28 Madrid, Spain*

and

K.C. SAHU and S.R. POTTASCH

*Kapteyn Astronomical Institute, Postbus 800, NL-970 AV Groningen, The Netherlands*

**Abstract.** During a time interval of only 2 years, a sudden change has been detected in the emission line spectrum of a young planetary nebula IRAS 0656 2-0337, whose nature has been recently discussed by several authors, was firstly observed in December 1987 at La Silla (Chile) and in October 1988 at La Palma (Spain), showing only the Balmer lines in emission. A third spectrum, obtained at La Silla during February 1990, shows a quite different appearance, with forbidden emission lines typical of a planetary nebula. The electronic density is very high, and there is a high infrared excess too. An expansion velocity of 45 km s$^{-1}$ has been estimated. It is shown that the effective temperature of the central star has significantly increased over this period of time. This is consistent with the evolution of a massive progenitor in the post-AGB phase, with $L = 7000 L_\odot$ at $D = 4$ kpc, in which the ionization of the neutral envelope, detected in CO, is now taking place.

# SEARCH FOR THE YOUNG PLANETARY NEBULAE.
# PRELIMINARY RESULTS

L.N. KONDRATJEVA

*Astrophysical Institute of Kazakh Academy of Sciences, Kamenskoje Plato,
480068 Alma-Ata Kazakhstan*

**Abstract.** Some criteria based on the model calculations and observational data were used to choose preplanetaries and young planetary nebulae. Observational results for the object TH4–4 are presented. The spectral and photometric variations of TH4–4 are discussed.

## 1. Introduction

The spectral survey of starlike planetary nebulae have been carried out to distinguish the objects being at the early stage of evolution. The low central star temperatures and as a result of that the low-excitation spectra of nebulae have been used as the main criterion to select objects for the further detail study.

## 2. Observations

The spectral survey have been taken out with the slit spectograph and with image-tube on $0.7m$ reflector. The spectrograms have been obtained with dispersions $16 - 300 A/mm$, resolution was equal to $1 - 20 A$ in dependence of dispersion. The full wavelength range covered $3700 - 8200 A$

## 3. Results

We present the spectral and photometric data for object TH4–4 from Catalogue of planetary nebulae. On the Palomar Survey Plates this object looks like the star $m_{pg} = 15.3$, $m_r = 13.0$. We have derived the very first spectrograms of TH4–4 in 1970. There were strong stellar continuum, some HeI, FeII, [FeII] emission lines and wide HI emissions in spectrum. $2\delta\lambda(H\alpha) = 5.5 \pm 0.2 A$. $T_{eff} = 22000 K$ (by HIZanstra method) $C(H\beta) = 1.2\pm0.2$ (by comparing of the star energy distribution curve with that of blackbody for $T = 22000 K$). At that time TH4–4 have been classified as Be-star. Beginning from 1975 the integral brightness of the object became to decline and some spectral variations have been marked, such as raise of HeI inensities (relatively to that of $H\beta$), appearance and following increasing of [OIII] lines, recently HeII emission has been recorded. All these changing indicate the increase of ionizing flux temperature.

## 4. Discussion

During about 20 years the integral light of TH4–4 has dropped from 13.4 to 15.3 mag. and $T_{eff}$ has raised more than twice. It is evidently that stellar radius had to decrease as a factor of 10 or so. The real compression and heating of star in so short period are quite doubtful. Perhaps we have watched the detachment of some dense envelope which surrounded the star at the early stage of our observations. This event would lead to variations like those remarked in case of TH4–4.

# ABOUT THE SUSPECTED VERY YOUNG PN IRAS 17516-2525

HANS ULRICH KÄUFL
*European Southern Observatory, D-8046 Garching, Germany*
and
LETIZIA STANGHELLINI
*Osservatorio Astronomico di Bologna, I-40126 Bologna, Italy*

Infrared photometry suggests that IRAS 17516-2525 is in transition from the AGB to a young PN (Van der Veen et al. 1989a,b, Manchado et al. 1989). IR spectra (van der Veen et al. 1989b, Käufl et al. 1992) revealed that the object has an ionized core. Optical observations show a weak $H_\alpha$ emitting unresolved object (Käufl et al. 1992) at the coordinates ($\alpha_{1950}$ : $17^h51^m37.8^s$, $\delta_{1950}$ : $-25°25'58''$) suggested by van der Veen et al. (1989a). The spectrum between 2.0 and 2.4$\mu m$ shows e.g. Br$_\gamma$, 2.0875$\mu m$ (unidentified) or the Na$_I$-doublet but no molecular Hydrogen (present in other compact PNs). The ratio of Pf$_\beta$ to Br$_\alpha$ (hardly affected by extinction, insensitive to electron density and temperature) has been studied in IRAS 17516-2525 and Hen1044. For Hen1044 we find this ratio to be in accordance with theoretical calculations (Hummer and Storey, 1987). For IRAS 17516-2525, however, Pf$_\beta$ is approximately twice as strong. Hence presumably Hen1044 is optically thin whereas IRAS 17516-2525 is optically thick at $\lambda \approx$4-5$\mu m$. Line fluxes appear to be constant over a time scale of 4 years. If the visual extinction $A_V$ in the object is $\geq 2^{mag}$ than the infrared luminosity equals the total luminosity (i.e. $L = 1400L_\odot * (\frac{distance}{kpc})^2$). A main-sequence object accidentally extincted by an intervening cloud can be excluded because the observed line-width of Br$_\alpha$ ($\approx 40\frac{km}{s}$, Van der Veen et al. 1989a) is too small as compared to that of O-stars ($\approx 1000\frac{km}{s}$, Käufl, 1992). Assuming an expansion-velocity of $\approx 20\frac{km}{s}$ the stellar appearance constrains dynamic age and size ($age \leq 360y * \frac{distance}{kpc}$, $size \leq 1500AU * \frac{distance}{kpc}$). A pre-main sequence object can also be excluded. The max. luminosity ($L_{PMS} \leq 5000L_\odot$) then requires the object to be closer than 2 kpc where it should be resolvable. In conclusion the observational evidence strongly suggests that IRAS 17516-2525 represents the searched 'missing link' between AGB stars and young compact Planetary Nebulae.

## References

Hummer, D.G., and Storey, P.J: 1987, *MNRAS* **224**, 801
Käufl, H.U. et al.: 1992, in Mass Loss on the AGB and Beyond, ed(s)., *H.E. Schwarz*, in press,
Käufl, H.U.: 1992, *Astron. Astrophys.* in press,
Manchado, A. et al.: 1989, *Astron. Astrophys.* **214**, 139
Van der Veen, W.E.C.J. et al.: 1989a, *Astron. Astrophys.* **216**, L1
Van der Veen, W.E.C.J. et al.: 1989b, *Astron. Astrophys.* **226**, 108

# A SPECTROSCOPIC SEARCH FOR HOT (B-TYPE) POST-AGB STARS

E.S. CONLON

*Department of Pure and Applied Physics, The Queen's University of Belfast, Belfast BT7 1NN, Northern Ireland*

**Abstract.** At Queen's University, we have been undertaking a spectroscopic programme to elucidate the nature of faint blue stars at high galactic latitudes. We have identified approximately 50 that appear to spectroscopically identical (even at high resolution and signal-to-noise) to normal young Population I B-type stars in the galactic disc. However, we have also found seven faint objects () that were previously classified as Population I on the basis of photometry and/or low resolution spectroscopy; careful model atmosphere analyses of high resolution spectra now indicate that they have non-Population I compositions. Their derived atmospheric parameters are coincident with theoretical post-AGB evolutionary tracks and thus together with their peculiar composition, they would appear to be hot evolved post-AGB objects.

A systematic search for other B-type post-AGB objects has been initiated. Criteria for target selection include high latitude, supergiant spectral classification and characteristic post-AGB infrared features. Spectroscopic data have been obtained at the Anglo-Australian Telescope and Mount Stromlo Observatory and analysed using model atmosphere techniques. Preliminary results are presented here.

# NEW CALCULATIONS OF THERMAL PULSES AND s-PROCESS NUCLEOSYNTHESIS IN AGB STARS

M. BUSSO[1], A. CHIEFFI[2], R. GALLINO[3], M. LIMONGI[2], C. M. RAITERI[1], O. STRANIERO[4]

[1] *Osservatorio Astronomico di Torino, 10025 Pino Torinese, Italy*

[2] *Istituto di Astrofisica Spaziale del CNR, Frascati, Italy*

[3] *Istituto di Fisica Generale dell'Università, via P. Giuria 1, 10125 Torino, Italy*

and

[4] *Osservatorio Astronomico di Teramo, Collurania (Teramo), Italy*

A set of thermal pulse models was computed, for initial stellar masses extending from low (M=1.5, 3 $M_\odot$) to intermediate (M=5, 7 $M_\odot$), using the FRANEC evolutionary code and assuming standard mass loss and solar metallicity. The main features are: i) the third dredge-up is naturally found, even for core masses below 0.7 - 0.8 $M_\odot$; ii) before the dredge-up occurrence, the main characteristics of the models (convective shell mass, interpulse duration, overlapping between adjacent pulses) are determined solely by the core mass $M_H$, well reproducing a behaviour which is typical in the current literature (see e.g. Schonberner, 1979): in particular, the shell mass is a decreasing function of $M_H$; iii) after dredge-up is started, the evolutionary track is modified and the strength of the pulses is enhanced; iv) the amount of dredge-up increases in time, from $\simeq 10^{-4}$ $M_\odot$ to $\simeq 10^{-3}$ $M_\odot$.

The s-process nucleosynthesis in the He-shell was computed using for the 3 $M_\odot$ model, assuming that a few $10^{-6}$ $M_\odot$ of $^{13}C$ are formed by proton mixing from the envelope after each pulse (see discussion in Busso et al., 1992). As the core mass $M_H$ grows with time, for any amount of $^{13}C$ burnt per cycle the decreasing convective shell mass leads to increasing neutron exposures per pulse, $\Delta\tau$. In the meantime, the pulse shape also changes: in particular, the *dilution factor*, D (ratio between the shell mass at $^{13}C$ ingestion and that at maximum convective expansion) decreases monotonically. The combined effects of increasing $\Delta\tau$ and of decreasing D cause the *effective* neutron exposure, $\tau_0$, to reach an asymptotic limit which depends only on the amount of $^{13}C$ burnt. This improved scenario does not modify the general results we obtained so far using average thermal pulse conditions. In particular, we confirm our previous suggestions that thermal pulses in low mass stars are the astrophysical site that simultaneously accounts for the *main* s-process component in the Solar System (Käppeler et al., 1990), for the abundance observations of s-enriched AGB stars (Busso et al., 1992) and for s-process anomalies in meteoritic SiC grains (see e.g. Gallino et al., 1992).

## References

Busso, M., Gallino, R., Lambert, D.L., Raiteri, C.M., and Smith, V.V.: 1992, *Ap. J.* (in press).
Gallino, R., Raiteri, C.M., and Busso, M.: 1992, *Ap. J.* (submitted).
Käppeler, F. Gallino, R., Busso, M., Picchio, G., and Raiteri, C.M.: 1990, *Ap. J.* **354**, 630.
Schonberner, D.: 1979, *A.&A.* **79**, 108.

# EVOLUTIONARY PROPERTIES OF POST-AGB AND POST-EAGB STARS

Marco Limongi[1,2], Amedeo Tornambe'[1], and Marco Castellani[1,3]

1) Istituto di Astrofisica Spaziale, CNR, CP67, I-00044 Frascati, Italy
2) Universita' di Roma "La Sapienza", Italy
3) Osservatorio Astronomico di Collurania, 64100 Teramo, Italy

*Subject headings:* galaxies: elliptical - stars: evolution - stars: interiors

**Summary.** *We discuss the evolutionary properties of shell burning star models which, due to the erosion of the external hydrogen envelope, leave the Hayashi track and evolve to the blue of the HR diagram. We analyze the properties of star models which evolve off the Asymptotic Giant Branch (AGB), the Early-AGB or the Red Giant Branch (RGB).*

We evolved several star models from the Zero Age Horizontal Branch to the cooling down phase, assuming different starting masses (from 0.48 $M_\odot$ to 0.54 $M_\odot$) and chemical compositions (Y=0.30, Z=$Z_\odot$ and Z=$2Z_\odot$). Mass loss has been allowed to start at different luminosities along the AGB, simulating in such a way a major variety of different evolutionary paths.

Our main findings are:

a) the structural parameters of stars evolving off the Hayashi track are dramatically different if the evolution starts from the thermally pulsing AGB (post-AGB) or if it starts from the Early-AGB. The first main difference is that post-EAGB models are endowed with a residual H envelope which is much larger than required by proportionally smaller He-cores (the physical reason for this occurrence is explained in CLT). In addition post-EAGB models are endowed with an additional energy source due to the actively burning He-shell.

b) We could define with a rather good degree of accuracy the border core mass which separates the post-EAGB evolution from the post-AGB evolution or from the RGB.

c) However, it exist one, and only one, correlation linking the critical H-envelope mass (at which the blueward excursion from the Hayashi track starts) and the luminosity, whatever being the evolutionary phase (RGB, E-AGB, AGB).

d) For low core masses (post-RGB stars) and large core masses (post-AGB stars) there exists one single correlation linking the critical H-envelope with the core mass while, for intermediate core masses (post-EAGB stars) the critical envelope correlates with the core in a quite different way, turning out to be much larger than in the two adjacent cases.

*References:*
Castellani, M.,Tornambe', A.: 1991, ApJ, **381**, 393
Castellani, M.,Limongi, M.,Tornambe', A.: 1992, ApJ, **389**, 227 (CLT)

# EVOLUTION OF A DUST SHELL ALONG A STELLAR POST-AGB TRACK

H. MARTEN[1], R. SZCZERBA[2] AND T. BLÖCKER[1]

[1] *Institut für Theoretische Physik und Sternwarte der Universität Kiel*
*Olshausenstr. 40, D-2300 Kiel, Germany*
[2] *Copernicus Astronomical Center, Laboratory of Astrophysics*
*ul. Chopina 12/18, 87-100 Toruń, Poland*

We study the spectral appearance of an evolving post-AGB object by means of two-fluid radiation hydrodynamics (gas and dust) in spherical symmetry. The stellar parameters as well as mass-loss rates are assumed to vary consistently with the stellar model calculations of Blöcker (1989, Diplom Thesis, University of Kiel) for a 3 $M_\odot$ sequence, resulting in a 0.605 $M_\odot$ remnant. The expelled mass is assumed to consist of spherical astronomical silicates with grain radii of 0.1 $\mu m$ and an *initial* dust-to-gas mass ratio of $7.5 \cdot 10^{-3}$. For further details on the computational method we refer to Szczerba & Marten (1992, in: Proc. of the ESO/CTIO Workshop, Mass Loss on the AGB and Beyond, La Serena 1992, *in press*).

Following the last 30000 yrs of AGB evolution along Blöckers 3 $M_\odot$ sequence with increasing mass-loss rates and an initial wind velocity of 5 km/sec, the star finally becomes totally obscured by a thick dust shell which expands with a final velocity of about 15 km/sec. In the calculations of Blöcker, the heavy mass loss of the red giant is eventually smoothly decreased between the stellar radial fundamental pulsation periods of 100 and 50 d (see also Blöcker & Schönberner, 1990, A&A 240, L11). During this short phase of only about 200 yrs the stellar continuum reappeares as a bump shortwards of the L photometric band. The decreasing mass loss (density) reduces the frictional coupling between dust and gas, and the dust is accelerated to about 150 km/sec ($v_{gas} \approx$ 50 km/sec). As a result, the contribution of newly formed (hot) dust from this high velocity/low density wind region to the overall spectra during the post-AGB phase is much smaller than expected from the quasi-stationary approach of other authors. Consequently, our model tracks with and without further dust condensation show a very similar behaviour in the $\log(\lambda F_\lambda(60)/\lambda F_\lambda(25))$ versus $\log(\lambda F_\lambda(25)/\lambda F_\lambda(12))$ plane with a maximum difference in their colors of only about 0.4 dex. At a post-AGB age of about 600 yrs ($T_{eff} \approx$ 7500 K), our model sequences in the IRAS diagram start to deviate from what is expected for a simply expanding and cooling shell. Instead of evolving downwards, describing a big loop until the colors of black bodies with the corresponding stellar temperature(s) are reached, the dust shells are re-heated due to the growing stellar temperatures and the models start to proceed to the upper right.

Both, the non-linear dynamical effects of the dust acceleration as well as consistent evolutionary changes of the stellar parameters and mass-loss rates *must* be taken into consideration in order to understand the exact course of the mass loss and its influence on the timescales for post-AGB evolution.

# DUST DRIVEN MASS LOSS FROM PULSATING AGB-STARS *

E.A. DORFI, M.U. FEUCHTINGER, S. HÖFNER
*Institut für Astronomie*
*A-1180 Wien, Austria*

ABSTRACT. A new numerical method allows an acurate calculation of the radiation hydrodynamics of time dependent stellar winds including also the radiation pressure on newly formed dust grains. The numerical procedure is based on an adaptive grid which distributes the grid points at locations of large gradients. All equations are written in conservation form and a monotonic 2. order transport scheme is used to advect the physical variables through the cell boundaries. We are able to resolve the shock waves running through the stellar atmospheres. These waves are generated by a pulsating star which is simulated by a moving piston. The following plots show the radial velocity and temperature structure of an extended atmosphere and several shock waves are clearly seen. Note that the innermost shock waves is a so-called supercritical shock where the radiative cooling zone behind the wave is clearly visible. The outer waves are almost isothermal because the material is optically thin in this region. The stellar parameters of this example are given by $M = 1.2\,M_\odot$, $L = 5315\,L_\odot$ and $R = 270\,R_\odot$ and the period of the moving piston is fixed at 350 days yielding a massive and slow wind with a mass loss rate of $\dot{M} = 1.24\,10^{-6}\,M_\odot\,\mathrm{yr}^{-1}$ and a final velocity of $v = 7.7\,\mathrm{km\,s}^{-1}$.

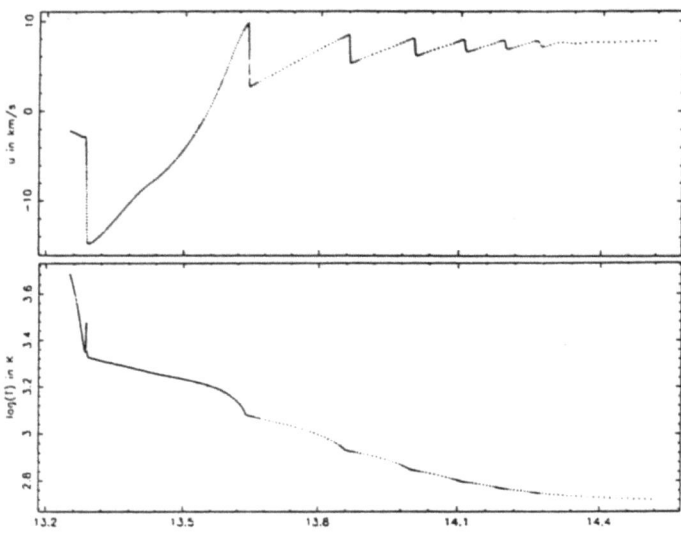

* supported by the FFWF under project number P8411

# IS THERE A CONNECTION BETWEEN THERMAL PULSES AND PNE HALOS: AN APPROACH TO AN ANSWER

ADAM FRANK and BRUCE BALICK
*Astronomy Department, FM-20, University of Washington, Seattle WA 98195, USA*
and
WILL VAN DER VEEN
*Astronomy Department, Columbia University, N.Y., N.Y. 10027, USA*

ABSTRACT. We wish to explore the hypothesis that the shells and halos of planetary nebulae are formed during the thermal pulsing of progenitor AGB stars. Using published data and model results we compare the AGB interpulse time $\Delta t_T$ and the time between shell ejections $\Delta t_D$ for a sample of PNe with halos. $\Delta t_T$ is derived from the Paczynski relation using the PNe central star luminosity. $\Delta t_D$ is calculated using the radii and velocity of PNe shells and halos.

**Conclusions.** 1. *5 out of 6 of the PNe in fig 1 have $\Delta t_D$ and $\Delta t_T$ that are within an order of magnitude of each other.* 2. The analysis is extremely sensitive to the assumed distances as the two points for NGC 7662 demonstrate. 3. One must be careful not to consider PNe central stars that are already on cooling tracks. The $\Delta t_T$-$L_*$ relation does not hold for these cases. This seems to apply to NGC 6720. 4. All the points lie close to or *above* the $\Delta t_D = \Delta t_T$ line but never below it.

After removing NGC 6720 out of the plot and considering the different luminosities and distances quoted for NGC 7662, *there is only one point more than 1 $\sigma$ from the line*$\Delta t_D = \Delta t_T$.

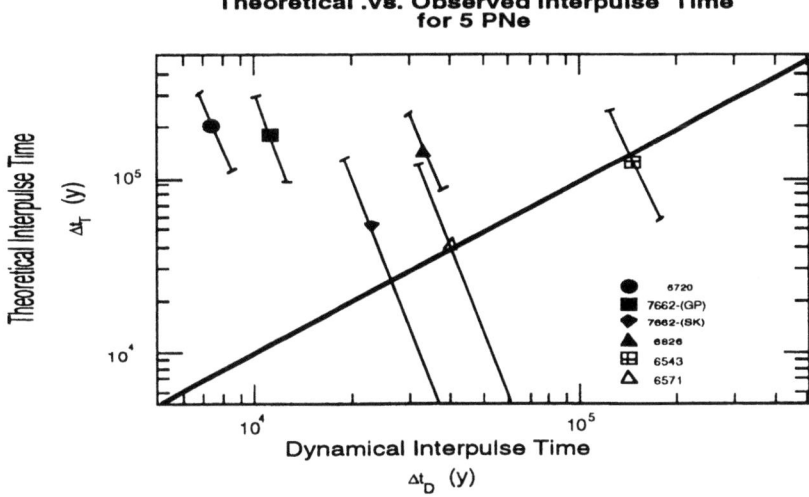

Fig 1. Data points are labeled with their NGC numbers. We used data from two authors for NCG 7662.

# LINEAR PULSATION PERIODS OF THE POST–AGB STARS

M.TAKEUTI, R. TAKANO and S.TAMURA

*Astronomical Institute, Tôhoku University, Sendai, Japan*

There may exist stars during and just after the asymptotic giant branch (AGB) stage. V441 (89) is a candidate star. Fernie and Sasselov (1989) have discussed the evolutionary changes of the pulsation periods based on those of the effective temperature, to compare those with observational data. They used Worell's theoretical results (Worell 1986) and have found that the observational changes of the period and the colors are too small compared with the theorical values. To examine the effect of opacities on the pulsations periods, we calculate the periods based on new opacities (OPAL, Iglesias and Rogers 1991), and compare them with the data reported previously.

We ignore convection through calculations. The OPAL opacities are used with the chemical composition of population II (Y,Z)=(0.299,0.001).

To compare the results with those of Fernie and Sasselov (1989), we choose the effective temperature of 5888K ($\log T_{eff} = 3.77$) for the models. the relation between the period, $\Pi$, and the radius, $R$, is calculated. The gradient of the $\log R - \log \Pi$ relation is determined for the period of 65 days: $\log \Pi = 1.813$ (Fernie 1991) The gradient of the $\log R - \log \Pi$ relation is less steep for pulsationally unstable models. For the convenience of comparison, $dT_{eff}/dt = 10\,\mathrm{K\,yr^{-1}}$ is chosen. Results are tabulated in table I. The rate labelled by Worrell are found in Fernie and Sasselov (1989). The changes of opacities affect the pulsation period only a little bit.

The combination of new results on stellar evolution and our pulsation properties will yield a new criterion to check the stellar evolution theory by using observational data on variable stars.

TABLE I
Theorical Period Change Rates

| $M/M_\odot$ | $d\Pi/dt$ (Worrell) | $d\Pi/dt$ (Present) |
|---|---|---|
| 0.644 | -0.24 | -0.24 |
| 0.600 | -0.11 | -0.13 |
| 0.565 | -0.07 | -0.20 |

## References

Fernie J.D. 1991, PASP, 105, 1087
Fernie J.D. and Sasselov D.D. 1989, PASP, 101, 513
Iglesias C.A. and Rogers F.J. 1991, Ap.J., 371, L73
Worell J.K. 1986 M.N.R.A.S., 223, 787

# EFFECTS OF NEW OPACITY ON THE POST-AGB EVOLUTION

M. KATO

*Department of Astronomy, Keio University, Yokohama 223, Japan*

and

I. HACHISU

*Department of Earth Science and Astronomy, University of Tokyo, Tokyo 153, Japan*

**Abstract.** Using the OPAL opacity, we have calculated the post-AGB evolutions of low mass stars. It is newly found that optically thick wind occurs for $M_{\rm core} > 0.55 M_\odot$ and the evolutionary time scale is drastically shortened.

**Key words:** post-AGB evolution, optically thick wind, mass loss, novae

## Optically thick winds on $M_{\rm core} > 0.55 M_\odot$

We have examined the effect of the new opacity (Rogers & Iglesias 1992, private communication) on the post-AGB evolution of low mass stars. The new opacity has a strong peak (2 or 3 times larger) around $\log T = 5.5$ due to the iron lines compared with the old opacity. The increase in the opacity leads to the decrease in the Eddington luminosity so that we expect strong wind mass loss even for low mass ($M_{\rm core} \sim 0.6 M_\odot$) post-AGB stars, which did show no wind mass loss for the old opacity (Kato & Hachisu 1989, ApJ, **364**, 424.).

Figure 1 shows our new results of the post-AGB evolution of low mass stars with the core mass ranging from $0.5 M_\odot$ to $1.377 M_\odot$. Optically thick wind occurs for $M_{\rm core} > 0.55 M_\odot$ while stars with the core mass of $0.5 M_\odot$ do not show any wind mass loss mainly because the luminosity does not exceed the Eddington luminosity.

The wind mass loss drastically shortens the time scale of the post-AGB evolution. For example, the evolutionary time from $\log T = 4$ to the end point of mass loss in Figure 1 is 12 yr and 24 yr for 0.7 $M_\odot$ and 0.6 $M_\odot$, respectively.

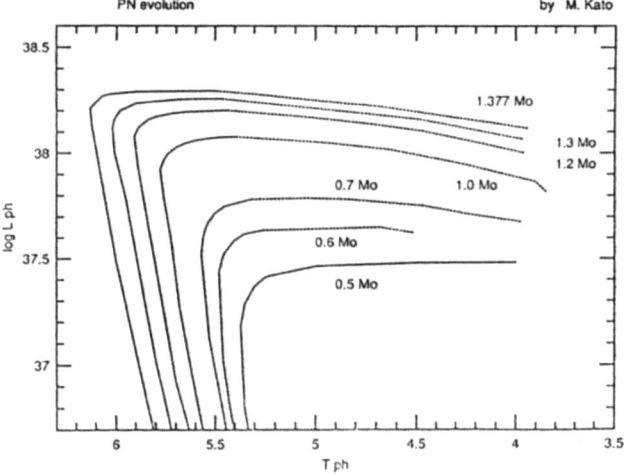

Fig. 1. Theoretical H-R diagram for $M_{\rm core} = 0.5, 0.6, 0.7, 1.0, 1.2, 1.3$, and $1.377\ M_\odot$. Dashed part of the lines denote the wind mass loss phase.

# ANGULAR MOMENTUM LOSS IN POST-MAIN SEQUENCE STELLAR EVOLUTION THROUGH THE PN STAGE

M. VILLATA

*Istituto di Fisica Generale, Università di Torino, Via Pietro Giuria 1, I-10125 Torino, Italy*

A simple analytical model can explain the large angular momentum loss which a star suffers in its post-main-sequence evolution up to the white dwarf stage.

Mass and angular momentum losses in the PN stage are a well-established fact. Indeed it is known that white dwarfs are very slow rotators compared with what would be expected if the initial angular momentum of the parent main-sequence stars were conserved (see, e.g., Hardorp 1974; Greenstein et al. 1977; Weidemann 1977; Pilachowski and Milkey 1987; Koester and Herrero 1988). Thus very effective transport of angular momentum from the deep layers towards the surface has to be postulated, and the star probably maintains solid-body rotation for a significant part of its post-main-sequence evolution.

The basic assumptions of the model are that main-sequence stars and white dwarfs rotate as rigid bodies and that the mass loss between the two stages is due to stellar winds. We neglect the possible braking effect of magnetic fields, so that our results for the angular momentum loss will be regarded as lower limits. Hence, at each stage of the post-main-sequence evolution, the angular momentum loss is given by (with the assumption of spherical symmetry)

$$-dJ \equiv -d(hMR^2\Omega) = \frac{2}{3}R^2\Omega(-dM), \qquad (1)$$

where $hMR^2$ is the moment of inertia of the star of mass $M$ and radius $R$, and $\Omega$ is its angular velocity.

A theoretical lower limit to the total angular momentum loss between the main sequence ("i") and the white dwarf stage ("f") is given by the simple formula (Villata 1992)

$$\frac{J_f}{J_i} = \left(\frac{M_f}{M_i}\right)^\delta, \quad \text{with} \quad \delta = \frac{2}{3}\frac{\ln(h_f/h_i)}{h_f - h_i}, \qquad (2)$$

and turns out to be very large, varying from 99% for a star which evolves into a 0.5-$M_\odot$ white dwarf to more than 99.99% for a final mass higher than 1.1 $M_\odot$.

We find that the theoretical upper limit to the rotational velocity of white dwarfs is $60\,\text{km}\,\text{s}^{-1}$, in very good agreement with observational values.

### References

Greenstein, J.L., Boksenberg, A., Carswell, R., and Shortridge, K. (1977), ApJ 212, 186.
Hardorp, J. (1974), A&A 32, 133.
Koester, D. and Herrero, A. (1988), ApJ 332, 910.
Pilachowski, C.A. and Milkey, R.W. (1987), PASP 99, 836.
Villata, M. (1992), MNRAS (in press)
Weidemann, V. (1977), A&A 59, 411.

# MODELLING PN FORMATION FROM HYDRODYNAMICS AND RADIATION

G. MELLEMA

*Sterrewacht Leiden, P.O. Box 9513, 2300 RA Leiden, The Netherlands*

The formation of planetary nebulae from the mass loss remnants of previous phases forms an interesting problem in both hydrodynamics and astrophysics. Not in the least because of the asphericity that most PNe show. The origin of this asphericity is still largely a mystery. Binarity is the most favoured explanation at the moment (Morris 1987)

By assuming that the asphericity is introduced by some cause in the AGB phase, the work of Soker & Livio 1989, Mellema et al. 1991, and Icke et al. 1992 has shown that even if the effects of radiation from the central star and the gas are neglected (adiabatic case), many interesting features resembling observed aspherical PNe are found. See the paper of Icke at this conference for an overview of this. This paper reports on the work of including radiation effects in these models.

The essence of the problem of PN formation is that both hydrodynamics and radiation play an important role. Furthermore, the radiation is what we observe. That is why we have modified our 2D hydrocode such that the hydrodynamics and the radiation and ionization state are solved at the same time. It is important to note that we treat the problem consistently this way. That is, the program calculates the radiation losses, which influence the flow, and which are also direct observables. This allows us to produce images of our models in the well-known emission lines of H$\alpha$, [OIII], and [NII]. Combined with the velocity information from the hydrodynamics, kinematic data can also be constructed.

This combined radiation and hydrodynamics code should be able to answer some of the questions regarding PNe formation. We plan to use it to simulate the emergence of PNe, especially concentrating on the role of the different shaping components. These are the (slow) mass loss on the AGB, and the (fast) one after that. It should be possible to place limits on the behaviour of the various winds, and separate influence of the wind interaction from the effects of ionization fronts. The possibility to construct both synthetic images and kinematic data should be a great help in comparing to the observational data.

All this is work done together with Adam Frank, Bruce Balick, and Vincent Icke whose contributions at these proceedings touch upon some other aspects of this project.

### References

Icke, V., Frank, A., Balick B. (1992) A&A 253, 244.
Mellema, G., Eulderink, F., Icke V. (1991) A&A 252, 719.
Morris, M. (1987) Publ. Astr. Soc Pac. 99, 1115.
Soker, N. and Livio, M. (1989) ApJ 339, 268.

# EVOLUTION OF PLANETARY NEBULAE ENVELOPES: AN EMPIRICAL APPROACH

V.V. GOLOVATY

*Astronomical Observatory of Lviv University, 290005 Lomonosov str. 8, Lviv, Ukraine*

and

YU. F. MALKOV

*Crimean Astrophysical Observatory, 334413 Nauchny, Crimea, Ukraine*

We carried out the empirical investigation of the evolution of gas density distribution in the envelopes of planetary nebulae (PN). For this purpose we analysed the isophotal maps of 10 PN in the lines H alpha, H beta or in the optically thin radio continuum. To obtain the spatial radial distribution of gas density $n(r)$ (where $n = n(H)+n(He)$) we used Abell's integral equation in the simplest, spherically-symmetric case. We found that $n(r)$ for all PN envelopes in our sample can be described by an approximative expression:

where $x=r/rc$, $A=M/(4pi\ V\ mH\ mH(3.086E18)2)$. Here M is the mass loss rate of the star-precursor, V is the terminal velocity of precursos's wind, mH is the mass of hydrogen atom; the value of rc (pc) characterizes the age of the envelope. The figure shows the agreement between the formula (1) (dots) and the empirical distributions $n(r)$ (solid curves). The straight line corresponds to precursor's wind with constant M/V. It is clearly seen that the formation of the envelope begins from the decrease of precursor's mass loss rate: young PN have the gas density which is less than the density of precursor's wind at the same distance from the center. This conclusion is insensitive to errors in distances to PN (as an example, the dashed curve shows the position of IC 5117 if we increase the distance to this PN by 3 times). Thus, the idea that PN are formed as a result of a rapid increase of precursor's mass loss rate ("superwind") must be ruled out.

# NUMERICAL STUDY OF THE SHAPING OF PLANETARY NEBULAE

I. V. IGUMENSHCHEV

*Institute of Astronomy, 48 Pyatnitskaya Str., 109017 Moscow, Russia*

An axisymmetric density distribution likely to be representative for many Planetary Nebulae (PN). We suggest that the axial symmetry of PN results from a predominant ejection of matter in the equatorial direction due to the duplicity of the central star (Livio *et al.*,1979, Ap.J.,**188**,1). We present an illustrative example of the formation of the bipolar PN Soutern Crab (He 2-104). In this model high velocity matter ejected by hot central star interacts with an outer oblate envelope located around a symbiotic binary star. The binary consists of a Mira variable, ejected matter forming a thick disk, and a hot component (Lutz *et al.*,1989, PASP 101,966). Accretion of some disk matter onto the hot component (dwarf) may lead to recurrent thermal shell flashes (Igumenshchev *et al.*,1990, Astrofizika,**30**,282), which result in the double shell nebula, observed in the Southern Crab (Burgarella *et al.*, 1991,A&A,**249**,199). The phase of forming a single shell stucture of the Southern Crab was simulated numerically. The results of a 2D hydrodynamical nonadiabatic calculation are illustrated in the Figure. The density contours with logarithmic spacing of 0.25 and the velocity field (arrows) of the model are shown at time $t \simeq$ 360 $yrs$ after the central star outburst with energy $10^{44}$ $erg$. The dense shell may be observed as a Crab-like nebula ("hour-glass" type). In this model the mass of the envelope is $7 \cdot 10^{-4} M_\odot$ and the maximal velocity is over of 200 $km/s$.

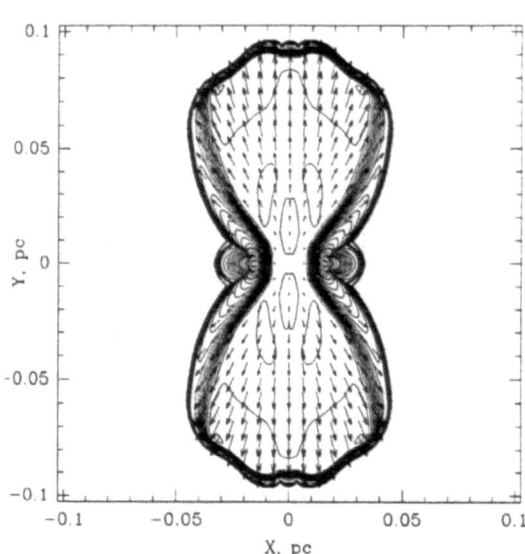

# DEPROJECTION OF PLANETARY NEBULA IMAGES

K. VOLK and D. A. LEAHY

*Department of Physics and Astronomy, University of Calgary, 2500, University Drive N.W., Calgary, Alberta, Canada T2N 1N4*

We present first results of the deprojection of some optical and radio images of planetary nebulae using an iterative technique based on the Lucy method, as discussed in Leahy (1991). This approach allows a greater range of possible geometries to be investigated and is less sensitive to the effects of noise compared with the direct matrix-inversion method which can be used for the spherical case. We assume that the emissivity function can be represented as

$$\epsilon(r,\theta) = f(r)g(\theta)$$

where $(r, \theta)$ are polar coordinates in a plane defined by our line of sight to the nebula and a radial cut of the sky image ($\theta = 0$ in the plain of the sky). By analyzing a number of cuts at different orientations we can approximate the three-dimensional emissivity function. In these models we assume that $g(\theta)$ is of the form $\cos^N(\theta - \theta_0)$ with $N = 0$ (symmetric case) or $N = 2$, and derive $f(r)$. So far we have always found that $N = 2$ gives a much better result than $N = 0$, which often cannot give any reasonable fit to the data. The $\theta_0$ parameter allows for a tilt of the plane of symmetry of the nebula with respect to the plane of the sky. So far we have done only a few trials with $\theta_0 \neq 0$.

The initial results suggest that the studied nebulae (NGC 40, NGC 6720, M3-35, NGC 6790, and IC 5117) have incomplete interacting-winds shells covering of order 25% of the sky as viewed from the star, and would appear more or less bi-polar if viewed in the plane of the sky. At intermediate orientations the nebulae would generally appear as incomplete rings. In some directions it appears that either no interacting-wind shell has formed or the shell has been disrupted. We suggest that the mass-loss process while the stars were on the AGB were quite asymmetric, so that when the second, fast wind develops the interacting-winds bubble quickly blows out and thereafter the sections of the shell which remain enter a snow-plow phase. Our results indicate that the nebulae may be better described in cylindrical coordinates $(r, \phi, z)$ with considerable $\phi$ dependence of the structure. We hope to develop a version of the deprojection program which uses cylindrical coordinates to investigate this possibility.

## References

Leahy, D. A. 1991, A&A, **247**, 584

# TWODIMENSIONAL AXIALSYMMETRICAL HYDRODYNAMICAL SIMULATIONS OF PN-EVOLUTION

J. ZWEIGLE and M. BREMER

*Astronomisches Institut der Universität Waldhäuserstrasse 64, D-7400 Tübingen, Germany*

M. GREWING

*Astronomisches Institut der Universität Waldhäuserstrasse 64, D-7400 Tübingen, Germany*

and

*Institut de Radioastronomie Millimtrique (IRAM), 300 Rue de la Piscine, F-38406 Saint Martin d'Hères*

In order to investigate the early evolution of planetary nebulae (PNe) we solved numerically the hydrodynamical equations in cylindrical coordinates (r,z) assuming azimutal symmetry. The numerical method used is described in detail by Mair et al. (1988). Our simulations model the interaction of a fast, tenuous, spherical symmetrical central star wind with a slow, dense, aspherical Red Giant Envelope (RGE) expelled from the progenitor star. For the aspherical RGE with a polar/equatorial density contrast we used the initial model given by Mellema et al. (1991) in cylindrical coordinates. We have investigated the influence of each initial model parameter upon the evolution of PNe. Thereby we confirm that the polar/equatorial density contrast in the RGE and the thickness of the RGE-disk play an important role for the morphology of PNe. In agreement with the results from Mellema et al. (1991). The polar/equatorial density contrast in the RGE influences the ratio of the distances of the bright inner rim to the central star in z- and r-direction. This ratio increases with decreasing polar/equatorial density contrast. We find the thickness of the RGE-disk to be a key parameter for getting an elliptical or a butterfly PN: thin RGE-disks produce the first type of nebulae, thick disks the latter. We thank G. Mair, E. Müller and W. Hillebrandt for making available to us a copy of the SADIE code.

### References

Mair, G. et al. 1988, Astron. Astrophys. 199,114
Mellema G. et al. 1991, Astron. Astropys. 252,718

# SPHERICALLY SYMMETRIC KINEMATIC MODELLING OF PLANETARY NEBULAE

C. DIESCH (1) and M. GREWING (1,2)
*(1)Astronomisches Institut der Universität
Waldhäuserstraße 64, D-7400 Tübingen
(2)Institute for Radioastronomy at Millimeter Wavelengths (IRAM)
300,rue de la Piscine,Domaine Universitaire,F-38406 St.Martin d'Hères, France*

**Abstract.** High resolution two-dimensional long slit spectra of NGC 7009, NGC 3242, IC 2448 and TC 1 in the H$\alpha$(6565 Å), OIII(5007 Å) and NII(6584 Å) emission lines have been used as input to determine radial velocity gradients in PNe by fitting calculated high resolution spatially resolved spectral structures of single emission lines. Contributions to the Doppler structure of emission lines taken into account are the global expansion of the PN shell and thermal line broadening due to temperatures of some $10^4$ K. This model is based on two parametrized empirically chosen spherical symmetric functions:
1. A empirical two parameter density function $\rho(r)$.
2. A linear radial velocity component function $v(r)$.
Fitting this model to the observed PN spectra yield the following results: All observed PNe show increasing radial velocity towards the outer rim of the shell. The velocity function $v(r)$ derived for NGC 7009 fits excellent to the former results for the outer regions of this PN (Weedman). In the case of TC 1 the velocity $v(r)$ function derived for different radii from OIII, H$\alpha$, and NII due to the ionisation structure in the PN are in very good agreement. The table below lists these results numerically:

|  | line | v(r=0) [km/s] | dv/dr [km/s/$10^{12}$ km] | appl. radii [$10^{12}$ km] |
|---|---|---|---|---|
| NGC 7009 | H$\alpha$ | -4.1 | 7.3 | 1.5 - 4.5 |
| NGC 3242 | H$\alpha$ | -28.2 | 11.4 | 2.5 - 4.5 |
| IC 2448 | H$\alpha$ | 4.5 | 3.6 | 0 - 5.0 |
| TC 1 | OIII | -1.2 | 3.7 | 0 - 2.5 |
|  | H$\alpha$ | -7.8 | 6.4 | 1.0 - 4.5 |
|  | NII | -9.1 | 6.5 | 1.5 - 5.5 |

### References

Weedman, D., Astrophys. J. **153**, 49
Weinberger, R., Astron. Astrophys. Suppl. Ser. **78**, 301

# SHOCK MODELLING AND HIGH RESOLUTION SPECTROSCOPY OF NGC 6905

L. CUESTA and J. P. PHILLIPS
*Instituto de Astrofísica de Canarias, E-38200 La Laguna, Tenerife, Spain*

NGC 6905 appears to be an unusual source in displaying conical outflow lobes superimposed upon a spheroidal inner shell.

We have recently undertaken high resolution spectroscopy of this nebula using the 2.5 m Isaac Newton Telescope (Observatorio del Roque de los Muchachos), and our results may be briefly summarised as follows:

a) The principal shell is expanding at velocities $\Delta V \sim 100$ km sec$^{-1}$, whilst spectra taken along the major axis reveal intense [NII] emission at the nebular limits, suggestive of ansae.

b) There is evidence to favour a change in extinction over the interior shell of $\Delta A_v \sim 1$ mag; presumably indicative of local extinction.

c) The spectral lines are well resolved, and yield 2D velocity maps which are well explained in terms of a tilted spheroidal inner shell, providing we assume velocity to be proportional to radius. The kinematics of the exterior cones appear to be radically different.

We have adopted a model similar to those of Canto (1980) in attempting to replicate the overall source morphology, whereby stellar winds are assumed to shock interact with ambient material, leading to distinct hour-glass configurations where the flow is collimated. In the present case, we suppose that either a) symmetrically disposed holes in the inner shell enable a stellar wind to interact with an exterior halo, or b) that the entire nebula is a consequence of such shock interaction, with the internal nebulosity constrained by a higher density disk.

The consequences are in both cases similar, and it is apparent that such models match the observed source morphology extremely closely. Comparison between observed line of sight velocities and model results are also reasonably good, whilst we find that structures of this kind would require wind velocities $V_w \sim 430$ km sec$^{-1}$, and typical mass-loss rates $\sim 10^{-7}$ M$_\odot$ yr$^{-1}$.

Finally, we note that such a model is not only capable of explaining the peculiar morphology of this source, and the observed kinematics, but also provides a ready explanation for the presence of ansae. In this case, we suppose such features to derive from the collision of opposing shock refracted streams, leading to the shock enhancement of lower excitation ionic species.

# A MODELLING OF EXPANSION VELOCITIES OF PLANETARY NEBULAE

K.GĘSICKI and R.SZCZERBA

*Nicolaus Copernicus Astronomical Center, Chopina 12/18, 87-100 Toruń, POLAND*

Recently written model of photoionization structure of planetary nebula has been used to calculate the emissivities of different nebular lines. With these values the line profiles have been modelled. The profiles can be obtained for different size and position of the spectrograph slit on the nebula image. The prepared computer code allows for estimation of the expansion velocities ($V_{exp}$) in the way consistent with the methods used by observers, i.e. defined as the half-width of the line profile (when nebula is not resolved, "whole neb." in the Figure) or defined as the half of the emission peaks separation (when the nebula can be resolved - "rectangle" in the Figure - and the line can be observed in the central part of the nebula). The real matter velocity at the radius where the observed surface brightness of $H\beta$ falls below 20 % of the maximal value is shown for comparison ("model" in the Figure).

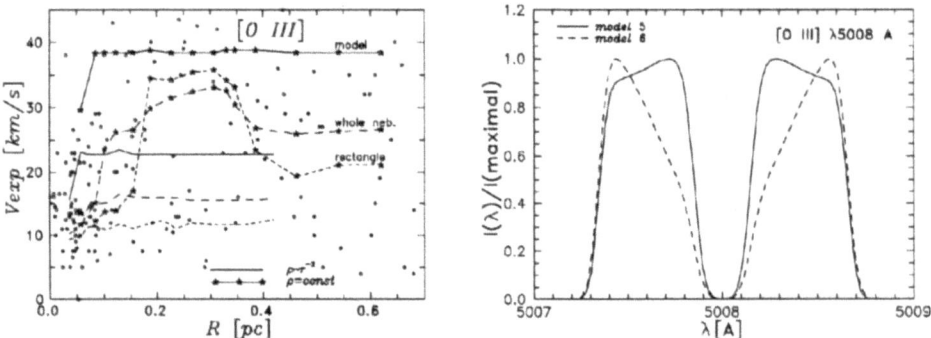

Figure: *The obtained evolutionary sequences (left) and an example of calculated profile (right)*

The simple models of planetary nebula evolution (shell expanding with $V_{exp}$ = 40 km/s at the outer radius and 10 km/s at the inner edge, when the central star follows Schönberner's model of 0.598 $M_\odot$) agree quite well with the observed correlations of $V_{exp}$ versus planetary radius, but rather do not follow the correlation of $V_{exp}$ versus excitation class of nebula as measured by line ratio I(HeII $\lambda 4686$)/I(H$\beta$). The presented line shapes show how they are influenced by ionization structure of the nebula, and movement of the ionization front explains the jump in measured velocities seen in the left Figure for model with constant density distribution.

# ECHELLE MEASUREMENTS OF THE EXPANSION VELOCITIES OF THE FAINT GIANT HALOES OF PLANETARY NEBULAE

M.BRYCE[1], J.MEABURN[1] and J.R.WALSH[2]

[1] Department of Astronomy, University of Manchester
Oxford Road, Manchester M13 9PL. UK
[2] Space Telescope European Coordinating Facility, ESO
Karl-Schwarzchild-Strasse 2, D-8046, Garching bei München, Germany

The faint, giant haloes observed in many Planetary Nebulae (PNe) are believed to be the remnants of the superwinds ejected by the progenitor asymptotic giant branch stars. Spatially resolved, high spectral resolution observations of [OIII]5007Å line profiles from the haloes of NGC 6543 and NGC 6826 have been obtained with the Manchester echelle spectrometer at the Isaac Newton Telescope. The widths of the observed profiles from NGC 6826 decrease towards the outer edge of the halo which implies a definite radial expansion of this halo. The expansion velocities $v_{exp}$ of the two haloes are derived using both thick and thin shell models.

In the thin shell model, the halo is assumed to be a spherically expanding thin shell of constant density. The model predicts line splitting (not seen in either data set) if $v_{exp} \geq$ the component widths. This model halo is limb brightened. In the thick shell model, the halo is assumed to be filled with a radially expanding wind with density inversely proportional to the square of the radius. The resultant profiles are found to be ~ Gaussian in shape. The surface brightness of this model halo falls off as $x^{-3}$ where $x$ is the angular distance from the centre of the nebula.

| Model | NGC 6543 | NGC 6826 |
|---|---|---|
| $v_{exp}$ Thin Shell (kms$^{-1}$) | 4.8±0.6 | 9.8 ± 0.6 |
| $v_{exp}$ Thick Shell (kms$^{-1}$) | 7 ± 1 | 13 ± 1 |

TABLE I
Expansion velocities of the two haloes derived using thick and thin shell models.

The values of $v_{exp}$ derived for these haloes using the two models are shown in Table I. The thick shell model predicts a larger value of $v_{exp}$. The halo of NGC 6543 is filamentary and appears to be limb brightened. Data from the bright knot has been included in the calculations of $v_{exp}$, the results are not significantly affected if this data is omitted. Both predicted values of $v_{exp}$ for NGC 6543 are lower than the expected superwind expansion velocity which is ~ 10kms$^{-1}$. The halo of NGC 6826 is more uniform although somewhat mottled in appearance. Limb brightening at [OIII]5007Å is not apparent, except for the bright knot. The decrease in observed profile width to larger angular radius indicates a systematic expansion of the halo. The value of $v_{exp}$ predicted by the thin shell model implies line splitting, which is not seen in the data. Together with the absence of limb brightening at [OIII]5007Å, this suggests that the thick shell model is more applicable to this halo than the thin shell model.

# KINEMATICAL STUDIES OF PLANETARY NEBULAE USING TAURUS+CCD

K.C. SAHU

*Kapteyn Laboratorium, University of Groningen, Groningen, The Netherlands*

J.R. WALSH

*Space Telescope European Coordinating Facility, ESO, Garching, Germany*

N.A. WALTON

*Department of Physics and Astronomy, University College London, London, England*

and

S.R. POTTASCH

*Kapteyn Laboratorium, University of Groningen, Groningen, The Netherlands*

TAURUS, the imaging Fabry-Perot spectrometer, using a CCD detector, has been commissioned on the 4.2m William Herschel Telescope, creating a powerful tool for studying the kinematics of emission line objects such as planetary nebulae. The mode of operation when using a CCD is different to that with IPCS (where rapid scanning is employed), so that account has to be taken of changes in atmospheric transmission between CCD readouts. The large dynamic range of the CCD ($>200$) allows the simultaneous study of the line profiles from the right and faint parts of the nebulae.

We have kinematic data on five planetary nebulae with very different properties: NGC 2440 (high temperature central star); NGC 3242 (multiple shell); Abell 30 (hydrogen deficient); NGC 1535 (bipolar); and IC 3568 (spherically symmetric). Observations in lines of different excitation: HeII 4686A, [OIII] 5007A, [NII] 6583A and [OI] 6300A, have been obtained in order to model the kinematics of the nebulae.

As an example we show results for NGC 2440 in the [OIII] 5007A line.

# HIGH-DISPERSION SPECTROSCOPY OF IC 351 AND NGC 3242. PLANETARIES WITH HIGH INTERNAL MOTION

Y.YADOUMARU and S.TAMURA

*Astronomical Institute, Tohoku University, Aoba-ku Sendai 980, Japan*

We made high-dispersion spectroscopic observations of IC 351 = PK 159−15°1, a compact high-excitation planetary nebula. We found the expansion velocity from H$\alpha$, [OIII] and HeII line, of $45.1 kms^{-1}$, $38.7 kms^{-1}$, and $33.3 kms^{-1}$, respectively. Moreover, for the H$\alpha$ line we detected an additional faint blue-shifted component escaping from the center of expansion with high velocity of $120 kms^{-1}$. This value is larger than the usual expansion velocity ($< 50 kms^{-1}$) but considerably smaller than that of stellar wind. This faint blue-shifted component of H$\alpha$ is also coincidently identified as the helium line, HeII$\lambda$6560 (Pickering 6). Using the emission coefficients of recombination lines (Case B, T= 10,000K; Osterbrock 1989), and the intensity ratio, I(HeII$\lambda$4686)/I(H$\beta$)=0.56±0.05 (Aller&Czyzak 1979), we can estimate the expected line intensity ratio, I(HeII$\lambda$6560)/I(H$\alpha$)=0.026±0.002. On the other hand, the intensity ratio of faint component to H$\alpha$ in the present observations is estimated as $I_{faintcomp.}$/I(H$\alpha$)=0.064±0.013, based up the H$\alpha$ profile in 1988. Comparing these two estimated intensity ratios it is clearly unable to emit the all intensity of the observed faint component by HeII$\lambda$6560 line only. Therefore, even if this component is partly polluted by HeII$\lambda$6560 line, we can consider the rest part as an emission from high velocity component of H$\alpha$. We can also recognize the bumped feature of [OIII]$\lambda$5007 line around the same velocity to the blue-shifted component of H$\alpha$. This fact supports the idea that these components can be attributed to the high velocity flow. We discuss the three ideas for the interpretation of such component; (a) colliding winds, (b) unresolved bipolar flow, and (c) secondary formation of an expanding shell. We have investigated whether the colliding winds model gives such high velocity or not. As the results of the calculation, if we use the usual mass-loss rate, the model can not give such high velocity. Therefore, it is doubtful that we can take this model as an explanation of the high velocity components. The high velocity component of IC 351 is not uncommon; such large internal motions are reported for other planetaries by Weinberger(1989) and compiled by Grewing(1989). The article on the results of IC 351 is submitted to **PASP** (Yadoumaru & Tamura 1992). We also report the other analysis on NGC 3242.

### References

Aller L.H. and Czyzak S.J.,1979,**Ap&SS**,62,397.
Grewing M.,1989,in "IAU Symposium No.131, *Planetary Nebulae*", ed. Torres-Peimbert S.,(Reidel, Dordrecht, Holland),p.241.
Osterbrock D.E.,1989,*Astrophysics of Gaseous Nebulae and Active Galactic Nuclei*,p.80 and p.85.
Weinberger R.,1989,in "IAU Symposium No.131, *Planetary Nebulae*", ed. Torres-Peimbert S.,(Reidel, Dordrecht, Holland),p.93.
Yadoumaru Y. and Tamura S.,1992,submitted to **PASP**.

# INTERACTION OF PLANETARY NEBULAE WITH PRENEBULAE DEBRIS

## JULIETA FIERRO
*Instituto de Astronomia, Unam*

In the present poster we suggest that some of the structures observed in the envelopes of planetary nebulae are caused by the interaction of central star wind and radiation with preplanetary nebula debris: planets, moons, minor objects, ring and ring arcs.

Recently considerable amount of planetary material has been reported to exist around solar type stars, this debris could be evaporated during the envelope ejection and alter the chemical abundance and produce some of the envelope inhomogenities.

If there are massive enough rings of material surrounding the progenitor and planets in their vicinity arc rings could be formed. If the rings are viewed pole on when the envelope is detached from the central star it will interact with the arc ring material and produce "ansae" and pedal and garden hose shape structures observed in some planetaries.

# THE MAGNETIC FIELDS IN THE ENVELOPES OF PROTO-PLANETARY NEBULAE

J.Y. HU

*Beijing Observatory, Academia Sinica, Beijing, 100080, China*

It is possible that the magnetic field plays important role in the formation of planetary nebulae(Poscoli,1992). In order to measure the strength of magnetic field in the envelope of proto-planetary nebulae(PPNe) we have used the Max-Planck-Institut fur Radioastronomie 100-m telescope at Effelsberg to obtain the high frequency resolution and high signal-to-noise ratio 1612 MHz spectra of PPNe, IRAS08005-2356, 18276-1431, and 20406+2953 in both circular polarization. The nature of PPN of these objects are confirmed by Slikhuis et al.(1991), Le Bertre(1987), and Hu et al.(1992) based on the extensive optical, infrared and radio molecular line observations.

The OH maser from IRAS08005 shows multi-peak structure and the Zeeman splitting is larger than width of individual peaks. We can easyly find the Zeeman split pairs with $\delta V=0.81$km/sec. It results the strength of magnetic field of 2.3mG. But we also found that some peaks are without counterparts. It means that in the masing region the magnetic field is not uniform.

In the case of IRAS18276 and 20406 the Zeeman splitting is smaller than width of OH maser peaks. The Stokes parameters of $I(\nu)$ and $V(\nu)$ are used to derive the strength of field. It results 0.33 and 0.26 mG. It is also clearly shown that not all of components of peaks are polarized. So the results are the low-limited values. Again this confirms that the magnetic field in the masing region in these two objects are not uniform.

The strength of magnetic fields in the masing region of PPNe are considerably stronger than that in the envelopes of OH/IR stars found by Zell and Fix(1991) on the order of 1-100$\mu$G.

Hu,J. Y. et al.,1992, in preparing.
Le Bertre, T., 1987, A&A, 180,160.
Pascoli, G.,1992, PASP,104, 350.
Slijkhuis, S. et al.,1991, A&A,248, 547.
Zell, P. J., Fix, J. D.,1991, ApJ,369,506.

## ON BIPOLAR JET FORMATION IN PLANETARY NEBULAE

### G. PASCOLI

*Faculté des Sciences, Department of Physics, Amiens, France*

Bipolar outflows or polar knots are rather ubiquitous phenomena. A wealth of observational data is now available on bipolar flows associated with young stellar objects or stellar formation regions (Snell, Proc.I.A.U.Symp.Nr. 115, 1987), or nuclei of active galaxies (Asseo and Sol, 1987, Phys.Rep. 148, 307).

In the Planetary Nebula (PN) context, the origin of bipolar jets has never been completely elucidated (Morris, Proc.Coll. "From Miras to PNe", Montpellier, France, 1990); even though interesting suggestions have been made in this sense (Soker and Livio, 1989, ApJ 339, 268).

We propose here a stationary magnetohydrodynamic model in order to explain the origin of bipolar jets or highly collimated outflows in PNe. Analytical calculations have been performed taking into account the effects of both a large scale azimuthal magnetic field and an anisotropic turbulent velocity field within the gas.

A comparison with the observational data (polar-to-equatorial density contrast, polar velocity magnitude and isophotal maps) allows us to expect the equatorial-to-polar density ratio to be typically in the range of 2-3. On the other hand, the present model does predict polar jet velocities $\sim$ 100 km(s (assuming an equatorial velocity $\sim$ 10 km/s), in good agreement with the observations. Stemming from our hypotheses, another interesting conclusion is that elongated turbulent cells, similar to filamentary structures, very likely appear within the polar regions. On the other hand a ring of magnetized matter is also formed in the equatorial plane. Theoretical isophotes are produced.

The model can be applied to objects such as He 2-104, Mz3, M2-9.

# STRIPPING OF A PLANETARY NEBULA FROM THE GLOBULAR CLUSTER M22

K.J. BORKOWSKI

*Department of Astronomy, University of Maryland, College Park, MD 20742, USA*

J.P. HARRINGTON

*Department of Astronomy, University of Maryland, College Park, MD 20742, USA*

and

Z. TSVETANOV

*Department of Physics and Astronomy, The Johns Hopkins University, Baltimore, USA*

High-spatial resolution imaging in the [O III] $\lambda 5007$ line of a planetary nebula (PN) in the globular cluster M22 reveals a strongly asymmetric (a half-moon shaped) nebular morphology. We confirm that this peculiar morphology is caused by the distortion of stellar ejecta by the ram pressure of the ambient interstellar medium (ISM) through which the cluster is moving with a high velocity of 200 km s$^{-1}$. Bright emission knots visible in the leading (upstream) nebular section confirm theoretical expectations that the shell should have been fragmented by the Rayleigh-Taylor instability. Stripping of the PN from M22 by the ambient ISM is the first direct evidence for removal of gas from globular clusters. M22 is located at a favourable location for the ram-pressure stripping to be effective, only 400 pc below the Galactic plane, well within a 2 kpc thick layer of ionized gas enveloping the plane. Ionized (possibly neutral) gas in this layer, with a hydrogen density of $\sim 0.1$ cm$^{-3}$, is responsible for the observed interaction.

# A DETAILED STUDY OF THE GALACTIC PLANETARY NEBULA G 258-15.7

P. LEISY and M. DENNEFELD
*Institut d'Astrophysique de Paris, IAP/CNRS*

The galactic Planetary Nebula G 258-15.7 is a large, bright nebula well suited for a detailed study. Known since Wray (1966), its morphology presents several blobs and ansae, generally associated with type I nebulae, and could be described as "late-butterfly" type according to the classification by Balick (1989). The central star has been classified as hydrogen-deficient by Mendez et al. (1985). Spectroscopy of the two main blobs shows a clear overabundance in He and N, with a marginally significant difference between the two sides. The most striking feature is the jet-like structure appearing on the [OIII]/Halpha picture (Fig. 1), the "jets" being located within the main blobs seen on the monochromatic images. A detailed appraisal of all the data will be presented in a subsequent paper.

## References

Balick, B., 1989, in IAU Symp. 131, p. 83, S. Torres-Peimbert, Edt
Mendez, R.H., Kudritzki, R.P. and Simon, K.P., 1985, Astron. Astrophys. **142**, 289
Wray, J.D., 1966, Ph.D Thesis, Northwestern University

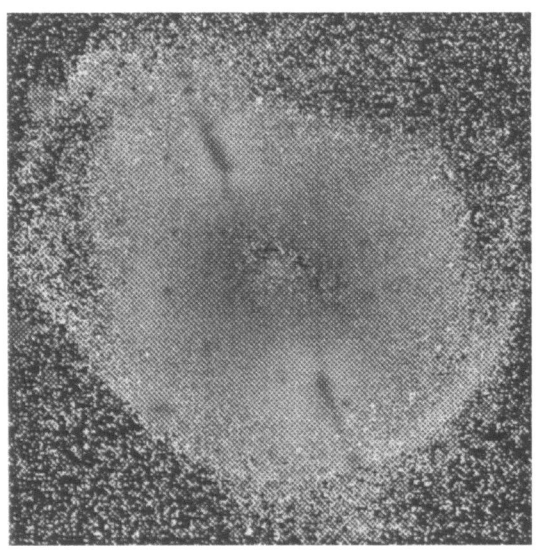

Fig. 1. NW blob : He/H = 0.15 ; N/O = 0.27.  SE blob : He/H = 0.14 ; N/O = 0.21.

# SHOCK MODELLING OF THE BIPOLAR OUTFLOW SOURCE NGC 6537

L. CUESTA and J. P. PHILLIPS

*Instituto de Astrofísica de Canarias, E-38200 La Laguna, Tenerife, Spain*

NGC 6537 is an unusual high excitation bipolar outflow source, with anomalous abundances indicative of a type I nebula. We have recently obtained a range of high resolution spectrophotometry for this source using the 2.5 m Isaac Newton Telescope (Observatorio del Roque de Los Muchachos, La Palma), together with narrow band optical imaging using the 2.6 m Nordic Optical Telescope (Observatorio del Roque de los Muchachos, La Palma). As a consequence, it is apparent that the source optical morphology is suggestive of the presence of an hyperbolic shock outflow surface, similar to those observed in other BPN (e.g. NGC 6302, Hb 5). Line ratio maps also indicate the presence of extremely strong [NII] emission at the periphery of the outflow, whilst expansion velocities are of order $\sim 400$ km sec$^{-1}$. These large shell expansion velocities may in turn be driven by an extremely high velocity wind, which in this case appears to extend over a range $\Delta V \geq 3000$ km sec$^{-1}$.

The morphology of this source is highly reminiscent of shock models in which stellar winds directly interact with the ambient medium, leading to shock refraction, appreciable emission in the post-shock zone, and the formation of bi-lobal cavities with typically hour-glass shapes. We have therefore applied such a model to our present results, assuming the ambient medium to vary in density as $R^{-\alpha}$, and the wind to be collimated by a toroidal disk.

The superposition of our model upon optical imaging reveals an extremely close correspondence, whilst a similar superposition of model and spectrophotometric results is also encouraging - with various minor differences attributable to non-symmetries in the source itself; perhaps a consequence of inhomogeneities within the ambient cloud.

Broadly speaking, we believe that few (if any) such models are capable of so closely replicating the morphology and kinematics of these sources - and in certain cases, it is clear that such structures would also explain the presence of ansae in nebulae such as NGC 3242 and NGC 6905. Similarly, it is clear that whilst care should be taken in the interpretation of our line ratio measures, the high peripheral ratios [NII]/H$\alpha$ would be anticipated under a broad variety of shock conditions.

For the present case, we conclude by noting that although a value $\alpha = 2$ has been assumed in the models described here, similar (if somewhat poorer) fits may be achieved for the values $\alpha = 0$ and 1. Similarly, we determine a source inclination i = 117 to the line of sight, and require wind velocities $V_w \simeq 900$ km sec$^{-1}$; although here again, these values are closely similar for all of our model fits.

# THE DUST IN THE HYDROGEN-POOR EJECTA OF ABELL 30

J.P. HARRINGTON and K.J. BORKOWSKI
*Department of Astronomy, University of Maryland*
W.P. BLAIR
*Department of Physics and Astronomy, The Johns Hopkins University*
J. BREGMAN
*NASA/Ames Research Ctr., Moffett Field, CA*

ABSTRACT: High-resolution images in [O III] $\lambda 5007$ of the hydrogen-poor knots of Abell 30 reveal comet-like structures which may be indicative of interaction with the stellar wind. In the near IR, new, higher-resolution, K-band images show an equatorial ring of hot dust that corresponds closely to optical knots 2 and 4 of Jacoby and Ford, while their polar knots 1 and 3 show no comparable IR emission. Both the thermal IR emission and the heavy internal extinction of the central star demands an extremely dusty ejecta. Greenstein showed that the UV extinction curve is fit by amorphous carbon. Our comprehensive dust models consider both the UV extinction and the IR emission from a population of carbon grains. The thermal emission from larger grains produces the far IR emission, while the stochastic heating of very small grains to high temperatures is essential to explain the near IR flux. We are able to reproduce the shape of the near IR spectrum with an $a^{-3.0}$ distribution of grain radii which extends down to a minimum grain radius of 8 Å.

We have constructed photoionization models of the dusty, hydrogen-poor gas which include the heat input due to photoelectrons ejected from the dust grains. The grain charge is important as is the enhanced photoelectric yield of small grains. We find that heating by grain photoelectrons dominates photoionization heating. This extra heat source is critical for the optical and UV line emission of the knots.

# THE CENTRAL REGION OF THE PLANETARY NEBULA A58

D.L. POLLACCO and P.W. HILL
*Department of Physics and Astronomy, North Haugh, St. Andrews,
Fife KY16 9SS, Scotland*

and

R.E.S. CLEGG
*Royal Greenwich Observatory, Madingley Road, Cambridge, CB3 0EZ, England*

We present images and high-resolution spectra of the hydrogen deficient knot at the centre of the old planetary nebula A58. The spectra confirm that this region contains essentially no hydrogen, as previously suspected. Emission lines from the knot are broad (FWHM $\sim$ 180 and 270 km/s for [NII] and [OIII] lines respectively) and are blue-shifted by $\sim$100 km/s relative to the systematic velocity.

We analyse these data in terms either of a one-sided collimated flow (the rear side flow being obscured or suppressed) or of dusty, spherical wind (where the rear side of the compact object is obscured by dust) with the data lending marginal support for the collimated flow model. The collimated flow or fast, H-poor wind is very likely to be associated with the 1919 nova-like event in Aquila.

New spectra of the central star are also presented.

# MORPHOLOGY & KINEMATICS OF THE 'BORN–AGAIN' PLANETARY ABELL 78

R.E.S. CLEGG and M.N. DEVANEY
*Royal Greenwich Observatory, Madingley Rd, Cambridge CB3 0EZ, UK.*

and

A.P. DOEL, C.N. DUNLOP, J.V. MAJOR, R.M. MYERS and R.M. SHARPLES
*Physics Dept., University of Durham, Durham DH1 3LE, UK.*

Abell 78 is one of a group of planetaries having an old, H-rich nebula surrounding a hot star which has more recently ejected highly-processed, H-deficient material. The A78 central star has O VI emission lines and a $3700\,\mathrm{km\,s^{-1}}$ hot wind, and is surrounded by knots of very dusty, H-deficient material. These objects are thought to have suffered a late helium shell flash which resulted in the central star (then a white dwarf) returning to the AGB and ejecting highly-processed material.

We obtained imaging and velocity data on the 4.2m William Herschel Telescope at La Palma, all in the [O III]5007Å line. Images of the central region were taken with the MARTINI image–sharpening device, with which the ambient seeing's FWHM of 1.6" was improved to 1.1". The velocity information was from the TAURUS Fabry-Perot Imaging Spectrometer, with which the central field was recorded with a a velocity step of $8.0\,\mathrm{km\,s^{-1}}$ per map channel over a free spectral range of $600\,\mathrm{km\,s^{-1}}$.

The deep images resolve new structures in the system of knots. They appear to form sets of radial, filament–like structures, which lie roughly in a plane which is almost in the line–of–sight. The (projected) expansion velocities of the knots are about $30\,\mathrm{km\,s^{-1}}$ in the radial direction.

The TAURUS data have also revealed a pair of fast-moving, diametrically opposed 'bullets', each located 13 arcsec from the star and at the two 'poles' of the disk system suggested by the knots. The bullets have (projected) expansion velocities of $+103$ and $-103\,\mathrm{km\,s^{-1}}$, but the deprojection factor is likely to be quite large, so the bullet 'ejection velocity' could be as high as $200\,\mathrm{km\,s^{-1}}$.

A model is suggested in which the hot central star has a main sequence companion. After the late thermal pulse, the born-again AGB star transferred matter onto an accretion disk around the dwarf, and the fast bullets represent condensations in a two–sided jet which emerges at each pole with ejection velocity $v(ej) \sim v(esc)$, the escape velocity from the dwarf companion's surface.

Such a model for A78, if correct, suggests that (a) fast bullet pairs seen in other PNs originate from near the stellar surface, and not from a two–wind hydrodynamic 'focussing' mechanism; and (b) such bullet pairs may enable us to determine the orientation of unseen binary systems at the centres of a few planetary nebulae.

# THE FORMATION OF SINGLE AND BINARY NUCLEI OF PLANETARY NEBULAE

L. R. YUNGELSON and A. V. TUTUKOV

*Institute for Astronomy, 48 Pyatnitskaya Str., 109017 Moscow, Russia*

The present birthrate of binaries may be written as

$$d^3\nu = 0.2 M_1^{-2.5} q^\alpha \, dM_1 \, d\log A \, dq \qquad \text{yr}^{-1},$$

where $M_1$ is the mass of the primary in $M_\odot$, $0 < q \leq 1$ - mass ratio of components, $A$ - semimajor axis of the orbit in $R_\odot$. The Eq.(1) implies that all stars are born as binaries with $10 \leq A/R_\odot \leq 10^6$, and that one binary with $M_1 \geq 0.8 M_\odot$ is formed annually in the Galaxy. We study numerically evolutionary scenarii of binaries within abovementioned range of $M_1$, $A$, $q$. As PN formation events we consider all ejections of common envelopes by close binaries and ejections of envelopes by red giants, after which one may expect a formation of a hot ($T_e \geq 30\,000 K$) star, surrounded by a nebula. Altogether about 20 different single and binary cores of PN can be formed. Combining the scenarii data with the Eq. (1) one can estimate the birthrates of most numerous kinds of PNN (single and with main-sequence, white dwarf, giant and relativistic companions) listed in the Table.

**The birth rate of PNN in the Galaxy** ($yr^{-1}$)

| Nucleus | Single | Companion | | | |
|---|---|---|---|---|---|
| | | MS | WD | Giant | Rel |
| CO, ONe dwarf | 0.17 | 0.62 | 0.08 | 0.0027 | 0.00082 |
| He dwarf | 0.0027 | 0.042 | 0.0042 | – | – |
| He star | – | 0.0014 | 0.01 | – | 0.0027 |

The single PNN within our concept of PN formation are mainly products of the merger of components of binaries inside common envelopes. The Figure shows the distribution over orbital periods of PNN with MS (thick line) and WD (thin line) companions, as well as positions of PNN with known orbital periods (dots). The theoretical distribution of PNN agrees with observational estimate of 10 per cent of PNN being close binaries with $P \preceq 10^d$. The absence of more wide, but still close, observed binary PNN is a result of observational selection, because they can be discovered mainly due to photometrical variability caused by the presence of close companion. The position of observed wide binary PNN is a result of the existence of lower angular separation limit for discovery of visual duplicity $\sim 1\,arcsec$.

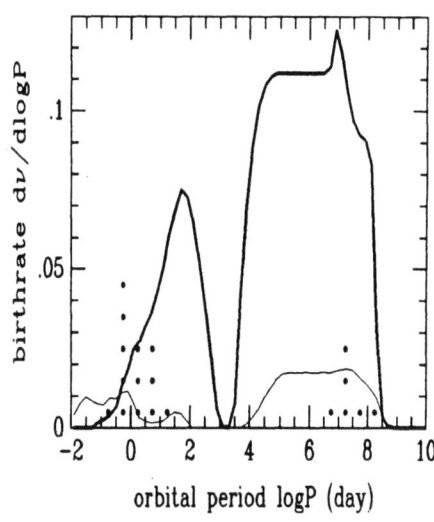

# A SPECTROSCOPIC STUDY OF BINARY STAR PLANETARY NEBULAE

## J.R. WALSH
*Space Telescope European Coordinating Facility, European Southern Observatory, Karl-Schwarzschild-Strasse 2, D8046 Garching bei München, Germany*

## N.A. WALTON
*Department of Physics & Astronomy, University College London, Gower Street, London WC1E 6BT, England*

and

## S.R. POTTASCH
*Kapteyn Laboratorium, Rijksuniversiteit Groningen, Postbus 800, 9700 AV Groningen, The Netherlands*

**Abstract.** A comprehensive spectrophotometric survey of the central stars of planetary nebulae with known or suspected binary star nuclei (BSPN) is underway. The aims of this programme are: to determine the nature of the companion to the ionising star of the nebula; to estimate the spectroscopic distance from the spectral type and magnitude of the companion and thence to determine the luminosity of the hot star; to compare the morphologies and excitation/abundance characteristics of BSPN with those of PN produced by single star evolution. The advantage of the distance determined from this simple method is that it is independent of any assumptions about the planetary nebula, in contrast to most other PN distance estimators.

To date 75 percent of known or suspected BSPN have been observed with the 2.5m Isaac Newton Telescope and the 1.5m ESO spectroscopic telescope at low and intermediate dispersions. A selection of preliminary results from the programme is presented, highlighting the binary central stars of NGC 246 and Abell 65.

## Preliminary results from the programme

Preliminary analysis of low (12Å) and higher (1.6Å) resolution spectra of the hot central star of NGC 246 and it's nearby visual companion, give the companion's visual magnitude, as $m_V = 14.4 \pm 0.1$ mags. The B-V colour is 1.09 mag. leading to an estimate of K2 V–K3 V. The identified stellar features imply a somewhat earlier spectral class of K0 V. Our preliminary classification of the cool star as type K0 V, with an absolute magnitude of $M_V = 5.9$ (Jaschek & Jaschek, 1990), leads to a distance to NGC 246 of 500 pc.

Our observations of the central star spectrum of Abell 65 show it to be quite remarkable, having an inverted Balmer decrement. It shows weak He I and He II lines with a strong C III,N III feature at 4645Å. Recently Acker & Stenholm (1990) have reported on their observations of the central star of HFG 1. On the basis of it's spectrum, containing an inverted Balmer decrement with lines of He I, He II, C III and N III in emission, they proposed that the central star of HFG 1 may be a member of the intermediate polar class of magnetic cataclysmic variables. Our spectrum of nucleus of Abell 65 shows it to be similar to that of HFG 1's nucleus, and hence it may also be related to the cataclysmic variables, having a high density, hot accretion disk surrounding the central star system.

## References

Acker, A., Stenholm, B. 1990, A&A, 233, L21
Jaschek, C., Jaschek, M. 1990, 'The Classification of Stars', Cambridge University Press

# THE PECULIAR LIGHT VARIATION OF THE PLANETARY NEBULA NGC 2346

XIANG-LIANG HAO

*Beijing Astronomical Observatory, Beijing, China*

Planetary nebula NGC 2346 is a well know butterfly bipolar nebula and the central star (AGK3-O°965) is known to be a spectroscopic binary with a period of about 16 days[1]. This object has been observed frequently and no bright variations were noticed from at least 1899 to 1981 Nov[2]. But in 1981 an unexpected large amplitude eclipse light variations were observed by Kohoutek[3]. Since then a lot of observations have been reported by many authors. They have revealed that the central star V651 Mon showed fast and complex light variations due to eclipse and the amplitudes of the eclipse varied rapidly.

We have observed this object using the 40/200 cm double astrograph and others at Beijing Observatory since 1981. From these observations we found that its drastic eclipse light variations wich started in 1981, decreased rapidly in 1986 and the eclipse nearly was unseen in the light curves of 1987[4].

From the observations during 1988-1991 we found that the brightness variations of V651 Mon have an obvious difference between 1987 and 1988 while we noticed that the eclipse events which nearly unseen in 1987, it reappeared in the light curves of 1988 and its amplitude is about 1.1 mag. and the maximum brightness in 1988 were brighter than 1987 and this result were supported by the observations during 1989-1991.

From these it is reasonable to think that this binary central star of planetary nebula NGC 2346 might undergoing another mass ejection from the sdO component during 1987-1988.

### References

(1) Mendez, R.H., 1980, IAU Symposium No. 88, p567
(2) Schaefer, B.E., 1983, IBVS, No. 2281
(3) Kohoutek, L., 1982, IBVS, No. 3271
(4) Hao, X-L, 1988, IBVS, No. 3271

# NEW ECLIPSING PHENOMENA OF THE CENTRAL STAR IN NGC 2346

R. COSTERO, M. PEÑA, W.J. SCHUSTER, M. TAPIA, J. ECHEVARRIA and J. FIERRO

*Instituto de Astronomía, UNAM, Mexico*

The central star of NGC 2346 is a well known binary with an A-type primary and a hot companion (Méndez and Niemela 1981). The star went through a series of periodic light variations which ceased in 1986 and were interpreted as an eclipse of a dust cloud passing in front of the binary system (e.g. Méndez et al. 1982, Costero et al. 1986). Recently, light variations reappeared with shallower minima compared to the previous eclipse (e.g. Kohoutek et al. 1992).

We have made photoelectric and CCD photometry of this star (V651 Mon) at the Observatorio Astronómico Nacional in San Pedro Mártir, Mexico, during April/May 1992. An eclipse of amplitude $\sim 0.25$ in the observed V magnitude was measured, similar to previous reports, but now it seems to be deeper and broader. We interpret this variations in terms of another fragmented dust cloud passing in front of the central star.

### References

Costero R., Tapia M., Méndez R.H., Echevarria J., Roth M., Quintero A., Barral J.F. 1986, Rev. Mexicana Astron. Astrof. 13, 149.
Kohoutek L., Mantegazza L., Hainaut, O. 1992, Inf. Bull. Var. Stars No. 3694.
Méndez R.H., Niemela V.S. 1981, ApJ 250, 240.
Méndez R.H., Gathier R., Niemela V.S. 1982, A&A 116, L5.

# OBSERVATIONAL STUDIES OF CLOSE BINARY CENTRAL STARS OF PLANETARY NEBULAE: HFG 1 AND A 63

HAKIM LUFTHFI MALASAN

*Department of Astronomy, Faculty of Science, University of Tokyo, Tokyo 113, Japan*

*Bosscha Observatory, Bandung Institute of Technology, Bandung 40391, Indonesia*

and

ATSUMA YAMASAKI

*Department of Geoscience, National Defense Academy, Hashirimizu, Yokosuka 239, Japan*

We have made spectroscopic and photometric observations of two close binary central stars of planetary nebulae HFG 1 (V 664 Cas) and A 63 (UU Sge), using the UH 2.2–m telescope (Mauna Kea Observatory) and the NAO 1.9–m and 0.9–m telescopes (Okayma Astrophysical Observatory).
Photometry (in $B_{42}$ and $V_{53}$) of V664 Cas reveals sinusoidal variations with a period of 0.58167 days, confirming the result of Grauer et al. (1987). No eclipses have been detected, indicating small orbital inclination. The spectrum in blue region shows strong emission lines of HI, HeI, HeII, CIII and NIII. In addition, the $H\beta$ emission line shows multiple peak. Variation along the orbital phase is found both in radial velocity (amplitude of 35 $km.s^{-1}$) and strength of all lines. We interpret that the emission lines originate from the illuminated part of the secondary atmosphere, since radial velocity motions of these lines coincide with the expected motion of the secondary component. Analysis of the light curves of V664 Cas results in orbital inclination of $50° - 60°$ and $T_{eff}$ (primary) of around 100,000 K. A model of V664 Cas based on spectroscopic observations includes: (a) Extreme UV radiation from the primary component, (b) Outer layer (optically thin) reprocessed component, and (c) Deeper layer (optically thin) reprocessed component.
Spectroscopic observations of UU sge reveal radial velocity variations with the period of 0.465 days and an amplitude of 84 $km.s^{-1}$. The deducted mass function is 0.028 $M_\odot$. Absolute parameters of UU Sge are obtained as:$M_1 = 0.55 - 1.44 M_\odot$ (assumed), $R_1 = 0.3 - 0.4 R_\odot$, $M_2 = 0.3 - 0.5 M_\odot$ and $R_2 = 0.5 - 0.7 R_\odot$. Both component are separated by 3.0 $R_\odot$. The nebular radial velocity has been determined to be 52 $km.s^{-1}$ from the [OIII] line.
A full account of this work will be published elsewhere.
HLM would like to thank Prof. M. Kondo for continuous encouragement and discussions, and the Hitachi Scholarship Foundation for travel grant. This work was supported in part by the Scientific Research Found of the Ministry of Education, Science and Culture of Japan (grant 63540189)

# IMAGING AND SPECTROSCOPY OF ABELL 63 (UU SGE)

N.A. WALTON
*Department of Physics & Astronomy, University College London,
Gower Street, London WC1E 6BT, England*

J.R. WALSH
*Space Telescope European Coordinating Facility, European Southern Observatory,
Karl-Schwarzschild-Strasse 2, D8046 Garching bei München, Germany*

and

S.R. POTTASCH
*Kapteyn Laboratorium, Rijksuniversiteit Groningen,
Postbus 800, 9700 AV Groningen, The Netherlands*

**Abstract.** UU Sge, the eclipsing binary central star (Bond et al, 1978) of the low-surface-brightness planetary nebula (PN) Abell 63, has been observed spectroscopically in the visible throughout its 11.2 hour period and especially during the minimum. A spectral determination of the binary system has been made. The primary hot central star is an 'O' type PN nucleus of temperature $\approx$40,000 K, consistent with the low excitation of the nebular spectrum (e.g. no He II 4686Å nebular emission detected). From the spectrum at minimum light, the secondary star appears to be a cool dwarf star around G7. Measurement of the magnitude of the secondary during the eclipse of the primary enabled the distance to the PN to be directly determined as 3.6 kpc. For this distance the luminosity of the hot star is approximately 4320 $L_\odot$, in good agreement with evolutionary tracks for (single) PN nuclei. Deep CCD images of Abell 63 show it has a 'butterfly' morphology implying that the close binary central system may have had a strong effect on the nebula shaping. The paper describing this work has been submitted (Walton et al, 1992).

**The cool star** : No strong features are seen in the dereddened spectrum of the secondary star at zero phase, but weak absorption at Ca II H and K, Fe I 4325Å, H$\gamma$, 5170Å(Mg I) and 5270Å(Ti II) is noted. Comparing with the Jacoby et al (1984) spectral library, the best fit to the secondary star spectrum is of a G7V spectrum with an uncertainty of three spectral subclasses. The dereddened V mag. of the secondary star (at zero phase), corrected for the hot star, is $m_V = 18.1 \pm 0.2$ mags. For an absolute mag., $M_V = 5.3$ for a G7V star (Lang 1992) the distance is 3.6 kpc (with uncertainty range of 2.9–5.0 kpc).

**The hot star** : The best fit N-LTE model atmosphere to the spectra of the primary together with IUE archive spectra, was of temperature 40 000 K, Log g=5.0. The dereddened V mag. of the hot primary of UU Sge, corrected for the cool secondary, is $m_V = 13.56 \pm 0.10$ mags. The luminosity of the central star of Abell 63 is then 4320 $L_\odot$ (with an uncertainty range of 2230–6610 $L_\odot$). The primary star is of spectral type 'O'. The location of the central star of Abell 63 on the H-R diagram, from comparison with theoretical single star evolutionary tracks (Schönberner, 1981), shows that it is consistent with the 0.6 $M_\odot$ model track.

## References

Bond, H.E., Liller, W., Mannery, E.J. 1978, ApJ, 223, 252
Jacoby, G.H., Hunter, D.A., Christian, C.A. 1984, ApJS, 56, 257
Lang, K.R. 1992, 'Astrophysical Data: Planets and Stars', Springer-Verlag, Berlin
Schönberner, D. 1981, A&A, 103, 119
Walton, N.A., Walsh, J.R., Pottasch, S.R. 1992, A&A, submitted

# NEW LIGHT ON UU SAGITTAE

S.A. BELL and D.L. POLLACCO

*Department of Physics and Astronomy, North Haugh, St. Andrews, Fife KY16 9SS, Scotland*

New V and I band CCD photometry and medium resolution spectroscopy are used to derive the masses, luminosities and radii accurate to < 10% for the individual components of the eclipsing central star of the planetary nebula A63-UU Sge ($M_1 = 0.63 \pm 0.06 M_\odot$, $R_1 = 0.33 \pm 0.01 R_\odot$, $M_2 = 0.29 \pm 0.04 M_\odot$ and $R_2 = 0.53 \pm 0.02 R_\odot$). Emission lines from the secondary component and HeII and NV absorption features from the primary component are used to determine the first radial velocity curves of the system. Ultra-violet and optical spectra show that the temperature of the primary compoment is $\sim 10^5$ K – much larger than previously suspected. As the techniques used are essentially independent this is probably the most accurately known mass for a planetary nebula central star and therefore allows meaningful comparison to be made with evolutionary tracks for these objects.

# PRECATACLYSMIC BINARIES IN THE CENTRE OF PLANETARY NEBULAE

G. JASNIEWICZ and A. ACKER
*URA 1280*
*Observatoire de Strasbourg*
*11, rue de l'Université*
*67000 STRASBOURG, France*

**Abstract.** We report at first on two old and large planetary nebulae (PN) of which central stars could satisfy the Ritter's criteria (1986) for being precataclysmic binaries: LoTr5 and Abell 35. Both nebulae have probably been ejected as a consequence of common-envelope evolution. A model of cataclysmic binary (CB) for the central star of Abell 35 has been tentatively attempted by Acker and Jasniewicz (1990). The nucleus of LoTr5 is a triple star (Jasniewicz et al., 1987; Malasan et al., 1991): action of a third body on the separation of the close binary could make this binary evolve into a CB (see Mazeh and Shaham, 1979). We report at second on the similarity between the spectrum of the central star of the extended PN HFG1 with that of a CB (Acker and Stenholm, 1990). The PN cited above could be fundamental objects just at the transition between the stage PN and the stage CB.

Abell 35 and LoTr5 are very old extended PN with a visible central star of late spectral type. The IUE satellite has revealed an extremely hot companion, discovered for Abell 35 by Grewing and Bianchi (1989), for LoTr5 by Feibelman and Kahler (1983). The central stars of LoTr5 and Abell 35 could be pre-CBs; according to Ritter (1986) a detached binary can be classified as pre-CB if it satisfies the properties: the primary component is a white dwarf or a white dwarf precursor; the secondary component is an essentially unevolved star of low mass ($M_2 \leq 1 M_\odot$); the orbital period is so short ($P \leq 2^d$) that the system can only have been formed via a common-envelope evolution; the binary is the central star of a PN.

The central star of the PN HFG1 presents some characteristics of an extreme member of the long-period intermediate polar class of magnetic CBs. Besides, two CBs, 063+71 and GK per, are surrounded by nebulae resembling old PNs (Bond, 1989). More, Krautter and Radons (1986) also wonder whether the CB 623+71 is the central star of a PN.

## References

Acker, A., Jasniewicz, G.: 1990, *A&A* **238**, 325
Acker, A., Stenholm, B.: 1990, *A&A* **233**, L1
Bond, H.E.: 1989, in *IAU Symp. No. 131, Planetary Nebulae*, ed. Torres-Peimbert (Dordrecht:Reidel), p.251
Feibelman, W.A., Kahler, J.B.: 1983, *ApJ* **269**, 592
Grewing, M., Bianchi, L.: 1989, in *IAU Symp. No. 131, Planetary Nebulae*, ed. Torres-Peimbert (Dordrecht:Reidel), p.314
Jasniewicz, G., Duquennoy, A., Acker, A.: 1987, *A&A* **180**, 145
Krautter, J., Radons, G.: 1986, *Mitt. Astron. Ges.* **67**, 308
Malasan, H.L., Yamasaki, A., Kondo, M.: 1991, *AJ* **101**, 2131
Mazeh, T., Shaham, J.: 1979, *A&A* **77**, 145
Ritter, H.: 1986, *A&A* **169**, 139

# THE ABELL 35-TYPE PLANETARY NUCLEI

HOWARD E. BOND
*Space Telescope Science Institute*
*Baltimore, Maryland USA*

ROBIN CIARDULLO
*Pennsylvania State University*
*University Park, Pennsylvania USA*

and

MICHAEL G. MEAKES
*Space Telescope Science Institute*
*Baltimore, Maryland USA*

The nuclei of the low-surface-brightness PNe A 35, LoTr 1, and LoTr 5 are binaries containing rapidly rotating late-type subgiants or giants and extremely hot ($T_{\rm eff} \gtrsim 100{,}000$ K) companions detected by the *IUE* satellite. All three objects show low-amplitude, periodic photometric variations in the optical band (with periods of 0.76 or 3.3, 6.6, and 5.9 days, respectively).

Since the photometric amplitudes change on timescales of weeks to months, we attribute the photometric variations to starspots on the cool components. The periodicities thus correspond to the rotational periods of the cool stars, and the true orbital periods remain unknown. However, for the very similar object HD 128220, which does not lie within a planetary nebula, the orbital period is known to be 872 days, suggesting that the orbital periods could be quite long.

The evolutionary origin of these systems remains puzzling, but we speculate that they have recently emerged from a common-envelope interaction in which the cool main-sequence component was spun up to rapid rotation, and still remains out of thermal equilibrium.

# THE IUE ULTRAVIOLET SPECTRUM OF PC 11

M. PARTHASARATHY
*Indian Institute of Astrophysics, Bangalore, India*

S.R. POTTASCH
*Kapteyn Astronomical Institute, Groningen, The Netherlands*

and

J. CLAVEL
*IUE Observatory, Madrid, Spain*

PC 11 (HD 149427, PK 331-5 1) is classified as a young planetary nebula with strong OIII 4363Å and a Zanstra temperature of $T_Z = 27000K$. It is also classified as (D' - type) yellow symbiotic star with A - F type companion. It is an IRAS source with detached cold dust with far intrared (IRAS) colours similar to planetary nebulae. The IUE short wavelength (SWP) spectra show emission lines due to OIII] (1661/1666Å). NIII] (1746/1754Å) CIII] (1907/1909Å). The OIII] and NIII] emission lines show significant variation. Variation in the strength of CIII] is not very significant. The strength of OIII] has decreased and NIII] has increased. The long wavelength (LWP) spectrum shows stellar continuum (A-F) and absorption lines due Mg II 2800Å feature. It also show emission lines at 2772Å (?) 3133Å - 3140Å (very strong) (OIII, [FeV], 3209Å (He II?) ([FEII]). The variation in the strength of emission line due OIII] and NIII] and the presence of stellar continuum (A-F) suggests that the central star of PC 11 is a binary.

# ON THE PHOTOMETRIC BEHAVIOUR OF THE CENTRAL STAR OF THE PLANETARY NEBULA Sh2-71

J. JURCSIK

*Konkoly Observatory of the Hungarian Academy of Sciences*

The central star of Sh2-71 was observed with the 1m telescope using $UBV(RI)_C$ photometric system at Konkoly Observatory in the seasons of 1990-92. A period of $68.^d06$ was found with variable amplitude and/or lightcurve shape. The amplitude is high, close to 1 magnitude. On the basis of old photographic observations (Kohoutek 1979) this period has not changed during the century. No significant colour variation was detected, but the nebular contamination should distort the colour curves. According to the five colour data the central object is a highly reddened B5 dwarf about 1700 pc distant. The nebular properties, however, indicate an evolved, higher mass, lower luminosity exciting source, therefore it is plausible to regard the central object as a binary. The light variation might be caused by obscuration of the B star by outflowing matter from the true nucleus. Bipolar outflow of 3" dimension has been detected on high dispersion $H\alpha$ spectra of the central star (Sabbadin et al. 1985; Cuesta et al. 1990). Further spectroscopic observations are highly needed in order to find the exact description of the system.

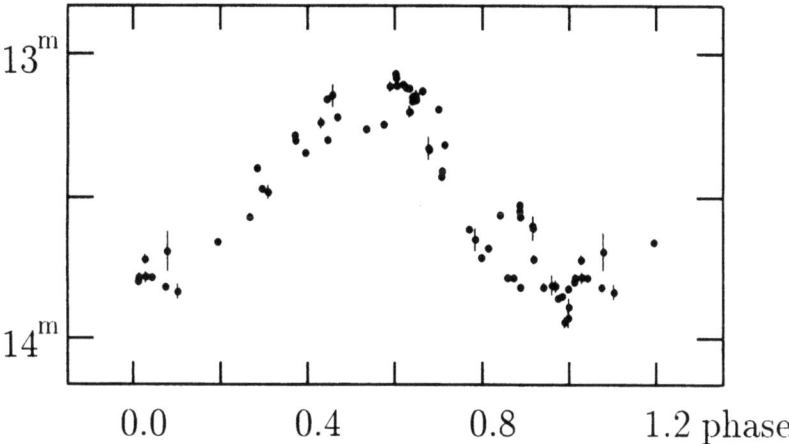

**Fig. 1.** The folded V lightcurve of the B star in the centre of Sh2-71 with the ephemeris Tmin=JDH2448190.0+$68.^d064$E. The scatter of the curve is larger than the observational error indicating real changes in the light variation.

### References

Cuesta, L., Phillips, J. P., Mampaso, A.: 1990, *Ap. Space. Sci.* **171**, 163.
Kohoutek, L.: 1979, *IBVS* **1672**.
Sabbadin, F., Ortolani, S., Bianchini, A.: 1985, *Monthly Not. Roy. Astron. Soc.* **213**, 563.

# ON SOME LINKS BETWEEN SYMBIOTIC STARS AND PLANETARY NEBULAE

LAURITS LEEDJÄRV

*Tartu Astrophysical Observatory, EE2444   Tõravere, Estonia*

**Abstract.** Some correlations found between different physical characteristics of symbiotic stars and planetary nebulae are discussed from the evolutionary point of view.

Symbiotic stars are considered as interacting binary stars, consisting of a red giant and a hot component which ionizes the stellar wind from the former, creating in this way physical conditions similar to those in planetary nebulae. Symbiotic stars are divided into types S and D, containing, respectively, a normal red giant or a Mira as the hot component. In present paper some physical characteristics of the gaseous nebulae in symbiotic systems have been estimated by fitting the observed continuous spectra of symbiotic stars to the computed ones in a wide wavelength range. Comparison of the obtained quantities to the compiled data for planetary nebulae demonstrates e.g. similar linear correlation between $\log M_{gas}$ and $\log n_e$ for both kind of objects (see Leedjärv, 1990 for more details). Also, it is found that the observed radio flux at 6 cm shows the tendency of decreasing when $n_e$ is increasing. When plotting these correlations, it turns out that D-type symbiotic stars occupy an intermediate position between S-type symbiotic stars and planetary nebulae. This can be explained by the different mass loss rates from Miras and from normal red giants. Miras in D-type symbiotic stars are quite near to the stage of planetary nebula, but probably it is not strictly correct to consider some symbiotic stars (V1016 Cyg, HM Sge) as protoplanetary nebulae. In planetary nebulae outer layers of the star are ionized by the core of the same star, while in symbiotic systems the source of ionizing radiation is another hot star. Further study of evolutionary links between symbiotic stars and planetary nebulae would yield a more detailed view in general evolutionary scheme of stars.

## References

Leedjärv, L., 1990, Astrophysics, **32**, 6.

# IS THERE ANY CONNECTION BETWEEN PLANETARY NEBULAE AND SYMBIOTIC STARS ?

MICHAEL FRIEDJUNG
*Institut d'Astrophysique, 98bis Boulevard Arago, 75014 Paris, France.*

Symbiotic stars have sometimes been misidentified as planetary nebulae, because their line spectra are similar to those of planetary nebulae particularly in the ultraviolet L.(Houziaux 1982 in "The Nature of Symbiotic Stars", M. Friedjung & R. Viotti eds., Reidel. p 229), while the infrared shows the presence of the spectrum of a cool star. The similarity is not complete; when H. E.Schwarz (1988 in "The Symbiotic Phenomenon", J.Mikolajewska et al eds., Kluwer, p 123) plotted a graph of two ratios derived from emission line fluxes, symbiotic stars did not occupy the regions of the diagram occupied by planetary nebulae and H II regions.

In fact almost all specialists believe for very good reasons that symbiotic stars are interacting binaries (M Friedjung 1982 in "The nature of Symbiotic Stars, M. Friedjung & R. Viotti eds., Reidel. p 253). A cool giant (surrounded by much dust when a mira variable) appears to transfer mass to a more compact companion, through its wind, or through Roche lobe overflow. The companion seems usually to be a white dwarf, but sometimes a main sequence star or even perhaps in a few cases a neutron star. Symbiotic stars go through stages of activity, believed to be associated with thermonuclear flashes of the companion if it is a white dwarf, or to an instability of the accretion process (mass losing star or more probably disk), if the companion is a main sequence star.

Optical and radio observations at high spatial resolution, show in a number of cases the presence of nebulosity, often with deviations from circular symmetry and signs of the presence of bipolar flows and jets. Perhaps the most sriking example is R Aqr, which has a nebula with double bipolar structure and a jet with several components, indicating ejections at different times. The measured ionized masses of such nebulosity are always much less than those of planetary nebulae, being around $10^{-4} M_O$ For R Aqr, HM Sge and RX Pup and much less for CH Cyg. It may be noted that the nebulosity seen around He 2-104, once claimed to be protoplanetary, appears also to belong to this class, resembling that around R. Aqr (D. Bugarella et al 1991: A&A, **249**, 199).

We can now ask whether there is any evolutionary link between planetary nebulae and symbiotic stars. When there is a white dwarf companion in the binary, previous evolution could in principle have led to the formation of a planetary nebula, but no signs of such a nebula have been seen up to now. In addition later evolution of the non Roche lobe filling cool giant of a widely separated symbiotic binary could lead to ejection of its envelope, with a white dwarf being left behind. In this last case if the companion was also a white dwarf, the planetary nebula would have a double white dwarf nucleus and unusual features. Separate bipolar structure would have been formed during each phase of activity of the symbiotic, leading to the existence of multiple bipolar features in outer regions.

# ELEMENTAL ABUNDANCES IN SYMBIOTIC STARS

H.M. SCHMID and H. NUSSBAUMER
*Institute of Astronomy, ETH Zentrum, CH-8092 Zurich*

Symbiotic objects are related to planetary nebulae in that they represent late stages of stellar evolution. They are interacting binary systems where a hot companion star ionizes the stellar wind of a red giant. This configuration offers the unique possibility for deriving elemental abundances for cool giants from a nebular spectrum with the diagnostic tools employed for HII regions. The analysis can be applied to different types of symbiotic systems having a G, K or M giant, a Mira variable or a carbon star as cool component. The great advantage of this technique is, that it does not depend on stellar parameters or molecular data, and that it can therefore be used as a test or an alternative for the traditional photospheric abundance determinations.

When abundances for symbiotic systems are compared with published abundances for cool giants and predictions from stellar evolution calculations we find:
(i) The abundances of the majority of symbiotic systems are typical for cool giants after CN cycle burning and the first dredge up phase. This is in agreement with previous abundance studies on symbiotic systems (Nussbaumer et al. 1988; Schmid & Schild 1990; de Freitas Pacheco & Costa 1992; Schild et al. 1992). Symbiotic Miras have enhanced carbon abundances. This is most likely the result of an ongoing third dredge up phase.
(ii) Strong nitrogen enrichment ($N/O > 0.67$), similar to type I planetary nebulae is only rarely observed in symbiotic systems (1 out of 32).
(iii) One of the analysed object, AS 210, is a symbiotic carbon Mira. Its relative nitrogen abundance supports the suspicion that in cool carbon stars N is underestimated and needs to be revised (see Lambert et al. 1986).
(iv) For some G type symbiotic systems (e.g. M1-2, HD 330036, AS 201), which are often considered to be young compact planetary nebulae with a binary core, an enhanced carbon abundance is derived. Two explanations are considered for the abundances of this special group. Either the ionized material was ejected form the present hot component (PN interpretation), or the material originates from the cool giant (symbiotic star interpretation). The first interpretation is supported if the abundances in the photosphere of the G giant are different from the abundances seen in the nebula. The latter interpretation implies that the G type giant in these systems must possess an abundance anomaly like a Ba star or an early R type carbon star. The relevant photospheric abundance determinations are still waiting to be done.

## References

de Freitas Pacheco J.A., Costa R.D.D.: 1992, *A&A* **257**, 619
Lambert D.L., Gustafsson B., Eriksson K., Hinkle K.H.: 1986, *ApJS* **62**, 373
Nussbaumer H., Schild H., Schmid H.M., Vogel M.: 1988, *A&A* **198**, 179
Schild H., Boyle S.J., Schmid H.M.: 1992, *MNRAS* , (in press)
Schmid H.M., Schild H.: 1990, *MNRAS* **246**, 84

# DIAGNOSTIC DIAGRAMS FOR PLANETARY NEBULAE AND SYMBIOTIC STARS

A. GUTIERREZ-MORENO, H. MORENO and G. CORTES
*Departamento de Astronomía, Universidad de Chile, Casilla 36-D, Chile*

ABSTRACT. One of us (Gutiérrez-Moreno 1988) presented some time ago a preliminary version of two diagrams for separating planetary nebulae (PN) from symbiotic stars (SS); these diagrams are based on the strength of [O III] lines $\lambda\lambda 4363, 5007$. Now a new version is presented, which includes more objects and uses measured total intensities corrected by reddening.

It is confirmed that the diagrams separate very well PN from SS. Besides, it is found that they also separate fairly well S- and D-type SS, with a certain overlap between the regions occupied by both kinds of objects. The diagrams also separate very young PN, which occupy in the diagram an intermediate region between SS and PN.

It is found that D-type SS with PN-like spectra for which we have intensities measured during a sufficiently large time interval seem to move down in the $F(4363)/F(H\gamma) - F(5007)/F(H\alpha)$ diagram, thus suggesting an evolution from the SS region to the PN region. The behaviour of two very young PN, IC 4997 and Hb 12, for which we have found observations made about 50 years ago, seem to confirm this suggestion, since in their first observed position, they were located in the D-type SS region.

An interesting object is V1016 Cyg, considered probably as a PN being formed. This objects is classified as a D-type SS, which suffered an eruption in 1965. From then on it started moving upwards in the diagram, arriving to a maximum $I(4363)/H(\gamma)$ value in 1974, when it began its downward motion until the 1987 observations, when it occupied again practically the same position than in 1965.

REFERENCES

Gutiérrez-Moreno, A. 1988, in "Progress and Opportunities in Southern Hemisphere Optical Astronomy". CTIO 25th Anniversary Symposium", eds. V.M. Blanco and M.M. Phillips (Utah: Brigham Young University), A.S.P. Conference Series, 1, 12.

# ON THE DEREDDENING OF SYMBIOTIC STARS

D. RAYKOVA

*Deparment of Astronomy, Bulgarian Academy of Sciences, Sofia, Bulgaria*

and

B. RAYTCHEV

*Observatoire de Strasbourg, Strasbourg, France*

In studying low dispersion spectra of symbiotic stars and following the usual dereddening procedure we obtain from different pairs of the Balmer lines $H\alpha$, $H\beta$ and $H\gamma$ considerably different values for the interstellar absorption $C$ at $H\beta$. There is a tendency the values to be $C_{\alpha\beta} > C_{\alpha\gamma} > C_{\beta\gamma}$. For one of the objects we obtained $C_{\alpha\beta} = 2.03$, $C_{\alpha\gamma} = 1.51$, $C_{\beta\gamma} = 0$. The mean values for a sample of 13 objects with measured $H_\gamma$ are $C_{\alpha\beta} = 1.82$, $C_{\alpha\gamma} = 1.63$ and $C_{\beta\gamma} = 1.13$. That could be caused by processes taking place in the emitting gas and neglected in the theory as well as by nonlinearity of the detector system. Selfabsorption in the Balmer lines and collisional excitation could affect in different ways the Balmer decrement.

The emitting gas formations in the symbiotic systems are of various densities and shapes. In different phases of activity and orbital motion the line intensities change and in some cases considerable selfabsorption in the Balmer and HeI lines has been observed (Tomov, 1991).

In order to find out what steepens the Balmer decrement we calculated the $C$ values for 25 well observed planetary nebulae from Kaler's catalogue (1976) and the same tendency was established. For 7 objects $\Delta C \leq 0.15$ and there one can safely use the average of $C_{\lambda_1\lambda_2}$. For 5 nebulae "optically thick" in the Balmer lines $\Delta C \geq 0.50$ and the means are $C_{\alpha\beta} = 1.20$, $C_{\alpha\gamma} = 1.02$, $C_{\alpha\delta} = 0.89$, $C_{\beta\gamma} = 0.60$, $C_{\beta\delta} = 0.42$. A priori one has no reason to consider either the average or $C_{\alpha\beta}$ as real interstellar absorption as it is the common practice.

On using the theorical Balmer decrement for planetary nebulae optically thick in the Balmer lines (Capriotti, 1964), we show that the $C$ values must have the observed behaviour when the selfabsorption is not taken into account. Limiting our estimates to $H\delta$, we obtained that $C_{\alpha\delta}$ is the value closest to the real absorption C at $H\beta$. If $H\delta$ is not measured, the most reliable is $C_{\alpha\gamma}$. This conclusion is supported by the symbiotic system PK 26 − 2.2. It has been observed two times a year apart and $C_{\alpha\gamma}$ is practically the same while $C_{\alpha\beta}$ changes by 0.4 and $C_{\beta\gamma}$ by 1.0.

We must note that there are several objects which show an opposite behaviour of $C_{\lambda_1\lambda_2}$ and our conclusions do not account for them.

### References

Capriotti E.R. 1964, Ap.J., 139, 225
Kaler J.B. 1976 Ap.J.Supp.Ser., 31, 517
Tomov N.A. 1991, Ph.D.thesis " Investigation of the gas dynamics in the symbiotic system AG Pegasi"

# ROSAT OBSERVATIONS OF SYMBIOTIC STARS

K.F.BICKERT
*MPI f. extraterrestrische Physik (MPE), Garching, FRG*

R.E.STENCEL
*Center f. Astron.& Astrophys. (CASA), Boulder, Colorado*

and

R.LUTHARDT
*Sternwarte Sonneberg, Sonneberg, FRG*

**Abstract.** The German X-ray astronomy satellite ROSAT (launched 1990 June 1) performed an all-sky survey (from 1990 Jul 30 till 1991 Jan 26) with the Position Sensitive Proportional Counter (2 deg FoV, .1-2.4 keV) covering 96% of the sky. Analysis of bright X-ray sources suggests PSPC positions can be accurately determined to within 30 arcsec, and 1..2 arcmin for soft weak ones. We have examined the survey data for detections from among a list of 189 symbiotic and symbiotic-like objects compiled by Vaidis (1988) with own extensions, chosen for completeness. For 178 objects, all-sky data was available. Standard Analysis Software System (SASS) and EXtended Standard Analysis Software packages (EXSAS) both use a three-step approach to detect sources. After quality screening, a sliding-window algorithm (LD) locates bright sources. These are excluded in the 2D-spline interpolated smoothed background map. A second sliding window MD runs on the background subtracted event files. All possible sources were then tested by maximum likelihood in the background-subtracted, exposure-corrected image. If LD, MD, or ML did not detected a source at the optical position, an upper limit ML program tried to find enhanced emission. For EXSAS, photon event files contain all photons within 20 arcmin radius around a source. Cut radii, background determination, and other parameters can be iteratively improved.

29 sources were flagged by SASS as detected with off-optical distances of up to 10 arcmin. 14 were accepted (w/in 3 arcmin). Source confusion was investigated using the ROSAT Master Source Catalog (X-ray source positions from SIMBAD, IUE, the Verron AGN catalog, Einstein, EXOSAT, and the (Edinburgh) scanned UK Schmitt blue (and some red) plate southern sky atlas (ca 200 million stars).

The 15 sources finally verified by EXSAS (including one not detected with SASS) are listed in Paper 1 (ROSAT Detections of Symbiotic Stars, Bickert et.al. 1992), Table 1. Re-detected were the symbiotic novae AG Dra, RR Tel, CH Cyg, AG Peg, V1016 Cyg, HM Sge, and at 3.5 $\sigma$ Z And (but not T CrB, V1017 Sgr, or RS Oph). We also see the near-by R Aqr, and the hard neutron star symbiotic GX1+4. New detections are NSV 4775 (which Kenyon suspects to be a VV Cep), the S-types Hen-1924=CD-43 14304 and AG Peg, the carbon stars Draco C-1 and LMC S63 (which has to be verified, possibly attitude problems), and the D- types RX Pup and H2-38 (containing Miras).

Paper 2 (ROSAT Upper Limits for Symbiotic Stars, Bickert et.al. 1992, preprint) results are summarized in Table 1: names, optical positions, stellar distances, galactic absorption, blue, visual, and K infrared magnitudes, spectral classes of the binary components, and the SASS and EXSAS derived values for 178 stars with distance between optical and maximum likelihood X-ray position, likelihoods, fluxes in cps (to avoid model- and fit-errors), exposure times and background (in count/sec/arcmin$^2$). Table 2 compares SASS/EXSAS with additionally hardness ratios (for EXSAS also background subtracted). All these are PRELIMINARY due to the data analysis learning curve: photon timing and diverse attitude problems as well as inconsistencies in vignetting and dead-time correction have been found and will be removed till the complete survey reprocessing beginning November 92.

Pointed observation data from AO phase 2 on Draco C-1 (PI:Bickert) and AG Dra (PI: Luthardt) are also shown (forthcoming papers). Table 3 summarizes pointings done, scheduled, or planned for AO3.2

# THE ACTIVE PHASE OF THE HOT COMPONENT OF Z ANDROMEDAE

T. FERNANDEZ-CASTRO
Planetario de Madrid. Spain

R. GONZALEZ-RIESTRA
IUE Observatory. Madrid, Spain

A. CASSATELLA
I.A.S, CNR. Frascati, Italy

A.R. TAYLOR
University of Calgary, Canada

and

E.R. SEAQUIST
University of Toronto, Canada

Z Andromeda is considered as the prototype of the symbiotic class. This system behaves essentially as the nearly uncontamined nucleus of a young planetary nebula. The parameters of its hot component ($R=0,07\ R_0$, $T > 10^5$ K) show that it lies close to the CSPN in the HR diagram (see Fernández-Castro et al. 1988). During the IUE lifetime (1978 onward), Z And experienced a phase of quiescence followed by an active phase which, starting in April 1984, lasted until about July 1986. Since then the star is recovering its quiescent appearance.

A multiwavelength study of the quiescence and activity phase of Z And shows that:

(a) In quiescence the energy distribution is typical of a very hot source ionizing an extended nebula which is fed by part of the wind lost by s cool giant companion, while accretion effects are negligible in the energetics of the system.

(b) In the outburst, the activity itself is superimposed to orbital variations. The ejection of an optically thick shell and the development of a moderately high velocity wind from the hot star are clearly seen at certain stages of the outburst. Those episodes could even have changed the normal behaviour of the system. A careful analysis of the available UV high resolution spectra will provide us with some more details about the nature of the energy sources of he system.

Theoretical models predict that further evolution of the system will turn Z And into a close binary consisting of two white dwarfs surrounded by an expanding planetary nebulae. A complete study of the system will be published elsewhere.

## References

Fernández-Castro, T. , Cassatella, A. , Giménez, A. and Viotti, R. 1988, Ap. J. 324, 1016.

# BZ CAMELOPARDALIS = 0623+71 : THE CATACLYSMIC VARIABLE INSIDE A BOW-SHOCK NEBULA

N.M.SHAKHOVSKOY and YU.S.EFIMOV
*Crimean Astrophysical Observatory, Nauchny 334413 Ukraine*

and

I.L.ANDRONOV and S.V.KOLESNIKOV
*Department of Astronomy, Odessa State University,*
*T.G.Shevchenko Park, Odessa 270014 Ukraine*

The object BZ Cam = 0623+71 (Krautter et al., 1987) was observed at the telescope AZT-11 of the Crimean Astrophysical Observatory by using the UBVRI photometer-polarimeter of the Helsinki University (Korhonen et al., 1984). No significant linear polarization was found, the mean values have the upper limit of 0.2 per cent.

The brightness variations in all five bands are highly correlated, with the amplitude increasing to the shorter wavelengths, and reaching 0.4 mag in U. The significant peaks at the periodogram occured at the frequences 12.8-17.6, 39.3-42.2, 66-70 (2-hour set at 12.12.91) and 10.2-10.7 (4- hour set at 23.02.92) cycles/day, far from the values 7.2 or 6.5 cycles/day, corresponding to the possible spectral periods suspected by Lu and Hutchings (1985). Such apparent variations of the 'periods' from night to night are similar to that observed in other cataclysmic variables, eg. MV Lyr.

For the high frequencies, we determined the parameter $\gamma$ for the power approximation of the periodogram : $S(f) \propto f^{-\gamma}$ (cf. Terebizh et al., 1989). Its value was found to be $\gamma = 1.08 \pm .05$ in U and $0.86 \pm 0.04$ in R, practically equal to unity. The values $\gamma = 0$ and $\gamma = 1$ were found for the inactive and active states of the magnetic binary AM Her (Shakhovskoy et al., 1992), and $\gamma = 2$ for the cataclysmic variable HQ And (Andronov et al., 1992). This parameter may give the additional information on the nature of the variability.

### References

Andronov I.L., Borodina I.G.,Kolesnikov S.V., Pavlenko E.P., Shakhovskoy N.M.: 1992a, in:
    'AM Her-type Stars and the Related Objects', Comm.Spec.Astrophys.Obs. **69**, 112
Korhonen T., Piirola V., Reiz A.: 1984, ESO Mess. **38**, 30
Krautter J., Klaas U., Radons G.: 1987, As.Ap. **181**, 373
Lu W., Hutchings J.B.: 1985, PASP **97**, 990
Shakhovskoy N.M., Kolesnikov S.V., Andronov I.L. : 1992, in: 'Magnetism in Stars', St.Peterburg,
    Nauka Publ.
Terebizh V.Yu., Terebizh A.V., Biryukov, V.V. :1989, Astrofizika **31**, 75

# THEORETICAL LIGHT CURVES OF RECURRENT NOVAE

M. KATO

*Department of Astronomy, Keio University, Yokohama 223, Japan*

**Abstract.** Theoretical light curves for the decay phase of recurrent novae are obtained using optically thick wind theory. With the OPAL opacity tables the decay time scale is drastically shortened.

Theoretical light curves for the decay phase of recurrent novae are presented. We have used OPAL new opacity tables ( Rogers and Iglesias 1992: private communication) which has a large peak about 2 or 3 times larger than the old opacity around $T = 2 \times 10^5$ K due to iron lines. This enhancement of the opacity causes strong mass loss (optically thick winds) and then the evolutional time-scale is drastically shortened. The decay time scale of recurrent novae strongly depends on the white dwarf mass and weakly depends on the hydrogen content of the envelope with solar composition of heavy element. From the fitting of theoretical light curve with observed data, I have found that the white dwarf mass is about 1.35 $M_\odot$ for U Sco, 1.0 $M_\odot$ for T Pyx, and 1.15 $M_\odot$ for RS Oph. As well as optical data, UV light curve is useful to determine the white dwarf mass from fitting of theoretical light curves with observational data.

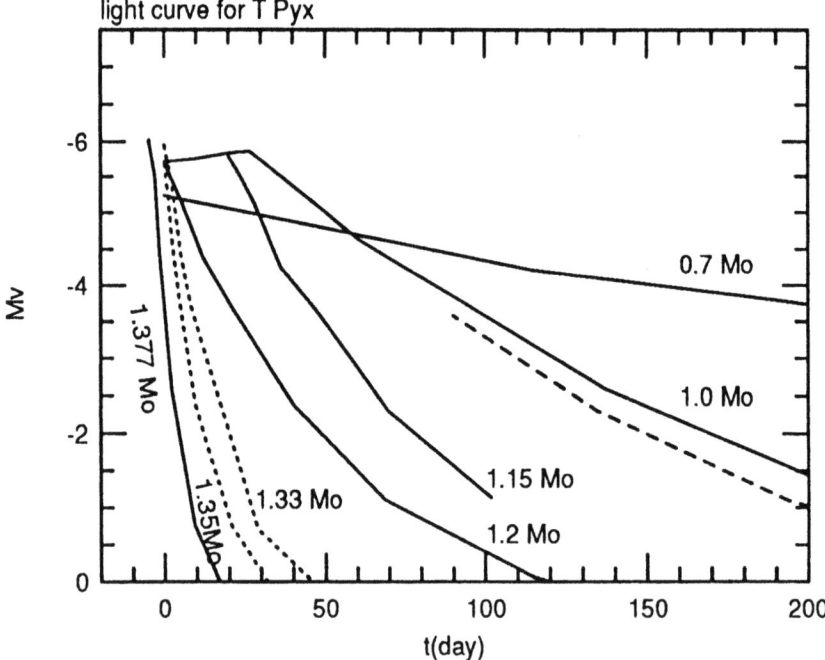

Fig. 1. Theoretical light curves for the decay phase of recurrent novae. White dwarf masses are given near each curve. Thick curve: $X = 0.7$, dotted: $X = 0.5$, and dashed: $X = 0.35$ while $Z = 0.02$ are assumed for all models.

# THE ENVIRONS OF SUPERNOVA PRECURSORS

O.A. TSIOPA

*Pulkovo Observatory, St.Petersbourg, 196140 Russia*

A high activity of a precursor during the last years of its existence can be connected with a close binary system orbit changing.
Some central binary stars in planetary nebulae are thought to be possible supernova precursors. The investigation of the supernova environs formed by the precursor matter before the explosion can help to verify this hypothesis.
Since the SN 1990M event we know that both types of supernovae can exhibit the existence of a rather strong stellar wind ejected prior to the explosion (Polcaro V.F. et al. 1991). The significantly inhomogenous structure of the wind witnesses a high activity of the precursor during the last years of its existence. The wind matter was thrown away as a thin ellipsoidal shell (Tsiopa O.A. 1992). The type Ia supernova precursor being a close white dwarf binary (Iben I.Jr. et al. 1984), the ejected matter can be accelerated by local thermonuclear explosion caused by the surface collision of the components in the periastrum point of the elliptical orbit. Such collisions can take place several times before the orbit becomes round with exentricity decreasing every time. In the case of 1990M one can estimate the time interval between the collision followed by the wind shell ejection and the supernova explosion of half a year. But it should be indicated that the presence of significant amount of hydrogen so close to the precursor is rather difficult to explain even in terms of a common shell surrounding the system of two white dwarfs.

### References

Iben I.Jr. and Tutukov A.U. (1984). "Supernovae of type I as end products of the evolution of binaries with components of moderate initial mass", Astrophys.J. Supplement, 54, 335–372
Polcaro V.F. and Viotti R. (1991), "H absorbtion in the type–Ia supernova 1990 M during its early phases", Astronomy and Astrophysics, 242, L9–L11
Tsiopa O.A. (1992), "The non–symmetric structure of the stellar wind of the supernova precursor", in "Bipolar outflows and stellar jets".

# CIRCUMSTELLAR NEBULAR LINES IN THE OPTICAL SPECTRUM OF SN 1987A*

IFTIKHARUDDIN KHAN and HILMAR W. DUERBECK

*Physics Department, University of Wuppertal, F.R.G., and Astronomical Institute, University of Münster, F.R.G.*

The nebulosities surrounding SN 1987A have a morphological structure strongly resembling some planetary nebulae (e.g. NGC 2392, NGC 3242).

Shock breakout from the photosphere of SN 1987A resulted in an intense UV flash of several $10^5$ K and a duration of 2 to 4 hours. This pulse ionized the circumstellar shell which has been formed by the interaction of the tenuous fast wind ($\approx 550$ km s$^{-1}$) from the blue supergiant phase sweeping up the dense slow wind (15 km s$^{-1}$) from the red supergiant phase of the progenitor.

Supernova spectra in the range 350 − 970 nm were obtained on days 605-607 after outburst with the ESO 1.52 m telescope and a Cassegrain spectrograph. The spectrum consists of narrow nebular lines from the ionized circumstellar shell, superimposed on broad emission lines of the supernova itself. Line fluxes of the broad lines of [C I], [O I], Mg I], [Ca II] are compared with theoretical predictions by Fransson (1987) and Fransson and Chevalier (1989); we suggest an 18 $M_\odot$ progenitor. Fluxes of the narrow nebular lines of [N II], [O II], [O III], [S II], [S III], [A III], [Fe II] and [Fe III] were also measured and a weak [Fe VII] line at $\lambda 5159$ was possibly detected. Other optical observations were made by Wampler and Richichi (1989) and Wang (1991).

An electron density $N_e = 2 \cdot 10^4$ cm$^{-3}$ is derived from the narrow circumstellar [S II] lines. Temperatures on days 605-607 obtained from the [N II] and [O III] lines differ ($1.22 \cdot 10^4$ K and $3 \cdot 10^4$ K, respectively) and show that these lines originate in different regions (clumps?) of the circumstellar shell, and thus meaningful abundance determinations are not feasible.

$N_e$ at different epochs (days 313 − 750) was $(1-3) \cdot 10^4$ cm$^{-3}$. The temperatures remained fairly constant between days 500 and 1285, the average derived from the [N II] lines is about 12500 K, and from [O III] about 26000 K.

The mass loss rate from the progenitor during its red supergiant phase was estimated from the [S II] lines. It is $\dot{M} = 4\pi R^2 N_e m_H v$, where $R$ is the radius of the spherical emitting volume of [S II]. The volume of the forbidden line emitting region can be obtained from the emission measure (EM) and $N_e$; EM is calculated from the [S II] line fluxes and atomic parameters. Then, $\dot{M}$ for the SN 1987A progenitor is $6 \cdot 10^{-6}$ $M_\odot$ yr$^{-1}$.

### References

Fransson, C.: 1987, in *ESO Workshop on the SN 1987A*, ed. I.J. Danziger, Garching: ESO, p. 467.
Fransson, C., Chevalier, R.A.: 1989, *ApJ*, **322**, L15.
Wampler, E.J., Richichi, A.: 1989, *A&A*, **217**, 31.
Wang, L.: 1991, *A&A*, **246**, L69.

* based on observations collected at the European Southern Observatory, La Silla, Chile

## SN 1987A DECONVOLVED BY MIM

GRATL H. and PFLEIDERER J.

*Institut für Astronomie der Leopold - Franzens - Universität, Technikerstraße 25,*
*A - 6020 Innsbruck, Austria*

Modelling of PNs from blurred images, as from the *Hubble Space Telescope*, needs good deconvolution, and the better that is, the more reliable are the results. As an example, we have MIM-deconvolved the HST image of SN 1987A taken in August 1990 by the FOC in [O III] $\lambda 5007$ Å (F501N). MIM (*minimum information method*) is a linear deconvolution method with a local smoothness constraint (Pfleiderer 1991). Our results are similar to those already published (Jakobson et al. 1991, Panagia et al. 1991) but we were able to decrease some of the uncertainties. The ring, centered on the SN, has diameter $1680 \pm 20$ mas ($\equiv 1.3$ ly) and tilt $44.5° \pm 1.5°$ (Fig 1.). The lumps deviate from the ellipse by up to $\leq 0.05$ ly, the ratio of brightest to faintest knots being $\approx 10$ (Fig.3).The SN is resolved (Fig.2), being *not* a uniformly illuminated disk but brighter in the center. Its FWHM is $100 \pm 10$ mas, i.e. an average expansion velocity since the explosion of $v \approx 3500$ km s$^{-1}$.

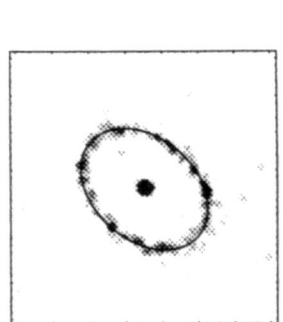
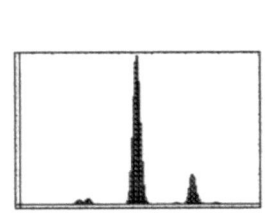
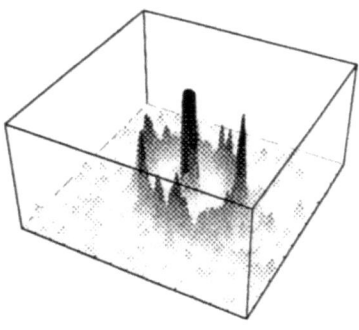

FIG. 1 MIM - processed image. The displayed area is a 141 pixel square submatrix.

FIG. 2 The flux (in arbitrary units) along a line centered on the SN.

FIG. 3 North is to the upper right corner, and east is to the upper left corner.

### References

Jakobson, P.,et al.: 1991, *ApJ* **369**, L63
Panagia, N.,Gilmozzi, R.,Machetto, F.,Adorf, H. M.,Kirshner, R.P.: 1991, *ApJ* **380**, L23
Pfleiderer, J.: 1991, in The Restoration of HST Images and Spectra, ed(s)., *R.L.White and R.J.Allen*, Space Telescope Science Institute, Baltimore, p. 50

We gratefully acknowledge the Space Telescope Science Institute, Baltimore, and the European Coordinating Facility, Garching, for providing us with the data. This work was supported by the *Fonds zur Förderung der wissenschaftlichen Forschung* under grant P8568-PHY and by the *Österreichische Akademie der Wissenschaften*.

# V. PLANETARY NEBULAE CONNECTION: EVOLUTION TO WHITE DWARFS

# EVOLUTIONARY TRACKS

DETLEF SCHÖNBERNER

*Institut für Theoretische Physik und Sternwarte, University of Kiel, Germany*

**Introduction**

It is now accepted without any doubts that the central stars of Planetary Nebulae (CPN) are rapidly evolving objects in the transition from the asymptotic giant branch (AGB) to the white-dwarf regime. After the pioneering study of Paczyński (1971) it has been demonstrated by Schönberner (1981) that Paczyński's calculations are too crude for the understanding of post-AGB evolution because the latter depends very sensitively on the detailed internal stellar structure, i.e. on the past AGB evolution. More precisely, the evolution of an AGB remnant is a function of the thermal-pulse cycle phase $\phi$ during which this remnant has been created by the planetary-nebula (PN) formation process. This has been shown by Schönberner (1979, 1983) and later fully been explored by Iben (1984).

In short, two major modes of post-AGB evolution towards the white-dwarf stage are possible, according to the two main modes of thermally pulsing AGB stars: the hydrogen-burning or helium-burning configuration. If, for instance, the PN formation, i.e. the removal of the stellar envelope by mass loss, occurs during the luminosity peak that immediately follows the helium shell instability, the remnant leaves the AGB while still only burning helium (Härm and Schwarzschild, 1975). On the other hand, PN formation may also occur during the quiescent hydrogen-burning phase, and the remnant then continues to burn hydrogen on its way to becoming a white dwarf.

Details and timing of the PN formation with respect to the thermal-pulse cycle phase are not known because of our poor knowledge of the mass-loss processes on the AGB. The high sensitivity of the post-AGB evolutionary tracks to the thermal-pulse phase $\phi$ allows, however, a distinction to be made between the helium-burning ($\phi = 0$) and hydrogen-burning mode ($\phi \geq 0.3$) by observations. An illustration is given in Fig. 1 where the temporal variations of different luminosity contributions of a 0.63 $M_\odot$ remnant for initial phases $\phi = 0$ and 0.85 resp. are shown (Blöcker, priv. comm.). The time is counted from a position close to the AGB (see below for more details). For hydrogen burning, the surface luminosity, $L$, remains constant ('plateau') until it drops rapidly by more than a factor of ten within a very short time span. Hydrogen burning becomes unimportant, and the further evolution is mainly controlled by the gravo-thermal energy release $L_g$. A helium-burning remnant behaves differently: helium burning decays only slowly, and the changeover to $L_g$ occurs on a larger timescale.

Observational evidences in favour of hydrogen-burning post-AGB models have already been collected by Schönberner (1989) and shall not be discussed again. I will only mention new results from PN progenitors, i.e. from cool post-AGB stars with detached, somewhat diluted dusty shells (v.d. Veen et al., 1989). The observed $T_{\text{eff}}$-post AGB age distribution is in excellent agreement with the predictions of hydrogen-burning post-AGB models with masses between 0.55 and 0.65$M_\odot$ (cf. also Schönberner, 1990).

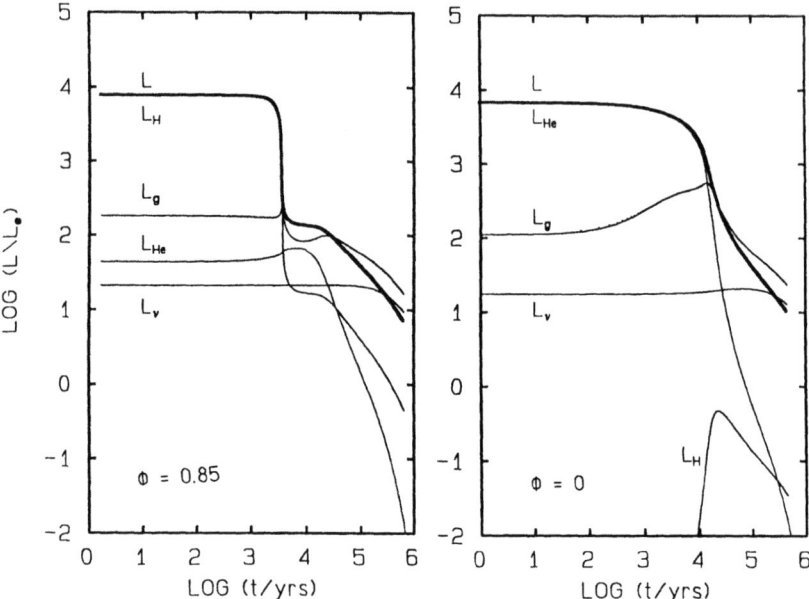

Fig. 1. Different luminosity contributions for a hydrogen-burning (left) and helium-burning remnant (right) of $0.63 M_\odot$ vs. post-AGB ages. Thick lines: surface luminosities $L$.

It must, however, be acknowledged that a substantial fraction ($\approx 10\ldots 30\%$) of the observed central-star population appears to have virtually hydrogen-free surfaces. Typical members belonging to this subclass are the WC, O VI and PG 1159 nuclei. Their origin is still quite uncertain but may be connected to a very 'late' helium shell flash that is accompanied by mixing and non-equilibrium burning (Iben et al., 1983). The reader is referred to Schönberner and Blöcker (1992) for a recent account of the problems in explaining hydrogen-free (i.e. helium-burning) CPN. We will consider here only hydrogen-burning central stars.

## Mass loss and evolution along the AGB

The structure of an AGB star is characterized by a very compact hydrogen-exhausted core with mass $M_H$ and a very dilute, fully convective envelope with mass $M_e$ containing the unprocessed matter. The hydrogen-burning shell is the interface between the core and the envelope, and the helium-burning shell sits further inside. The inert, helium-exhausted inner part of the stellar core consists of carbon and oxygen and is electron-degenerated. The stellar radius is about $10^4$ times larger than that of the core.

The luminosity of a giant star at the Hayashi limit depends practically only on the mass of its core (Paczyński, 1970; Kippenhahn, 1981). This statement holds, in general, also for stars on the AGB since it is mainly hydrogen burning that determines the overall course of evolution, i.e. virtually $L = L_H$ for more than 80% of the evolutionary time. This luminosity, $L_H$, in turn determines the growth rate,

$\dot{M}_H$, of the core according to

$$\dot{M}_H = L_H/XE_H,$$

where $E_H (= 6.3 \cdot 10^{18}\,\text{erg/g})$ is the energy released per gram of hydrogen and $X$ the hydrogen mass fraction of the stellar envelope. For a typical core mass $M_H$ of $0.6 M_\odot$, $\dot{M}_H$ is about $10^{-7} M_\odot/\text{yr}$. The core mass-luminosity relationship does not hold for massive stars where envelope convection cuts into the hydrogen-burning shell. The surface luminosity is substantially larger in these cases and depends also on the envelope mass. For more details see Blöcker and Schönberner (1991). Close to the tip of the AGB where the envelope mass is already small ($< 0.1 M_\odot$), the core mass-luminosity relation is always valid.

The stellar core can be thought of as being a very hot (pre) white dwarf which accretes nuclearly processed matter from the envelope. It grows by mass at the expense of the envelope at a rate given above while its radius shrinks. The core evolves independently of the envelope as long as the latter contains sufficient mass as to keep up the burning temperature in the hydrogen shell source by contraction. This also means that the evolution is independent of mass loss as long as thermal equilibrium is maintained. The evolutionary track of an AGB star in the HR diagram is entirely due to the envelope's response to the (masswise) growing core: expansion of the outer parts along the Hayashi-limit in order to accommodate the increasing luminosity, but overall contraction at nearly constant luminosity when the envelope mass drops below $\approx 0.05 M_\odot$. If hydrogen burning cannot be sustained anymore ($M_e < 10^{-4} M_\odot$), the luminosity drops and the envelope shrinks finally to white-dwarf dimensions (see also discussion in the next Section).

While the nuclear evolution of AGB stars is controlled by the core $\dot{M}_H$, the total lifetime on the AGB is determined by mass loss from the surface of the star. This is a consequence of the observed stellar winds with mass-loss rates up to $\dot{M}_w \approx 10^{-4}\,M_\odot/\text{yr}$ for the most luminous AGB stars (e.g. Knapp, 1985). Even if many stars may not reach such large rates, it is clear that always $\dot{M}_w \gg \dot{M}_H$ on the upper AGB (at $50000\,L_\odot$, $\dot{M}_H \simeq 7 \cdot 10^{-7} M_\odot/\text{yr}$ only).

Defining an evolutionary timescale by

$$-\frac{M_e}{\dot{M}_e} = \frac{M_e}{\dot{M}_H + \dot{M}_w} \quad \text{with } \dot{M}_H > 0,\ \dot{M}_w > 0,\ \dot{M}_e < 0,$$

we immediately conclude that stellar evolution is determined by the wind term, $\dot{M}_w$, on and in the vicinity of the AGB. Thus, the lifetime at the tip of the AGB is very short (planetary-nebula formation!), and the following post-AGB evolution is completely determined by the internal stellar structure at the AGB tip, which in turn is a function of the thermal-pulse cycle phase $\phi$ (cf. Iben 1984, also Schönberner 1979, 1983, or previous Section).

A certain mass-loss law has to be specified in evolutionary calculations, with the constraint that the total amount of mass lost from the star, i.e. $\int \dot{M}_{\text{agb}} dt$, must be consistent with a reasonable initial-final mass relation. A simple Reimers law (Reimers, 1975) will not work since it predicts a too small integrated mass loss which is not consistent with the rather flat initial-final mass relationship of Fig. 2 (see also Weidemann, 1992). Instead, Blöcker and Schönberner (1990) employed a formula for $\dot{M}_{agb}$ adapted from Bowen's (1988) calculations of pulsating Mira atmospheres. In terms of the Reimers rate, $\dot{M}_R$, they got $\dot{M}_{agb}/\dot{M}_R \sim L^{2.7}$ which yields $10^{-3}$ to $10^{-4} M_\odot/$yr at the tip of the AGB. The resulting combinations $(M_i, M_f) = 3M_\odot, 0.605M_\odot$ and $5M_\odot, 0.836M_\odot$ resp., are in fair agreement with the empirical galactic initial-final mass relation of Weidemann (1987), as is illustrated in Fig. 2. There are also the sequences of Wood and Faulkner (1986) plotted, and one realizes immediately that only their combination $(M_i, M_f) = 2M_\odot, 0.6M_\odot$ is consistent with the initial-final mass relation! Their more massive remnants have been computed from too small initial masses, which has consequences for the late central-star evolution (next Section).

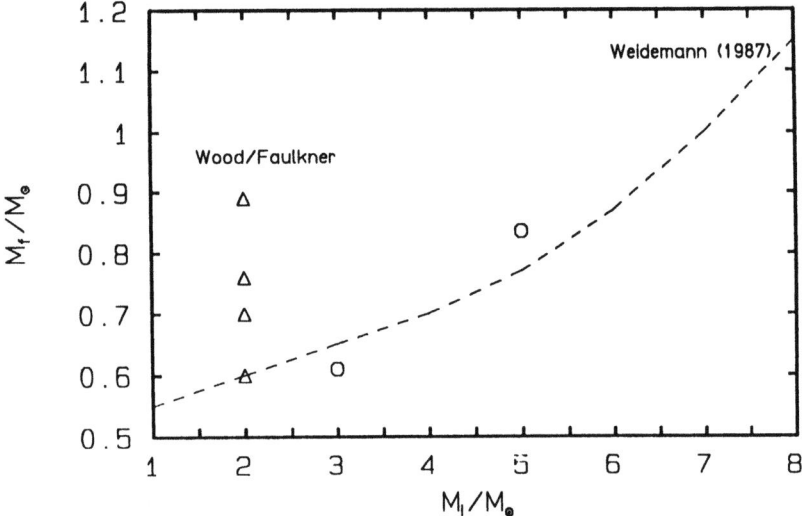

Fig. 2. Empirical initial-final mass relationship from Weidemann (1987) with different evolutionary sequences as explained in the text.

### Evolution through the central-star region

The remnant's transition towards the hotter parts of the HR diagram is also influenced by mass loss. The situation is, however, worse than in the case of the AGB evolution because neither reliable observations nor a theory exists. Furthermore, we do not exactly know where (i.e. at which effective temperature) the strong AGB mass loss ceases. It is normally assumed that this happens when $M_e \approx 10^{-3} M_\odot$, corresponding to an effective temperature between 5000...6000 K (Schönberner, 1983; Wood and Faulkner 1986). For the post-AGB evolution mass loss has either been neglected (Wood and Faulkner 1986) or set equal to the Reimers rate (Schönberner,1979,1983). Since the latter predicts $\dot{M}_w = \dot{M}_R \sim T_{eff}^{-2}$, such a wind term is only important for $T_{eff} < 10000$ K.

A more physical approach has recently been described by Blöcker and Schönberner (1990): the end of heavy AGB mass loss is coupled to the pulsational properties of the models, because the periods of radial pulsations decrease as the effective temperature becomes larger when the model is slowly departing from the Hayashi line. The pulsational periods are calculated from the stellar parameters according to Ostlie and Cox (1986). The mass-loss rate is reduced from its AGB value to the Reimers rate while the period (fundamental mode) changes from 100 to 50 d. A period of 50 d corresponds to $T_{\text{eff}} \approx 6000\ldots6500\,\text{K}$, depending on the remnant mass, and thus gives very similar zero points for the post-AGB evolution as used by Schönberner (1983). The Reimers rate is kept until it falls below the rate predicted by radiatively driven winds (Pauldrach et al., 1988) which is then used for central stars. This rate scales like $\dot{M}_\text{w} = \dot{M}_\text{P} \sim L^{1.9}$, roughly independent of effective temperature. $\dot{M}_\text{P}$ is well below the burning rate, $\dot{M}_\text{H}$, and does not afflict the evolutionary speed of central stars (cf. Perinotto, 1989).

The two post-AGB models with 0.605 and 0.836 $M_\odot$, resp., calculated by Blöcker and Schönberner (1990), behave on the 'plateau' as expected: the 0.836 $M_\odot$ model heats up about ten times faster than the 0.605 $M_\odot$ model. For the final dimming on the 'cooling' track, however, the results are different: at $L \approx 10^{2.1} L_\odot$ the massive remnant is *twice* as old as the lighter one (cf. also Fig. 3 in the next section). This result is in variance with the computations of Wood and Faulkner (1986) which predict a steadily increasing fading speed with increasing remnant mass.

An explanation is possible with Fig. 1 (left part). When hydrogen-burning is extinct, $L \simeq L_\text{g}$. Then the temporal evolution of the photon luminosity depends mainly on the thermo-mechanical structure of the core, modified somewhat by the neutrino losses. The structure of the core depends, on the other hand, on its evolutionary history: With increasing age (or thermal-pulse number) the core becomes smaller and denser, thereby steadily reducing its available potential-energy reservoir on its way to a fully electron degenerate configuration. We expect, in general, an increasing fading speed with increasing core age (core mass) or thermal-pulse number. Now more massive stars start their AGB evolution with already quite massive cores which are less condensed and substantially hotter than old cores of the same mass evolved from stars with smaller initial mass. These young and massive cores have a large reservoir of potential and thermal energy which enables them to fade with a considerably slower pace. With increasing age on the AGB such a massive core contracts and cools (neutrino losses!) until it converges to the limiting configuration for which heating by compression and cooling by neutrinos is nearly balanced. Only central stars which possess cores with such a 'converged' structure are expected to fade independently of their past evolution. For $M_\text{i} = 5 M_\odot$ about 40 thermal pulses are necessary, leading to $M_\text{H} \simeq 0.89\,M_\odot$ (Blöcker, priv. comm.). It is questionable whether mass loss does allow such a long lifetime ($\approx 2.5 \cdot 10^5 \text{yr}$) at very high luminosities ($M_\text{bol} < -6$).

The above discussion of fading properties did not consider the influence of the thermal-pulse cycle phase $\phi$. Indeed, over a complete pulse cycle $L_\text{g}$ does not remain constant. Large amounts of heat are converted into potential energy during the helium-shell instability by expanding the outer core regions. These regions contract after the shell flash and give rise to a slowly decreasing release of gravo-thermal energy. The core is smallest immediately before the onset of the next helium shell flash. This is the reason why Wood and Faulkner (1986) found a slight increase of the fading speed with $\phi$.

The different fading speeds of three selected AGB remnants are illustrated in Blöcker and Schönberner (poster, these proceedings). The tracks of the two remnants with the relatively young cores show clearly the importance of further contraction on the upper part of the 'cooling' track. The plateau evolution of the remnants with equal masses but different core ages, i.e. different initial masses, is virtually identical. This behavior corresponds with the earlier discussion above, since the evolution along the 'plateau' is controlled by the actual values of envelope mass, nuclear burning and mass loss. The influence of the earlier evolutionary history is only small. Note that the $0.836 M_\odot$ remnant with the older and smaller core is slightly more luminous.

### Comparison with observations

For comparison with observations we need a larger set of theoretical tracks to cover the expected mass spectrum of CPN. We thus combined the results of Blöcker and Schönberner (1990) with the earlier calculations of Schönberner (1983) and determined post-AGB isochrones (Fig. 3). These isochrones bend sharply downwards because of the fast luminosity drop when hydrogen burning extinguishes, and then upwards because of the slow fading of our massive AGB remnant. These isochrones suggest that the least luminous central stars should have about $0.65\,M_\odot$!

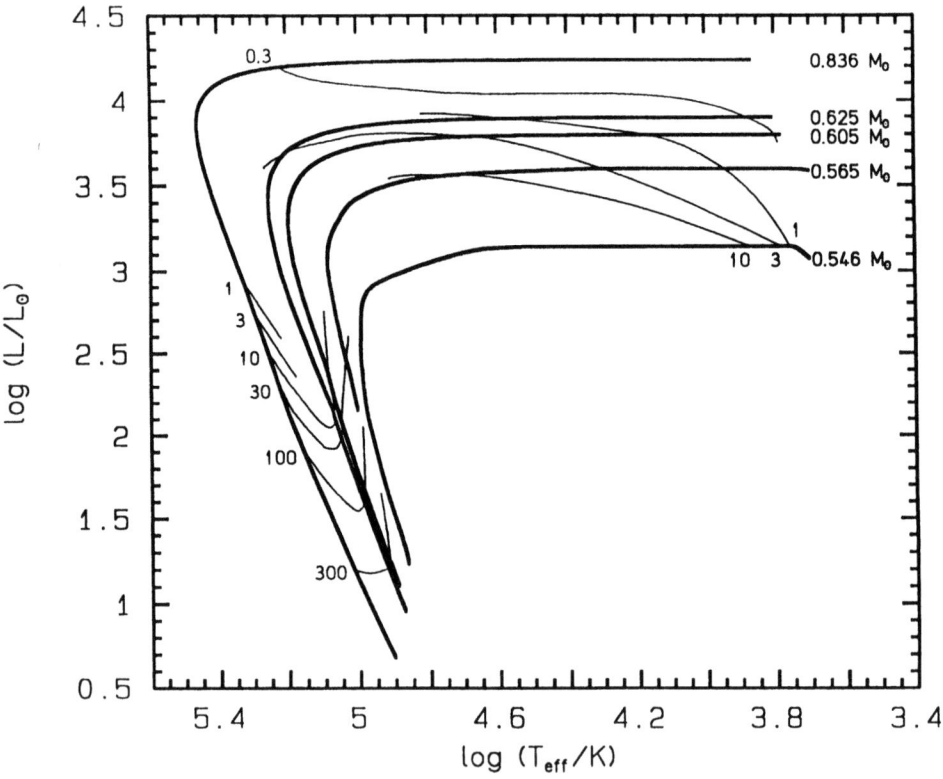

Fig. 3. HR diagram with hydrogen-burning post-AGB tracks and isochrones.

The 'plateau' luminosities have very successfully been determined by observing and analysing Magellanic Cloud PN (cf. Walton et al., 1991). Upon comparison with the tracks they found a very narrow range of CPN masses, $0.58\ldots0.64 M_\odot$, with a mean mass of $M = 0.59 \pm 0.2\,M_\odot$. In respect with the isochrones depicted in Fig. 3 it is, however, more interesting to study the late CPN evolution. For this purpose we used the fact that many PN with hot, intrinsically faint central stars appear to be optically thick even in the hydrogen Lyman continuum, making a temperature determination by the Zanstra method possible. Also model PN coupled to hydrogen-burning central-star models show that recombination caused by the rapid luminosity drop leads temporarily to optically thick nebular shells (Schönberner and Tylenda, 1990; Marten and Schönberner, 1991). The ionized masses are a few tenth of a solar mass, and the well-known Shklovsky method for distance determinations is in these cases applicable! We thus selected all clear-cut cases of optically thick galactic PN from the lists of Jacoby and Kaler (1989) and Kaler and Jacoby (1989), resulting in a sample of 29 very hot central stars (out of 82 listed objects). Their loci in the HR diagram are estimated from the Zanstra temperatures ($T_\mathrm{HI} \simeq T_\mathrm{HeII}$) and Shklovsky distances ($M_\mathrm{ion} = 0.2 M_\odot$) and compared to the theoretical tracks in Fig. 4. The agreement with our calculations is striking: practically all objects accumulate where the fading speed of the models decreases, with the least luminous objects having indeed only modest masses of, say, a little bit above $0.63\,M_\odot$. The hottest (and most massive) nuclei have comparatively larger luminosities, just as our calculations predict. A few objects appear to be rather young, with still high luminosities. They are likely also more massive CPN which evolve so quickly along the 'plateau' that the usual optically-thin PN phase does not occur.

The ionized nebular mass, $M_\mathrm{ion}$, for the oldest objects plotted in Fig. 4 may have been underestimated because of accretion of matter from the old AGB wind (cf. Marten and Schönberner (1991), Fig. 12). Thus distances to hot, low-luminous CPN are very likely *underestimated*, contrary to the general belief (cf. Kaler and Jacoby, 1989). Increasing the distances to these objects would not change our conclusion.

In conclusion we can summarize that the late evolution of more massive central stars depends on their history and is not a unique function of their core mass as normally assumed in the past. Moreover, the investigation of a larger sample of low-luminosity CPN indicates that relatively massive AGB stars experience only few thermal pulses before PN ejection, and that the resulting remnants have young cores which fade rather slowly along the upper part of the 'cooling' track. It would be very important for the future to concentrate on the hottest central stars and to improve their temperature and luminosity determinations.

## References

Blöcker, T., Schönberner, D.: 1990, A&A 240, L11
Blöcker, T., Schönberner, D.: 1991, A&A 244, L43
Bowen, G.H.: 1988, ApJ 329, 299
Härm, R., Schwarzschild, M.: 1975, ApJ 93, 200
Iben, I.,Jr.: 1984, ApJ 277, 333
Iben, I.,Jr., Kaler, J.B., Truran, J.W., Renzini, A.: 1983, ApJ 264, 605
Jacoby, G.H., Kaler, J.B.: 1989, AJ 98, 1662
Kaler, J.B.: 1983, ApJ 271, 188
Kaler, J.B., Jacoby, G. H.: 1989, ApJ 345, 871
Kippenhahn, R.: 1981, A&A 102, 293
Knapp, G.: 1985, ApJ 293, 273
Marten, H., Schönberner, D.: 1991, A&A 248, 590
Ostlie, D.A., Cox, A.N.: 1986, ApJ 311, 864

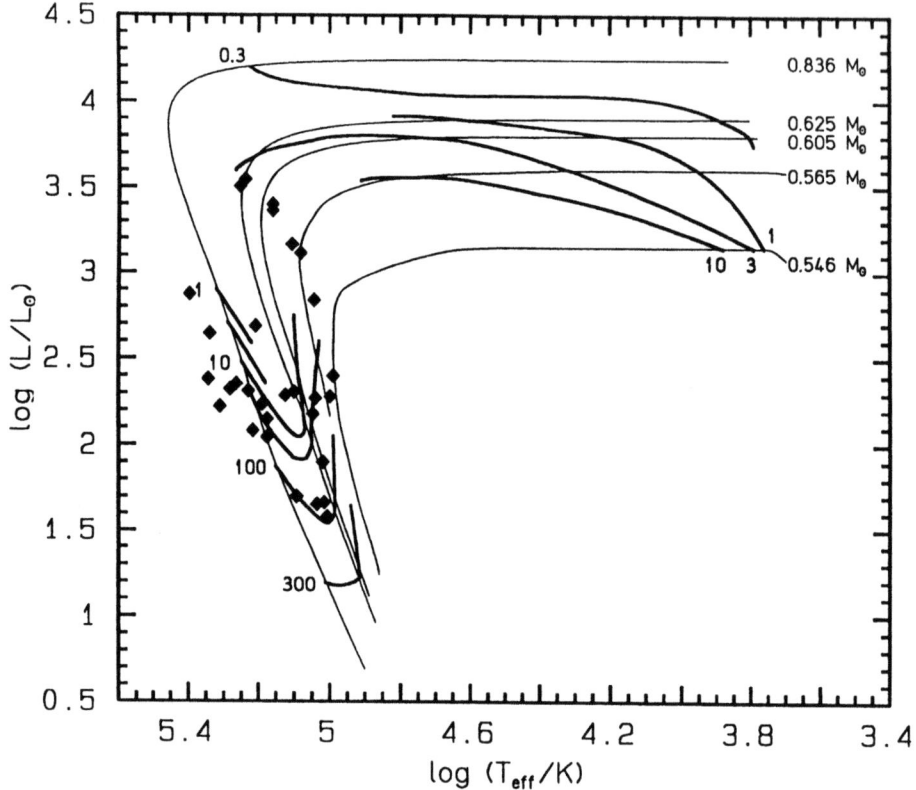

Fig. 4. HR diagram of hot central stars with optically thick planetaries.

Paczyński, B.: 1970, Acta Astron. 20, 47
Paczyński, B.: 1971, Acta Astron. 21, 417
Pauldrach, A., Puls, J., Kudritzki, R.P., Méndez R., Heap, S.R.: 1988, A&A 207, 123
Perinotto, M.: 1989, IAU Symp. No. 131, *Planetary Nebulae*, S. Torres-Peimbert, Ed., Kluwer, p. 293
Reimers, D.: 1975, *Problems in Stellar Atmospheres and Envelopes*, B. Baschek, W.H. Kegel, G. Traving, Eds., Springer, p. 229
Schönberner, D.: 1979, A&A 79, 108
Schönberner, D.: 1981, A&A 103, 119
Schönberner, D.: 1983, ApJ 272, 708
Schönberner, D.: 1989, Proc. IAU Symp. No. 131, *Planetary Nebulae*, S. Torres-Peimbert, Ed., Kluwer, p. 463
Schönberner, D.: 1990, Proc. Int. Coll. *From Miras to Planetary Nebulae: Which Path for Stellar Evolution?*, M.O. Mennessier and A. Omont, Eds., p. 355
Schönberner, D., Tylenda, R.: 1991, A&A 248, 590
Schönberner, D, Blöcker, T.: 1992, Proc. Workshop *Atmospheres of Early-Type Stars*, U. Heber and C.S. Jeffery, Eds., Springer, p. 305
Van der Veen, W.E.C.J., Habing, H.J., Geballe, T.R.: 1989, A&A 226, 108
Walton, N.A., Barlow, M.J., Clegg, R.E.S., Monk, D.J.: 1991, Proc. IAU Symp. No. 148, *Magellanic Clouds*, P.R. Haynes and D. Milne. Eds., Kluwer, p. 334
Weidemann, V.: 1987, A&A 188, 74
Weidemann, V.: 1992, Proc. ESO-CTIO Workshop *Mass Loss on the AGB and Beyond*, in press
Wood, P.R., Faulkner, D.J.: 1986, ApJ 307, 659

# DIAGRAMS FOR OBSERVATIONAL TESTING OF EVOLUTION OF PLANETARY NEBULA NUCLEI

R. TYLENDA
*Copernicus Astronomical Center, Laboratory of Astrophysics*
*ul. Chopina 12/18, 87-100 Toruń, Poland*

**Abstract.** Confrontation of theoretical models with observations is the main way of progress of our understanding of the nature of astrophysical objects. This paper discusses different diagrams used for observational testing of theoretical evolutionary models of planetary nebula nuclei (PNNi). Particular attention is paid in indicating and discussing sources of errors and uncertainties which may result in wrong conclusions drawn from the diagrams. Finally, results of recent studies comparing the theory and observations for the Galactic PNNi are critically reviewed.

## 1. H-R diagram

This is the most classic and most widely applied diagram for observational testing of the theory of stellar evolution and of the PNN evolution, in particular. The most often used version of this diagram plots the PNN luminosity, $L_\star$, versus the PNN effective temperature, $T_\star$. Thus the main advantage of the H-R diagram is that it displays fundamental parameters of a PNN, parameters whose physical meaning and interpretation are obvious and clear. Moreover, these parameters can be reliably determined from theoretical model calculations.

Unfortunately, from an observational point of view the problem is subject to substantial uncertainties. In the case of a typical PNN the bulk of energy is emitted in the EUV region which cannot be directly observed. As a result, determination of $T_\star$ of a PNN is usually difficult and uncertain. This subject is discussed in detail in another review (Preite-Martinez in this volume). As shown by Schönberner & Tylenda (1990) errors in $T_\star$ can be important, especially if the observational material is poor or if one uses not entirely correct assumptions (e.g. that all nebulae are optically thick). As a result the obtained positions in the H-R diagram may be very different from the true ones leading to incorrect conclusions.

The errors in $T_\star$ have also important consequences on the PNN luminosity. This is because $L_\star$ is often calculated from an observed PNN magnitude and a bolometric correction; the latter being a strong function of the adopted $T_\star$.

Another way of calculating $L_\star$ is to use nebular $H\beta$ or radio fluxes. These quantities can be measured with a relatively high precision which is not often the case for the PNN magnitudes (e.g. problems with nebular contamination). Another advantage is that apart from the coolest PNNi the relation between the $H\beta$ (or radio) flux and the PNN luminosity is only weakly dependent on $T_\star$. However, this method gives reliable results only if the nebula completely absorbs the ionizing radiation. Otherwise one gets only a lower limit to $L_\star$.

Observational verification of a question whether a particular planetary nebula (PN) completely absorbs the PNN ionizing flux or not, is not a straightforward problem. Strong lines of low excitation ([NII], [OII]) are sometimes taken as a criterion for optically thick PNe. However, for a non-spherical PN - which is often the case - strong low-excitation lines may indicate that the PN is indeed optically thick but only in some directions. In some cases - especially for old, low luminosity objects - when the so-called ionizing parameter is low, the nebular gas is mildly ionized. Then even in innermost PN regions there is an important proportion of neutral or singly ionized species and the strength of low-excitation lines is a measure of the ionizing parameter rather than of the optical thickness.

Some authors interpret an observed increase of the ionized PN mass, $M_{ion}$, with the nebular radius, $R_{PN}$, as an argument that the PNe are opaque (e.g. Pottasch & Acker 1989; Zijlstra & Pottasch 1989). However, this correlation has to be interpreted with a great caution since $M_{ion}$ and $R_{PN}$ are *not* independent. This fact in combination with observational selection effects and uncertainties in observed quantities can produce an apparent trend of increasing $M_{ion}$ with $R_{PN}$ no matter whether the PNe are optically thick or thin (Stasińska et al. 1991a; Zijlstra 1990; Stasińska & Tylenda in this volume).

As usually in the case of stellar luminosities a crucial problem lies in the PN distances. Another review is devoted to this subject (Terzian in this volume). The present situation in this field is far from being satisfactory and is a source of substantial errors and uncertainties. Hence importance of investigations of PN samples with fairly accurate distances, i.e. Galactic bulge, Magellanic Clouds.

Another source of uncertainties is the interstellar extinction. A common opinion seems to be that in the case of PNe the extinction can be determined quite accurately, e.g. from an observed Balmer decrement. Then the observations can be corrected without particular problems, at least in the optical. However, it has recently been found that the extinction from Balmer decrement, $C_{opt}$, is often larger than that from the radio/$H\beta$ flux ratio, $C_{rad}$ (Stasińska et al. 1991a; Cahn et al. 1992; Tylenda et al. 1992). A thorough discussion of the problem in Stasińska et al. (1992) reveals that there is a systematic trend between the two values with an average relation $C_{opt} \simeq 1.17\, C_{rad}$. The origin of this effect is not clear. Nevertheless, this shows that depending on the method for determining extinction one can obtain a systematic shift in luminosity by factor 2 for heavily reddened PNe.

In conclusion, placing an observed PNN on the H-R diagram is not a straightforward or easy problem. There are numerous sources of errors while determining $T_\star$ and $L_\star$. If this is combined with the fact that theoretical tracks are usually close one to another in the $\log L_\star - \log T_\star$ plane (cf. dashed curves in Fig.1a) it is clear that an estimation of PNN masses from observed positions in the H-R diagram can be subject to large uncertainties. In addition, some objects may appear in a region where no theoretical track passes through. Therefore, conclusions have to be drawn with a great caution while interpreting an observed PNN H-R diagram

## 2. $M_v$ - $t_{ev}$ diagram

Theoretical models of the PNN evolution predict that the evolutionary time scale strongly decreases with the PNN mass (Schönberner 1979, 1983; Wood & Faulkner 1986). This fact offers an important possibility for observational testing of PNN models. Schönberner (1981) was first to explore this possibility using a plot of absolute visual PNN magnitude, $M_v$, versus PN radius, $R_{PN}$, originally proposed by Abell (1966).

Principal advantages and disadvantages of the $M_v$ - $t_{ev}$ diagram can be seen if one plots theoretical tracks on it (cf. dashed curves in Fig. 1b). First of all, the tracks for different masses are well separated for $M_v < 5$, i.e. during the initial, constant luminosity phase. Thus in this phase different theoretical models can be well tested against observations. However, during the cooling phase the tracks approach one another in the $M_v$ - $t_{ev}$ plane. Thus nothing sure can be said about old, low-luminosity objects from this plot.

Another advantage of the discussed diagram is that it uses parameters directly available from observations ($M_v$, $R_{PN}$). In other words, the $M_v$ - $t_{ev}$ diagram stays closely to observations.

Much of uncertainty when confronting theoretical models to observed PNNi in the $M_v$ - $t_{ev}$ plane is due to the fact that the theoretical age, $t_{ev}$ has to be compared to the observed $R_{PN}$ or expansion time, $t_{exp} = R_{PN}/v_{exp}$ ($v_{exp}$ - expansion velocity). Observational determinations of $R_{PN}$ can be subject to significant uncertainties, especially for PNe having ill defined outer rims or highly irregular structures, as well as, for compact, badly resolved objects. Errors in $v_{exp}$ are not negligible either. For many objects there is no measurement and one is forced to use a sort of typical $v_{exp}$, usually taken as 20 km/sec.

The theoretical evolutionary time, $t_{ev}$, depends on three factors: nuclear burning rate, zero-age point (moment when the PN formation is finished) and mass loss rate. The first process is rather well known from the stellar evolution theory. Unfortunately, the other two factors are taken from simplified and somewhat arbitrary considerations which results in significant uncertainties in the theoretical tracks. This problem is discussed in a poster of Górny (see this volume). It appears that within a reasonable range of parameters defining zero-age point and PNN wind rate the uncertainty in $\log t_{ev}$ is typically within 0.3.

Similarly as in the H-R diagram, the PN distances introduce substantial uncertainties in the $M_v$ - $t_{ev}$ plane. It is, however, interesting to note that systematic errors in distances affect the observed positions in the $M_v$ - $t_{ev}$ plot in an opposite sense than in the H-R diagram. For instance, increasing distances shifts the observed points in the $M_v$ - $t_{ev}$ diagram towards tracks of lower masses whereas in the H-R diagram the result is to get higher $L_\star$ and, usually, higher PNN masses. This effect can be used as a test of different distance scales (Tylenda & Stasińska 1989).

## 3. Effects of the Zanstra method and the Shklovsky distances

The Zanstra method is the simplest and most widely used method for estimating $T_\star$, similarly as the Shklovsky method for the PN distances. It is therefore instructive to investigate possible uncertainties and systematic effects introduced by these two methods while constructing observational diagrams.

For this purpose let us consider a simple scenario in which a model PNN is surrounded by a 0.2 $M_\odot$ nebula expanding with 20 km/sec and having a form of a spherically symmetrical shell of a constant thickness $\Delta R/R = 0.3$. At a given time moment the H ionization structure of the nebula can be calculated from simple (Strömgren) considerations adopting that the PNN radiates as a blackbody. This allows to estimate the $H\beta$ flux and, subsequently, the HI Zanstra temperature and luminosity. Form the $H\beta$ flux and the extension of the $H^+$ zone in the model PN one can simulate observational determination of the Shklovsky distance and its effect on various parameters, such as $L_\star$, $M_v$, $t_{ev}$.

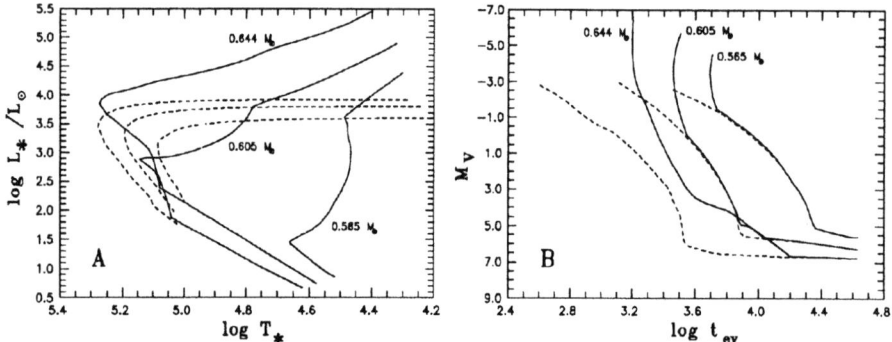

Fig. 1. *Effects of the HI Zanstra method and the Shklovsky distances on the observed evolution of PNNi in the H-R diagram (a) and in the $M_v-t_{ev}$ diagram (b). Dashed curves - original Schönberner's tracks. Full curves - observed evolution. Curves are labeled with PNN masses.*

The results are displayed in Fig. 1. Dashed curves show the original Schönberner's tracks (Schönberner 1979, 1983; Blöcker & Schönberner 1990). Full curves present the "observed" evolution from our simulations.

The curves for lower mass PNNi (particularly $0.565 M_\odot$) in Fig. 1a illustrate quantitatively what has been said in Sect. 1 about the danger of using the observed $H\beta$ (or radio) flux as an indicator of $L_\star$. A low mass, slowly evolving PNN leaves enough time for its PN to expand considerably. When the PNN starts emitting ionizing photons the diffuse PN quickly becomes totally ionized and the $H\beta$ line measures only a fraction of the ionizing flux. Consequently, the PNN appears as a very low $T_\star$ and very low $L_\star$ object, i.e. at a position where no PNN is expected according to the stellar evolution theory.

Systematic errors due to the Shklovsky method are stronger for more massive PNNi (see particularly $0.644 M_\odot$ in Fig. 1), especially in initial phases. These PNNi evolve fast and they start ionizing the nebular gas very early when their PNe are compact and dense. Consequently, $M_{ion}$ is much smaller than $0.2\ M_\odot$ and the Shklovsky method considerably overestimates the distance. As a result the tracks of more massive PNNi are moved upwards in the H-R diagram (Fig. 1a) and right-upwards in the $M_v$ - $t_{ev}$ plot (Fig. 1b).

An interesting observation can be made upon comparing Figs. 1a and 1b. As a result of the Zanstra method and the Shklovsky distances the "observed" positions of PNNi in the H-R diagram are much more dispersed than the purely theoretical tracks (cf. full and dashed curves in Fig. 1a). Consequently, the "observed" PNN mass distribution derived from Fig. 1a would be *wider* than the real one. The effect in the $M_v$ - $t_{ev}$ plane is, however, qualitatively different. Due to the Shklovsky distances more massive PNNi move towards low mass tracks. Thus in this case the "observed" PNN mass distribution would show *less* high mass objects than in reality. Perhaps this is the main reason why the PNN mass distributions derived from the $M_v$ - $t_{ev}$ diagram (e.g. Schönberner 1981; Weidemann 1989) are usually narrower and peaked at lower masses than those obtained from the H-R diagram (e.g. Kaler 1983; Kaler et al. 1990).

As can be seen from Fig. 1, observational methods can substantially modify PNN positions in the diagrams. It is not easy to correct for these effects in a real situation. Hence an idea, so far explored by Stasińska & Tylenda (1990) and Tylenda et al. (1991), to apply exactly the same methods of deriving parameters for both, observed objects and theoretical evolving PNNi surrounded by expanding model PNe. In this way the theoretical tracks are brought much closer to the observations and systematic errors due to observational methods are minimized. Unfortunately this procedure introduces uncertainties due to the model PNe.

## 4. Distance-independent diagrams

As mentioned above, investigations of PNNi very often suffer from uncertainties in the distances. Hence importance of analyses using observational parameters independent of the distances.

Tylenda & Stasińska (1989) have proposed a distance-independent diagram which plots $T_\star$ against a parameter $f = L_v/(4 \pi t_{ev}^2)$ where $L_v$ is the PNN luminosity in the $V$ band. For observed objects, adopting that $t_{ev} = t_{exp}$ one gets $f = (F_v v_{exp}^2)/\theta^2$ where $F_v$ denotes the PNN flux observed (corrected for extinction) in the $V$ band and $\theta$ is the observed angular radius of the PN.

The $T_\star$ - $f$ plane combines observational parameters used in the H-R and $M_v$ - $t_{ev}$ diagrams. Hence many remarks from Sect. 1 and 2 are valid here, as well. Similarly as in the H-R diagram, uncertainties in $T_\star$ affect the observed positions in the $T_\star$ - $f$ plot. Problems of the $M_v$ - $t_{ev}$ diagram (zero-age point, mass loss, observational measuring of $\theta$ and $v_{exp}$) introduce uncertainties while comparing the $f$ values for observed objects and theoretical tracks. The $T_\star$ - $f$ diagram, similarly as $M_v$ - $t_{ev}$, is particularly suitable for investigating PNNi with active shell sources. During the cooling phase the tracks of different masses follow similar paths and an analysis becomes very uncertain.

Several distance-independent diagrams have recently been investigated by Zhang & Kwok (1992). Most useful for testing of the PNN evolution is a plot of $T_b$ versus $T_\star$, where $T_b$ is the PN radio brightness temperature. This diagram greatly reduces problems due to extinction. There are, however, several sources of uncertainties while analyzing this diagram. Firstly, as mentioned in Sect. 1, determination of $T_\star$ is far from being a simple problem. Secondly, uncertainties and errors while measuring PN radii (ill defined PN rims, irregular objects, small, badly resolved PNe) can seriously affect the derived brightness temperature as $T_b \sim \theta^{-2}$. Thirdly, when calculating $T_b$ for a theoretical track one has to adopt a PN model. Bearing in mind numerous problems while modelling PN formation and evolution one realizes that the positions of theoretical tracks in the $T_b$ - $T_\star$ plane are uncertain.

The PNN evolution can also be studied using other diagrams which I cannot discuss because of lack of space. Let me mention that Schönberner (1986) has attempted to constrain PNN models from a plot of the nebular excitation versus $M_v$.

Szczerba has applied a $HeII\lambda4686$ line intensity histogram (Szczerba 1987) and a plot of the nebular ratio $HeII\lambda4686/H\beta$ versus $F(H\beta)/F_{PNN}(\lambda4861)$ where $F_{PNN}(\lambda4861)$ is the PNN continuum flux at 4861Å (Szczerba 1990). A PN luminosity function in $H\beta$ and [OIII] can also be used as shown by Jacoby (1989) and Stasińska et al. (1991b). Finally, studies of PN line intensities can provide constraints to PNN models as well, as in Vilkoviskii et al. (1983) and Stasińska (1989).

## 5. Observed PNN evolution and masses

This section summarizes the results obtained by different authors concerning the evolution and masses of the Galactic PNNi. Results for other galactic systems are discussed elsewhere (e.g. Dopita in this volume).

At the very beginning it should be mentioned that it is rather difficult and somewhat dangerous to compare and draw conclusions from results obtained in different papers. The reason is that different authors use different diagrams, different distance scales, different PN samples in which observational selection effects can also be different. My feeling is that much of difference between individual determinations is indeed due to these factors.

Schönberner (1981) has investigated the problem of PNN masses using the $M_v$ - $t_{ev}$ diagram and the Shklovsky distances. He has concluded that most of PNNi burn hydrogen quiescently and found a very narrow distribution with an average PNN mass of $0.58M_\odot$ and 85% of the objects within $\pm0.03M_\odot$.

Heap & Augensen (1987) have used the same type of diagram but with $UV$ PNN magnitudes and Daub's (1982) distances. They have obtained a PNN mass distribution much wider than Schönberner with a mean value of $\sim0.60M_\odot$ and an important high mass tail extending up to $\sim0.8M_\odot$.

Weidemann (1989) has argued that the Daub's distance scale is too small. Using the same observational material as Heap & Augensen but with modified distances he has derived a distribution essentially the same as Schönberner.

Kaler (1983) has explored the H-R diagram for investigating the PNN masses. His study has been limited to large PNe for which the Shklovsky method is expected to give a correct distance scale. The Kaler's distribution is wider than the Schönberner's with $\sim40$% of objects above $0.6M_\odot$.

The Kaler's study has been extended in Kaler et al. (1990) using new observational data. They have obtained a still wider distribution with $\sim40$% above $0.7M_\odot$.

The H-R diagram has also been investigated by Pottasch (1984) and Gathier & Pottasch (1989) who adopted the so-called individual distances. They have obtained a large number of cool PNNi at low luminosities ($< 10^3 L_\odot$) which would imply masses of $\sim0.5M_\odot$. This is in conflict with the theoretical models which predict that PNe with central stars $<0.55M_\odot$ should not be observed because of too a long time scale of their post-AGB evolution (Schönberner 1983). Gathier & Pottasch conclude that low-mass PNNi evolve much faster than theoretically predicted.

Tylenda & Stasińska (1989) have analyzed the PNN evolution in three diagrams, i.e. H-R, $M_v$ - $t_{ev}$ and $T_\star$ - $f$. They have found that the individual distances lead to inconsistent results suggesting that these distances are often underestimated.

Instead, the Shklovsky distances have given consistent results from different diagrams. The authors conclude that most of PNNi have masses between $0.55 M_\odot$ and $0.65 M_\odot$ and that the H-burning PNN models better reproduce the observations than the He-burners.

Stasińska & Tylenda (1990) have used the same diagrams as in Tylenda & Stasińska (1989) but with theoretical tracks modified by the presence of a PN and observational effects due to the Zanstra and Shklovsky methods. They have obtained the PNN masses between $0.55 M_\odot$ and $\sim 0.7 M_\odot$ and a peak mass at $0.61 M_\odot$.

Zhang & Kwok (1992) have derived a strongly asymmetric PNN mass distribution from their distance-independent $T_b$ - $T_\star$ diagram. The distribution peaks at $0.60 M_\odot$ and has a high mass tail extending up to $0.85 M_\odot$.

From a $HeII\lambda 4686$ line intensity histogram and a plot of $HeII\lambda 4686/H\beta$ versus $F(H\beta)/F_{PNN}(\lambda 4861)$ Szczerba (1987, 1990) has concluded that the PNN mass distribution is strongly peaked at $0.60 M_\odot$ and that the PNNi mostly burn hydrogen.

### 5.1. GALACTIC BULGE

Investigations of PNe in the Galactic bulge are important because of two reasons. Firstly, this is the only sample of PNe in the Galaxy for which we have fairly well known distances. Secondly, it is expected that these objects might be somewhat different from the rest of the PNe in the Galaxy.

Webster (1988) compared the excitation classes of PN samples in the Galactic bulge and in the Magellanic Clouds. She has concluded that the bulge has more low-mass progenitor stars than the LMC and the solar neighbourhood.

Zijlstra & Pottasch (1989) and Pottasch & Acker (1989) have constructed H-R diagrams for the Galactic bulge PNNi. In both studies the PNN luminosities have been calculated from the $H\beta$ (or radio) fluxes as the authors argue that the PNe are opaque. The resulting diagrams show most of the objects to be of low $L_\star$ and low $T_\star$ which cannot be explained by the existing PNN models. Pottasch & Acker conclude that high mass PNNi evolve much slower and low mass PNNi much faster than predicted from the theory.

Tylenda et al. (1991) have analyzed several diagrams with theoretical tracks modified for observational effects due to the presence of a nebula. They have obtained a mean PNN mass of $0.593 M_\odot$ with a standard deviation of $0.025 M_\odot$. This can be compared to the value of $0.615 \pm 0.036$ $M_\odot$ found for a sample of Galactic disc PNe by Stasińska & Tylenda (1990) using a similar procedure of analysis. This would suggest that the bulge PNNi are less massive. However, the conclusion is not sure as the selection effects occurring in the two samples are not the same.

From an analysis of [OIII] and $H\beta$ luminosity functions for the bulge PNe Stasińska et al. (1991b) have obtained a mean PNN mass of 0.57 - 0.60 $M_\odot$ and a standard deviation of 0.04 - 0.05 $M_\odot$.

Mendez (1992) has estimated the PNN luminosities in the bulge using observed PNN magnitudes and bolometric corrections derived from the Stoy method. He has obtained a mean $\log L_\star/L_\odot = 3.86$ which corresponds to $0.625 M_\odot$ as a mean PNN mass (according to the Schönberner's models). Comparison of the results of

Mendez with those of Tylenda et al. (1991) shows the effect of different methods for deriving extinction (Balmer decrement in the former, radio/$H\beta$ in the latter).

## 5.2. PNN MASSES FROM SPECTROSCOPIC STUDIES

The most sophisticated method for determining PNN masses is the one elaborated by Mendez et al. (1988), based on a model atmosphere analysis of observed PNN absorption line profiles (see also Kudritzki in this volume). The method has a great advantage of being independent of distances. However, it can be applied only to bright central stars showing little nebular contamination for which accurate line profiles can be obtained spectroscopically.

Mendez et al. (1988, 1990) have analyzed 23 PNNi using non-LTE plane-parallel model atmospheres. They have obtained PNN masses substantially higher from those found in other studies summarized above. A mean value of their masses is $0.70 M_\odot$ with a standard deviation of $0.09 M_\odot$.

Recently, Mendez et al. (1992) have reanalyzed their observational material using more realistic non-LTE model atmospheres including spherical extension and stellar winds. As a result the PNN gravities have been somewhat reduced leading to slightly lower - but still quite high - masses, i.e. $0.65 \pm 0.05\ M_\odot$.

It is often argued that the reason why Mendez et al. find high PNN masses is that they picked up the most luminous and thus the most massive PNNi. However, this argument does not seem to be convincing. The fact that a sample contains objects apparently brightest does not necessarily imply that they are intrinsically most luminous. In fact it can be argued on theoretical grounds that within luminous PNNi it is more probable to find a low mass object than a high mass one.

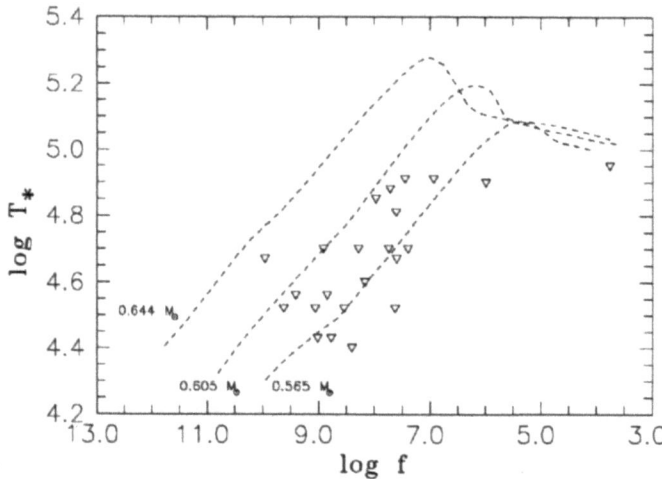

Fig. 2. $T_\star$ - $f$ diagram for the PNN sample of Mendez et al. (1992). Dashed curves - theoretical tracks of Schönberner labeled with PNN masses. Triangles - observed objects.

Fig. 2 shows the sample of Mendez et al. in the $T_\star$ - $f$ diagram. The values of $T_*$ have been taken from Mendez et al. (1992). Observational data necessary to calculate $f$ have been taken from Acker et al. (1992).

As can be seen from Fig. 2, there is nothing exceptional in the positions of PNNi from the sample of Mendez et al. They are fairly typical for other PNNi (see e.g. Fig. 3 in Tylenda & Stasińska 1989). A similar conclusion, although from a different diagram, has been drawn by Szczerba (1990). An estimation of the PNN masses from Fig. 2 gives $0.58 \pm 0.02\ M_\odot$.

The consequences of the above result are the following. If the PNN masses derived by Mendez et al. are correct than all the PNNi are more massive and more luminous than derived in other works. In addition, as concluded by McCarthy et al. (1990), the theoretical evolutionary time scales are wrong. The PNNi, after having formed their nebulae, evolve much slower than predicted. If, however, as found in other studies, the present PNN models are roughly correct than there is something wrong or missing in the present model atmosphere analyses which leads to a systematic overestimate of the PNN masses.

**Acknowledgement.** This work was supported in part from grant no. 2-2114-92-03 financed by Polish Committee for Scientific Research.

**References**

Abell G.O., 1966, ApJ 144, 259
Acker A., Ochsenbein F., Stenholm B., Tylenda R., Marcout J., Schohn C., 1992, Strasbourg-ESO Catalogue of Galactic Planetary Nebulae (ESO publ.)
Blöcker T., Schönberner D., 1990, A&A 240, L11
Cahn J.H., Kaler J.B., Stanghellini L., 1992, A&AS 94, 399
Daub C.T., 1982, ApJ 260, 612
Gathier R., Pottasch S.R., 1989, A&A 209, 369
Heap S.R., Augensen H.J., 1987, ApJ 313, 268
Jacoby G.H., 1989, ApJ 339, 39
Kaler J.B., 1983, ApJ 271, 188
Kaler J.B., Shaw R.A., Kwitter K.B., 1990, ApJ 359,392
McCarthy J.K., Mould J.R., Mendez R.H., Kudritzki R.P., Husfeld D., Herrero A., Groth H.G., 1990, ApJ 351, 230
Mendez R.H., 1992, A&A in press
Mendez R.H., Kudritzki R.P, Herrero A., Husfeld D., Groth, H.G., 1988, A&A 190, 113
Mendez R.H., Herrero A., Manchado A., 1990, A&A 229, 152
Mendez R.H., Kudritzki R.P., Herrero A., 1992, A&A in press
Pottasch S.R., 1984, Planetary Nebulae, Reidel, Dordrecht
Pottasch S.R., Acker A., 1989, A&A 221, 123
Schönberner D., 1979, A&A 79, 108
Schönberner D., 1981, A&A 103, 119
Schönberner D., 1983, ApJ 272, 708
Schönberner D., 1986, A&A 169, 189
Schönberner D., Tylenda R., 1990, A&A 234, 439
Stasińska G., 1989, A&A 213, 274
Stasińska G., Tylenda R., 1990, A&A 240, 467

Stasińska G., Tylenda R., Acker A., Stenholm B., 1991a, A&A 247, 173
Stasińska G., Fresneau A., da Silva Gameiro G.F., Acker A., 1991b, A&A 252, 762
Stasińska G., Tylenda R., Acker A., Stenholm B., 1992, A&A in press
Szczerba, R., 1987, A&A 181, 365
Szczerba, R., 1990, A&A 237, 495
Tylenda R., Stasińska G., 1989, A&A 217, 209
Tylenda R., Stasińska G., Acker A., Stenholm B., 1991, A&A 246, 221
Tylenda R., Acker A., Stenholm B., Köppen J., 1992, A&AS in press
Vilkoviskii E.V., Kondrateva L.N., Tambotseva L.V., 1983, Sov. Astron. 27, 194
Webster B.L., 1988, MNRAS 230,377
Weidemann, V., 1989, A&A 213, 155
Wood P.R., Faulkner D.J., 1986, ApJ 307, 659
Zhang C.Y., Kwok S., 1992, A&A in press
Zijlstra A.A., 1990, A&A 234, 387
Zijlstra A.A., Pottasch S.R., 1989, A&A 216, 245

# THE EVOLUTION OF THE PLANETARY NEBULAE IN THE MAGELLANIC CLOUDS AND THE GALACTIC BULGE

M.A. DOPITA

*Mt. Stromlo and Siding Spring Observatories, Institute of Advanced Studies,*
*The Australian National University*

ABSTRACT. From self-consistent photoionisation modelling of 147 Magellanic Cloud PN, we have constructed the H-R Diagram for the central stars, and have derived both the chemical abundances and the nebular parameters. We find that the central stars have core masses generally between 0.55 and 0.7 $M_\odot$, and find strong evidence to support the model that they leave the AGB as helium-burning stars following the final shell flash. Type I PN have more massive cores, up to near the Schwartzschild limit, and show clear evidence for the Third dredge-up episode. From HST images, younger PN are very compact. The expansion velocity of the nebula is closely correlated with the position of the central star on the H-R Diagram.

## 1. Introduction

The study of the nearby Galactic Bulge and Magellanic Clouds (MC) populations of planetary nebulae provides a vital key to our understanding of the details of post-asymptotic Giant Branch evolution. These PN are near enough to be studied in detail, are at known distance, and details of their internal structure can be resolved. A comparison of the relatively young, metal poor MC population with the PN in the Galactic Bulge would be expected to allow the evolutionary effects of both age and metallicity of the parent stars to be distinguished and understood.

The MC population has been the subject of a systematic and detailed study by us and by the University College group in recent years, and by the time of IAU Symposium #131, data on the diameters, fluxes, expansion velocities and kinematics had all been accumulated (see the review by Barlow, 1989, also Monk, Barlow and Clegg 1988,Dopita *et al.* 1985; Dopita, Ford, and Webster 1985; Meatheringham *et al.*, 1988 ; Meatheringham, Dopita, Morgan 1988 ; Wood, Bessell, and Dopita, 1986; Wood *et al.*, 1987). This has led to a general understanding of the outlines of an evolutionary sequence (Dopita and Meatheringham, 1990), a necessary prerequisite to detailed modelling of individual PN. Since then, the acquisition of high quality spectrophotometry for some 147 PN in the Magellanic Clouds has permitted rapid progress to be made in the detailed modelling of these PN and their central stars.

Likewise the Galactic Bulge population has been subjected to intense study in recent years. A number of new objects have been discovered by optical techniques (Kinman, Feast and Lasker 1988). However, the most productive technique has been to identify PN candidates by their IRAS colours, and then to subsequently search for the radio continuum emission from these objects (Pottasch *et al.* 1988; Pottasch, Ratag and Olling 1990; Ratag *et al.*1990; Ratag 1990). Absolute H$\beta$ fluxes have been measured for many PN by Acker *et al.* 1991). Extensive Spectrophotometric data of high quality has been obtained by Webster (1988), Ratag (1990) and Acker *et al.* (1991). This raises the number of objects observed to almost 300, permitting detailed comparison of the Bulge and the Disk populations of PN.

## 2. Dynamical and Size Evolution.

For the MC PN, dynamical evidence (Dopita *et al.* 1987) leads us to the adoption of a two-wind model (Kwok *et. al.* 1978; Kwok 1982) to describe the outlines of the evolutionary sequence for PN. In this model, the ionised gas of the PN will be trapped between the compressed AGB wind and a hot pad of shocked stellar wind. In the particular case that the total energy content can be represented by a simple power law in time, $E(t) = E_o t^\alpha$, and if the radial density distribution in the undisturbed AGB wind is given by a power law in radius, $\rho(r) = \rho_o r^\beta$, then, from dimensional considerations, the radius of the outer shock $R$, and the velocity of expansion $V_{exp}$, are given by:

$$R = A. (E_o/\rho_o)^{1/(5+\beta)}. t^{(2+\alpha)/(5+\beta)}; \qquad V_{exp} = B. (E_o/\rho_o)^{1/(5+\beta)}. t^{(\alpha-\beta-3)/(5+\beta)}$$

A and B being dimensionless constants. If we assume that the nebula is evolving into a AGB wind that has been blown at a steady mass-loss rate and velocity($\beta = -2$), then the observational material on the Magellanic Cloud PN limits $\alpha$ to lie in the bounds $1 < \alpha < 2$. This result is in accord with radiative-driven wind theory since the PNn evolutionary models of Wood and Faulkner (1986) predict $\alpha$ to lie in the theoretical range $1 < \alpha < 1.7$. In the limiting cases, $\alpha = 1$ and $\alpha = 2$, we have, for $\beta = 2$:

$\alpha = 1.0;$ $\qquad V_{exp} = const.$ ; $\qquad R = const.\ t$
$\alpha = 2.0;$ $\qquad V_{exp} = const.\ R^{1/4}$ ; $\qquad R = const.\ t^{4/3}$

The velocity of expansion therefore depends only very weakly on radius during the optically thick evolution, a result supported by observations (Sabbabin and Hamzaoglu, 1982; Phillips, 1984; Hippelein and Weinberger 1990). The weakness of this dependence makes it a poor test of theoretical evolutionary scenarios, despite its fairly extensive use in the literature (e.g.Sabbadin *et al.*, 1984; Okorokov *et al.*, 1985).

From the MC sample, we find that the effective temperature, luminosity, and the velocity of expansion of the nebula are well correlated through

$$(V_{exp} / km.s^{-1}) = -128 \pm 4 + 38 \pm 2 \ [\ \log(T_{eff}) - 0.25 \pm 0.05 \log(L/L_\odot)\ ]$$

It is clear that this relationship is fundamental to the understanding of the dynamical evolution of PN, showing that the PN shells are accelerated continuously during the evolution of the PNN towards the blue. This acceleration occurs early for low-mass PN, but higher terminal velocites are reached in the case of the high-mass PN. This is consistent with a lower shell mass in the low mass PN, and more energetic stellar winds in the high-mass PN.

In this model, the stellar wind provides the pressure which confines the ionised material to a (somewhat) thin shell. Gathier *et al.* (1983), and, more recently, Pottasch and Acker (1989) and Dopita and Meatheringham (1990) have pointed out that the strong relationship between nebular mass and nebular radius is strong evidence to support the idea that these PN are optically thick. However, for the Galactic Bulge PN, both Zijlstra (1990) and Stasinska *et al.* (1991) have argued that this relationship may simply due to selection effects, or artefacts due to error propagation, and that most PN are density bounded. However, this conclusion seems to be at variance with spectral data, for which the relative strength of the low ionisation lines appears to exclude the possibility that the nebulae are optically thick in the majority of cases.

For the Galactic Bulge PN, there appears to be room for improvement in the measurement of their diameters. For PN smaller than about 5 arc sec. in diameter, the radio continuum and optical diameters do not agree, in the sense that the optical diameters are larger. However, where carefully calibrated optical observations have been made (Dopita *et al.* 1990), the optical diameters agree very well with the radio sizes. It appears therefore that optical sizes for Galactic Bulge PN should be treated with caution, and conclusions concerning the mass-radius relation are premature.

In the last year, HST images have been obtained for a number of PN in the Magellanic Clouds (see Dopita *et al.*, this conference). This data clearly shows that the smallest PN are more compact than this expected, so that their dynamical ages are smaller than would be expected from theory. This implies that the final phase of mass ejection was denser, and occurred at a lower velocity (~3km.s$^{-1}$) than the simple AGB wind model would predict. It is tempting to associate this mass ejection with a "superwind" generated during the final helium shell flash. If this were the case, PN would leave the AGB as helium-burners (see below). For these objects, the dynamical age is simply the time taken for the newly-ionised gas to flow out of the central region. For optically thick PN, simple Strömgren theory predicts that the ionised mass $M_{neb} \propto (L_*R^3)^{1/2}$. Hence, we may define a reduced mass which is the nebular mass corrected to unit luminosity of the central star. This is shown in Figure 1. The slope defined by these points is close to unity, rather than the value of 3/2 that would obtain for constant filling factor. The implication is that the total thickness of the ionised shell remains approximately constant during the expansion of the nebula, *i.e.* the filling factor of the ionised shell varies as $R^{-1}$. This is clear evidence that the PNn wind has a major rôle in shaping and accelerating the nebular shell in the later phases of evolution (see Marten and Schönberner 1991).

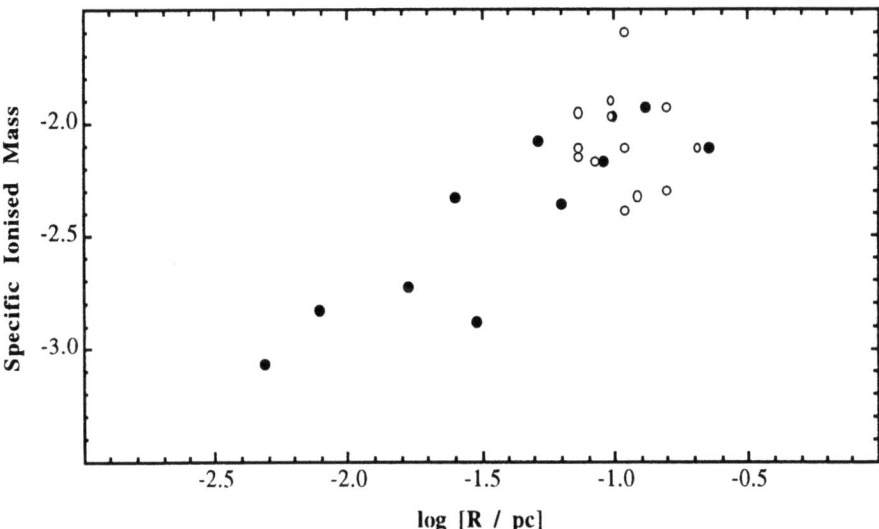

Figure 1: The reduced mass: radius relation for the LMC PN for which the [O III] sizes have been directly measured. The filled circles are from HST PC images (see Dopita *et al.*, this volume), and the open circles are from the imaging of Jacoby, Walker and Ciardullo (1990).

## 3. The Observed Hertzsprung-Russell Diagram.

For the Magellanic Clouds we use the spectrophotometric results presented by Meatheringham and Dopita (1991a,b) and of Vassiliadis *et al.* (1992). The objects are drawn from the Sanduleak, MacConnell and Philip (1978) list; a fairly uniform magnitude limited sample. Fainter PN such as the Jacoby objects studied by Henry, Leibert and Boronson (1989), are not well represented. However, recently a number of new planetary nebulae (PN) have been found the outer fields of the LMC ( Morgan and Good, 1992; Morgan 1992). Many of these objects have been confirmed as PN and spectrophotometry has been obtained (Dopita, Vassiliadis and Morgan, this volume).

In order to determine the PN nebular abundances, and the position of the central star on the H-R diagram we require to know the absolute Hβ flux, the nebular density, and to have accurate spectrophotometry over as wide a wavelength as possible. This is possible because, with the aid of a photoionisation code, the ionisation temperature can be determined from the nebular excitation, the luminosity of the central star can be determined from the absolute Hβ flux, and the chemical abundances can be determined from the electron temperature of the nebula and from its detailed emission line spectrum. We have used the generalised modelling code MAPPINGS (Binette, Dopita and Tuohy, 1985) and its major upgrade MAPPINGS 2 (Sutherland and Dopita 1992) to compute the emission line spectra of isobaric nebulae in photoionisation equilibrium with central stars having a Black-Body photon distribution.

The "observed plane" of the H-R Diagram for the Magellanic Cloud PN is the Excitation Class - Hβ Flux plane, since Excitation Class is closely related to $T_{eff}$, and Hβ Flux is closely related to $L_*$. The challenge for the modeller is to discover the transformation.

The detailed definition of excitation class, $E$, differs somewhat from author to author (c.f. Aller 1956; Feast 1968; Webster 1975; and Morgan 1984). We will use here the classification given by Dopita and Meatheringham (1990), which, since it was defined in terms of two line ratios, is a continuous variable. Our modelling allows us to derive an excitation temperature, related fundamentally to these two line-ratios, but involving many other diagnostic line ratios. Since our models are self- consistent, abundance effects are largely eliminated. This method is closely related to the so-called energy balance method developed by Preite-Martinez and Pottasch (1983) from an original idea of Stoy (1933). In general, a temperature derived by a global model which allows for abundance variations should be more accurate. From our models, at MC abundances, we find that the correlation between E.C. and log $(T_{eff})$ is:

$$log [ T_{eff} ] = 4.489 + 0.112.E - 0.0017.E^2$$

Given the effective temperature, the luminosity of the central star is determined by the absolute Hβ flux. For the case of optically thick objects, our models show that reddening - corrected Hβ flux can be represented as a function of luminosity and excitation class:

$$log [ L_*/L_\odot ] = log [ L_{H\beta} ] - 31.262 - 0.179E + 0.035E^2 - 0.00166E^3$$

This relationship is only weakly metallicity dependent (Dopita, Jacoby and Vassiliadis 1992). Optically-thin PN may also be placed on the H-R diagram, though with lower accuracy. The effective temperatures and luminosities of the central stars, as derived from the detailed photoionisation modelling of only the optically thick PN, are given in Figure 2. A distance modulus to the LMC of 18.5 (corresponding to a distance of 50 kpc) has been assumed. Figure 2 multiplies the sample of Magellanic Cloud PNn which have been placed on the H-R Diagram by a factor of five, previous studies being those of Aller et al.(1987), and Monk, Barlow and Clegg (1988). The bright optically thick objects are found in a range of core masses between 0.56-0.70 $M_\odot$, with a mean of about 0.62 $M_\odot$. This is consistent with the results of Barlow (1989), or Dopita and Meatheringham (1990). However, it is very evident that the mean mass is higher, and the range of core masses wider, than those derived by Schönberner (1981) for a group of nearby, evolved PNn. Note that Type I PN are found preferentially at high $T_{eff}$, lower luminosities and, in general, at larger values of the core mass ($M_{core} \geq 0.7 M_\odot$). This is consistent with the assumption that these PN represent the more massive progenitor stars. Optically-thin PN occupy a broader region of the diagram, of somewhat lower mean mass, from 0.546 to 0.65$M_\odot$, implying that the efficiency of mass loss during the AGB is a very strongly decreasing function of the core mass and luminosity.

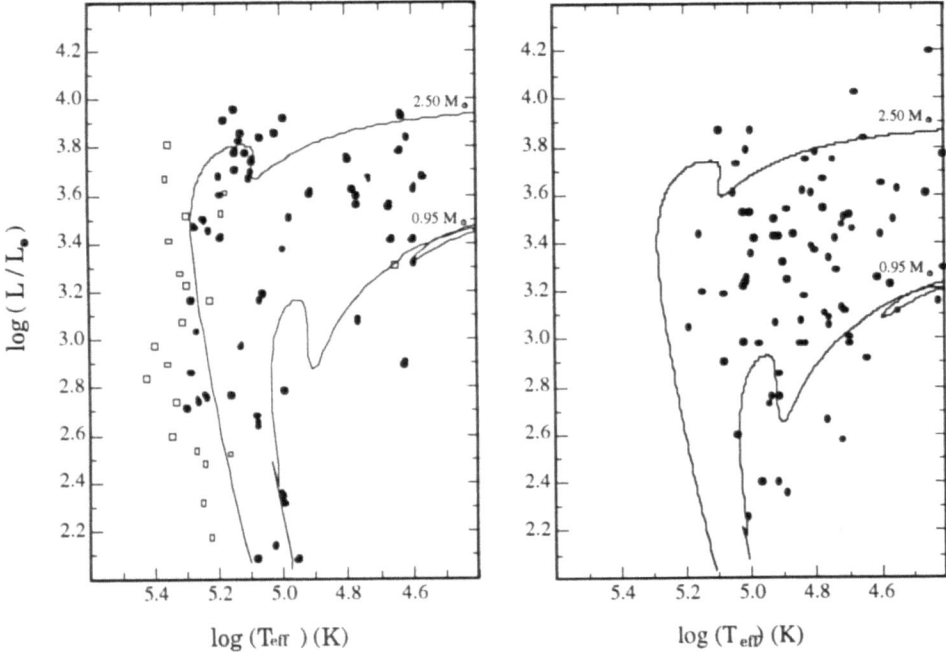

Figure 2: The observed H-R Diagram for the optically-thick PN in the LMC (left), and the corresponding diagram for the Galactic Bulge PN, taken from Ratag (1991), compared with theoretical tracks for He-burning stars from Vassiliadis and Wood (1992). These correspond to core masses of 0.57 and 0.68$M_\odot$. These tracks appear to be able to better describe the observed distribution of PN. In particular, the Galactic Bulge PN in the region log L<3.0, log $T_{eff}$<5.0 would be difficult to account for in the case of hydrogen-burning models. It is clear that the differences between these two H-R Diagrams are mainly due the the different ages of the two populations. The LMC Type I PN are shown as open squares, and clearly represent a high-mass population, regardless of the burning status of the central star.

The positions of the Galactic Bulge PN on the H-R Diagram has been best determined by Ratag (1991) using the energy-balance technique, fully accounting for energy losses in the IR. The agreement he finds between this technique and the measured Zanstra temperatures gives credence to the results, and lends further support that most of the optically-observed PN are optically thick to the ionising radiation. These results are also shown in Figure 2. Note the absence of luminous, high-temperature stars, corresponding to young, massive PNN. Also, the relatively large numbers of objects in the region log $L/L_\odot$ < 3.0; log $T_{eff}$ < 5.0. Both of these results are consistent with the older PN population in the Bulge.

## 4. Comparision of the Observed and the Theoretical Hertzsprung-Russell Diagrams

The availability of evolutionary tracks for the PNn has been somewhat limited (Paczynski 1971; Harm and Schwartzschild 1975; Schönberner 1981; Iben 1984; Wood and Faulkner 1986).

Recently however, Vassiliadis and Wood (1992) presented fully self-consistent tracks fom the Main Sequence though to the White Dwarf stage. This work demonstrated that the most important factor determining the relationship of the PNN to the main-sequence star from which it had evolved is the treatment of mass-loss on the AGB. In the Vassiliadis and Wood work, this was treated in an internally consistent manner which gives the observed period-luminosity relationship for the long period variables. For the first time, it is possible to relate the initial mass $M$ and metallicity $[Z]$ of the star to the age of the star when it reaches the PN phase of evolution $\tau$, its core mass $M_{core}$, and its (hydrogen-burning) luminosity (Dopita, Jacoby and Vassiliadis 1992). Empirically;

$$(\tau / \text{Gyr}) = 14.35 \, [Z]^{0.195} \, (M/M_\odot)^{-(3.4 + \log[Z])}$$

$$(M_{core}/M_\odot) = 0.493 \, [Z]^{-0.035} \{1 + 0.147(M/M_\odot)\}$$

and $\quad\quad\quad (L/L_\odot) = 57340\{(M_{core}/M_\odot) - 0.507\}.$

Usually, the observed H-R diagram for PNN is compared with hydrogen-burning tracks. However, the observed distribution of points in Figure 2 is difficult to reconcile with this hypothesis for a number of reasons. First, the rapidly accelating rates of evolution as we go to higher core masses, and the rapidly decreasing number of stars of higher mass resulting from the slope of the IMF means that the luminosity function PNN should be very rapidly decreasing above $\log(L/L_\odot) \sim 3.7$ (Shaw 1989). This is not seen. Second, the clump of LMC points in the vicinity of $4.0 > \log(L/L_\odot) > 3.7$; $5.2 > \log(T_{eff}) > 5.05$ has no explanation. Hydrogen-burning models evolve through here in less than 1000 yr. Third, the Type I PN are highly unlikely to be seen in the region of the H-R diagram above $\log(L/L_\odot) \sim 2$, since their fading times are so rapid, if hydrogen burning. Finally, the existence of PN in the region $\log(L/L_\odot) < 3.0$; $\log(T_{eff}) < 5.0$ found by Ratag, Pottasch and Waters (1992) cannot be explained. These would have to have core masses in the range 0.54-0.58 $M_\odot$, and such objects would have ages greater than the Universe, according to the above equations.

All these results can be reconciled if the PNN eject a large amount of matter in the final shell flash event, causing them to leave the AGB as helium-burners. The PNN fade until hydrogen re-ignition takes place, causing the characteristic dog-leg in the tracks of Figure 2. At this time the rate of evolution across the H-R diagram is slowed, allowing a greater probability for objects to be observed in this phase. In the helium-burning tracks, the rate of evolution blueward is very much less mass-dependent that for the hydrogen-burning tracks, and the fading of more massive objects is considerably delayed, allowing the Type I objects to be observed while still relatively bright.

The distribution of the brighter PNN in Figure 2, if interpreted as helium-burners implies that the rate of star formation in the LMC has not been uniform over time. The faintest PNn are consistent with the lower track, for which the stellar age ~15 Gyr. This is similar to the Bulge and to the older globular clusters in our Galaxy, suggesting that both the Galaxy and the LMC formed at about the same epoch. For $\log(T_{eff}) > 5.0$, there is a concentration of points near the 1.5$M_\odot$ track (0.62$M_\odot$ core mass). This would imply a burst of star formation about 3-4 Gyr ago. This age estimate is similar to that arrived at by Meatheringham et al. (1988) from a study of the kinematics of the PN (2.5-3.6 Gyr), where it was assumed that the current day vertical velocity dispersion has been achieved by orbital diffusive processes operating over the lifetime of the precursor star. Finally, the large concentration of points near and above the 2.5$M_\odot$ track implies that a burst of star formation took place within the last $\sim 3 \times 10^8$ yr. Both these major bursts coincide reasonably well with those estimated from the distribution of ages of open clusters (van den Bergh 1991), and are coincident with the major bursts identified in the field star Giants (Frogel and Blanco 1983; Stryker 1984, and references therein).

## 5. The Evolutionary Status of the Type I PN in the Magellanic Clouds.

It has been frequently assumed, on the basis of evolutionary models with dredge up, that the N-rich Type I PN are derived from a more massive population than ordinary PN. The position of the LMC Type I PN in Figure 2 clearly proves that this assumption is correct. In fact, it is likely that for a number of Type I PN, the luminosity is underestimated because UV radiation can escape (poleward) through the characteristic ansae which these PN display. Indeed, images taken with the HST (Dopita *et al.*, this volume) confirm this to be the case in at least two Type I objects (SMP #83 and 96). In the case of SMP#83, the equatorial ring of nebular gas and the optically-thick lobes together intercept only ~10% of the ionising radiation, meaning that the true luminosity of the central star is ~$4 \times 10^4$ $L_\odot$. Likewise the nucleus of the Type I PN N67 (SMP#22) in the SMC has been detected at X-ray wavelengths, placing it at $\log (L/L_\odot) = 4.6 \pm 0.7$; $\log(T_{eff}) \sim 5.5$ (Wang, 1991). These figures imply core masses of order $1.2 M_\odot$ for both objects.

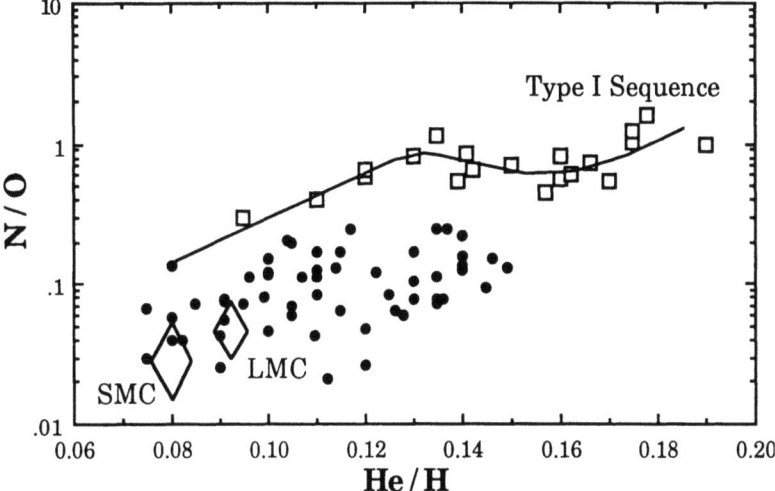

Figure 3: The N/O ratio as a function of He/H ratio for the luminous optically-thick PN in both the LMC and the SMC (from Dopita and Meatheringham, 1991b). The mean abundances in the ISM of each galaxy is shown as an error diamond from Russell and Dopita (1992). The Type I PN show as a distinct sequence on this diagram.

The abundance differences between the Type I objects and the sample as a whole is clearly exemplified in figure 3, in which we plot the N/O ratio as a function of helium abundance. Many points are found near the mean points for the LMC and the SMC found by Russell and Dopita (1990), and Type I objects stand out as having both high He/H ratios and high N/O ratios. Such a diagram has been extensively used by Kaler (1983; 1985) as a diagnostic of the importance of the various dredge-up processes occuring during the first Giant and AGB phase of evolution (Becker and Iben 1980; Renzini and Voli 1981; Iben and Renzini 1983). It is apparent from figure 3 that some objects show appreciable He enhancements without any corresponding N / O enhancement; a signature of the third dredge up phase without C/N conversion. Since the Type I objects have been shown to have more massive core masses, and hence more massive precursors, we can conclude that the C to N conversion is mass-dependent. Lastly, for a number of Type I PN, there appears to be a depletion in the absolute abundance of O, relative to other $\alpha$-process elements such as Ne (Aller *et al.* 1987; Dopita and Meatheringham 1991). This appears to be evidence for the dredge up of ON-processed material, which as yet has no clear theoretical explanation.

## 6. Conclusions

Since the last IAU Symposium #131, the study of the H-R Diagrams of PN has finally reached the stage where a detailed confrontation of observation and theory has become possible. In this review, we have shown how the dynamical / radius evolution is determined by AGB mass loss from the central star and by the high speed wind during the PN phase of evolution. An excellent agreement between the observed and the theoretical evolutionary tracks is obtained only provided that the PN leave the AGB as helium-burning objects following a shell flash. On this interpretation, details of the star formation history of the LMC may be distiguished, which agree with other techniques. The Type I PN, are found to have high core masses, and to lie on the descending branch of the evolutionary tracks. They show evidence for a 3rd. dredge-up episode resulting in correlated He and N abundance enhancements. We also find clear evidence, most notably in the Type I objects for dredge-up and the ejection of ON processed material into the PN envelopes.

## References

Acker, A., Stenholm, B., Tylenda, R., and Raytchev, B. 1991a, Ast. Ap., 90, 89.
Acker, A., Köppen, J., Stenholm, B., and Raytchev, B. 1991b, Ast. Ap. Suppl. Ser., 89, 237.
Aller, L.H. 1956 Gaseous Nebulae (New York:Wiley).
Aller, L.H., Keyes, C.D., Maran, S.P., Gull, T.R., Michalitsianos, A.G.
    and Stecher, T.P., 1987 Ap. J., 320, 159.
Barlow, M.J. 1987, M. N. R. A. S., 227, 161.
_____. 1989 in IAU Symp. #131"Planetary Nebulae", Ed. S.Torres-Peimbert
    (Kluwer:Dordrecht), p319.
Becker, S.A., and Iben, I. Jr. 1980, Ap. J., 237, 111.
Binette, L., Dopita, M.A., and Tuohy, I.R. 1985, Ap. J., 297, 476.
Dopita, M.A., Ford, H.C., Lawrence, C.J., and Webster, B.L. 1985, Ap. J., 296, 390.
Dopita, M.A., Ford, H.C., and Webster, B.L. 1985, Ap. J., 297, 593.
Dopita, M.A., Henry, J.P, Tuohy, I.R., Webster, B.L., Roberts, E.H., Y-I. Byun, Cowie, L.L.
    and Songaila, A., 1990, Ap. J., 365, 640.
Dopita, M.A., Jacoby, G.H., and Vassiliadis, E. 1992, Ap. J., 389, 27.
Dopita, M.A., and Meatheringham, S.J. 1990 Ap. J., 357, 140.
_____. 1991a, Ap. J., 367, 115.
_____. 1991b, Ap. J., 377, 480.
Dopita, M.A., Meatheringham, S.J., Webster, B.L., and Ford, H.C. 1988,
    Ap. J., 327, 639.
Dopita, M.A., Meatheringham, S.J., Wood, P.R., Webster, B.L., Morgan, D.H.,
    and Ford, H.C.,1987, Ap. J. (Lett.), 315, L107.
Feast, M.W. 1968, M. N. R. A. S., 140, 345.
Frogel, J.A., and Blanco, V.C., 1983, Ap. J., 274, L57.
Gathier, R., Pottasch, S.R., Goss, W.M.,and van Gorkom, J.M. 1983, Ast. Ap., 128, 325.
Harm, R., and Schwartzschild, M., 1975, Ap.J., 200, 324.
Henry, R.B.C., Liebert, J., and Boroson, T.A., 1989, Ap. J., 339, 872.
Hippelein, H., and Weinberger, R. 1990, Ast. Ap., 232, 129.
Iben, I., Jr. 1984, Ap. J., 277, 333.
Iben, I., Jr., and Renzini, A. 1983, Ann. Rev. Astr. Ap., 21, 271.
Jacoby, G.H., Walker, A.R., and Ciardullo, R. 1990, Ap. J., 365, 471.
Kaler, J.B. 1983, "Planetary Nebulae", ed. D.R. Flower (Reidel:Dordrecht),p245.

_____. 1985 Ann. Rev. Ast. Ap., 23, 89.
Kaler, J.B., and Jacoby, G.H. 1990, Ap. J., 362, 491.
Kinman, T.D., Feast, M.W., and Lasker, B.M.. 1988 AJ 95, 804.
Kwok, S. 1982, Ap. J., 258, 280.
Kwok, S., Purton, C.R., and FitzGerald, P.M. 1978, Ap. J. (Lett), 219, L125.
Martin, H., and Shönberner, D. 1991, Ast. Ap., 248, 590.
Meatheringham, S.J., and Dopita, M.A., 1991a, Ap. J. Suppl. Ser., 75, 407.
_____., 1991b, Ap. J. Suppl. Ser., 76, 1085.
Meatheringham, S.J., Dopita, M.A., Ford, H.C., and Webster, B.L. 1988, Ap. J., 327, 651.
Meatheringham, S.J., Dopita, M.A., and Morgan, D.H., 1988 Ap. J., 329, 166.
Monk, D.J, Barlow, M.J., and Clegg, R.E.S. 1988 M.N.R.A.S., 234, 583.
Morgan, D.H. 1984, M.N.R.A.S., 208, 633.
Morgan, D.H., andGood, A.R., 1985, Mon Not. R.A.S., 213, 491.
Morgan, D.H. 1992, (in prep).
Okorokov, V.A., Shustov, B.M., Tutukov, A.V., and Yorke, H.W., 1985, Ast. Ap., 142, 441.
Paczynski, B. 1971, Acta Astron., 21,417.
Phillips, J.P., 1984, Astr. Ap., 137, 92.
Pottasch, S.R., and Acker, A., 1989, Ast. Ap., 221, 123.
Pottasch, S.R., Bignell, C., Olling, R., and Zijlstra, A.A. 1988, Ast. Ap., 205, 248.
Pottasch, S.R., Ratag, M. A., and Olling, R. 1990, in "From Miras to PN - Which path for Stellar Evolution?", eds M.O. Mennessier and A. Omont, Editions Frontières, France, p381.
Preite-Martinez, A., and Pottasch, S.R., 1983, in IAU Symp#103, "Planetary Nebulae",
    ed. D.R. Flower, (Reidel:Dordrecht), p547.
Ratag, M.A. 1990, Thesis, U.of Groningen.
Ratag, M.A., Pottasch, S.R., and Waters, L.B.F.M., 1992, Ast. Ap. (in press).
Ratag, M.A., Pottasch, S.R., Zijlstra, A.A., and Menzies, J. 1990, Ast. Ap., 233, 181.
Renzini, A. and Voli, M. 1981, Ast. Ap., 94, 175.
Sabbadin, F., Gratton, R.G., Bianchini, A., and Ortolani, S., 1984, Ast. Ap., 136, 181.
Sabbabin, F. and Hamzaoglu, E., 1982, Ast. Ap., 110, 105.
Sanduleak, N., MacConnell, D.J., and Philip, A.G.D., 1978, P. A. S. P., 90, 621.
Schönberner, D., 1981, Astr. Ap., 103, 119.
Shaw, R.A. 1989, in IAU Symp. #131"Planetary Nebulae", Ed. S.Torres-Peimbert
    (Kluwer:Dordrecht), p319.
Stasinska, G., Tylenda, R., Acker, A., and Stenholm, B. 1991, Ast. Ap., 247, 173.
Stoy, R.W., 1933, M.N.R.A.S., 93, 588.
Stryker, L.L., 1984, IAU Symp#108, *"Structure and Evolution of the Magellanic Clouds,*
    eds. S. van den Bergh and K.S. de Boer, Dordrecht:reidel, p79.
Sutherland, R.E., and Dopita, M.A. 1992, Ap. J. Suppl., (in press).
Vassiliadis, E, Dopita, M.A., Morgan, D.H., and Bell, J.F. 1992, Ap. J. Suppl. Ser, (in press).
Vassiliadis, E., and Wood, P.R. 1992, (in prep).
Wang, Qingde 1991, Mon. Not. R. Astr. Soc., 252, 47p.
Webster, B.L. 1975, M. N. R. A. S., 173, 437.
Webster, B.L. 1988, M.N.R.A.S., 230, 377.
Wood, P.R., Bessell, M.S., and Dopita, M.A. 1986, Ap. J., 311, 632.
Wood, P.R., and Faulkner, D.J. 1986, Ap. J., 307, 659.
Wood, P.R., Meatheringham, S.J., Dopita, M.A., and Morgan, D.H. 1987, Ap. J., 320, 178.
van den Bergh, S., 1991, IAU Symp.#148, *"The Magellanic Clouds"*, eds. R.Haynes and D. Milne,
    Kluwer:Dordrecht.
Zijlstra, A.A., 1990, Ast. Ap. 234, 387.

# WHITE DWARF CENTRAL STARS

JAMES W. LIEBERT
*Steward Observatory, University of Arizona*
*Tucson, Arizona USA*

Planetary nebula nuclei on lower luminosity, diagonal tracks in the HR Diagrams are degenerate dwarf stars close to their final radii. For several reasons, these stars have until recently been difficult to identify and study. With the advent of new techniques and technologies, both hydrogen–rich and hydrogen–poor atmospheric sequences have been found.

## 1. Introduction

It has long been realized that the nuclei of planetary nebulae (PNN) are becoming white dwarfs when their luminosities start to fall. Nonetheless, it has traditionally been very difficult to study central stars which fall on diagonal portion of a post–asymptotic giant branch (AGB) track in an HR Diagram. These are stars of rather low luminosity, and hence quite faint in apparent magnitude at the typical distances of planetary nebulae. Also, they are quite hot, so that spectral absorption features are likely to be weak and shallow. As they grow dimmer, these nuclei are more easily masked by the nebular radiation, though this is counterbalanced by the fact that the nebulae are likely to be old and low in surface brightness. Planetary nebulae are traditionally far enough away and concentrated to the Galactic plane, so that reddening may be a problem. Until recently, therefore, only a few planetary central stars were shown to have white dwarf gravities from direct studies of their spectra.

Due to some exciting developments the past several years, many more white dwarf central stars have been discovered and studied in some detail. The first reason for this is the employment of CCD detectors to obtain accurate, high signal-to-noise (S/N) ratio line profiles and spectrophotometry. Secondly, these can now be analyzed using improved, NLTE, line–blanketed model atmospheres and synthetic spectrum codes. Finally, a most important development has been the discovery of larger planetary nebulae of very low surface brightness – the "senile" planetaries (Ishida and Weinberger 1987). It comes as no surprise that this oldest group of nebulae often harbors white dwarf central stars. In fact, several weak nebulae have now been found around previously–discovered hot white dwarfs.

What can we learn from studying the white dwarf central stars and their nebulae? First, these various discoveries have in effect extended the planetary nebula luminosity function and age distribution. We want to match up the formation rates of the PNs with those of the white dwarfs in order to find out what subset of the latter pass through a PN phase. The potential problem with the respective birthrates is reviewed in Phillips (1988). In addition, it is of interest to see whether the well–established H–rich and H–poor PNN sequences evolve directly into the DA and non–DA white dwarf sequences. Does each composition group evolve as a continuous sequence, or do they change from one dominant atmospheric composition to another? A critical question for establishing the PN formation rate is whether the post–AGB evolutionary ages of the PNN can be reconciled with the expansion ages of the nebular shells: for cooling–dominated stars, this may be a severe test.

At this point, only very preliminary and partial answers to these questions are possible. The main reason for this is that – despite the great advances in atmospheric modelling and in the spectrophotometric data – it is not clear in many cases that accurate enough atmospheric and stellar parameters can be determined, for example, to place the stars in an HR Diagram for quantitative comparison with evolutionary tracks. Ironically, the temperature determinations for the H–poor sequence with their complicated and mixed atmospheric abundances of He and CNO elements may presently be more accurate than those of the H–rich sequence. We shall therefore discuss the two groups separately.

## 2. High Gravity Planetary Nuclei with Hydrogen-poor Atmospheres

The most extreme, hydrogen-poor nuclei are generally luminous objects with Wolf-Rayet (WC) spectra featuring broad emission due to helium and CNO ions. Because of the prominence of an O VI doublet in the optical ultraviolet, they are sometimes called "O VI" nuclei. The sequence apparently includes extremely hot stars, some of which may have dropped from their maximum luminosities, for which the emission lines are much weaker. The spectra are dominated by absorption features due to carbon, oxygen, usually helium and sometimes nitrogen, often with the cores reversed into emission. In the PNN spectral classification system of Mendez et al. (1986), these would be called O(C) or hgO(C), designations which denote the appearance of strong helium and carbon features. These stars range in temperature from 100,000 K to perhaps 170,000 K, as determined from the analyses cited below.

Some stars showing this kind of spectrum are pulsationally-unstable, with complicated mode structures of non-radial g modes. The prototype of this class of pulsators is PG 1159–035 or GW Vir (McGraw et al. 1979). Though the early discoveries belonging to this group were field stars in the Palomar Green (PG) Survey – and these appear to lack even "senile" nebulae – Grauer and Bond (1984) discovered that the nucleus of K 1–16 is a similar pulsator. That it's similar absorption lines were sharper indicated a higher luminosity and the implication that the apparent pulsational instability strip extends to somewhat higher luminosities. Several pulsating PNN, all with similar spectra, have subsequently been found. Several key features of the spectra are illustrated in Werner (1992) which includes a table of known objects of this "PG 1159" spectroscopic class.

The magnificent work by the Kiel and Munich groups has given us a quantitative understanding of this post–AGB, H–poor sequence. High S/N line profiles of He, CNO, and most recently Fe ions have been analyzed using model atmospheres which treat over 100 NLTE levels and hundreds of individual transitions. Temperatures, surface gravities, and abundances for eight PG 1159 stars including two nuclei are summarized in Werner (1993). A major surprise was the Kiel result (Werner, Heber and Hunger 1991; WHH) that carbon and oxygen mass fractions are competitive with helium in the atmospheres of these stars, while N has a significant abundance apparently only in a minority of objects. For example, the prototype PG 1159–035 has abundances of 33% He, 50% C and 17% O by mass. Even more extreme are those of the very hot object H 1504+65 – <1% He, C~O~50%. These abundances imply that most of the helium envelope which exists at the end of the AGB phase is lost during post–AGB evolution. WHH hypothesized that much of this mass is lost during the WC phase, at which time a high-velocity wind is observed; these stars would then be predecessors of those with PG 1159 or O(C) spectra, as first suggested by Sion, Liebert and Starrfield (1985).

The discovery last year of a spectacular mass loss event in the nucleus of Longmore 4 cements the relationship of the WC and PG1159 spectral types. This star, found to pulsate by Bond and Meakes (1990), normally exhibits an absorption–line PG 1159 spectrum. Werner et al. (1992) found that showed broad WC emission features early in 1992; these faded in strength over a period of days.

There is also a growing case that at least some of the H–deficient PNN and PG 1159 sequence lose their envelopes in a late helium shell flash (Iben et al. 1983). Abell 78 is one of three objects known to have inner nebular knots which are extremely H–deficient, and enriched in such elements as helium and carbon, but surrounded by a large nebula of "normal" abundances (cf. Jacoby and Ford 1983). The inner nebula is assumed in this hypothesis to be due to a second AGB–like phase, resulting from the ejection of the hydrogen and much of the helium envelopes due to the shell flash. Such a "born again" event may actually have been observed in V605 Aql, which was observed to have a novalike outburst in 1917–1921; it has now faded and exhibits a WC–like spectrum. This object is the central star of the old PN Abell 58. Hubble Space Telescope imaging shows what may be the ejecta of this explosion, knots of H–poor nebulosity not unlike those of Abell 78 (Bond, Liebert, Renzini and Meakes 1993, these Proceedings).

It is generally assumed that the H–deficient nuclei and PG 1159 stars cool down into the H–deficient DO white dwarfs, whose atmospheres are predominantly helium–dominated. However, the sequence can only be continuous down to a temperature of about 45,000 K, since no white dwarfs are known with He-rich spectra between 30,000 K and 45,000 K (Liebert 1985). There may also be a discrepancy between the luminosity functions of the hottest DA and DO white dwarfs, as discussed in the next section. Thus, the possibility remains that some stars evolve from one sequence to the other (see Shipman 1988 and Liebert 1988 for discussions of various hypotheses).

## 3. Hydrogen–rich White Dwarf Central Stars

Spectra of H–rich composition are generally dominant among both the PNN and the hot white dwarfs. Do these form a continuous sequence? It is not clear that they do. In particular, it has been difficult to identify high gravity, H–rich examples of similar temperatures (100,000–170,000 K) to the H–poor PG 1159 sequence discussed previously, or to even the hottest white dwarfs of type DO at $T_{eff}$ 80,000 K.

Among the H–rich white dwarf nuclei are the well–studied cases of Abell 7 and NGC 7293, which may be close to or in the DO temperature range. Several H–rich central stars are known with $T_{eff}$ ~70,000 K, with one faint, poorly–observed star with an estimated temperature near 100,000 K (Napiwotzki and Schönberner 1991). One of the hottest, recently–reported examples is the central star to S 216 (or LS V +46 21) which has a rough temperature estimate of 90,000 K (Tweedy and Napiwotzki 1992).

The apparent paucity of H-rich objects of very hot temperatures is more dramatic for complete samples of field white dwarfs. Best studied of these is the PG Survey (Fleming, Liebert and Green 1986). Some 353 DA stars were studied in this sample to a B magnitude of typically 16.2 over 10,000 square degrees of sky at high Galactic latitude. Yet the hottest of these appear to be cooler than 80,000 K (Holberg et al. 1989); to my knowledge, there are no other field DA stars (ie. non PNN) with reported temperatures above this value. As already noted, a sequence of DO and PG 1159 stars – most lacking nebulae – extends to 170,000 K (Werner 1993; Wesemael, Green and Liebert 1985).

Some selection effects might be envisioned for these samples which might be responsible for the apparent paucity of very hot, H-rich cases. First of all, if these stars have thick hydrogen envelopes and active hydrogen shell sources, their rate of evolution down to log L $\sim$2 could be considerably faster than for those lacking hydrogen shells. It is controversial whether either the H-rich PNN or the DA white dwarfs have such thick shells (cf. the reviews in Schönberner 1988 and Liebert 1988).

One might then expect that any rapidly evolving H-rich stars are more likely to be enveloped in a bright PN, which could prevent the parameters from being determined by a standard photospheric analysis. For the PG Sample, it is possible to answer this concern by asking how many PNs appear in the area of sky that was surveyed, and how many of these could possibly harbor very hot nuclei. The answer to the first question is that there are only nine nebulae in the relevant field (NGC 6058, NGC 7094, A 30, A 33, A 39, H4-1, Sn-1, EGB 6 and 61+41°1). Of these, none harbor poorly-observed central stars and the only hot H-rich nucleus is that of EGB 6 / PG 0950+139; this is a 70,000 K white dwarf with unique properties (Liebert et al. 1989), but not an extremely hot star. Such extremely hot PNN, imbedded in bright and dusty nebulae do exist (cf. NGC 7027), but they are likely to be quite rare in a volume-limited sample.

We must also note that white dwarf and PNN samples may not be sampling the same stellar population. The former come from a volume very near the Sun, with a scale height of 275 pc (Boyle 1989). The scale height estimates for the nuclei are generally much smaller (see M. Peimbert, this Conference). The difference in these samples should be kept in mind in attempts to reconcile the differing formation rates.

Finally, we come to what is currently the most difficult problem of all – that the $T_{eff}$ estimates for the H-rich white dwarfs and central stars may not currently be accurate enough to draw reliable comparisons with the H-poor sequence. No problem was apparent in prior studies of the hottest DA white dwarfs. Wesemael et al. (1980) compared LTE and NLTE, pure-H models and found the former to be adequate. Line profile fits for the series of Balmer lines and for Lyman alpha appeared to yield consistent temperatures, using just LTE models. The physics of the hydrogen atom is understood well enough to conclude that temperatures determined for the DA white dwarfs should be more accurate over a wide range than for He-rich cases.

Napiwotzki and Schönberner (1992) first showed that, above a temperature of perhaps 50,000–60,000 K, some hydrogen–rich PNN line profiles sometimes do not yield self–consistent fits. When there was a discrepancy, the temperature of the fit would increase with an increasing upper level of a Balmer transition. That is, H$\alpha$ might yield a temperature as low as 50,000 K, while the H$\delta$ fit gives 90,000 K for a given central star (Napiwotzki 1992). Most disturbing is the fact that this discrepancy appears most likely to affect the PNN and DA stars of highest temperature – so that we may not really know how hot the sequence extends! Perhaps such stars in S 216 and NGC 7293 are much hotter than the published estimates.

This lack of self–consistency in fitting a sequence of hydrogen lines for hot stars – which might be dubbed the Napiwotzki effect – can apparently extend to lower temperatures. Lamontagne et al. (1993) faced the same result in fitting the star Feige 55, for which they got a best fit temperature of about 54,000 K. Yet most of the field PG white dwarfs yield self–consistent Balmer line temperature fits up to 70,000 K, so that it is unlikely their true temperatures are much higher than the estimates.

Since this problem arises only in some stars within what appears to be a similar temperature range, there clearly is another parameter to the problem. It was initially believed that the effect was related to the appearance of helium in the atmosphere (a DAO spectrum), though He II is usually seen at least weakly in the spectra of the hottest H–rich PNN and white dwarfs. While helium might yet be part of the answer, the dominant characteristic of Feige 55 is the very strong spectrum of heavier elements – ions of carbon, nitrogen and especially iron – observed in ultraviolet spectra obtained with the it International Ultraviolet Explorer. It is argued by Lamontagne et al. (1993) that a properly treatment of the extensive blanketing in the ultraviolet might lead to better self-consistency in the hydrogen line fits and more accurate temperatures for such stars. A similar investigation of the effects of line blanketing is underway at Kiel (Dreizler and Werner 1993).

## 4. Conclusions

Our goal was to attempt to answer some of the interesting questions posed in the Introduction, but we are currently stopped in our tracks. Until the Napiwotzki effect is properly taken into account, no proper comparison of the H–rich and H–poor sequences is possible, and the luminosity functions and formation rates of the PNN and white dwarf samples may undergo some revision.

## References

Bond, H.E. and Meakes, M.G. 1990, AstronJ, 100, 788
Boyle, B.J. 1989, MonNotRAS, 240, 533
Dreizler, S. and Werner, K. 1993, in White Dwarfs: Advances in Observation and Theory, NATO ASI Series, ed. M.A. Barstow (Dordrecht, Kluwer), in press
Fleming, T.A., Liebert, J. and Green, R.F. 1986, Ap.J., 308, 176
Grauer, A. D. and Bond, H.E. 1984, ApJ, 277, 211
Holberg, J.B., Kidder, K., Liebert, J. and Wesemael, F. 1989, in White Dwarfs, Proc. IAU Coll. 114, ed. G. Wegner (Berlin, Springer–Verlag), p. 188
Iben, I., Jr., Kaler, J.B., Truran, J.W. and Renzini, A. 1983, ApJ, 264, 605
Ishida, K. and Weinberger, R. 1987, AstrAp, 178, 227
Jacoby, G.H. and Ford, H.C. 1983, ApJ, 266, 298

Lamontagne, R., Wesemael, F., Bergeron, P., Liebert, J., Fulbright, M.S. and Green, R.F. 1993, in White Dwarfs: Advances in Observation and Theory, NATO ASI Series, ed. M.A. Barstow (Dordrecht, Kluwer), in press

Liebert, J. 1985, in Hydrogen Deficient Stars and Related Objects, Proc. IAU Coll. 87, , eds. K. Hunger, D. Schönberner and N.K. Rao (Dordrecht, D. Reidel Publ.), p. 367

Liebert, J. 1988, in Planetary Nebulae, Proc. IAU Symp. 131, ed. S. Torres–Peimbert (Dordrecht, Kluwer Academic Publ.), p. 545

Liebert, J., Green, R.F., Bond, H.E., Holberg, J.B., Wesemael, F., Fleming, T.A. and Kidder, K. 1989, ApJ, 346, 251

McGraw, J.T., Starrfield, S.G., Liebert, J. and Green, R.F. 1979, in, White Dwarfs and Variable Degenerate Stars, Proc. IAU Coll. 93, eds. H. M. Van Horn and V. Weidemann (Rochester, Univ. of Rochester Press), p. 377

Mendez, R.H., Miguel, C.H., Heber, U. and Kudritzki, R.P. 1986, in Hydrogen Deficient Stars and Related Objects, Proc. IAU Coll. 87, eds. K. Hunger, D. Schönberner, and N.K. Rao (Dordrecht, D. Reidel Publ.), p. 323

Napiwotzki, R. 1992, in Atmospheres of Early–Type Stars, Proc. of a Workshop at the University of Kiel (Berlin, Springer– Verlag), in press

Napiwotzki, R. and Schönberner, D. 1992, in White Dwarfs, eds. G. Vauclair and E.M. Sion (Dordrecht, Kluwer Academic Publ.), p. 39

Phillips, J.P. 1988, in Planetary Nebulae, Proc. IAU Symp. 131, ed. S. Torres–Peimbert (Dordrecht, Kluwer), p. 425

Schönberner, D. 1988, in Planetary Nebulae, Proc. IAU Symp. 131, ed. S. Torres-Peimbert (Dordrecht, Kluwer Academic Publ.), p. 463

Shipman, H.L. 1988, in Planetary Nebulae, Proc. IAU Symp. 131, ed. S. Torres–Peimbert (Dordrecht, Kluwer Academic Publ.), p. 555

Sion, E.M., Liebert, J. and Starrfield, S.G. 1985, ApJ, 292, 471

Tweedy, R.W. and Napiwotzki, R. 1992, MonNotRAS, 259, 315

Werner, K. 1992, in Atmospheres of Early-Type Stars, Proc. of a Kiel / CCP7 Workshop at the University of Kiel (Berlin, Springer–Verlag), in press

Werner, K. 1993, in White Dwarfs: Advances in Observation and Theory, NATO ASI Series, ed. M.A. Barstow (Dordrecht, Kluwer), in press

Werner, K., Hamann, R., Heber, U., Rauch, T. and Wessolowski, U. 1992, AstrAp, 259, L69

Werner, K., Heber, U. and Hunger, K. 1991, AstrAp., 244, 437

Wesemael, F., Green, R.F. and Liebert, J. 1985, ApJS, 58, 379

Wesemael, F., Auer, L.H., Van Horn, H.M. and Savedoff, M.P. 1980, ApJS, 43, 159

# ON THE RELATION OF CORE MASS WITH CHEMICAL COMPOSITION IN PN

S.R. POTTASCH
*Kapteyn Laboratory*
*P.O. Box 800*
*9700 AV Groningen*
*The Netherlands*

Abstract. The attempts in the literature to find a relation between planetary nebula core mass and the nebular abundance ratios N/O and He/H are described. A critical discussion concludes that the evidence for such a relation is weak.

## 1. Introduction

It is clear that in the course of their evolution to planetary nebulae various elements have been produced in the nuclear burning stages, brought to the surface of the central star, and ejected so as to produce nebular overabundances. The evidence for this is quite direct.[1] The oxygen abundance in PN is usually somewhat less than solar (see Fig. 1a), while the nitrogen to oxygen ratio varies from the solar value to an order of magnitude higher (see Fig. 1b). Since most stars must have begun their 'life' with approximately the solar value, it is inferred that the nitrogen was formed during the course of evolution of the central star. Similarly, helium and carbon are also found to have been enriched in the course of evolution of some, but not all, central stars. The other elements which have been well studied in PN, neon, and argon, are apparently not enriched. The helium and nitrogen enrichment are related to each other in a general way (see Fig. 2) and probably have been formed in the same or similar processes.

It is thought that the processes which lead to the enrichment of these elements are understood at least in a rudimentary way. These are discussed by Renzini and Voli (1981) and Iben and Renzini (1983) and many others (e.g. Groenewegen and De Jong, this volume). Three 'dredge-up' events are identified. The first occurs in all stars on the red giant branch following exhaustion of hydrogen in the core. Convection extending into the interior brings material processed in the CN cycle to the surface, resulting in an increase of nitrogen (by a factor of 2) and a decrease of both carbon (by 30%) and the $^{12}C/^{13}C$ ratio. The second 'dredge-up' occurs only in higher mass stars (from about 2.3 $M_\odot$ to 8 $M_\odot$), and results in an increase of both He and N abundances. The third 'dredge-up' occurs along the AGB: Following each helium shell flash helium and carbon are convected to the surface. In these theories, oxygen is the least affected element and its abundance is expected to reflect that of the interstellar medium.

---

[1] Solar abundance is accepted as being similar to the abundance of the (interstellar) material out of which stars were formed.

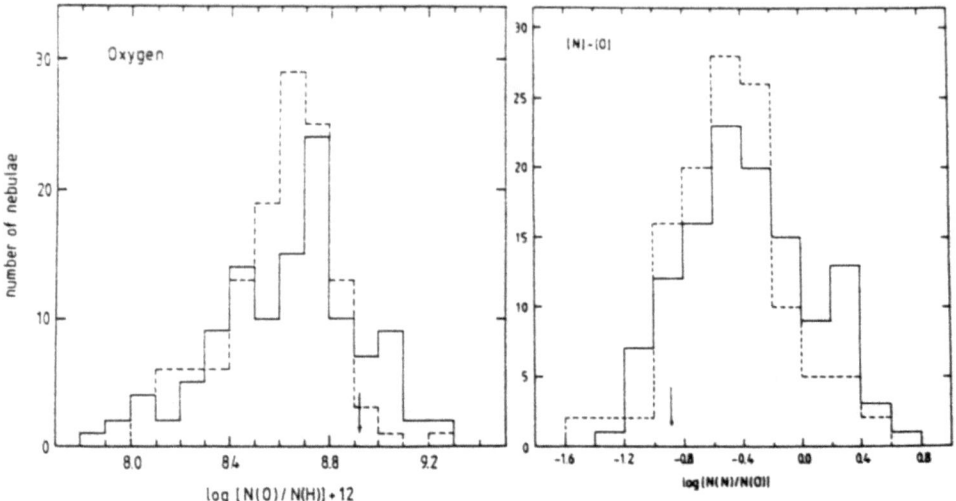

Fig. 1.
(a) Histogram of the oxygen abundance of galactic PN. Solid line bulge PN, dashed line 'nearby' PN. The arrow indicates the solar abundance. (b) same as (a) except that the nitrogen to oxygen ratio is varying.

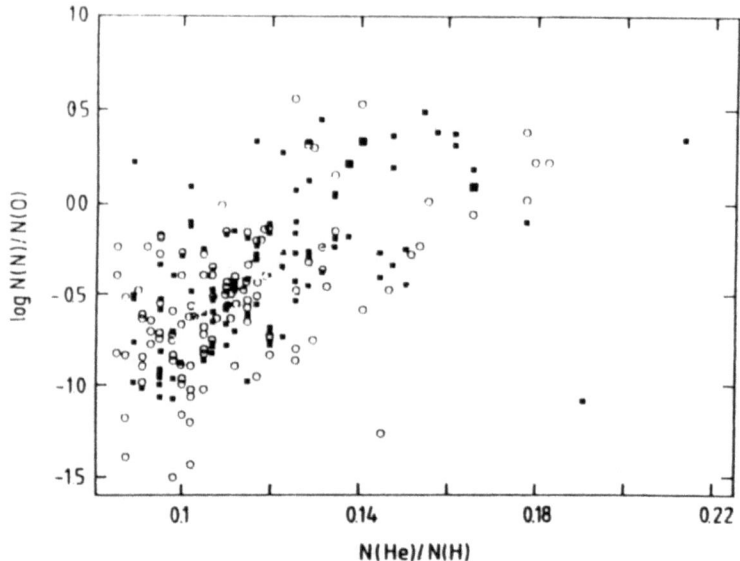

Fig. 2.
The nitrogen to oxygen ratio plotted against the He/H ratio. The filled squares are the bulge PN, the open circles are the 'nearby' PN.

There is a large uncertainty in the prediction of the enrichment. There are several reasons for this. The mixing length theory introduces parameters which are difficult to estimate. The amount of mass loss and its ejection are poorly known. Estimates of these quantities can be made which lead to predictions of the abundance ratio's N/O, He/H and C/O as functions of the initial mass of the star undergoing evolution. These predictions cannot be directly checked because the initial mass (or progenitor mass) that the central star had when on the main sequence cannot be found. One must therefore assume that there is a one-to-one relationship between the present central star mass (core mass) and the initial mass of the star.

The task of the present review is to summarize the present status as to whether there is an observed relationship between the core mass and the abundance ratios mentioned above. We will limit ourselves to the N/O and He/H ratio's because these are more widely observed and better known than the carbon abundance.

## 2. Determining the Core Mass

In a sense the determination of the core mass is straightforward. Paczynski (1971) demonstrated from models of stellar evolution that a relationship exists between the luminosity of the central star and its mass. The exact form of this relationship is somewhat uncertain. Besides the form given by Paczynski, forms are given by Schönberner (1981), Wood and Zarro (1981) and Boothroyd and Sackmann (1988) and others. For the present purpose it is not important which (if any) of these relations is the correct one. It is important that always the same relationship is used. This is not always the case in the literature so that caution is desirable.

The problem of determining the core mass is now shifted to the problem of determining the stellar luminosity. This is not so straightforward. First of all, there is the problem of determining the distance which for galactic PN is not easy to do accurately (see Terzian, this volume). Therefore a great deal of emphasis is placed of measurements of PN in the galactic bulge and the Magellanic Clouds, where distances are probably known to within 15%. The luminosity can then be determined from measurements of the nebular lines if the nebula is optically deep in the ionizing radiation, i.e. all the stellar emission is absorbed by the nebula. There are various ways of estimating whether this is true or not (see Pottasch, 1993). We will not discuss this further, but use the estimates of the individual authors for the optical depth. If one has a doubt as to whether the nebula is optically deep, the luminosity can be determined from the stellar temperature and the observed visual magnitude assuming that the star radiates as a blackbody (or some known model atmosphere). This last procedure applies only to the galactic (bulge) nebulae, since it is usually very difficult (but not impossible) to see stellar radiation from the Magellanic Clouds.

Before the literature results of the luminosity vs abundance are given it is useful to try to obtain some insight as to what one would expect to see.

## 3. Some Insight

What general results would one expect to see if:
1) the Schönberner (1981) tracks describe the evolution correctly
2) there is a monotonic relation between the core mass and the N/O ratio.

In Fig. 3 the evolutionary tracks of Schönberner (1981) and Wood and Faulkner (1986) for various core masses have been plotted. The heavy lines indicate the ages of 3000 and 10000 years since the star had a surface temperature of 5000 K. This is thought to be approximately when the last thermal pulse and expulsion of the nebula took place. Thus most of the observed PN will be expected to be found between the heavy lines or close to them.

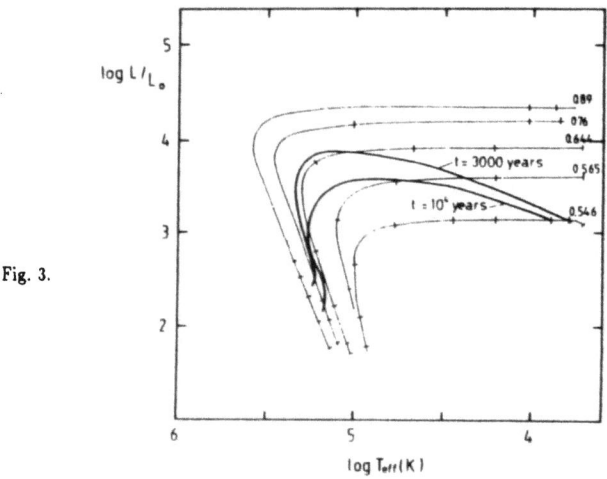

Fig. 3.

Theoretical evolutionary tracks (Schönberner, 1981; Wood and Faulkner, 1986). The thick lines indicate the ages of 3000 and 10000 years since the star had a surface temperature of 5000 K.

Assume that the higher core mass stars have higher N/O ratios. Then one would expect that:
(1) for central stars below $10^5$ K there will be a relation between central star temperature and N/O ratio such that the higher temperatures have the higher N/O ratio.
(2) that for central stars between $10^5$ K and $2 \times 10^5$ K (and with luminosities greater than $10^2$ $L_\odot$) a wide range of N/O will be found, but because those PN with higher N/O will have central stars which evolved more quickly, these PN will have higher densities. One would therefore expect that for these nebulae there will be a relation between nebular density and N/O ratio such that the higher densities correspond to higher N/O ratio.
(3) PN with central stars having temperatures greater than $2 \times 10^5$ K will all have higher N/O ratio's.

Points (2) and (3) will be considered in the following section. Point (1) is illustrated in Fig. 4. The top figure is a plot of N/O against the central star temperature as taken from the work of Ratag et al (1991) for a large selection of galactic bulge PN. The middle diagram is for a selection of 'nearby' PN where the N/O ratio is taken from a variety of sources (Henry, 1990; de Freitas Pacheco et al, 1992; Perinotto, 1991; Stasinska and Tylenda, 1991) and the central star temperature is taken from a compilation of Zanstra temperatures by Zhang and Kwok (private communication, 1991). A diagram similar to this has been published by Stasinska and Tylenda (1991) for a somewhat different selection of galactic PN. The bottom diagram in the figure is for the LMC and is taken from the work of Dopita and Meatheringham (1991a,b). In this case other sources of data could have been used (e.g. Monk et al, 1988) but this was chosen because of the large numbers of PN measured in a uniform way. The temperature determination by Dopita and Meatheringham above $10^5$ K is questionable as will be discussed below.

Fig. 4.

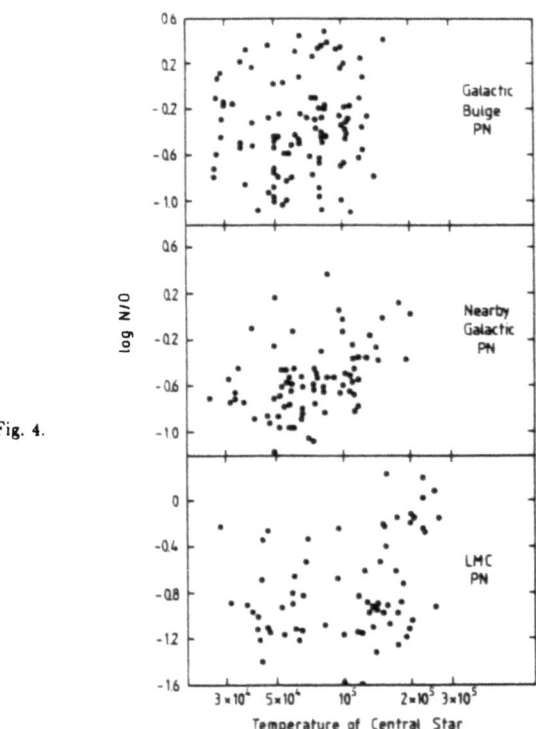

(a) N/O ratio plotted against central star temperature for galactic bulge PN (Ratag et al, 1991); (b) same as (a) but for a 'nearby' PN; (c) same as (a) but for LMC nebulae.

If only central stars with temperatures below $10^5$ K are considered, it appears very difficult to find any correlations in these diagrams. The diagrams for the galactic bulge and the LMC are clearly scatter diagrams, while any correlation in the diagram for the 'nearby' nebulae is at best marginal. This can also be said of the diagram of Stasinska and Tylenda (below $10^5$ K). This negative result indicates that caution should be exercised in claims of correlations between N/O and core mass.

## 4. Results from the Literature

The discussions in the literature are usually based on attempts to find a relation between the luminosity and the abundance ratios. We will discuss separately the LMC and the Galactic Bulge PN, where the distance in known, and the 'nearby' PN, where the unknown distance is a further complication.

### 4.1. PN IN THE LMC

A large sample of LMC PN have been studied by Dopita and Meatheringham (1991a,b) using spectra taken by themselves, and by Kaler and Jacoby (1990, 1991) using spectra taken from several sources, especially Monk et al (1988) and the first paper by Dopita and Meatheringham. Both of these pairs of authors derive a temperature for the central star and a luminosity for each PN. A spectral criterion is given to distinguish between nebulae which are optically thick and thin in the Lyman continuum, and most are found to be optically thick.

The central star temperature derived in these studies are found from model fits to the spectrum assuming the star radiates as a blackbody. This is probably a reasonable approximation for temperatures less than 80,000 K. For temperatures above $10^5$ K, the temperature is essentially determined by the HeII $\lambda$ 4686 to H$\beta$ ratio.[2] Very high stellar temperatures (sometimes above $2 \times 10^5$ K) are obtained, which are probably unreliable for the following reasons:
1) Koppen and Preite-Martinez (1991) show that high ratios of $\lambda$ 4686/H$\beta$ can be obtained even when the stellar temperature is as low as 50,000 K.
2) If the method were applied to NGC 2392 a stellar temperature of almost $2\times10^5$ K is obtained while 50,000 K is the more accepted value. Likewise for NGC 4361 a value close to $3 \times 10^5$ K would be found, compared to the more accepted value of 80,000 to 100,000 K (Mendez et al, 1988; Torres-Peimbert et al, 1990).

The temperature thus found has no effect on the derived luminosities which depend mainly on the measured H$\beta$ flux. It does strongly affect the core mass which is determined from the position on the HR diagram. These temperatures place the central star on the cooling curves of very high core mass models. This can be seen in Fig. 5 where the filled circles represent central stars of 'low temperatures' and the open circles are from the 'high temperature' central stars. While the former

---

[2] Kaler and Jacoby use only this ratio. Dopita et al use the NeV line as well.

do not have core masses higher than 0.64 $M_\odot$, the latter extend to as high as 1.2 $M_\odot$.

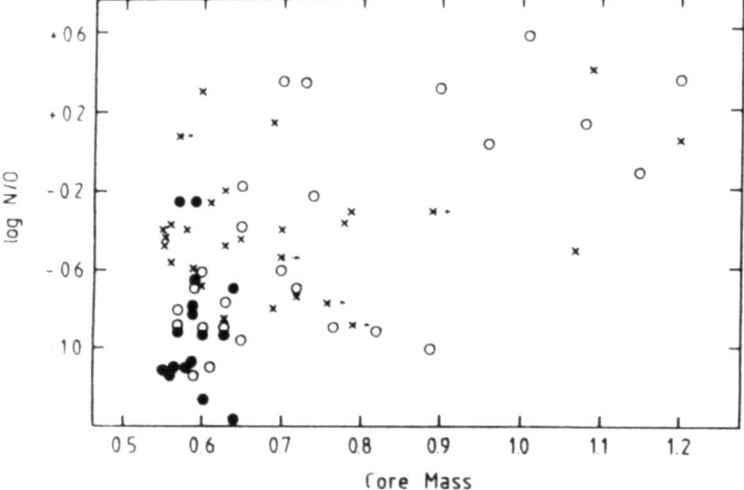

Fig. 5.
Core mass vs. nitrogen to oxygen ratio. The open circles are high temperature LMC central stars (Kaler and Jacoby, 1990) while the filled circles are LMC central stars with $T < 10^5$ K (Kaler and Jacoby, 1991). Results from Dopita and Meatheringham (1991a,b) are also included. The crosses are galactic central stars whose temperature is determined by the same method (Kaler et al, 1990).

Fig. 5 is the kind of plot presented by Kaler and Jacoby (1990, 1991) and Kaler et al (1990) as evidence for relation between core mass and nitrogen enrichment. There is clearly no correlation below core masses of 0.9 $M_\odot$. The correlation above 0.9 $M_\odot$ rests on the reliability of these high core masses and certainly needs further confirmation.

### 4.2. PN IN THE GALACTIC BULGE

A large sample of PN in the galactic bulge has been studied by Ratag et al (1991; 1993) using spectral measurements of Ratag et al (1992), Webster (1988) and Acker et al (1991). The central star temperatures are determined by a combination of the Energy Balance method, the Zanstra method and nebular models. The luminosity is then found by combining the stellar temperature and observed magnitude assuming a distance to the bulge of 7.8 kpc. The luminosity can also be determined from the $H\beta$ flux for optically deep nebulae. Their resultant plot of luminosity against the nitrogen to oxygen ratio is shown as Fig. 6. No correlation can be seen in this plot. The highest luminosity in Fig. 6 is less than $2 \times 10^4$ $L_\odot$ which corresponds to a core mass of less than 0.75 $M_\odot$ Fig. 6 is therefore consistent with Fig. 5 which is also a scatter diagram below 0.75 $M_\odot$ But there appears no evidence from the Energy Balance or Zanstra methods for the high temperatures found from the HeII λ 4686/Hβ ratio, and further brings into question the existence of core masses above 0.9 $M_\odot$.

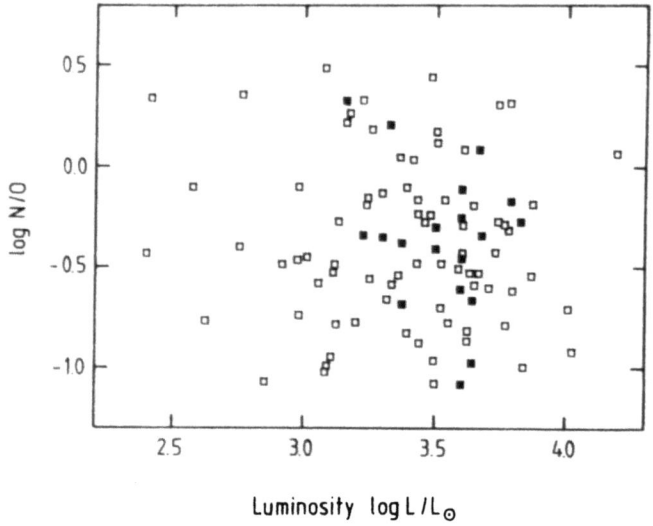

Fig. 6.
N/O ratio plotted against central star luminosity. Open squares are PN believed to be in 'horizontal' evolution; filled squares are PN already on the cooling track.

### 4.3. NEARBY PLANETARY NEBULAE

This is the most difficult group to discuss because the distances are most uncertain. Furthermore a selection effect enters because a limited group of PN are discussed at one time. Kaler et al (1990) select a group of large PN with high temperature central stars. They obtain distances from the Shklovski method. Central star temperatures are found essentially from the HeII λ 4686/Hβ ratio, although sometimes the HII Zanstra temperature is considered. Because the method of determining the temperature is so similar to that of Kaler and Jacoby (1990), it is not surprising that the resultant core masses are also very similar. In fact, some of the results are also plotted in Fig. 5.

Stasinska and Tylenda (1991) consider a sample with very little overlap with that of Kaler et al. Again distances are from the Shklovski method and central star temperatures are Zanstra determinations. Furthermore the theory plays an unclear role in their considerations, since they obtain three different core masses for a given nebula with a specified distance and central star temperature and luminosity. They find a possible correlation between the N/O ratio and core mass when one set of core masses is used. However when the core mass is determined from the position of the star on the HR diagram, the possible correlation disappears.

One of the few PN common both to the discussion of Stasinska and Tylenda (1991) and Kaler et al (1990) is NGC 650. The former authors find a temperature of 107,000 K and a core mass of 0.59 $M_\odot$ for the central star, while the latter find a temperature of 175,000 K and a core mass greater than 0.89 $M_\odot$. This gross discrepancy is not due to the distance determination, which is the same in both cases.

Another argument against the high core masses derived by Kaler et al (1990) has been given by Napiwotzki and Schönberner (1991). Using high resolution spectra of the central star of PW 1 and a non LTE analysis, they derive a temperature (60000–70000 K), effective gravity (log g $\approx$ 7.5) and mass (0.5 to 0.6 $M_\odot$) fot this object. This is quite different than given by Kaler et al, who assign a core mass of 1.0 $M_\odot$ to PW 1 (N/O $\approx$ 1). Napiowotzki and Schönberner state that a hot white dwarf of 1 $M_\odot$ is expected to have a log g $\approx$ 8.5, which is in contradiction with the observed spectrum.

Gathier and Pottasch (1989) consider a sample of nebulae with independent distance determinations. They point out that there is a tendency that those with the highest luminosity also have the highest N/O ratio. Their discussion is somewhat incomplete however since they make no distinction between high He/H and high N/O.

The above approach, to consider individual objects, has advantages and disadvantages. The latter is that selection effects play an important role. For example the PN NGC 2440 and NGC 6537 have independent distance determinations. Their temperature is undoubtedly higher than $2 \times 10^5$ K (Heap et al, 1990). The luminosities suggest core masses of between 0.7 and 0.8 $M_\odot$. Both are known to have very high N/O and He/H ratios and it is tempting to cite these cases as evidence that high core mass objects have high N/O ratios. The argument is even further strengthened since high stellar temperature is what is expected for high core mass objects. But counter examples exist: 2–3.3, 3+3.1, 351+5.1, 355–4.2. These are all galactic bulge objects with N/O between 1 and 3 and high He/H as well (in 3 cases). But they all have low temperature central stars and low to average luminosity (Ratag et al, 1993a). The nebulae are of average size. All of this indicates that they are low core mass objects having high N/O and He/H ratios.

It has also been suggested that NGC 2392 is a high core mass object, both from its high luminosity (Mendez et al, 1988) and from its suspected temperature increase in the past decade (Heap, 1992). But its N/O ratio is about average and its He/H ratio does not show a significant increase. Mention can also be made of the newly discovered PN, SAO 244567 (Parthasarthy et al, 1992), which has shown a strong increase in central star temperature over the past 40 years, yet it has only average N/O and He/H ratios.

## 5. Discussion

The evidence that a relation exists between core mass and N/O ratio is very weak. Fig. 7 is a histogram of all the available N/O ratios for PN in either the galactic bulge or in the LMC. In both groups distance uncertainties are unimportant. The central stars were divided in two temperature groups: Those below 75,000 K and those above 80,000 K. Furthermore the low temperature group is subdivided into two luminosity categories: $L < 3000\ L_\odot$ and $L \geq 3000\ L_\odot$. The values of temperature and luminosity are as given by the authors cited earlier.

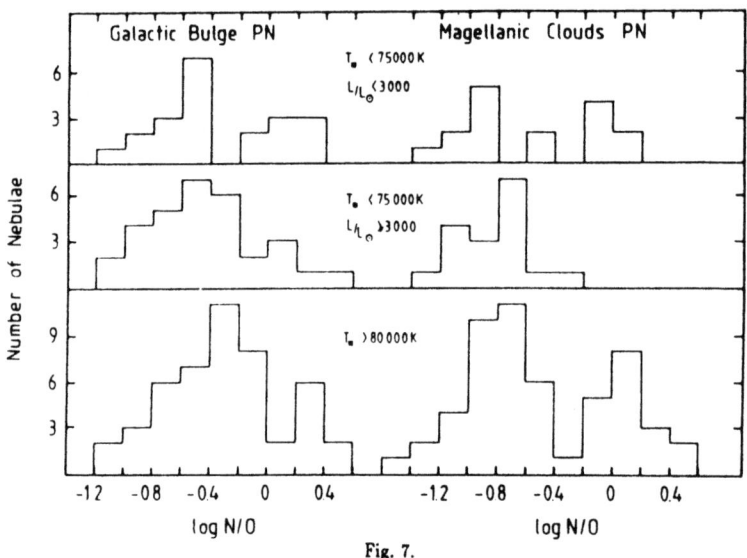

Fig. 7.
Histogram of the N/O ratio in the Galactic Bulge and the LMC for various central star temperatures and luminosities.

For the low temperature objects, shown in the upper two diagrams in Fig. 7 both high and low values of N/O are found. If the high core masses are found in the high luminosity group, as expected from the core mass-luminosity relation, we would expect to see a predominance of high N/O ratios in the high luminosity group. This is definitely not seen, either in the LMC or in the galactic bulge objects.

In the lower diagram in Fig. 7 the high temperature objects are shown. While we have expressed doubts earlier about the values of the temperature found from the HeII $\lambda$ 4686/H$\beta$ ratio, it is still likely that all these objects have temperatures higher than 80,000 K and fall in this group. As can be seen from the figure, a wide range of N/O is also found for this group. If there is a relation between core mass and N/O, those nebulae with high N/O would be expected to have evolve much more quickly and therefore have higher densities than the PN with low N/O ratios (as discussed above). To see if this is true, the densities found from the [SII] or [OII] ratios have plotted in a histogram (Fig. 8) separately for those nebulae with high N/O ratios and those with low ratios. As can be seen, there is no indication in Fig. 8 that the PN with higher N/O ratios also have higher densities, either in the galactic bulge or in the LMC.

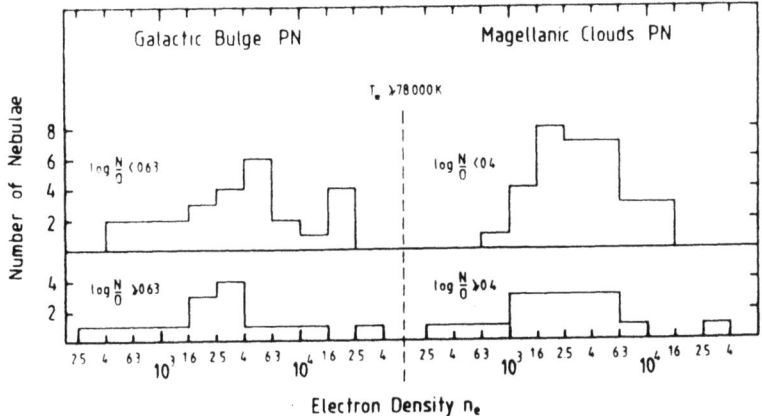

Fig. 8.
Histogram of the electron density for PN in the Galactic Bulge and the LMC. The PN are divided into groups with low (upper) and high N/O ratio (lower).

This lack of evidence for a relation between N/O and core mass makes it rather unlikely that a clearcut relation actually exists. It is possible that the luminosity has been incorrectly calculated, but this seems unlikely since it has been done by many groups independently. It is also unlikely that errors in the abundance ratio are large enough to cause such confusion.

It is possible that the assumption that the core mass is uniquely determined by the initial (progenitor) mass is wrong. Weidemann and Koster (1983), based on a study of white dwarfs in open clusters, show a trend that massive cores correspond to large initial masses. However, their relation shows a large scatter in the low mass range. It may be argued that this scatter is due to observational errors, but alternative explanations in terms of variation in mass loss efficiency have been advanced (Van der Veen and Habing, 1990).

Another possible source of confusion is the implicit assumption that hydrogen shell burning determines the evolutionary tracks. If there were a mixture of hydrogen and helium burning PN (which cannot at present be distinguished), an initial-final mass relation could exist, but we would be unable to see it.

## References

Acker, A.A., Köppen, J., Stenholm, B., Raytchev, B. 1991, *Astron. Astrophys. Suppl.* **89**, 237.
Boothroyd, A.I., Sackmann, I.J. 1988, *Astrophys. J.* **328**, 641.
Dopita, M.A., Meatheringham, S.J. 1991a, *Astrophys. J.* **367**, 115.
Dopita, M.A., Meatheringham, S.J. 1991b, *Astrophys. J.* **377**, 480.
de Freitas Pacheco, J.A., Maciel, W.J., Costa, R.D.D. 1992, *Astron. Astrophys.*..

Gathier, R., Pottasch, S.R., 1989, *Astron. Astrophys.* **209**, 369.
Groenewegen, M.A.T., De Jong, T. 1992, *Astron. Astrophys.* submitted.
Heap, S.R., Corcoran, M., Hintzen, P., Smith, E. 1990, in: *From Mira to PN*, ed. Mennessier M.-O. and Omont, A.
Henry, R.C. 1990, *Astrophys. J.* **356**, 229; **363**, 728.
Iben, I.J., Renzini, A. 1983, Ann. Rev. Astron. Astrophys. **21**, 271.
Kaler, J.B., Jacoby, G.H. 1990, *Astrophys. J.* **362**, 491.
Kaler, J.B., Jacoby, G.J. 1991, *Astrophys. J.* **382**, 134.
Kaler, J.B., Shaw, R.A., Kwitter, K. 1990, *Astrophys. J.* **359**, 392.
Köppen, J., Preite-Martinez, A. 1991, *Astron. Astrophys.* **248**, 191.
Mendez, R.H., Kudritzki, R.P., Herrero, A., Husfeld, D., Groth, H.G. 1988, *Astron. Astrophys.* **190**, 113.
Monk, D.J., Barlow, M.J., Clegg, R.E.S. 1988, Mon. Not. Roy. Astr. Soc. **234**, 583.
Napiwotzki, R., Schönberner, D. 1991, in: *White Dwarfs*, ed. G. Vauclair, E. Sion, p. 39
Paczynski, B. 1971, *Acta Astr.* **21**, 417.
Parthasarathy, M., Garcia-Lario, P., Pottasch, S.R., Manchado, A., Clavel, J., de Martino, D., v.d. Steene, G.S., Sahu, K.C. 1992 *Astron. Astrophys.*.
Perrinoto, M. 1991, *Astrophys. J. Suppl. Ser.* **76**, 687.
Pottasch, S.R. 1993, *Astr. Astrophys. Rev.*
Ratag, M.A. 1991, Thesis, Univ. of Groningen.
Ratag, M.A., Pottasch, S.R. Dennefeld, M., Menzies, J.W. 1992, *Astron. Astrophys. Suppl. Ser.*
Ratag, M.A., Pottasch, S.R. Dennefeld, M., Menzies, J.W. 1992, *Astron. Astrophys.*
Ratag, M.A., Pottasch, S.R., Water, L.B.F.M. 1993a, *Astron. Astrophys.*
Ratag, M.A., Waters, L.B.F.M., Pottasch, S.R. 1993b, *Astron. Astrophys.*
Renzini, A., Voli, M. 1981, *Astron. Astrophys.* **94**, 175.
Schönberner, D. 1981, *Astron. Astrophys.* **103**, 119.
Stasinska, G., Tylenda, R. 1990, *Astron. Astrophys.* **240**, 467.
Torres-Peimbert, S., Peimbert, M., Pena, M. 1990, *Astron. Astrophys.* **233**, 540.
V.d. Veen, W.E.C.J., Habing, H.J. 1990, *Astron. Astrophys.* **231**, 404.
Webster, B.L. 1988, Mon. Not. Roy. Astr. Soc. **230**, 377.
Weidemann, V., Koester, D., 1983 *Astron. Astrophys.* **121**, 77.
Wood, P.R., Faulkner, D.J. 1986, *Astrophys. J.* **307**, 659.
Wood, P.R., Zarro, D.M. 1981, *Astrophys. J.* **247**, 247.

# SIMULATIONS OF A POPULATION OF PLANETARY NEBULAE

G. STASINSKA
*DAEC, Observatoire de Paris-Meudon, France*
and
R. TYLENDA
*DAEC, Observatoire de Paris-Meudon, France*
*Copernicus Astronomical Center, Torun, Poland*

ABSTRACT. We present a simulation of the population of Galactic bulge planetary nebulae (GBPN), which matches the diagrams obtained from VLA radio observations. This simulation may not be the only one fitting the observed data, but it helps understanding the role of observational uncertainties and selection effects in the interpretation of observational diagrams.

## 1. INTRODUCTION

Statistical properties of planetary nebulae (PN) are derived from diagrams in which selection effects and observational uncertainties enter in a complicated manner. In order to understand how they operate, we have performed a numerical experiment, in which we have simulated a population of PN, assumed to be lying in the Galactic bulge. We have then simulated the observing conditions. This enabled us to construct diagrams which are directly comparable to observational ones. The results of such an experiment is presented below.

## 2. PROCEDURE FOR SIMULATING THE GBPN POPULATION

### 2.1. The model for a single PN

The schematic model adopted consists in a constant density shell expanding uniformly around a post AGB star which evolves according to the tracks of Schönberner (1979, 1983) and Blöcker and Schönberner (1991). At any time, the ionized mass is computed analytically by balancing the number of ionizing photons and the number of hydrogen recombinations to excited levels.

### 2.2. Simulation of the GBPN population

Individual model PN are randomly selected out of a uniform distribution for the ages, and a Gaussian distribution for the central star masses as well as for the logarithms of nebular masses and expansion velocities The distance to each nebula is randomly selected from a Gaussian distribution, centered at 7.8Kpc with a standard deviation of 0.8Kpc, appropriate for GBPN.

### 2.3. Simulation of the observing conditions

In order to understand diagrams obtained from 6cm radio measurements, we must simulate the observing conditions. We assume that logarithmic errors in the fluxes are Gaussian, with a dispersion of 0.05dex, and that the dispersion of errors in the diameters φ ranges from 0.15 dex for φ=1" to 0.05 dex for φ > 10". Since, in order to exclude most of foreground PN, the observed sample of GBPN is defined by eliminating objects having φ > 20", the same is done for simulations. Finally, observed and simulated populations must be compared using complete subsamples. This requires taking only PN with F(6cm) > 10mJy (81 objects in the observed sample).

## 3. AN EXAMPLE OF A MODEL GBPN POPULATION MATCHING THE VLA DATA

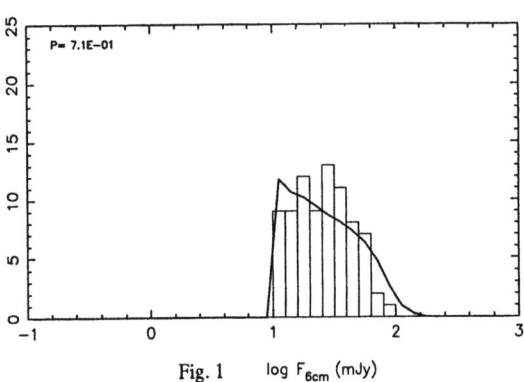

Fig. 1    log $F_{6cm}$ (mJy)

Fig. 2    φ (arc sec)

We present a few figures relative to a specific model population leading to diagrams that are in agreement with the VLA observations of GBPN published in Zijlstra et al. (1989). In this theoretical population, the central star masses are distributed around 0.57 $M_\odot$ with a standard deviation of 0.02 $M_\odot$; the dispersion of the nebular masses is 0.3 dex around 0.1 $M_\odot$, and that of the expansion velocities $v_{exp}$ is 0.3 dex around 20 km/s.

Figs. 1, 2 and 3 show respectively the distributions of the 6cm fluxes, of the angular diameters, and of the Shklovsky distances (computed assuming an ionized mass of 0.2$M_\odot$). Boxes are histograms relative to observations. Curves represent the simulated distributions. The agreement is good (confirmed by thesignificance of the Kolmogorov-Smirnov test, respectively found to be 0.71, 0.80 and 0.19).

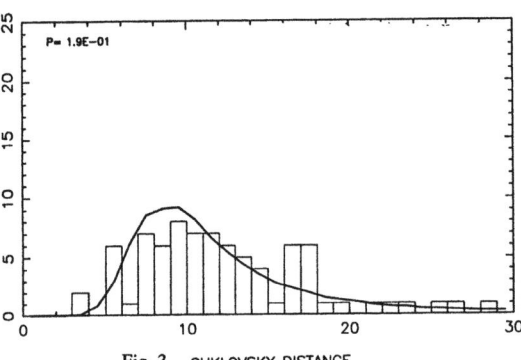

Fig. 3    SHKLOVSKY DISTANCE

Fig. 4a shows the classical nebular mass-radius diagram for the observed sample, where $M_\phi$ is calculated using the formula $M_\phi = 3.12 \; 10^{-3} \; F(6cm)^{0.5} \phi^{1.5}$. The observed points gather along a line of slope 1.5. This is often interpreted as the consequence of the gradual ionization of the expanding nebulae. In

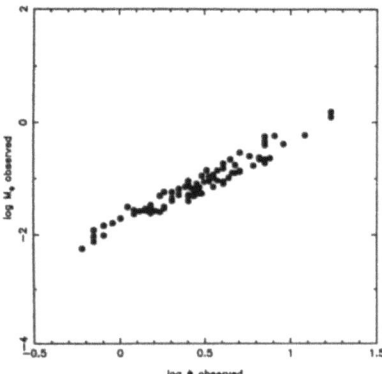

Fig. 4a

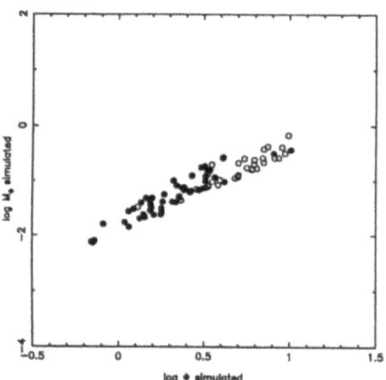

Fig. 4b

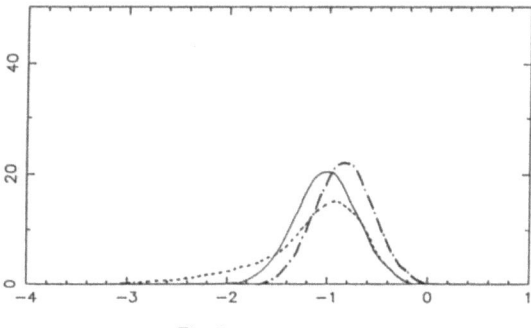

Fig. 5   log $M_{neb}$

fact, it has already been shown by Stasińska et al (1989) that the aspect of this diagram is mainly the result of the formal dependance of $M_\phi$ on $\phi$. Fig. 4b shows the same diagram for the simulated population. White circles represent density bounded model PN, and we see that the observational diagram is well reproduced, while more than 50% of the model PN are density bounded.

Fig. 5 shows the effect of observational selection on the derivation of the PN mass distribution. The thin continuus line represents the original distribution of total nebular masses - i.e. Gaussian in the log, with the parameters given above-. The thick dot-dashed line corresponds to the total nebular masses for those PN with F(6cm) < 10 mJy, while the thick dotted line represents the ionized masses for the same subsample. All three distributions are normalized to 81 objects. The dot-dashed curve is shifted rightwards with respect to the original nebular mass distribution, because the less massive nebulae are generally too faint to be observed. The mean nebular mass corresponding to the observable sample is 0.19 $M_\odot$ while it was 0.13 $M_\odot$ in the original distribution. The distribution of the ionized masses, which are the masses that can be observed, has a long tail towards small masses, constituted of the compact, ionization bounded nebulae.

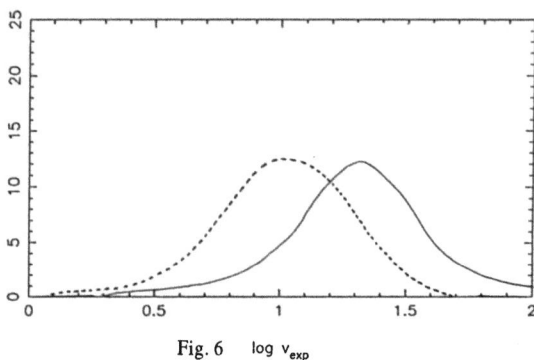

Fig. 6 log $v_{exp}$

Fig. 6 shows the simulated distribution of log $v_{exp}$, normalized to 81 PN; thin curve: original population, thick dashed curve: PN with $\phi < 20"$ and $F(6cm) > 10mJy$. Because PN with large $v_{exp}$ quickly reach a diameter of 20", the latter distribution contains a smaller proportion of rapidly expanding PN than the original distribution. The peak in the distribution lies at ~ 11km/s, instead of 20km/s.

## 4. DISCUSSION

This very schematic model population is only to be taken as an instructive example. It reproduces quite well the classical observational diagrams shown above, including the $M_\phi$ vs $\phi$ relation. This example shows that, because of selection effects, the apparent properties of a sample of PN may be quite different from the characteristics of the underlying PN population.

Real nebulae are neither spherical, nor ionized by blackbodies, and the Schönberner tracks may not be appropriate for at least some PN nuclei (because of the rôle of helium flashes, different mass loss, binary stars). This introduces a scatter which was not taken into account in the simulations. The true nebular mass and expansion velocity distributions should therefore be narrower than assumed in any simulation.

In thesimulated population presented above, 68% of the PN have nebular masses within $\pm 0.3$dex of $0.1 M_\odot$. If the true population corresponds to this solution, this would put a strong constraint on dynamical models of the formation of nebular envelopes. The distribution of the PN nuclei masses is also very narrow, with a mean of 0.57 $M_\odot$ in the original distribution (0.58 $M_\odot$ after including selection effects), quite below the value admitted up to now.

However, we stress that the simulated population discussed here may not represent the real one, and we are pursuing this work to explore all the possible solutions. In order to obtain a more reliable picture of the GBPN population, is crucial to extend the sample of GBPN radio observations down to 6 cm fluxes of 1mJy. Indeed, the aspect of the histogram of F(6cm) between 1 and 10mJy is very sensitive to the number of PN ionized by stars of masses higher than $0.61 M_\odot$ as can be seen in diagrams presented by Stasińska et al. (1991). Also, accurate angular diameters are needed for the whole sample. Finally, measurements of the expansion velocities for GBPN, which are presently known for only a few objects, would be most helpful. Indeed, they provide an important constraint, since the expansion velocity determines the epoch when a given PN becomes optically thin.

A complete version of this work will be submitted to Astronomy and Astrophysics.

References
Blöcker T., Schönberner D. (1990) A&A 240, L11
Schönberner D. (1979) A&A 79, 108, (1983) APJ 272, 708
Stasińska G., Tylenda, R., Acker A., Stenholm B. (1989) A&A 247, 173
Stasińska G., Fresneau A., Gameiro, G.F., Acker A. (1991) A&A 252, 762
Zijlstra A.A., Pottasch S.R., Bignell C. (1989) A&A S 79, 329

# HYDROGEN AND HELIUM BURNING EVOLUTIONARY TRACKS

P.R. WOOD and E. VASSILIADIS
*Mount Stromlo and Siding Spring Observatories, Private Bag, Weston Creek P.O., Canberra, ACT 2611, Australia*

ABSTRACT. We discuss the effect of metallicity on the luminosities of planetary nebula nuclei (PNNi), and we use theoretical stellar evolution calculations to predict the fraction of PNNi that enter the planetary nebula domain burning helium rather than hydrogen. Both these factors will clearly influence the planetary nebula luminosity function.

## 1. Introduction

Stellar evolution calculations from the main sequence to the white dwarf stage have recently been made for stars of initial masses 1-5 $M_\odot$ and metal abundance Z from $Z_\odot$ to 0.25 $Z_\odot$ (Vassiliadis and Wood 1992). These tracks allow us to predict the effect of metal abundance on the luminosities of PNNi produced by stars of a given initial mass. Of course, the luminosity of a PNN during its high luminosity transition across the HR diagram from the AGB is determined almost entirely by AGB mass loss. In these calculations, the empirical mass loss rate varies exponentially with pulsation period on the AGB up to a maximum value corresponding to the radiation pressure driven limit $\dot{M} = L/cv_\infty$ (Wood 1990; Vassiliadis and Wood 1992). There is no explicit abundance dependence in the mass loss formulae used, but the abundance is involved implicitly since lower abundance gives a hotter AGB which in turn leads to smaller pulsation periods at a given luminosity. Within the limits of current observational data, the maximum observed period for optically-visible, long-period variables on the AGB appears to be similar in the solar vicinity, the Galactic bulge, the LMC and the SMC. If we assume that a star will not be optically-visible once "superwind" mass loss starts, this result indicates that the high superwind mass loss rate begins at a similar pulsation period in each of these stellar systems, each of which has a different characteristic abundance. Furthermore, if we identify the superwind phase with the radiation pressure driven limit, then the result indicates that the mass loss rate depends mostly on the pulsation period, without further explicit abundance dependence. Our mass loss formula should therefore reasonably reproduce the transition to the superwind mass loss regime at different abundances.

The second topic we wish to discuss is the likelihood that a star entering the planetary nebula region of the HR diagram will be burning helium or hydrogen. Since we have evolved a relatively large number of tracks (22) from the main-sequence to the PN domain without any deliberate attempt to control the phase of the shell flash cycle at which the stars leave the AGB, we can use the statistics of our H and He burning tracks to roughly estimate the relative numbers of observed PNNi that burn H or He on entering the PN domain of the HR diagram.

## 2. Comparison of hydrogen and helium burning tracks

The most detailed study of the possible PNN track morphologies that can result from stars leaving the AGB at different phases of the helium shell flash cycle is that of Iben (1984), this study being done for a 0.6 $M_\odot$ star. Basically, PNN tracks fall into two categories. Firstly, there are tracks where the star evolves smoothly through the PN region of the HR diagram burning hydrogen before the decline into the white dwarf state. Some examples of these tracks, which we will call hydrogen burning tracks, are shown in Figure 1. The second class of track is exhibited by stars that enter the PN domain of the HR

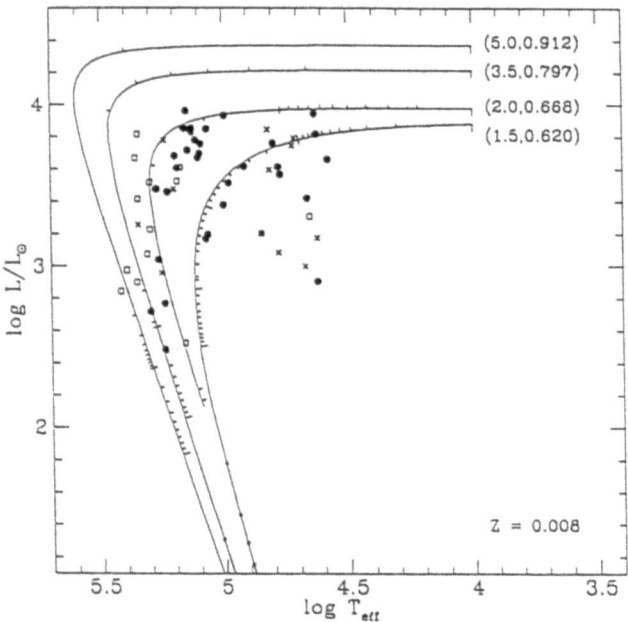

Figure 1. Evolutionary tracks for PNNi that cross the HR diagram burning hydrogen. Each track is labelled by (Initial mass, Mass at log $T_{eff}$ = 4). Tick marks are initially at intervals of 100 yr (above track) then at 1000 yr (below track). Dots mark time steps of $10^5$ yr from log $T_{eff}$ = 4. Points represent LMC PN from Dopita and Meatheringham (1991a,b): open squares are type I PN, filled circles are optically thick objects, and crosses are optically thin objects.

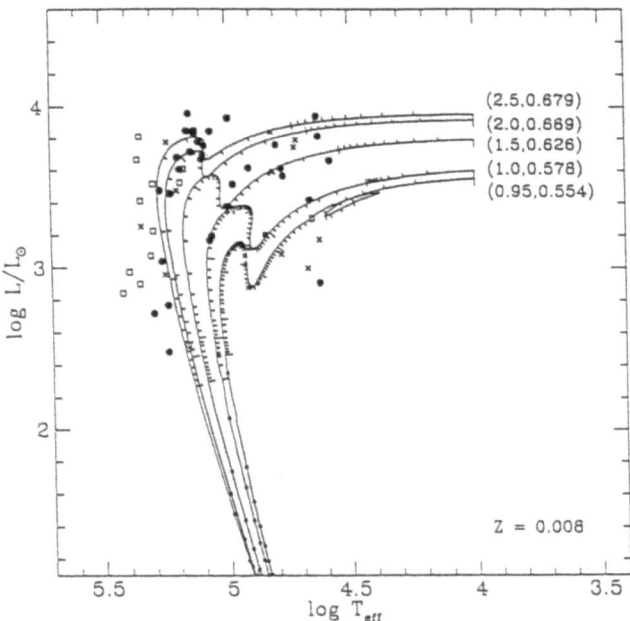

Figure 2. Same as Fig. 1 but for PNNi that enter the PN domain burning helium.

diagram burning helium (as a result of a helium shell flash) and which re-ignite the hydrogen-burning shell at high luminosity and high $T_{eff}$ (log $T_{eff} \approx 5$). Such stars we will call helium burners, although this is a misnomer since the stars spend most of their PN lifetimes burning hydrogen. Examples of these tracks can be seen in Figure 2.

## 3. The metal abundance dependence of PNN luminosity

As can be seen in Figure 1, PNNi have their maximum luminosity when evolving across the HR diagram from the AGB. The luminosity L(4) when log $T_{eff} = 4$ is given as a

Table 1

| $M_i$ | Z | log L(4)/$L_\odot$ | $M_i$ | Z | log L(4)/$L_\odot$ |
|---|---|---|---|---|---|
| 1.0 | 0.016 | 3.541 | 2.5 | 0.016 | 3.979 |
| 1.0 | 0.004 | 3.641 | 2.5 | 0.004 | 4.152 |
| 1.0 | 0.001 | 3.776 | ... | ... | ... |
| 3.5 | 0.016 | 4.138 | 5.0 | 0.016 | 4.360 |
| 3.5 | 0.008 | 4.216 | 5.0 | 0.008 | 4.370 |
| 3.5 | 0.004 | 4.296 | 5.0 | 0.004 | 4.399 |

function of initial mass $M_i$ in Table 1 for PNNi of various metal abundances. All tracks have initial helium abundance Y = 0.25.

It can be seen from Table 1 that $d\log L(4)/d\log Z \sim -0.25$, at least for $M \lesssim 3.5\ M_\odot$. The increase in maximum PNN luminosity with decreasing Z is as expected, since a higher metallicity gives a cooler giant branch and the period of ~500 days required for superwind mass loss rates occurs at lower luminosity.

## 4. The fraction of hydrogen and helium burning PNNi

Whether a star ends up as a hydrogen or helium burning PNN depends on (1) the phase of the shell flash cycle at which the star leaves the AGB (which in turn depends on how the mass loss rate varies throughout the shell flash cycle) and (2) the duration of the transition from the AGB to $\log T_{eff} = 4.5$ where nebula ionization becomes significant. Factors (1) and (2) together determine the phase of the shell flash cycle the star finds itself in when in the PN domain of the HR diagram.

In the present calculations, the very rapid increase of mass loss rate with luminosity means that, at least for the lower mass stars with $M \lesssim 2.5\ _\odot$, mass loss is concentrated in the latter half of the quiescent (hydrogen burning) phase of the shell flash cycle (Vassiliadis 1992; see also Wood 1992). If $\phi$ is the phase of the shell flash cycle (with the shell flash occurring at $\phi = 0.0$), then the star is therefore most likely to leave the AGB during the interval $0.5 < \phi < 1.0$. Assuming a transition time of ~7000 years and a shell flash cycle length of ~ 70000 years, those PNN with $0.9 \lesssim \phi \lesssim 1.0$ will suffer a shell flash during post-AGB, pre-PN evolution and will therefore become helium burners (as defined above). In this case, we would expect ~20% of low mass stars to become helium burning PNNi. In our calculations, we found that 6 out of 22 stars (27%) became helium burners. If this factor is combined with the fact that the high luminosity phase of PN evolution is at least twice as long for helium burners as for hydrogen burners, then we would expect that at least 40% of observed high luminosity ($\log L/L_\odot \gtrsim 2.5$) PNNi would be helium burners. This estimate is similar to that of Iben's (1984) guesses 1 and 2.

## References

Blöcker, T. and Schönberner, D. 1990, A&AL, 240, L11.
Iben, I. 1984, ApJ, 277, 333.
Paczynski, B. 1971, Acta Astr., 21, 417.
Vassiliadis, E. 1992, Thesis, Australian National University.
Vassiliadis, E. and Wood, P.R. 1992, in preparation.
Wood, P.R. 1990, in *From Miras to Planetary Nebulae: Which Path for Stellar Evolution*, eds. M.O. Mennessier and A. Omont (Editions Frontières), p.67.
Wood, P.R. 1992, in IAU Symposium 155 *Planetary Nebulae*, eds. R. Weinberger and A. Acker (Kluwer: Dordrecht), these proceedings.
Wood, P.R. and Faulkner, D.J. 1986, ApJ, 307, 659.

# FURTHER MODELS OF PLANETARY NEBULA SPECTRAL EVOLUTION

K. VOLK

*Department of Physics and Astronomy, University of Calgary, 2500 University Drive N.W., Calgary, Alberta, Canada T2N 1N4*

**Abstract.** Studies of the total spectra of planetary nebulae (PN) show that they usually emit significant fractions of their total energy at infrared wavelengths. Many PN have IRAS observations but we do not understand their broad-band colours. To model these observations requires a combination of a photoionization treatment of the ionized region and a radiative transfer treatment of any neutral dust shell. Including the effects of the central star evolution and an assumed galactic distribution allows many observables to be simulated. I find that the Schönberner 0.64 $M_\odot$ evolutionary track is too slow in the transition to high temperature to match the IRAS colours of PN if a typical AGB wind expansion speed is 10 km s$^{-1}$. Accelerating the evolution track by a factor of 2 produces reasonable agreement. An equal mixture of carbon-rich and oxygen-rich sources with mass loss rates of $2 \times 10^{-5}$ $M_\odot$ yr$^{-1}$ seems to reasonably match the general properties of the PN data from IRAS although the details are still unclear. These models strongly indicate that the post-AGB luminosity decline deduced by Knapp (1986) from radio CO line observations is not correct.

## 1. Introduction

The IRAS satellite observed hundreds of planetary nebulae (PN) at wavelengths from 12 to 100 $\mu$m. In addition LRS spectra are available for some of the brighter PN in the 8 to 23 $\mu$m wavelength range. It was surprising to observe so many PN because the dust was thought to be relatively short-lived within the ionized region due to ionic sputtering. The IRAS observations are not well understood: the LRS spectra and broad-band colours do not seem to indicate a clear evolutionary path and there is a tremendous diversity in the infrared character of PN, much more so than for compact HII region sources. Zhang and Kwok (1991) have shown that many PN emit a large fraction of their energy in the far-infrared, as had previously been observed for smaller groups of PN (i.e. Cohen and Barlow 1974).

To understand these observations better requires a spectral evolution model which includes the evolution of both the nebular spectrum and the thermal dust emission. After the asymptotic giant branch (AGB) evolution ends, the remnant dust shell expands away from the star at constant speed while the star evolves to higher temperature. There may also be a second, fast wind which produces an interacting winds shell. This circumstellar shell is ionized from within by the star. All of these factors must be considered as functions of time. The basic evolution of the spectrum is determined by competition between the central star evolution and the expansion of the dust shell.

## 2. The Model

A multi-component modelling process is required because of the wide range of physical processes involved in forming the energy distribution. I used the Schönberner central star tracks to follow the central star evolution (although I ended up accelerating the evolution by a factor of 2 as discussed below). For any time after the AGB the Schönberner tracks give the temperature and luminosity of the central star which is assumed to be a blackbody. There were 4 runs for different initial dust shells, corresponding to $\dot{M}$ values of $2.1 \times 10^{-5}$ and $5.2 \times 10^{-6}$ $M_\odot$ yr$^{-1}$ and either silicate or graphite grains. The dust density distribution was inverse-square until a fast wind begins, when $T_* = 10^4$K, after which a shell + wind density distribution was used.

Once the star becomes hot enough to ionize the surrounding gas, the photoionization program CLOUDY (kindly provided by Gary Ferland) was used to model the ionized gas including an appropriate amount of dust. If the nebula is predicted to be ionization bounded the radius of the ionization front and the output spectrum from CLOUDY were used as inputs to a dust radiative transfer program (DUSTCD, originally by C. M. Leung). There is therefore an inconsistency in the modelling because no dust radiative transfer calculation is done for the ionized region. This type of calculation was carried out for about 120 times during the post-AGB evolution, until the star has passed onto the white dwarf cooling curve at an age of $\approx 10^4$ years.

These model runs differ from those in Volk (1992) because the newer version of CLOUDY includes dust in the ionized region, rather than having to carry out the dust radiative transfer independently of the photoionization calculation. A discrepancy between CLOUDY and the DUSTCD radiative transfer program mentioned in Volk (1992) has been resolved, but there still remain some difficulties in combining CLOUDY and DUSTCD results which produce odd effects in certain parts of the model tracks. The process used to convert the spectral evolution results to simulate observations – photometry, 5 GHz radio flux density, angular radii, infrared excess values, and line ratios – is discussed in Volk (1992).

## 3. Results Of The New Runs

Early trial runs indicated that the Schönberner (1983) model tracks evolve too slowly to match the IRAS colours of PN even for the 0.64 $M_\odot$ case. This effect is quite pronounced and seems to be a very robust conclusion for an expansion speed of 10 km s$^{-1}$. It was found that the 0.64 $M_\odot$ track had to be accelerated by a factor of 2, at least during the initial temperature rise to $> 3 \times 10^4$K, for the model colour track to "turn back" at the proper 12/25 $\mu$m colour. This can be explained in several ways: the expansion speed could be only 5 km s$^{-1}$, the typical PN central star could be slightly more massive than 0.64 $M_\odot$, or possibly there is stronger post-AGB mass loss from the residual envelope than was assumed by Schönberner. As observations of suspected post-AGB stars often show signs of continuing mass loss, I consider the last possibility to be the most likely. A wider range of post-AGB stellar evolution tracks is needed to study this problem, because this difficulty with the original Schönberner tracks has been found before from other studies.

One new result of these models was that the graphite dust runs consistently predict more infrared emission than did the silicate ones. The predicted IRAS 12 $\mu$m fluxes were generally 1 to 2 magnitudes higher at any given time for the graphite shell. Thus, even if there are not as many carbon-rich PN and PPN as oxygen-rich ones there is a strong observational bias towards the carbon-rich sources using IRAS. This may explain why most of the brighter PN and PPN observed by IRAS appear to be carbon-rich. By contrast optically selected nebulae are then strongly biased against young, carbon-rich nebulae due to the strong dust extinction from the shell, so objects such as IRAS 21282+5050 or 07027−7934 were discovered by IRAS.

The runs for the lower $\dot{M}$ values using either type of dust were found to be strongly matter-bounded for most of the PN evolution, and the predicted optical line strengths do not match well with what is generally observed. Thus these tracks were rejected. For the higher $\dot{M}$ values the nebulae are predicted to be matter-bounded during the phase of maximum ionizing photon flux but ionization-bounded both as young nebulae and once the luminosity of the central star falls below about $10^3$ $L_\odot$. Together the graphite and silicate dust tracks cover the centre of the observed IRAS colours of PN, but the predicted colours do not show nearly as much scatter as is observed. Part of the cause of the scatter may be PAH emission in the carbon-rich nebulae, but clearly more work needs to be done to identify why some observed PN have unusual colours compared with the model tracks.

To show the degree to which the models match the IRAS data, Figure 1 gives the observed brightness distribution of optically identified PN in the 4 IRAS bands along with the model predictions for an equal mix of carbon-rich and oxygen-rich nebulae. A scale factor of 0.73 has been applied to the model curves. A good match in shape and level is obtained at 60 $\mu$m. At 100 $\mu$m the data probably suffers completeness problems even at relatively high flux densities, so the observed curve rolls off from the model curve at $\approx$ 10 Jy. At 25 $\mu$m a good match is obtained for PN brighter than $\approx$ 2 Jy but thereafter there are no model sources. At 12 $\mu$m the shape of the model curve is correct but the number of predicted sources is $\approx$ 10% too low which may simply mean that the 12 $\mu$m luminosities are $\approx$ 10% too low, due to PAH emission in real carbon-rich nebulae for example. Together the agreement at 12 and 60 $\mu$m and the disagreement at 25 $\mu$m are rather peculiar and I do not know what the cause of this is. Overall the match is fairly good. The fit varies somewhat for different mixtures of the carbon-rich and oxygen-rich nebulae.

## 4. The $\beta$ Problem For PN

There is a direct conclusion from Figure 1 which has a bearing upon papers in the literature on the radio CO line observations of PN and PPN. For both AGB and post-AGB sources observed in CO it is customary to calculate a $\beta$ value which is the ratio of the (scalar) momentum carried in the mass-loss to the photon momentum from the star. This value is found to be of order 1 for AGB stars but rises to $\approx$ 100 in PN. Knapp (1986) interprets this as a direct result of a luminosity drop by a large factor soon after the AGB, whereas the central star models indicate a fairly constant luminosity during the rise to high temperature. In the context of this paper, that the Schönberner tracks produce a reasonable fit to the IRAS observations for an assumed galactic population of 18000 PN makes Knapp's interpretation difficult to sustain. For if post-AGB sources do indeed fade by a factor of 50, say, soon after the AGB then these models have the young PN too bright by a factor of 50. To produce the general agreement in infrared brightness – and optical and radio brightness too – with observations would be impossible unless my assumed galactic population is 50 times too low (or the Schönberner AGB luminosity values are a factor of 50 too low, but that would cause further conflicts with observations). However, as bad as PN distance determinations are it is impossible that there could be a total galactic population of about $10^6$ PN and that all those we observe are quite local. Thus I conclude that the change in $\beta$ values do not indicate a luminosity effect even though this is the simplest interpretation.

## References

Cohen, M., and Barlow, M. J. (1974) ApJ, 193, 401.
Knapp, G. R. (1986) ApJ, 311, 731.
Schönberber, D. (1983) ApJ, 272, 708.
Volk, K. (1992) ApJSupp, 80, 347.
Zhang, C.-Y., and Kwok, S. (1991) A&A, 250, 179.

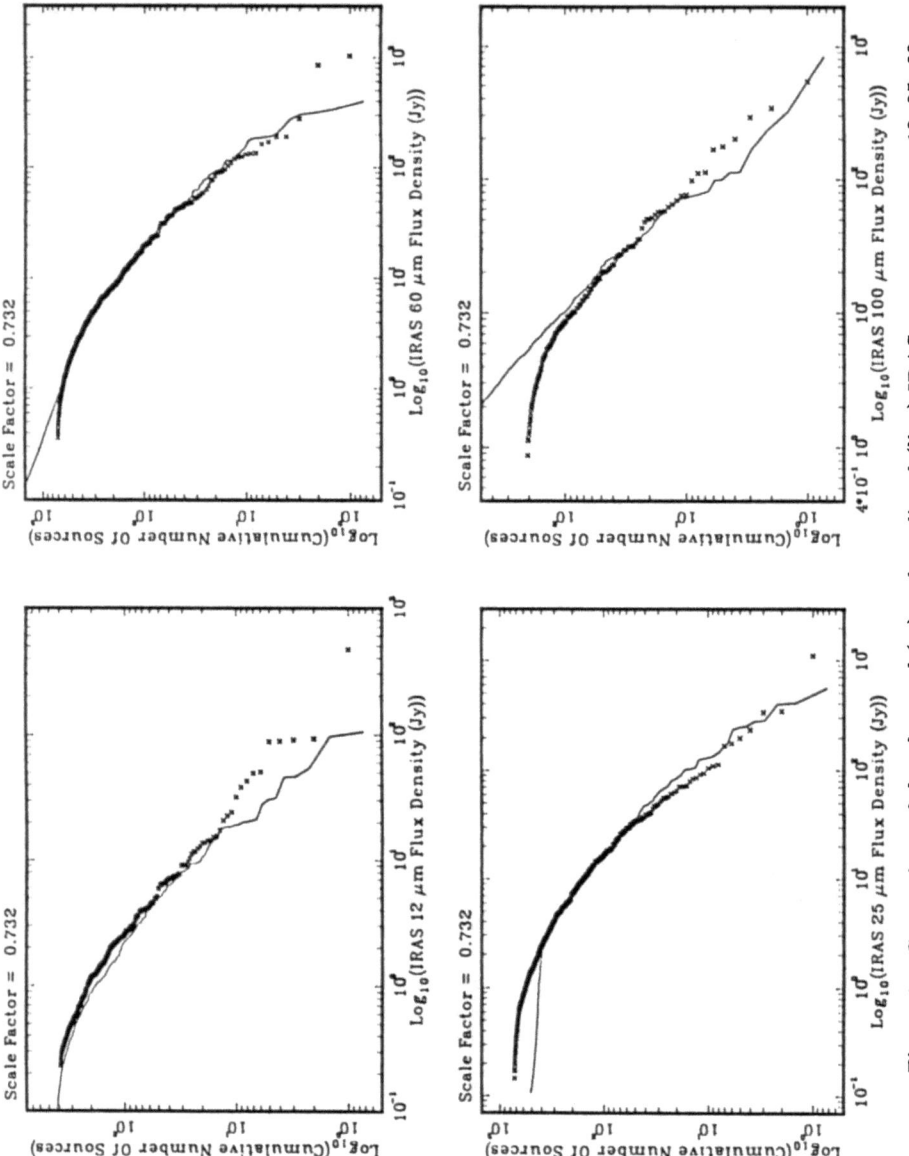

Figure 1 – Comparison of the observed (×) and predicted (line) IRAS source counts at 12, 25, 60 and 100 μm. The model curves have been scaled by factor 0.73 to best match the observations.

# SYNTHETIC P-AGB EVOLUTION

LETIZIA STANGHELLINI
Osservatorio Astronomico di Bologna, via Zamboni 33, I-40126 Bologna, Italy
and
ALVIO RENZINI
Dipartimento di Astronomia, Università di Bologna, via Zamboni 33, I-40126 Bologna, Italy

ABSTRACT. *Extensive Montecarlo simulations of Post-Asymptotic Giant Branch (P-AGB) populations have been constructed, exploring the effects of various assumptions on synthetic H-R diagrams, luminosity functions, and inferred mass distributions. Such assumptions include the IMF, the initial mass-final mass relation, the AGB to PN transition time, the duration of the planetary nebula (PN) stage, etc. We have also investigated how the observational errors in luminosity and temperature propagate into the inferred mass distribution of the P-AGB stars.*

In recent years many authors have compared theoretical evolutionary tracks for P-AGB stars to observational data of Planetary Nebulae Nuclei (PNni) in order to infer several properties of these stars (e.g. Schönberner 1981; Tylenda et al. 1991; Stasińska et al. 1991). While several interesting results have been achieved, it became also clear that some of these comparisons may be affected by intrinsic limitations (cf. Acker 1989; Schönberner & Tylenda 1990) some of which are further explored in this paper. We have developed a versatile method to produce Montecarlo simulations of P-AGB stars based on the evolutionary tracks of Paczyński (1971) and Schönberner (1983). The method is able to produce a synthetic population of PNni for specified IMF, AGB to PN transition time, and initial mass-final mass relation. Only a very restricted selection of our simulations is presented here, while a more extensive set will be published elsewhere in the near future.

In principle, the mass of PNni can be determined from their location in the H-R diagram, making use of P-AGB evolutionary tracks. However, both the luminosity and the effective temperature of PNni are affected by systematic and random errors that affect the derived mass distribution. Here we explore the effect of systematic errors in temperature. The temperatures are in general estimated using the Zanstra method, which assumes the nebula to be optically thick to the ionizing radiation. The method gives different results depending on the considered ion; in particular, the H I Zanstra temperature is almost always lower than the He II one, being the nebulae generally more optically thick to the radiation able to ionize the latter ion. Since in some cases the nebula is thin to He II ionizing radiation too, it appears clear how uncertain these temperature determinations may be. For these reasons, we have constructed two P-AGB stellar populations with completely identical premises, except that in one case we arificially introduce a 30% systematic error in the effective temperature. We extract synthetic P-AGB populations of 200 objects each, using a Salpeter's IMF, and Weidemann's (1987) initial mass-final mass relation. A $\sim 30\%$ error in temperature is rather common for the Zanstra method.

Figure 1 shows the effect of the temperature error on the resulting H-R diagram (left panel) and on the inferred mass distribution (righ panel). One can see how different the two mass distributions are; the average PNN mass is respectively

$< M >= 0.596 M_\odot$ for the *true* distributions and $< M >= 0.669 M_\odot$ for the *observed* one. We should nonetheless note that the peak in the mass distributions is *stable* against effective temperature errors, while only the high mass tail of the distribution is affected. This particular behavior follows from the low mass objects being predominantly on the horizontal ($L \simeq const$) part of the tracks, while higher mass PNNi are found predominatly in the fading sections of them. Therefore, temperature errors do not appreciably affect the inferred mass of the former, while do so for the latter ones. With the same line of reasoning we can also easily realize that errors in the PN distance scale will have the opposite effect. We conclude that this method of deriving the mass distribution of PNNi is prone to systematic (and random) errors in the effective temperature of individual objects, in particular for what concerns the high mass tail of the distribution.

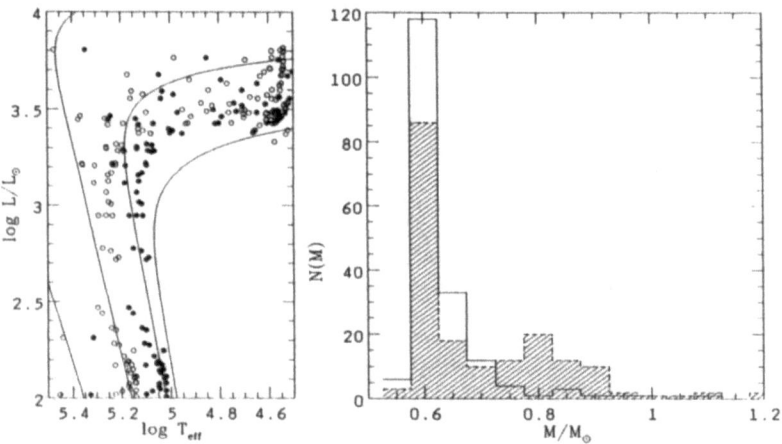

*Fig. 1. Left: P-AGB populations on the H-R diagram; filled circles: true distribution; open circles: observed distribution. The tracks correspond to $M/M_\odot = 0.55, 0.6, 0.8, 1.2$. Right: corresponding mass histograms; solid line: input or true distribution; broken line and shaded histogram: derived or observed distribution.*

In Figure 2 we show a further example of how systematic errors in the temperature scale can interfere with other aspects of the problem and potentially lead to erroneous conclusions. Figure 2 shows three mass histograms of synthetic populations that differ only for the initial mass-final mass relation. We can see that the *bimodality* in the mass distribution of Figure 1 referring to the case with temperature error included mimics that shown in Figure 2 when the initial mass-final mass relation adopted by Ciotti et al. (1991) is used instead of that of Weidemann (1987).

The last test concerns the problem of the AGB to PN transition time. It is well known that the planetary nebula formation depends on the delicate interplay between the time scales of the dynamical evolution of the shell and the evolution of the nucleus. The AGB-PN transition time is controlled by the exact value of the envelope mass ($M_e^R$) that survives the *superwind* envelope ejection at the tip of the AGB (Renzini 1989), a quantity that can hardly be theoretically predicted.

Recent observations associated to simple modelling (Käufl, Renzini, & Stanghellini

1992) have shown that on average the transition time is of the order of 1000 yrs or more, i.e., a non negligible fraction of the *nebular* time scale. In Figure 3 we show the effects of the adopted $M_e^R$ on the luminosity function of a systhetic distribution of PNNi, via the implied transition time. The first synthetic distribution is derived under the assumption of a constant residual envelope mass ($M_e^R = 10^{-3} M_\odot$), irrespective of the stellar mass; for the second simulation (following Paczyński 1971) we have adopted a linear $M_e^R(M)$ relation, with $M_e^R = 10^{-2}$ and $10^{-3} M_\odot$ respectively for $M = 1.2$ and $0.6$ $M_\odot$. The resulting luminosity functions are clearly very sensitive to the adopted transition times. We should caution however that for these simulations we have not taken into account the effects of the duration of the PN phase, which is also expected to be a function of the stellar mass. The luminosity function in Figure 3 referes to all P-AGB objects with luminosity down to Log $L = 1.5$, which in the case of low mass stars is reached in a time much longer than the PN phase. The effect of the duration of the PN phase will be thoroughly examined in our forthcoming paper, along with other explorations.

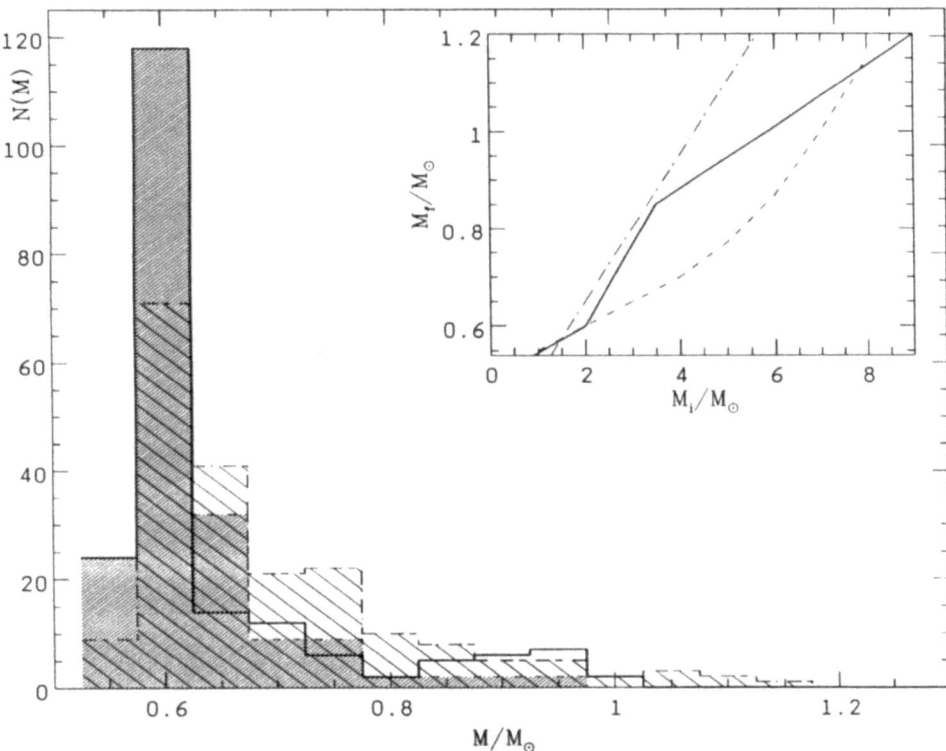

*Fig. 2. Effects of different initial mass-final mass relations on the PNN mass distribution. The initial mass-final mass relations (shown in the insert) are from Iben & Renzini (1983), with $\eta = 2$ and $b = 1$, dot-dash line and lightly shaded histogram, Ciotti et al. (1991), solid line and unshaded histogram, and Weidemann (1987), dashed line and heavily shaded histogram.*

In conclusion, errors in the effective temperature can produce strong variations in the inferred mass distribution, and can mask the effect of the initial mass-final mass relation. Also, current uncertainties affecting the AGB to PN transition times can have serious effects on the predicted luminosity function of PNNi.

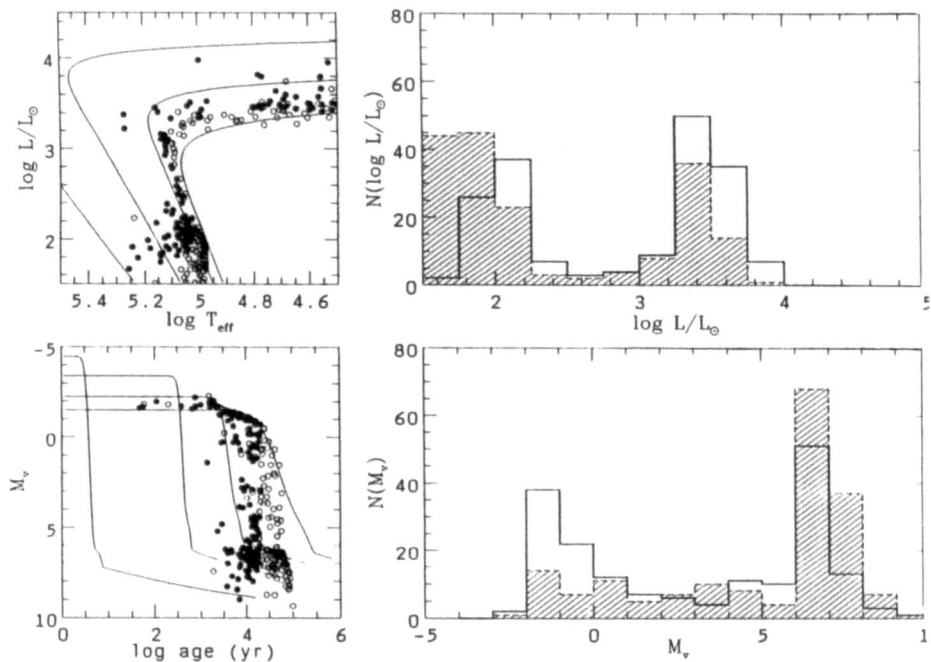

*Fig. 3. Transition time effects on luminosity distribution. Left panels: distribution of 200 PNNi on the H-R and $M_v - t$ diagrams; solid symbols: case A with $M_e^R = 10^{-3} M_\odot$; open symbols: case B with $M_e^R(M)$. Right panels: corresponding luminosity functions, with solid/unshaded histograms for case A, and dashed/shaded histogram for case B.*

### References

Acker A., 1989, in *Planetary Nebulae*, ed. A. Acker (Starsbourg Obs.), p. 77

Ciotti, L., D'Ercole, A., Pellegrini, S., & Renzini, A. 1991, ApJ, 376, 380

Käufl H. U., Renzini A., & Stanghellini L., 1992, ApJ *submitted*

Paczyński B., 1971, Acta Astron. 21, 417

Renzini A., 1989, in *Planetary Nebulae*, ed. S. Torres-Peimbert (Dordrecht: Kluwer), p. 391

Schönberner D., 1981, A&A 103, 119; 1983, ApJ 272, 708

Schönberner D. & Tylenda R., 1990, A&A 234, 439

Stasińska G., *et al.* 1992, in *Stellar Populations of Galaxies*, ed. B. Barbuy & A. Renzini (Dordrecht: Kluwer), p. 492

Tylenda R., Stasińska G., Acker A., & Stenholm B., 1991, A&A 246, 221

# FURTHER MODELS OF PLANETARY NEBULA SPECTRAL EVOLUTION

K. VOLK

*Department of Physics and Astronomy, University of Calgary, 2500 University Drive N.W., Calgary, Alberta, Canada T2N 1N4*

**Abstract.** Four new calculations of planetary nebulae spectral evolution are presented, as in Volk (1992). These models use the 0.64 $M_\odot$ central star evolution track of Schönberner (1983) but with the rate of evolution accelerated by a factor of 2, as the original models evolve to the planetary nebula phase too slowly to match the observations. Models were calculated for mass-loss rates of $2.1 \times 10^{-5}$ and $5.2 \times 10^{-6}$ $M_\odot$ yr$^{-1}$ using solar composition and silicate dust, and using the average observed planetary nebula composition and graphite dust. An interacting winds shell was assumed to form. The model results were combined with an assumed Galactic distribution of 25000 planetary nebula to simulate a variety of observables including V magnitudes, H$\beta$ fluxes, the IRAS colours, the 5 GHz radio flux densities, and the nebular radii.

The models show that even for the higher mass-loss rate the nebulae are matter-bounded during most of the high luminosity phase of the evolution. For the nebulae to stay ionization-bounded mass-loss rates of order $4 \times 10^{-5}$ are required for steady mass-loss. If there is no interacting winds shell or if the mass-loss rate is increasing with time this limit will be higher. The low mass-loss rate models do not produce typical planetary nebulae optical spectra because of extreme matter-bounded conditions. The carbon-rich nebulae are found to be considerably brighter in the infrared than the solar composition nebulae, which explains the observed preponderance of carbon-rich planetary nebulae and proto-planetary nebulae with IRAS low resolution spectra.

A 50%/50% mixture of carbon-rich and oxygen-rich planetary nebulae from the higher $\dot{M}$ model runs is able to reasonably match the IRAS photometry of planetary nebulae at 12, 60, and 100 $\mu$m but fails for 25 $\mu$m at low flux densities. Among other things, this implies that the rapid post-AGB luminosity decline for PPN and PN deduced by Knapp (1986) from CO observations is incorrect.

## References

Knapp, G. R. 1986 ApJ, **311**, 731
Schönberber, D. 1983, ApJ, **272**, 708
Volk, K. 1992, ApJSupp, **80**, 347

# EVOLUTION OF 1-5 $M_\odot$ STARS WITH MASS LOSS

E. VASSILIADIS and P.R. WOOD
*Mount Stromlo and Siding Spring Observatories, Private Bag, Weston Creek P.O., Canberra, ACT 2611, Australia*


Stars of mass 1-5 $M_\odot$ and composition Y=0.25 and Z=0.016 have been evolved from the main-sequence to the white dwarf stage with an empirical mass loss formula based on observations of mass loss rates in AGB stars. This mass loss formula (Wood 1990) causes the mass loss rate to rise exponentially with pulsation period on the AGB until *superwind* rates are achieved, where these rates correspond to radiation pressure driven mass loss rates. The formula was designed to reproduce the maximum periods observed for optically-visible LPVs and it also reproduces extremely well the maximum AGB luminosities observed in star clusters in the Magellanic Clouds (see Vassiliadis and Wood 1992 for details).

During AGB evolution, deep envelope convection exits in the 5 $M_\odot$ star while the envelope mass is > 1.6 $M_\odot$. The deep envelope convection has several consequence: (1) the hydrogen-exhausted core mass $M_c$ is effectively constant with time, since the increase in $M_c$ during the hydrogen-burning phases is eliminated at the next shell flash when almost all the burnt hydrogen is dredged up by envelope convection ($\lambda \approx 1$), (2) the burning of the helium shell at a shell flash is quenched very rapidly by the cooling effect of the deep envelope convection, so that the helium exhausted core mass does not increase significantly with time, and (3) shell flashes have very little effect on the surface luminosity because of the small amount of energy released by helium burning and the quick recovery of hydrogen burning.

When mass loss has reduced the envelope mass below ~1.6 $M_\odot$, convective penetration by the envelope is no longer deep and 'normal' helium shell flash behaviour returns - the core mass increases with time, the hydrogen and helium shells burn outward at the same average rate, and the surface luminosity varies by large amounts during the shell flash cycle.

The 5 $M_\odot$ star produces a 0.90 $M_\odot$ remnant which evolves through the planetary nebula region of the HR diagram very rapidly until log $L/L_\odot$ has declined to ~2.5 where the evolution begins to slow down, in agreement with the calculations of the evolution of massive PNN by Paczynski (1971) and Faulkner and Wood (1986). The results contrast with those of Blöcker and Schönberner (1990) who find a 0.836 $M_\odot$ remnant which is ~3 times more luminous at an age of $10^4$ years than similar models from our calculations.

**References**

Blöcker, T. and Schönberner, D. 1990, A&AL, 240, L11.
Paczynski, B. 1971, Acta Astr., 21, 417.
Vassiliadis, E. and Wood, P.R. 1992, in preparation.
Wood, P.R. 1990, in *From Miras to Planetary Nebulae*, eds. M.O. Mennessier and A. Omont (Editions Frontières), p. 67.
Wood, P.R. and Faulkner, D.J. 1986, ApJ, 307, 659.

# ON THE FADING OF AGB REMNANTS

T. BLÖCKER AND D. SCHÖNBERNER
*Institut für Theoretische Physik und Sternwarte, Kiel, FRG*

July 8, 1992

We investigated the question how the evolution of post-AGB models depends on their history, i.e. on their initial mass and AGB evolution. Therefore, we calculated the evolution of a 3 and $5 M_\odot$ star from the main sequence towards the stage of white dwarfs. These models suffered from 9 and 17 thermal pulses on the AGB, resp., and the common mass-loss law led to final masses of 0.61 and $0.84 M_\odot$, resp., which are consistent with reasonable initial-final mass relationships. It was found that more massive AGB remnants fade much more slower than hitherto assumed. Thus, we conclude that only a reliable combination of initial and final mass yields the right fading time scales for more massive post-AGB models. To prove that we have re-calculated the evolution of the $3 M_\odot$ model with another mass-loss law leading to 86 thermal pulses and a remnant mass of $0.84 M_\odot$, a combination which, however, does not comply with initial-final mass relations. Comparing now the post-AGB evolution of the two massive models of *equal* remnant mass ($0.84 M_\odot$) but *different* initial masses (3 and $5 M_\odot$, resp.) yields completely different fading time scales. Thus we confirm by direct calculations the suggestion of Blöcker and Schönberner (1990, A&A 240, L11) that not only the remnant mass but also the initial mass determines the time scales of more massive central stars.

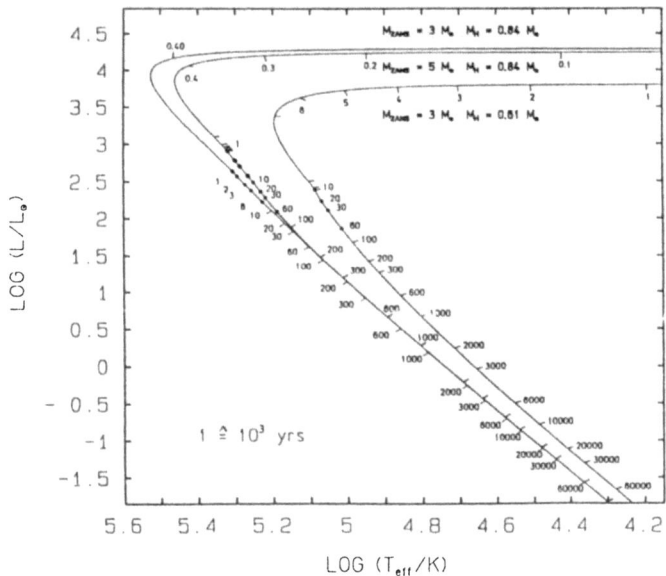

Fig. 1: Evolutionary tracks of three hydrogen-burning post-AGB models (pulse phase $\phi = 0.5$) of $(M_{\mathrm{ZAMS}}/M_\odot, M_{\mathrm{H}}/M_\odot) = (3, 0.605), (5, 0.836)$ and $(3, 0.836)$. Note that the latter combination of initial and final mass are not consistent with reasonable initial-final mass relationships. Timemarks are in units of $10^3$ yrs.

# PLANETARY NEBULA EVOLUTION TRACED BY DISTANCE-INDEPENDENT PARAMETERS

C.Y. ZHANG

*Department of Astronomy, The University of Texas at Austin, Austin, TX 78712*

and

S. KWOK

*Department of Physics and Astronomy, The University of Calgary, Canada*

ABSTRACT. Making use of the results from recent infrared and radio surveys of planetary nebulae, we have selected 431 nebulae to form a sample where a number of distance-independent parameters (e.g., $T_b$, $T_d$, $I_{60\mu m}$, and IRE) can be constructed. In addition, we also made use of other distance-independent parameters $n_e$ and $T_*$ where recent measurements are available. We have investigated the relationships among these parameters in the context of a coupled evolution model of the nebula and the central star. We find that most of the observed data in fact lie within the area covered by the model tracks, therefore lending strong support to the correctness of the model. Most interestingly, we find that the evolutionary tracks for nebulae with central stars of different core masses can be separated in a $T_b$-$T_*$ plane. This implies that the core masses and ages of the central stars can be determined completely independent of distance assumptions. The core masses and ages have been obtained for 302 central stars with previously determined central-star temperatures. We find that the mass distribution of the central stars strongly peaks at 0.6 $M_\odot$, with 66% of the sample having masses <0.64 $M_\odot$. The luminosities of the central stars are then derived from their positions in the HR diagram according to their core masses and central star temperatures. If this method of mass (and luminosity) determination turns out to be accurate, we can bypass the extremely unreliable estimates for distances, and will be able to derive other physical properties of planetary nebulae.

This work was supported by a grant to SK from the Natural Sciences and Engineering Research Council of Canada. CYZ also acknowledges support by the NASA grant NAG 2-67.

# EXCITATION CLASS OF NEBULAE AS AN EVOLUTION CRITERION

G.A. GURZADYAN and A.G. EGIKYAN
*Garny Space Astronomy Institute, Armenia*

A principally new, quantitative system of the classification of spectra of planetary nebulae is proposed. The excitation class of a nebula, p, is determined according to the relative intensities of emission lines, $N_1+N_2/4686$ HeII and $N_1+N_2/H\beta$. The excitation classes are obtained for 177 PN with known distances and sizes of all classes – low (p=1-3), middle (p=4-8) and high (p=9-12$^+$). An empirical relationship between the excitation class p and the mean radius of the nebulae $R_n$ is discovered – the largest sizes occur for highly excited nebulae, and the smallest for low excitation ones. This relationship as well as excitation class p, as an independent parameter admit an evolutionary interpretation (Gurzadyan, Egikyan 1991).
It is shown that after reaching the highest class of excitation p=12$^+$ some part of nebulae again decreases their class of excitation with the further increases of sizes. The general diagram of this relationship, p$\sim R_n$, has two nearly symetric branches – rising and descending with the apogee on p=12$^+$. The whole number of nebulae in the descending part of this diagramm is more than an order smaller than their number in the rising part of the diagram.
The distribution of the number of PN versus the excitation class shows that the "Marathone" of nebulae from p=1 to p=12$^+$ takes place with a rate too far to be constant: a nebula "lives" longer in the lowest excitation classes, p=1-3, after which very quickly gallops the classes p=4-6, and again the longer duration of life in classes p$\sim$9-11.
Moreover, there is also a tendancy that the mean expansion velocities increase when passing to high excitation nebulae.
The relationship p$\sim R_n$ admits an evolutionary interpretation, and can be taken to some degree as an analogon of the Hertzsprung-Russel diagram for planetary nebulae.
At present we can establish that PN may be of low excitation class, from 1 to 6, at their birth, and of classes from 12$^+$ up to 1-2 in their destruction or disappearing period.

### References

Gurzadyan G.A., Egikyan A.G. 1991, Astrophys.Space.Sci.,181,73

# MORPHOLOGY AND EVOLUTION OF PLANETARY NEBULAE

L. STANGHELLINI

*Osservatorio Astronomico, via Zamboni 33, I-40126 Bologna, Italy*

and

R. L. M. CORRADI and H. E. SCHWARZ

*European Southern Observatory, Casilla 19001, Santiago 19, Chile*

A large set of narrow-band images of planetary nebulae (PNe) have been studied together with the location of their nuclei (PNNi) on the $\log T_{\text{eff}} - \log L/L_\odot$ plane, in order to disclose possible correlations between the morphological class of the PNe and the evolutionary stage of their PNNi.

We have used a new set of about 180 H$\alpha$ images of galactic PNe, mostly taken with the NTT telescope at ESO (Schwarz *et al.* 1992, A&AS, *in press*). We used the He II Zanstra temperatures for the PNNi, and the Cahn *et al.* (1992, A&AS, *in press*) distance scale to derive the absolute luminosities.

Although the investigation is still at a preliminary stage, and the errors on distances and Zanstra temperatures do not allow sound conclusions, there seem to be differences on the distributions of PNNi hosted by the different morphological types (the morphological classification is described in Schwarz *et al.*, these proceedings). In particular, we have found that:

1. PNNi hosted by single and multiple shells PNe occupy different loci on the $\log T_{\text{eff}} - \log L/L_\odot$ plane, these differences are likely to be related to different evolutionary paths (Iben *et al.* 1983, ApJ 264, 605);
2. bipolar PNe host nuclei that are likely to have different mass distribution than those hosted by elliptical PNe;
3. PNe of the different morphological types form a optical thickness sequence that goes from the thick stellar to the thin elliptical PNe, going through bipolar and pointsymmetric nebulae;

These and other findings reveal possible differences in the evolutionary paths of central stars of nebulae with different morphologies.

# INFLUENCE OF THE STELLAR WINDS ON THE EVOLUTION OF THE PLANETARY NEBULA NUCLEI.

S.K.GÓRNY

*Nicolaus Copernicus Astronomical Center, Chopina 12/18, 87-100 Toruń, POLAND*

A grid of homogeneous models of evolution of hydrogen burning planetary nebulae nuclei, assuming different stellar winds and the zero points for the post-AGB evolution, have been constructed from original Schönberners tracks. Following a simplified line-driven wind theory the mass loss rate has been adopted to be

$$\dot{M} = \alpha \frac{L}{c \cdot V_\infty}$$

with $\alpha = 0.50, 0.25, 0.12, 0.06$ and $V_\infty$ changing from $V_\infty = 1 \cdot V_{esc}$ at $T_{eff} = 10^{3.6}$K to $V_\infty = 4 \cdot V_{esc}$ at $T_{eff} = 10^{5.1}$K. The zero points for evolution ("superwind" cessation) have been chosen at the moments when the fundamental mode period of pulsations is $P_0 = 100, 75, 50, 25$ days.

Using a simple model of a coupled evolution of a planetary nebula and its central star the observational characteristics and the various diagrams usually used for investigation of the planetaries have been calculated (e.g. $M_V - t_{exp}$ - see figures).

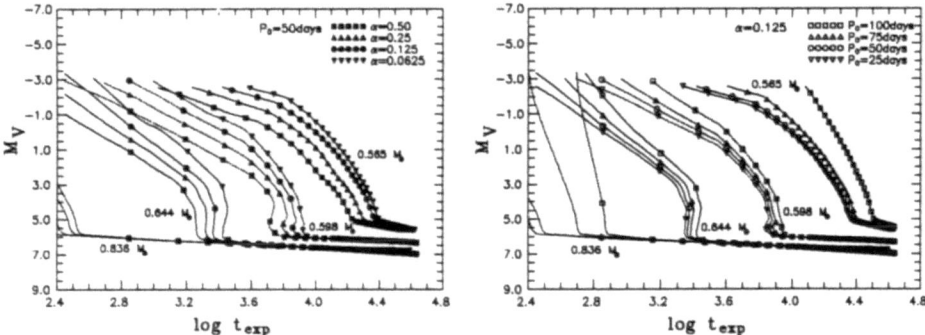

Figure: *The influence of stellar winds (left) and starting points (right) on the PNN evolution.*

From diagrams it can be seen that modifications connected with different stellar winds are generally more significant than the moment of "superwind" cessation. The zero point is important only if the "superwind" ends at low temperatures but even for this case a high mass loss may significantly speed up the evolution.

It is worth to notice that although the mass loss and the zero points have substantially been modified the new time scale of evolution for each nuclei of a given mass is generally in agreement with already existing models.

The sensitivity of the tracks positions to the adopted wind scenario has been investigated on the observational diagrams constructed applying the standard Zanstra and Shklovski methods for deriving stellar parameters. An extensive discussion of the problem will be published in near future.

# DETECTION OF EVOLUTION OF THE NUCLEUS OF NGC 2392

SARA R. HEAP

*Laboratory for Astronomy and Solar Physics, Code 681, Goddard Space Flight Center, Greenbelt MD 20771, U.S.A*

**Abstract.** Ultraviolet spectra taken with the IUE satellite indicate that the central star of the planetary nebula, NGC 2392, has declined in brightness by 7% per decade. We interpret this fading as a consequence of evolution toward higher temperature. With this interpretation, the rate of optical fading suggests that the stellar mass is 0.73 $M_\odot$.

Recently, two independent studies concluded that the nucleus of NGC 2392 is a relatively massive central star. From a non-LTE spectroscopic analysis, McCarthy *et al.* (1990) derived a stellar mass, $M = 0.77$ (+.10,-.03) $M_\odot$. In the second study, Pauldrach *et al.* (1988) showed that the anomalously low terminal velocity of the wind from the central star ($V_\infty$=500 km s$^{-1}$, Heap 1986) can be matched by theoretical models only if the star has a mass as high as $0.87 \pm 0.15$ $M_\odot$.

Stars this massive should evolve to higher temperatures quite rapidly. Blöcker and Schönberner's (1990) evolutionary models indicate that the central star should increase in temperature by 243 °K per year if $M$=0.77 $M_\odot$ and 928 °K per year if $M$=0.87 $M_\odot$. Are such changes detectable by observation, and if so, how? Looking for direct signs of a temperature increase is a relatively coarse approach. A more sensitive test is optical fading. Since a central star evolves to higher temperatures at a constant bolometric luminosity, a larger fraction of its luminosity is emitted in the unobservable extreme ultraviolet, and a smaller fraction is emitted in the optical region of the spectrum.

Unfortunately, published estimates of stellar magnitude have been obtained in a variety of ways and are thus, not inter-comparable. Comparing ultraviolet fluxes measured by the IUE satellite is presently the most reliable way to detect optical fading in the central star of NGC 2392. The sensitivity of the short-wave spectrograph has been relatively stable and is well calibrated over its 14-year lifetime. IUE observations of NGC 2392 indicate that the flux from the central star has decreased by about 7% per decade in the wavelength interval, 1400-1900Å. The decline shows no significant wavelength dependence, which is to be expected, since this spectral region is on the Rayleigh-Jeans tail of the blackbody curve. The decline in apparent brightness over a decade corresponds to a 0.73 $M_\odot$ star.

## References

Blöcker, T. & Schönberner, D. 1990 A & A, **240**, L11
Heap, S. R.: 1986 ESA SP-263, p. 291
McCarthy, J. *et al.*: 1990 Ap.J., **351**, 230
Pauldrach, A. *et al.*: 1988 A & A, **207**, 123

# A SEARCH FOR OPTICAL-UV FADING OF CENTRAL STARS

B. ALTNER
Applied Research Corporation (ARC), Landover, MD USA
and
S.R. HEAP
Laboratory for Astronomy and Solar Physics
NASA Goddard Space Flight Center, Greenbelt, MD USA

**Abstract.** We derive estimates of the masses of planetary nebulae central stars (CSPN), based on their rate of optical-UV fading.

As a central star evolves to higher temperatures, a decreasing fraction of its light is emitted in the optical-UV region of the spectrum. The rate of fading is extremely sensitive to the stellar mass. After reprocessing early IUE spectra and correcting later spectra for the known rate of camera sensitivity degradation (Garhart 1992, *IUE Newsletter*, **47**), we are able to detect evidence of fading in a number of CSPN at a level greater than the minimum detection threshold ($\sim 4\%$). The first five columns in the following Table list the star and its type, photometric variability characteristics (Bond 1992, priv. comm.), temperature (in kK), and number of IUE spectra used. The sixth column lists the the mean rate of fading in selected wavelength bands free of nebular emission and camera defects. The last two columns show the corresponding stellar masses, calculated by interpolating within the hydrogen-burning sequences of Wood and Faulkner (1986, *Ap. J.*, **307**, 659: W & F) and Blöcker and Schönberner (1990, *Astron. Astrophys.*, **240**, L11: B & S), respectively.

| Central Star | Type | Var? | $T_3$ | $N_{SWP}$ | Fading Rate (%/decade) | Mass ($M_\odot$) W & F | Mass ($M_\odot$) B & S |
|---|---|---|---|---|---|---|---|
| BD +30 3639 | WC9 | — | 26: | 4 | $+3.3 \pm 1.6$ | — | — |
| He 2-131 | O7f-eq | — | 27 | 2 | $-4.1 \pm 2.4$ | $0.74^{+0.05}_{-0.05}$ | $0.71^{+0.03}_{-0.05}$ |
| He 2-138 | BC OIa | — | 27 | 3 | $-8.1 \pm 1.2$ | $0.78^{+0.04}_{-0.04}$ | $0.74^{+0.02}_{-0.02}$ |
| NGC 40 | WC8 | Irreg. | 31: | 3 | $-4.1 \pm 1.5$ | $0.73^{+0.02}_{-0.03}$ | $0.70^{+0.02}_{-0.03}$ |
| IC 418 | O6f | " | 36 | 5 | $-8.9 \pm 3.3$ | $0.77^{+0.03}_{-0.03}$ | $0.73^{+0.02}_{-0.03}$ |
| IC 4593 | O7f | Irreg. | 40 | 4 | $-8.5 \pm 1.1$ | $0.76^{+0.01}_{-0.01}$ | $0.72^{+0.01}_{-0.01}$ |
| NGC 2392 | O6f | Const. | 47 | 6 | $-5.5 \pm 1.5$ | $0.74^{+0.02}_{-0.02}$ | $0.70^{+0.01}_{-0.02}$ |
| NGC 6891 | O3f | Irreg. | 50 | 5 | $-1.3 \pm 1.3$ | — | — |
| NGC 6826 | O4f | " | 50 | 4 | $-0.2 \pm 4.5$ | — | — |
| IC 3568 | O5f | " | 50 | 3 | $-0.3 \pm 1.8$ | — | — |
| NGC 1535 | O3 | Irreg? | 70 | 3 | $-0.6 \pm 1.3$ | — | — |
| NGC 1360 | sdO | Const? | 80 | 7 | $-1.1 \pm 1.5$ | — | — |
| NGC 7009 | cont. | — | 82 | 6 | $-2.1 \pm 2.2$ | — | — |
| NGC 4361 | sdO | — | 82 | 3 | $+2.1 \pm 1.1$ | — | — |
| NGC 7293 | wD | " | 90 | 8 | $-1.5 \pm 1.1$ | — | — |
| NGC 6853 | wD | " | >100 | 2 | $-0.6 \pm 2.5$ | — | — |
| NGC 246 | C3/OVI | Const. | >130 | 4 | $-3.5 \pm 3.8$ | — | — |
| K1-16 | PG 1159 | Puls. | >150 | 10 | $+4.9 \pm 1.9$ | — | — |

# THE CENTRAL STARS OF He 2-131 AND He 2-138: PHOTOMETRIC VARIATIONS

R. G. HUTTON

*Observatorio Félix Aguilar, Univ. de San Juan, Argentina, and Yale Southern Observatory, San Juan, Argentina*

and

R. H. MÉNDEZ

*University Observatory, Munich, Federal Republic of Germany*

The central stars of the planetary nebulae He 2-131 and He 2-138 show variations in their visual magnitudes, with amplitudes of about 0.1 mag. and time scales of a few hours. This behavior appears to be very similar to that exhibited by the central stars of IC 418 and IC 4593. These four central stars have several other characteristics in common: a relatively low effective temperature, between 27000 and 40000 K; clear spectroscopic evidences of mass loss, both in the ultraviolet (IUE) and visible spectral regions; and short-term spectroscopic variability, in the form of radial velocity variations and/or of substantial changes in emission and P-Cygni-type line profiles. None of these central stars has shown convincing evidence of binarity; we attribute their behavior to variations in the stellar winds.

It is interesting to contrast these variations against those exhibited by massive early-type stars. These massive and luminous counterparts are well known to show spectroscopic wind variations with similar time scales. However, at least sometimes these variations are not accompanied by photometric variations in the Paschen continuum. The LBV's (luminous blue variables) do show photometric variations of similar amplitudes, but apparently with much longer time scales. Since the physical mechanism responsible for the variations is not unambiguously identified, the photometric variations in central stars of planetary nebulae may become a very useful constraint for the hydrodynamic modeling of hot star winds. A satisfactory theory must explain the photometric behavior of all hot stars, massive or not.

Suggestions of observational programs which may help to better define the problem are, for example, to verify the existence or absence of similar variations in low-temperature central stars with weaker winds, and in central stars with surface temperatures in the range from 60000 to 100000 K. Multi-color photometry would permit to check if there are detectable temperature variations in the cooler variables, like He 2-138.

One of the main difficulties in this kind of studies is the lack of observations covering a sufficiently long, uninterrupted time interval. Observations with the "whole Earth telescope" or similar consortia might be useful to learn more about the time behavior of the variations. Of course it would be desirable to obtain, simultaneously, high-resolution spectrograms; this complement would be very difficult to organize, but it would provide important information for a better description of the phenomenon.

# TIME-RESOLVED CCD-PHOTOMETRY OF PLANETARY NEBULA NUCLEI

M. M. ROTH, T. SOFFNER and W. MITSCH
Universitäts-Sternwarte München, Scheinerstr. 1, 8000 München 80, FRG

The enormous potential of direct imaging CCDs with respect to stellar photometry was soon discovered after these detectors became available for standard instrumentation at modern telescopes. Among favourable properties like high quantum efficiency, linearity, and large dynamic range the multiplexing advantage of the 2-dimensional detector has permitted to perform quantitative work that otherwise would have been impossible with classical photoelectric photometry. Striking examples for this are the progress in constructing colour magnitude diagrams for globular clusters (see e.g. Hesser et al. 1987, their Fig.9) or the high precision (down to the m-mag level) achieved in differential photometry on ensembles of stars (Gilliland and Brown 1988). Several groups have also used CCDs for time series measurements employing specialized instruments and/or proper observing strategies (Dunham et al. 1985, Howell and Jacoby 1986, Stover and Allen 1987). If high time resolution (on the order of seconds or less) is required, however, the photoelectric method is still superior to CCDs mainly because of problems like data rates and readout time overhead (Barwig 1987).

Motivated by the participation in a global photometric campaign on the central star of NGC1501 (Bond et al., 1992) and by specific interest from two working groups studying stellar atmospheres of hot stars and cataclysmic variables we have started to experiment with time-resolved photometry using a CCD camera system which has lately become available for the 0.8m telescope on Wendelstein Observatory (Roth 1990, Roth et al. 1991).

We present three examples for preliminary results that have been obtained recently. Despite the disadvantages mentioned above the CCD is particularily useful for measuring stars embedded in crowded fields or nebulosities where aperture photometry is likely to fail.

## References

Barwig, H., Schoembs, R., Buckenmayer, C., 1987, A&A **175**, 327
Bond, H.E. et al., 1992, in preparation
Dunham, E.W. et al., 1985, PASP **97**, 1196
Gilliand, R.L. and Brown, T.M., 1988, PASP **100**, 754
Hesser, J.E. et al., 1987, PASP, **99**, 739
Howell, S.B. and Jacoby, G.H., 1986, PASP **98**, 802
Roth, M.M., 1990, in: **CCDs in Astronomy**, ed. G.H. Jacoby, ASP Conf. Ser., p. 380
Roth, M.M., Soffner, T., Hoffmüller, J., 1991, in: Proc. **3rd ESO/ST-ECF Data Analysis Workshop**, ed. P.J. Grosbøl and R.H. Warmels, p. 107
Stover, R.J. and Allen, S.A., 1987, PASP **99**, 877

# VARIABLE SPECTRA OF IC 4997 AND NGC 6572

SIEK HYUNG and LAWRENCE H. ALLER

*University of California, 8979 Math Science bldg., 405 Hilgard Ave.*
*Los Angeles, California 90024-1562*

and

WALTER A. FEIBELMAN

*NASA/Goddard Space Flight Center, Code 684, Greenbelt, Maryland 2077*

Variability of the [OIII]4363/4340 H $\gamma$ ratio in IC 4007 was established in 1956 by William Liller and L.H. Aller who attributed the changes to a gradual decrease of electron density with time. Subsequent 4363/4340 ratio fluctuations negated this explanation. Ferland pointed out that small changes in the radiative flux of the Planetary nebula nucleus (PNN) could explains the variations. Our pervious study emphasized IUE observations, here we compare high dispersion spectra obtained with the Hamilton Echelle Spectrograph with previous measurements to asses line intensity variations. Emission line variability in PNN spectra as noted by Mendez *et al.* (1988) and by other for HeII 4686 in NGC 6572 may offer significant clues. PNN 4686 appeared by 1990 in IC 4997. Possibly both of these PNN may be evolving into Wolf-Rayet objects, but this development does not necessarily imply that the nebular excitation will increase with time.

Attempts to interpret these nebulae with the aid of theoretical models encountered troubles. Many trials for IC 4997 resulted in a model consisting of a dense inner shell ($N_H \sim 10^7$ atoms/$cm^3$) of radius 0.0005 pc and outer radius 0.000565 pc surrounded by a shell of lower density ($N_H \sim 10,000$ atoms/$cm^3$) with an outer radius of 0.013 pc. This model, however, encounters a paradox in that with an expansion velocity of 20 $km/sec$, the inner shell would have left the photosphere of the PNN about 30 years ago. These is no evidence for any unusual events in the star and nebula at that time. If a greater distance is assumed for IC 4997, the date of ejection can be pushed backward but then difficulties are encountered if we try to reproduce the observed nebular line intensities.

For NGC 6572 a toroid of density $N_H = 25,000$ atoms/$cm^3$, inner radius of 0.01pc and thickness $\sim 0.014$ pc was combined with a conical segment of density $N_H = 10,000$ atoms/$cm^3$, inner radius 0.04 pc, and thickness 0.0102 pc. T(PNN)= 55,000 and 50,000 for IC 4997 and NGC 6572 respectively. While we cannot assert that long term changes are taking place in either of these PNs, their continued monitoring to check on spectral variations is recommended. Gurzadyan has suggested that the [OIII] 4363/H$\gamma$ ratio varies with a 50 year cycle upon which smaller fluctuations are superposed, but the most recent (1992) ratio measurements do not seem to support this conclusion.

# A SEARCH FOR PULSATIONS IN O VI PLANETARY NUCLEI

HOWARD E. BOND

Space Telescope Science Institute
Baltimore, Maryland USA

and

ROBIN CIARDULLO

Pennsylvania State University
University Park, Pennsylvania USA

The first two pulsating central stars of planetary nebulae to be discovered were those of K 1-16 (Grauer & Bond 1984) and Lo 4 (Bond & Meakes 1990). They are nonradial, multiperiodic $g$-mode pulsators, with typical periods near 25–31 min. They are O VI nuclei or related objects, with extremely high temperatures ($T_{\mathrm{eff}} \gtrsim 100,000$ K), hydrogen deficiency, and high abundances of C and O.

We have used CCD time-series photometry to search for pulsational variability in 20 additional planetary nuclei with O VI or "PG 1159"-type spectra, using the 0.9-m and 1.5-m telescopes at KPNO and CTIO. Four new pulsators have been discovered and observed more intensively: NGC 1501, 2371-2, and 6905, and Sanduleak 3. A few details are given below.

**NGC 1501** shows pulsation amplitudes of up to 0.1 mag (peak-to-peak). Power spectra from four observing runs show considerable changes in the mode structure. (Such changes in pulsation amplitudes and frequencies, on time scales of a few months, appear to be a general property of pulsating PNNs.) A 1524-sec (25.4-min) mode was present during all four runs. **NGC 2371-2** showed very low-amplitude variations (if any) in October 1989, but obvious pulsations (amplitude up to $\sim$ 0.07 mag) in April 1990. The strongest mode in the April 1990 data is at a period of 983 sec (16.4 min). **NGC 6905** shows pulsation amplitudes of up to $\sim$0.1 mag. Power spectra calculated from data taken only 4 months apart are very different. The strongest pulsation modes have periods of 875 sec (14.6 min) and 710 sec (11.8 min). **Sanduleak 3** is a 13th-mag field star, classified as a "WO"-type Wolf-Rayet star. Our discovery of pulsations similar to those of O VI nuclei establishes Sand 3 as a low-mass pre-white dwarf, rather than a high-mass W-R star. The power spectra show a rich mode spectrum, dominated by a peak at 932 sec (15.5 min).

Pulsations have not been detected in the following O VI or PG 1159 planetary nuclei: NGC 246, 2452, 2867, 5189, 5315, and 6751; IC 1747, Abell 30 and 78, Ba 1, He 2-55, IW 1, Jn 1, M 3-30, PB 6, and VV 47.

### References

Bond, H.E., and Meakes, M.G. 1990, AJ, 100, 788
Grauer, A.D., and Bond, H.E. 1984, ApJ, 277, 211

# GLOBAL PHOTOMETRIC CAMPAIGNS ON PULSATING PLANETARY NUCLEI

ROBIN CIARDULLO
Pennsylvania State University
University Park, Pennsylvania USA

and

HOWARD E. BOND
Space Telescope Science Institute
Baltimore, Maryland USA

As reported in a contributed paper at this symposium, six "O VI"-type planetary nuclei are known to be pulsating variables. Through the techniques of asteroseismology, it should be possible to explore the interior structures of these stars, determine their masses and rotation rates, and measure the stellar evolutionary timescales through observations of changes in the pulsation periods.

In 1991 November and 1992 May, we and worldwide networks of collaborators carried out CCD photometry of the pulsating nuclei of NGC 1501 and Sanduleak 3, respectively, for continuous 10-day intervals.

The data analysis for both campaigns is still underway, so this paper represents a progress report.

Collaborators for the NGC 1501 campaign (1991 November 14-27) were as follows: **Kitt Peak National Observatory 0.9-m:** H. E. Bond; **Lick Observatory 1.0-m:** R. Stover; **Nishiharima Observatory 0.6-m:** T. Kuroda, T. Ishida, et al.; **Okayama Observatory 0.9-m:** H. Malasan, S. Tamura, A. Yamasaki, E. Kambe, et al.; **Ouda Observatory 0.6-m:** T. Kato, R. Hirata, et al.; **Beijing Observatory 2.2-m:** J.-S. Chen; **Wise Observatory 1.0-m:** E. Leibowitz; **Wendelstein Observatory 0.8-m:** M. Roth, T. Soffner; **Data Reduction:** R. Ciardullo; **Theory:** S. Kawaler (Iowa State University).

Due to the favorably high declination of NGC 1501, the coverage was good: 218 hours over the interval 1991 Nov 14.5-27.5 (70% coverage).

The light curve of the central star was very irregular, with pulsation amplitudes of up to $\sim 0.1$ mag (peak-to-peak), but usually smaller. The power spectra are dominated by two peaks at 0.758 and 0.866 mHz (periods of 1318 and 1154 sec), but there are many lower-amplitude modes as well. The amplitude of the 1154-sec mode appears to vary with a period of $\sim$8-9 days, suggesting it is composed of at least two closely spaced components. Data reduction and analysis for the complete set of observations are underway.

Collaborators for the Sanduleak 3 campaign (1992 May 4-15) were as follows: **Cerro Tololo Interamerican Observatory 0.9-m:** H. E. Bond; **Mt. John Observatory 1.0-m:** D. Sullivan, W. Tobin, M. Clark, J. Pritchard; **Kavalur Observatory 1.0-m:** M. Parthasarathy; **South African Astr. Obs. 1.0-m:** D. Buckley; **Data Reduction:** R. Ciardullo; **Theory:** S. Kawaler.

During 1992 May 4.5-15.5, we obtained 155.5 hours of data, or 59% coverage. Detailed analysis of the data was just beginning at this writing.

# PHOTOELECTRIC PHOTOMETRY OF FIVE PNNi

R. SILVOTTI, C. BARTOLINI, F.R. BOFFI, G. COSENTINO, A. GUARNIERI and A. PICCIONI

*Dipartimento di Astronomia, Università di Bologna, via Zamboni 33, I-40126 Bologna, Italy*

and

L. STANGHELLINI

*Osservatorio Astronomico di Bologna, via Zamboni 33, I-40126 Bologna, Italy*

In september 1991 we started a photoelectric monitoring of O VI PNNi that are candidate to be non radial pulsators. The observations were obtained using the two head photometer described by Piccioni et al. (1979, Acta Astron. 29,463), mounted on the 1.5 m telescope of the Bologna Observatory. We generally used the B and sometimes the U filter; only BA 1 was always observed without filters because of its faintness. The reductions have been pursued by means of a program that yields the moving averages of the counts of the variable and comparison star. For the search of periodicities the methods of Deeming (Kurtz 1985, MNRAS 213,773) and Scargle (1982,ApJ 263,835) were principally used. The list of observations is reported in the following Table, where $\sigma$ is the mean value of the standard deviations obtained by computing the moving averages.

### Log of observations

| PNNi | nights | lenght | ampl. | $\sigma$ | princ. periods |
|---|---|---|---|---|---|
| BA 1 | 2 | $4^h.1$ | $0^m.4$ | $0^m.05$ | $23.3^{min}$, 18.4 |
| IC 2003 | 4 | 8.6 | 0.02 | 0.01 | 35.1, 16.0 |
| NGC 1501 | 1 | 2.0 | 0.05 | 0.01 | 2.9, 31.9 |
| NGC 6905 | 3 | 9.3 | 0.04 | 0.02 | 12.9, 17.6 |
| NGC 7026 | 4 | 8.6 | 0.01 | 0.01 | 22.6, 8.9 |

**BA 1** Beside to the quoted periods, other high frequency pulsations at 7.2,8.5,9.4 and 11.6 min have been detected in both nights.
**IC 2003** The periodicities at 35.1, 36.9, 29.4 and 16.0 min are present in the periodogram of three close nights. Several other oscillations with periods from 120 to 780 sec, registered in at least two nights, confirm the variability of this object.
**NGC 1501** Further peaks have been detected at 9.3, 7.1, 11.2 and 18.4 min.
**NGC 6905** The periodicities at $(772\pm6)$ and $(1058\pm15)$ sec are present in all three nights. A very prominent peak at 888 sec appears in the U observation of september 18 and also on september 6, with remarkable precision. Other low frequency pulsations seems to be real at 22.9, 23.8, 28.8 and 36.3 min. Finally we have noted a number of coincidences among different nights at about 193, 331, 435, 486, 551, 641 and 702 sec. The results confirm the variability of NGC 6905, announced by Bond and Ciardullo (1990, in "Confrontation between stellar pulsation and evolution" C. Cacciari and G. Clementini (eds.) A.S.P. Conf. Series vol.11, p. 529).
**NGC 7026** The results are uncertain, we can not rule out variations with amplitude of one hundredth of magnitude or less.

# O VI CENTRAL STARS OF PLANETARY NEBULAE: NGC 2371

L. STANGHELLINI

*Osservatorio Astronomico di Bologna, via Zamboni 33, I-40126 Bologna, Italy*

and

J.B. KALER

*Dept. of Astronomy, University of Illinois, 1002 W. Green Street, Urbana, IL 61801, USA*

and

R.A. SHAW

*Space Telescope Science Institute, 3700 San Martin Drive, Baltimore, MD 21218, USA*

We performed detailed spectral analysis of the planetary nebula NGC 2371 and its nucleus. The central star of NGC 2371 is a member of the O VI PNNi class, and it shows luminosity variations (Bond & Ciardullo 1990, ASP Conf. Ser. vol 11, *Confrontation between Stellar Pulsation and Evolution*, C. Cacciari & G. Clementini (eds.), p. 529) that could be associated to nonradial pulsations. From the spectrum of NGC 2371 we calculate the nebular parameters and the abundance of the most prominent ions. The nucleus is hot ($T_{\text{eff}} \geq 120,000$K) and luminous (V=15.4), it is probably close to the blue *bend* of the post-AGB evolutionary sequence, and shows prominent O VI emission with (so far) unique double narrow-broad feature, as shown in the Figures. This feature can not be related to the shock front of the fast wind into the planetary nebula. This star has been analyzed together with the other O VI PNNi known. A preliminary correlation between the stellar parameters and the total O VI strength have been derived, and the locus of NGC 2371 on the $\log T_{\text{eff}} - \log L/L_\odot$ plane have been studied in relation to the locus of the post-AGB nonradial instability strips. These results, together with a study on the other O VI central stars of planetary nebulae, will be published in the near future.

# PRECISION ASTEROSEISMOLOGY OF PRE-WHITE DWARFS AND PN CENTRAL STARS

STEVEN D. KAWALER

*Department of Physics and Astronomy, Iowa State University, Ames IA 50011 USA*

and

PAUL A. BRADLEY

*Department of Astronomy, University of Texas, Austin TX 78712 USA*

The nonradially pulsating GW Vir (a.k.a. DOV, or pulsating PG1159) stars and the K1-16 type variable stars provide an unparalleled opportunity to probe the internal structure of planetary nebula central stars and their immediate descendants, hot pre-white dwarfs. Recent global photometry campaigns have provided data of sufficient quality to severely constrain evolutionary models of these stars.

This paper reports on a sequence of new structural models of pre-white dwarfs to explore their seismological properties. In the asymptotic limit, the periods of $g$-modes with consecutive radial overtone number are spaced equally in period. The value of this period spacing is a property of the equilibrium structure of the star (Kawaler 1991 and references therein). We explored how the period spacings of models in our grid depend on the stellar mass, luminosity, and helium layer thickness. The mean spacing depends principally on mass, with much weaker dependence on luminosity and surface helium layer thickness. We also explored how mode trapping at the He/C-O composition transition zone causes regular departures from uniform period spacing.

Observations of the pulsation spectrum of PG 1159 -035 with the Whole Earth Telescope (Winget et al. 1991) provide rich detail on the $g$-modes in this hot pre-white dwarf star. Using our models, the best fit to all the periods seen in PG1159 is for a model mass of $0.58 \pm 0.01 M_\odot$, $L = 200 L_\odot$, $T_{eff} = 143,000K$, $M_{helium} = 0.003 M_*$, and $Y_{surf} = 0.30$. Analysis is currently in progress on W.E.T. data on the DOV stars PG 1707 and PG 2131, and on the CCD network data on NGC1501 and Sand 3 (Bond and Ciardullo, these proceedings). A paper in preparation (Kawaler & Bradley 1993) extends this work considerably using truly evolutionary models, and discusses more fully the implications for the formation and future evolution of pre-white dwarf stars.

### References

Kawaler, S.D. (1991), C. Cacciari and G. Clementini (eds.), *Confrontation Between Stellar Evolution and Pulsation*, (PASP Conference Series: Provo) pp. 494-512.

Kawaler, S.D. and Bradley, P.A. (1993), in preparation.

Winget, D., et al. (1991), Ap. J., **378**, 326.

# A SPECTACULAR MASS-LOSS EVENT OF THE CENTRAL STAR OF LONGMORE 4

K. WERNER, W.-R. HAMANN, U. HEBER[1], R. NAPIWOTZKI, T. RAUCH, U. WESSOLOWSKI

*Institut für Theoretische Physik und Sternwarte, Kiel, FRG*

and

[1] *Dr.-Remeis-Sternwarte Bamberg, FRG*

The nucleus of the PN Lo4 is very hot and H-deficient and may be classified as a PG 1159 star (Werner 1992). Besides K 1-16, it is the second known pulsating central star. New spectra were taken between May 1991 and Febr. 1992. Within 18 days we witnessed a rapid decline of an emission line phase, which has begun less than 7 months ago. During this time Lo4 has changed its spectral type from PG 1159 to early WC (WC2 or WC3) and back to PG 1159. This phenomenon has never been observed before in hot post-AGB stars. The event is interpreted as an intrinsic phenomenon. Because of the short time scales, the variations are probably confined to the outermost stellar layers. It is known that the pulsation driving zones are close to the stellar surface and we speculate about a causal relation between enhanced mass-loss and possible variations in the pulsation behaviour. Spectral analysis was performed using NLTE model codes for spherically expanding atmospheres (Hamann et al. 1991) to analyze the WC-type spectrum and for plane-parallel static atmospheres (Werner 1992) to analyze the PG 1159 type spectrum. We find $T_{\rm eff}$=120kK and $\log g$=5.5 during the PG 1159 phase. For the WC phase we obtain the same $T_{\rm eff}$ and the mass-loss rate $\log (\dot{M}/(M_\odot/{\rm yr}))$=-7.3. The element abundances are typical for PG 1159 stars (He:C:O=46:43:11, by mass) and $v_\infty$=4000 km/s. More details can be found in Werner et al. (1992).

Hamann, W.-R., Koesterke, L., Wessolowski, U. 1991, NATO ASI Series C, 341, 69
Werner, K. 1992, Lecture Notes in Physics 401, Springer, Berlin, p. 273
Werner, K., Hamann, W.-R., Heber, U., Napiwotzki, R., Rauch, T.,
    Wessolowski, U. 1992, A&A 259, L69

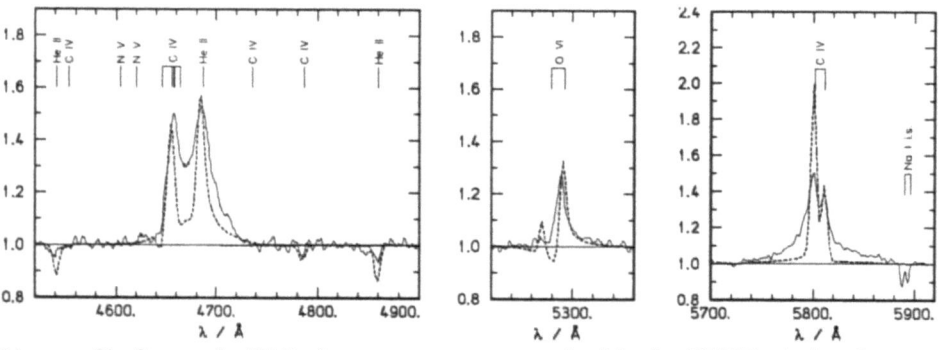

Line profile fits to the WC-phase spectra computed with the NLTE wind-code

# DISCOVERY OF A PLANETARY NEBULA ASSOCIATED WITH THE WHITE DWARF GD 561

R. NAPIWOTZKI and D. SCHÖNBERNER
*Institut für theoretische Physik und Sternwarte der Universität, Olshausenstr. 40,
W-2300 Kiel, Germany*

The search for old planetary nebulae around hot white dwarfs, PG 1159 stars, and sdO stars is subject of intensive studies (e.g. Kwitter et al. 1989). We report here on the discovery of such a nebula around the white dwarf GD 561.

GD 561 (WD 2342+806) was discovered by Giclas et al. (1970). It was classified as DAO by Bergeron et al. 1992. A comparison of a spectrogram of GD 561, taken by us, with the DAO central stars of S 216 and NGC 7293 shows striking similarity. From a preliminary analysis of the Balmer lines we got $T_{\text{eff}} = 50,000\,\text{K} \ldots 60,000\,\text{K}$. But according to the discussion in Napiwotzki (1992 and these proceedings) this determination needs further scrutiny.

In the background of our exposure the forbidden O III lines at 4959 Å and 5007 Å, typical for a PN, are clearly visible. The associated nebula was listed before in the catalogue of Sharpless (1959) as H II region (S 174). The distance of S 174 was determined by Fich & Blitz (1984) from its radial velocity using the galactic rotation curve to be 220 pc. The distance is in agreement with the estimated distance of GD 561. The resulting radius (0.3 pc) is consistent with the interpretation of S 174 as beeing an old PN. The position of GD 561 is not at the center but near the outer rim of the PN. But this is not unexpected: the central stars of very old PN are very often off-center. We conclude that GD 561 is the central star of S 174.

*References*
Bergeron, P., Saffer, R.A., Liebert, J., 1992, ApJ, in press
Fich M., Blitz, L., 1984, ApJ 279, 125
Giclas, H.L., Burnham, R., Thomas, N.G., 1970, Lowell Obs. Bull., 7, 183
Kwitter, K.B., Massey, P., Congdon, C.W., Pasachoff, J.M., 1989, ApJ 97, 1423
Napiwotzki, R., 1992, Proc. Kiel/CCP7 workshop on Atmospheres of Early Type
    Stars, eds. U. Heber, S. Jefferey, Springer, p. 310
Sharpless, S., 1959, ApJS 4, 257

# A NEW PG 1159-TYPE CENTRAL STAR DISCOVERED IN THE ROSAT XRT ALL SKY SURVEY: NON-LTE ANALYSIS OF X-RAY AND OPTICAL SPECTRA

K. WERNER
*Institut für Theoretische Physik und Sternwarte, Kiel, FRG*

C. MOTCH
*Max-Planck-Institut für Extraterrestrische Physik, Garching, FRG*

and

M. PAKULL
*Landessternwarte Heidelberg, FRG*

We report on the discovery of a new PG 1159 star in the ROSAT XRT all sky survey and give results of a model atmosphere analysis. The X-ray source RX J2117.1+3412 is relatively faint (0.33 cnt s$^{-1}$) and extremely soft. Ground based optical follow-up spectroscopy (OHP, France) proofs its PG 1159 nature: It belongs to the "low gravity emission" spectral subtype. Optically, it is the second brightest PG 1159 star. CCD [O III] imagery reveals that the star is surrounded by an old arc-shaped planetary nebula of faint surface brightness. The spectral analysis of the central star was performed with non-LTE line blanketed model atmospheres (Werner 1992). We find a complete agreement between the atmospheric parameters determined at optical wavelengths and in the ROSAT PSPC energy range.
A detailed paper was submitted to A&A (Motch, Werner, Pakull 1992).

Werner, K. 1992, in *Atmospheres of Early-Type Stars*, eds. U.Heber and C.S.Jeffery, Lecture Notes in Physics 401, Springer, Berlin, p. 273

| | | |
|---|---|---|
| $T_{\rm eff}$ | = | $150\,000 \pm 15\,000$ K |
| log g | = | 5.6 – 6.3 |
| C/He | = | $0.5^{+1.1}_{-0.34}$ by number |
| O/He | = | $0.05^{+0.15}_{-0.03}$ |
| N/He | < | $10^{-3}$ |
| log $n_H$ | = | 20.41 – 20.51 |
| E(B-V) | = | 0.05±0.01 |
| M/M$_\odot$ | = | $0.65^{+0.2}_{-0.1}$ |
| log L/L$_\odot$ | = | $3.95 \pm 0.5$ |
| d [kpc] | = | $1.4^{+0.7}_{-0.5}$ |
| z [pc] | = | $250^{+120}_{-80}$ |
| V | = | $13.2^m$ |

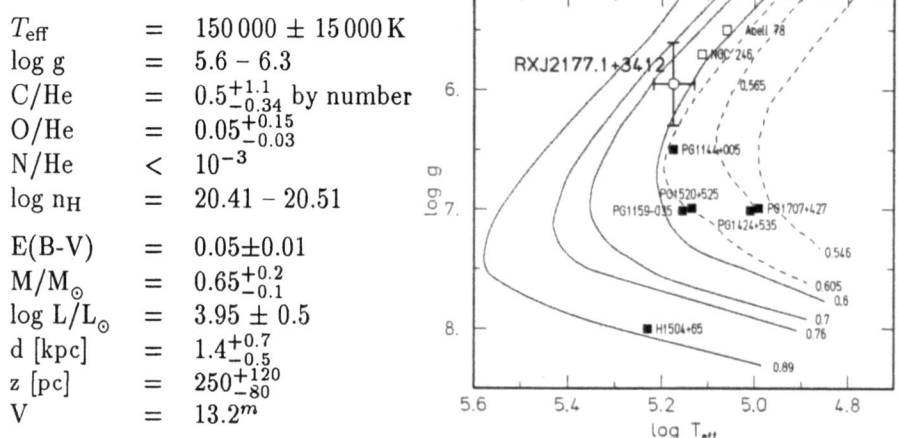

Fig. 1: Position of the new PG 1159- and other H-deficient stars in the log g–log Teff diagram. Open squares: central stars, solid lines: He burners, dashed: H burners.

# ROSAT STUDIES OF THE COMPOSITION AND STRUCTURE OF DA WHITE DWARF ATMOSPHERES

C.J. DIAMOND[1], M.A. BARSTOW[2], A.E. SANSOM[2], M.C. MARSH[2], S.R. ROSEN[2], T.A. FLEMING[3], D. KOESTER[4], D.S. FINLEY[5], J.B. HOLBERG[6], K. KIDDER[6]

1. School of Physics and Space Research, University of Birmingham, UK
2. Dept. of Physics and Astronomy, Univeristy of Leicester, UK
3. Max-Planck-Institut Für Extraterrestrische Physik, Garching, GER
4. Dept. of Physics and Astronomy, Louisiana State University, Baton Rouge, USA
5. Center for EUV Astrophysics, University of California, Berkeley, USA
6. Lunar and Planetary Laboratory, University of Arizona, Tucson, USA

We have made a detailed study of a sample of 28 hot DA white dwarfs detected in the ROSAT EUV and soft X-ray all-sky-survey.

The survey data have been combined with subsequent ROSAT pointed observations, earlier EXOSAT results and optically determined temperatures and gravities (from Kidder/Holberg and Finley). These data have been analysed using theoretical model atmosphere grids (hydrogen with trace helium homogeneously mixed or stratified with a hydrogen layer over helium).

We have found that below $\approx$40,000K, DA atmospheres are well described by a nearly pure H composition with a small opacity due to trace helium (H/He$< 3 \times 10^{-5}$) or a thick (mH $> 6 \times 10^{-13} M_\odot$) hydrogen layer. Above 40,000K however, H+He models (homogeneous or stratified) cannot in general explain the observed X-ray/EUV fluxes.

Analysis of the results shows that additional opacity must be present in the photospheres of these hotter stars and we believe that it is due to the presence of trace metals. In support, other observations have detected trace metal lines in spectra of some of the hottest white dwarfs[1], and in two of stars of this subsample, Feige 24[2] and G191-B2B[3], no significant He 228Å feature was detected. The disappearance of trace metals in DA's when they cool below 40,000K is in agreement with theoretical radiative levitation calculations (see for example[4]).

Thus we propose that these results are direct observational evidence that the atmospheric composition of DA white dwarfs is dominated by the balance between gravitational and radiative forces. Secondly, upper limits on the amount of helium in the atmosphere obtained from the modelling indicate that it makes a minimal contribution to the opacity.

## References
1. Vennes, S., Thejll, P., & Shipman, H.L., in 'White Dwarfs', ed. G. Wegner, Springer-Verlag, 176.
2. Paerels et al., 1986, *Ap.J.Lett*,309,L33
3. Wilkinson, E., Cash, W. & Green, J.C., 1992,*Ap.J*, submitted.
4. Chayer et al., 1989, in 'White Dwarfs', ed. G. Wegner, Springer-Verlag, 176.

# HST FOS OBSERVATIONS OF KPD0005+5106: A SUBLUMINOUS WN-WC DESCENDANT WITH ONGOING MASS OUTFLOW?

EDWARD M. SION

Dept. of Astron. & Astrophys., Villanova University, Villanova, PA 19085, USA

and

RONALD A. DOWNES

STScI, 3700 San Martin Drive, Baltimore, MD 21218, USA

**Abstract:** We report the results of spectroscopic observations of the ultra-high excitation, helium-rich, pre-white dwarf KPD0005+5106 obtained with the Hubble Space Telescope Faint Object Spectrograph (FOS) in 1991 May and July with the red (FOS/RD) and blue (FOS/BL) Digicon detectors. The data reveal a rich line spectrum both in absorption and in emission with ultra-high excitation species present including O VIII, N V, possibly C V, Fe VI, Fe VII and numerous weaker high n, low l, transitions of C IV and O VI as well as the predominant He II (3 $\rightarrow$ n) and He II (Balmer $\alpha$) absorption lines. There is a strong emission complex at 2981A which we identify primarily as three transitions of N V commonly seen in WN Wolf-Rayet spectra. We present evidence that high ionization species in emission (O VIII, N V, C IV, Si IV) and in absorption (He II, Fe VI, Fe VII) are *longward-shifted* relative to the far UV resonance (circumstellar) absorption lines by 25-50 km/s. Based upon the detected species, line velocities, line widths and emission features, we conclude that (1) KPD0005 is the very likely the evolutionary descendant of a WN-WC subluminous Wolf-Rayet progenitor and (2) has ongoing, possibly episodic, mass outflow.

If KPD0005 is the progeny of a post-AGB star which has suffered a late helium thermal pulse, then the nitrogen should have been rapidly destroyed in the triple-$\alpha$ runaway. Werner and Heber (1992) have recently summarized theoretical mechanisms (based upon work by Iben and Renzini) whereby nitrogen could survive this event. KPD0005 could indeed be a descendant of the WN-WC progenitor Abell 78 and PG1144+005 but at higher gravity. Thus the strong He II spectrum would be due to the surviving He from a late thermal pulse, which has diffused upward (and is possibly undergoing radiative acceleration leading to ejection), giving eventually a canonical DO photosphere. On the other hand, it cannot be ruled out that KPD0005 is an evolved, WN-WC Wolf-Rayet central star with its nebula no longer optically detectable, and which has left the AGB for the *first time*. It has not yet begun its re-trace of the AGB for the second time via a late He thermal pulse but is losing mass due to radiative or mechanical driving at a significant rate. This issue could be settled with NLTE modelling and wind analysis of the peculiar emission/absorption spectrum of KPD0005 along the lines of those advanced by the K. Werner and collaborators at Kiel.

This research was supported in part by NSF grant AST90-16289 to Villanova University.

## References

Werner, K., and Heber, U. 1992, in Atmospheres of Early Type Stars, eds. U. Heber and C.S. Jeffrey, (Springer: Berlin), in press.

# HST OBSERVATIONS OF THE NUCLEI OF EGB 6 (0950+139) AND ABELL 58 (V605 Aql)

HOWARD E. BOND AND MICHAEL G. MEAKES
*Space Telescope Science Institute*
JAMES W. LIEBERT
*Steward Observatory, University of Arizona*
and
ALVIO RENZINI
*Universita di Bologna*

This paper deals with the central stars of two large, low-surface-brightness planetary nebulae: **V605 Aquilae**, central star of **Abell 58**, and **0950+139**, central star of **EGB 6**. Both of these nuclei are associated with compact emission-line nebulosities, which are unresolved from the ground. We obtained images with the Faint Object Camera (FOC) on the *Hubble Space Telescope* of both objects, in order to determine the nature of the compact nebulae.

**V605 Aquilae** experienced an outburst during 1917-1921 to ~11th mag, but the star is currently fainter than 20th mag. In 1921 it showed the spectrum of a hydrogen-deficient carbon red giant. The R CrB-like spectrum at maximum suggests that the outburst may have represented a born-again episode following a final helium shell flash in a hot white dwarf.

Ground-based images of V605 Aql reveal an unresolved nebular knot, which is extremely hydrogen deficient. An *HST* image, obtained with the FOC in the light of [O III] 5007 Å, shows that the knot is resolved into a patchy nebula with a diameter of about $0''.5$. No star is visible, presumably because of dust in our line of sight.

The present diameter of the compact nebula is consistent with ejection around 1920 at approximately a red-giant escape velocity, and thus is consistent with the suggestion that V605 Aql experienced a final helium shell flash around 1920 that temporarily turned it into a "born-again," hydrogen-deficient red giant. The remarkably short timescale probably requires a rather high stellar mass, $\sim 1 M_\odot$.

**0950+139** is the 16th-mag central star of the large, very faint PN EGB 6. Ground-based spectra of the star reveal a hot (70,000 K) DA white dwarf, with superposed [O III] and Balmer emission. The emission-line component might arise from (a) ongoing mass loss from the white dwarf; (b) ejection during a "born-again" phase, as in V605 Aql; or (c) ablation from a close (few AU) substellar companion.

In order to test these explanations, we obtained FOC images in H$\beta$ and [O III] 5007 Å. The images unexpectedly show a point-like companion, at a projected distance of $0''.18$ or ~80 AU, which appears to be the source of the entire emission-line component. We speculate that the emission-line companion is associated with a dM5 component discovered in the IR by Zuckerman et al. in 1991 and recently confirmed by Fulbright & Liebert. However, the object is much too far from the white dwarf to be losing mass by ablation, and it remains unclear why it should be surrounded by a compact nebula.

# THE LOW LUMINOSITY CENTRAL STAR OF THE PN ESO166-21

M.T.RUIZ

*Depto. de Astronomia, Universidad de Chile, Casilla 36-D, Santiago, Chile*

and

M.PEÑA and TORRES-PEIMBERT

*Instituto de Astronomia, Universidad Nacional Autónoma de México, Apdo. Postal 70 264, México D.F. 04510, México*

We present low dispersion UV and optical spectrophotometry of the central star of the PN ESO166-21. The stellar spectrum, from 1200 to 6600 Å, is a featureless continuum. The energy distribution is consistent with a black body of $120,000 \pm 20,000$K. The observed visual magnitude is 18.1.

Ruiz et al. (1989) showed that the nebula is very extended ($\Phi = 160"$) with a spherical shape and bright knots. The emission lines indicate a high ionization degree and the chemical composition shows He and N enrichment, typical of PNe with massive progenitors. We found a nebular expansion velocity of 28km/s. From these parameters we derived a distance of $1.0 \pm 0.2$kpc and we estimated a stellar luminosity of $L/L_\odot \cong 20$ and a radius of $R/R_\odot \cong 0.01$. From the theorical evolutionary tracks in the H-R diagram (see Shaw & Kaler 1989) we obtain a mass of about $1 M_\odot$ and log g=8.4 for this star.

These parameters correspond to one of the most evolved central stars of PNe indicating that this object is already in the white dwarf cooling sequence. Only a few objects have been reported with these characteristics, among them NGC7293, A21 and A31. These rare objects are very important in the study of the link between white dwarfs and their precursors.

### References

Ruiz M.T., Heathcote S.R. and Weller W.G. 1989, IAU symp. 131, 192
Shaw R.A. & Kaler J.B. 1989, Ap.J.Supp. 69, 495

# VI. PLANETARY NEBULAE IN GALACTIC SYSTEMS

G. JACOBY

G.A. TAMMANN

M. PEIMBERT

X. HUI

H. DEJONGHE

R.E.S. CLEGG

J. KÖPPEN

I. IBEN, JR.

# LUMINOSITY FUNCTIONS OF PLANETARY NEBULAE

GEORGE JACOBY

*Kitt Peak National Observatory, National Optical Astronomy Observatories, P.O. Box 26732, Tucson, Arizona 85726, USA*

and

ROBIN CIARDULLO

*Department of Astronomy and Astrophysics, Penn State University, 525 Davey Lab, University Park, Pennsylvania 16802, USA*

ABSTRACT. Luminosity functions of planetary nebulae contain information about the central star mass distributions, nebular, central star, and progenitor evolution, stellar death rates, and a galaxy's star formation and chemical evolution histories. Appropriate observing strategies can be used in combination with various models to extract some of the parameters of these functions. The principal results from these studies are that the central star mass distribution is narrow ($\sigma \sim 0.02 - 0.04 M_\odot$), the number of PN in a galaxy depends on galaxy color, and the number of PN in the Galaxy is $\sim 10^4$.

The most extensive application of luminosity function studies has been exploiting the bright end cutoff as a distance indicator. Distances for 25 galaxies have been measured using the methodology outlined by Jacoby, Ciardullo, and collaborators. The PNLF method compares extremely well with other techniques, and is accurate to $\sim 5\%$. In fact, there is no evidence for systematic effects of any kind, although a small (5-10%) metallicity correction needs to be applied for metal-poor systems.

## 1. Introduction

Luminosity function studies of planetary nebulae (PN) have recently received a great deal of attention. The increased awareness derives primarily from using the [O III] planetary nebula luminosity function (PNLF) to derive distances to galaxies. These distances, in turn, have important consequences for the Hubble Constant and the age of the Universe.

The first discussions of the PNLF can be traced back 30 years to Henize and Westerlund (1963) who measured the bright end of the SMC PNLF. Those authors stated quite clearly that they found an upper limit to the PN luminosities, and this limit was very similar to that found by O'Dell (1962) for the Galaxy and predicted by Shklovsky (1956). They stopped short, however, of suggesting that this fiducial may be used as an extragalactic distance indicator, perhaps because Galactic PN luminosities have always been plagued by inaccurate distances. Consequently, most PNLF studies have targeted large collections of PN with a single distance (*e.g.*, those in the Galactic Bulge, the Magellanic Clouds, M31) rather than those in the solar neighborhood. It is a remarkable irony that distances to far-off galaxies can be measured quite accurately ($\sim 5\%$) using PN, but distances to the much closer Galactic PN usually cannot be measured to better than a factor of 2.

In addition to the extragalactic distance scale, motivation to study the PNLF derives from interest in 1) estimating the total number of PN in galaxies (including ours), 2) deriving central star mass distributions, 3) testing central star and nebula evolution theory, and 4) testing progenitor evolution and dredge-up theory. In addition, the various parameters describing the star formation history and chemical enrichment are also somehow convolved into the physical processes that determine the PNLF, but these represent a weak dependence and cannot be extracted easily.

In the following discussion, we define the PNLF to be the number of PN as a function of magnitude in a particular emission line. An example is shown in Figure 1. For distant galaxies, [O III] λ5007 is the only reasonable line to explore because it is generally the brightest line in the visible spectrum. Furthermore, the [O III] PNLF exhibits much lower sensitivity to metallicity variations than the Balmer line PNLF (Dopita et al. 1992), and so it has greater value as an extragalactic distance indicator. The use of [O III] does, however, introduce a selection effect against low excitation nebulae that may affect estimates of the total PN population. Other lines, such as H$\beta$, suffer less from this effect and have been used extensively for nearby PN samples (e.g., Pottasch 1990; Stasińska et al. 1991).

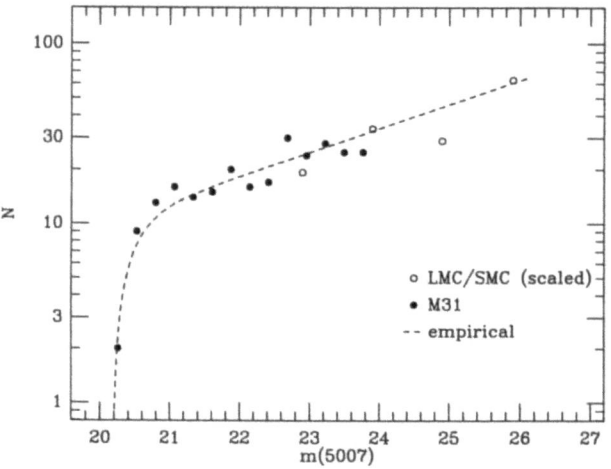

**Figure 1.** The [O III] PNLF of the bulge of M31 (solid points) extending 3.5 mag below $M^*$. Open points show the Jacoby (1980) Magellanic Cloud PNLF (minus the invalid identifications noted by Boroson and Liebert [1989]) after scaling to match the sample size of the M31 data. The overall PNLF spans 6 magnitudes. Equation (2) is shown as the smooth curve.

## 2. Results From PNLF Studies

The first observational determination of the [O III] PNLF was made by Jacoby (1980) for PN in the Magellanic Clouds. His primary interest was to derive the number of PN in Local Group galaxies by comparing complete samples of objects in the brightest few magnitudes of the PNLF with the PNLF of the LMC and SMC. Jacoby derived the V-band luminosity-specific PN density for 10 Local Group galaxies and used their mean PN density to estimate a Galactic PN population of $10,000 \pm 4000$. A more detailed accounting of the problem has been carried out by Peimbert (1990; see also this volume) who derived a total Galactic PN population of $7200 \pm 1800$. Since these estimates are based on extrapolations of [O III] PNLFs, they may be subject to the aforementioned selection effect.

Pottasch (1984) compared the H$\beta$ PNLF in the Galactic bulge with the Magellanic Cloud PNLF. The functions exhibit a similar range of luminosities, which Pottasch took as a strong indication that the adopted distances to the Magellanic Clouds, the Galactic Bulge, and the solar neighborhood sample were credible. The shapes of the PNLFs were not directly comparable, however, due to drastically different selection criteria, and so it was not possible to argue for the use of PN as a distance indicator. Although the possibility had been suspected much earlier (Hodge 1966), the first serious proposal outlining the PNLF technique for deriving extragalactic distances was presented many years later (Jacoby, Ciardullo, and Ford 1988). Jacoby (1989) and Ciardullo et al. (1989a) described the underlying astrophysics and details of the method (see §3).

In principle, the shape of the PNLF can be computed from stellar and nebular evolution theory. Henize and Westerlund (1963) made the initial attempt, by assuming a non-evolving central star and a uniformly expanding nebular shell. Despite the apparent simplicity of this model, it predicts the faint end of the PNLF rather well.

Jacoby (1989) simulated the [O III] PNLF to a much finer precision by using improved nebula models and central star evolutionary tracks to produce a grid of models having central star age and mass as independent variables. For a given set of central star evolutionary tracks (e.g., Wood and Faulkner 1986), these 2 parameters define the luminosity and temperature of the star, while age alone determines the size and density of the surrounding nebula. The grid represents the time history of the emission-line fluxes escaping the nebula as a function of central star mass. By selecting central star masses according to some distribution function, and selecting ages randomly for many hypothetical PN, a PNLF in any emission line can be simulated. Jacoby found that it was necessary to invoke a rather small central star dispersion (0.02 $M_\odot$) to match the rapidity of the bright end cutoff in the observed PNLF of 5 different galaxies. Combined with the fast evolutionary time scales for high mass central stars, the 0.02 $M_\odot$ high mass Gaussian width serves to truncate the PNLF. It is worth emphasizing that the small dispersion applies only to the very brightest PN as measured in $\lambda 5007$ and that the dispersion width depends on the adopted core mass-luminosity relationship (that of Schönberner [1979] in this case). Furthermore, the possible convergence of evolutionary tracks due to continuing mass loss after the AGB (Vassiliadis and Wood 1992) would reduce the high mass cutoff rate since stars with high initial masses will approach the low mass tracks later in their lifetimes.

Stasińska et al. (1991) used a similar Monte Carlo approach to determine that the ionized mass in the Galactic bulge nebulae is $\sim 0.2 M_\odot$ and that the dispersion in central star masses is 0.04-0.05 $M_\odot$. Although the latter value is higher than that derived by Jacoby (1989), the two results cannot be compared directly because Stasińska et al. adopted the Schönberner (1981, 1983) central star evolutionary tracks. In addition, selection effects and line-of-sight depth to the Galactic bulge act to smooth the observed PNLF such that the apparent dispersion in core mass is larger than the intrinsic dispersion. Consequently, the estimate of 0.04-0.05 $M_\odot$ should be viewed as an upper limit.

Dopita et al. (1992) investigated the effects of metallicity variations on the PNLF. Jacoby's (1989) limited effort to probe the sensitivity of the PNLF to metallicity was concerned only with the effect on the nebula. Because the central star characteristics also change with metallicity, Dopita et al. included that effect in assessing the impact of metallicity variations on the total system (see §3.2).

## 3. Extragalactic Distances Using the PNLF

Early attempts to use PN as distance indicators (Ford and Jenner 1978; Jacoby and Lesser 1981; Lawrie and Graham 1983; Ford et al. 1989) were based on the brightest few objects. Any method that relies on the extremes of a population is subject to systematic sampling errors; consequently, those early studies received relatively little attention.

The systematic errors can be minimized by incorporating the shape of the luminosity function into the analysis. Along these lines, Ciardullo et al. (1989a) described a robust and objective method for deriving distances using the method of maximum likelihood. The procedure accounts for all known observational uncertainties (such as photometric errors and filter calibration) in a rigorous manner without resorting to the deleterious effects of histogram sampling. Given the [O III] magnitudes for a few dozen PN, the approach is to find the magnitude shift relative to a reference PNLF such that the probability of observing the PN is maximized. Figure 2 illustrates a typical PNLF and probability curve.

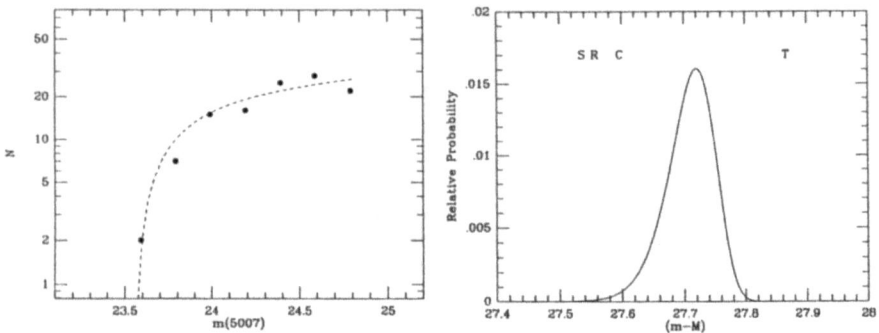

**Figure 2.** The [O III] PNLF for the bulge of M81, extending ~ 1.2 mag below $M^*$ (left panel) with equation (2) shown as the smooth curve, and the maximum likelihood probability distribution (representing the formal errors only) for the distance modulus (right panel). Symbols S, R, T, C refer to the distance moduli derived using SBF, brightest red stars, Tully-Fisher, and Cepheids. Uncertainties for each method are typically ±0.3 mag, corresponding to the total extent of the figure.

Ciardullo et al. adopted the PNLF from the bulge of M31 as a reference because its distance is well-determined (van den Bergh 1991), it has a large PN population (see Figure 1), and it has the color and metallicity typical of the giant ellipticals for which the method is most applicable and provocative. For ease of computation, the M31 PNLF has been approximated by the empirical law (solid curve in Figure 1),

$$N(M) \propto e^{0.307M}(1 - e^{3(M^*-M)}) \qquad (1)$$

where $M$ is related to the $\lambda 5007$ flux of a PN by

$$M = -2.5 \log F_{5007} - 13.74 \qquad (2)$$

and $M^*$ is the absolute magnitude of the bright end cutoff. From the M31 observations, $M^* = -4.48$. The revised M31 distance (770 kpc) of Freedman and Madore (1990) and the extinction ($A_{5007} = 0.28$) from Burstein and Heiles (1984) would yield an $M^*$ that is 0.06 mag brighter. Distances quoted in this review refer to the original estimate for $M^*$.

Table 1 lists all galaxies with an observed PNLF and their derived distances.

### 3.1. TESTING PROCEDURES

One reason for the quick acceptance of PNLF distances is that its proponents did not present their results until after they had verified that the method works. They performed numerous tests, both internal and external, designed to assess the accuracy of their results; in fact, the PNLF method has been tested more carefully than any other general purpose distance technique.

The easier tests to perform were the "internal" tests which compared the distances to several galaxies within a single cluster. This was done for the Leo I galaxies NGC 3377, NGC 3379, and NGC 3384 (Ciardullo et al. 1989b). Distances to these galaxies demonstrated unprecedented internal accuracy (2.5% rms) for a galaxy distance technique. Furthermore, the method was tested over a small, but non-negligible, range in both Hubble type (E6-SB0) and metallicity (25%). Following these encouraging results, Jacoby et al. (1990) pushed the method further and derived distances to 6 galaxies in the Virgo Cluster core. Here, a wider range of metallicities was encompassed (55%), yet the distances all agreed (5% rms) within the range expected for a large galaxy cluster. Another internal test compared the distance of an edge-on Sb galaxy (NGC 891) to that of its neighbor, an SB0 galaxy with relatively recent star formation (NGC 1023) (Ciardullo et al. 1991). The purpose of this test was to identify any systematic error introduced by using an Sb galaxy (e.g., M31) as a reference for earlier galaxy types. The resulting distances were identical, and the relatively sparse sample of PN identified above the disk of NGC 891 (33) compared to NGC 1023 (110) did not compromise the results.

TABLE 1. Summary of PNLF Results to Date

| Name | Type | Nr. PN | $(m - M)_0$ | $\alpha_{2.5}(\times 10^9)$ |
|---|---|---|---|---|
| Local Group | | | | |
| LMC | SBm | 42 | $18.44 \pm 0.18$ | 32 |
| SMC | Im | 8 | $19.09 \pm 0.29$ | 48 |
| 185 | dE3p | 4 | ... | ... |
| 205 | S0/E5p | 12 | $24.68 \pm 0.35$ | $54 \pm 18$ |
| 221 | E2 | 9 | $24.58 \pm 0.60$ | $38 \pm 17$ |
| 224 | Sb | 104 | $24.26 \pm 0.04$ | $11.0 \pm 2.0$ |
| NGC 1023 Group | | | | |
| 891 | Sb | 34 | $29.97 \pm 0.16$ | ... |
| 1023 | SB0 | 97 | $29.97 \pm 0.14$ | $22.3 \pm 3.7$ |
| Fornax Cluster | | | | |
| 1399 | E1 | 53 | $31.01 \pm 0.08$ | ... |
| 1404 | E2 | 53 | $31.17 \pm 0.07$ | ... |
| Leo I Group | | | | |
| 3377 | E6 | 22 | $30.07 \pm 0.17$ | $38.3 \pm 8.8$ |
| 3379 | E0 | 45 | $29.96 \pm 0.16$ | $21.4 \pm 3.3$ |
| 3384 | SB0 | 43 | $30.03 \pm 0.16$ | $39.7 \pm 6.4$ |
| Virgo Cluster | | | | |
| 4374 | E1 | 37 | $30.98 \pm 0.18$ | $17.5 \pm 3.1$ |
| 4382 | S0 | 59 | $30.79 \pm 0.17$ | ... |
| 4406 | S0/E3 | 59 | $30.98 \pm 0.17$ | $13.9 \pm 2.0$ |
| 4472 | E1/S0 | 26 | $30.71 \pm 0.19$ | $6.7 \pm 1.4$ |
| 4486 | E0 | 36 | $30.81 \pm 0.17$ | $8.8 \pm 1.5$ |
| 4649 | S0 | 16 | $30.76 \pm 0.19$ | $6.5 \pm 2.0$ |
| NGC 5128 Group | | | | |
| 5128 | S0p | 224 | $27.73 \pm 0.04$ | $26.8 \pm 5.7$ |
| 5253 | Amor | 16 | $28.08 \pm 0.29$ | ... |
| Other | | | | |
| 3031 | Sb | 88 | $27.72 \pm 0.25$ | $16.2 \pm 2.0$ |
| 3109 | Sm | 7 | 26.00 | 145 |
| 3115 | S0 | 52 | $30.11 \pm 0.20$ | $25.0 \pm 7.4$ |
| 4594 | Sa | 204 | $29.76 \pm 0.04$ | $17.6 \pm 1.7$ |
| Bulge | Sbc | 22 | $14.54 \pm 0.20$ | ... |

Notes:

1. No metallicity corrections (Ciardullo and Jacoby 1992) have been applied to the distances.

2. All results are from Ciardullo, Jacoby, and collaborators except: the Galactic Bulge (Pottasch 1990), NGC 3109 (Richer and McCall 1992), and NGC 4594 (Hui et al. 1993).

More important than these internal tests were the tests for external errors between the PNLF method and distances derived by other reliable indicators. Only the RR Lyrae and Cepheid variables are considered "unassailable" techniques. This is unfortunate because the former cannot be used beyond ~1 Mpc, and the latter are not found in early-type systems where the PNLF method is most applicable. Nevertheless, the PNLF method can be used in the nearer spirals where HII region discrimination is possible based on spatial resolution. Three Cepheid galaxies have been observed: M81 (Jacoby et al. 1989), the LMC, and the SMC (Jacoby et al. 1990). The PNLF and Cepheid distances are 3.5 and 3.3 Mpc, 49 and 51 kpc, and 66 and 56 kpc, respectively. The agreement is exceptionally good for the first 2 cases. The SMC has a metallicity ~0.1 that of M31's bulge which, according to the models of Dopita et al. (1992) and the observational test by Ciardullo and Jacoby (1992), requires a correction (see §3.2) of 12% downward to 59 kpc. (The SMC PN sample represents about 30% of the data used in the latter experiment, so this distance correction is not completely independent of the empirical calibration.) Thus, the method reproduces accepted values whether sample sizes are large (88 PN in M81) or small (35/8 in the LMC/SMC), thereby demonstrating the robustness of the maximum likelihood luminosity function fitting technique. In addition, Pottasch (1990) derived the distance to the Galactic bulge using the PNLF approach. His result of 8.1 kpc compares well with that of other indicators that yield 7.7 kpc (Reid 1989).

Another way to search for systematic effects is to compare distances to individual galaxies using different methods. Until recently, this approach provided little new information because most methods suffer from large random errors that mask the external effects. With the advent of the PNLF and surface brightness fluctuation (SBF) techniques, we can make comparisons for the first time that are capable of resolving systematic effects at the 5% level. Jacoby et al. (1992) performed a cross-comparison for the 7 primary methods of extragalactic distance determination currently in use: PNLF, SBF, globular cluster luminosity functions (GCLF), novae, Type Ia supernovae, $D_n - \sigma$, and Tully-Fisher. The review by Jacoby et al. (1992) describes each of these methods in detail: their advantages, disadvantages, strengths, and weaknesses. The comparisons demonstrate that each method yields distances having accuracies very close to what their proponents predict. The PNLF method, for instance, agrees with the SBF distances on a galaxy-by-galaxy basis to within 8% rms, suggesting that the PNLF contribution to the error is ~ 5%. Figure 3 illustrates the excellent agreement among the best methods.

Based on direct observational tests, it appears that the internal consistency and external accuracy of the PNLF technique is excellent. Bottinelli et al. (1991), however, raised a specter of doubt about the PNLF distance to Virgo. In particular, they suggested that (1) there is a correlation between the distance moduli and apparent galaxy magnitudes in Virgo in the sense that brighter galaxies are found to be closer, (2) there is a trend in the comparison between the PNLF distances and the SBF distances presented by Tonry (1991), (3) the brightest few tenths of the PNLF is similar to a power law for which the tradeoff between distance and sample size represents a degenerate solution, and (4) there is a correlation between PN identification rates and parent galaxy luminosity.

Mendez et al. (1992) review these issues. They demonstrate quantitatively that items (1) and (3) are insignificant or erroneous and item (2) is most likely the result of a small (5%) systematic error in the SBF distances rather than in the PNLF method. Item (4) is quite real, as was first discussed by Peimbert (1990), and further supported by Ciardullo et al. (1991) and Richer and McCall (1992). However, rather than being an artifact of improperly applying the PNLF technique, the drop-off in the bright PN population as galaxy luminosity rises can be understood as a consequence of stellar evolution (see §3.3).

## 3.2. EFFECTS OF METALLICITY

Every distance indicator (other than geometric methods) is affected by metallicity, but most indicators are too imprecise to discern the effects (e.g., $D_n - \sigma$). Because the PNLF technique is among the most precise, it is possible to distinguish effects as small as 5%. Dopita et al. (1992) predicted that the distances derived using the PNLF method would be affected by less than $\sim$ 10% for metallicities within a factor of $\sim$ 3 of solar, and this trend has been confirmed observationally (Ciardullo and Jacoby 1992).

It seems remarkable that the net effect on the [O III] PNLF due to changes in progenitor metallicity is so small, but it can be understood easily as a fortuitous, near-perfect balance between 2 metallicity sensitive processes. While the nebula is affected in a direct manner (higher abundances produce higher [O III] fluxes thanks to the greater availability of oxygen atoms), the central star's UV luminosity is affected in the reverse. The latter behavior is a consequence of the mass loss history of the progenitor: higher metallicity implies greater losses over the lifetime of the star, resulting in a lower central star mass. The two effects offset each other to first order so that the PNLF bright end cutoff remains nearly constant (to within 0.1 mag) until the metallicity of the PN ensemble deviates significantly from solar abundances. Under extreme conditions, one or the other of these non-linear effects overwhelms the balance and a metallicity correction becomes necessary.

Another way to estimate the effects of metallicity is to measure the distance to a single galaxy using different PN populations. Hui (1992) has done this for the nearby galaxy NGC 5128 by dividing a large sample of PN into 4 groups with average galactocentric radii of 2, 5, 7, and 13 arcmin. The abundance gradient in the galaxy serves to create metallicity subsamples, ranging from about twice solar to about half solar. Hui finds the distances to these groups to be 3.65, 3.55, 3.55, and 3.42 Mpc. Furthermore, she finds that the PN production rate increases with radius by 50%, in good agreement with the following section. For a galaxy as close as NGC 5128, the PN production rate variations cannot be an artifact of sampling the PNLF inadequately as Bottinelli et al. (1991) have suggested. Also, note that the 5% decrease in the derived distance with radius (i.e., decreasing metallicity) is in excellent agreement with the experiment of Ciardullo and Jacoby (1992).

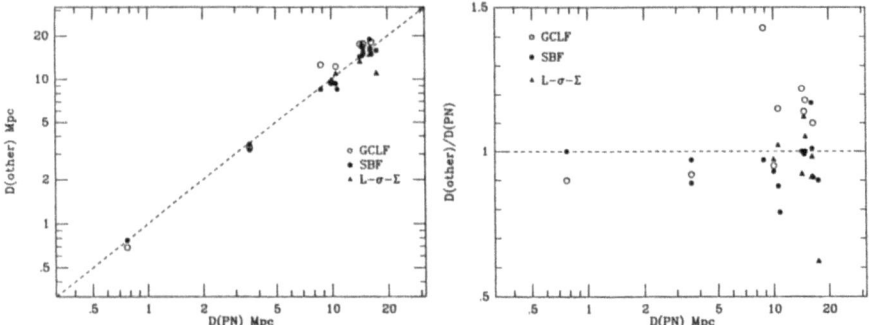

**Figure 3.** A comparison between the PNLF distances to individual galaxies and those derived using SBF (Tonry 1991), GCLF (Harris 1992, priv. comm.), and the fundamental plane relation (L-$\sigma$-$\Sigma$) for ellipticals (Pierce 1989). PNLF distances have been increased by 3% to account for the recently revised M31 distance (see text). The (zero-point offsets, rms dispersions) are (-4%,9%), (+11%,17%), and (-6%,14%) for the comparison of the 3 methods respectively. These values become (-3%,8%), (+7%,13%), and (-1%,8%) if the most discrepant galaxy is removed from each sample of 14, 9, and 9 galaxies, respectively. Obviously, the PNLF distances are in very good agreement with distances derived using other methods.

## 3.3. THE PN PRODUCTION RATE: A VALUABLE BY-PRODUCT

The procedure for deriving distances requires that the PNLF of a target galaxy be matched to that of the reference galaxy; *i.e.*, the sample sizes must be normalized. In fact, there are 2 variables that enter the solution: distance and number of PN. We define the luminosity-specific PN rate, $\alpha_{2.5}$, as the number of PN in the first 2.5 mag of the PNLF relative to the bolometric luminosity of the sampled region in the host galaxy. In practice, $\alpha_{2.5}$ may have to be estimated from a survey which does not extend 2.5 mags and a correction factor, obtained by extrapolating the integral PNLF, is applied. Figure 4 illustrates M81's 2-dimensional solution for distance modulus and $\alpha_{2.5}$.

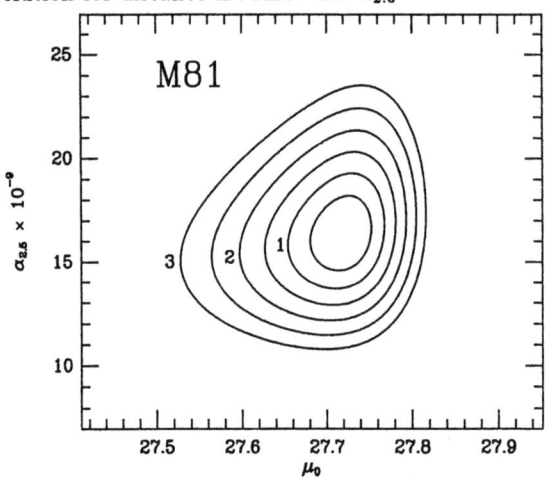

**Figure 4.** The complete maximum likelihood solution for M81, illustrating the 0.5, 1, 1.5, 2, 2.5, and 3 $\sigma$ confidence levels in the 2 variables, distance modulus and luminosity specific PN density, $\alpha_{2.5}$. The right panel of Figure 2 is recovered if the vertical span of Figure 4 is collapsed to 1 dimension.

Peimbert (1990) showed that $\alpha_{2.5}$ varies by nearly an order of magnitude, and that the variations correlate extremely well with host galaxy color and absolute magnitude. This result would seem to be disparate with the prediction of Renzini and Buzzoni (1986) that a population's stellar death rate should be nearly independent of its age or IMF. However, the stellar death rate and the PN birth rate are not necessarily identical. Peimbert (1992) proposed that the [O III] bright PN form from young stars and the proportions of the mix of young and old stars correlates with galaxy luminosity and color.

Ciardullo *et al.* (1991) suggested another possibility: the progenitor-core mass relationship is "noisy" so that a given turnoff mass is capable of producing a distribution of core masses, possibly as a consequence of rotation (Weidemann 1990). Weidemann and Koester (1983) showed that this possibility occurs in the Galaxy. Statistically then, an older, redder system produces fewer high mass central stars than a younger one. The [O III] PNLF is truncated, not by the non-existence of high mass central stars, but by their fast evolutionary rates and by their enhanced nebular abundances.

Following Peimbert (1990), Ciardullo *et al.* (1991) and Richer and McCall (1992) confirmed that $\alpha_{2.5}$ correlates with host galaxy color, and showed that it may correlate with UV (1550$\lambda$) excess (*e.g.*, Burstein *et al.* 1988) in the sense that the greater the UV excess, the fewer bright PN in the galaxy. A possible explanation is that the source of UV excess originates, at least in part, from those stars that lose so much mass prior to climbing the AGB that they bypass the PN stage to become hot horizontal branch stars (Greggio and

Renzini 1990). Since AGB mass loss depends on metallicity, these stars may be the high metallicity tail of a galaxy's population. It may therefore be possible for a galaxy's mean metallicity to be so high that the PN represent only a tiny fraction of the dying stars.

Figure 5 illustrates the relationship between $\alpha_{2.5}$ and galaxy color (U-V), metallicity, and UV color. The number of PN in our Galaxy can be estimated from these relations given the color $(U - V = 0.45)$ and luminosity $(M_{bol} = -21.2)$ (de Vaucouleurs and Pence 1978; de Vaucouleurs 1977). Peimbert's (1990) determination of 7200 is 3-4 times smaller than other recent estimates (Phillips 1989), but the latter generally depend on Galactic PN distance measurements. This matter is still being debated, since $\alpha_{2.5}$ may be subject to excitation level selection effects and Galactic PN distances are uncertain.

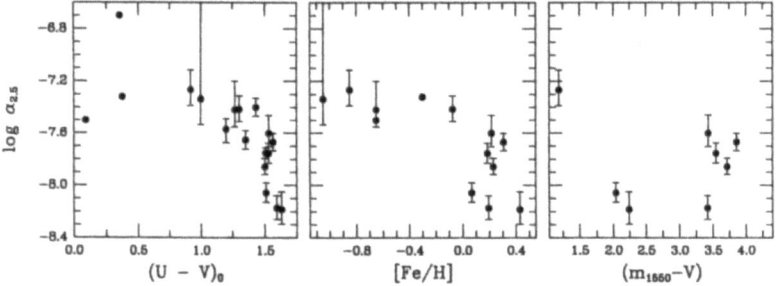

**Figure 5.** The relationship between $\alpha_{2.5}$ and galaxy color, metallicity, and UV color. A trend is evident for the two leftmost panels. There is only a weak trend, if any, with UV color. The bluest galaxy shown in the right hand panel is NGC 205, a galaxy known to have a rich population of young blue stars, and is therefore very different than the giant ellipticals. If excluded, a more respectable correlation appears.

### 3.4. THE BRIGHT TAIL OF THE PNLF

A bright extension to the PNLF has been seen in a small number of galaxies. One possible cause for this bright tail arises when galaxies more distant than $\sim 10$ Mpc are surveyed. The spatial scale becomes so compressed that 2 individual PN can coincide within the seeing disk. The measured brightness of the "double" can exceed $M^*$, extending the PNLF up to 0.75 mag. It is very unlikely that 2 PN having an $M^*$ luminosity will coincide, and so the magnitude of this effect is typically small ( $\lesssim 0.3$ mag; Jacoby et al. 1990). Since the superposition process is well understood, it can be modeled easily and the effects removed during the maximum likelihood fitting process. Generally, there are so few objects in this category ($\sim$ 1%) that this is unnecessary. Additional sources of "PN" having $M_{5007} < M^*$ can be HII regions (Ciardullo et al. 1991) and supernova remnants. Since both of these objects are rare or absent in early-type galaxies, they do not impact the method.

Although the mass distribution of PN central stars tends strongly toward low masses (Stasińska et al. 1991; Tylenda et al. 1991), a few high mass stragglers appear in the the Magellanic Clouds (Kaler and Jacoby 1991) and possibly in the Galactic disk (e.g., Kaler and Jacoby 1989; Mendez et al. 1988). Due to rapid evolutionary rates, these are unlikely to be found while bright, and should not participate in the bright end of the PNLF (Kaler and Jacoby 1990, 1991). Furthermore, most of the massive central stars are identified with Type I PN which have enhanced abundances. A simple numerical experiment shows that the nitrogen enhancement competes for collisional energy and therefore serves to diminish the [O III] $\lambda$5007 line by 0.4 mag. Thus, PN having high mass central stars are pushed down the PNLF causing a de-selection (Figure 3 of Kaler and Jacoby 1991) when bright [O III] PN are being collected. The possibility cannot be dismissed, however, that some of the objects on the bright tail of the PNLF derive from PN having high mass central stars.

## 4. Further Issues

The PNLF distance method, while among the best techniques, could be better. The principal areas for improvement are (1) refining the adopted PNLF shape, (2) deriving distances to additional Cepheid galaxies for calibrators, and (3) examining the metallicity sensitivity over a wider range and for more galaxies, and (4) clarifying the theoretical rationale for the bright end cutoff. The technique is currently limited to distances $\lesssim$ 25 Mpc, but this can be pushed to $\sim$ 40 Mpc with large telescopes on excellent sites.

What creates the correlation between $\alpha_{2.5}$ and galaxy luminosity and color? The answer to this question blends stellar evolution, galaxy formation, and PN theories.

An important by-product from PNLF studies is that several hundred PN may be identified in a galaxy. Kinematics of these objects provide a unique sampling of the galaxian gravity field to test for mass-to-light variations and dark matter. The unprecedented study by Hui (1992; also this volume) based on 433 PN in NGC 5128 illustrates the enormous value that extragalactic PN offer.

The measurement of abundances for individual stars in distant galaxies is another by-product with considerable potential. With large telescopes and fiber spectrographs, we are no longer limited to the Magellanic Clouds. M31 and its neighbors are easily within reach, and it will soon be possible to determine the chemical compositions for stars in M81, NGC 5128, and other galaxies at distances up to 4 Mpc.

## References

Bottinelli, L., Gouguenheim, L., Paturel, G, and Teerikorpi, P. 1991, A&A, 252, 550.
Boroson, T.A., and Liebert, J. 1989, ApJ, 339, 844.
Burstein, D., and Heiles, C. 1984, ApJS, 54, 33.
Burstein, D., Bertola, F., Buson, L.M., Faber, S.M., and Lauer, T.R. 1988, ApJ, 328, 440
Ciardullo, R., and Jacoby, G.H. 1992 ApJ, 388, 268.
Ciardullo, R., Jacoby, G.H., Ford, H.C., and Neill, J.D., 1989a ApJ, 339, 53.
Ciardullo, R., Jacoby, G.H., and Ford, H.C. 1989b ApJ, 344, 715.
Ciardullo, R., Jacoby, G.H., and Harris, W.E. 1991 ApJ, 383, 487.
de Vaucouleurs, G.A. 1977, in The Evolution of Galaxies and Stellar Population, eds. B. Tinsley and R. Larson (Yale Univ. Obs.: New Haven), p.43.
de Vaucouleurs, G.A., and Pence, W.D. 1978, AJ, 84, 1163.
Dopita, M.A., Jacoby, G.H., and Vassiliadis, E. 1992, ApJ, 389, 27.
Ford, H.C, Ciardullo, R., Jacoby, G.H., and Hui, X. 1989, in IAU Symposium 131, Planetary Nebulae, (Dordrecht:Reidel), ed. S. Torres-Peimbert, p. 335.
Ford, H.C., and Jenner, D.C. 1978, BAAS, 10, 665.
Freedman, W.L., and Madore, B.F. 1990, ApJ, 365, 186.
Greggio, L., and Renzini, A. 1990, ApJ, 364, 35.
Henize, K.G., and Westerlund, B.E. 1963, ApJ, 137, 747.
Hodge, P.W. 1966, in Galaxies and Cosmology, (New York: McGraw Hill), p. 130.
Hui, X. 1992, PhD. thesis, Boston University.
Hui, X., Ford, H.C., Ciardullo, R., and Jacoby, G.H. 1993, in preparation.
Jacoby, G.H. 1980, ApJS, 42, 1.
Jacoby, G.H. 1989, ApJ, 339, 39.
Jacoby, G.H., Branch, D., Ciardullo, R., Davies, R.L., Harris, W.E., Pierce, M.J., Pritchet, C.J., Tonry, J.L., and Welch, D. 1992, PASP, 104, in press.
Jacoby, G.H, Ciardullo, R., and Ford, H.C. 1990, ApJ, 356, 332.
Jacoby, G.H, Ciardullo, R., and Ford, H.C., and Booth, J. 1989, ApJ, 344, 704.

Jacoby, G.H, Ford, H.C., and Ciardullo, R. 1988, in *The Extragalactic Distance Scale*, A.S.P. Conference Series No. 4, ed. S. van den Bergh and C.J. Pritchet (Provo: Brigham Young University Press), p. 42.
Jacoby, G.H., and Lesser, M.P. 1981, AJ, 86, 185.
Jacoby, G.H, Walker, A.R., and Ciardullo, R. 1990, ApJ, 365, 471.
Kaler, J.B., and Jacoby, G.H. 1989, ApJ, 345, 871.
Kaler, J.B., and Jacoby, G.H. 1990, ApJ, 362, 491.
Kaler, J.B., and Jacoby, G.H. 1991, ApJ, 382, 134.
Lawrie, D.G., and Graham, J.A. 1983, BAAS, 15, 907.
Mendez, R.H., Kudritzki, R.P., Ciardullo, R., and Jacoby, G.H. 1992, A&A, 00, submitted.
Mendez, R.H., Kudritzki, R.P, Herrero, A., Husfeld, D., Groth, H.G. 1988, A&A, 190, 113.
O'Dell, C.R. 1962, ApJ, 135, 371.
Peimbert, M. 1990, Rev. Mex. A. A. 20, 119.
Phillips, J.P. 1989, in IAU Symposium 131, Planetary Nebulae, (Dordrecht:Reidel), ed. S. Torres-Peimbert, p. 425.
Pottasch, S.R. 1984, in Planetary Nebulae, A Study of Late Stages of Stellar Evolution, (Dordrecht: D. Reidel), p. 117.
Pottasch, S.R. 1990, A&A, 236, 231.
Reid, M.J, 1989, in IAU Symposium 136, The Center of the Galaxy, ed. M. Morris (Dordrecht: Reidel), p. 195.
Renzini, A., and Buzzoni, A. 1986, in Spectral Evolution of Galaxies, ed. C. Chiosi and A. Renzini (Dordrecht: Reidel), p. 195.
Richer, M.G., and McCall, M.L. 1992, AJ, 103, 54.
Schönberner, D. 1979, A&A, 79, 108.
Schönberner, D. 1981, A&A, 103, 119.
Schönberner, D. 1983, A&A, 272, 708.
Stasińska, G., Fresnau, A., da Silva Gameiro, G.F., and Acker, A. 1991 A&A, 252, 762.
Shklovsky, I.S. 1956, Astr. J. Soviet Union 33, 315.
Tonry, J.L. 1991, ApJ, 373, L1.
Tylenda, R., Stasińska, G., Acker, A., and Stenholm, B. 1991, A&A, 246, 221.
Weidemann, V. 1990, ARA&A, 28, 101.
Weidemann, V., and Koester, D. 1983, A&A, 121, 77.
Wood, P.R., and Faulkner, D.J. 1986, ApJ, 307, 659.
van den Bergh, S. 1991, PASP, 103, 1053.
Vassiliadis, E., and Wood, P.R. 1992, ApJ, 00, submitted.

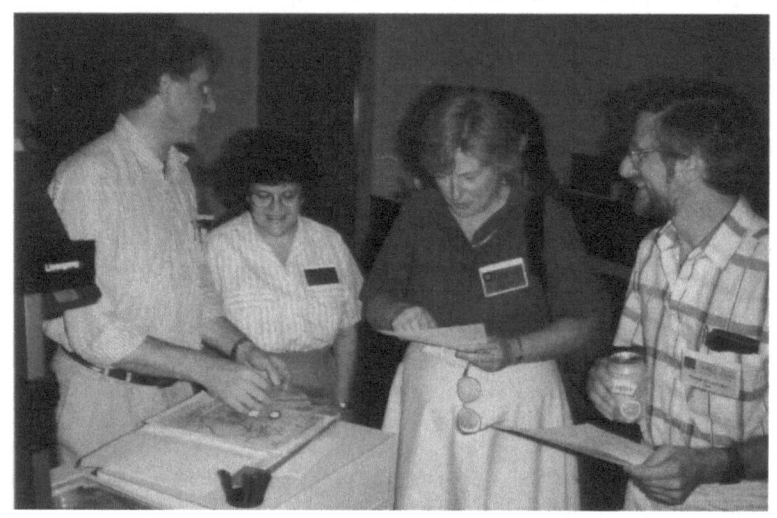

J. KÖPPEN, K.B. KWITTER, S.R. HEAP, G. JACOBY

# WHY ARE PLANETARY NEBULAE POOR DISTANCE INDICATORS?

G.A. TAMMANN

*Astronomisches Institut der Universität Basel, Venusstraße 7, CH-4102 Binningen*


ABSTRACT. It is shown from multiple evidence that the λ 5007 Å luminosity of the brightest shells of planetary nebulae has intrinsic scatter, which is well approximated for the brightest objects by an exponential luminosity function, as proposed by Bottinelli et al. (1991). The resulting open-ended luminosity function weakens the usefulness of planetary nebulae as distance indicators.

## 1. Introduction

The decisive requirement for any blind application of a distance indicator is that its luminosity (or size) can be determined without any intrinsic scatter. In the presence of intrinsic scatter the actual luminosity (or size) becomes a function of sample size. The problem is particularly severe, if the brightest (or largest) objects of a class are used as distance indicators. Clearly, the brightest objects (with intrinsic scatter) in a giant galaxy are brighter than in a dwarf galaxy, simply because the extreme wings of the luminosity distribution are more completely populated in the former case. The effect is well known for the largest HII regions (Sandage and Tammann 1974a) and for the brightest stars (Sandage and Tammann 1974b; Sandage and Carlson 1988) of late-type galaxies. Therefore the brightest and largest objects are generally expected to be particularly extreme in giant galaxies. Neglect of this effect will always lead to an underestimate of the distance of large galaxies. Since the latter are rare and therefore typically distant, they necessarily lead to a compressed distance scale. In fact, a compressed distance scale can be taken as an unfailing indicator of internal scatter of the distance indicator under consideration.

It has been proposed to use those PNe of a galaxy, which are brightest in the [O III] λ 5007 Å line, as distance indicators (Jacoby 1989). The situation is here somewhat more favorable, because instead of using the *one* brightest PN the method relies on the luminosity function (LF) of the brightest tens of PNe. This luminosity function is claimed to drop to zero at a critical luminosity $M^*_{5007}$ = -4.48 (Ciardullo et al. 1989). But still the question arises whether a universal luminosity function, which is defined by only a handful of brightest objects, can hold regardless of sample size.

The very steep decrease of PN lifetimes with increasing mass makes it reasonable to postulate an upper cutoff luminosity of the *nuclei* of PNe. However, the λ 5007 Å luminosity measures the *shell*, not the nucleus. And there cannot be a one-to-one relation between the nuclear and the nebular luminosities. The relation must be modulated by non-sphericity and other geometrical effects of the shells and, above all, by the very strong evolution of the shell luminosity - particularly of the most massive PNe - of any given PN (cf. Jacoby 1989). Therefore, even if the luminosity of the central star is sharply bounded towards high values, the luminosity of the brightest shells is expected to vary with time and from galaxy to galaxy, and hence to depend also on the sample size, i.e. galaxy size.

## 2. A Comparison of Distances

Distances derived by Jacoby and collaborators on the assumption of a constant shape of the λ 5007 Å LF and a unique value of the cutoff magnitude M*, are compiled in Table 1. They are confronted with modern independent distance determinations. Within 4 Mpc the PNe can reproduce the high-accuracy distances from Cepheids to within ~ 10%. But beyond the PN distance scale is progressively compressed. At the Virgo cluster the distance discrepancy amounts to a factor of 1.4.

TABLE 1. Distances from PNe and Other Methods

| (1) | $(m-M)^\circ$ from PNe (2) | Source (3) | $(m-M)^\circ$ from others (4) | Method (5) | $\Delta(m-M)$ (6) |
|---|---|---|---|---|---|
| LMC | 18.44 | 1 | 18.50 | Cepheids (7) | -0.06 |
| SMC | 19.09 | 1 | 18.87 | Cepheids (7) | +0.22 |
| M31 | 24.26 | 2 | 24.44 | Cepheids (7) | -0.18 |
| M81 | 27.72 | 3 | 27.59 | Cepheids (7) | +0.13 |
| Leo | 30.02 | 4 | 30.50 | Source 8 | -0.48 |
| NGC 1023 NGC 891 | 29.97 | 5 | 30.96: | Source 9 | -0.99: |
| Virgo | 30.90 | 6 | 31.64 | Source 10 | -0.74 |

Sources. 1. Jacoby et al. (1990). 2. Ciardullo et al. (1989a). 3. Jacoby et al. (1989). 4. Ciardullo et al. (1989b). 5. Ciardullo et al. (1991). 6. Jacoby et al. (1990a). 7. Madore and Freedman (1991). 8. The difference between the Leo Group and the Virgo cluster is determined from metallicity-corrected surface brightness fluctuations (Tammann 1992), PNe corrected for sample size effects (Bottinelli et al. 1991), globular clusters (Harris 1990), and the $D_n$-$\sigma$ relation (Faber et al. 1989). 9. The distance is determined from the recession velocity $v_{220}$ = 855 km s$^{-1}$ (corrected for a Virgocentric infall model) of the group and $H_o$ = 55 km s$^{-1}$ Mpc$^{-1}$. Due to peculiar motions the modulus may be off by ~ 0.$^m$4. 10. From globular clusters, novae, supernovae, the $D_n$-$\sigma$ relation, 21cm line widths, and the scale length of the Galaxy and M31 (Tammann 1992, and references therein).

The significance of the PN distance scale being too short has recently been enhanced by new determinations of the Virgo distance. Expanding-photosphere models of two supernovae of type II provide a Virgo modulus of (m-M) = 31.71 ± 0.26 (Schmidt et al. 1992). A direct luminosity calibration of the type I supernova 1937C through the Cepheids in the parent galaxy IC 4182 (Sandage et al. 1993), if applied to the 10 supernovae of that type in the Virgo cluster, provides a Virgo modulus of (m-M) = 31.64 ± 0.22. These new determinations are again much higher than the distance from PNe.

In a Critical Review (Jacoby et al. 1992), which defies its name, much emphasis is given to the fact that the Virgo distance from PNe agrees with that from the surface brightness fluctuation method (Tonry 1991). Unfortunately the latter method is dominated by metallicity effects and does not yield useful distances (Tammann 1992).

The expectation that the compression of the PN distance scale is due to an increase of the λ 5007 Å luminosity of the brightest PNe with the sample (galaxy) size has first been substantiated by Bottinelli et al. (1991), who have shown that the PN distance modulus μ of individual Virgo galaxies depends on the apparent (and hence absolute) magnitude of the galaxy (Fig. 1).

To elaborate the dependence of the brightest PNe on the galaxy size, the data in Table 2 were compiled. Column 2 lists the absolute bolometric magnitude of the galaxy, or in cases where only a fraction of the galaxy was surveyed, the corresponding magnitude of that fraction. The data were taken from the sources 1-6 in Table 1 and adjusted to the distances adopted in col. 4 of Table 1.

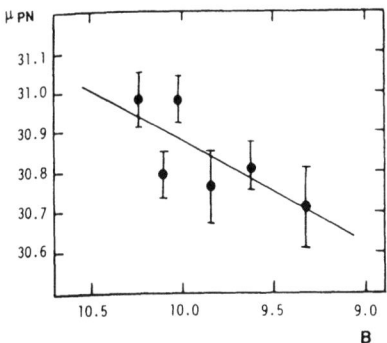

Figure 1. PN distance moduli μ of individual Virgo galaxies against their apparent (B) magnitude. (From Bottinelli et al. 1991).

TABLE 2. Bolometric Magnitudes of Surveyed Galaxy Populations, Absolute Magnitudes of the Brightest PNe, and Nominal Deathrates of Stars

| (1) | $M°_{bol}$ (2) | $M°_{5007}(3)$ (3) | $\log \alpha_{25}+9$ (4) |
|---|---|---|---|
| LMC | - | -4.33 | - |
| SMC | - | -3.79 | - |
| M31 | -19.74 | -4.41 | 1.19 |
| NGC 185 | -15.93 | -3.49 | 1.38 |
| NGC 205 | -16.18 | -3.67 | 1.70 |
| M32 | -16.81 | -3.77 | 1.32 |
| M81 | (-20.95) | -4.54 | (1.26) |
| NGC 3377 | -19.36 | -4.98 | 1.38 |
| NGC 3379 | -20.67 | -5.13 | 1.13 |
| NGC 3384 | -20.04 | -4.91 | 1.39 |
| NGC 1023 | (-19.98) | -5.50 | 0.94 |
| NGC 891 | - | -5.34 | - |
| NGC 4374 | -21.25 | -5.44 | 0.92 |
| NGC 4382 | -21.58 | -5.61 | 1.00 |
| NGC 4406 | -22.28 | -5.75 | 0.83 |
| NGC 4472 | -22.19 | -5.24 | 0.52 |
| NGC 4486 | -21.95 | -5.74 | 0.62 |
| NGC 4649 | -22.08 | -5.11 | 0.51 |

The mean absolute λ 5007 Å magnitude of the three brightest PNe in column 3 includes *all* objects found by Jacoby and collaborators. These authors had excluded some of the brightest objects for the sole reason that they do not fit under their adopted LF. But because this LF is in question, all objects, which are bright in the λ 5007 Å line, must be considered as shells of PNe, unless proven to the contrary.

The data of Table 2 are plotted in Fig. 2. They expose a pronounced correlation of the mean absolute magnitude $M°_{5007}(3)$ of the three brightest planetry shells with the bolometric magnitude of the surveyed population. The correlation is illustrated in the Figure by a line with slope 0.25. This slope necessarily follows if the *bright end* of the PN LF has the form $N(M) \propto 10^{1.6M}$. This form of the LF approximates indeed the available data well, as shown in the next Section.

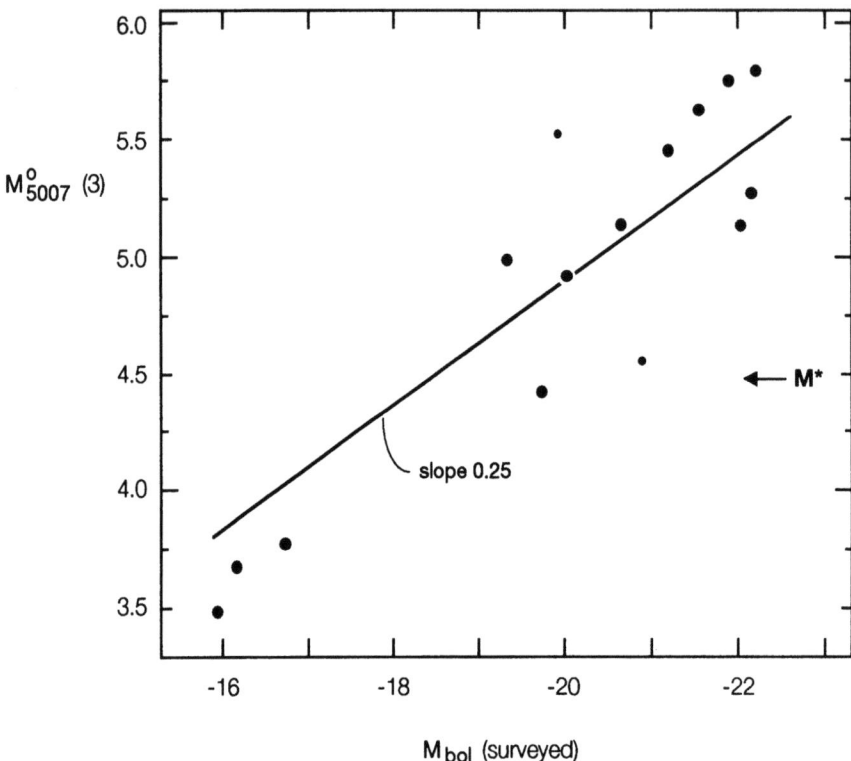

Figure 2. The dependence of the mean absolute magnitude of the three brightest PN shells on the sample size, i.e. $M°_{bol}$ (surveyed). The slope of 0.25 of the drawn line follows from log $N(M) \propto$ 1.6 M. Small symbols have uncertain $M°_{bol}$.

## 3. The Luminosity Function of Bright PNe

The largest available body of data to determine the bright end of the (differential) λ 5007 Å LF of planetary shells is provided by the six Virgo ellipticals which were searched for PNe, and which are assumed to lie at the same distance. The combined data are shown in Figure 3.

Two fits are shown to the data in Fig. 3. The one is the luminosity function as proposed by Jacoby and collaborators with a cutoff magnitude M*, the other one is an exponential function of the form $N(M) \propto 10^{1.6M}$ as proposed by Bottinelli et al. (1991). Clearly the latter is more realistic by giving a finite probability to the *observed* objects brighter than M*. The finite, albeit small probability of overluminous shells reaches order of unity in large samples. In fact the LF of Bottinelli et al. (1991) very well explains the observed correlation between the brightest PN shells and the sample size (cf. Fig. 2).

The data are too sparse to allow an *exact* determination of the shape of the LF. The exponential function of Bottinelli et al. (1991) is a useful working model at *least for the extreme end* of the luminosity function, because it allows for the obvious intrinsic luminosity scatter of the brightest planetary shells. If the exponential LF is adopted here, this is not to say, that it holds over a wider luminosity interval nor that it reflects the final shape of the LF.

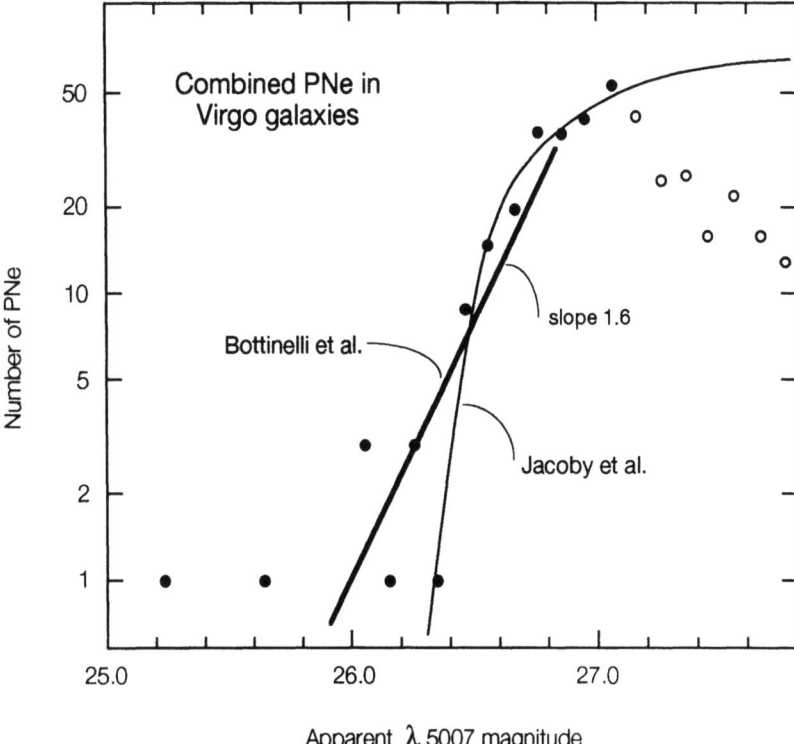

Figure 3. The observed differential λ 5007 Å luminosity function of planetary shells for a combined sample of six elliptical galaxies in the Virgo cluster. Open symbols are affected by incompleteness.

The great impact of the shape of the extreme end of the luminosity function on the distance scale comes from the fact, that the surveyed populations are widely different and increase in size with increasing distance. At the distance of the Virgo cluster the surveyed samples are on average seven times larger than that of the calibrator M31.

The expectation that the luminosity of the three brightest objects decreases, when the population size decreases, can be checked in the Virgo galaxies by defining random subsamples, which agree in population size, for instance, with that of the calibrator M31. Monte Carlo calculations show that $M°_{5007}(3)$ becomes $0.^m6$ fainter on average for the subsamples of appropriate size than for the total surveyed population of the six Virgo ellipticals. This immediately explains why the use of the artificial cutoff magnitude M* must lead to a compressed distance scale.

Another puzzle of the compressed distance scale is the large galaxy-to-galaxy variation of the specific PN number $\alpha_{2.5}$, where $\alpha_{2.5}$ is the number of PNe per $10^9$ $L_\odot$ within the brightest 2.5 magnitude interval. The variation of $\alpha_{2.5}$, which is proportional to the specific stellar death rate, is expected to be roughly constant in elliptical galaxies, but Jacoby and collaborators derived values which differ by a factor of 6. The log $\alpha_{2.5}$ values derived from the presently adopted distance scale are listed in Table 2, column 4, and are plotted in Fig. 4 against $M°_{bol}$ of the surveyed population. Also here is a clear variation of $\alpha_{2.5}$ with $M°_{bol}$, but the trend is well represented by a line of slope 0.3, which is exactly expected if $N(M) \propto 10^{1.6M}$ (cf. Bottinelli et al. 1991). Only the dwarf

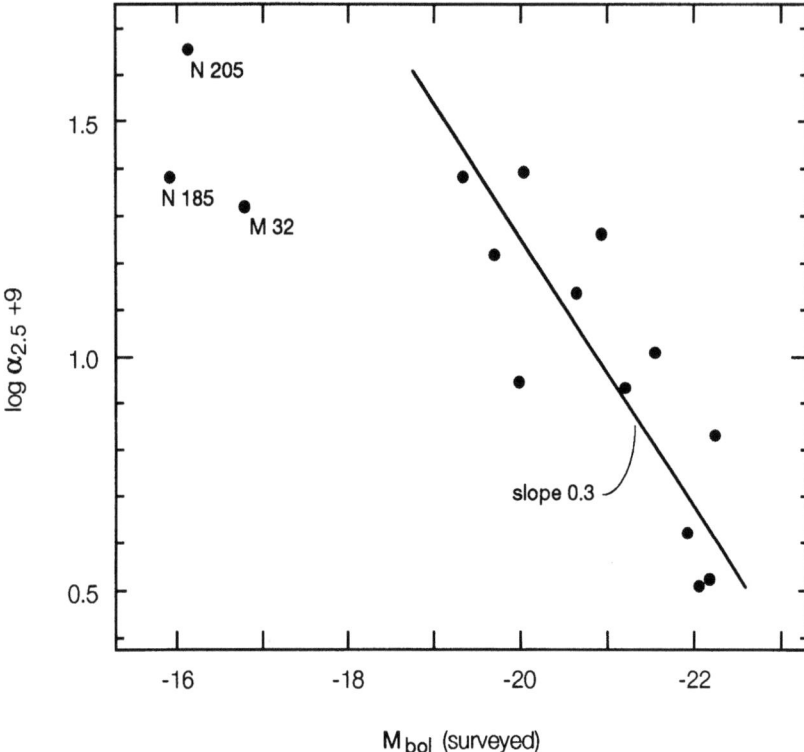

Figure 4. log $\alpha_{2.5}$ versus $M°_{bol}$ of the surveyed population. The slope of 0.3 of the drawn line follows from log $N(M) \propto 1.6 M$.

ellipticals fall off the line, since their PN numbers are so small that they do not populate the bright exponential tail of the luminosity function. In the case of an exponential LF the stellar death rates as measured by $\alpha_{2.5}$ cannot be compared directly. In large galaxies $\alpha_{2.5}$ measures the small formation rate of overluminous shells, whereas in smaller galaxies it measures the (high) formation rate of more average shells.

Ferguson and Davidsen (1992) have pointed out a correlation of the original $\alpha_{2.5}$ values with the far-UV flux of elliptical galaxies. This correlation is maintained also with the new values of $\alpha_{2.5}$ in Table 2. In either case the correlation is surprising because $\alpha_{2.5}$ has no clear physical meaning. Unfortunately the integration of the total PN numbers down to a uniform limiting luminosity is not possible, because the exponential shape of the luminosity shape holds only for the brightest part.

## 4. Conclusions

The assumption of a cutoff magnitude M* of the λ 5007 Å LF of PN shells leads to a Virgo cluster distance of ~ 15 Mpc, which is to be compared with all independent evidence requiring 21 ± 2 Mpc. The obvious reason for the compressed distance scale is the *intrinsic scatter* of the λ 5007 Å luminosity of the brightest shells. This makes the distances from PNe sensitive to the population size, and leads to systematic distance underestimates for distant and hence large galaxies.

Even if the masses of planetary nuclei obeyed a delta function, some luminosity scatter of the brightest *shells* would be expected simply because of their strong luminosity evolution. The probability of catching a PN during its brightest shell phase increases, of course, with increasing sample size. This effect is strongly enhanced if the planetary masses have a finite, albeit narrow range. Jacoby and collaborators assumed a Gaussian mass function with $\mathfrak{m} = 0.61 \pm 0.02\ \mathfrak{m}_\odot$. A central star, with three standard deviations above the mean mass, produces a short-lived, extremely bright shell (cf. Jacoby 1989), whose detection probability clearly depends on population size.

The intrinsic scatter of the luminosity of the brightest shells produces a LF which - as Bottinelli et al. (1991) have shown - is well approximated at the extreme tail by $N(M) \propto 10^{1.6M}$. This form indeed explains the observed distance dependence of Virgo ellipticals on their luminosity (Fig. 1), the luminosity increase of the three brightest shells with $M^\circ_{bol}$ of the surveyed population (Fig. 2), and the wide variation of the specific PN numbers $\alpha_{2.5}$ with $M^\circ_{bol}$ (Fig. 4).

Planetary nebulae as distance indicators are still plagued by other problems. The calibration of the (artificial) cutoff magnitude M* rests on the PNe in the bulge of an Sb galaxy (M31), whereas it is applied to E galaxies. Jacoby et al. (1990b) have argued that M* is independent of Hubble type, because they could recover the distances of LMC and SMC, and Ciardullo et al. (1991) found the same distance for the neighboring galaxies NGC 1023 (SB0) and NGC 891 (Sb). But this may be the result of the interplay of three varying parameters: population size, metallicity, and evolutionary history. In the case of the Magellanic Clouds and NGC 891 not even M*$_{bol}$ of the surveyed population has been specified. Metallicity variations alone affect M* by ~ $0.^m3$ according to Ciardullo and Jacoby (1992). Moreover Dopita et al. (1992) have shown from theoretical models that the evolutionary history has a very strong influence on the shell luminosity. A variation of the ages of bulge stars in spirals from 0.8 to 5.0 Gyr affects the $\lambda$ 5007 Å luminosity by $0.^m44$, and the age variation from 5.0 to 10.0 Gyr of the stars in (Virgo) ellipticals causes an additional luminosity change by $0.^m59$!

No stringent observational test of the claimed stability of M* has so far been attempted. This would involve about ten E galaxies in the Virgo cluster of widely different luminosity. The expectation is that their brightest PN shells follow the same relation as in Fig. 2. If indeed the relation is confirmed, PNe are weakened as distance indicators, because their brightness *and* the sample size depend on distance (cf. Bottinelli et al.). Moreover, the problem of the local calibration remains, because no fundamental distance is known to any normal E galaxy.

As a final note it is added that the observation of PN shells through the $\lambda$ 5007 Å filter is even at the distance of the Leo group by no means a simple matter. Ongoing work on sub-arcsec observations of NGC 3379 by S. Wagner and collaborators recovers - as a preliminary result - only about 80% of the objects of Ciardullo et al. (1989b), but finds roughly 50% of additional objects. Independent checks of the existing lists of PNe are in any case desirable.

Acknowledgement: Support of the Swiss National Science Foundation is gratefully acknowledged.

## References

Bottinelli, L., Gouguenheim, L., Paturel, G., and Teerikorpi, P. 1991, A.A. 252, 550.
Ciardullo, R., and Jacoby, G.H. 1992, Ap.J., in press.
Ciardullo, R., Jacoby, G.H., and Ford, H.C. 1989b, Ap.J. 344, 715.
Ciardullo, R., Jacoby, G.H., Ford, H.C., and Neill, J.D. 1989a, Ap.J. 339, 53.
Ciardullo, R., Jacoby, G.H., and Harris, W.E. 1991, Ap.J. 383, 487.
Dopita, M.A., Jacoby, G.H., and Vassiliadis, E. 1992, Ap.J. 389, 27.
Faber, S.M., Wegner, G., Burstein, D., Davies, R.L., Dressler, A., Lynden-Bell, D., and Terlevich, R.J. 1989, Ap.J.Suppl. 69, 763.
Ferguson, H.C., and Davidsen, A.F. 1992, preprint.
Harris, W.E. 1990, P.A.S.P. 102, 966.
Jacoby, G.H. 1969, Ap.J. 339, 39.
Jacoby, G.H. et al. 1992, P.A.S.P. 104, 599.
Jacoby, G.H., Ciardullo, R., and Ford, H.C. 1990a, Ap.J. 356, 332.
Jacoby, G.H., Walker, A.R., and Ciardullo, R. 1990b, Ap.J. 365, 471.
Jacoby, G.H., Ciardullo, R., Ford, H.C., and Booth, J. 1989, Ap.J. 344, 704.

Jacoby, G.H., Walker, A.R., and Ciardullo, R. 1990, Ap.J. 365, 471.
Madore, B.F., and Freedman, W.L. 1991, P.A.S.P. 103, 933.
Sandage, A., and Carlson, G. 1988, A.J. 96, 1599.
Sandage, A., Saha, A., Tammann, G.A., Panagia, N., and Macchetto, F. 1993, Ap.J.Letters, in press.
Sandage, A., and Tammann, G.A. 1974a, Ap.J. 194, 559.
Sandage, A., and Tammann, G.A. 1974b, Ap.J. 191, 603.
Schmidt, B.P., Kirshner, R.P., and Eastman, R.G. 1992, Ap.J., in press.
Tammann, G.A. 1992, Physica Scripta T 43, 31.
Tonry, J.L. 1991, Ap.J.Letters 373, L1.

# PLANETARY NEBULA BIRTH RATES IN THE GALAXY AND OTHER GALAXIES

MANUEL PEIMBERT
Instituto de Astronomía, UNAM, Apartado Postal 70-264, México 04510 D.F., México

ABSTRACT. A review is presented of PN birth rates. The observational determinations for other galaxies are compared with theoretical predictions based on stellar evolution models. The galactic birth rate is derived for different distance scales and is compared with the predicted one from galactic models of the solar neighborhood and with those determined for other galaxies; to reach agreement between observations and predictions long distance scales to PN have to be adopted. The total number of PN is estimated for the Galaxy and other galaxies.

## 1. Introduction

Phillips (1989) presented an excellent review on the estimates of: the PN formation rate, $\dot{\rho}(\text{pc}^{-3}\text{yr}^{-1})$, and the total number of PN in the Galaxy, $N_T$. Estimates of $N_T$ and $\dot{\rho}$ comprised a range of about two orders of magnitude. In this review I will try to estimate the constraints in the PN birth rate per unit luminosity, $\dot{\xi}(\text{yr}^{-1}L_\odot^{-1})$, and $N_T$ provided by stellar evolution models of a stellar ensemble (Renzini and Buzzoni 1986, Chiosi 1992) and by the seminal work of Jacoby and collaborators on the determinations of $\dot{\xi}$ and $N_T$ for extragalactic systems (Jacoby 1980, 1989; Jacoby et al. 1989, 1990a, 1990b; Ciardullo et al. 1989a, 1989b, 1991; Ciardullo and Jacoby 1992). An earlier account of some of these matters was presented by Peimbert (1990a).

## 2. Birth rates

The determinations of galactic birth rates of PN are important to: a) find out which is the fraction of the stars in the $0.8 \leq M_i(M_\odot) \leq 8$ range that undergo the PN phase, b) to study the chemical enrichment of the interstellar medium, and c) to test the different statistical distance scales proposed for galactic PN.
The local PN birth rate is given by

$$\dot{\rho} = \rho/\Delta t = \rho \langle v \rangle/(R_f - R_i), \qquad (1)$$

where $\rho$ is the density, in the solar vicinity, of PN with $R_i < R < R_f$, $\Delta t$ is the time needed for $R$ to increase from $R_i$ to $R_f$, and $\langle v \rangle$ is the average velocity of expansion from $R_i$ to $R_f$. Usually $\rho$ is given in pc$^{-3}$ and $\Delta t$ in years; $R$ denotes the radius of ionized hydrogen.
In the optically thick (ionization bounded) phase $\langle v \rangle$ denotes the average velocity of the ionization front relative to the central star, $v_{ion}$, while in the optically thin (matter bounded) phase $\langle v \rangle$ denotes the average velocity of expansion of matter, $v_{exp}$,

given by the Doppler effect.

There are several sources of error associated with the use of equation (1) (see Phillips 1989, Peimbert 1990b). The main uncertainty is due to the adopted distance scale: since $\dot{\rho}$ is proportional to $d^{-4}$ the $\dot{\rho}$ estimates vary to a first approximation like $k^{-4}$ (see Table 3).

We will use the birth rate per unit luminosity, $\dot{\xi}$, instead of $\dot{\rho}$ because it is a more convenient quantity to compare with results for other galaxies and with model predictions for stellar ensembles. The galactic PN birth rate, $\dot{N}_T(\text{yr}^{-1})$, has been estimated by multiplying $\dot{\xi}$ by the bolometric luminosity of the Galaxy. The total number of PN has been derived by multiplying the galactic birth rate by the mean lifetime of a PN.

## 3. Stellar death rate

It is possible to compare $\dot{\xi}$ with the stellar death rate per solar bolometric luminosity, $\dot{S}$. Renzini and Buzzoni (1986) computed $\dot{S}_b$ values as a function of time from models with a single burst of star formation, without subsequent star formation or star accretion, and three widely different initial mass functions, IMF, independent of time and chemical composition. Their assumed IMF by number are given by: $\phi(M_i) = AM_i^{-\alpha}$, with $\alpha$ equal to 1.5, 2.35 and 3.5 for $M_i \geq 0.57 M_\odot$ and $\alpha = 2.35$ for $M_i < 0.57 M_\odot$ for the three IMF.

Peimbert (1990a) has shown that to a very good approximation $\dot{S}_b(t_1)$, where $t_1$ is the age of the system, corresponds to the stellar death rate per unit luminosity for systems with a constant rate of star formation and for systems with a decreasing rate of star formation. This result is very powerful and implies that for galaxies where star formation started more than $10^9$ years ago and that are not suffering from an extremely violent burst of star formation, $\dot{S}(10^{-12}\text{yr}^{-1}L_\odot^{-1})$ should be in the 18 to 22 range; moreover it also implies that if in these systems all the stars in the $0.8 \leq M(M_\odot) \leq 8$ range go through the PN phase then $\dot{\xi} = \dot{S}$. Chiosi (1992) from an independent study of the stellar death rate obtains very similar results to those of Renzini and Buzzoni (1986).

## 4. Extragalactic systems

Table 1 presents the absolute bolometric magnitudes of 17 galactic systems. Also Table 1 presents the $\dot{\xi}$ and $N_T$ values based on the following assumptions: a) the PN luminosity function established by Jacoby (1989), Jacoby et al. (1989) and Ciardullo et al. (1989b) applies to all the systems and b) the lifetimes of the PN are of 25,000 years.

In Figures 1a and 1b we present the $(B-V)_0$ versus $\dot{\xi}$ and $M_{bol}$ versus $\dot{\xi}$ diagrams respectively. In these figures NGC 205 and NGC 185 were not plotted due to the large estimated errors. The errors quoted correspond to the square root over the total number of detected PN but they do not include errors in the estimated lifetimes nor in the PNLF. The $\dot{\xi}$ values for the SMC and the LMC are smaller than those presented by Peimbert (1990a) mainly due to the study made by Boroson and Liebert (1989) who found that a considerable fraction of the PN candidates in the list by Jacoby (1980) were not PN.

The decrease of $\dot{\xi}$ with $(B-V)_0$ and with increasing luminosity has been discussed by Peimbert (1990a) and by Ciardullo et al. (1991).

For all the galaxies in Table 1 a value of $\dot{S}$ in the range of $15 \leq 10^{-12}\text{yr}^{-1}L_\odot^{-1} \leq 22$ is predicted; slightly higher $\dot{S}$ values for the more luminous galaxies are expected due to their older stellar population (see Peimbert 1990a). Therefore the decrease of $\dot{\xi}$ with $(B-V)_0$ and with $M_{bol}$ is not due to a decrease in $\dot{S}$.

TABLE 1. Distance moduli, absolute bolometric magnitudes, birth rates and total number of PN.

| Object | $(m-M)_0$ | $M_{bol}$ | $\dot{\xi}$ $(10^{-12}\mathrm{yr}^{-1}L_\odot^{-1})$ | $N_T$ $(10^3)$ | References |
|---|---|---|---|---|---|
| NGC 4472 | 30.71 | -23.29 | 2.7 ± 0.6 | 11.1 ± 2.5 | 1,2,3,4,5 |
| NGC 4486 | 30.81 | -23.08 | 3.4 ± 0.6 | 11.5 ± 2.0 | 1,3,4,5,6 |
| NGC 4649 | 30.76 | -22.85 | 2.6 ± 0.8 | 7.1 ± 2.2 | 1,2,3,4,5 |
| NGC 4406 | 30.98 | -22.78 | 5.4 ± 0.8 | 13.9 ± 2.1 | 1,2,3,4,5 |
| NGC 4374 | 30.98 | -22.60 | 6.8 ± 1.3 | 14.8 ± 2.8 | 1,2,3,4,5 |
| NGC 4382 | 30.79 | -22.31 | 8.0 ± 1.2 | 13.3 ± 2.0 | 1,3,4,5,6 |
| M31 | 24.26 | -21.96 | 6.6 ± 1.2 | 8.0 ± 1.5 | 5,7,8,9 |
| M81 | 27.72 | -21.88 | 8.4 ± 1.8 | 9.4 ± 2.0 | 5,10,11 |
| NGC 3379 | 29.96 | -21.57 | 8.5 ± 1.7 | 7.2 ± 1.4 | 3,5,8,12 |
| NGC 1023 | 29.97 | -21.51 | 9.1 ± 1.5 | 7.3 ± 1.2 | 3,8,13,14,15 |
| NGC 3384 | 30.03 | -21.25 | 15.0 ± 3.0 | 9.4 ± 1.9 | 3,5,8,12 |
| NGC 3377 | 30.07 | -20.69 | 15.0 ± 3.0 | 5.6 ± 1.4 | 3,5,8,12 |
| LMC | 18.44 | -19.03 | 8.1 ± 1.8 | 0.64 ± 0.14 | 8,15,16,17,18,19,20 |
| SMC | 18.76 | -17.20 | 7.7 ± 2.3 | 0.12 ± 0.04 | 8,15,16,17,18,19,21 |
| M32 | 24.26 | -17.14 | 10.1 ± 2.5 | 0.14 ± 0.04 | 5,7,8 |
| NGC 205 | 24.26 | -16.86 | 23.5 ± 6.8 | 0.26 ± 0.08 | 7,8,15,22 |
| NGC 185 | 23.78 | -15.89 | 14.5 ± 7.3 | 0.07 ± 0.04 | 7,8,15,23 |

1) Burstein and Heiles 1984, 2) Poulain 1988, 3) Sandage and Tammann 1981, 4) Jacoby et al. 1990a, 5) Peimbert 1990a, 6) Michard 1982, 7) Ciardullo et al. 1989b, 8) de Vaucouleurs et al. 1976, 9) Freeman 1970, 10) Brandt et al. 1972, 11) Jacoby et al. 1989, 12) Ciardullo et al. 1989a, 13) Frogel et al. 1978, 14) Ciardullo et al. 1991, 15) This work, 16) Hindman 1967, 17) Kruit 1990, 18) Bothun and Thompson 1988, 19) Jacoby et al. 1990b, 20) Boroson and Liebert 1989, 21) Feast 1988, 22) Ciardullo and Jacoby 1992, 23) Saha and Hoessel 1990.

Most of the PNLF used for the determinations of $\dot{\xi}$ in Table 1 are complete only at the high luminosity end; the completeness limit of the observations extends only from 0.8 to 2.5 mag below the bright end cutoff. The total number of PN is obtained by adjusting the observed upper end of the PNLF with the PNLF derived by Jacoby and collaborators. Since the PNLF spans about 8 magnitudes (Ciardullo et al. 1989b), the $\dot{\xi}$ values represent the upper end of the PNLF. Therefore strictly speaking, it can only be said that the number of bright PN decreases with increasing luminosity of the galaxy for $(B-V)_0 > 0.90$ or $M_{bol} < -22.0$.

The decrease in the number of bright PN with the increase of $(B-V)_0$ and luminosity could be due to: a) an age effect in galaxies with stellar populations of different ages (Peimbert 1990a) or b) to a spread in the masses of the central stars of PN, $M_c$, produced by stars with similar initial masses in the main sequence (Ciardullo et al. 1991). In both cases larger ages will produce fainter and fewer PN due to the following reasons: a) lower luminosity of the central star, b) smaller mass of the shell and c) longer stellar evolutionary times that might prevent the star to become hot enough to ionize the nebula before it has dissipated.

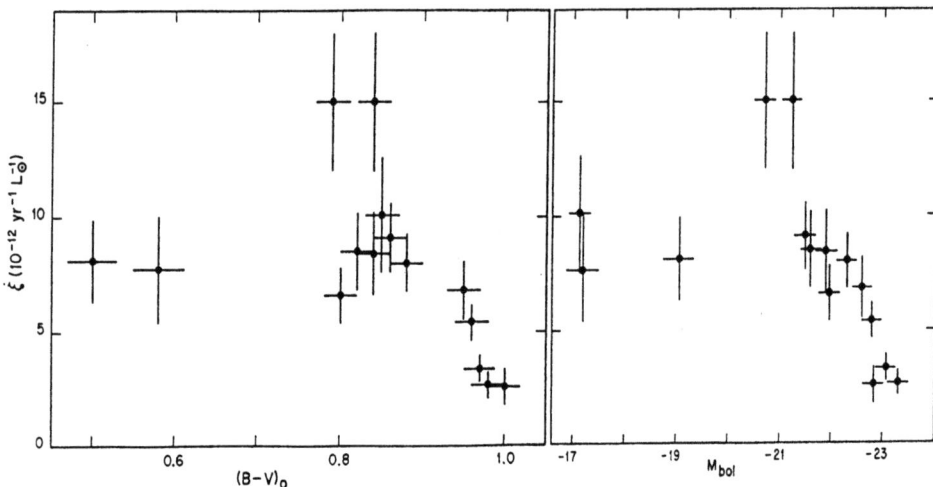

Figure 1. a) PN birth rate per solar luminosity, $\dot{\xi}$, versus intrinsic color index, $(B-V)_0$, for a group of fifteen galaxies. b) $\dot{\xi}$ versus bolometric magnitude, $M_{bol}$.

## 5. The Galaxy

### 5.1 GALACTIC MODELS

To derive $\dot{\xi}$ and $N_T$ for the Galaxy we need to know the luminosity of the solar neighborhood and the $M_{bol}$ for the whole system. Table 2 presents galactic properties based on three different sources, they are very similar; in what follows we will adopt $M_{bol} = -21.8$ for the Galaxy.

Other authors have determined $N_T$ based on the specific number of PN per unit mass (e.g. Alloin et al. 1976, Phillips 1989). We consider the determinations of $N_T$ based on the number of PN per unit luminosity to be more accurate than those derived from the number of PN per unit mass. All the $N_T$ values presented in this review are based on the luminosity of the system involved.

TABLE 2. Galactic properties

| Solar Neighborhood | | | | Whole Galaxy | | | References |
|---|---|---|---|---|---|---|---|
| L $(L_\odot pc^{-3})$ | $L_v$ $(L_\odot pc^{-3})$ | $\mu_v$ $(L_\odot pc^{-2})$ | B.C. | B-V | $M_v$ | $M_{bol}$ | |
| $1.15 \times 10^{-1}$ | $5.34 \times 10^{-2}$ | ... | -0.84 | ... | -20.5 | -21.3: | 1 |
| ... | ... | ... | ... | $0.53 \pm 0.05$ | -20.6 | -21.4: | 2 |
| ... | ... | $22.2 \pm 2.8$ | ... | $0.85 \pm 0.15$ | -21.15 | -21.95: | 3,4 |

1) Allen 1973, 2) de Vaucouleurs and Pence 1978, 3) Kruit 1986, 4) Kruit 1990.

TABLE 3. A comparison of distance scales for PN with $0.10 \leq R(\text{pc}) \leq 0.30$.

| $k = d/d_{\text{Seaton}}$ | M(rms) $(M_\odot)$ | Scale |
|---|---|---|
| 1.43 | 0.50 | Cudworth (1974) |
| 1.41 | 0.49 | Kingsburgh and Barlow (1992) |
| 1.41 | ... | Mallik and Peimbert (1988) |
| 1.40 | 0.47 | Schneider and Terzian (1983) |
| 1.30 | 0.39 | Weidemann (1977) |
| 1.06 | ... | Maciel and Pottasch (1980) |
| 1.00 | 0.20 | Seaton (1968) |
| 1.00 | 0.20 | Cahn and Kaler (1971) |
| 1.00 | 0.20 | Milne and Aller (1975) |
| 0.86 | 0.14 | Daub (1982) |

5.2 DISTANCE SCALES

The density of PN in the solar neighborhood depends on the distance scale to the fourth power, see equation (1). The distance scale is still a controversial issue and widely different distance scales are used by different authors. To compare different distance scales it is possible to introduce a relative scale factor, $k$, with the normalization $k = 1$ for Seaton's (1968) distance scale. In Table 3 we present a comparison in the $0.10 \leq R(\text{pc}) \leq 0.30$ range for some of the most frequently used distance scales. For optically thin PN samples $k \propto M(\text{rms})^{2/5}$, where $M(\text{rms})$ is the envelope mass derived from the root mean square density. For optically thick PN samples $k$ increases with $R$ and an average value for the considered range is presented in Table 3. If objects smaller than 0.1 pc are considered the spread in $k$ values is even larger (Gathier 1987). In this comparison the $M(\text{rms})$ values have been computed under the assumptions that $N(\text{He})/N(\text{H}) = 0.11$ and $N(\text{He}^+)/N(\text{H}^+) + 2N(\text{He}^{++})/N(\text{H}^+) = 0.13$.

5.3 BIRTH RATES AND TOTAL NUMBER OF PN

Table 4 presents four determinations of $\dot{\xi}$. For the first determination we took the surface density of PN by Mallik (1991), $\sigma = 5.8 \pm 1.2$ kpc$^{-2}$, the $\mu_v$ value in Table 2, a lifetime of 25,000 years and a B.C. of 0.84 mag. For the second value we took the volume density of PN by Mallik, $\rho = 19 \pm 4$ kpc$^{-3}$, the $L$ value in Table 2 and a lifetime of 25,000 years. The errors come from the PN densities. The difference between the first two determinations is mainly due to the different scale heights of PN and stellar luminosity in the solar neighborhood (see §5.4). The third $\dot{\xi}$ value was derived from Figure 1a assuming a $(B-V)_0 = 0.85$ for the Galaxy and the last $\dot{\xi}$ value from Figure 1b and assuming $M_{bol} = -21.8$ for the Galaxy (see Table 2).

Table 5 presents $\dot{\xi}$ determinations derived from $\rho$ values by different authors, that were obtained based on different samples and different distance scales, and the L value by Allen (1973) for the solar neighborhood presented in Table 2. We have assumed a lifetime of 25,000 years for the PN, with the exception of the determination by Ishida and Weinberger (1987) for which we used a lifetime of 40,000 years, given by them, because they are considering larger and fainter nebulae than the other authors. For all the $N_T$ determinations we assumed a lifetime of 25,000 years for the PN and an $M_{bol} = -21.8$ for the Galaxy. Most of the differences in Table 5 are due to the distance scale adopted.

TABLE 4. Galactic PN birth rate

| $\dot{\xi}$ $(10^{-12}\text{yr}^{-1}L_\odot^{-1})$ | Method | References |
|---|---|---|
| 4.8 ± 1.0 | Surface density of PN and surface brightness at the solar galactocentric distance | 1,2,3,4 |
| 6.6 ± 1.4 | Local density of PN and solar neighborhood luminosity function | 1,2,3,5 |
| 9.0 ± 3 | $(B-V)_0$ versus $\dot{\xi}$ (Fig. 1a and Table 2) | 3 |
| 9.0 ± 3 | $M_{bol}$ versus $\dot{\xi}$ (Fig. 1b and Table 2) | 3 |

1) Mallik and Peimbert 1988, 2) Mallik 1991, 3) this work, 4) Kruit 1986, 5) Allen 1973.

TABLE 5. WD and PN birth rates in the solar neighborhood and the total number of PN in the Galaxy

| $\dot{\xi}$ $(10^{-12}\text{yr}^{-1}L_\odot^{-1})$ | | $N_T(\text{PN})$ $(10^3)$ | $k$ | References |
|---|---|---|---|---|
| 5.4 ± 1.1 | WD | ... | ... | 1 |
| 6.3 ± 2.2 | WD | ... | ... | 2 |
| 6.6 ± 1.4 | PN | 6.9 | 1.41 | 3 |
| 10.1 ± 2.6 | PN | 10.5 | 1.43 | 4,5,6 |
| 12.2 ± 2.2 | PN | 12.7 | 1.30 | 7 |
| 16.6 ± 2.2 | PN | 17.3 | ... | 8 |
| 16.7 | PN | 17.4 | ... | 9 |
| 20.0 | WD | ... | ... | 10 |
| 42.3 ± 11 | PN | 44.1 | 1.00 | 5,6,11 |
| 43.5 ± 17 | PN | 45.3 | 0.86 | 6,12 |
| 69.6 | PN | 72.6 | ... | 13 |

1) Fleming et al. 1986, 2) Downes 1986, 3) Mallik 1991, 4) Cudworth 1974, 5) Alloin et al. 1976, 6) Peimbert 1990a, this work, 7) Weidemann 1977, 8) Phillips 1989, 9) Acker 1978, 10) Weidemann 1991, 11) Cahn and Kaler 1971, 12) Daub 1982, 13) Ishida and Weinberger 1987.

5.4 SCALE HEIGHTS

To compare the $\dot{S}$ values from galactic evolution models with the observed $\dot{\xi}$ values we need to know if all the stars go through the PN phase, this can be tested by comparing the scale heights of the PN and of the stellar luminosity of the solar neighborhood.

The scale heights derived by different authors are: 90 pc (Cahn and Kaler 1971), 100 pc (Ishida and Weinberger 1987), 115 pc (Cahn and Wyatt 1976), 125 pc (Daub 1982), 130 pc (Amnuel et al. 1984) and 153 pc (Mallik 1991). In general the larger the distance scale adopted the larger the derived scale height. These scale heights can be compared with those derived for the different populations in the solar neighborhood. The model of Kruit (1986) has a scale height of 80 pc for population I stars with $L_v(z=0) = 39.5\%$ and a scale height of 325 pc for old dwarfs and disk giants with $L_v(z=0) = 60.5\%$.

From the model of Kruit (1986) and Table 2 I assume that the luminosity distribution in the solar neighborhood can be approximated by

$$L(z) = 1.15 \times 10^{-1}(0.40e^{-z/80} + 0.60e^{-z/325})L_\odot \text{pc}^{-3}, \qquad (2)$$

where $z$ is the distance to the plane in pc, the first term corresponds to the population I component and the second to the old disk population. The scale height of $L(z)$ is approximately 227 pc. The population I component with $M < 8M_\odot$ has $\dot{S} = 18 \times 10^{-12}\text{yr}^{-1}L_\odot^{-1}$ while the old disk component has $\dot{S} = 21 \times 10^{-12}\text{yr}^{-1}L_\odot^{-1}$ (Renzini and Buzzoni 1986). By assuming that all the stars of the population I component with $M < 8M_\odot$ produce PN and that only a fraction of the old disk stars produce PN we have computed Table 6. From this table we obtain that approximately 28% of the old disk stars go through the PN stage for the scale height derived by Mallik (1991). In Figure 2 the region to the left of the model is permitted and implies that long distance scales with $k \geq 1.3$ and $h \geq 140$ pc should be used. In Table 7 we present $\dot{\xi}$, $h$, $N_T$ and $\dot{N}_T$ as a function of $k$ based on the local density of PN determined by Mallik.

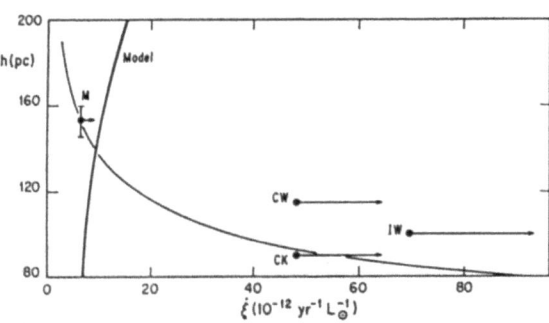

Figure 2. $\dot{\xi}$ versus scale height, $h$, for the solar neighborhood. The four points correspond to: Mallik ($k = 1.41$), Ishida and Weinberger, Cahn and Kaler ($k = 1$) and Cahn and Wyatt ($k = 1$). The model values come from Table 6. The other curve indicates the values derived from Mallik (1991) assuming a variable $k$ factor. The arrows indicate the growth in $\dot{\xi}$ under the assumption that the lifetime of PN is of 18,750 years instead of 25,000.

TABLE 6. Birth rate for $z = 0$ pc as a function of the PN scale height.

| $\dot{\xi}$ $(10^{-12}\text{yr}^{-1}L_\odot^{-1})$ | h (pc) | % of old disk stars that become PN |
|---|---|---|
| 7.2 | 80 | 0 |
| 7.9 | 100 | 6 |
| 9.1 | 125 | 15 |
| 10.7 | 150 | 27 |
| 12.5 | 175 | 42 |
| 15.3 | 200 | 64 |
| 19.4 | 225 | 97 |
| 19.8 | 227 | 100 |

TABLE 7. $\dot{\xi}$, h, $N_T$ and $\dot{N}_T$ as a function of the galactic distance scale, $k$, based on the local volume density of PN determined by Mallik (1991). Hubble parameter based on the consistency of the galactic and extragalactic PNLF with $k = 1.20$ and $H_0 = 74$ km s$^{-1}$ Mpc$^{-1}$ (Méndez et al. 1992a, 1992b, Jacoby and Ciardullo 1992).

| $k$ | $\dot{\xi}$ $(10^{-12}\text{yr}^{-1}L_\odot^{-1})$ | $h$ (pc) | $N_T$ $(10^3)$ | $\dot{N}_T$ (yr$^{-1}$) | $H_0$ (km s$^{-1}$ Mpc$^{-1}$) |
|---|---|---|---|---|---|
| 0.8  | 64.5 | 87  | 67   | 2.7  | 111 |
| 1.00 | 27.2 | 109 | 28   | 1.13 | 89  |
| 1.20 | 12.7 | 130 | 13.2 | 0.53 | 74  |
| 1.40 | 7.1  | 152 | 7.4  | 0.30 | 63  |
| 1.50 | 5.4  | 163 | 5.6  | 0.22 | 59  |
| 1.75 | 2.9  | 190 | 3.0  | 0.12 | 51  |
| 2.00 | 1.7  | 217 | 1.8  | 0.07 | 44  |

## 6. Optical thickness

The assumption of 25,000 years for the PN lifetime, combined with a velocity of expansion of 25 km s$^{-1}$ (Phillips 1989), imply a limiting size of $R = 0.64$ pc for optically thin PN. If we assume that the samples are made of optically thin PN with $R < 0.64$ pc, then the $\dot{\xi}$ values in Tables 1, 4, 5 and 7 are correct. On the other hand if we assume that $R_{max} = 0.64$ pc and that during part of their lifetimes the PN are optically thin and the rest optically thick, the lifetimes become smaller because during the optically thick phase $v_{ion}$ should be used instead of $v_{exp}$.

From the distance independent sample by Mallik and Peimbert (1988) it follows that most PN in the solar neighborhood are optically thick, in at least some directions, a similar result was obtained by Gathier (1987). A complementary result was obtained by Méndez et al. (1992a) who find that most PN in their sample are optically thin in at least some directions. Consequently a large fraction of PN are optically thick in some directions and thin in others. Marten (1992) finds for an optically thick model of PN that $v_{ion}/v_{exp} \approx 1.5$. By assuming that during part of their evolution PN are optically thick, then their lifetimes will get reduced and their $\dot{\xi}$ values will be increased by $v_{ion}/v_{exp}$; this effect will affect the $\dot{\xi}$ values in Tables 1, 4, 5, and 7 but obviously will not affect the $N_T$ values in all tables nor the $\dot{\xi}$ values in Table 6. In Figure 2 the arrows represent an increase of 1.33 in the $\dot{\xi}$ values.

PN with $R > 0.64$ pc present several problems and I consider that it is a good idea to avoid them in the determination of $\dot{\xi}$: a) the PN samples suffer from severe incompleteness of intrinsically large PN due to their faintness, b) their expansion could be decelerated by the interstellar medium (Ishida and Weinberger 1987), c) in the late stages of evolution PN could become optically thick again, and if the number of ionizing photons decreases rapidly with time the ionized size could also decrease with time.

## 7. Discussion

By comparing the distances of Cudworth (1974) and Méndez et al. (1992a) I find that $k = 1.20 \pm 0.2$ for the latter sample, the large error comes from the small number of PN involved. The Méndez et al. sample has a selection effect in favor of high latitude

PN, which presumably have on the average fainter stars than the samples used by other authors; notice however that Cudworth (1974) does not find a significant difference in the distance scales for B and C nebulae. Méndez et al. (1992b) find that the galactic PNLF derived from the Méndez et al. (1992a) sample is similar to the extragalactic PNLF derived by Jacoby and collaborators. If the galactic PNLF derived from the sample of Méndez et al. (1992a) agrees with the extragalactic one, the $H_0$ value of 74 km s$^{-1}$ Mpc$^{-1}$ derived by Jacoby and Ciardullo (1992) corresponds to $k = 1.2$; the error in this equivalence is of a factor of $\sim 1.2$. In Table 7 we present the $H_0$ values for different $k$ values.

The decrease of $\dot{\xi}$ with increasing $M_{bol}$ for galaxies brighter than $M_{bol} = -22.0$ or redder than $(B - V)_0 = 0.90$ could be due to a smaller fraction of the older stars producing PN in agreement with the scale height determination for the Galaxy where we find that only $\sim 28\%$ of the old disk population objects go through the PN stage.

The $\dot{\xi}$ determinations derived from: a) the $(B - V)_0$ and $M_{bol}$ values for the Galaxy (Figure 1 and Table 2) for $v_{ion}/v_{exp} = 1.3$, b) the distance scales by Cudworth (1974) and Mallik and Peimbert (1988) (Table 5) for $v_{ion}/v_{exp} = 1.3$ and c) the scale height determinations by Mallik (1991) (Table 6), imply that $\langle\dot{\xi}\rangle = (11 \pm 3) \times 10^{-12} \text{yr}^{-1} L_\odot^{-1}$ for the Galaxy. This value compared with the results by Renzini and Buzzoni (1986) implies that about half of the IMS go through the PN phase.

Similarly from the same sources it is found that $N_T = 9000 \pm 3000$. From a) and c) above it is found that $k = 1.4 \pm 0.1$; where the upper limit comes mainly from the $\dot{\xi}$ extragalactic results, and the lower limit from the stellar death rate predictions and the PN scale height.

From the $\langle\dot{\xi}\rangle$ value for the Galaxy it is found that $\dot{N}_T(\text{PN}) = 0.46$ yr$^{-1}$. Alternatively from Renzini and Buzzoni (1986) if all the stars go through the PN phase then $\dot{N}_T = 0.83$ yr$^{-1}$. These numbers can be compared with the white dwarf formation rate: $\dot{N}(\text{WD}) = 0.25$ yr$^{-1}$ (Fleming et al. 1986, Downes 1986), $\dot{N}(\text{WD}) = 0.83$ yr$^{-1}$ (Weidemann 1991); and the OH/IR star formation rate $\dot{N}(\text{OH/IR}) \sim 0.9$ yr$^{-1}$ (Herman and Habing 1985).

It is a pleasure to acknowledge several fruitful discussions with G.H. Jacoby, D.C.V. Mallik and R. Méndez.

## References

Acker, A. 1978, *Astr. and Ap. Suppl.*, **33**, 367.
Allen, C.W. 1973, *Astrophysical Quantities*, (London: Athlone).
Alloin, D., Cruz-González, C., and Peimbert, M. 1976, *Ap. J.*, **205**, 74.
Amnuel, P.R., Guseinov, O.H., Novruzova, H.I., Rustamov, Yu. S. 1984, *Ap. Sp. Sci.*, **107**, 19.
Boroson, T.A. and Liebert, J. 1989, *Ap. J.*, **339**, 844.
Bothun, G.D. and Thompson, I.B. 1988, *A.J.*, **96**, 877.
Brandt, J.C., Kalinowski, J.K., and Roosen, R.G. 1972, *Ap. J. Suppl.*, **24**, 421.
Burstein, D., and Heiles, C. 1984, *Ap. J. Suppl.*, **54**, 33.
Cahn, J.H. and Kaler, J.B. 1971, *Ap. J. Suppl.*, **22**, 319.
Cahn, J.H. and Wyatt, S.P. 1976, *Ap. J.*, **210**, 508.
Chiosi, C. 1992, private communication.
Ciardullo, R. and Jacoby, G.H. 1992, *Ap. J.*, in press.
Ciardullo, R., Jacoby, G.H., and Ford, H.C. 1989a, *Ap. J.*, **344**, 715.
Ciardullo, R., Jacoby, G.H., Ford, H.C., and Neill, J.D. 1989b, *Ap. J.*, **339**, 53.
Ciardullo, R., Jacoby, G.H., and Harris, W.E. 1991, *Ap. J.*, **383**, 487.

Cudworth, K.M. 1974, *A.J.*, **79**, 1384.
Daub, C.T. 1982, *Ap. J.*, **260**, 612.
de Vaucouleurs, G., de Vaucouleurs, A., and Corwin, H.G. Jr. 1976, *Second Reference Catalogue of Bright Galaxies*, (Austin: University of Texas Press).
de Vaucouleurs, G. and Pence, W.D. 1978, *A.J.*, **84**, 1163.
Downes, R.A. 1986, *Ap. J. Suppl.*, **61**, 569.
Feast, M.W. 1988, in *The Extragalactic Distance Scale*, A.S.P. Conference Series No. 4, ed. S. van den Bergh and C.J. Pritchet (Provo: Brigham Young University Press), p. 9.
Fleming, T.A., Liebert, J., and Green, R.F. 1986, *Ap. J.*,, **308**, 176.
Freeman, K.C. 1970, *Ap. J.*, **160**, 811.
Frogel, J.A., Persson, S.E., Aaronson, M. and Matthews, K. 1978, *Ap. J.*, **220**, 75.
Gathier, R. 1987, *Astr. and Ap. Suppl.*, **71**, 245.
Herman, J. and Habing, H.J. 1985, *Physics Reports*, **124**, No. 4, p. 255.
Hindman, J.V. 1967, *Australian J. Phys.*, **20**, 147.
Ishida, K. and Weinberger, R. 1987, *Astr. and Ap.*, **178**, 227.
Jacoby, G.H. 1980, *Ap. J. Suppl.*, **42**, 1.
Jacoby, G.H. 1989, *Ap. J.*, **339**, 39.
Jacoby, G.H. and Ciardullo, R. 1992, this Symposium.
Jacoby, G.H., Ciardullo, R., and Ford, H. 1990a, *Ap. J.*, **356**, 332.
Jacoby, G.H., Ciardullo, R., Ford, H.C., and Booth, J. 1989, *Ap. J.*, **344**, 704.
Jacoby, G.H., Walker, A.R. and Ciardullo, R. 1990b, *Ap. J.*, **356**, 332.
Kingsburgh, R.L. and Barlow, M.J. 1992, *MNRAS*, **257**, 317.
Kruit, P.C. van der 1986, *Astr. and Ap.*, **157**, 230.
Kruit, P.C. van der 1990, in *The Milky Way as a Galaxy*, eds. G.F. Gilmore, I.R. King and P.C. van der Kruit (University Science Books, California) p. 331.
Maciel, W.J. and Pottasch, S.R. 1980, *Astr. and Ap.*, **88**, 1.
Mallik, D.C.V. 1991, *Proc. Astron. Soc. Australia*, **9**, 15.
Mallik, D.C.V. and Peimbert, M. 1988, *Rev. Mexicana Astron. Astrof.*, **16**, 111.
Marten, H. 1992, this Symposium.
Méndez, R.H., Kudritzki, R.P. and Herrero, A. 1992a, *Astr. and Ap.*, **260**, 329.
Méndez, R.H., Kudritzki, R.P., Ciardullo, R. and Jacoby, G.H. 1992b, in preparation.
Michard, R. 1982, *Astr. and Ap. Suppl.*, **49**, 591.
Milne, D.K. and Aller, L.H. 1975, *Astr. and Ap.*, **38**, 183.
Peimbert, M. 1990a, *Rev. Mexicana Astron. Astrof.*, **20**, 119.
Peimbert, M. 1990b, *Reports on Progress in Physics*, **53**, 1559.
Phillips, J.P. 1989, in *IAU Symposium No. 131, Planetary Nebulae*, ed. S. Torres-Peimbert (Dordrecht: Kluwer), p. 425.
Poulain, P. 1988, *Astr. and Ap. Suppl.*, **72**, 215.
Renzini, A. and Buzzoni, A. 1986 in *Spectral Evolution of Galaxies*, ed. C. Chiosi and A. Renzini (Dordrecht: Reidel), p. 195.
Saha, A. and Hoessel, J.G. 1990, *A.J.*, **99**, 97.
Sandage, A. and Tammann, G.A. 1981, *A Revised Shapley-Ames Catalog of Bright Galaxies*, (Washington: Carnegie Institute of Washington).
Schneider, S.E. and Terzian, Y. 1983, *Ap. J.*, **274**, L61.
Seaton, M.J. 1968, *Ap. (Letters)*, **2**, 55.
Weidemann, V. 1977, *Astr. and Ap.*, **61**, L27.
Weidemann, V. 1991, in *Proc. $7^{th}$ European Workshop on White Dwarfs*, eds. G. Vauclair and E.M. Sion (Dordrecht: Kluwer), p. 67.

# PLANETARY NEBULAE AND HALO DYNAMICS IN EARLY TYPE GALAXIES

X. HUI

*Astronomy Department, California Institute of Technology, USA*

and

H.C. FORD

*Department of Physics and Astronomy, The Johns Hopkins University
and Space Telescope Science Institute, USA*

## 1. Introduction

The study of extragalactic planetary nebulae has made rapid progress in recent years with the help of high quantum efficiency detectors. A brief but distinctive phase in the late stage of stellar evolution, planetary nebulae (PNe) are not only interesting objects in their own right, but also are extremely valuable and unique tools for probing the host galaxies. Recent studies have used planetary nebulae as test particles to investigate the dynamics and mass distributions in the halos of early type galaxies.

In spiral galaxies HI rotation curves are measured far beyond the optically bright disks and are usually flat at large radii, indicating a dark halo whose mass is often several times that of the luminous matter. Because most early type galaxies contain very little gas, and individual stars are too faint to observe, information on the mass distributions and stellar kinematics in these galaxies is usually derived from integrated absorption spectra. However, the rapid falloff of surface brightness severly limits such measurements, and high signal-to-noise spectra are only possible in the bright central regions. Very little is known about the halos of early type galaxies.

Planetary nebula surveys have been carried out in over 20 early type galaxies (Ciardullo *et al.* 1989, Jacoby *et al.* 1990, Ciardullo *et al.* 1991, Hui *et al.* 1992a and Ford *et al.* 1992). By comparing the CCD images taken on the characteristic emission line [O III] $\lambda 5007$ and the adjacent continuum, a large number of planetary nebulae can be identified in individual galaxies up to 15 Mpc. The radial velocities of these planetaries can then be measured efficiently with a multifiber spectrograph. As an example, in NGC 5128, a nearby giant elliptical galaxy, over 400 planetary nebula (PN) velocities were measured out to a galactic radius of 20 kpc (Hui *et al.* 1992b). Planetary nebulae thus provide a unique way to study the stellar dynamics and mass distributions in galactic halos. If the globular clusters are also observed, comparison of the stars and clusters will help clarify how the halos formed.

In this paper we focus on the PN kinematics and the halo mass distribution in NGC 5128, the best studied galaxy to date. We then briefly describe the continuing efforts to measure PN velocities in more distant galaxies such as NGC 4594, NGC 3379, and elliptical galaxies in the Virgo Cluster.

## 2. NGC 5128

NGC 5128 (also known as the radio source Centaurus A), which has a dust lane lying along its photometric minor axis, is the nearest (3.5 Mpc, Hui et al. 1992a) example of a galaxy merger. Graham (1979) used observations of the HII regions along the edge of the dust lane to show that the gas and dust are in a highly inclined, rotating disk. Graham proposed that the gas and dust came from a gas cloud or small galaxy which merged with NGC 5128. Subsequent observations of the disk $H_\alpha$ emission (Bland, Taylor & Atherton 1987; Nicholson, Bland-Hawthorn & Taylor 1992) and neutral hydrogen 21-cm emission (van Gorkom et al. 1990) confirmed the earlier findings and revealed that the dust lane is "the projection of a severely warped, thin disc of gas and dust". Wilkinson et al. (1985) used absorption spectra to show that the rotation of the main stellar body is around the photometric minor axis of the elliptical component, perpendicular to the disk rotation. These papers demonstrate the prescience of Baade and Minkowski's belief that NGC 5128 is two galaxies in collision (cf. Sandage 1961).

However, because of the low surface brightness in the halo, previous studies were largely restricted to the observations of the bright stellar envelope and the ionized gas associated with the conspicuous dust lane. To obtain stellar kinematic information in the halo, we surveyed NGC 5128 for planetary nebulae. The observations were made on the CTIO 4-m telescope using the prime focus CCD camera (Hui et al. 1992a). Figure 1 shows the locations of our CCD fields. Along the photometric major and minor axes, the survey frames were placed on both sides of the galaxy in order to measure the halo rotation. The outermost halo fields are approximately 20 kpc from the center of the galaxy.

Spectroscopic observations were made using multifiber spectrographs on both the AAT and CTIO 4-m (Hui et al. 1992b). Over 400 velocities were measured from the [O III] $\lambda 5007$ line. The PN velocities are typically accurate to 4 km s$^{-1}$ for spectra obtained on AAT and 30 km s$^{-1}$ at CTIO. The observed PNe cover the entire galaxy to a radius of 10 kpc and extend along the photometric major axis to 20 kpc.

### 2.1 KINEMATICS

2.1.1 *A schematic PN velocity field* The PN spatial distribution as projected on the sky is shown in Figure 2, where the symbols correlate with the PN velocities (relative to the systemic velocity of the galaxy): a circle is used if a PN is approaching and a cross is used if it is receding; the size of the symbol scales with the velocity. To reduce the scatter of the intrinsic velocity dispersion, each nebular velocity is replaced by the mean velocity of its neighboring PNe within 1.5 kpc radius. Therefore, the figure is in fact a representation of the mean PN velocity field. Also shown in the figure are the approximate isophotes of the galaxy (Dufour, et al. 1979). The semi-major axes of the ellipses vary from 1 to 4 effective radius ($r_e$) with a step size of $1r_e$.

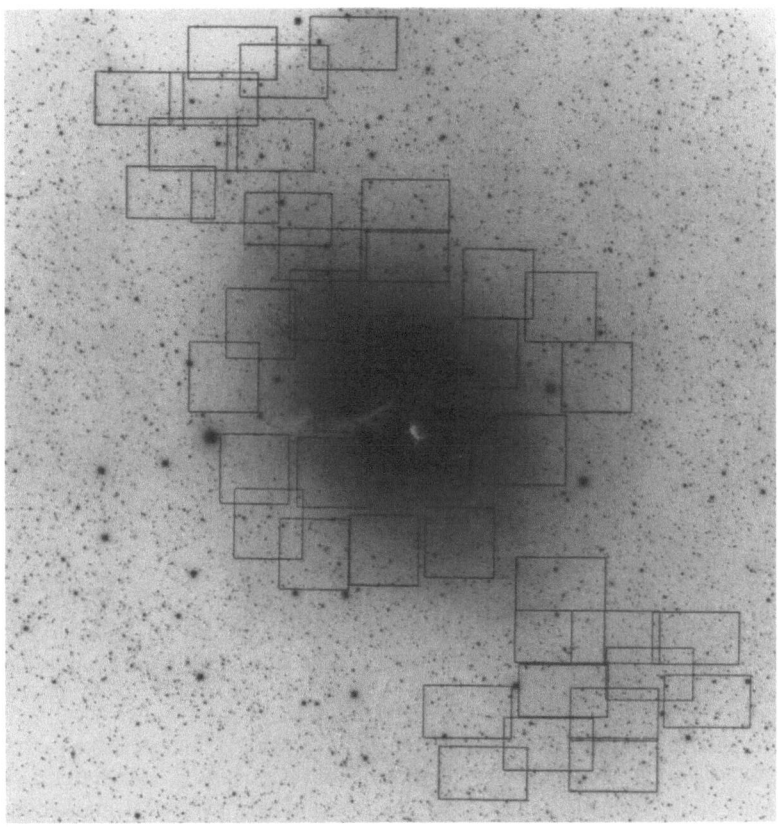

**Figure 1.** Each rectangle or square shows a CCD survey field. The center of the most distant fields are approximately 20 kpc from the nucleus. North is at the top and the east at the left.

Several things can be seen in Figure 2. First, the nebulae apparently rotate around both the photometric major and minor axes. Second, the PN velocities extend to as far as $4r_e$, a factor of 4 increase in radius compared to the observations of ellipticals made through integrated light. Finally, virtually no planetary nebula is found in the central $3 \times 6$ kpc region due to obscuration by the dust lane. The other blank areas in the distribution are due to the incomplete spatial coverage of our survey (*cf.* Figure 1).

2.1.2 *Rotation Axis* Given the two dimensional PN velocity field, the rotation axis can be readily obtained by examining the velocity variation as a function of azimuthal angle. We fit the function

$$V = A \times sin(\Phi - \Phi_0) + V_0 \qquad (1)$$

to individual PN velocities in a ring between 5 and 10 kpc radius where the PN spatial coverage is most complete (*cf.* Figure 2). In equation (1), $\Phi$ is the azimuthal angle on the

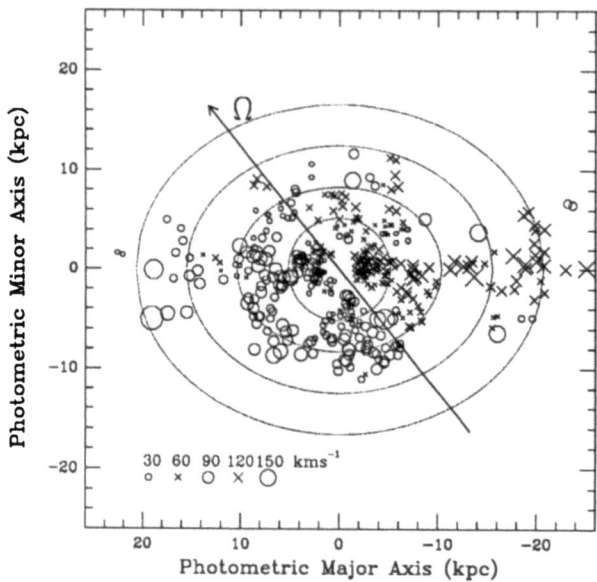

**Figure 2.** A schematic PN velocity field. The legend at lower left relates the symbol size to velocities (see text). The ellipses are the approximate isophotes and the letter $\Omega$ marks the position of the rotation axis.

sky with respect to the photometric minor axis (P.A. = 305°), and positive eastward. The best fit yields $\Phi_0 = 39° \pm 10°$. Thus, the rotation axis is at P.A. = 344° ± 10°, a significant misalignment of 39° from the minor axis. The rotation axis position is marked in Figure 2 with the letter $\Omega$.

The large offset between the photometric and dynamical axes provides convincing evidence that NGC 5128 is *not* an axisymmetric system. Otherwise the rotation axis would fall either along the photometric minor axis (an oblate spheroid) or the photometric major axis (a prolate spheroid). Instead, the galaxy must be triaxial, in which case the observation can be explained naturally. Two effects can contribute to the misalignment. Generically, stars can rotate around both the intrinsic short axis or long axis in a triaxial potential (*cf.* Statler 1987). Thus, the net angular momentum vector could be anywhere in the plane defined by the two. Alternatively, because the observed photometric axes are generally not the principal axes in projection for a triaxial ellipsoid, projection could also introduce misalignment even if the short axis is intrinsically the rotation axis (Stark 1977; Binney 1985).

2.1.3 *Major and Minor Axis Velocity Profiles* The major axis rotation is in the sense of NE approaching and SW receding, while the minor axis rotation is SE approaching and NW receding, the same as the dust lane. We construct the major axis rotation curve and velocity dispersion profile by folding the planetary nebulae in the NE onto the SW side. The PNe are binned in $2'$ intervals inside $10'$ and in $5'$ intervals at larger radii. The mean velocity and its standard deviation yield a rotation velocity and a velocity dispersion for each bin. Similarly, the minor axis rotation curve and velocity dispersion profile are

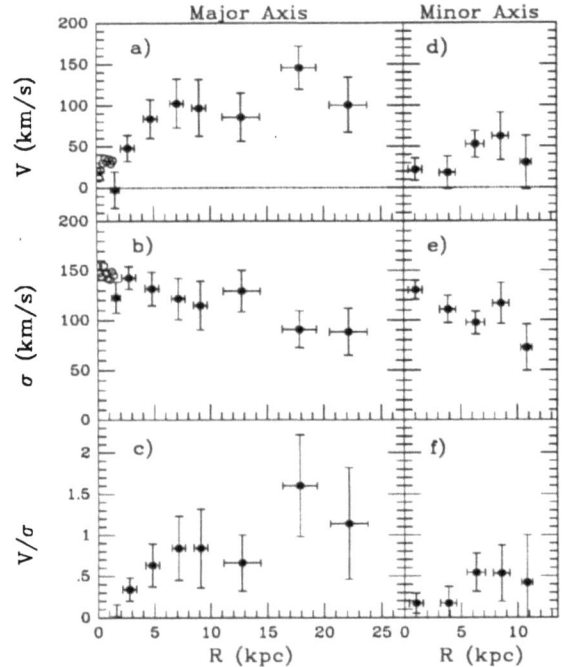

**Figure 3.** a) PN major axis rotation curve (black dots). Circles are the long slit data (Wilkinson et al. 1985). b) PN major axis velocity dispersion profile. c) Local $V/\sigma$ ratio along the major axis. d), e) and f) are the same but for the minor axis.

obtained by reflecting the SE side onto the NW side. The PNe are grouped in 2.5′ intervals. The resulting rotation curves and velocity dispersion profiles are shown in Figure 3.

Along the major axis, the rotation increases with radius in the inner galaxy and reaches approximately 100 km s$^{-1}$ at about 7 kpc. Beyond that, the rotation is nearly flat. The velocity dispersion is about 143 km s$^{-1}$ at 2.8 kpc, then declines slowly through out the halo to $\sim 90$ km s$^{-1}$ at 20 kpc. The streaming motion becomes more important with increasing radius. The local $V/\sigma$ ratio is approximately unity beyond 10 kpc (Figure 3c). The minor axis rotation is about 50 km s$^{-1}$ between 5 and 10 kpc. By comparing with the velocity dispersion, we find that the $V/\sigma$ ratio is about 0.5 for the minor axis rotation (Figure 3f). Along the major axis, the PN rotation and velocity dispersion are consistent with those measured from the stellar absorption spectra (Wilkinson et al. 1985). However, planetary nebulae extend the measurement much further into the halo of the galaxy.

## 2.2 MASS DISTRIBUTION OF THE GALAXY

To derive the mass distribution of the galaxy, we apply a spherical Jeans equation to the major axis rotation and velocity dispersion:

$$\frac{d(\rho\sigma^2)}{dr} = -\frac{GM(r)\rho}{r^2} + \frac{\rho V_{rot}^2}{r} \quad (2)$$

where the velocity dispersion is assumed to be isotropic. If the velocity dispersion in fact becomes more tangential with increasing radius, the isotropic model will overestimate the halo mass; on the other hand, the mass will be underestimated if the stellar orbits become more radial.

Because the light distribution of the galaxy follows a de Vaucouleurs' law, we adopt the Hernquist mass model

$$\rho(r) = \frac{M_l a}{2\pi} \frac{1}{r(r+a)^3} \qquad (3)$$

where the scale length $a = r_e/1.8153$. This model closely resembles de Vaucouleurs' law, yet allows dynamical properties of a galaxy to be evaluated analytically (Hernquist 1990; de Zeeuw 1990, private communication). The mass distribution corresponding to equation (3) can be written as

$$M(r) = \frac{M_l r^2}{(r+a)^2} + \frac{M_d r^2}{(r+d)^2} \qquad (4)$$

with $M_l$ being the total mass of the potential. The second term is included to allow for dark matter. Although the Hernquist model has a finite total mass, it can approximate an isothermal halo when the scale length $d$ is sufficiently large (Dubinski & Carlberg 1991).

To solve equation (2), we further assume that the intrinsic rotation curve is

$$V_{rot}(r) = \frac{v_0 r}{(r^2 + r_0^2)^{1/2}} \qquad (5)$$

which is asymptotically flat at large radii ($r > r_0$). We determine $v_0 = 133 \pm 14$ km s$^{-1}$ and $r_0 = 5.5$ kpc by fitting equation (5) to the PN major axis rotation curve. No effort is made to correct for the inclination of the galaxy since we are seeing it edge-on (Hui 1992).

*2.2.1 Central Mass-to-light Ratio* We first assume that the luminous matter is self-gravitating. The scale length of the mass distribution is determined by the effective radius of de Vaucouleurs' law. Because $r_e = 5.18$ kpc, we have $a = 2.85$ kpc. The only free parameter in the model, M/L, is then adjusted to fit the velocity dispersion inside 1.5 kpc. The total mass resulting from the best fit is $M_l = 1.56 \pm 0.04 \times 10^{11} M_\odot$. Adopting the luminosity of the galaxy to be $B_0 = 6.71$, the mass-to-light ratio is $M/L_B = 3.9$. The value follows the M/L − L relationship comfortably (Figure 7, Kormendy 1987) for the galaxy's absolute magnitude $M_B = -20.80$. Kormendy argued that the dark matter is probably negligible in the center of elliptical galaxies as implied by the M/L − L relation. The fact that the M/L ratio of NGC 5128 follows the general relationship suggests that dark matter is of minor importance in the inner region. The predicted velocity dispersion is plotted in Figure 4 as dashed line. Although the fit is excellent in the central region (upper panel), in the halo the model decreases much faster than the observed PN velocity dispersion (lower panel).

*2.2.2 Halo Mass Distribution* We adopt the two component mass model of equation (4), which has four parameters: two scale lengths, the total luminous mass $M_l$ and the parameter $M_d$. While the scale length of the luminous component is set by de Vaucouleurs' law, the others are determined by fitting the model to the combined velocity dispersion profile of the long slit spectra in the inner 1.5 kpc and the PN data between 2 and 22 kpc. The best fitting model yields $M_l = 1.24 \times 10^{11} M_\odot$, $M_d = 9.49 \times 10^{11} M_\odot$, and $d = 27.5$ kpc. The total luminous mass here is smaller than the value $M_l = 1.56 \times 10^{11} M_\odot$ derived in the self-consistent model. This is because we allow the dark mass density to vary continuously towards the center of the galaxy so that it does not have a hollow core. The total mass distribution, its two components, and the mass-to-light ratio are plotted in Figure 5.

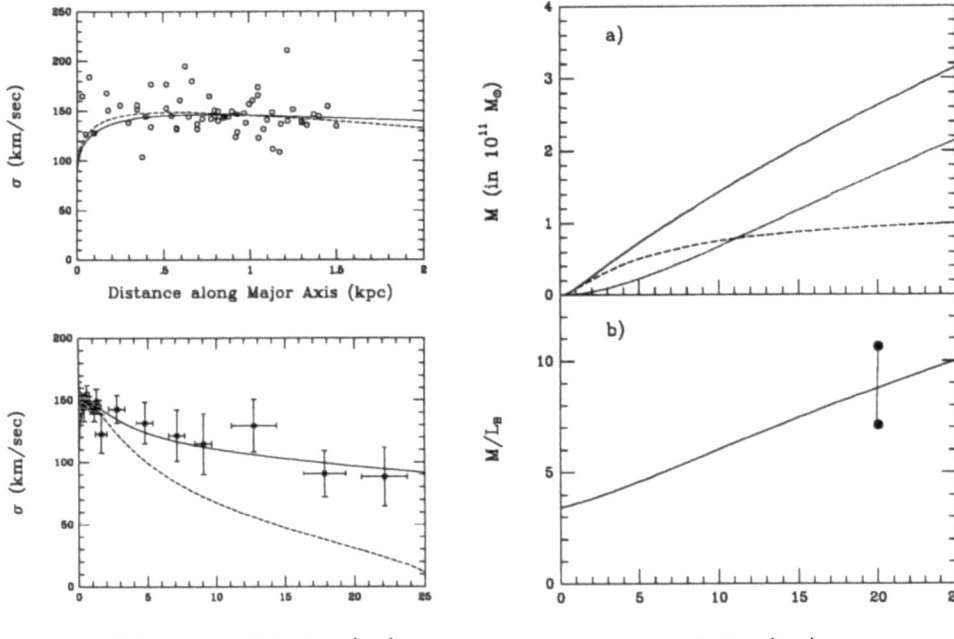

**Figure 4.** Comparison between the models and the velocity dispersion of integrated light (upper panel) and of planetary nebulae (lower panel). The dashed line is for the self-consistent model and the solid line is for the model with dark matter.

**Figure 5.** a) Distribution of the total mass (solid curve), the luminous component (dashed curve) and the dark component (dotted curve) as a function of radius. b) Mass-to-light ratio. The two dots linked by a dotted line are $M/L_B$ derived using globular clusters.

Within 25 kpc, the total mass of the galaxy is $3.1 \times 10^{11} M_\odot$, and $M/L_B = 10$. The model velocity dispersion is shown in Figure 4 as the solid line.

We used the projected mass estimator (Heisler, Tremaine & Bahcall 1985) to calculate the total mass of the galaxy within 20 kpc from 81 globular cluster velocities (Harris et al. 1988; Sharples 1988). The mass is $3.6 \times 10^{11} M_\odot$ if the GCs are on isotropic orbits, or $2.4 \times 10^{11} M_\odot$ if they are on tangential orbits. The two dots linked by a dotted line in Figure 5b are the corresponding mass-to-light ratios. These two values bracket the mass-to-light ratio given by the PN velocity dispersion. This result suggests that globular clusters are on partially tangential orbits. This is not surprising given the fact that globular clusters beyond 10 kpc are identified by their slightly fuzzy appearance. The resulting sample thus consists of extended and bright globular clusters almost exclusively. These clusters are expected to be on low eccentricity orbits which result in little tidal stripping.

## 3. Current and Future Works

Planetary nebula systems are under study in more distance galaxies. We have made an extensive PN survey of NGC 4594, the Sombrero galaxy (Ford et al. 1992). Over 300 PNe were identified in the galaxy to a radius of 18 kpc. In collaboration with K. Freeman, velocities of 150 PNe were measured on AAT. Detailed dynamical study is underway. Similar work has been carried out on NGC 3379 by Ciardullo and Jacoby (this volume). Interestingly, they found that the PN velocity dispersion decreases to 65 km s$^{-1}$ at about $3r_e$ in good agreement with a constant mass-to-light ratio model. Both work demonstrate that PN velocities can be measured to at least 10 Mpc with 4-m telescopes. With the new generation telescopes of 8-m or 10-m class, one can expect to apply the same method to galaxies in the Virgo Cluster and beyond. A significant increase in the sample size promises to make important contributions to the dynamical studies of early type galaxies.

## Reference

Binney, J. 1985, *M.N.R.A.S.*, **212**, 767
Bland, J., Taylor, K., & Atherton, P.D. 1987, *M.N.R.A.S.*, **228**, 595
Ciardullo, R., Jacoby, G.H. & Ford, H.C. 1989, *Ap. J.*, **344**, 1989
Ciardullo, R., Jacoby, G.H. & Harris, W.E., 1991, *Ap. J.*, **383**, 487
Dubinski, J & Carlberg, R.G. 1991, *Ap. J.*, **378**, 496
Dufour, R.J., van den Berge, S., Harvel, C.A., Martins, D.H., Schiffer, III, F.H., Talbot, Jr., R.J., Talent, D.L. & Wells., D.N. 1979, *A. J.*, **84**, 284
Ford, H.C., Hui, X., Ciardullo, R., Jacoby, G.H. & Freeman, K.C. 1992, *submitted to Ap.J*
Graham, J.A. 1979, *Ap. J.*, **232**, 60
Harris, H.C., Harris, G.L.H., & Hesser, J.E. 1988, *in IAU Symposium 126 , Globular Cluster Systems in Galaxies*, eds. J.E. Grindlay & A.G.D. Philip, (Kluwer Academic), p 205
Heisler, J., Tremaine, S. & Bahcall J.N. 1985, *Ap. J.*, **298**, 8
Hernquist, L. 1990, *Ap. J.*, **356**, 359
Hui, X. 1992, *Ph.D Thesis, Boston University*
Hui, X., Ford, H.C., Ciardullo, R. & Jacoby, G.H. 1992a, *submitted to Ap.J*
Hui, X., Ford, H.C., Freeman, K.C. & Dopita, M.A. 1992b, *in preparation*
Jacoby, G.H., Ciardullo, R. & Ford, H.C. 1990, *Ap. J.*, **356**, 332
Kormendy, J. 1987, in *IAU Symposium No. 127, Structure and Dynamics of Elliptical Galaxies*, ed. T. de Zeeuw (Dordrecht: Reidel), p17
Nicholson, R.A., Bland-Hawthorn, J. & Taylor, K. 1992, *Ap. J.*, **387**,503
Sandage, 1961, in *The Hubble Atlas of Galaxies*
Sharples, R.M. 1988, *in IAU Symposium 126 , Globular Cluster Systems in Galaxies*, eds. J.E. Grindlay & A.G.D. Philip, (Kluwer Academic), p 545
Stark, A.A. 1977, *Ap. J.*, **213**, 368
Statler, T.S. 1987, *Ap. J.*, **321**, 113
van Gorkom, J.H., van der Hulst, J.M., Haschich, A.D. & Tubbs, A.D., 1990, *A. J.*, **99**, 1781
Wilkinson, A., Sharples, R.M., Fosury, R.A.E. & Wallace, P.T. 1986, *M.N.R.A.S.*, **218**, 297

# DYNAMICS OF AGB STARS AND PLANETARY NEBULAE IN THE GALAXY

H. DEJONGHE

*Sterrenkundig Observatorium, Universiteit Gent, Belgium*

**Abstract.** The available kinematical data on OH/IR stars and PNs are reviewed. Dynamical models for the OH/IR stars are presented.

**Key words:** Dynamics - Equilibrium models - the Galaxy

## 1. Introduction

It is proverbial that one picture says more than a thousand words. Pictures shape our ideas about all kinds of astronomical objects, from planets to the universe itself. Moreover, the inquiring mind transcends the (visual) appearance of the things with theory, only to find itself trying to visualise abstract concepts back again (graphs,...). A good example thereof is the H-R diagram, which is a picture that compactly summarizes stellar evolution.

In this contribution, I will try to introduce a particular way of visualising the dynamical state of a galaxy. This doesn't seem to fit in well with the subject of this symposium, were it not that I will use tracer populations of AGB stars (mainly OH/IR stars and PNs) to reach that goal.

## 2. Elements of Dynamical Modelling

The ultimate dynamical knowledge is to know for every star, at all times, its location in phase space, which is the product of configuration space (normal 3-space) and velocity space. On the other hand, such a complete knowledge is impossible and (as of now) unmanageable; this is why both theory and assumptions are needed to simplify this ultimate picture.

When studying the dynamics of a tracer population, it is obvious that self-consistency is not required, i.e. the gravitational potential is not generated by the tracer population. This means that the gravitational potential can be decoupled from the original set of equations, and this potential therefore must be a given. The specification of the gravitational potential is the first important decision one has to make when building an equilibrium model. In particular, one has to decide on the prevailing geometry, i.e. whether the potential is spherical, axisymmetric or triaxial.

The potential generates structure in phase space, because it creates orbits. The quintessential orbit is the linear harmonic oscillator, e.g. a spring. At every moment it has a length $z$ which changes at a rate $v_z$. Phase space is the 2-dimensional space $(z, v_z)$. If we do not know the dynamical state of the spring completely, it is natural to ask what information we can single out as particularly important. The maximum length $z_m$ must be such a quantity, because then, at least, we can confine the length of the spring, though we've lost the ability to predict its actual length at any particular moment. This $z_m$ is an example of an integral of the motion [1]: it is a function of phase space coordinates that remains a constant along the orbit, and therefore it can be used as a label for that orbit. Hence, we can now describe the linear harmonic oscillator with $z_m$ ( a constant) and only one rapidly changing coordinate ($z$, or $v_z$, or something else).

In 3 dimensions, one would expect 3 constants of the motion and 3 rapidly changing variables for every orbit. This is true for integrable potentials, by definition. Most potentials however are not integrable, but it is likely that for most astrophysical purposes there exist good integrable fits (Goodman and Schwarzschild 1981, Dejonghe & de Zeeuw 1988), safe possibly for tumbling triaxial figures. A very elegant class of integrable potentials are the Stäckel potentials, which have the nice property that the integrals of the motion are quadratic functions of the velocities.

In order to better understand the significance of these integrals, let's consider the following experiment. We affix many springs to a flat surface. The springs only vibrate in the $z$-direction (perpendicular to the surface), and hence their $x$ and $y$ coordinates which are markers on the surface are constants of the motion. Now we disturb the springs, for example by pushing them down simultaneously by hand. The imprint of the hand will be lost very quickly, and in the analysis of the resulting dynamical state, it will certainly not matter very much to focus on the description of the rapidly changing coordinates. If we only knew $z_m(x,y)$, then we would know the profile of the perturber, which is everything that there is to know in this experiment. Consider next the somewhat different situation that at every location $(x,y)$ there are a lot of springs (for example molecules), which may or may not start to vibrate due to infalling light, then the number of excitations $N$ will be proportional with the intensity, while the degree of excitation (the $z_m$) will tell us something about the wavelength of the infalling light. The function $N(x,y,z_m)$ we call a distribution function. It is written here as a function in integral space. It cannot exist without a medium for which it is a probability density, though it may provide us with important information on something else (the infalling light). This function is very analogous to the concept with the same name in stellar dynamics; the medium there is called a tracer population. On a photographic plate, the distribution function is a faithful representation of the perturbing radiation, and in stellar dynamics it is hoped that the distribution function will teach us similarly important things about the formation of galaxies.

Though the distribution function $F$ is defined as a probability density in phase space, we will, according to Jeans' theorem, write it as a function of the integrals of the motion. Hence, it is, for all purposes, a function in integral space. For time independent potentials, the specific binding energy $E$ is always one of them, and according to the geometry and the potential, there may be 2 more integrals. The big difference with the previous examples is that all these integrals are highly non-local, just as an orbit is.

In order to determine this distribution function, we must write down its relation with observable quantities. This relation can almost always be written in the form

---

[1] The concept is actually rather complicated, and a precise definition is far beyond the scope of this contribution

of an average of the distribution function. As already indicated, the body of data can be very inhomogeneous, including star counts, mean velocities and velocity dispersions, line profiles, proper motions, etc... No general theorem exists that would enable us to decide on the uniqueness of this inversion, let alone analytic procedures to perform such an inversion. Hence, a pragmatic approach is in order.

One way to proceed is with quadratic programming (QP, Dejonghe 1989). In this method we assume that the distribution function can be written as a linear combination of (preferably analytically simple) components, with coefficients $c_i$. A $\chi^2$-type function (quadratic in the $c_i$) is then mimimized, subject to the constraint that the distribution function must be positive everywhere (linear constraints in the $c_i$).

Only numerical experience at this point can give us an idea to what degree we can have confidence in the computed distribution function. It is obvious that the more the data cover phase space, the more the distribution function will be constrained. Also, the more restrictive we are in the functional form of $F$ (function of one, two, or three integrals), the less indeterminacy we will encounter when trying to determine a distribution function, but also the less realistic our results may be. Only partial results on these issues are available (Dejonghe & Merritt 1992), and there still remains a lot to do.

## 3. Some models for tracer populations

As an obvious consequence from the preceding section, a tracer population which may be used to stake some claims about the global dynamical state of our Galaxy must be present in a substantial volume of the Galaxy, and must also be detectable there. This inevitably puts infrared astronomy in a privileged place, and the IRAS satellite certainly has earned her marks in that respect.

### 3.1. OH/IR STARS

The OH/IR stars have the advantage of being old, strong infrared emitters with a characteristic spectrum around the 1612 Mhz maser peak. Their mostly old age and AGB status make them rather useful for equilibrium dynamical modelling, since it helps that we can assume that somehow older objects must have undergone some form of relaxation, which is presumably sufficient (but not necessary) for equilibrium. Their strong infrared emission makes them shine right through the dusty galactic plane (GP), a property which is needed for a sufficient spatial coverage. The characteristic double maser peak, caused by an expanding shell, is easily recognizable, and provides a simple way of measuring the line-of-sight velocity. In addition, the velocity of the expanding circumstellar shell may indicate an age, in the (statistical) sense that larger expansion velocities are associated with younger stars. This provides an interesting test on the models, since it can be expected that the younger population is more confined to the disk.

All this is reason enough to systematically search for them, and this has been done by Eder *et.al.* (1988), Sivagnanam *et.al.* (1989) and te Lintel Hekkert *et.al.* (1991a), hereafter tLH, and further references therein. The IRAS PSC provides a $F_\nu(12\mu m)/F_\nu(25\mu m)$ versus $F_\nu(25\mu m)/F_\nu(60\mu m)$ color diagram in which the OH/IR stars occupy a fairly well defined place. A list of candidate OH/IR stars results, and these are then individually radio-checked for the 1612 Mhz emission. About 1500 OH/IR stars are now known.

The IRAS satellite was severely confused in the GP, because of the high density of sources. In order to find OH/IR stars there, one must resort to mapping type surveys. The first such survey close to the galactic center (GC) was done by Habing et.al. (1983), and yielded 34 stars within a radius of 1° around the GC. This survey has been substantially improved in a more limited region by Lindqvist et.al. (1992a), hereafter L, yielding 134 stars. None of these surveys have been completely satisfactory in their velocity coverage, for technical and feasibility reasons. a window of ±217 km/s, the tLH sample that (a few) high velocity stars can be expected to turn up when searching for them (van Langevelde 1992).

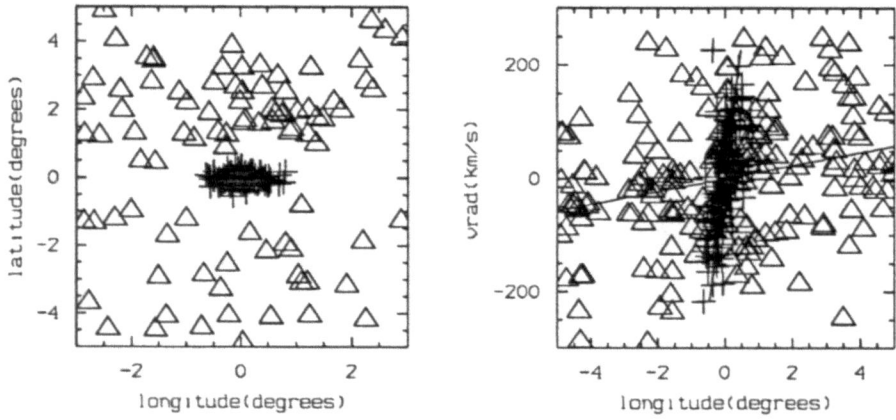

Fig. 1. (a) The L sample (crosses) and the truncated tLH sample (triangles) in $(\ell, b)$ on the sky. (b) The L sample (crosses) and the truncated tLH sample (triangles) in $(\ell, v_r)$ space.

I will now summarize the first results from the tLH and L surveys. Some of these results can be found in te Lintel Hekkert et.al. (1991b), and Lindqvist et.al. (1992b). Figure 1a shows the $(\ell, b)$ diagrams for the L and the tLH samples, the latter being truncated to ±3° in longitude and ±5° in latitude. It is obvious that both samples are complementary, but certainly not enough so. Currently a consortium headed by Habing and te Lintel Hekkert is working on surveys at the VLA and the AT to fill in a few gaps. The tLH survey is believed to be fairly complete up to 3 Jy at $12\mu m$, for $|b| > 2.5°$, the latter limitation due to IRAS. This is very clear from the "zone of avoidance" in Fig. 1 for the tLH sample. In future surveys, considerable attention and care will have to be taken to treat completeness properly, in order to link the different surveys together.

Figure 1b shows the $(\ell, v_r)$ plot for both samples (heliocentric velocities). The regression lines are the linear approximations to the rotation curves. The L rotation curve (crosses) in this plot has a slope of about 500 km/s/degree or about 3.7 km/s/pc, using $R_\odot = 7.5$kpc. These values are about right as is obvious from the plot, but are a factor of three higher than quoted by Lindqvist et.al. (1992b). They note however that the slope is very uncertain, depending on the elimination of a few stars, as must be since the slope is very steep. The mass of the point source in the GC is therefore not so well constrained, and values are given in Lindqvist et.al. (1992b). The slope of the tLH sample is 11 km/s/degree or about 82 km/s/kpc. Such a rotation curve reaches its presumed peak value of about 220 km/s at about 2.5 kpc, which is very reasonable. These slopes are so different because the tLH sample ignores the GC and the GP.

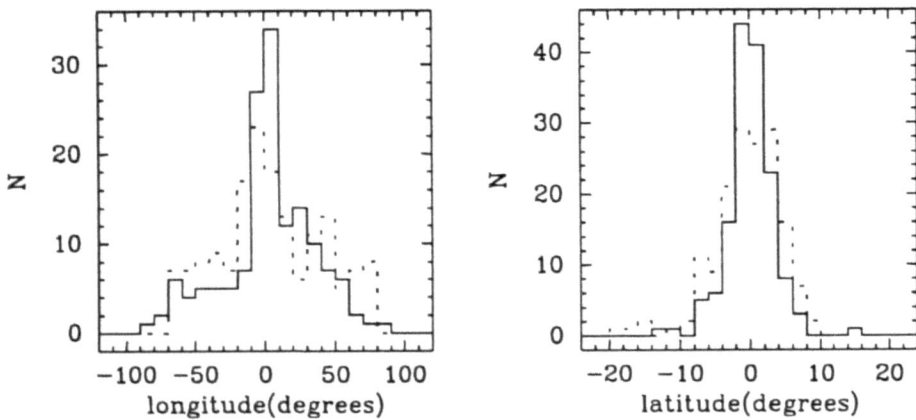

Fig. 2. The histograms of old stars in the tLH sample (dashes) and the young stars (solid) as a function of galactic longitude and latitude

From the analysis by Lindqvist et.al. (1992b) it would seem that indeed the projected velocity dispersion of the younger stars is smaller than the dispersion for the older stars, as one could expect if stars are born on primarily circular orbits. In Fig. 2 we see histograms as a function of longitude and latitude for a selection of the tLH sample, divided into two groups defined by the "oldest" and "youngest" stars, each containing about 150 stars. It is clear that, a few dissenters notwithstanding, the younger population is more confined to the disk, and somewhat more bulgy.

As explained in the first section, equilibrium dynamical modelling needs an assumption on the potential. There is no obvious sign of triaxiality in the tLH data, and therefore it seems unnecessary to waste ones effort at this point in producing dynamical models in a triaxial potential. This does not mean that, sooner or later, the potential of the galaxy will turn out to be triaxial, especially in the central regions. In fact, it is hard to see how it could not be! But, in any case, triaxiality in a tracer population (stars or gas) does not imply that the underlying gravitational potential is triaxial, since there is certainly no self-consistency requirement.

Similarly, since the radial velocity does not give any information on the $z$-component of the velocity for stars in the GP, it is natural to try two integral models first (based on the specific binding energy $E$ and the $z$-component of the angular momentum, which are both integrals of the motion in an axisymmetric potential). Such models have the property that $\sigma_r = \sigma_z$. Only when the projected velocity dispersion turns out to be much too small (since for two–integral models $\sigma_r$

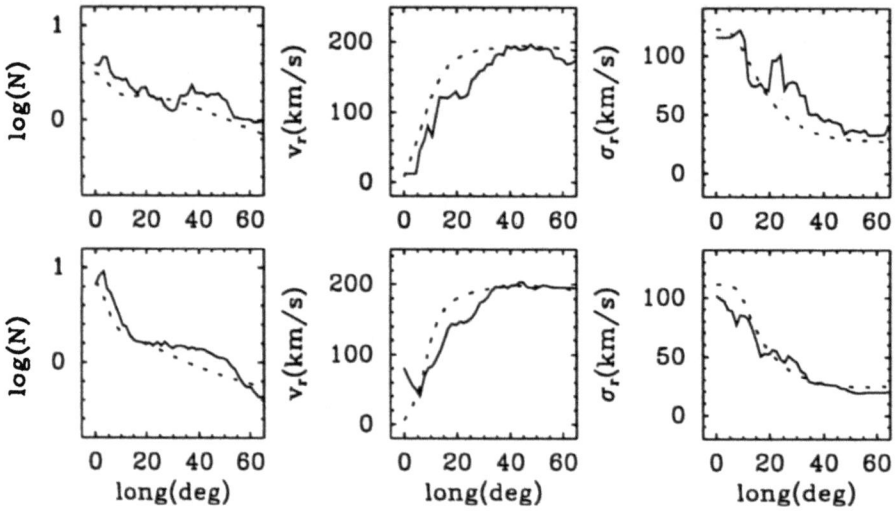

**Fig. 3.** The data for the tLH sample at $b = 2°$ (solid lines) compared with a 2-integral model (dashes) in an approximation to the BS potential. Left panels: logarithm of projected star counts per square degree, middle panels: projected mean velocity, right panels: projected velocity dispersions. Top panels: the old stars, bottom panels: the young stars.

is determined by the thinness of the disk), will we be able to rule out two-integral models.

Figure 3 shows the result of the fit of two-integral models embedded in an approximation to the Bahcall-Soneira potential (BS) for the Galaxy. For details, see te Lintel Hekkert *et.al.* (1991b). The data where smoothed by averaging at every point with the 15 closest neighbours, and the fit was produced on the basis of the projected star counts, fluxes and pressures. The fit looks good, to the degree that it cannot be expected to reproduce all details in the data, which may not be real anyway. Problem areas may be the projected velocity dispersion for the old stars, which is somewhat too low, and the mean velocity which rises a bit too fast for both samples. This may indicate the possible need for a third integral. In Fig.4, the spatial number density, mean rotation and radial velocity dispersion are plotted

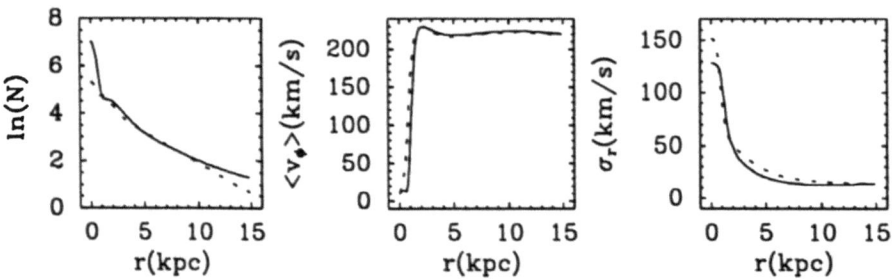

**Fig. 4.** The logarithm of the spatial density in the plane, the mean rotation and the radial velocity dispersion for the young (solid) and old (dashed) stars

for both groups of stars. The old stars show a nice exponential disk (no artifact of the components used in QP!) with scale factor 3.5 kpc. The mean rotation follows very closely the rotation curve, and the old stars have overall a somewhat higher velocity dispersion, which clearly shows a bulge component. Finally, Fig. 5 shows the distribution function in turning point space (the color version is much nicer!). All well-known components are present. The dynamic range is very large: the highest value is about $10^7$ stars/kpc$^3$/(km/s)$^3$ in the thin disk, but, clearly, such values are very uncertain.

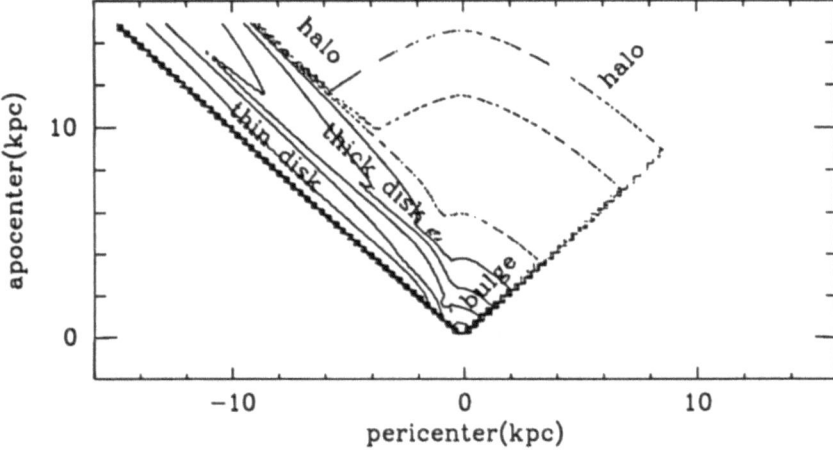

Fig. 5. The distribution function in turning point space. Contours are logarithmically spaced. Dotted contours correspond to values that are smaller than one. The dynamic range is of the order $10^{12}$.

3.2. PLANETARY NEBULAE

About 1500 PNs are known. An ancient but standard reference is the catalogue by Perek & Kohoutek (1967). It has been (and still is) the basis for many surveys. This list however contains quite a few objects that later turned out to be something else, like M stars, symbiotic stars, or worse, HII regions. Subsequent cataloguing includes the work of Acker et.al. (1983), which is a good reference to the literature prior to 1983, and Acker et.al. (1991,1992).

A different approach uses the IRAS PSC, which, again, has been essential in the search for PN's. Just as is the case with the OH/IR stars, the PNs are strong infrared emittors, and occupy a fairly well defined place in the $F_\nu(12\mu m)/F_\nu(25\mu m)$ versus $F_\nu(25\mu m)/F_\nu(60\mu m)$ color diagram (Pottash et.al. 1988, Ratag et.al. 1990). Subsequent radio interferometry can be used to decide on the true nature of the candidates (see also Zijlstra et.al. 1989). This method up to now yielded about 50 new PNs within 15° from the galactic center, on a total of about 400 in roughly the same region (Acker et.al. 1991). In any case, the final number of detected PNs in the Galaxy is in reasonable agreement with the total number of OH/IR stars known.

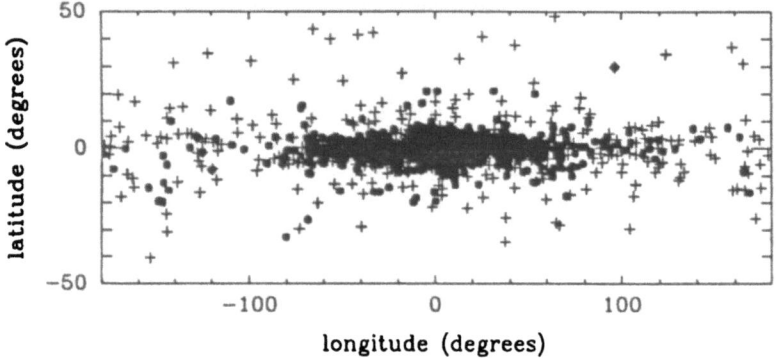

Fig. 6. The tlH sample (dots) ant the Acker et.al. sample (crosses)

No extensive dynamical modelling has been done as of now for the PNs. A preview of what is in store there can be seen by simply comparing the tLH sample and the Acker et.al. (1991) sample, such as if Fig. 6. The PN's extend towards higher latitudes then the OH/IR's. In the GP the coverage of the PNs is poor.

## Acknowledgements

It is a pleasure to thank H. Habing for providing the data of the OH/IR stars near the GC in electronic form, and A. Acker for doing the same for the PNs.

## References

Acker, A., Marcout, J., Ochsenbein, F., Lortet, M.C., 1983, *Astron. Astrophys. Suppl.*, **54**, 315
Acker, A., Köppen, J., Stenholm, B., Raytchev, B., 1991, *Astron. Astrophys. Suppl.*, **89**, 237
Acker, A., Ochsenbein, Stenholm, B., Tylenda, R., Marcout, J., Schohn, C., 1992, *Strasbourg-ESO Catalogue of Galactic Planetary Nebulae*, Part I, ESO publication
Dejonghe, H., 1989, *Astrophys. J.*, **343**, 113
Dejonghe, H. & de Zeeuw, P.T., 1988, *Astrophys. J.*, **329**, 720
Dejonghe, H. & Merritt, D., 1992, *Astrophys. J.*,
Eder, J., Lewis, B.M., Terzian, Y., 1988, *Astrophys. J. Suppl.*, **66**, 183
Goodman, J. & Schwarzschild, M., 1981, *Astrophys. J.*, **245**, 1087
Habing, H.J., Olnon, F.M., Winnberg, A., Matthews, H.E., Baud, B., 1983, *Astron. Astrophys.*, **128**, 230
Lindqvist, M., Winnberg, A., Habing, H.J., Matthews, H.E., 1992a, *Astron. Astrophys. Suppl*, **92**, 43
Lindqvist, M., Habing, H.J., Winnberg, A., 1992b, *Astron. Astrophys.*, **259**, 118
Perek, L., & Kohoutek, L., 1967, *Catalogue of Galactic Planetary Nebulae*, Prague Academic Press
Pottash, S.R., Bignell, C., Olling, R., Zijlstra, A.A., 1988, *Astron. Astrophys.*, **205**, 248
Ratag, M.A., Pottash, S.R., Zijlstra, A.A., Menzies, J., 1990, *Astron. Astrophys.*, **233**, 181
Sivagnanam, P., Braz, M.A., Le Squeren, A.M., Tran Minh, F., 1989, *Astron. Astrophys.*, **211**, 341.
te Lintel Hekkert, P., Caswell, J.L., Habing, H.J., Norris, R.P., Haynes, R.F., 1991a, *Astron. Astrophys. Suppl.*, **90**, 327
te Lintel Hekkert, P., Dejonghe, H., Habing, H.J., 1991b, *Proc. of Astron. Soc. of Austr.*, **9**, 20
van Langevelde, H.J., Brown, A.G.A., Lindqvist, M., Habing, H.J., de Zeeuw, P.T., 1992, *Astron. Astrophys. Lett.*, submitted
Zijlstra, A.A., Pottash, S.R., Bignell, C., 1989, *Astron. Astrophys. Suppl.*, **79**, 329

# PN ABUNDANCES IN DIFFERENT GALACTIC SYSTEMS

R.E.S. CLEGG

*Royal Greenwich Observatory, Madingley Road, Cambridge CB3 0EZ, U.K.*

**Abstract.** Developments in the measurement of abundances in planetary nebulae since IAU Symposium 131 are reviewed. Present uncertainties and outstanding problems in the interpretation of emission-line measurements are discussed. Results for different samples of PN are reviewed, in terms of the galactic site. Results for the Galactic disk, Galactic Bulge, Galactic Halo, LMC & SMC, Fornax Dwarf galaxy and M31 are discussed. Attention is drawn to the importance of Magellanic Cloud samples for testing dredge-up theory, and to the strong variation of PN C/O ratios with the local site's metallicity and age. New results for carbon in Cloud PNs are presented. Important problems in the interpretation of the metallicity of Bulge PNs are highlighted. Attention is drawn to the idea that metal-rich stellar populations do not produce PNs, such stars leaving the lower AGB early.

## 1. Introduction & Overview

In this review I will concentrate on developments in the field of PN abundances since the last IAU Symposium (No 131, in Mexico City) was held. In section 1 I will cover the general abundance pattern, recent surveys and compilations, and some current uncertainties in deriving abundances. In Section 2 I discuss briefly results for the nearby Galactic disk. Sections 3, 4, 5 & 6 cover the Halo, Bulge, Magellanic Clouds (hereafter, 'Clouds') and other extragalactic systems, respectively, with new results for Cloud PNs being presented in Sec. 5.

Some recent reviews on abundances and related properties were given by Clegg (1989, 1991), Peimbert (1990, 1992) and Henry (1990). The systematics of a large set of Galactic PN abundances were studied by Perinotto (1992).

Three interesting examples of how PN properties vary (or not!) as a function of 'metallicity' or galactic site are given here. Note that different metallicity regions usually have different *chemical evolution histories* and *ages* (and hence different *AGB star masses*).

Firstly, Henry (1989) compiled O/H and Ne/H ratios for PNs in the disk, Bulge, Halo and Clouds, *i.e.* over a large spread in metallicity. The Ne/O ratio is remarkably constant at 0.17 by number, and Henry showed that this puts a significant constraint on the constancy of the ratio of the numbers of massive stars above $15 M_\odot$ (since there the relative yields of O and Ne vary with mass).

A second example is provided by the idea that metal-rich stellar populations may not produce many PNs. This was suggested to explain the UV-excess seen in elliptical galaxies (Greggio & Renzini 1989, Brocato *et al.* 1990) and justified by later stellar evolutionary models (Castellani & Tornambe 1991, 1992). The suggestion is that, in old systems of $\sim 2\times$ Solar metallicity, high mass-loss results in stars only reaching the early AGB (non-thermally-pulsing) before the envelope mass tends to zero, and the stars move to the left in the HR diagram without ejection of a PN. A population of hot 'post-early-AGB' stars rather than PN central stars would contribute to the far-UV light in the most metal-rich galaxy cores, such as NGC1399 (Castellani & Tornambe 1990, Ferguson & Davidson 1993). Ferguson & Davidsen show that the number of PNs per unit bolometric luminosity in early-type galaxies *decreases* as the (1550Å–V) colour get bluer, a trend which supports the above hypothesis. Peimbert (1990) noticed that the PN number and the PN birthrate per unit luminosity correlated with galaxian B–V colour and with $M_{BOL}$, which may be closely related effects.

If correct, this idea would affect the interpretation of mean PN abundances in an old, metal-rich (Solar or above) galactic site: the mean planetary nebula O/H value would be lower than the mean stellar value, for example.

Lastly, the variation of nebular C/O ratio with galactic site is particularly strong. The percentage of PNs having C/O > 1 varies from 18 (Bulge); 50 ('Southern' PNs, Kingsburgh & Barlow these proceedings); 65 (local disk, Zuckerman & Aller 1987); 60 (Halo objects) and almost 100 (Magellanic Clouds; see below). Although this is partly due to a 'titration effect' (less C needed when the O/H ratio is low in a stellar envelope) it is also due to the 'efficiency' of the 3rd dredge-up (which can include opacity, mixing, and envelope-mass effects).

## 2. Abundance Determinations

The main current uncertainties arise from uncertainties in:

• Depletion of atoms in grains (especially, for O and C). However, recent determinations of PN dust/gas ratios by mass (reviewed in article by Barlow) are so low ($\sim 10^{-4}$) that the required depletions of C & O are below 20%.
• Ionization correction factors for sulphur & argon.
• Temperature fluctuations in nebulae – is the fluctuation parameter $t^2$ significant? A fresh stimulus for this is the finding from new CCD data that measured Balmer continuum temperatures are lower than [O III] values (Liu & Danziger 1993). The difference corresponds to a parameter $t^2 = 0.035$.
• Whether localized shock-excitation is common or significant. Peimbert *et al.* (1991) described situations in which PNs & H II regions might have shock-excitation. Middlemass *et al.* (1991) presented a shock model for PN giant halos; it was found that derived O/H & Ne/H ratios could be a factor 2 too low. But, they also showed that such an effect would not operate in the main, bright part of the nebulae. Several Type I PNs in the SMC (discussed below) have $T_e(O^{++}) \sim 25,000$ K, which cannot arise from photo-ionization but could be due to shock-excitation.
• The still un-resolved discrepancy between $C^{++}/H^+$ ratios derived from the [C III] 1908 and C II 4267Å lines. From IUE & optical spectra of Cloud PNs, Walton *et al.* (1991) found that $T_e(C^{++})$ is much less than $T_e(O^{++})$, disagreeing with simple expectations and with photo-ionization models. The problem is not confined to Galactic PNs! The $C^{++}/H^+$ ratios would be significantly increased for the Cloud PNs if the $T_e(C^{++})$ values were used to interpret the C III]1908Å fluxes. Torres-Peimbert *et al.* (1990) made a photo-ionization model of NGC 4361 with a gradient in the C/H ratio, and obtained good fits to C line strengths. This provides a possible answer to this old problem.

From analysis of near-IR HF lines, Jorissen *et al.* (1992) found fluorine to be over-abundant in many red giants, with F/H up to 30× Solar in carbon stars. The F/O & C/O ratios were well-correlated. The suggested mechanism for $^{19}$F production was $\alpha$-particle capture on $^{14}$N in the He-burning shell at the AGB stage. Aller & Czyzak (1983) do quote a mean PN log F abundance of 4.6 (the same as the Solar value adopted by Jorissen *et al.*), but presumably this is based on guessed atomic data. It would be useful to improve the accuracy of PN F/H ratios. The strongest (!) line seen in PNs is [F IV]4059Å (analogue of [O III]5007); new deep CCD spectra are required, together with atomic data for the FIV lower levels. Even rough abundances accurate to a factor 2 would be of value, in view of the large reported enhancement factors in red giants.

## 3. The Galactic Disk

Fresh abundance sets have been given by Aller & Keyes (1987), Henry (1989), de Freitas Pacheco *et al.* (1991, 1992), Koeppen *et al.* (1991) & Kingsburgh & Barlow (Poster VI-225) - the last three studies concentrate on Southern PNs which may be systematically closer to the Galactic Centre than other samples of bright nebulae. A large compilation of Galactic PN abundances was given by Perinotto (1992). Abundances in large PNs were studied by Kaler *et al.* (1990), and abundance gradients were studied by Faundez-Abans & Maciel (1988) & Maciel (1991) (see also Posters VI-232 & 233).

The situation of 1987 still stands: the average PN has abundances of He, C & N (*only*) enhanced over the values for H II regions in the same galactic site. The increases are attributed to the 1st, 2nd & 3rd 'dredge-up' episodes (*e.g.*, Becker & Iben 1979,1980; Iben & Renzini 1983). Ca, Si, Al, Mg and Fe have ratios to oxygen much lower than Solar, which is taken as due to depletion in grains.

The mean O/H ratio for local PNs, $4.4 \times 10^{-4}$, agrees well with the value for H II regions (*e.g.*, Orion) and M, S & N-type red giants (Smith & Lambert 1990a). However, the (well-determined) Solar photospheric value is a factor two higher. Are many O atoms in grains in nebulae, or is the nebular abundance scale wrong because large temperature fluctuations really exist? An interesting clue comes from analysis of B stars in the Orion cluster: Cunha & Lambert (1992) found them to have the *same* O/H ratio as the Orion nebula. It seems that the Sun may really be 'metal-rich' relative to the local ISM.

One issue still not resolved is whether, from patterns of CNO & He, there is evidence that the O-N part of the CNO cycle has operated. All papers disagree, some samples producing a finding that it has and others that it has not. The answer is that the evidence *is still marginal and sample-dependent*.

## 4. The Galactic Bulge

Webster (1988) analyzed 65 Bulge PNs, and found 'normal' abundances, and a ratio of Type I to other nebulae the same as in disk and Cloud samples. In a large sample of Bulge PNs, Ratag *et al.* (1992; and Ratag 1991) found most elements had abundances similar to local disk PNs, with *marginal* evidence for slightly-higher He/H and N/O ratios. Walton *et al.* (Poster VI-237) also report 'disk—like abundances and a lack of C-rich PNs in the Bulge. The most important result from these three studies is that the mean O/H ratio in the Bulge samples is the same as for local disk PNs. This is suprising, since for K giants in the Bulge, Rich (1988) found a large spread of abundances - from 0.1 to 10× Solar - with a mean which was 2× Solar. Why is this not reflected in the PN oxygen value?

Two answers spring to mind. The chemical evolution of the Bulge could have been quite different from the local disk, with a different O–Fe relation versus time being set up. However, in a recent model Matteucci & Brocato (1990) predict that the effect will go *the wrong way* in galactic bulges, with O expected to be overabundant compared with Fe at about 'Solar' metallicity.

A second idea is that mentioned in the Introduction: that metal-rich stars tend not to make PNs. Ferguson & Davidsen (1993) suggest that this could explain the PN–K giant discrepancy in the Bulge: choosing a PN sample 'selects' the low end of any range in metallicity.

Because the discrepancy rests on only one stellar study, higher- dispersion spectroscopy of K giants is needed to confirm Rich's findings.

## 5. The Galactic Halo

Abundance patterns in these very-low abundance objects are important. The sample of *extreme* halo objects has been increased by four: IRAS1833–2357 (Gillett *et al.* 1989) in the globular cluster M22, PRMG-1 & PRTM-1 (Peña *et al.* 1989, 1990), and M2-29 is a very low-abundance system (Webster 1988, Peña *et al.* 1991). Peimbert (1991a,b) discussed the abundances of He, C, N, O, Ne, S & Ar in a sample of 10 such 'halo' PNs, including less extreme objects such as the metal-deficient NGC 4361 & NGC 2242. IRAS1833 seems to be a 'born-again' object like A30, A58 & A78, with highly-processed nebular material. M2-29 is a metal-poor object in the Bulge, presumably 'passing through' at high velocity.

As Peimbert shows, a few extreme halo objects show unusual abundance ratios. For example, although Henry (1990) showed the Ne/O ratio quite constant in PNs, the two glaring exceptions are the halo objects BB-1 (Ne/O high) and H4-1 (Ne/O low). It is difficult to explain these ratios just through galactic chemical evolution effects, and I suggest that rather they reflect pollution by different individual massive stars in an inhomogeneous early halo.

## 6. The Magellanic Clouds

Reviews of Cloud PN ('MCPN') abundances are given by Dopita (1991; and these proceedings) and Barlow (1989, 1991). Fuller information is in Dopita's text.

Recent work includes three papers on optical line fluxes (Meatheringham & Dopita 1991a,b; Vassiliadis *et al.* 1992) with the derived abundances and discussion given by Dopita & Meatheringham (1991a,b). In this large programme the optical spectra of about 150 MCPN have now been measured. Torres-Peimbert *et al.* and de Freitas & Costa (Posters VI-240 & 242) report new abundance surveys of MCPNs. Kaler & Jacoby (1990, 1991) compared the correlations between abundances and core mass in MCPN samples with dredge-up theory. Groenewegen & de Jong (Poster VI-241) used parametrised, 'synthetic' AGB evolution models to predict MCPN abundance patterns - results from this useful work should be compared with the new observed abundance sets.

553

**TABLE 1: Elemental Abundances (logarithmic, scale H=12.0)**

|    |              | Ref | He | C | N | O | Ne | S | Ar |
|----|--------------|-----|------|------|------|------|------|------|------|
| 14 | Non Type I   | a   | 10.99 ± 0.03 | 8.86 ± 0.27 | **7.51 ± 0.20** | 8.24 ± 0.13 | 7.40 ± 0.20 | 6.44 ± 0.27 | 5.61 ± 0.24 |
| 3  | Type I       | a   | 11.18 ± 0.08 | 7.45 ± 0.17 | **8.38 ± 0.12** | 8.27 ± 0.11 | 7.45 ± 0.10 | 6.62 ± 0.00 | 5.68 ± 0.05 |
| 17 | SMC PN       | a   | 11.03 ± 0.10 | 8.78 ± 0.42 | **7.84 ± 0.35** | 8.24 ± 0.12 | 7.41 ± 0.18 | 6.48 ± 0.24 | 5.62 ± 0.21 |
|    | SMC H II     | b   | 10.92 ± 0.05 |             | 6.55 ± 0.13 | 8.13 ± 0.10 | 7.69 ± 0.19 | 6.83 ± 0.08 | 5.84 ± 0.08 |
|    | SMC H II     | c   | 10.90 ± 0.02 | 7.28 ± 0.04 | 6.46 ± 0.12 | 8.02 ± 0.08 | 7.22 ± 0.12 | 6.49 ± 0.14 | 5.78 ± 0.12 |
|    | ...          |     |              | C+N →       | **7.34 ± 0.05** |             |             |             |             |
| 12 | Non Type I   | a   | 11.00 ± 0.05 | 8.80 ± 0.21 | **7.66 ± 0.25** | 8.42 ± 0.13 | 7.60 ± 0.17 | 6.47 ± 0.18 | 5.86 ± 0.19 |
| 5  | Type I       | a   | 11.08 ± 0.07 | 8.55 ± 0.83 | **8.47 ± 0.24** | 8.47 ± 0.08 | 7.86 ± 0.14 | 6.61 ± 0.17 | 6.05 ± 0.08 |
| 17 | LMC PN       | a   | 11.02 ± 0.07 | 8.74 ± 0.30 | **8.07 ± 0.34** | 8.44 ± 0.12 | 7.70 ± 0.22 | 6.51 ± 0.20 | 5.93 ± 0.18 |
|    | LMC H II     | b   | 10.96 ± 0.06 |             | 7.07 ± 0.20 | 8.37 ± 0.22 | 7.68 ± 0.11 | 6.87 ± 0.14 | 6.07 ± 0.25 |
|    | LMC H II     | c   | 10.93 ± 0.02 | 7.93 ± 0.15 | 6.97 ± 0.10 | 8.43 ± 0.08 | 7.64 ± 0.10 | 6.85 ± 0.11 | 6.20 ± 0.06 |
|    | ...          |     |              | C+N →       | **7.97 ± 0.15** |             |             |             |             |
|    | Gal PN       | d,e | 11.00 ± 0.04 | 8.74 ± 0.21 | **7.96 ± 0.32** | 8.68 ± 0.14 | 7.98 ± 0.20 | 6.88 ± 0.41 | 6.31 ± 0.21 |
|    | Gal H II     | f,g | 11.00 ± 0.02 | 8.46 ± 0.20 | 7.57 ± 0.04 | 8.70 ± 0.04 | 7.90 ± 0.10 | 7.06 ± 0.06 | 6.42 ± 0.04 |
|    | ...          |     |              | C+N →       | **8.51 ± 0.05** |             |             |             |             |
|    | Solar        | h   | 10.99 ± 0.03 | 8.60 ± 0.05 | 8.00 ± 0.04 | 8.93 ± 0.03 | 8.09 ± 0.10 | 7.21 ± 0.06 | 6.56 ± 0.10 |

Notes: MCPN: (a) This work.
H II regions: (b) Russell & Dopita (1990), (c) Dufour (1984, 1990)
Galactic PN: (d) Torres-Peimbert & Peimbert (1977), (e) Aller & Czyzak (1983)
Galactic H II: (f) Shaver et al (1983), (g) Dufour et al. (1982: carbon)
Solar: (h) Grevesse & Anders (1989), Grevesse et al. (1990:nitrogen; 1991:carbon)

Walton, Barlow & Clegg (1991, and in prep.) analysed 34 Cloud PN abundances, concentrating on C/H ratios (which need IUE spectra). We update the work of Monk, Barlow & Clegg (1988) with new AAT optical spectra and fresh, (plus all archived) IUE spectra. New ionization correction factors were derived from photo-ionization models of PNs with Cloud abundances.

Results are summarized in Table 1. Mean abundances for Type I and non-Type I PNs in each Cloud, plus the Galactic disk, are compared with H II region abundances at the same galactic site. Detailed comparisons with results of Dopita and collaborators has not been made yet, so a summary of our results is given here.

As discussed by Barlow (1991) & Clegg (1991), the mean CNO abundances can be interpreted simply by using the fact that the 1st and 2nd dredge-up episodes recycle the original C,N & O in the stellar envelope. We take the initial composition of progenitor stars to be those of the relevant H II regions [this neglects any possible change through chemical evolution of the region, and also a current discrepancy between Cloud C/H ratios for H II regions and stars, which disagree (Russell & Dopita 1990, Pagel 1993)].

In the Table the sum of 'C+N' gives the available nitrogen abundance after CN-cycle processing. Further N can only come from the ON-cycle or from fresh $^{12}$C created in the 3rd dredge-up. However, since the mean O/H ratios for all types of PNs and the H II regions in each system are the same within the errors, there is no evidence for ON-cycle processing (which would reduce O). The conclusion is that the enhancement of N over the listed 'C+N' values is due to the latter effect, and thus that most of the Type I PNs suffered a 3rd dredge-up. Since many now have C/O <1 and high N, their fresh $^{12}$C was largely converted to $^{14}$N.

Such analyses shows that the 'efficiency' of the both the 1st & 3rd dredge-ups *increases* at lower metallicities. Thus, while it is 'easier' to make a carbon star on the AGB at a given envelope mass when the O/H ratio is low (a simple 'titration effect'), in fact this is added-to by the increased efficiency, which has also been noted in a number of theoretical studies (see article by Lattanzio).

Full details will be published. However, it should be noted that in our sample of 34 MCPNs, *virtually all* non-Type I PNs have C/O>1.

Attention is drawn to 3 SMC Type I PNs for which a 'standard analysis' yields very high values of $T_e(O^{++})$ ($\sim$25,000 K) and low abundances of all heavy elements (SMC N67, L305 & L536; Walton *et al.*). Meatheringham *et al.* (1990) discovered this too for SMC L536 (=SMP 28), noting the apparently–low abundances, especially of carbon (log C < 6.0!). However, the high $T_e(O^{++})$ is hard to explain from photo–ionization alone, and moreover for the same nebulae $T_e(N^+)$ is *very* much lower. Either these objects have high-density ($\geq 10^5$ cm$^{-3}$) central regions where the O$^{++}$ is located (with the N$^+$ further out), or some other process affects the [O III] lines. Perhaps, shock-excitation enhances the 4363Å line – one object (N67) is a strong X-ray source (Wang 1991). [The data of Table 1 are derived assuming these nebulae to have the average value of $T_e(O^{++})$ from LMC Type I PNs].

It would be of great interest to increase the links between known types and PNs and AGB stars in the Clouds. Luminous AGB stars may provide one such link. Wood, Bessell & Fox (1983) showed that at the bright end of the AGB ($M_{BOL} \leq -6$) in both Clouds, there were hardly any carbon stars and the (few) O-rich stars here commonly had MS or S characteristics. Smith & Lambert (1990b) measured high lithium abundances in such AGB stars in the LMC & SMC. The suggested interpretation was that they made Li during 'envelope burning'. This will also result in a hefty conversion of $^{12}C$ into $^{14}N$ and $^{13}C$, and should result, after envelope ejection, in a Cloud Type I PN. The abundance patterns of these two sets of objects should be compared in more detail.

## 7. Beyond the Magellanic Clouds

Abundance analyses have been made for one nebula in the dwarf Fornax galaxy in the Local Group (Danziger *et al.* 1978, Maran *et al.* 1984) and of 3 PNs in M31 (Jacoby & Ford 1986). The Fornax PN confirms the trend we noted earlier - in a metal-poor dwarf galaxy, it has a high carbon abundance ($\log O=8.38$, $\log C=8.95$). Jacoby & Ford (1986) analysed 3 PNs in M31, at projected distances of 3.5, 18 & 33kpc from the centre. In 3-5hr exposures they were able to detect the important [O III]4363Å line and determine $T_e$ values, but were unable to measure $N_e$. Derived O/H values were from $(1-5) \times 10^{-4}$, one nebula being a high-velocity (and thus halo-population) object, but still with disk-type O and N abundances.

Further advances in abundance studies within the Local Group (*i.e.*, out to about 1 Mpc) will surely come from multi-object spectrographs (mainly, fibres and to a lesser extent, multi-slits). With $4m$ telescopes and efficient, thinned CCD detectors, good progress can now be made. A greater challenge is to measure abundances in early-type galaxies beyond the Local group, say out to 3 Mpc distance. Very long exposures with fibres and multi-slits may just be able to obtain a useful S/N ratio for the crucial diagnostic lines of [O III]4363 or [N II]6584Å.

In conclusion, studies of PN abundances as a function of site and metallicity have yielded some very important information; future advances can be expected mainly from Galactic Bulge, Magellanic Clouds, and other Local group galaxy samples. Observational tests of the possible lack of metal-rich PNs (and AGB stars) need to be developed.

## References

Aller, L.H. & Czyzak, S.J..: 1983, *Ap. J. Suppl.*, **51**, 211
Aller, L.H. & Keyes, C.D.: 1987, *Ap. J. Suppl.*, **65**, 405
Barlow, M.J.: 1991, *IAU Symp 148*, eds. R. Haynes & D.K. Milne, p291
Becker, S.A. & Iben, I.: 1979, *Ap. J.*, **232**, 831
Becker, S.A. & Iben, I.: 1980, *Ap. J.*, **237**, 111
Brocato, E., Matteucci, F., Mazzitelli, I. & Tornambe, A.: 1990, *Ap. J*, **349**, 458
Castellani, M. & Tornambe, A.: 1991, *Ap. J*, **381**, 393
Castellani, M., Limongi, M. & Tornambe, A.: 1992, *Ap. J*, **389**, 227
Clegg, R.E.S.: 1987, *MNRAS*, **229**, 31P
Clegg, R.E.S.: 1989, *IAU Symp 131*, ed. S. Torres-Peimbert, p139
Clegg, R.E.S.: 1991, *IAU Symp 145*, eds. G. Michaud & A. Tutukov, p387
Cunha, & Lambert, D.L.: 1992, *Ap. J.*, **399**, 586
Dopita, M.A.: 1991, *IAU Symp 148*, eds. R. Haynes & D.K. Milne, p299
Dopita, M.A. & Meatheringham, S.J.: 1991a, *Ap. J.*, **367**, 115
Dopita, M.A. & Meatheringham, S.J.: 1991b, *Ap. J.*, **377**, 480
Dufour, R.J.: 1984, *IAU Symp.* 108, p353
Dufour, R.J., Shields, G.A. & Talbot, R.J.: 1982, *Ap. J.*, **252**, 461

Faundez-Abans, M & Maciel, W.: 1988, *Rev. Mex. Astr. Astrof.*, **16**, 105
Ferguson, H.C. & Davidsen, A.F.: 1993, *Ap. J*, in press
de Freitas Pacheco, J.A., Maciel, W.J. & Costa, R.: 1992, *Astr. Astroph.*
de Freitas Pacheco, J.A., Maciel, W.J., Costa, R. & Barbuy, B.: 1991, *Astr. Astroph*, **250**, 159
Gillett, F.C., et al.: 1989, *Ap. J.*, **338**, 862
Greggio, L. & Renzini, A.: 1989, *Ap. J*, **364**, 35
Henry, R.B.C.: 1989, *MNRAS*, **241**, 453
Henry, R.B.C.: 1990, *Ap. J.*, **356**, 229
Iben, I. & Renzini, A.: 1983, *Ann Rev Astr. Astrophys.*, **21**, 272.
Jacoby, G.H., & Ford, H.C.: 1986, *Ap. J.*, **304**, 490
Jorissen, A., Smith, V.V. & Lambert, D.L.: 1992, *Astr. Astroph.*, **261**, 164
Kaler, J.B. & Jacoby, G.H.: 1990, *Ap. J.*, **362**, 491
Kaler, J.B. & Jacoby, G.H.: 1991, *Ap. J.*, **382**, 134
Kaler, J.B., Shaw, R. & Kwitter, K.: 1990, *Ap. J.*, **359**, 392
Koeppen, J., Acker, A. & Stenholm,: 1991, *Astr. Astroph*, **248**, 197
Liu, X-W. & Danziger, I.J.: 1993, *MNRAS*, in press
Maciel, W.J.: 1991, in *Elements and the Cosmos*, eds. M.G. Edmunds & R.J. Terlevich (CUP)
Maciel, W, de Freitas Pacheco, J.A. & Landaberry, S.J.C.: 1990, *Rev. Mex. Astr. Astrof.*, **21**, 517
Manchado, A. & Pottasch, S.R., 1989, *Astr. Astroph*, **222**, 219
Maran, S.P., Gull, T.R., Stecher, T.P., Aller, L.H. & Keyes, C.D.: 1984, *Ap. J.*, **280**, 615
Matteucci, F. & Brocato, E.: 1990, *Ap. J.*, **365**, 539
Meatheringham, S.J. & Dopita, M.A.: 1991a, *Ap. J. Supp.*, **75**, 407
Meatheringham, S.J. & Dopita, M.A.: 1991b, *Ap. J. Supp.*, **76**, 1085
Meatheringham, S.J., et al.: 1990, *Ap. J*, **361**, 101
Middlemass, D., Clegg, R.E.S., Walsh, J.R. & Harrington, J.P.: 1991, *MNRAS*, **251**, 284
Monk, D.J., Barlow, M.J. & Clegg, R.E.S..: 1988, *MNRAS*, **234**, 583
Peimbert, M.: 1990, *Rep. Prog. Phys.*, **53**, 1559
Peimbert, M.: 1991a, in *Elements and the Cosmos*, eds. M.G. Edmunds & R.J. Terlevich (CUP)
Peimbert, M.: 1991b, in *Highlights in Astronomy*, Vol 9, (ed. J. Bergeron), (Dordrecht: Kluwer)
Peimbert, M. & Torres-Peimbert, S.: 1987a, *Rev. Mex. Astr. Astrof.*, **14**, 540
Peimbert, M. & Torres-Peimbert, S.: 1987b, *Rev. Mex. Astr. Astrof.*, **15**, 117
Peimbert, M., Sarmiento, A. & Fierro, J.: 1991, *PASP*, **103**, 815
Peña, M., Ruiz, M.T., Maza, J. & Gonzalez, L.E.: 1989, *Rev. Mex. Astr. Astrof.*, **17**,
Peña, M., Torres-Peimbert, S. & Ruiz,:1991, *Rev. Mex. Astr. Astrof.*,
Peña, M., Torres-Peimbert, S. & Ruiz, M.T.: 1991, *PASP*, **1**, 865
Perinotto, M.: 1992, *Ap. J. Suppl.*, **76**, 687
Ratag, M.A.: 1991, *Ph.D. Thesis*, University of Groningen
Ratag, M.A., Pottasch, S.R., Dennefeld, M. & Menzies, J.: 1992, *Astr. Astroph*, **255**, 255
Renzini, A. & Voli, M.: 1981, *Astr. Astroph*, **94**, 175
Rich, R.M.: 1998, *A. J.*, **95**, 828
Russell, S.C. & Dopita, M.A.: 1990, *Ap. J. Supp.*, **74**, 93
Russell, S.C. & Dopita, M.A.: 1992, *Ap. J.*, **384**, 508
Shaver, P.A., McGee, R.X., Danks, A.C. & Pottasch, S.R.: 1983, *MNRAS*, **204**, 53
Smith, V.V. & Lambert, D.L.: 1990a, *Ap. J. Supp.*, **72**, 387
Smith, V.V. & Lambert, D.L.: 1990b, *Ap. J.*, **361**, L69
Torres-Peimbert, S. & Peimbert, M.: 1991, *Rev. Mex. Astr. Astrof.*, **2**, 181
Torres-Peimbert, S., Peimbert, M. & Peña, M.:1991, *Rev. Mex. Astr. Astrof.*, **15**, 117
Torres-Peimbert, S., Peimbert, M. & Peña, M.:1991, *Astr. Astroph.*, **233**, 540
Vassidialis, E., Dopita, M.A., Morgan, D.H. & Bell, J.F.: 1992, *Ap. J. Supp.*, **83**, 87
Walton, N.A., Barlow, M.J. & Clegg, R.E.S.: 1991, *IAU Symp 148*, eds. R. Haynes & D. Milne, 291
Wang, Q.: 1991, *MNRAS*, **252**, 47p
Webster, B.L.: 1988, *MNRAS*, **230**, 377
Wood, P.R.: 1981, *Physical Processes in Red Giants*, ed. A. Renzini, p205
Zuckerman, B. & Aller, L.H.: 1987, *Ap. J.*, **301**, 772

# HOW TO MODEL THE CHEMICAL EVOLUTION OF GALAXIES

JOACHIM KÖPPEN

*Institut für Theoretische Physik und Sternwarte, Olshausenstr. 40, D-W-2300 Kiel, F.R.G.*

## 1. The First Glimpse: The Simple Model

For a first interpretation of the comparison of observational data, the crude "Simple Model" of chemical evolution is quite useful. Since it has well been described in the literature (e.g. Pagel and Patchett 1975, Tinsley 1980), let us here just review the assumptions and whether they are satisfied:

1. The galaxy is a closed system, with no exchange of matter with its surroundings: For the solar neighbourhood this probably is not true (the infamous G-dwarf-"problem", Pagel 1989b). For the Magellanic Clouds this is most certainly wrong, because of the presence of the Inter-Cloud Region and the Magellanic Stream, and evidence for interaction with each other and the Galaxy as well (cf. e.g. Westerlund 1990).
2. It initially consists entirely of gas (without loss of generality of primordial composition): This is good approximation also for models with gas infall, as long as the infall occurs with a time scale shorter than the star formation time scale.
3. The metal production of the average stellar generation (the yield $y$) is constant with time: Initially, it is reasonable to make this assumption. For tables of the oxygen yield see Köppen and Arimoto (1991).
4. The metal rich gas ejected by the stars is completely mixed with the ambient gas. To neglect the finite stellar life times ("instantaneous recycling approximation") is appropriate for elements synthesized in stars whose life time is much shorter than the star formation time scale, such as oxygen, neon, sulphur, and argon.
5. The gas is well mixed at all times: We don't know. The dispersion of H II region abundances may give an indication. In the Magellanic Clouds Dufour (1984) finds quite a low value ($\pm 0.08$ dex for oyxgen).

Then the metallicity $Z$ of a primarily produced element (such as O, Ne, S, Ar) at every time $t$ is a simple function of the gas fraction (by mass) $f_g$ at that moment:

$$Z(t) = y\ln(1/f_g(t)) \tag{1}$$

where stellar nucleosynthesis, evolution, and the initial mass function all are concentrated into a **single** number (the yield $y$). The (hydro)dynamic evolution of the galaxy and its star formation history are described **separately** by the 'astration parameter' $\ln(1/f_g)$, also a **single** number. Note that the metallicity does not depend on the actual temporal history of the star formation rate (how complicated it may be) or on its dependence on the gas fraction. Also, the metallicity is always proportional to the yield of the stellar population, and thus, in the same galaxy, the abundance ratios essentially give information on stellar nucleosynthesis!

## 2. A panorama of types of chemical models

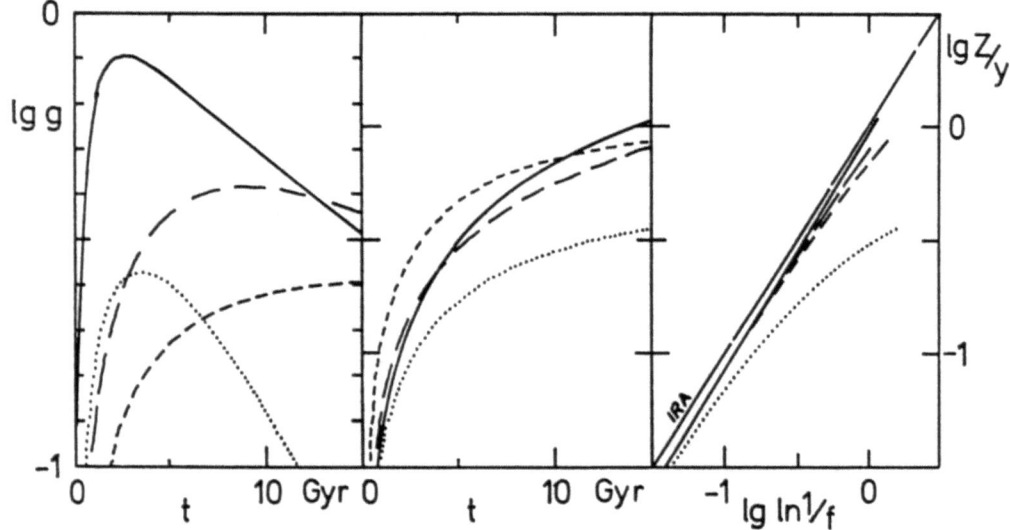

**Fig. 1.** Chemical evolution of infall models: time evolution of gas density $g$, metallicity $Z$, and the $\lg Z$ vs. $\lg\ln(1/f_g)$ diagram for models with yield 1, a star formation rate proportional to $g$. An approximative treatment of finite stellar lifetimes is used. If they are neglected, the line labelled IRA is obtained. Infall time scales are $0.1\,\tau_{SFR}$ (full lines), $0.7\,\tau_{SFR}$ (long dashes) and $50\,\tau_{SFR}$ (short dashes). Also shown (dots) is a model with continuous galactic wind (mass loss rate = 3∗SFR).

In the following let's look at various types of chemical evolution models: Fig. 1 shows the temporal evolution of the gas density $g$ and metallicity $Z$ and the $\lg Z$ vs. $\lg\ln(1/f_g)$ diagram for models with exponential infall of zero metallicity gas. If infall occurs quickly — i.e. with a time scale shorter than for star formation — one essentially gets the "closed-box" Simple Model. We note that in the metallicity-gas fraction diagram this corresponds to a straight line of slope 1. All models were computed with an approximate treatment of the finite stellar lifetimes, and so avoid the "instantaneous recycling approximation", but as can be seen in the figure, this only leads to a fairly small (less than 0.1 dex) deviation from Eq. 1. Infall gives rise to lower metallicities, i.e. a lower effective yield

$$y_{\text{eff}} = Z/\ln(1/f_g) \tag{2}$$

but this is only noticeable with very slow infall at large ages. So, in a way, all these models — despite their rather different evolution of gas density or metallicity — behave very closely as a Simple Model.

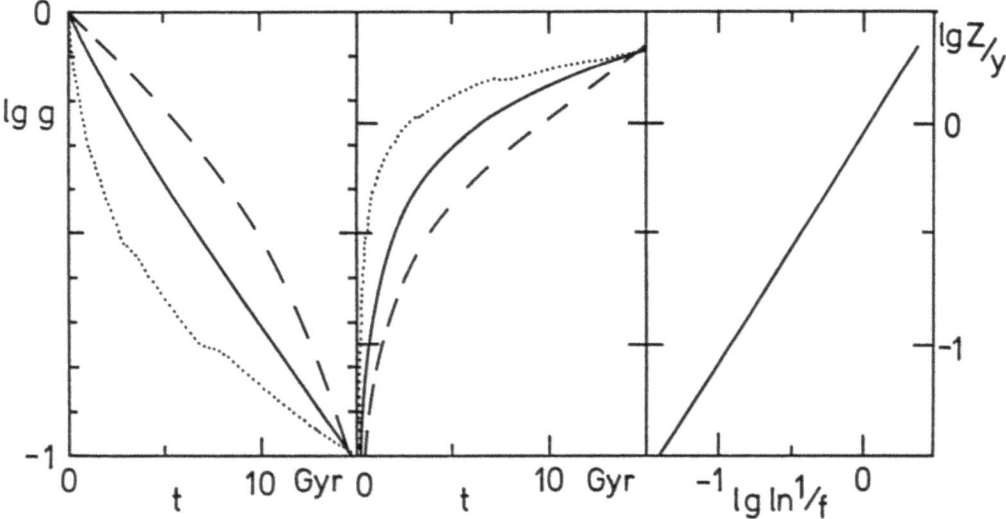

**Fig. 2.** Chemical evolution of the Simple Model with different dependences of the star formation rate on gas density $g$: linear (full line), $g^{0.25}$ (dashed), and quadratic (dotted). The constant of proportionality is adjusted to give the same gas density ($g = 0.1$) after 15 Gyr. The kinks in the dotted curves as due to our representing the spectrum of stellar lifetimes by only 5 values.

The exact dependence of the star formation rate on the gas density, if it exists at all, is unknown. The previous models were done with a linear dependence. In Fig. 2 we see that this dependence may alter the gas density evolution and thus the star formation history, and the age-metallicity relation, but no change at all is seen in the $Z$ vs. $f_g$ diagram.

That star formation must be a simple function of time is not a necessity of the physics of galaxies, but rather a simplifying assumption to compute models. For irregular galaxies the idea of stochastical self-propagating star formation (SSPSF, Gerola et al. 1980) is often discussed: star formation may be induced in the vicinity of previous star formation activity. This gives rise to a strongly fluctuating star formation rate. Similar fluctuations are also possible due to the nonlinear behaviour of the equations for the evolution of interstellar clouds (Struck-Marcell and Scalo 1987), or of the energy balance in detailed multi-phase models of the interstellar medium (Theis et al. 1992). Matteucci and Chiosi (1983) simulated the SSPSF process by computing the chemical evolution of models with a rapidly fluctuating star formation rate. Since perfect mixing in the interstellar medium is assumed, the abundance of oxygen is still closely linked to the gas fraction as in a Simple Model (see our Fig. 3 and their Fig. 8).

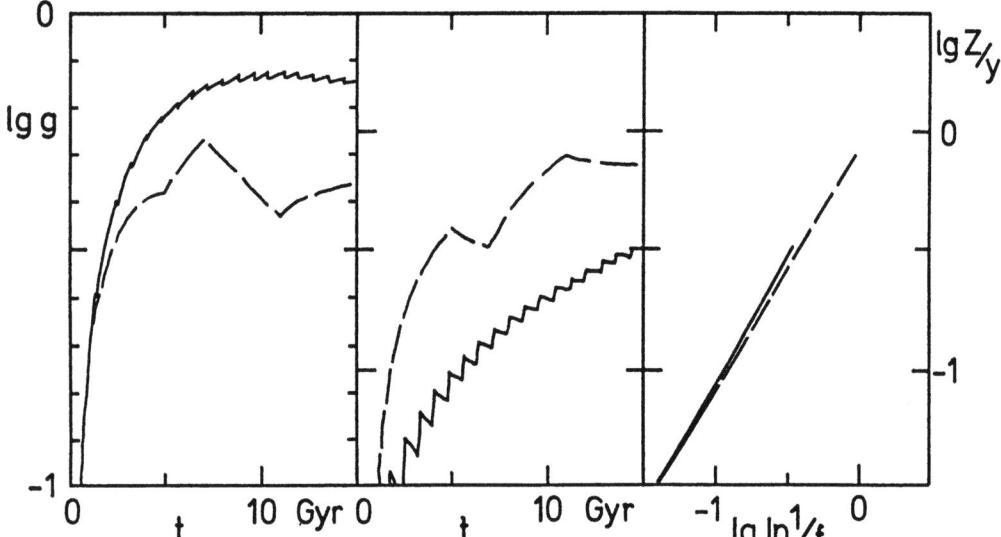

**Fig. 3.** Chemical evolution of models with very short star formation bursts every 1 Gyr (full line) and long ones (from 1 ... 5 and 7 ... 11 Gyr; dashed line).

However, these simulations are rather simplified: They do not take into account the multi-phase structure of the ISM (and thus the various delay times of mixing and enrichment). Also, they assume that chemical enrichment is a purely local process: gas flows in e.g. the hot phase, distribution of metals over a finite volume by galactic fountains, etc. are ignored. Calculations of the evolution with incomplete mixing (Wilmes and Köppen 1991) show that non-local effects do not affect the average abundance values, but the correlation between secondary and primary elements (e.g. N/O ratio).

Many galaxies are not isolated: The Magellanic Clouds are connected with the Inter-Cloud Region consisting of gas and embedded stars, which contains as much H I gas as the SMC. The Magellanic Stream forms a long, gaseous tail to the Clouds, and is regarded as an indication of global interaction in the Magellanic system (Wayte 1990).

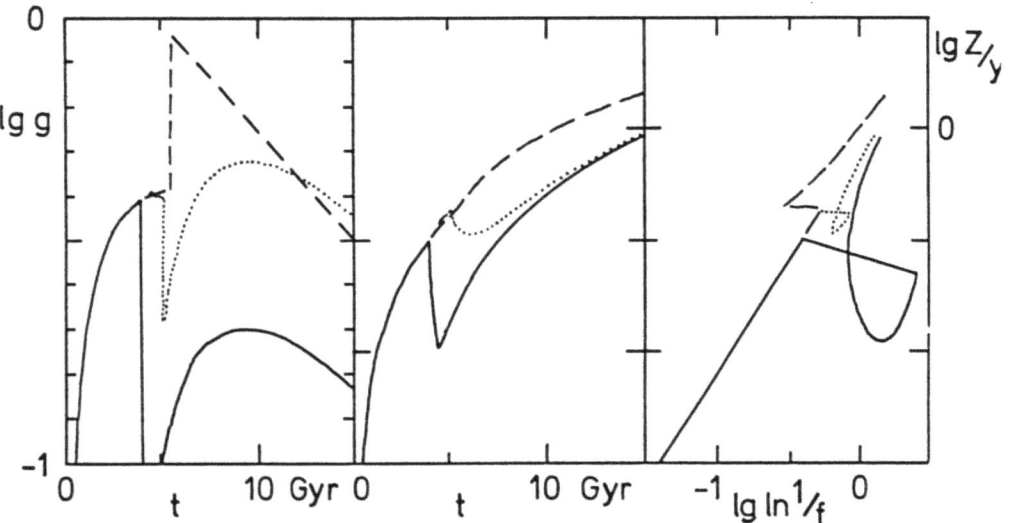

**Fig. 4.** Effects of possible processes caused by galactic interactions: sudden 99 percent gas loss at 4 Gyr (full lines), sudden gas inflow (of additional 50 percent mass) at 6 Gyr (dashes), star burst (3*SFR) at 4.5 Gyr and 50 percent gas loss at 5 Gyr followed by this gas flowing back with time scale 1 Gyr (dots).

What could happen during an interaction? While we do not claim to produce a physically consistent list of processes and event, let us consider the effects of the following items on chemical evolution (as shown in Fig. 4):

1. Sudden gas loss by tidal interaction: Within a few Gyr the gas that had been bound in stars of intermediate life time (and low metal production) is released and — depending on the IMF — a fair fraction of the lost gas is replenished, the metallicity drops. If star formation continues, the system recovers to an almost normal Simple Model after few $\tau_{SFR}$ (full lines).

2. Sudden inflow of gas or loss of stars: Both processes increase the gas fraction for some time. After few $\tau_{SFR}$ the model evolves again as a Simple Model (dashed lines).

3. Partial exchange of mass: Combination of the previous two events.
4. Star formation burst: The effects can be seen in Fig. 3: a quick increase of the metallicity followed by a decrease when the intermediate mass stars die. Essentially like a Simple Model.
5. SN-driven galactic wind following an intense starburst: The dotted curves in Fig. 4 show that the evolution is dominated by the gas loss effects.

Thus, substantial deviations from the Simple Model can only be caused by really drastic changes in the parameters.

Pure chemical evolution has one weak spot: one can only determine the "chemical age" of a system, not the absolute one: e.g. the more gas-rich SMC could be a galaxy genuinely younger than the more metal-rich LMC, if they had a similar star formation time scale. On the other hand, it could be of the same age, but with a lower star formation time scale.

## 3. Observational constraints

If one wants to make a chemical evolution model for a galaxy, what observational constraints are valuable?

- Present metallicity: this can be obtained from H II region and SNR emission line spectra. While one gets e.g. the oxygen abundance, the iron abundance — so often taken as a metallicity indicator from stellar photometry and spectra — cannot be obtained, because of weakness or lack of suitable lines, poorly known collisional rates, and probable lock-up into dust grains.
- Present gas fraction: the neutral gas mass comes from H I measurements; molecular mass could be estimated from CO. For the galaxy's total mass which takes part in the chemical evolution, one often uses the dynamical mass (from the velocity dispersion seen in the H I 21 cm line widths). This is a problematic approach, since this could well be affected by dark matter. Ideally, one should use the total mass in gas and stars. The latter is obtained from the absolute luminosity of the stars by using stellar population models, thus it is model-dependent. A grid of Simple Models can be found in Arimoto and Tarrab (1990) who also compute the photometric colours.

These two data may already tell important things: Pagel et al. (1978) find that that the average oxygen abundances of both Magellanic Clouds can well be explained with a simple model with the **same** metal yield $y = 0.003 = 0.15 Z_\odot$ (it is also compatible with regions in M 33 and M 101). This model also explains the chemical compositions and gas fractions of 6 irregular and blue compact galaxies (Lequeux et al. 1979).

- Age metallicity relation: This is a more powerful constraint to distinguish between different model types (see our figures). However it is very difficult to obtain, since both metallicities and ages must be derived (cf. Twarog 1980). In the Magellanic Clouds the use of the star clusters does not seem to be a proper way (Richtler 1992).
- Stellar metallicity distribution function: This involves only the determination of metallicity, but in a statistically complete sample of stars of the same type (G-dwarfs in the solar neighbourhood (Pagel and Patchett 1975, Pagel 1989b), K-giants in the Bulge (Rich 1988)). If known, this function is an extremely useful constraint on infall (e.g. Köppen and Arimoto 1990a,b).
- Photometric colours: Since the colours are sensitive to the ratio of present and past integrated star formation rate (e.g. Searle et al. 1973, Rocca-Volmerange et al. 1981), they can help to fix the star formation history, however, not in great detail, especially in the far past.
- Abundances from planetary nebulae: with planetaries one can probe into the past (as with stars), and also take advantage of the better abundance diagnostics by emission lines. The difference of mean abundances of planetaries and H II regions gives some constraints on the age-metallicity relation. Of course, several problems remain: accurate abundances, determination of ages, spatial dispersion of the star from its place of birth ought to be mentioned.

## 4. Abundance gradients in spiral galaxies

The trouble with abundances gradients seen with H II regions is that there are far too many theories to explain them (Pagel 1989a, Götz and Köppen 1992): radial variations of the yield or of the star formation rate, various forms of gas infall, including a radial variation of the infall time scale, radial gas flows in the disk (caused by various ways). This ambiguity may be resolved by looking at the temporal evolution of the gradients: Here one can distinguish between different models (Köppen 1992), and the information about the gradient, say at about 10 Gyr, would be very helpful to put constraints on the star formation rate and on dynamical processes (infall, radial flows). Planetaries may be very useful here (see Maciel and Köppen, this conference).

**Acknowledgements:** I thank H.Meusinger and N.Arimoto for helpful discussions. Financial support from the Deutsche Forschungsgemeinschaft (SFB 328) is gratefully acknowledged.

**References:**

Arimoto N., Tarrab I.: 1990, *Astron.Astrophys.* **228**, 6
Dufour R.J.: 1984, *IAU Symposium* **108**, 353
Gerola H., Seiden P.E., Schulman L.S.: 1980, *Astrophys.J.* **242**, 517
Götz M., Köppen J.: 1992, *Astron.Astrophys.* , in press
Köppen J.: 1992, *Astron.Astrophys.* , submitted
Köppen J., Arimoto N.: 1990a, *Chemical and Dynamical Evolution of Galaxies*, Proc.Elba Intern.Phys.Centre, ed. F.Ferrini, J.Franco, F.Matteucci, ETS editrice, Pisa
Köppen J., Arimoto N.: 1990b, *Astron.Astrophys.* **240**, 22
Köppen J., Arimoto N.: 1991, *Astron.Astrophys.Suppl.* **87**, 109
    Erratum: *Astron.Astrophys.Suppl.* **89**, 420
Lequeux J., Peimbert M., Rayo J.F., Serrano A., Torres-Peimbert S.: 1979, *Astron.Astrophys.* **80**, 155
Matteucci F., Chiosi C.: 1983, *Astron.Astrophys.* **123**, 121
Pagel B.E.J.: 1989a, *Rev.Mex.Astron.Astrof.* **18**, 161
Pagel B.E.J.: 1989b, *Evolutionary Phenomena in Galaxies*, Proc. Advanced Study Inst., eds. J.E.Beckman and B.E.J.Pagel, Cambridge University Press, Cambridge, p. 201
Pagel B.E.J., Patchett B.E.: 1975, *Mon.Notices Roy.Astron.Soc.* **172**, 13
Pagel B.E.J., Edmunds M.G., Fosbury R.A.E., Webster B.L.: 1978, *Mon.Notices Roy.Astron.Soc.* **184**, 569
Rocca-Volmerange B., Lequeux J., Maucherat-Joubert M.: 1981, *Astron.Astrophys.* **104**, 177
Rich R.M.: 1988, *Astron.J.* **95**, 828
Richtler T.: 1992, *New Aspects of Magellanic Cloud Research*, eds. B.Baschek, G.Klare, J.Lequeux, Springer, Heidelberg, in press
Searle L., Sargent W.L.W., Bagnuolo W.G.: 1973, *Astrophys.J.* **179**, 427
Struck-Marcell C., Scalo J.M.: 1987, *Astrophys.J.* **64**, 39
Theis C., Burkert A., Hensler G.: 1992, *Astron.Astrophys.* , submitted
Tinsley B.M.: 1980, *Fund.Cos.Phys.* **5**, 287
Twarog B.A.: 1980, *Astrophys.J.* **242**, 242
Wayte S.R.: 1991, *IAU Symposium* **148**, 447
Westerlund B.E.: 1990, *Astron.Astrophys.Rev.* **2**, 29
Wilmes M., Köppen J.: 1991, *Evolution of Interstellar Matter and Dynamics of Galaxies*, CTS Workshop 1, Prague, ed. J.Palouš, Cambridge University Press, Cambridge, in press

## SMOTCH - Simple MOdel waTCH
The dedicated hardware solution for DIY chemical evolution of galaxies
© 1992, Joachim Köppen, Kiel, pat.pend. etc. pp.

The three scales near the bottom solve the basic equation of the Simple Model of chemical evolution $Z = y \ln(1/f_g)$, in convenient units and with proper yields $y$:

- **Gas fraction** (fixed scale) is read off or set at the pointer (moving black triangle). Note: if the gas fraction is below about 10 percent, a large portion of the gas is returned with appreciable delay times by intermediate mass stars. Thus, the Simple Model can no longer give correct results.
- Exponent for power law **IMF**s (fixed scale) (Salpeter: 1.35, marked by **S**) and a mass range from 0.05 to 60 $M_\odot$. This is basically the oxygen yield, as computed by Köppen and Arimoto (1991) A&A Suppl. **87**, 109 and **89**, 420:
  - The actual yield can be read from the moving abundance scale, when the dial is set in the position as shown (gas fraction 37%): so an IMF with a slope of 2.0 gives an oxygen yield of $\lg y/Z_\odot = -1.35$ dex.
  - The outer fixed scales $M_u$ and $M_\ell$ indicate how the yield changes, if upper and lower mass limits are varied: e.g. using $M_u = 100\ M_\odot$ raises the yield by 0.2 dex. Using also $M_\ell = 0.2\ M_\odot$ gives a further increase by 0.4 dex.
  - The yields of the IMFs of Tinsley (**T**), Miller/Scalo (**MS**), and Scalo (**Sc**) are marked, with the original mass ranges given by the authors.
- The gas **metallicity** $\lg Z/Z_\odot$ (moving) is found opposite the selected value of the IMF. Note that this is exact only for $O^{16}$; all other 'metals' are O.K. within $\pm 0.5$ dex (except He and secondary nuclei such as $C^{13}$ and $N^{14}$).

Any of the three quantities can be computed from the other two: Given a gas faction, the metallicity can be read off for any yield (IMF), or the yield is obtained from an observed Z. Likewise, any combination of IMF and Z gives the gas fraction.

IF one makes the additional assumption of a linear star formation law $SFR = f_g/\tau_{SFR}$, one can estimate the actual age of a galaxy or the star formation time scale $\tau_{SFR}$. This is done with the top supplementary scales which solve $f_g(t) = \exp(-\alpha t/\tau_{SFR})$. The locked-up mass fraction $\alpha$ depends only weakly on the IMF.

- **Time** $t$ (fixed scale) in Gyr. The black dot marks 15 Gyr.
- **Star formation timescale** $\tau_{SFR}$ (moving scale) in Gyr. The solar symbol (at 5 Gyr) indicates a reasonable value for the solar neighbourhood: the gas fraction after 15 Gyr would be 2%. Since the gas is depleted more slowly by the factor $1/\alpha$ (the moving short scale **IMF** gives $\approx 0.2$ dex for Salpeter IMF), the true gas fraction is higher (10%). For steep IMFs, this correction is rather small.

Note that these top scales assume a specific SFR-law. However, for other laws, they can be used as a guideline.

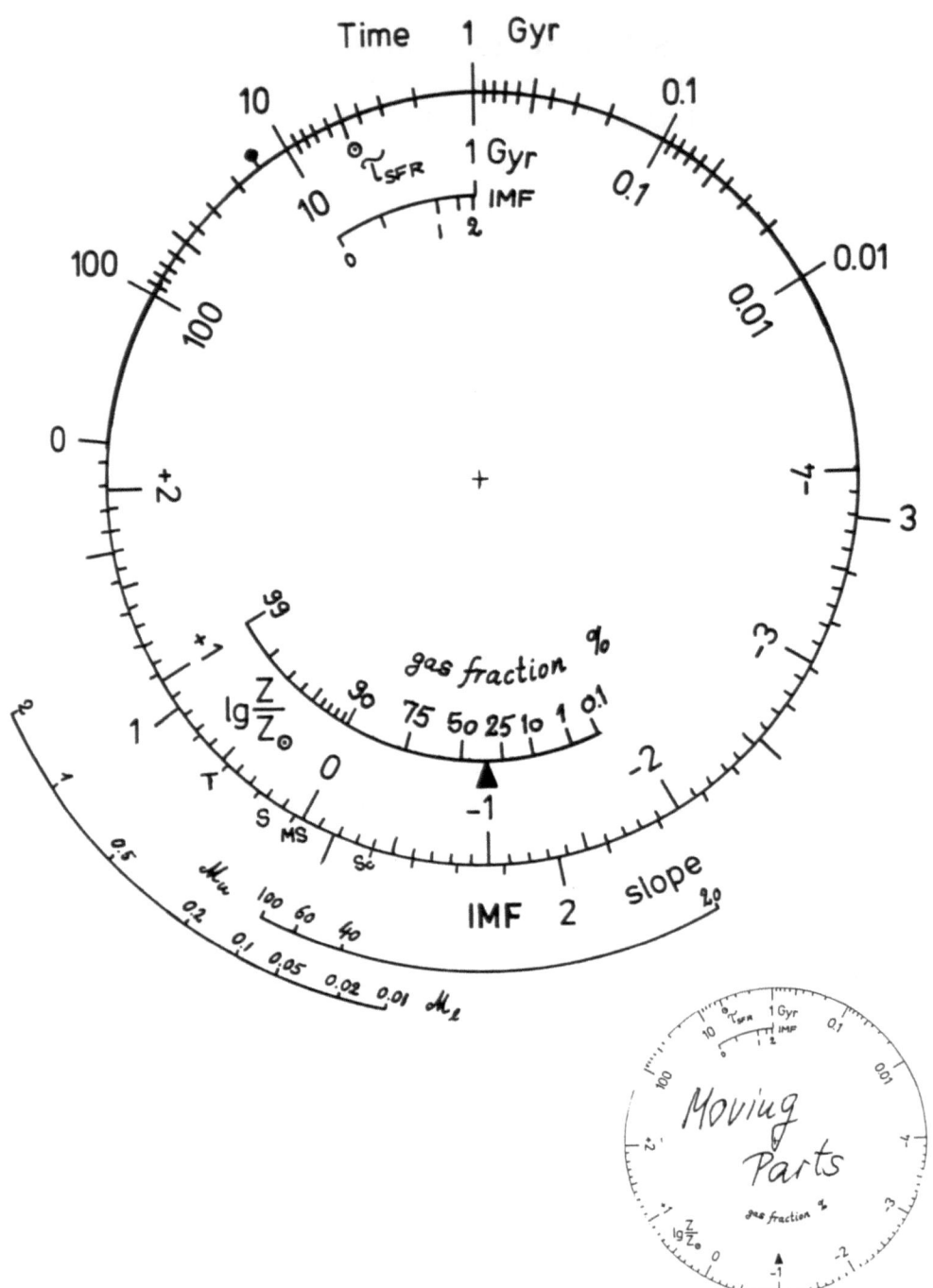

# DISTRIBUTION OF PLANETARY NEBULAE PERPENDICULAR TO THE DISK

D.C.V. MALLIK and S. CHATTERJEE
*Indian Institute of Astrophysics, Bangalore 560034, India*

From an analytical solution to the Boltzmann-Poisson equations for a thin, vertically isothermal self-gravitating disk with the assumptions that the mass spectrum of stars is expressible as an inverse power-law and the velocity dispersion perpendicular to the plane follows a law of the form $< V_z^2(m) > = V_o^2$ for $m < m*$ and $< V_z^2(m) > = V_o^2(m*/m)^\theta$ for $m \geq m*$, we have obtained a vertical height distribution of planetary nebulae : $n_{PN}(z) = \int_m^{m_u} n(m,z) \tau_{PN}/\tau(m) dm$, where $m_u = 7.0 m_\odot$, $m_l = 1.0 m_\odot$ and $n(m,z) = n(m,o) \exp(-\phi(z)/< V_z^2(m) >)$, $\phi(z)$ being the potential at $z$. Figure 1 shows a normalised height distribution for various values of $V_o$ where we have assumed a Salpeter slope, a $\rho(o) = 0.10 M_\odot pc^{-3}$ and set $m* = m_l = 1.0 m_\odot$. Figure 2 shows filled circles obtained from the observational sample of Daub (1982, ApJ **260**, 612) superposed on the theoretical distributions. Although there is agreement for large values of $z$, closer to the plane the observational sample falls below the theoretical curves. Since the observational sample is size-limited, we may be missing a larger fraction of small nebulae closer to the plane. It is also possible a single value of $V_o$ is not a correct representation of reality as these nebulae originate from stars of greatly differing ages and $V_o$ may have changed over the lifetime of the Galactic disk.

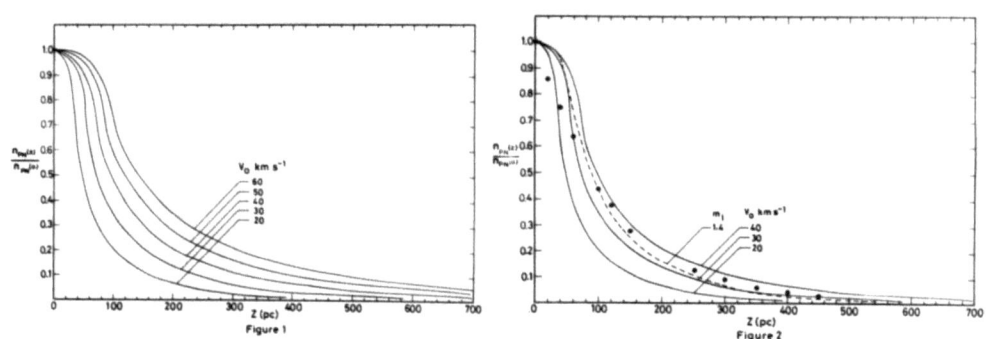

Figure 1

Figure 2

# KINEMATICS OF DISK PLANETARY NEBULAE

W.J. MACIEL and C.M. DUTRA

*Instituto Astronômico e Geofísico da USP, Av. Miguel Stefano 4200, 04301-904 São Paulo SP, Brazil*

In the past few years, it has become clear that planetary nebulae (PN) are a true stage in stellar evolution, especially regarding their chemical composition. This fact led to the introduction of the classification scheme developed by Peimbert, which has been applied to a sample of galactic objects for which a detailed amount of data exists.

On the other hand, the kinematical properties of PN make them interesting objects to study the kinematics of the disk population. In particular, there are controversies regarding the rotation curve of the Galaxy outside the solar circle, and it is not clear whether the curve decreases, remains constant or even increases thereafter. Since PN are luminous objects that can be observed to relatively large distances and have accurately measured velocities, it is interesting to investigate their possible contribution to this problem.

In the present work, we will explore the kinematical properties of the different types of PN in comparison with classical population I objects. We will concentrate on the space distribution and kinematics of a sample containing 150 disk nebulae, especially considering the determination of the galactic rotation curve and of Oort's constants.

It is shown that planetary nebulae of Types I, IIa, IIb, III, and IV form an approximately continuous sequence in terms of these properties, confirming similar conclusions based on their chemical composition. Some kinematical consequences regarding the connection between planetary nebulae and HII regions are explored, leading to the determination of the galactic rotation curve and Oort's constants. It is found that the rotation curve presents a flattening near the solar circle and a moderate increase for larger galactocentric distances.

*Work partially supported by CNPq and FAPESP.*

# SPECTROPHOTOMETRY AND KINEMATICS OF THE NEWLY DISCOVERED PN IN THE OUTHER FIELD OF THE LMC

M.A. DOPITA and E. VASSILIADIS

*Mt. Stromlo and Siding Springs Observatories, Institute of Advanced Studies, The Australian National University, Canberra*

and

D.H. MORGAN

*The Royal Observatory, Blackford Hill, Edinburgh*

Recently a number of new planetary nebulae (PN) have been found in the outer field of the LMC using deep objective–prism J plates taken with a Schmidt Telescope at the AAO. For a number of these PN, we have obtained spectrophotometry on the 2.3m ATT and for a sub–sample, determined the radial velocities using the Echelle Spectrograph of the 2.3m Telescope. All this data is to appear in the ApJ.
Since the PN are concentrated in the outer portions of the face of the LMC, they allow an improved determination of a rotation solution given by:

$$V(\theta, r) = V_m(r)(1 \pm (\tan(\theta - \theta_0)\sec i)^2)^{-0.5} + V_0 \ (0 \leq \theta \leq 2\pi)$$

where $V(\theta, r)$ is the rotational velocity projected onto the line of sight at position angle $\theta$ and at radial coordinate $r$. The rotation solutions derived are as follow:

| Sample | No. Objects | $V_0$(km/s) | $\theta_0$ |
|---|---|---|---|
| This sample | 11 | 40 | 164 |
| All PN outside 4deg. | 30 | 46 | 170 |
| All PN in the LMC | 106 | 42 | 168 |

The faint PN fall into three distinct groups: first Type I PN with high electron temperatures, very strong lines in HeII, [NII], and often, [Ne V]; second, optically thick PN displaying unusually strong [OII] lines for a PN; and third, optically thin PN of intermediate class. Optically thin objects are comparatively rare in this sample, whereas one would have expected a large number among the faint PN. The optically thick non Type I objects are mainly of excitation class 2–6. However, this appears to be due mainly to low ionisation parameter, rather than to cool nuclei. The predominance of faint optically thick PN appear to be a result of the evolution of central star. Normally a PN ionises more and more of the surrounding gas as it expands, thanks to the decreasing density. If a nebula becomes optically thin, its $H\beta$ luminosity will decrease as $t^{-3}$ and it will rapidly become very faint as a PN. Suppose now that the PN is optically thick when it commences its rapid fading. Models of the evolution of PNN along the fading tracks show that the fading of the central star is likely to be so rapid $\sim t^{-4\,to\,5}$ as to overcome the effects of the nebular expansion, and thus the nebula may actually start to recombine until the luminosity of the central star falls below $\sim 100 L_\odot$. After this the rate of the decline slows considerably ($\sim t^{-1}$), and the nebula becomes optically thin to the ionising radiation. All the Type I PN were found to lie near or to the high mass side of the $2.5 M_\odot$ track, confirming that these objects originate from high mass precursors.

# THE RADIAL VELOCITIES OF PLANETARY NEBULAE IN NGC 3379

ROBIN CIARDULLO

Department of Astronomy and Astrophysics, Penn State University, 525 Davey Lab, University Park, Pennsylvania 16802, USA

and

GEORGE JACOBY

Kitt Peak National Observatory, National Optical Astronomy Observatories, P.O. Box 26732, Tucson, Arizona 85726, USA

Several authors have analyzed the kinematics of elliptical galaxies using surface photometry in combination with absorption line velocity dispersion measurements. However, these analyses never explore the halos of galaxies, since the best absorption line measurements extend only ~1 $r_e$. The only way to extend our knowledge of stellar kinematics to larger radii is to use the emission lines of planetary nebula for radial velocity measurements.

In May 1991 we placed Kitt Peak's multi-fiber spectrograph (NESSIE) at the R-C focus of the Mayall 4 m telescope and measured the radial velocities of 29 PN in the outer envelope of the normal elliptical galaxy NGC 3379. These observations achieved an internal precision of ~ 17 km s$^{-1}$, and allow us to measure the galaxy's mass out to a radius of $3'.5$ (3.5 $r_e$, or 11 kpc). Assuming that our PN trace the mass, then the projected mass estimator of Heisler, Tremaine, and Bahcall (1985) gives a value of ~ $1.4 \times 10^{11} M_\odot$. The implied mass-to-light ratio, $M/L_B \sim 18$, is in agreement with that found from absorption line analyses and suggests that dark matter is not a major constituent of NGC 3379's halo.

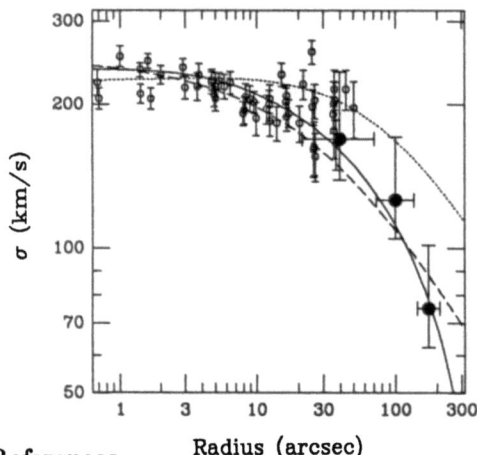

Figure 1. Displayed is the velocity dispersion profile for NGC 3379. The open circles are the absorption line measurements, mostly from Davies and Birkinshaw (1988); the filled circles show our PN data. For comparison, 3 constant mass-to-luminosity models are shown: an $r_t/r_c \sim 2.2$ King (1966) model (solid line), an isotropic orbit Jaffe (1983) model (dashed line), and a circular orbit Jaffe model (dotted line). A Jaffe model with radial orbits is ruled out by the data. Note that in general, isotropic orbits fit the data reasonably well.

### References

Davies, R.L., and Birkinshaw, M. 1988, Ap. J. Suppl., **68**, 409.
Heisler, J., Tremaine, S., and Bahcall, J.N. 1985, Ap. J., **298**, 8.
Jaffe, W. 1983, M.N.R.A.S., **202**, 995.
King, I.R. 1966, A. J., **67**, 471.

# UV AND OPTICAL ABUNDANCES FOR A SAMPLE OF SOUTHERN GALACTIC PLANETARY NEBULAE

ROBIN L. KINGSBURGH and M.J. BARLOW

*Dept. of Physics & Astronomy, U.C.L., Gower Street, London WC1E 6BT, U.K.*

We present abundances for a sample of 57 southern hemisphere galactic planetary nebulae (PN). Optical spectra covering the 3100-7400 Å range were obtained at the AAT. Low resolution UV spectra obtained with the IUE satellite were available for half of these objects and were accessed via the IUE Uniform Low-Dispersion Archive. Additionally, new low resolution IUE SWP observations of Fg 1, M 3-1 and M 3-3 were obtained.

The abundance analysis includes a derivation of He/H, O/H, N/H, C/H, Ne/H, Ar/H and S/H ratios (by number). Helium abundances were derived following Clegg (1987, MNRAS 221, 31p). In the cases where unseen stages of ionization are present, we use ionization correction factors which are not only based on the similarity of ionization potentials of various ions, but also incorporate results from 10 detailed ionization structure models (Walton et al., 1992, in prep.). We classify five PN in this sample as Type I: He 2-111, NGC 2440, He 2-15, He 2-112 and NGC 5189; all have log(N/O)$\geq$-0.3 and He/H$\geq$0.125 and are bipolar in appearance. The mean oxygen abundance for 51 non-Type I PN is found to be 12+log(O/H)=8.66$\pm$0.31. We find one object with high oxygen abundance, PC 14, whose 12+log(O/H)=9.16.

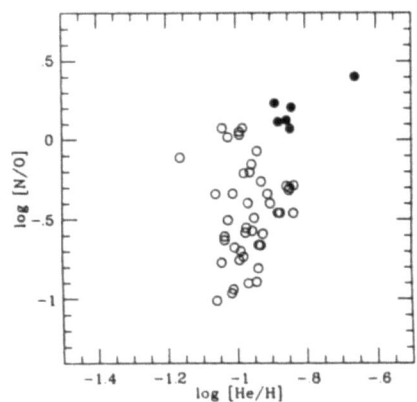
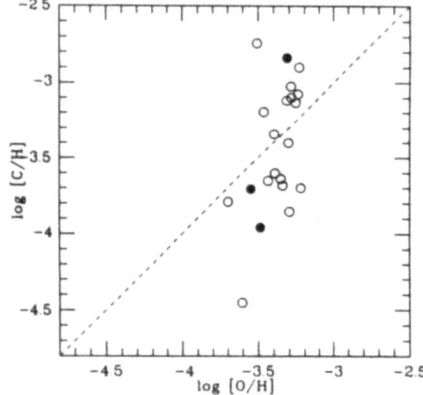

FIGURE 1 plots log(N/O) vs. log(He/H). Filled circles are Type I PN. N/O is strongly correlated with He/H for only the Type I PN. For the remainder of the sample a continuous range of N/O is found, with He/H ranging from $\sim$ 0.09 to 0.14. For the 52 non-Type I PN, we find a mean He/H ratio of 0.111$\pm$0.017 by number.

FIGURE 2 plots log(C/H) vs. log(O/H). Filled circles are Type I PN. For this sample of 21 PN, we find the ratio of carbon-rich to oxygen-rich PN to be $\sim$50%. This value is somewhat lower than, but consistent with, the ratio of 62% found by Zuckerman & Aller for a sample of 68 galactic PN (1986, ApJ 301, 772).

# CHEMICAL ENRICHMENT AND CENTRAL STAR PROPERTIES

C.Y. ZHANG

*Department of Astronomy, The University of Texas at Austin, Austin, TX 78712*

ABSTRACT. We have selected a sample of planetary nebulae, for which the core masses are determined using distance-independent parameters (Zhang and Kwok 1992). The chemical abundances of He, N, O, and C are taken from the literature for them. Relationships of the ratios of He/H, N/O, and C/O with various stellar parameters of planetary nebulae (PN), such as the core mass, the mass of the core plus the ionized nebular gas, the stellar age and temperature, are examined. It is found that the N/O increases with increasing mass, while the C/O first increases and then decreases with the core mass. No strong correlation seems to exist between the He/H and the core mass. A correlation of the N/O and He/H with the stellar temperature exists. The current dredge-up theory for the progenitor AGB stars cannot satisfactorily account for these patterns of chemical enrichment in PN. Furthermore, the correlations of the N/O and He/H with the stellar age and temperature indicate that besides the dredge-ups in the RG and AGB stages, physical processes that happen in the planetary nebula stage may also play a role in forming the observed patterns of chemical enrichment in the planetary nebulae.

CYZ was supported by the NASA grant NAG 2-67.

## References

Zhang, C.Y., and Kwok, S. 1992, A&A (in press)

# DETERMINATION OF ELEMENT ABUNDANCES IN PLANETARY NEBULAE FROM RECOMBINATION LINE SPECTRA

A.A. NIKITIN and A.F. KHOLTYGIN
*S.-Petersburg Univ. Astron. Obs., 198904, Petergof, Russia*

and

A.A. SAPAR and T.KH. FEKLISTOVA
*Tartu Astrophysical Observatory, EE2444, Tõravere, Estonia*

**Abstract.** Carbon, nitrogen and oxygen abundances have been found from the recombination lines for 63 planetary nebulae.
Intensities of the recombination lines give important information about ion abundances in planetary nebulae. The results of our calculations of the effective recombination coefficients and the relative line intensities for ions CII, CIII, CIV, NIII, NIV, OIV and OV [1] are presented. The abundance ratio of the ions $A^+$ and $H^+$ is defined by [1]:

$$\frac{N(A^+)}{N(H^+)} \equiv \{A^+/H^+\} = \frac{\lambda_{ki}}{\lambda(H_\beta)} \frac{\alpha^{eff}(H_\beta)}{\alpha_{ki}^{eff}} = X(T_e) \frac{I_{ki}}{I(H_\beta)},$$

where $I_{ki}$ is the observed line intensity for transition $k \to i$ corrected for interstellar extinction and $I(H_\beta)$ is the same for $H_\beta$, the quantities $\alpha_{ki}^{eff}$ and $\alpha^{eff}(H_\beta)$ are the corresponding effective recombination coefficients. Both the radiation and dielectronic recombination contribution have been taken into account in $X(T_e)$. The values of $X(T_e)$ for ions CII, CIII, CIV, NII, NIII, NIV, OII, OIII, OIV and OV are given in table. Using the ionization correction factors given in [1] the total ion abundances for C, N and O have been obtained. We have determined the abundances of CNO elements in 63 planetary nebulae. The abundances found from the recombination line intensities are often much higher than those from collisionally excited lines.

### References

Golovatyj V.V., Sapar A.A, Feklistova T.Kh., Kholtygin A.F. (1991), Atomic data for the spectroscopy of rarefied astrophysical plasma, Tartu Astr. Obs. Teated No 109, Tallinn

# CLUMPS IN THE PLANETARY NEBULAE

A.F. KHOLTYGIN

*S.-Petersburg Univ. Astron. Obs., 198904, Petergof, Russia*

and

T.KH. FEKLISTOVA

*Tartu Astrophysical Observatory, EE2444, Tõravere, Estonia*

**Abstract.** The effect of density and temperature fluctuations on the intensity of the recombination and collisionally excited lines is investigated.

The abundances of CNO ions in planetary nebulae found from the recombination lines often exceed those from collisionally excited lines. We have investigated the dependance of the ratio $\{X_{rec}\}/\{X_{coll}\}$ versus $\{X_{coll}\}$ where $\{X_{rec}\}$ and $\{X_{coll}\}$ are respectively the abundancies of the ion X determined from the recombination and collisional excited lines. The negative correlation between the values $\{X_{rec}\}/\{X_{coll}\}$ and $\{X_{coll}\}$ for planetaty nebulae type II is found. It is proposed to use the ratio of the abundances determined using the lines dominated by different line formation mechanisms as the measure of the electron temperature and ion density fluctuations in planetary nebulae. The effect of the fluctuations on the intensities of the recombination lines and collisionally excited lines are investigated. Dimensions of dense clumps in the planetary nebulae are estimated. We have proposed that the clumps are the relic of the maser condensations in the OH/IR progenitors. Discrepancy in the C, N and O abundances determined from recombination lines and collisionally excited lines is a real fact. This discrepancy is due to the thermal and density inhomogenities in planetary nebulae.

# THE CHEMICAL FEATURES OF GALACTIC PLANETARY NEBULAE

P. R. AMNUEL

*School of Physics and Astronomy, Tel Aviv University, 69978 Tel Aviv, Israel*

The chemical composition of 218 galactic planatary nebulae is investigated, all the nebulae are divided into four classes according to the masses of the nebulae and progenitor stars. The values of local abundances, galactic abundances and electron temperature gradients are found for each class of nebulae. The correlations between element abundances are also investigated. The results are compared with theorical predictions.

Data on correlations between elements allow to conclude that:

- Only one dredge–up process takes place, obviously, in L– and In–PNe progenitor stars (masses below $3M_\odot$). In M–PNe progenitors (masses $\sim 3$–$8M_\odot$) three dredge–up processes could occur;
- CN–cycle can, possibly, play a role in progenitor for all PNe classes. In particular, this leads to low (in comparison to predicted values) C/O in M–PNe. On the other hand, there are some arguments that ON–cycle would exist not in massive ($\sim 3$–$8M_\odot$) progenitors only, but can also play a role in stars with masses below $\sim 3M_\odot$.

As for other elements our data allow to conclude that:

- When second PN envelope arises (A–PN), some enrichment of Ne takes place while abundances of other elements do not vary essentially.
- The most essential cooling factor in PNe is oxygen. Abundances of nitrogen and carbon are less essential.
- Argon, perhaps, plays a certain role which is more essential than it was earlier suggested, in nucleosynthesis processes for the stars with masses below $\sim 8M_\odot$.
- Although total number of PNe used reaches 218, subdivision of PNe into the masses classes decreases the number of PNe used in the different samples. This tells on the reliability of conclusions.

# O, S, Ar FROM PLANETARY NEBULAE DATA AND THE CHEMICAL EVOLUTION OF THE GALACTIC DISK

J.A. DE FREITAS PACHECO

*Instituto Astronômico e Geofísico - USP, Av. Miguel Stefano, 4200 - CEP 04301, S. Paulo, Brasil*

The O, S, Ar abundances for a sample of 122 planetary nebulae (merging LNA data and those by Köppen, Acker and Stenholm 1991) were analysed. Average abundances were calculated for progenitors having different metallicities (ages). Our study suggests that type I planetaries, whose progenitors are not older than 1-2 Gyr, have average oxygen abundances 0.2 dex lower than the solar value. This agrees with O-abundance determinations in HII regions, intermediate mass supergiants and B stars in young associations. S and Ar show a different behaviour. We suggest that such a paucity of O in the ISM is produced by recent infalling gas from the halo, having abundance ratios similar to those expected from type Ia supernovae.

# EVOLUTION OF RADIAL ABUNDANCES GRADIENTS FROM PLANETARY NEBULAE

W.J. MACIEL

*Instituto Astrônomico e Geofísico da USP, Av. Miguel Stefano 4200, 04301-904 São Paulo SP, Brazil*

and

J.KÖPPEN

*Inst.f.Theor.Physik u. Sternwarte, Olshausenstr. 40, DW 2300 Kiel, Germany*

The presence of radial abundance gradients is now well established, both in our Galaxy and in other spirals. These gradients were originally derived from oxygen abundance in HII regions, but in the past few years, other elements produced by massive stars have also been shown to display radial abundance gradients similar to the oxygen gradient, which amounts to -0.07 dex kpc$^{-1}$. Moreover, planetary nebulae (PN), belonging to the intermediate disk population (the so-called Peimbert "type II" PN) also exhibit gradients of the same order of magnitude as the HII regions.

The origins of the gradients is not yet clear, although several possibilities have been proposed in the literature, ranging from the adequate gas flow to variations of the SFR/IMF or of the chemical yields. Also connected with the origin of the gradients is the question of their time evolution. In this respect, planetary nebulae are likely to give a significant contribution. Originated from progenitor stars in a large range of main sequence masses (0.8 to 8 $M_\odot$), PN include objects of different populations, whose properties are reflected in their chemical compositions. Therefore, by determining the radial gradients of the main heavy elements in a sample of disk PN, one would be able in principle to study the time evolution of the interstellar medium out of which the PN progenitor stars have been formed.

In this paper, an effort is made to determine radial gradients of the elements O,S,Ne and Ar for PN of types I,II and III, according to the classification scheme originally proposed by Peimbert. The obtained gradients and their variations along the PN sequence are then interpreted in terms of a model for the chemical evolution of the Galaxy.

*Work partially supported by CNPq, FAPESP, and DFG (SFB 328)*

# DEPENDANCE OF THE METALLICITY OF PLANETARY NEBULAE WITH THE GALACTIC HEIGHT ABOVE THE DISK

F. CUISINIER and A. ACKER

*Observatoire de Strasbourg, URA 1280, 11 rue de l'Université, F 67 000 Strasbourg, France*

and

J. KÖPPEN

*Inst.theor.Astroph., Im Neuenheimer Feld 561, D-6900 Heidelberg 1, Germany*

**Abstract.** Abundances of O,N,He,S,Ar have been derived from the observations of PN at various heights above the plane with the plasma diagnosis code HOPPLA (Köppen et al. 1991). A gradient in O and N appears in the first 1000 pc. No correlation is found above.

We used the "Strasbourg–ESO catalogue of Galactic PN" (Acker et al., 1992) as a basis to select homogeneously PN above 300pc using Shklovskii distances calculated from the de–reddened $H\beta$ fluxes. 23 PN could be observed at the ESO 1.52m telescope with a B&C long slit spectrograph. Abundances have been deduced via the plasma diagnosis code "HOPPLA".

The remarkable features are an absence of variation above 1000pc and a gradient below. The first point could be compatible with a fast collapse of the Thick Disk. The second can not be explained if we assume that metallicity is a marker of time. PN spend the most of their time near their maximum $|z|$. Thus, the galactic structures where they are originating from should be hollow, which is hard to concieve. Therefore we prefer to explain it through a variation of the age–metallicity relationship with $|z|$ for low $|z|$ PN.

## References

Acker A., Ochsenbein F., Tylenda R., Stenholm B., Marcout J. : 1992, *"The Strasbourg–ESO catalogue of Galactic Planetary Nebulae" publ. ESO* ,
Barker T.: 1978, *ApJ* **220**, 193
Faundez–Abans M., Maciel W.J.: 1988, *Rev.Mex.Astron.Astrof.* **16**, 105
Köppen J., Acker A., Stenholm B.: 1991, *A&A* **248**, 197

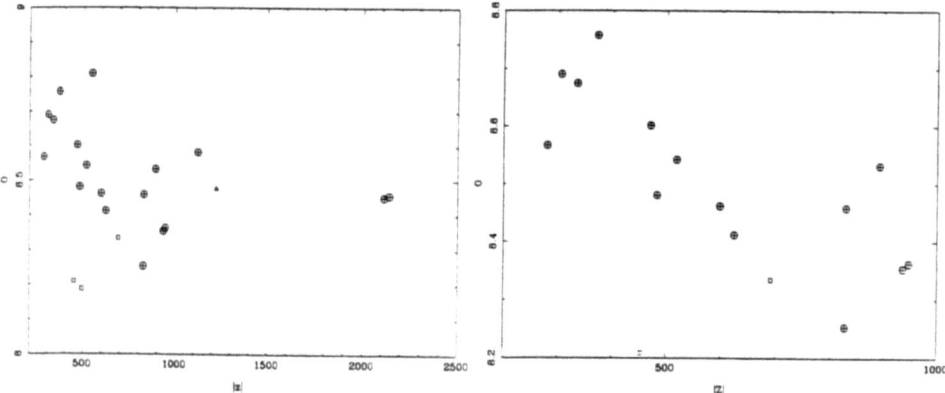

Fig. 1. oxygen abundance vs. $|z|$          Fig. 2. lookup at the first 1000pc

# CHEMICAL COMPOSITION OF PLANETARY NEBULAE: A NEW DETERMINATION

V.V. GOLOVATY

*Astronomical Observatory of Lviv University, 290005 Lomonosov str. 8, Lviv, Ukraine*

and

YU. F. MALKOV

*Crimean Astrophysical Observatory, 334413 Nauchny, Crimea, Ukraine*

A new method of the determination of planetary nebulae abundances is proposed. Unobserved ionization stages are taken into account with aid of the correlations between relative abundances of various ions which had been obtained from the grid of the photoionization models of planetary nebulae luminescence calculated by us. Simple approximative expressions for the determination of He/H, C/H, N/H, O/H, Ne/H, Mg/H, Si/H, S/H, and Ar/H are found. The chemical composition of 130 galactic planetary nebulae is revised. The observational data were compiled from 73 papers of many authors published in 1972-1991. Our mean abundances of C, N, O, Ne, S, Ar are O.1-O.3 dex lower than the mean abundances of these elements found previously by other authors. Such a discrepancy may be due to an overestimation of "empirical" ionization correction factors in previous works. It is shown that the abundance of oxygen in massive stars-precursors may be reduced by 0.2 dex on average due to the ON-cycle, but the abundance of neon remains practically unchanged. A comparative analysis of the abundances in the galactic disk, bulge and halo nebulae is carried out. We found that helium is enhanced in the galactic bulge nebulae relative to the disk ones, but the mean nitrogen abundance and mean ratio N/O are lower in the bulge. Our data suggest that the second dredge-up did not take place in the stars-precursors of the bulge planetaries, and the helium and nitrogen enhancement in these nebulae is due to other mixing processes. The mean abundances of O, Ne, S, Ar in the galactic disk and galactic bulge nebulae are quite similar, and we must conclude that the stars-precursors of the bulge planetaries had been formed during the same burst of star formation as the stars-precursors of the disk nebulae, while the halo nebulae correspond to the previous burst.

# GALACTIC BA ENRICHMENT FROM TP-AGB STARS

C. M. RAITERI[1], M. BUSSO[1], F. MATTEUCCI[2], R. GALLINO[3]

[1] *Osservatorio Astronomico di Torino, 10025 Pino Torinese, Italy*

[2] *ESO, Karl-Schwarzschild-Str.2, D-8046 Garching bei München, Germany*

and

[3] *Istituto di Fisica Generale, Università di Torino, Via P.Giuria 1, 10125 Torino, Italy*

The production of the bulk of barium has long been ascribed to the *main* component of the *s*-process, whose astrophysical site has been envisaged in the convective He shell of Thermally Pulsing Asymptotic Giant Branch (TP-AGB) stars of low mass (1–3 $M_\odot$; see Käppeler et al. 1990). The main neutron source is the $^{13}C(\alpha,n)^{16}O$ reaction, operating at the thermal energy of $kT = 12\,\mathrm{keV}$. We have calculated neutron captures in such environment with an updated nuclear physics, adopting the neutron capture cross sections of Beer, Voß, & Winters (1992) together with their temperature-dependence. Stellar models producing a mean neutron exposure of $\tau_0 \simeq 0.30\,\mathrm{mb}^{-1}$ are able to reproduce the solar distribution of the *s*-abundances satisfactorily, but the Ba isotopes show some overproduction. Such a strong indication suggests a revision of the Ba cross sections (see Gallino, Raiteri, & Busso 1992). Once that a suitable choice of $\sigma_{n,\gamma}(\mathrm{Ba})$ is made, it is found that a *r*-contribution to solar Ba of the order of 10% can be expected.

We also investigated the production of Ba in stellar models of different neutron exposures, corresponding to stars of various metallicities. A constant amount of the $^{13}C$ neutron source was assumed. The results of our calculations are shown in Table I: the overabundances of the Ba isotopes are given with respect to the initial (solar-scaled) ones. From these numbers the Ba yields from TP-AGB stars can be derived; by inserting them into a detailed model for the chemical evolution of the Galaxy (Matteucci & François 1989), we can predict the behaviour of barium as a function of metallicity. The results, that must be compared with the observation of [Ba/Fe] vs. [Fe/H], will be presented in a forthcoming paper.

TABLE I

| [Fe/H] | -1.3 | -0.82 | -0.51 | -0.35 | -0.22 | -0.12 | -0.05 | 0.0 |
|---|---|---|---|---|---|---|---|---|
| $^{134}$Ba | 6.75e3 | 3.98e3 | 1.72e3 | 9.15e2 | 4.88e2 | 2.73e2 | 1.60e2 | 1.14e2 |
| $^{135}$Ba | 1.73e3 | 9.91e2 | 4.04e2 | 2.07e2 | 1.07e2 | 5.86e1 | 3.35e1 | 2.36e1 |
| $^{136}$Ba | 7.29e3 | 4.19e3 | 1.74e3 | 8.96e2 | 4.65e2 | 2.54e2 | 1.45e2 | 1.02e2 |
| $^{137}$Ba | 6.24e3 | 2.51e3 | 9.28e2 | 4.59e2 | 2.31e2 | 1.24e2 | 7.01e1 | 4.91e1 |
| $^{138}$Ba | 1.25e4 | 5.30e3 | 1.80e3 | 8.21e2 | 3.83e2 | 1.92e2 | 1.02e2 | 6.90e1 |
| $\tau_0(\mathrm{mb}^{-1})$ | 1.19 | 0.47 | 0.29 | 0.23 | 0.19 | 0.17 | 0.15 | 0.14 |

## References

Beer, H., Voß, F., & Winters, R.R. 1992, ApJS, 80, 403
Gallino, R., Raiteri, C.M., & Busso, M. 1992, ApJ, (submitted)
Käppeler, F., Gallino, R., Busso, M., Picchio, G., & Raiteri, C.M. 1990, ApJ, 354, 630
Matteucci, F. & François, P. 1989, MNRAS, 239, 885

# CHEMICAL ABUNDANCES IN GALACTIC BULGE PN

N.A. WALTON and M.J. BARLOW

*Department of Physics & Astronomy, University College London,
Gower Street, London WC1E 6BT, England*

and

R.E.S. CLEGG

*Royal Greenwich Observatory, Madingley Road, Cambridge CB3 0EZ, England*

**Abstract.** We present abundance determinations, in particular of carbon, and C/O ratios, for 11 Galactic bulge planetary nebulae (PN) based on our low resolution UV data from IUE observations and optical spectrophotometry from the Anglo-Australian Telescope. We compare the observed abundances with those predicted by dredge-up theory for the high metallicity Galactic bulge. The sample abundances are also contrasted with the abundances found for PN in the Galactic disk. The mean C/O ratio for the bulge PN is significantly lower than that found for Galactic disk PN. Further, we present an abundance analysis of the very metal-poor bulge PN M2-29. From an analysis of the differential extinction found from the observed ratios of the He II 1640,4686Å lines, we find that the ultraviolet reddening law towards the bulge is steeper than in the solar neighbourhood.

**Table:** Our abundances determined for the bulge PN and the values found for local neighbourhood PN, taken from Torres-Peimbert & Peimbert (1977, Rev Mex Astron Astrofis, 2, 181) and Aller & Czyzak (1983, ApJS, 51, 211), are shown. M2-29 is excluded from the means as it is a low metallicity halo PN located in the bulge.

∗ The only one of our bulge PN with C/O clearly in excess of unity (Cn 1-5) has a WC4 Wolf-Rayet central star.

## Bulge PN Elemental Abundances

| Object | He | C ×10⁴ | N ×10⁴ | O ×10⁴ | Ne ×10⁴ | C/O | N/O |
|---|---|---|---|---|---|---|---|
| M 3-21 | 0.118 | 2.28 | 2.98 | 7.04 | 1.57 | 0.32 | 0.42 |
| H 1-42 | 0.107 | – | 0.97 | 5.34 | 0.74 | – | 0.18 |
| M 2-23 | 0.084 | – | 1.34 | 2.93 | 0.45 | – | 0.46 |
| M 1-42 | 0.194 | 0.72 | 3.21 | 1.94 | 0.63 | 0.37 | 1.65 |
| Cn 1-5∗ | 0.143 | 13.3 | 3.98 | 7.67 | 2.44 | 1.74 | 0.52 |
| M 2-30 | 0.112 | 0.99 | 1.08 | 4.65 | 1.05 | 0.21 | 0.23 |
| Hb 8 | 0.104 | 0.35 | 0.32 | 3.29 | 0.60 | 0.11 | 0.096 |
| Al 1 | 0.103 | 5.36 | 2.31 | 4.75 | 1.03 | 1.13 | 0.49 |
| Vy 2-1 | 0.105 | 4.35 | 1.94 | 5.96 | 1.38 | 0.73 | 0.33 |
| M 3-33 | 0.112 | 0.55 | – | 2.61 | – | 0.21 | – |
| M 2-29 | 0.093 | 0.01 | 0.17 | 0.28 | 0.039 | 0.036 | 0.62 |
| Local PN | 0.100 | 5.50 | 0.91 | 4.79 | 0.96 | 1.15 | 0.19 |
| Bulge mean | 0.118 | 3.49 | 2.01 | 4.62 | 1.10 | 0.60 | 0.49 |
| ± | 0.029 | 4.10 | 1.14 | 1.83 | 0.59 | 0.53 | 0.44 |

# A REANALYSIS OF C/O RATIOS IN PLANETARY NEBULAE

C. ROLA

*DAEC, Observatoire de Paris-Meudon, France*
*Centro de Astrofísica, Universidade do Porto, Portugal*

and

G. STASINSKA

*DAEC, Observatoire de Paris-Meudon, France*

It is often mentioned in the literature that, in planetary nebulae, carbon abundances derived from optical lines are systematically much larger than derived from UV lines. Various explanations have been proposed: line blends, inaccurate atomic physics, temperature fluctuations, non uniform distribution of carbon inside nebulae... Here, we study the effect of random errors in the line intensity measurements.

First, we compiled all the published UV and optical line intensities which allow to estimate C/O. We then derived the C/O ratios by classical empirical methods, assuming that the electron temperature $T_e$ is given by the [OIII]4363/5007 ratio, and that C/O is equal to $C^{++}/O^{++}$ (abundances derived from CIII]1909/[OIII]5007, CIII]1909/OIII]1661 and CII 4267/[OIII]5007 are denoted by $C_{uv}/O_{opt}$, $C_{uv}/O_{uv}$ and $C_{opt}/O_{opt}$ respectively). We find that both $C_{uv}/O_{opt}$ and $C_{uv}/O_{uv}$ are systematically smaller than $C_{opt}/O_{opt}$.

In order to understand to what extent observing conditions alone affect observational diagrams, we have performed the following numerical experiment. We have randomly selected a sample of 100 objects out of a population of fictive planetary nebulae having C/O distributed around 0.8, with a dispersion of 0.3 dex. Temperatures were assumed uniform, but varying from object to object, with a mean of 12000K and a dispersion of 1400K. The observational errors in the line intensities were simulated by randomly selecting the intensities out of a Gaussian distribution centered on the true intensity, with a dispersion chosen so as to reproduce the typical uncertainties stated in observational papers. A line was considered to be detected when its intensity was at least 1.5 times greater than the noise. To these simulated intensities, we have applied the same empirical methods as above to derive the C/O ratios.

The simulated diagrams comparing $C_{uv}/O_{opt}$, $C_{uv}/O_{uv}$ and $C_{opt}/O_{opt}$ show qualitatively the same patterns as the observational ones. There is no systematic bias in C/O when derived from $C_{uv}/O_{opt}$ because these lines are very strong. On the contrary, $C_{opt}/O_{opt}$ tends to overestimate the true C/O, especially at small C/O ratios. Indeed, there is a selection effect against nebulae in which random realizations of the CII 4267 intensities are below the true value, because these weak lines are often lost in the noise.

We conclude that random uncertainties in the line intensity measurements are an important factor in explaining the discrepancy between UV and optical determinations of C/O ratios.
It is unwise to combine $C_{uv}/O_{opt}$ with $C_{opt}/O_{opt}$ for statistical studies of planetary nebulae, because the systematic differences depend on the signal-to-noise of the observations.

# THE HELIUM-TO-METALS ENRICHMENT RATIO IN PLANETARY NEBULAE

C.M.L. CHIAPPINI and W.J. MACIEL

*Instituto Astronômico e Geofísico da USP, Av. Miguel Stefano 4200, 04301-904 São Paulo SP, Brazil*

The helium-to-metals enrichment ratio $(dY/dZ)$ and the pregalactic helium abundance $(Y_p)$ are important parameters for the understanding of the galactic chemical evolution, and also as a cosmological test to the Big Bang theory. In this work, we present a new determination of these parameters based on a sample of galactic HII regions, HII galaxies, and planetary nebulae.

The determination of $dY/dZ$ depends mainly on planetary nebulae, while $Y_p$ is essentially determined by the low metallicity objects. For the objects with measured abundances of O, N, C, Ne, S, Ar, and Cl, the correlation between the total metal abundance $(Z)$ and the abundances of oxygen, nitrogen, and carbon can be studied in detail, so that the slope $dY/dZ$ can be determined from plots of the helium abundance by mass $Y$ as a function of $Z$, O/H and N/H. We have taken into account the contamination of the observed abundances of He, N, and C in planetary nebulae due to the fresh material produced and dredged up by their central stars.

The results show that $dY/dZ \geq 5$, which is higher than previously reported. We obtained further evidence that this value depends on the PN type, decreasing for high metallicities. Therefore, the simple linear model of chemical evolution is limited to low metallicities.

*Work partially supported by CNPq and FAPESP.*

# SPECTROPHOTOMETRY OF SELECTED PLANETARY NEBULAE OF TYPE I IN THE MAGELLANIC CLOUDS

S. TORRES-PEIMBERT and M. PEIMBERT
*Instituto de Astronomía, UNAM, Apartado Postal 70-264, México 04510 DF, México*

M.T. RUITZ
*Observatorio Astronómico Nacional, U. de Chile, Casilla 36-D, Santiago, Chile*

and

M. PEÑA
*Instituto de Astronomía, UNAM, Apartado Postal 70-264, México 04510 DF, México*

We carried out spectroscopic observations of N67 (in the SMC), and N66, N97 and N102 (in the LMC) with the 4-m telescope of CTIO. The wavelength range is $\lambda\lambda$ 3500 – 7400. From these we obtained physical conditions and chemical abundances of these objects.

In N66 we found a broad feature centered at $\lambda 4686$ present in our August 1990 spectrum that was not present in a similar dispersion spectrum taken in January 1985 (Peña and Ruitz 1988, RevMexAA, 16, 55). The FWHM of the feature is of 33 Å, which amounts to 2100 km s$^{-1}$, and indicates that the central star is developing WR features in a short time scale.

In N67, N97 and N102 the O/H ratio is 0.25 dex smaller, while the (N + O)/H is similar, to those of the HII regions of their corresponding galaxy. These results are strong arguments in favor of ON cycling in these objects.

# SYNTHETIC AGB EVOLUTION IN THE LMC:
# THE ABUNDANCES OF PN

MARTIN GROENEWEGEN

*Astronomical Institute, Kruislaan 403, NL-1098 SJ Amsterdam, Holland.*

TEIJE DE JONG

*Astronomical Institute, Kruislaan 403, NL-1098 SJ Amsterdam, Holland.*

and

*SRON, Space Research Groningen, P.O. Box 800, NL-9700 AV Groningen, Holland.*

We have developed a model to calculate the evolution of AGB stars in a synthetic way. The evolution is started at the first thermal pulse (TP) and ends when the envelope mass has been lost due to mass loss or when the core mass reaches the Chandrasekhar mass.

Our model is more realistic than previous synthetic evolution models in that more physics has been included. The variation of the luminosity during the interpulse period is taken into account as well as the fact that, initially, the first few pulses are not yet at full amplitude and that the luminosity is lower than given by the standard core mass-luminosity relations. The effects of first, second and third dredge-up are taken into account. Hot Bottom Burning (HBB) is included in an approximate way. Mass loss on the AGB is included through a Reimers Law. We also take into account mass loss prior to the AGB. Most of the relations used are metallicity dependent.

The free parameters in our calculations are the minimum core mass for dredge-up ($M_c^{min}$), the dredge-up efficiency ($\lambda$) and the Reimers mass loss scaling parameter ($\eta_{AGB}$).

The model has been applied to the LMC using a recent determination of the age-metallicity and Star Formation Rate history for the LMC. A model with $M_c^{min} = 0.58~M_\odot$, $\lambda = 0.75$, $\eta_{AGB} = 5$, including HBB fits the observed carbon star luminosity function, the C/M-ratio of AGB stars and the initial-final mass relation.

For this model the abundances of PN are calculated and compared to the observation in the N/O-He/H, C/O-He/H, C/O-C/N and N/O-N/H diagrams. The agreement in all four diagrams is good. There is a relation between the abundance of a PN and its main sequence mass. Our model predicts the high N/O and He/H ratios observed in some PN. To get agreement in the N/O-N/H diagram we have to assume that the main sequence oxygen abundance varies like $(O/Z) = 1~/~(1 + 0.838~(Z/Z_0)^{0.7})$, where $Z_0$ is the present metallicity in the LMC. The fact that the relative main sequence oxygen abundance was higher in the past is consistent with a recent chemical evolution model for the LMC and obervations of the oxygen abundance in our Galaxy.

# CHEMICAL ABUNDANCES OF PLANETARY NEBULAE IN THE LMC

J.A. DE FREITAS PACHECO, R.D.D. COSTA

*Instituto Astronômico e Geofísico - USP, S. Paulo, Brasil*

We report new spectroscopic data on a sample constituted of 21 well observed planetary nebulae in the LMC. The observations were performed at the National Laboratory for Astrophysics (Brazópolis - Brasil) using the 1.6m telescope and a CCD detector. Extinction, electron temperature and densities were derived for all the planetaries and a comparison is made with results obtained by other surveys, including common objects. Chemical abundances of helium, nitrogen, oxygen, sulphur and argon were also derived for all sample objects. Enrichment of the progenitor stars in He and N due to mixing episodes as well as the relation with the chemical evolution of the LMC are discussed in terms of our data and other observations.

# THE EVOLUTION OF PLANETARY NEBULAE, THEIR PRECURSORS AND THEIR PROGENY — A COMMENTARY

ICKO IBEN, JR.
*Departments of Astronomy and Physics, University of Illinois 1002 W. Green St.,*
*Urbana, IL 61801*

## 1. The Legend

There appears to exist in the minds of most of the participants of this conference a highly stylized picture of how the majority of all stars evolve after reaching the asymptotic giant branch. This is a fascinating development as, just three decades ago, no one understood what an asymptotic giant branch (AGB) star is or does, either theoretically or observationally, and, just two decades ago, no one understood the nature of the transition from AGB to planetary nebula, other than that it happened. And, yet, today, the evolution from AGB to planetary nebula and to white dwarf is described by a very beguiling picture whose outlines have become substantially fixed.

The standard picture is: (1) The AGB star alternately burns hydrogen and helium in shells above an electron-degenerate core. Helium burning always begins as a thermonuclear runaway, and elements processed by partial helium burning during the flash make their way to the surface after each flash in a dredge-up episode. (2) As the AGB star brightens, the rate of mass-loss from its surface accelerates, and eventually there is a complete detachment of a contracting stellar remnant from an expanding ejected shell. The mechanism of mass loss involves first inflation of the stellar envelope by shocks driven by pulsations and then radiation pressure on grains formed in the outer part of the envelope. (3) The remnant continues to contract as it burns hydrogen or helium in a shell. Whether helium or hydrogen burns depends on where in a flash cycle of the AGB precursor final detachment of the expanding shell and contracting remnant occurs. (4) The fraction of nebular mass in molecular form relative to the fraction in ionized form depends on, among other things, the surface temperature of the remnant and the amount of mass in the nebula which is self shielding. When the surface temperature of the remnant becomes high enough, photons from the remnant cause a portion of the nebula to fluoresce in the optical. (5) A fast wind from the remnant forms a hot bubble which helps shape the nebula. If the remnant has a companion, this companion will also contribute to shaping. Bipolarity is common and the symmetry axis could coincide with the spin angular momentum or magnetic moment axis of the remnant, if single, or with the orbital angular momentum vector, if in a binary. (6) Eventually, nuclear burning dies out and the remnant cools as a white dwarf. The nebula disperses. (7) Depending on where in its flash cycle the AGB precursor leaves the AGB, the remnant white dwarf may experience a final helium shell flash to become a "bornagain" AGB star, or a final hydrogen shell flash to become a "self-induced nova". In the born-again case, the central star burns helium and essentially retraces its steps as a hydrogen burner on a similar time scale; when the nebula fluoresces a second time, it is substantially larger then it was the first time. During the final nuclear-burning phase, further mass is lost from the surface of the remnant, possibly exposing highly processed matter. (8) The fact that there are different evolutionary channels leading to the white dwarf state, even for single stars, may be responsible for some of the differences among white dwarfs with regard to surface composition. The fact that some are formed in close binaries can be responsible for additional differences.

In the following, articles referenced by author, but not by year, appear in this volume; they are not cited in the list of references at the end.

## 2. Thermally Pulsing Asymptotic Giant Branch Interiors

An AGB star possesses an electron-degenerate core composed of carbon and oxygen (if the mass of its progenitor is $\sim$ 1-8 $M_\odot$) or of oxygen, neon, and magnesium (if the mass of its progenitor is $\sim$ 8-10 $M_\odot$). In both instances, the core has the dimensions of a hot white dwarf $\sim 10^9$ cm and the hydrogen-rich envelope has the dimensions of a giant of radius a few times $10^{13}$ cm. During the early AGB (E-AGB) phase, hydrogen does not burn as the core grows in mass until it nearly reaches the hydrogen-rich envelope, whereupon hydrogen is reignited. In the case of a CO core, hydrogen and helium then burn alternately in shells situated between the core and a convective envelope. Hydrogen burning is quiescent and lasts approximately 90% of the thermally pulsing AGB (TP-AGB) lifetime.

Helium burning begins as a thermonuclear runaway and both carbon and s-process elements are produced, as detailed by Lattanzio. During the helium flash, the surface luminosity first dips, as hydrogen burning is extinguished, and then increases, as the energy produced by helium burning leaks out of the production region (see Wood). During this phase, carbon and s-process elements are dredged to the surface. Then the luminosity drops as helium continues to burn quiescently for about 10% of the time between flashes. Helium stops burning and another phase of quiescent hydrogen burning ensues. The excursions in luminosity during a complete cycle can be substantial.

In the case of an ONeMg core, carbon-burning and helium-burning shells alternate (Nomoto 1984, 1987), but *no one* has yet followed the ignition and extinction of hydrogen burning in a *complete* model. Yet, unless they are "Blöcker-Schönberner" (1990) stars, the most luminous LPV's in the Magellanic clouds have ONeMg cores. They are surely experiencing helium shell flashes and dredge-up (Wood, Bessell, and Fox 1983), and, because of their superabundances of lithium (Smith and Lambert 1989, 1990), they must also experience phases of quiescent hydrogen burning, interrupted by helium shell flashes.

## 3. Pulsation, Dust, and Mass Loss

The mechanism for mass loss from AGB stars and whether or not a "superwind" terminates the AGB phase in all cases have been debated for years. It is definitely clear from the statistics in the Magellanic clouds that thermally pulsing AGB stars with small core masses (0.5-0.6 $M_\odot$) exist for of the order of 5-20 $\cdot 10^5$ yr, whereas those with large core masses ($\gtrsim 0.8$-$0.9\ M_\odot$) exist only for of the order of $10^5$ yr or less. From this, one can infer average mass-loss rates of $\dot M \sim 10^{-7} - 10^{-6}\ M_\odot$ yr$^{-1}$ for stars of small core mass and $\sim 10^{-5} - 10^{-4}\ M_\odot$ yr$^{-1}$ for those of large core mass. Thus, there has long been evidence that mass-loss rates increase with luminosity.

A promising theoretical approach has been adopted by Bowen (1988) and Bowen and Willson (1991). The models show that, beyond a certain radius, shocks inflate the envelope beyond that of a corresponding hydrostatic model, producing an extended region of much higher density than in a hydrostatic model. Radiation pressure on grains formed in the envelope accelerates the dust grains and collisions between grains and ambient atoms and molecules accelerate the gas. The nearly periodic oscillatory motion near the photosphere gradually develops into an outflowing wind. Applying the results of the hydrodynamic calculations to construct synthetic AGB model evolutionary sequences, Bowen and Willson (1991) show that the mass loss rate of the evolving AGB star accelerates exponentially, growing indeed into a "superwind" (Renzini 1981).

## 4. Superwinds and Pre-Planetary Nebulae

Both large OH/IR sources and carbon-rich systems such as IRC+10216 could rightfully be called proto-planetary nebulae or pre-planetary nebulae (PPNe) in the sense of being precursors of what has traditionally been called a PN, namely a very hot central star and an ionized nebular shell. However, the term has come to mean different things to different people. I think it is probably wise to follow Sun Kwok's lead in insisting that the name be reserved for systems in which pulsation driven mass loss has ceased once and for all and in the spectrum of which two components can clearly be identified as coming from, respectively, a central star and a circumstellar shell (or shells) no longer connected by high density matter to the central star. The consensus definition appears to include all systems for which the central star has a surface temperature between 5,000K and 30,000K. By demanding that there be no large amplitude light variations, one insures that the mass of the remnant hydrogen-rich envelope of the central star is small enough that pulsation can no longer drive mass loss at superwind rates, and physical detachment of central star and expanding shell has taken place.

## 5. Classical Planetary Nebulae and Their Central Stars

### 5.1. THE INTERACTING WINDS MODEL

The zero-order standard model of a classical PN is characterized by a set of spherically symmetric layers surrounding a hot, compact central star. The central star emits ionizing radiation (surface temperature between 30,000K and 300,000K) and emits a fast ($\sim$ 3000 km s$^{-1}$) particulate wind with a momentum flux comparable with or larger than that of the circumstellar shell (Perinotto). The fast wind interacts with the matter in the shell (or shells) previously ejected in the superwind(s), causing this matter to be swept up into a new shell compressed between the inner, more rapidly moving, wind and the outer, less rapidly moving, wind (Kwok, Purton, and FitzGerald 1978; Kwok 1982; Kahn 1983, 1989; Dyson).

The Kahn model assumes that the fast wind flowing from the central star is converted into a shock heated gas bubble which is confined from below by the ram pressure of the wind as it escapes the star and from above by the back pressure from matter in the layer formed by the earlier superwind.

It is disappointing that there is as yet no satisfactory theory which, once parameters have been normalized to observational estimates, might allow one to predict the mass-loss rate of the central star as a function of mass and position in the H-R diagram. Such a theory would enable us to make progress in understanding the mystery presented by the surface composition of hydrogen-deficient central stars and white dwarfs.

## 5.2. CENTRAL STAR EVOLUTION

The first attempt to model the evolution of the central star is that of Paczyński (1971). This work has been extended by Schönberner (1979, 1981), whose classic tracks have been the standard reference for observers for over a decade. Wood and Faulkner (1986) have also provided a set of tracks in wide use.

A series of imaginative papers by Renzini (1979, 1981, 1982, 1983, 1989) provides a comprehensive framework for comparing theory with the observations, raising issues that are still being debated and acted upon. Heap emphasizes the fact that the rate at which a model central star of constant mass evolves to the blue during the high-luminosity plateau phase varies with about the $10^{th}$ power of the mass of the model (e.g., Iben and Renzini 1983), and that this fact can be used to estimate the masses of PNNi which are known to be in the plateau phase.

The growing wealth of information about PNNi in the Magellanic clouds (Dopita) is very gratifying to see. Combining their work (Kaler and Jacoby 1990, 1991) with that of Dopita and Meatheringham (1991a,b) and that of Aller et al. (1987), Kaler and Jacoby (1991) show that estimated masses of 80 studied PNNi lie in the range 0.56-1.22 $M_\odot$. The masses have been estimated on comparison with theoretical tracks of hydrogen-burning PNNi. The bulk of the brightest 33 central stars have estimated masses in the range 0.56-0.66 $M_\odot$, but there appears to be a secondary peak with masses $\sim$ 0.7-0.74 $M_\odot$. The masses of the other 47 stars appear, in the main, to follow a distribution similar to that defined by the bright sample, but roughly a dozen have masses extending from 0.74 $M_\odot$ to 1.22 $M_\odot$ in what appears to be a flat distribution. The absence, in the bright sample, of PNNi with large masses is consistent with the fact that the rate of evolution through the plateau phase increases rapidly with mass.

The clustering of PNN masses about 0.6 $M_\odot$ is due to the facts that (1) stars of initial mass $\lesssim 2$ $M_\odot$ all develop electron-degenerate helium cores which grow to $\sim$ 0.5 $M_\odot$ before a helium flash lifts degeneracy, and (2) $\sim$ 0.05 $M_\odot$ is added to the helium core during the quiescent core helium-burning (horizontal branch or red giant clump) phase. Thus, when they reach the TP-AGB phase, all stars initially less massive than $\sim 2$ $M_\odot$ develop a CO core of approximately the same mass $\sim$ 0.55 $M_\odot$. The spread in final core mass about 0.6 $M_\odot$ may be interpreted to be a consequence of the fact that initially more massive stars (in the 1-2$M_\odot$ range) must lose more mass than lighter ones and therefore live longer and grow larger core masses before departing from the AGB as PNNi. The $\sim 0.05$ $M_\odot$ mass difference between the common initial mass and the final mean mass can be used in conjunction with the theoretical fuel-consumption rate to derive a mean lifetime of $\sim 5 \cdot 10^5$ yr for TP-AGB stars originating from low mass main sequence progenitors. This estimate is quite consistent with the estimate of $\sim 3 \cdot 10^5$ yr given by Whitelock and Feast for Miras in the inner part of our Galaxy.

Dopita makes the interesting statement that all of the PNNi in the Magellanic Clouds are helium burners, and this is consistent with Kawaler's (1988) statement that Galactic PNNi cannot be hydrogen burners. Kawaler's argument is based on a prediction that hydrogen burners should pulsate at a detectable amplitude, contrary to observations of galactic PNNi in the plateau phase. My feeling is that, on the basis of analyses thus far performed, it is not possible to tell whether the majority of cloud PNNi are either helium burners or hydrogen burners, but a definitive check is potentially possible using an analysis such as that performed for Galactic PNNi by Schönberner (1986), who concludes that most of the PNNi in his sample are hydrogen burners.

5.3. NEBULAR MASS, CHEMISTRY, AND COMPOSITION

Quantitative estimates of the amount of matter in ionized form are now being complemented by quantitative estimates of the amount of matter in molecular form (Huggins, Bieging 1988), giving total masses of detected matter which in many cases are substantially larger than the $\sim 0.2$-$0.3$ $M_\odot$ of ionized matter that is often used in obtaining statistical parallaxes.

Huggins and Healy (1989) find that the ratio of molecular mass to ionized mass is very tightly anti-correlated with nebular radius R, and this is a nice demonstration of how the ionization front passes through superwind matter. A reasonable fit to the Huggins-Healy results over the range R = 0.002-0.6 pc is $M_i/M_m \sim 10$ $(R/0.3$ $pc)^{1.5}$. Complementary to this result is the existence of a correlation between the mass of the ionized nebular layer and nebular radius.

At this conference and elsewhere, Clegg (1989, 1991) has extensively reviewed the element abundance distributions in PNe, showing how departures from the solar system distribution and from the distribution in nearby HII regions provide clues to the nature of nucleosynthesis and mixing in the parent star during its nuclear-burning phases. Other recent reviews include those of Peimbert (1991), Barlow (1991), and Henry (1990). One of the most dramatic departures from the solar system distribution forms the basis for the "Type I PNe" classification of Peimbert (1978). In Galactic nitrogen-rich Type I PNe, the mean value of He/H is $\sim 0.13$ versus $\sim 0.085$ for both non-Type I nebulae and HII regions. The mean ratio of nitrogen to oxygen depends on the sample, but appears to be in the range 0.5-1 (Henry 1990). Large helium and nitrogen abundances could be evidence for the occurrence of the first and second dredge-up episodes in fairly massive intermediate-mass stars prior to the onset of thermal pulses (Kaler, Iben, and Becker 1978; Becker and Iben 1980). These dredge-up episodes mix into the hydrogen-rich envelope results of hydrogen burning in the interior prior to the TP-AGB phase.

Barlow gives an intriguing interpretation of the fact that emission from the nebula NGC 6302 has characteristics of both carbon-rich and oxygen-rich gas. Analysis of the optical spectrum shows that (C+N+O)/H $\gg$ solar, a clear signature of the third dredge-up, and that C/O < 1 and N/O $\gg$ 1, a clear indication of envelope burning. He points out that, because the envelope is losing mass while dredge-up and burning are progressing, the abundances in the ejected matter will be continuously changing. One might expect in the final ejecta an outer O-rich region emitted before the third dredge-up has done much to alter the composition, an intermediate C-rich region composed of matter in which dredge-up increases the carbon abundance faster than it can be converted into nitrogen, and an innermost O-rich region made of matter in which envelope burning has converted C into N faster than C can be dredged up. After the system has evolved into a PNe with a hot PNN, instabilities produced by the fast wind as it strikes the neutral layers could cause some mixing between the different layers.

As usual, the Magellanic clouds provide additional insights and conundrums. The fact that the Galaxy, the LMC, and the SMC are an ordered sequence with regard to metallicity might be expected to lead to some systematic differences in the mean properties of various components in passing from one aggregate to another in the same order.

Clegg emphasizes that *all* Magellanic cloud PNe which are not of type I have C/O > 1, whereas only about 60% of Galactic PNe have C/O > 1 (Zuckerman and Aller 1986). This would suggest that all AGB stars in the clouds experience third dredge-up episodes before departing the AGB. However, AGB carbon stars have not been identified in the oldest Magellanic cloud globular clusters (Frogel, Mould, and Blanco 1990). Could it be that, for AGB stars of very small mass and metallicity, the superwind is triggered almost at once by a large increase in the abundance of carbon due to dredge-up? Models show that dredge-up does not occur until the core mass exceeds a critical value and that one dredge-up episode in a low mass, low metallicity star is enough to establish carbon star characteristics. If carbon is increased, the formation rate of dust grains is also enhanced, leading perhaps to a sufficiently large mass-loss rate that departure from the AGB occurs long before another flash can take place. In this picture, the absence of AGB carbon stars in old clusters is simply due to the fact that the lifetime of such stars is extremely abbreviated by the sudden triggering of a superwind. Or, could it be that the central stars of the very lowest mass ($\lesssim 0.55\ M_\odot$) are "lazy" (Renzini 1981), evolving so slowly during the plateau phase that the ejected matter is dispersed over too large a volume before it can be ionized?

## 6. PNN Chemistry and the Born-Again Phenomenon

The superwind terminates once the mass of hydrogen-rich material left near the surface decreases below a critical value. What happens next depends on precisely where in the flash cycle this occurs (see Iben 1984, 1987, 1989). Of particular interest is the evolution of a model which departs from the AGB late during the quiescent hydrogen-burning phase with a mass of helium-rich matter which is close to the mass necessary for helium ignition during the AGB phase. In such a model, a final helium shell flash can take place after the star has departed the AGB. The model returns to the AGB as a "born-again" AGB star (Iben et al. 1983, Iben 1984). After the star departs from the AGB a second time and evolves to high temperatures, further mass loss via a fast wind may remove all of the remaining hydrogen, exposing first the He-N buffer layer and, thereafter, layers that have experienced partial helium burning.

An inspiration for the early quantitative work exploring the born-again scenario was the discovery that the central stars of Abell 30 (Hazard et al. 1980) and of Abell 78 (Jacoby and Ford 1983) appear to have ejected hydrogen-deficient matter rich in He and N long after the major portion of the nebula was formed. Another instance of the born-again phenomenon is V605 Aql, the central star of Abell 58, which underwent a nova-like explosion in 1919. The inner part of the nebula has developed some characterists of the Abell 30 and 78 systems. A clumpy nebulosity of dimensions $\sim 10^{16}$ cm hides the central star (Bond, Liebert, and Renzini) and an H-deficient knot with a speed of $\sim 100$ km s$^{-1}$ and dimensions $\sim 2 \cdot 10^{16}$ cm has been detected (Pollacco et al.). Bond et al. point out that at maximum visual light, the spectrum of Nova Aql 1919 resembled that of an R CrB star, and this is indicative of hydrogen reignition during the post-AGB helium shell flash.

For the central star of A78, Werner and Koesterke (1992) estimate abundances by mass (He, C, N, O) $\sim$ (0.33, 0.5, 0.02, 0.15). For the central star of NGC 6751, Hamann and Koesterke estimate abundances by mass (He, C, N, O) $\sim$ (0.615, 0.27, 0.015, 0.10). These abundance mixes bear some resemblance to the mix predicted by the born-again scenario, except that the O/C abundance ratios, at $\sim 1/6 - 1/3$, are much larger than obtained in the theoretical calculations which use cross section for the $^{12}C(\alpha,\gamma)^{16}O$ reaction given by Fowler, Caughlan, and Zimmerman (1975). Thus, the estimated abundance ratios could be demonstrating that a much larger value for this cross section is in order. The ratio N/C can be used to estimate how much matter which has experienced partial helium burning is mixed with hydrogen-rich matter in the convective shell which is driven by hydrogen burning. For the PNN in A78, the ratio of the two forms of matter is $\sim 15$, and for the PNN in NGC 6751 it is $\sim 10$. For a 0.6 $M_\odot$ model, the mass of hydrogen-rich material remaining at the surface when the first phase of hydrogen-burning is completed is $\sim 10^{-4}$ $M_\odot$, giving 0.001-0.0015 $M_\odot$ for the mass of He, C, and O incorporated into the convective shell driven by hydrogen burning. This is only $\sim 5$-10% of the mass of the original convective shell driven by helium burning, and this seems quite reasonable.

## 7. Common-Envelope PNe and Binary Central Stars

As Livio has emphasized, at least half of all stars are born in binaries with short enough periods that the very presence of a secondary will have some *influence* on the shape of the nebula produced by the primary. Perhaps half of all binaries are in tight enough orbits that the formation of a shell of nebular material is a consequence of Roche-lobe overflow with the formation of a common envelope. Using an algorithm based on the orbital characteristics of main-sequence binaries and on theoretical considerations of the outcome of common-envelope evolution, Yungelson and Tutukov estimate the frequency of PNNi of various types. In their scenario, essentially *all* stars are initially in binaries (with separations up to $10^6$ AU), and it is therefore not surprising that their model predicts that the majority of PNNi have a main sequence companion. A more startling prediction is that *most single* PNNi (20% of all PNNi in observable PNe) are actually a result of mergers which take place during a common envelope event, with the merged product eventually going on to become an AGB star.

The shapes of many PNe with bipolar symmetry (especially the "butterfly" types in Balick's classification of morphological types) may be a consequence of the drag forces that are set up as the ejected wind passes by the companion. The degree of density contrast between matter along the polar axis and matter in the equatorial plane is the major distinguishing feature of the observationally based classification scheme, and Bond and Livio (1990) argue that a moderate-to-high density contrast is a natural result of the dynamics of the common envelope process, as it is modelled in theoretical calculations. However, among the set of $\sim 13$ PNe known to contain binary PNNi in tight enough orbits that they must have passed through a common envelope phase, only about half are clearly identified as of the butterfly or elliptical variety, corresponding, respectively, to high and moderate density contrasts. Thus, a common envelope phase may not be a guarantee that bipolarity will arise during the PNe phase.

Since fully 80% of all PNe exhibit bipolarity (Zuckerman and Aller 1986), it seems reasonable to suppose that many PNe in wide binaries (orbital separation $3 \cdot 10^{-5}$ pc $< A < .01$ pc) develop bipolarity even though they are not an ejected common envelope. If azimuthal symmetry is to be imposed on the PNe, the orbital period of the binary must be substantially less than the time scale over which the circumstellar shell has been ejected by the primary AGB star. For a mass-ejection rate of $\sim 10^{-5} M_\odot$ yr$^{-1}$ and an ejected mass of 1 $M_\odot$, this means that $P_{orb} \ll 10^5$ yr, or $A \ll 2 \cdot 10^3$ AU $\sim 0.01$ pc.

Common-envelope evolution will lead to a detectable PN only if the compact remnant of the star which fills its Roche lobe develops a surface temperature hotter than $\sim 30{,}000$K before the ejected material becomes too dispersed. Systems which will develop evolutionary characteristics most similar to those achieved by single stars (here we include as "single" those stars which are in wide enough binaries that Roche-lobe overflow never occurs) are, of course, those in which the Roche-lobe filling star is an AGB star. If the mass-losing star is of the TP-AGB variety, Roche-lobe filling will in general occur during the large luminosity, large radius portion of a helium shell flash, and the result will be a PNN of the helium-burning variety. Depending on the degree of orbital shrinkage, which determines how much matter remains above the burning shell when mass loss ceases, and on the efficacy of the fast wind, the surface composition of the PNN and its white dwarf progeny could be either: H, He, and CNO elements; He and N; or some combination of C, O, and Ne. In the first instance, a self-induced nova event (Iben and MacDonald 1986) might be expected to occur.

In building up an understanding of close binary star evolution, it is extremely important to have reliable information about as many real PNe as possible which have close binary central stars. It is equally important to have reliable information about as many binaries as possible which are likely to evolve into PNe containing close binary central stars. Armed with this information, one may make a beginning at constructing a reasonable evolutionary scenario for systems of various types and estimating the efficiency of drag forces in reducing orbital separation during common envelope events.

UU Sge, the central star of Abell 63, is a fascinating example not only of close binary evolution, but of the evolution of our understanding of an observable system. Four different observational studies of the central star have been undertaken in recent years. The three most recent studies suggest sharply different characteristics for both components (see Walton, Walsh, and Pottash; Pollacco and Bell). Theoretical scenarios can be constructed to match each of the three sets of estimated characteristics (see Iben and Tutukov 1989 for one scenario), and a consideration of the expansion age of the nebula allows one to discriminate among the scenarios (Iben and Tutukov 1992).

Preparation of the manuscript has been supported by the NSF grant AST91-13662.

## References

Aller, L. H., Keyes, C. D., Maran, S. P., Gull, T. R., Michalitsianos, A. G., & Stecher, T. P. 1987, ApJ, **320**, 159
Barlow, M. J. 1991, in The Magellanic Clouds, ed. R. Haynes and D. Milne (Dordrecht: Kluwer), p. xx
Becker, S. A., and Iben, I. Jr. 1980, ApJ, **237**, 111.
Bieging, J. H. 1988, PASP, **100**, 97
Blöcker, and Schönberner, D. 1990, A&A, **240**, L11
Bond, H. E., and Livio, M. 1990, ApJ, **355**, 1990
Bowen, G. H. 1988, ApJ, **329**, 299
Bowen, G. H., and Willson, L. A. 1991, ApJ, **375**, L53
Clegg, R. E. S. 1989, in Planetary Nebulae, ed. S. Torres-Peimbert (Dordrecht: Kluwer), p. 139
―――― 1991, in The Photospheric Abundance Connection, ed. G. Michaud, and A. Tutukov (Dordrecht: Kluwer), p. 387
Dopita, M. A., Meatheringham, S. J. 1991a, ApJ, **367**, 115
―――― 1991b, ApJ, **377**, 480
Hazard, C., Terlevich, B., Morton, D. C., Sargent, W. L. W., and Ferland, G. 1980, Nature, **285**, 463
Henry, R. B. C. 1990, ApJ, **356**, 229
Huggins, P. J., and Healy, A. P. 1989, ApJ, **346**, 201
Iben, I. Jr. 1984, ApJ, **277**, 333
―――― 1987, in Late Stages of Stellar Evolution, ed. S. Kwok, and S. R. Pottasch (Dordrecht: Reidel), p. 175
―――― 1989 in Evolution of Peculiar Red Giant Stars, ed H.R. Johnson and B. Zuckerman (Cambridge: Cambridge University Press), p. 205
Iben, I. Jr., Kaler, J. B., Truran, J. W., and Renzini, A. 1983, ApJ, **264**, 605
Iben, I. Jr., and MacDonald, J. 1986, ApJ, **301**, 164
Iben, I. Jr., and Renzini, A. 1983, ARAA, **264**, 271
Iben, I. Jr., and Tutukov, A. V. 1989, in Planetary Nebulae, ed. S. Torres-Peimbert (Dordrecht: Kluwer)
―――― 1992, in preparation
Jacoby, G., and Ford, H. 1983, ApJ, **266**, 298
Kahn, F. D. 1983, in Planetary Nebulae, ed. D. R. Flower (Dordrecht: Reidel), p. 305
―――― 1989, in Planetary Nebulae, ed. S. Torres-Peimbert (Dordrecht: Kluwer), p. 411

Kaler, J. B., Iben, I. Jr., and Becker, S. A., ApJL, **224**, L63
Kaler, J. B., & Jacoby, G. H. 1989, ApJ, **345**, 871
—— 1991, ApJ, **382**, 134
Kawaler, S. 1988, ApJ, **334**, 220
Kwok, S. 1982, ApJ, **258**, 280
Kwok, S., Purton, C. R., and FitzGerald, P. M. 1978, ApJL, **219**, L127
Nomoto, K. 1984, ApJ **277**, 791
—— 1987, ApJ, **322**, 206
Paczyński, B. 1971, Acta Astron., **21**, 417
Peimbert, M. 1978, in <u>Planetary Nebulae, Observations and Theory</u>, ed. Y. Terzian (Dordrecht: Reidel), p. 215
—— 1990, Reports on Progress in Physics, **53**, 1559
Renzini, A. 1979, in <u>Stars and Star Systems</u>, ed B. Westerlund (Dordrecht: Reidel), p. 155
—— 1981, in <u>Physical Processes in Red Giants</u>, ed I. Iben, Jr., and I. Iben, Jr. 1981 (Dordrecht: Reidel), p. 431
—— 1982, in <u>Wolf Rayet Stars</u>, ed. C. W. H. de Loore and A. J. Willis (Dordrecht: Reidel), p. 413
—— 1983, in <u>Planetary Nebulae</u>, ed. D. R. Flower (Dordrecht: Reidel), p. 267
—— 1989, in <u>Planetary Nebulae</u>, ed. S. Torres-Peimbert (Dordrecht: Kluwer), p. 391
Schönberner, D. 1979, A&A, **79**, 108
—— 1981, D. 1981, A&A, **103**, 119
—— 1986, A&A, **169**, 189
Smith, V. V., and Lambert, D. L. 1989, ApJL, **345**, L75
—— 1990, ApJL, **361**, L69
Werner, K., and Koesterke, L. 1992, in <u>Atmospheres of Early-type Stars</u>, ed. U. Heber and C. S. Jeffery, (Berlin: Springer), p. 305
Wood, P. R., Bessel, M. S., and Fox, M. W. 1983, ApJ, **272**, 99
Wood, P. R., and Faulkner, D. J. 1986, ApJ, **307**, 659
Zuckerman, B., and Aller, L. H. 1986, ApJ, **301**, 772

# Author's index

| | | | |
|---|---|---|---|
| ACKER A. | : 33, 174, 177, 396, 578 | DIESCH C. | : 197, 207, 374 |
| ALLER L.H. | : 1, 193, 488 | DIMITRIJEVIC M.S. | : 94, 189 |
| ALTAMORE A. | : 217 | DOEL A.P. | : 388 |
| ALTNER B. | : 86, 485 | DOPITA M.A. | : 212, 219, 433, 569 |
| AMNUEL P.R. | : 575 | DORFI E.A. | : 364 |
| ARENS J.F. | : 211, 320, 342 | DOWNES R.A. | : 209, 498 |
| ANDRONOV I.L. | : 323, 324, 325, 407 | DREIZLER S. | : 83 |
| BACHILLER R. | : 224, 227 | DUERBECK H.W. | : 410 |
| BALICK B. | : 131, 365 | DUFOUR R.J. | : 198 |
| BALL J.R. | : 211 | DUFTON P.L. | : 356 |
| BARKER T. | : 181 | DUNLOP C.N. | : 388 |
| BARLOW M.J. | : 92, 163, 185, 213, 321, 340, 341, 571, 581 | DUTRA C.M. | : 568 |
| | | DYSON J.E. | : 299 |
| BARSTOW M.A. | : 497 | ECHEVARRIA J. | : 392 |
| BARTHES D. | : 322 | EFIMOV Y.S. | : 407 |
| BARTOLINI C. | : 491 | EGIKYAN A.G. | : 481 |
| BÄSSGEN M. | : 207, 215 | ENGELS D. | : 330 |
| BECKER S.R. | : 84 | ERMOLAEV A.M. | : 189 |
| BELL S.A. | : 395 | ERICKSON E.F. | : 183 |
| BIANCHI L. | : 85 | ESCALANTE V. | : 220 |
| BICHERT K.F. | : 405 | FAZIO G.G. | : 210, 344 |
| BLADES J.C. | : 213 | FEAST M.W. | : 251 |
| BLAIR W.P. | : 386 | FEIBELMAN W.A. | : 204, 488 |
| BLÖCHER T.G. | : 479 | FEKLISTOVA T.KH. | : 573, 574 |
| BLÖKNER T. | : 363 | FERLAND G.J. | : 123 |
| BLOMMAERT J.A.D.L. | : 243 | FERNANDEZ-CASTRO T. | : 406 |
| BOFFI F.R. | : 491 | FERRARI-TONIOLO M. | : 37 |
| BOHIGAS J. | : 206 | FEUCHTINGER M.U. | : 364 |
| BOHLIN R.C. | : 212 | FIERRO J. | : 380, 392 |
| BOND H.E. | : 397, 489, 490, 499 | FINLEY D.S. | : 497 |
| BORKOWSKI K.J. | : 307, 383, 386 | FLEMING T.A. | : 497 |
| BRADLEY P.A. | : 493 | FLOWER D.R. | : 221 |
| BREGMAN J. | : 386 | FORD H.C. | : 212, 533 |
| BREMER M. | : 215, 218, 373 | FORVEILLE T. | : 224 |
| BRYCE M. | : 377 | FRANK A. | : 311, 365 |
| BUCKLEY D. | : 179 | FRIEDJUNG M. | : 401 |
| BUSSO M. | : 361, 580 | GABLER A. | : 82 |
| BUTLER K. | : 73, 84 | GABLER R. | : 82 |
| CASSATELLA A. | : 217, 406 | GAIGALAS G. | : 95, 96 |
| CASTELLANI M. | : 362 | GALLINO R. | : 361, 580 |
| CERNICHARO J. | : 347 | GARCIA-BURILLO S. | : 347 |
| CHARNLEY S.B. | : 329 | GARCIA-LARIO P. | : 330, 331, 332, 348, 352, 357 |
| CHATTERJEE S. | : 567 | GEBALLE T.R. | : 335, 342 |
| CHIAPPINI C.M.L. | : 583 | GESICKI K. | : 315, 376 |
| CHIEFFI A. | : 361 | GEZARI D.Y. | : 230 |
| CHU Y.H. | : 139, 209, 219 | GLEIZES F. | : 322 |
| CIARDULLO R. | : 397, 489, 490, 503, 570 | GOLOVATYJ V.V. | : 187, 370, 579 |
| CLAVEL J. | : 398 | GONGORA- T. A. | : 220 |
| CLEGG R.E.S. | : 229, 231, 232, 387, 388, 549, 581 | GONZALEZ-RIESTRA R. | : 406 |
| | | GORNY S.K. | : 483 |
| COLGAN S.W.J. | : 183 | GOULDSWORTHY S.N. | : 221 |
| CONLON E.S. | : 356, 360 | GRATL H. | : 411 |
| CORRADI R.L.M. | : 214, 216, 482 | GREWING M. | : 197, 207, 218, 228, 373, 374 |
| CORTES | : 196, 403 | GROENEWEGEN M. | : 585 |
| COSENTINO G. | : 491 | GRUENWALD R. | : 89, 188 |
| COSTA R.D.D. | : 217, 586 | GUARNIERI A. | : 491 |
| COSTERO R. | : 392 | GUELIN M. | : 347 |
| COX P. | : 224 | GUILLOTEAU S. | : 227, 347 |
| CRUZ-GONZALEZ G. | : 202 | GURZADYAN G.A. | : 481 |
| CUESTA L. | : 199, 375, 385 | GUTIERREZ-MORENO A. | : 196, 403 |
| CUISINIER F. | : 578 | HAAS M.R. | : 183 |
| DAHN C.C. | : 175 | HABING H.J. | : 243 |
| DAMINELLI NETO A. | : 217 | HACHISU I. | : 367 |
| DANZIGER J. | : 186, 190 | HAMANN W.R. | : 87, 494 |
| DE FRANCESCO G. | : 85 | HAO X.L. | : 41, 391 |
| DE FREITAS PACHECO J.A. | : 217, 576, 586 | HARRINGTON J.P. | : 212, 383, 386 |
| DE JONG T. | : 333, 585 | HARRIS H.C. | : 175 |
| DEJONGHE H. | : 541 | HAWKINS G. | : 320, 343 |
| DEGUCHI S. | : 225 | HEAP S.R. | : 23, 86, 205, 230, 484, 485 |
| DENNEFELD M. | : 384 | HEBER U. | : 494 |
| DEUTSCH L.K. | : 210, 344 | HILL P.W. | : 387 |
| DEVANEY M.N. | : 388 | HIRANO N. | : 225 |
| DIAMOND C.J. | : 497 | HOARE M.G. | : 232 |

| | | | |
|---|---|---|---|
| HOFFMANN W.K. | : 210, 344 | MALLIK D.C.V. | : 182, 567 |
| HÖFNER S. | : 364 | MAMPASO A. | : 202 |
| HOLBERG J.B. | : 497 | MANCHADO A. | : 330, 331, 332, 348, 357 |
| HOPFENSITZ W. | : 228 | MANTEIGA M. | : 202 |
| HORA J.L. | : 210, 344 | MARAN S.P. | : 212, 230 |
| HRIVNAK B.J. | : 335, 336, 339, 342, 354 | MARCOUT J. | : 35 |
| HRON J. | : 326, 327 | MARENZI A.R. | : 37, 217 |
| HU J.Y. | : 328, 333, 334, 336, 338, 381 | MARSH M.C. | : 497 |
| HUBENY I. | : 86 | MARTEN H.G. | : 315, 363 |
| HUGGINS P.J. | : 147, 224, 227 | MATTEUCI F. | : 580 |
| HUI X. | : 533 | MEABURN J. | : 377 |
| HUTTON R.G. | : 486 | MEAKES M.G. | : 397, 499 |
| HYUNG S. | : 193, 488 | MEATHERINGHAM S.J.: 212, 230 | |
| IBEN I. Jr | : 587 | MEIXNER M. | : 211, 223, 320, 343 |
| ICKE V. | : 346 | MELLEMA G. | : 369 |
| IGUMENSHCHEV I.V. | : 371 | MENDEZ R.H. | : 42, 47, 81, 82, 486 |
| JACOBY G.H. | : 176, 503, 570 | MENNESSIER M.O. | : 322 |
| JAIN S.K. | : 353 | MERKELIS G. | : 95, 96 |
| JASNIEWICZ G. | : 396 | MIHAJLOV A.A. | : 189 |
| JERNIGAN J.G. | : 211, 320, 343 | MILLER C.O. | : 191 |
| JIANG B.W. | : 328, 333, 336 | MITSCH W. | : 487 |
| JURCSIK J. | : 399 | MODOGLIANI A. | : 93 |
| JUSTTANONT K. | : 321, 341 | MONET D.G. | : 175 |
| KALER J.B. | : 492 | MONT A.O. | : 227 |
| KAMESWARA RAO N. | : 200 | MORENO H. | : 196, 403 |
| KAMEYA O. | : 225 | MORGAN D.H. | : 569 |
| KASUGA T. | : 225 | MORRIS M. | : 349 |
| KATO M. | : 367, 408 | MOTCH C. | : 496 |
| KÄUFL H.U. | : 359 | MYERS R.M. | : 388 |
| KAWALER S.D. | : 493 | NAPIWOTZKI R. | : 88, 494, 495 |
| KEENAN F.P. | : 356 | NERI R. | : 347 |
| KERSCHBAUM F. | : 326 | NIKITIN A.A. | : 573 |
| KHAN I. | : 410 | NUSSBAUMER H. | : 402 |
| KHOLTYGIN A.F. | : 573, 574 | OCHSENBEIN F. | : 33 |
| KIDDER K. | : 497 | O'DELL C.R. | : 191 |
| KIENEL C. | : 39 | OSMER S. | : 213 |
| KIMESWENGER S. | : 39 | PAKULL M. | : 496 |
| KINGSBURGH R.L. | : 185, 571 | PARKER Q.A. | : 38 |
| KISIELIUS R. | : 95, 96 | PARTHASARATHY M. | : 352, 353, 398 |
| KHOLTYGIN A.F. | : 192 | PASCOLI G. | : 382 |
| KOESTER D. | : 497 | PATI A.K. | : 182 |
| KOESTERKE L. | : 87 | PATRIARCHI P. | : 93 |
| KOHOUTEK L. | : 36 | PEIMBERT M. | : 523, 584 |
| KOLESNIKOV S.V. | : 407 | PENA M. | : 90, 392, 500, 584 |
| KONDRATJEVA L.N.: 195, 358 | | PERSI F. | : 37, 217 |
| KÖPPEN J. | : 174, 557, 577, 578 | PERINOTTO M. | : 57, 93 |
| KREYSING H.C. | : 197 | PFLEIDERER J. | : 411 |
| KUDASHKINA L.S. | : 323, 324 | PHILLIPS J.P. | : 199, 375, 385 |
| KUDRITKKI R.P. | : 42, 47, 82 | PICCIONI A. | : 491 |
| KWITTER K.B. | : 209 | PIER J.R. | : 175 |
| KWOK S. | : 263, 335, 336, 342, 480 | PISMIS P. | : 202 |
| LAME N.J. | : 194 | POGGE R.W. | : 194 |
| LANGILL P.P. | : 336, 342 | POLCARO V.F. | : 217 |
| LATTANZIO J.C. | : 235 | POLLACCO D.L. | : 180, 201, 387, 395 |
| LEAHY D.A. | : 372 | PORRO I. | : 222 |
| LEBRE A. | : 322 | POTTASCH S.R. | : 40, 331, 332, 348, 352, 357, |
| LEEDJÄRV L. | : 400 | | 378, 390, 394, 398, 449 |
| LEISY P. | : 384 | PREITE-MARTINEZ A.: 37, 65 | |
| LIEBERT J.W. | : 443, 499 | QUIGLEY R. | : 198 |
| LIMONGI M. | : 361, 362 | RAITERI C.M. | : 361, 580 |
| LIU X.W. | : 186, 190 | RAMSAY G. | : 180 |
| LIVIO M. | : 279 | RAUCH T. | : 494 |
| LONSDALE C.J. | : 38 | RAYKOVA D. | : 404 |
| LOPEZ J.A. | : 203, 208 | RAYTCHEV B. | : 404 |
| LORD S.D. | : 183 | READ M.A. | : 38 |
| LOZINSKAYA T.A. | : 219 | REID N. | : 319 |
| LUCAS R. | : 347 | RENZINI A. | : 473, 499 |
| LUTHARDT R. | : 405 | RIERA A. | : 345, 348 |
| LUTZ J.H. | : 19 | ROCHE P.F. | : 232 |
| MAC CAUSLAND R.J.H.: 356 | | ROLA C. | : 582 |
| MACIEL W.J. | : 568, 577, 583 | ROSEN S.R. | : 497 |
| MAGAZZU A. | : 226 | ROSSI C. | : 217 |
| MAJOR J.V. | : 388 | ROTH M.M. | : 42, 203, 208, 487 |
| MALASAN H.L. | : 393 | RUBIN R.H. | : 183 |
| MALKOV Yu. F. | : 187, 370, 579 | RUDNISKIJ G.M. | : 323 |

| | |
|---|---|
| RUDZIKAS Z. | : 96 |
| RUIZ M.T. | : 91, 500, 584 |
| SAGER R. | : 182 |
| SAHAI R. | : 229 |
| SAHAL-BRECHOT S. | : 94 |
| SAHU K.C. | : 271, 357, 378 |
| SAMLAD M. | : 174 |
| SANSOM A.E. | : 497 |
| SAPAR A.A. | : 573 |
| SASSELOV D.D. | : 259 |
| SAURER W. | : 43, 178 |
| SEAQUIST E.R. | : 406 |
| SCARROTT S.M. | : 3 |
| SCHMID H.M. | : 402 |
| SCHNEIDER S.E. | : 179 |
| SCHÖNBERNER D. | : 415, 479, 495 |
| SCHUSTER W.J. | : 392 |
| SCHWARZ H.E. | : 214, 216, 482 |
| SHAKHOVSKOY N.M. | : 407 |
| SHARPLES R.M. | : 388 |
| SHAW R.A. | : 492 |
| SHIBATA K.M. | : 225 |
| SHIVANANDAN K. | : 210, 344 |
| SHORE S.N. | : 230 |
| SILVESTRO G. | : 222 |
| SILVOTTI R. | : 491 |
| SIMPSON J.P. | : 183 |
| SION E.M. | : 498 |
| SKINNER C.J. | : 211, 320, 321, 341, 343 |
| SLIJKHUIS S. | : 333, 337 |
| SMITH R.G. | : 329 |
| SOFFNER T. | : 487 |
| STANGHELLINI L. | : 184, 214, 359, 473, 482, 491, 492 |
| STASINSKA G.F. | : 177, 461, 582 |
| STAUBERT R. | : 197 |
| STECHER T.P. | : 212 |
| STENCEL R.E. | : 405 |
| STENHOLM B. | : 11, 174, 177 |
| STOREY P.J. | : 92 |
| STRANIERO O. | : 361 |
| STRAZZULLA G. | : 226 |
| SURENDIRANATH R. | : 200 |
| SYLVESTER R.J. | : 321 |
| SZCZERBA R. | : 315, 350, 363, 376 |
| TAKANO R. | : 366 |
| TAKEUTI M. | : 366 |
| TAMURA S. | : 34, 225, 355, 366, 379 |
| TAMMAN G.A. | : 515 |
| TAPIA M. | : 203, 208, 392 |
| TAYLOR A.R. | : 406 |
| TERZIAN Y. | : 109 |
| THORNLEY M.D. | : 230 |
| TIELENS A.G.G.M. | : 155 |
| TORNAMBE A. | : 362 |
| TORRES-PEIMBERT S.T. | : 90, 500, 584 |
| TSIOPA O.A. | : 409 |
| TSVETANOV Z. | : 383 |
| TUTUKOV A.V. | : 389 |
| TWEEDY R.W. | : 91 |
| TYLENDA R. | : 177, 423, 461 |
| VAN BLERKOM D. | : 179 |
| VAN DE STEENE G.C. | : 40 |
| VAN DER VEEN W. | : 365 |
| VAROSI F. | : 230 |
| VASSILIADIS E. | : 291, 465, 478, 569 |
| VIEGAS S.M. | : 89, 188 |
| VILKAS M. | : 95, 96 |
| VILLATA M. | : 368 |
| VIOTTI R. | : 217 |
| VOLK K. | : 372, 469, 477 |
| WALSH J.R. | : 231, 377, 378, 390, 394 |
| WALTON N.A. | : 340, 378, 390, 394, 581 |
| WANG J.J. | : 334 |
| WANG L. | : 351 |
| WATERS L.B.F.M. | : 271 |
| WEINBERGER R. | : 34, 43 |
| WELCH W.J. | : 223 |
| WERNER K. | : 83, 494, 496 |
| WESSOLOWSKI U. | : 494 |
| WHITELOCK P.A. | : 251 |
| WOOD P.R. | : 212, 291, 465, 478 |
| WOOTTEN A. | : |
| WOLSTENCROFT R.D. | : 38 |
| WOODSWORTH A.W. | : 354 |
| YADOUMARU Y. | : 379 |
| YAMASAKI A. | : 393 |
| YATSYK O.S. | : 44 |
| YUNGELSON L.R. | : 389 |
| ZHANG C.Y. | : 99, 173, 480, 572 |
| ZHOU X. | : 334 |
| ZWEIGLE J. | : 197, 373 |

# OBJECTS INDEX

## Planetary Nebulae (true and possible)

| Name | PN G | page |
|---|---|---|
| A 7 | 215.5-30.8 | : 109 |
| A 12 | 198.6-06.3 | : 23, 197 |
| A 16 | 153.7+22.8 | : 131 |
| A 21 | 205.1+14.2 | : 109, 175, 307, 500 |
| A 24 | 217.1+14.7 | : 175 |
| A 29 | 244.5+12.5 | : 109, 175 |
| A 30 | 208.5+33.2 | : 123, 131, 139, 378, 386, 443, 489, 587 |
| A 31 | 219.1+31.2 | : 109, 175, 307, 500 |
| A 33 | 238.0+34.8 | : 131, 443 |
| A 34 | 153.7+22.8 | : 131 |
| A 35 | 303.6+40.0 | : 38, 109, 279, 307, 396, 397 |
| A 36 | 318.4+41.4 | : 23, 109 |
| A 39 | 047.0+42.4 | : 131, 443 |
| A 41 (MT Ser) | 009.6+10.5 | : 279 |
| A 46 | 055.4+16.0 | : 279 |
| A 58 (V 605 Aql) | 037.5-05.1 | : 387, 443, 499, 587 |
| A 62 | 047.1-04.2 | : 109 |
| A 63 (UU Sge) | 053.8-03.0 | : 279, 393, 394, 395, 587 |
| A 65 | 017.3-21.9 | : 279, 390 |
| A 74 | 072.7-17.1 | : 109, 175 |
| A 78 | 081.2-14.9 | : 1, 23, 87, 57, 139, 388, 443, 489, 498, 587 |
| Al 1 | 006.8-08.6 | : 581 |
| AS 201 | 249.0+06.9 | : 402 |
| AS 267 | 006.3-00.4 | : 35 |
| Ba 1 | 171.3-25.8 | : 489, 491 |
| BD+30 3639 | 064.7+05.0 | : 1, 23, 99, 109, 147, 163, 187, 197, 232, 340, 344, 485 |
| Cn 1-5 | 002.2-09.4 | : 581 |
| CPD -56°8032 | 332.9-09.9 | : 92, 229 |
| CRL 618 | 166.4-06.5 | : 19, 147, 223, 225, 299, 343, 347 |
| CRL 2688 | 080.1-06.5 | : 340 |
| CW Leo (IRC +10216) | | : 243, 320 |
| DdDm 1 | 061.9+41.3 | : 443 |
| DeHt 1 (LoTr 1) | 228.2-22.1 | : 279, 397 |
| DHW 5 | | : 109 |
| DS 1 | 283.9+09.7 | : 279 |
| EGB 4 (BZ Cam) | 143.6+23.8 | : 407 |
| EGB 5 | 211.9+22.6 | : 279 |
| EGB 6 (0950+139) | 221.5+46.3 | : 109, 443, 499 |
| ESO 166-21 (Wray 17-31) | 277.7-03.5 | : 500 |
| ESO 215-04 (DS 1) | 283.9+09.7 | : 279 |
| Fg 1 | 290.5+07.9 | : 203, 299 |
| GJJC 1 | 009.8-07.5 | : 90 |
| H 1-42 | 357.2-04.5 | : 581 |
| H 2- 1 | 350.9+04.4 | : 47 |
| H 2-17 | 003.1+03.4 | : 449 |
| H 4- 1 | 049.3+88.1 | : 443 |
| HaTr 4 | 335.2-03.6 | : 279 |
| HaWe 4 | 149.4-09.2 | : 109, 178 |
| Hb 5 | 359.3-00.9 | : 163, 202, 226 |
| Hb 8 | 003.8-17.1 | : 581 |
| Hb 12 | 111.8-02.8 | : 1, 99, 193 |
| HD 165970 | 010.6-00.0 | : 35 |
| HD 330036 (Cn 1-1) | 330.7+04.1 | : 402 |
| HDW 3 | 149.4-09.2 | : 178 |
| He 2-15 | 261.6+03.0 | : 571 |
| He 2-55 | 286.3+02.8 | : 489 |
| He 2-77 | 298.1-00.7 | : 109 |
| He 2-99 | 309.0-04.2 | : 57 |
| He 2-104 | 315.4+09.4 | : 345, 371, 382 |
| He 2-108 | 316.1+08.4 | : 23, 65, 109 |
| He 2-109 | 315.4+05.2 | : 196 |
| He 2-111 | 315.0-00.3 | : 19, 147, 208, 571 |

| Name | PN G | page |
|---|---|---|
| He 2-112 | 319.2+06.8 | : 571 |
| He 2-114 | 318.3-02.0 | : 147 |
| He 2-119 | 317.1-05.7 | : 208 |
| He 2-123 | 323.9+02.4 | : 348 |
| He 2-131 | 315.1-13.0 | : 23, 109, 485, 486 |
| He 2-138 | 320.1-09.6 | : 23, 91, 109, 485, 486 |
| He 2-145 | 331.4+00.5 | : 196 |
| He 2-152 | 333.4+01.1 | : 196 |
| He 2-163 | 327.8-07.2 | : 196 |
| HEN 1578 | 009.3+00.3 | : 35 |
| Hf 48 | 290.1-00.4 | : 147 |
| HFG 1 | 136.3+05.5 | : 109, 279, 390, 393 |
| Hu 1-2 | 086.5-08.8 | : 193 |
| Hu 2-1 | 051.4+09.6 | : 23, 57, 99 |
| IC 351 | 159.0-15.1 | : 193, 379 |
| IC 418 | 215.2-24.2 | : 23, 57, 65, 99, 109, 147, 163, 193, 210, 211, 231, 485, 486 |
| IC 1297 | 358.3-21.6 | : 204 |
| IC 1747 | 130.2+01.3 | : 109, 489 |
| IC 2003 | 161.2-14.8 | : 491 |
| IC 2149 | 166.1+10.4 | : 93, 99, 131, 193, 204 |
| IC 2165 | 221.3-12.3 | : 193 |
| IC 2448 | 285.7-14.9 | : 65, 99, 109, 374 |
| IC 3568 | 123.6+34.5 | : 23, 47, 65, 139, 378, 485 |
| IC 4406 | 319.6+15.7 | : 1, 147, 229 |
| IC 4593 | 025.3+40.8 | : 23, 57, 65, 85, 307, 485, 486 |
| IC 4634 | 000.3+12.2 | : 193, 216 |
| IC 4776 | 002.0-13.4 | : 99 |
| IC 4997 | 058.3-10.9 | : 99, 147, 193, 488 |
| IC 5117 | 089.8-05.1 | : 99, 211, 370, 372 |
| IC 5217 | 100.6-05.4 | : 193 |
| IRAS 06562-0337 | 217.0-00.0 | : 35, 271, 357 |
| IRAS 07027-7934 | 291.3-26.2 | : 163, 469 |
| IRAS 07131-0147 | 217.3+04.5 | : 35, 38 |
| IRAS 08355-4027 | 260.7+00.9 | : 35, 38 |
| IRAS 08574-5011 | 270.1-02.9 | : 35 |
| IRAS 11339-6004 | 293.7+01.1 | : 35 |
| IRAS 12238-4907 | 298.7+13.2 | : 35 |
| IRAS 12262-6147 | 300.5-01.7 | : 35, 37 |
| IRAS 12302-6317 | 300.8-00.7 | : 35, 37 |
| IRAS 14132-5839 | 313.7+02.1 | : 35 |
| IRAS 14150-6718 | 311.1-06.0 | : 35 |
| IRAS 17423-1755 | 009.3+05.7 | : 35, 348 |
| IRAS 17514-1555 | 012.2+04.9 | : 201 |
| IRAS 18333-2357 | 009.8-07.5 | : 163 |
| IsWe 1 | 149.7-03.3 | : 109, 489 |
| IsWe 2 | 107.7+07.8 | : 109 |
| J 900 | 194.2+02.5 | : 163 |
| Jn 1 | 104.2-29.6 | : 489 |
| K 648 (Ps 1) | 065.0-27.3 | : 57, 173 |
| K 1-2 | 253.5+10.7 | : 279 |
| K 1-16 | 094.0+27.4 | : 485, 493 |
| K 2-2 | 204.1+04.7 | : 109 |
| K 3-68 (We 2-21) | 178.3-02.5 | : 139 |
| Lo 4 | 274.3+09.1 | : 87, 494 |
| LoTr 5 | 339.9+88.4 | : 23, 109, 197, 279, 396, 397 |
| M 1-2 | 133.1-08.6 | : 402 |
| M 1-7 | 189.8+07.7 | : 147, 225 |
| M 1-11 | 232.8-04.7 | : 99 |
| M 1-16 | 226.7+05.6 | : 147, 229, 348 |
| M 1-26 | 358.9-00.7 | : 47, 99 |
| M 1-30 | 355.9-04.2 | : 449 |
| M 1-37 | 002.6-03.4 | : 449 |
| M 1-42 | 002.7-04.8 | : 581 |
| M 1-67 (209 BAC) | 050.1+03.3 | : 139, 279 |

601

| | | | | | | |
|---|---|---|---|---|---|---|
| M 1-78 | 093.5+01.4 | : 109 | | NGC 6565 | 003.5-04.6 | : 109 |
| M 2-2 | 147.8+04.1 | : 139 | | NGC 6567 | 011.7-00.6 | : 99, 193 |
| M 2-5 | 351.2+05.2 | : 449 | | NGC 6572 | 034.6+11.8 | : 1, 65, 99, 109, 163, 193, 344, 488 |
| M 2-9 | 010.8+18.0 | : 204, 211, 228, 344, 345, 348, 382 | | NGC 6578 | 010.8-01.8 | : 109 |
| M 2-13 | 011.1+11.5 | : 99 | | NGC 6620 | 005.8-06.1 | : 139 |
| M 2-23 | 002.2-02.7 | : 581 | | NGC 6629 | 009.4-05.0 | : 23 |
| M 2-29 | 004.0-03.0 | : 90, 581 | | NGC 6720 | 063.1+13.9 | : 109, 131, 139, 147, 175, 181, 198, 209, 299, 365, 372 |
| M 2-30 | 003.7-04.6 | : 581 | | | | |
| M 2-51 | 103.2+00.6 | : 147 | | | | |
| M 3-6 | 253.9+05.7 | : 65 | | NGC 6741 | 033.8-02.6 | : 193 |
| M 3-21 | 355.1-06.9 | : 581 | | NGC 6751 | 029.2-05.9 | : 87, 131, 307, 489, 5 |
| M 3-28 | 021.8-00.4 | : 19 | | NGC 6772 | 033.1-06.3 | : 147, 224 |
| M 3-30 | 017.9-04.8 | : 489 | | NGC 6781 | 041.8-02.9 | : 147, 224 |
| M 3-33 | 009.6-10.6 | : 581 | | NGC 6790 | 037.8-06.3 | : 211, 372 |
| M 3-35 | 071.6-02.3 | : 372 | | NGC 6803 | 046.4-04.1 | : 99, 109 |
| M 4-9 | 024.2+05.9 | : 147 | | NGC 6804 | 045.7-04.5 | : 109 |
| M 4-18 | 146.7+07.6 | : 200, 211 | | NGC 6826 | 083.5+12.7 | : 1, 23, 47, 57, 65, 85 86, 181, 259, 377, 4 |
| Mz 1 | 322.4-02.6 | : 147 | | | | |
| Mz 3 | 331.7-01.0 | : 345, 382 | | NGC 6853 | 060.8-03.6 | : 23, 109, 131, 139, 147, 175, 181, 209, 307, 485 |
| NGC 40 | 120.0+09.8 | : 1, 57, 85, 109, 209, 372, 485 | | | | |
| NGC 246 | 118.8-74.7 | : 91, 109, 209, 390, 485, 489 | | NGC 6884 | 082.1+07.0 | : 109 |
| | | | | NGC 6886 | 060.1-07.7 | : 109 |
| NGC 650-51 | 130.9-10.5 | : 209, 231, 449 | | NGC 6891 | 054.1-12.1 | : 23, 57, 65, 85, 109, 485 |
| NGC 1360 | 220.3-53.9 | : 23, 109, 209, 485 | | | | |
| NGC 1501 | 144.5+06.5 | : 487, 489, 490, 491, 493 | | NGC 6905 | 061.4-09.5 | : 57, 375, 385, 489, 4 |
| | | | | NGC 7009 | 037.7-34.5 | : 1, 23, 57, 65, 109, 181, 193, 204, 229, 374, 485 |
| NGC 1514 | 165.5-15.2 | : 109, 209 | | | | |
| NGC 1535 | 206.4-40.5 | : 23, 57, 65, 109, 181, 231, 378, 485 | | | | |
| NGC 2022 | 196.6-10.9 | : 209 | | NGC 7026 | 089.0+00.3 | : 109, 491 |
| NGC 2242 | 170.3+15.8 | : 90 | | NGC 7027 | 084.9-03.4 | : 1, 19, 47, 109, 147, 155, 163, 191, 205, 221, 225, 227, 230, 231, 232, 340, 443 |
| NGC 2346 (V651 Mon) | 215.6+03.6 | : 109, 147, 221, 225, 231, 279, 391, 392 | | | | |
| NGC 2371-72 | 189.1+19.8 | : 131, 489, 492, 493 | | | | |
| NGC 2392 | 197.8+17.3 | : 23, 65, 109, 131, 139, 181, 186, 231, 410, 449, 484, 485 | | NGC 7094 | 066.7-28.2 | : 23, 139, 443 |
| | | | | NGC 7139 | 104.1+07.9 | : 19 |
| | | | | NGC 7293 | 036.1-57.1 | : 1, 23, 47, 88, 91, 1C 147, 187, 209, 299, 307, 443, 485, 495, 500 |
| NGC 2440 | 234.8+02.4 | : 1, 23, 109, 131, 180, 181, 193, 207, 378, 449, 571 | | | | |
| NGC 2452 | 243.3-01.0 | : 109, 182, 489, 571 | | NGC 7354 | 107.8+02.3 | : 109, 209 |
| NGC 2792 | 265.7+04.1 | : 109 | | NGC 7662 | 106.5-17.6 | : 109, 123, 131, 163, 181, 191, 193, 209, 365 |
| NGC 2818 | 261.9+08.5 | : 109, 147 | | | | |
| NGC 2867 | 278.1-05.9 | : 109, 489 | | | | |
| NGC 2899 | 277.1-03.8 | : 147, 229 | | NVS 8355 | 004.1+11.2 | : 35 |
| NGC 3132 | 272.1+12.3 | : 109, 147, 207, 229 | | PB 6 | 278.8+04.9 | : 489 |
| NGC 3211 | 286.3-04.8 | : 109 | | PC 11 | 331.1-05.7 | : 279, 398 |
| NGC 3242 | 261.0+32.0 | : 23, 65, 109, 131, 163, 181, 186, 374, 378, 379, 385, 410 | | PC 14 | 336.2-06.9 | : 571 |
| | | | | PHL 932 | 125.9-47.0 | : 279, 307 |
| | | | | PuWe 1 | 158.9+17.8 | : 109, 175, 449 |
| NGC 3587 | 148.4+57.0 | : 1, 209, 307 | | Sa 4-1 | 075.7+35.8 | : 57 |
| NGC 3918 | 294.6+04.7 | : 109, 163 | | Sh 2-68 | 030.6+06.2 | : 109 |
| NGC 4071 | 298.3-04.8 | : 147 | | Sh 2-71 | 035.9-01.1 | : 199, 279, 399 |
| NGC 4361 | 294.1+43.6 | : 23, 90, 197, 449, 485 | | Sh 2-176 | 120.2-05.3 | : 109 |
| NGC 5189 | 307.2-03.4 | : 109, 131, 147, 489 | | Simeiz 22 (Sh 2-188) | 128.0-04.1 | : 109, 178, 307 |
| NGC 5315 | 309.1-04.3 | : 109, 489 | | Sn 1 | 013.3+32.7 | : 443 |
| NGC 5882 | 327.8+10.0 | : 65 | | Sp 1 | 329.0+01.9 | : 279 |
| NGC 6058 | 064.6+48.2 | : 443 | | SwSt 1 | 001.5-06.7 | : 1 |
| NGC 6072 | 342.1+10.8 | : 147 | | TaWe a | 065.4+03.1 | : 35 |
| NGC 6210 | 043.1+37.7 | : 23, 57, 65, 85, 109, 191 | | TaWe b | 070.8+03.7 | : 35 |
| | | | | TaWe c | 071.5+04.9 | : 35 |
| NGC 6302 | 349.5+01.0 | : 1, 57, 89, 123, 147, 163, 202, 206, 220, 229, 279, 385 | | TaWe d | 208.9-07.8 | : 35 |
| | | | | Tc 1 | 345.2-08.8 | : 65, 93, 374 |
| | | | | Tc 233 | 354.4+04.0 | : 35 |
| NGC 6309 | 009.6+14.8 | : 209 | | Th 4-4 | 008.3+03.7 | : 358 |
| NGC 6369 | 002.4+05.8 | : 109, 229 | | VV 47 | 164.8+31.1 | : 224, 489 |
| NGC 6445 | 008.0+03.9 | : 147 | | Vy 2-1 | 007.0-06.8 | : 581 |
| NGC 6537 | 010.1+00.7 | : 1, 89, 109, 193, 205, 385,449 | | Vy 2-2 | 045.4-02.7 | : 1, 19, 109, 147 |
| | | | | We 1-6 | 224.9+01.0 | : 178 |
| NGC 6543 | 096.4+29.9 | : 1, 19, 23, 57, 85, 131, 139, 163, 193, 197, 209, 259, 299, 377 | | WeDe 1 (WDHS 1) | 197.4-06.4 | : 109 |
| | | | | WMP a | 080.3-10.4 | : 35 |
| | | | | Wray 17-1 | 258.0-15.7 | : 384 |
| NGC 6563 | 358.5-07.3 | : 147 | | V744 Sgr | 000.6+00.6 | : 35 |

## OTHER OBJECTS

| Name | page |
|---|---|
| AC Her | : 322 |
| AFGL 2343 | : 321, 355 |
| AFGL 3068 | : 320 |
| AG Car (PK 289-00 1) | : 139, 217 |
| AG Dra | : 405 |
| AG Peg | : 405 |
| BD +39 4926 | : 352 |
| BD +75 325 | : 23, 47 |
| CB 623+71 | : 396 |
| CD -41 13967 | : 47 |
| CH Cyg | : 401, 405 |
| CIT 6 | : 223 |
| CRL 2688 (PK 080-06.1) | : 19,343 |
| DEM 39 | : 139 |
| DEM 231 | : 139 |
| Draco C-1 | : 405 |
| Dzeta Puppis | : 47, 82 |
| Feige 24 | : 497 |
| G 333.6-0.2 | : 183 |
| GD 2 | : 91 |
| GD 394 | : 91 |
| GD 561 | : 495 |
| GL 618 | : 163 |
| GW Vir | : 443 |
| GX1+4 (PK 1+ 4 1) | : 405 |
| HD 44179 | : 271, 352 |
| HD 46703 | : 271, 355 |
| HD 52961 | : 271, 352 |
| HD 101584 | : 271 |
| HD 112374 | : 355 |
| HD 128220 | : 397 |
| He 2-38 (PK 280-02 1) | : 405 |
| Hen 1924 | : 405 |
| HM Sge (PK 53-03 2) | : 400, 401, 405 |
| HR 4049 | : 271, 352 |
| IRAS 04296+3429 | : 163, 263, 339, 341 |
| IRAS 05113+1347 | : 263,339, 342 |
| IRAS 05343+0852 | : 163, 342 |
| IRAS 05381+1012 | : 263 |
| IRAS 06297+4045 | : 330 |
| IRAS 06530-0213 | : 263 |
| IRAS 07134+1005 (HD 56126) | : 263, 341, 342, 352, 354, 355 |
| IRAS 08005-2356 | : 381 |
| IRAS 09371+1212 | : 163 |
| IRAS 10215-5916 (DM -58 3221) | : 263 |
| IRAS 12175-5338 (SAO 239853) | : 263 |
| IRAS 12316-6401 | : 37 |
| IRAS 14122-5947 | : 333 |
| IRAS 1429-4539 | : 338 |
| IRAS 16037+4218 | : 330 |
| IRAS 16279-4759 | : 333 |
| IRAS 16552-3050 | : 333 |
| IRAS 17150-3224 | : 263 |
| IRAS 17436+5003 (HD 161796) | : 163, 263, 321, 343, 355 |
| IRAS 17441-2411 | : 263 |
| IRAS 17475+3119 (BD +31 3797) | : 263 |
| IRAS 17516-2525 | : 359 |
| IRAS 18025-3906 | : 263 |
| IRAS 18095+2704 | : 263, 354, 355 |
| IRAS 18276-1431 | : 381 |
| IRAS 18401+2854 | : 334 |
| IRAS 19114+0002 (HD 179821, GL 2343) | : 163, 263, 343 |
| IRAS 19454+2920 | : 263, 338 |
| IRAS 19477+2401 | : 263, 338 |
| IRAS 19480+2504 | : 263 |
| IRAS 19500-1709 (HD 187885) | : 263, 355 |
| IRAS 20000+3239 | : 263, 338, 339, 342 |
| IRAS 20004+2955 (V 1027 Cyg) | : 263 |
| IRAS 20028+3910 | : 263 |
| IRAS 20406+2953 | : 338, 381 |

| Name | page |
|---|---|
| IRAS 21282-5050 | : 147, 211, 225, 344, 469 |
| IRAS 22223+4327 (DO 41288) | : 263, 339, 342, 354 |
| IRAS 22272+5435 (HD 235858) | : 163, 263, 335, 339, 341, 343, 354 |
| IRAS 22574+6609 | : 263 |
| IRAS 23304+6147 | : 263, 338, 339 |
| IRAS 23321+6545 | : 263 |
| K 4-7 (PK 026-02 2) | : 404 |
| KPD0005+5106 | : 498 |
| LS IV-12°111 | : 356 |
| LS V +46 21 | : 443 |
| M 5 | : 499 |
| M 15 | : 57, 173 |
| M 22 | : 163, 307, 383 |
| M 31 | : 1, 19, 47, 109, 515, 523 |
| M 32 | : 515, 523 |
| M 81 | : 1, 515, 523 |
| M 82 | : 155 |
| M 1-92 | : 147 |
| Magellanics Clouds | : 1, 11, 19, 23, 47, 109, 184, 185, 212, 213, 219, 235, 243, 251, 291,319, 405, 433, 449, 465,503,515, 523, 569, 584, 585, 586 |
| NGC 185 | : 515, 523 |
| NGC 205 | : 515, 523 |
| NGC 891 | : 515 |
| NGC 1023 | : 515, 523 |
| NGC 2359 | : 139 |
| NGC 2453 | : 139 |
| NGC 3199 | : 139, 182 |
| NGC 3377 | : 515, 523 |
| NGC 3379 | : 515, 523, 533, 570 |
| NGC 3384 | : 515, 523 |
| NGC 4374 | : 515, 523 |
| NGC 4382 | : 515, 523 |
| NGC 4406 | : 515, 523 |
| NGC 4472 | : 515, 523 |
| NGC 4486 | : 515, 523 |
| NGC 4594 | : 533 |
| NGC 4649 | : 515, 523 |
| NGC 5128 | : 533 |
| NGC 6164/65 (PK 336-00 1) | : 139 |
| NGC 6397 | : 47 |
| NGC 6522 | : 251 |
| NGC 6888 | : 405 |
| Nova Her 1991 | : 321 |
| OH 12.3-0.2 | : 329 |
| OH 12.8-1.9 | : 329 |
| OH 13.1+5.0 | : 329 |
| OH 16.1-0.3 | : 329 |
| OH 17.7-2.1 | : 329 |
| OH 20.2-0.1 | : 329 |
| OH 21.5+0.5 | : 329 |
| OH 26.4-1.9 | : 329 |
| OH 26.5+0.6 | : 329 |
| OH 30.1-0.7 | : 329 |
| OH 30.7+0.4 | : 329 |
| OH 32.0-0.5 | : 329 |
| OH 32.8-0.3 | : 329 |
| OH 53.8+20.2 | : 321 |
| OH 127.8+0.0 | : 163 |
| OH 231.8+4.2 | : 163 |
| OH 0739-1435 | : 343 |
| ORION | : 155 |
| PG 1144+005 | : 87, 498 |
| PG 1159 | : 47, 83, 87, 92, 415, 443, 489, 493, 494, 495, 496 |

| | | | |
|---|---|---|---|
| PG 1707 | : 493 | Sanduleak 3 | : 489, 490 |
| PG 2131 | : 493 | Sgr I | : 251 |
| R Aqr | : 401 | Sh 2-174 | : 35, 495 |
| R Cen | : 322 | Sh 2-216 (PK 158+00 1) | : 23, 109, 175, 307, 443, 495 |
| R CrB | : 355, 499 | | |
| RCW 58 | : 139, 299 | SK -69 202 | : 139 |
| RCW 104 | : 139 | SN 1987a | : 139, 279, 410, 411 |
| R Nor | : 322 | T CrB | : 405 |
| ROB 162 | : 47 | T Pyx | : 408 |
| RR Tel | : 405 | U CMi | : 322 |
| R Sct | : 322 | U Mon | : 322 |
| RS Oph | : 405, 408 | U Sco | : 408 |
| RV Tau | : 263, 322 | UU Her | : 263, 355 |
| RX Pup | : 401, 405 | VV Cep | : 405 |
| SAO 163075 | : 341 | V 1016 Cyg (PK 75+05 1) | : 400, 402, 405 |
| SAO 173329 | : 271 | V 1017 Sgr | : 405 |
| SAO 244567 | : 449 | Z And | : 405, 406 |
| | | 89 Her | : 263, 271, 355 |

MIX
Papier aus verantwortungsvollen Quellen
Paper from responsible sources
FSC® C105338

If you have any concerns about our products,
you can contact us on
**ProductSafety@springernature.com**

In case Publisher is established outside the EU,
the EU authorized representative is:
**Springer Nature Customer Service Center GmbH
Europaplatz 3, 69115 Heidelberg, Germany**

Printed by Libri Plureos GmbH
in Hamburg, Germany